Architecture and Mathematics from Antiquity to the Future

Birkhäuser

Kim Williams • Michael J. Ostwald
Editors

Architecture and Mathematics from Antiquity to the Future

Volume II: The 1500s to the Future

Birkhäuser

Editors
Kim Williams
Kim Williams Books
Torino, Italy

Michael J. Ostwald
School of Architecture and Built Environment
The University of Newcastle
Callaghan, Australia

Iidabashi Subway Station, Makoto Sei Watanabe, architect (2000), entrance at street level.
Photograph © Andrew I-kang Li, reproduced by permission.

ISBN 978-3-319-00142-5 ISBN 978-3-319-00143-2 (eBook)
DOI 10.1007/978-3-319-00143-2
Springer Cham Heidelberg New York Dordrecht London

Library of Congress Control Number: 2014958159

Printed on acid-free paper

Springer International Publisher is part of Springer Science+Business Media
(www.birkhauser-science.com)

Preface to Architecture and Mathematics from Antiquity to the Future

In June of 1996, in his keynote address at the conference 'Nexus'96: Relationships Between Architecture and Mathematics', the founding international conference of what would become an international community for research in a new interdisciplinary field, eminent engineer Mario Salvadori asked, '[c]an there be any relationship between architecture and mathematics?' Over the next 18 years, the Nexus community came together for a series of bi-yearly conferences in Italy, Portugal, Mexico, Turkey and the USA to examine, debate and celebrate the relationships that exist between architecture and mathematics. The conferences were hosted in locations where important historic connections had been proposed between architecture and mathematics: in Europe these locations include Fucecchio (1996), Mantua (1998), Ferrara (2000), Óbidos (2002), Genoa (2006), Porto (2010) and Milan (2012). Further afield, conferences were held in Mexico City (2004), San Diego (2008) and Ankara (2014). Conference venues were chosen to permit participants to visit local sites of historic importance for architecture and mathematics in post-conference workshops, such as Pompeii and Herculaneum in 1996, the villas of Palladio in 1998 and Teotihuacan in 2004. The speakers at these events include some of the most influential people in architecture, art, mathematics and engineering. Lionel March, Robert Tavenor, Alberto Pérez-Gómez, Marco Frascari, Michele Emmer, Leonard Eaton and Mario Salvadori, amongst many other luminaries, have all presented at the Nexus conferences and taken part in round-table discussions, forums and visits to some of the great architecture of these regions.

The first Nexus conference was actually conceived out of the frustration caused by the difficulty of finding a venue for publishing interdisciplinary research: papers in architecture and mathematics were seen as too mathematical for architectural journals, but not mathematical enough for mathematics journals. At best, such research was viewed as a curiosity, too far from the mainstream to garner much interest. Because there was no single journal that encouraged such research, when authors were fortunate enough to have an article accepted, publications were scattered, and authors seldom knew about the work of others examining similar topics. The Internet was in its infancy at that time, leaving far-flung scholars to work in isolation. One journal, *The Mathematical Intelligencer*, and its particularly open-minded editor-in-chief, Chandler Davis, had accepted papers by three of the participants at the first conference, Kim Williams, Benno Artmann and Heinz Götze, who subsequently began to correspond. The 23 people who met in 1996 at the first conference knew of each other's work by word of mouth: friends sending their work to friends. But already by the second conference, 2 years later, the growing group felt the need for a publishing venue, and it was decided to found the *Nexus Network Journal*—*Nexus*, from the name chosen for the first conference to represent the idea of interweaving ideas from two disciplines, and *Network*, to describe the group of people whose acquaintances and collaborations were continuing to expand. The first issue of the journal, with Kim Williams as editor-in-chief, was introduced online in 1999, was added to at trimester intervals of the course of that year and was produced in print at its end. The journal continued in that way for its first 2 years, but by volume 3 in 2001, submissions had grown so much that it was published in two issues per year, until with volume 9 in 2010, it grew to three issues per year.

Across 15 volumes, 35 issues and over 500 refereed papers, the international reputation and impact of the journal have grown considerably. Now published jointly in the Birkhäuser programme of Springer-Basel and Kim Williams Books of Torino, Italy, the journal is highly respected and has a growing readership. Beginning with volume 16 in 2014, the *NNJ* will be overseen jointly by the editors of these present two volumes.

Foreseen along with the conferences was the publication of the proceedings. The series entitled 'Nexus: Architecture and Mathematics' comprised seven volumes from the first seven conferences. At the beginning, the conference books were seen as separate from the journal. This changed with the eighth conference, when speakers voiced the desire to see their papers published in the *NNJ*, which was by that time mature and esteemed. Thus, since 2010, papers presented in the Nexus conferences have been published in special issues of the journal and are available online. However, the research presented at the early conferences was only available in a series of limited edition books. With many of these being out of print there has been growing pressure to make the most highly cited works from the early years of the Nexus conferences available. Rather than simply republishing selected works in the order in which they were written, such was the scope of these early Nexus publications that an alternative proposition presented itself.

We, the editors, have assembled almost a hundred papers from the early years of the Nexus conferences, and arranged them both thematically and chronologically to trace key moments in the history and theory of architecture and mathematics, from antiquity to the present day, along with predictions for the future. These chapters describe over 60 major buildings and architectural works, consider more than twenty major theories of geometry and design and cover themes and ideas arising from five continents and spanning over four millenia.

Having said this, the present two-volume work does not pretend to be a comprehensive encyclopaedia of the history and theory of every facet of the relationship between architecture and mathematics. Being works by more than one hundred authors with backgrounds in not only architecture and mathematics but also engineering, physics, chemistry, philosophy, music and more, there is a rich diversity of approaches to the topic, along with some insightful synergies and informative disagreements. All of the chapters have undergone minor editorial revisions, including, in some cases, updated bibliographies. In a few cases authors have chosen to make more substantial revisions, to bring their chapters up to date, or direct the reader to advances that are currently occurring in their areas. In addition to this, we have provided an overview chapter for each volume (Chap. 1 in vol. I and Chap. 48 in vol. II), to frame the sequence and structure of the whole as well as a chapter entitled 'Mathematics *in*, *of* and *for* Architecture: A Framework of Types' (Chap. 3) which seeks to classify, and thereby make more accessible, the myriad connections proposed across this work.

Each of the chapters in the present work have become crucial landmarks in the scholarly landscape of architecture and mathematics. Some represent pioneering research, the first studies of the relationships between architecture and mathematics in a specific period, or in the oeuvre of a given architect. They serve as both points of departure for new voyages of discovery and as destinations for people entering unfamiliar terrain. For the novice researcher these works provide a grounding for their explorations, and for seasoned scholars these chapters offer a critical record of the efforts of fellow travellers. We, the editors, hope that through this two-volume work, these chapters can continue to inspire and guide future generations.

We wish to thank Maria Roberts, Valentina Filemio and Marco Giorgio Bevilacqua for assistance with editing and proofing, and Michael Dawes for support with image preparation and research assistance. We thank all authors for permission to reuse their material, and for their help in updating texts and references. Finally, we thank Anna Mätzener, Editor for Mathematics and History of Science, and Thomas Hempfling, Executive Editor for Mathematics, Birkhäuser, for their support of the Nexus conferences and the *Nexus Network Journal* throughout the years, and especially for their support of this present work.

Torino, Italy — Kim Williams
Newcastle, Australia — Michael J. Ostwald
January 2015

Contents for Volume I

Contents for Volume II

Contributors to Volume II

Franca Caliò Department of Mathematics, Politecnico di Milano, Milan, Italy

Paul A. Calter Vermont Technical College, Randolph, VT, USA

Alessandra Capanna Dipartimento di Architettura e Progetto, Università di Roma "La Sapienza", Rome, Italy

John Clagett Engelwood, NJ, USA

Henry Crapo Centre de Recherche Les Moutons Matheux, La Vacquerie, France

Claude Le Conte De Poly-Barbut Centre d'analyse et de mathématique sociales, EHESS, Paris, France

Sylvie Duvernoy Politecnico di Milano, Milan, Italy

Leonard K. Eaton (1922–2014)

Eugenia Victoria Ellis Department of Civil, Architectural and Environmental Engineering, College of Engineering, Drexel University, Philadelphia, PA, USA

Michele Emmer Dipartimento di Matematica, Sapienza Università di Roma, Rome, Italy

Steven Fleming School of Architecture and Design, University of Tasmania, Launceston, TAS, Australia

Rachel Fletcher New York School of Interior Design, NY, USA

Tomás García-Salgado Facultad de Arquitectura, UNAM, Ciudad Universitaria, Coyoacan, Mexico

Ning Gu School of Architecture and Built Environment, University of Newcastle, Callaghan, NSW, Australia

Rumiko Handa College of Architecture, University of Nebraska-Lincoln, Lincoln, NE, USA

John G. Hatch Department of Visual Arts, The University of Western Ontario, London, ON, Canada

Saori Hisano Miyagi, Sendai-shi, Japan

Jay Kappraff Department of Mathematics, New Jersey Institute of Technology, University Heights, Newark, NJ, USA

Ulrich Kortenkamp Martin-Luther-Universität, Institut für Mathematik, Halle (Saale), Germany

Kristina Luce Department of Art, Western Washington University, Bellingham, WA, USA

Lionel March The Martin Centre, University of Cambridge, Cambridge, UK

Elena Marchetti Department of Mathematics, Politecnico di Milano, Milan, Italy

Rosirene Mayer Faculdade de Arquitetura, Universidade Federal do Rio Grande do Sul, Porto Alegre, RS, Brazil

Eugene McGovern Dublin Institute of Technology, Dublin 1, Ireland

Tessa Morrison The School of Architecture and Built Environment, The University of Newcastle, Callaghan, NSW, Australia

Maurice Murphy Dublin Institute of Technology, Dublin 1, Ireland

Ben Nicholson School of the Art Institute of Chicago, Chicago, IL, USA

Michael J. Ostwald School of Architecture and Built Environment, University of Newcastle, Callaghan, NSW, Australia

Richard Padovan Richmond upon Thames, Surrey, UK

Sara Pavia Trinity College, Dublin, Ireland

Alberto Pérez-Gómez School of Architecture, McGill University, Montreal, QC, Canada

Graham Pont Balmain, NSW, Australia

John Poros College of Architecture, Art, and Design, Mississippi State University, Mississippi State, MS, USA

Peter Proudfoot Roseville, NSW, Australia

Mark A. Reynolds Mill Valley, CA, USA

Paul L. Rosin School of Computer Science and Informatics, Cardiff University, Roath, Cardiff, UK

Zafer Sağdiç Faculty of Architecture, Department of Architecture, Branch of History of Architecture, Yildiz Technical University, Istanbul, Besiktas, Turkey

Arzu Gönenç Sorguç Department of Architecture, Middle East Technical University, Ankara, Turkey

David Speiser Université catholique de Louvain, Louvain-La-Neuve, Belgium

Randy S. Swanson Randy Swanson Architect, St. Petersburg, FL, USA

Richard Talbot Fine Art, The Quadrangle, Newcastle University, Newcastle Upon Tyne, UK

Chris Tucker School of Architecture and Built Environment, University of Newcastle, Callaghan, NSW, Australia

Benamy Turkienicz Faculdade de Arquitetura, Universidade Federal do Rio Grande do Sul, Porto Alegre, RS, Brazil

Josephine Vaughan School of Architecture and Built Environment, University of Newcastle, Callaghan, NSW, Australia

Stephen R. Wassell Department of Mathematics and Computer Science, Sweet Briar College, Sweet Briar, VA, USA

Donald J. Watts The College of Architecture, Planning and Design, Kansas State University, Manhattan, KS, USA

Kim Williams Kim Williams Books, Turin (Torino), Italy

João Pedro Xavier Faculdade de Arquitectura da Universidade do Porto, Porto, Portugal

Maria Zack Department of Mathematical, Information and Computer Sciences, Point Loma Nazarene University, San Diego, CA, USA

Andrzej Zarzycki New Jersey Institute of Technology, College of Architecture and Design, University Heights, Newark, NJ, USA

Chapter 48
The Revolutionary, The Reactionary and The Revivalist: Architecture and Mathematics After 1500

Michael J. Ostwald and Kim Williams

revolutionary |ˌrevəˈloō sh əˌnerēl *adjective*
Involving or causing a complete or dramatic change.

reactionary |rēˈak sh əˌnerēl *adjective*
Opposing a person, a set of views, or political or social practice.

revivalist |riˈvīvəˌlistl *adjective*
A tendency or desire to resurrect a former custom or practice.

Premise

A common conceit in historical analysis is to view the world and its history as following an essentially forward, progressive or evolutionary trajectory (Foucault 1972; Turchin 2003). This way of conceptualizing the past permits the construction of a narrative sequence that commences with a time dominated by superstitions and primitive practices and then moves towards an era of reason and civilization. Even the naming of historic periods—from the "Dark Ages" to the age of "Enlightenment"—serves to reinforce the message that advances in science, philosophy and art illuminate and thereby clarify the world around us (Adorno 1998). As a corollary to this idea, if history is following an essentially forward trajectory towards a more advanced or aware state, then movements or events that

M.J. Ostwald (✉)
School of Architecture and Built Environment, University of Newcastle, Callaghan, NSW 2308, Australia
e-mail: michael.ostwald@newcastle.edu.au

K. Williams
Kim Williams Books, Corso Regina Margherita, 72, 10153 Turin (Torino), Italy
e-mail: kwb@kimwilliamsbooks.com

K. Williams and M.J. Ostwald (eds.), *Architecture and Mathematics from Antiquity to the Future*, DOI 10.1007/978-3-319-00143-2_1,

either retard or undermine this progress are regarded as following an inferior or retrograde path. One example of this view is found in Edward Gibbon's argument, in *The History of the Decline and Fall of the Roman Empire* (1996) that a loss of civic virtue, coupled with wanton immorality and decadence, set a once great civilisation on a regressive path, culminating in its collapse and the consequent spread of barbarism across Europe (Womersley 1994).

A variation of this way of thinking about history as a progressive trajectory is found in the work of the German philosopher Theodore Adorno. Adorno developed a school of thought, known as Critical Theory, which used techniques from the social sciences and humanities to interpret or analyse history (Benzer 2011). One feature of Adorno's critique of the twentieth century is his division of various cultural, artistic and scientific theories into two progressive types: the reactionary and the revolutionary (Adorno 1981). Reactionary works are those motivated by failures of the recent past to argue for the introduction of changes to improve the situation. Adorno is mildly critical of such works because, through their process of rejection, they inadvertently reinforce and sustain the continued presence of some trace of a past, flawed, system. In contrast, revolutionary works are motivated by a desire for something completely new. They are not intellectually framed in opposition to the present or the past. Instead, the revolutionary agenda is relentlessly forward-looking and both unfettered and uncontaminated by the past (Adorno 2003). While both reactionary and revolutionary developments continue the forward trajectory of civilization, there is also a category of behaviour which theorists of the Frankfurt school, like Adorno, suggest might reverse this progress. These regressive behaviours could be described as revivalist or revisionary practices; they either stall or reverse progress, no matter which field they occur in. This way of thinking about history as a trajectory towards (or away from) enlightenment is useful for interpreting political, social, and aesthetic movements, but there are also parallel theories for examining developments in science and mathematics.

Thomas Kuhn (1962), Karl Popper (2002) and Bruno Latour (1987) have all observed that knowledge in science and mathematics is developed through a series of revolutionary stages (paradigm shifts), which are then followed by successive reactionary sub-stages, each of which refine or test knowledge. Such cyclical revolutionary and reactionary processes are progressive, in the sense that they are motivated by the will to develop new knowledge, even if they accept past knowledge as part of a larger evolutionary trajectory. Although this way of thinking might describe the majority of recent research in science and mathematics, throughout history there have been moments when revivalist notions have deterred this forward progress. For example, prior to the fifteenth century scientists and mathematicians were frequently involved in debates about esoteric and hermetic principles as well as being engaged in alchemical and occult practices. In the sixteenth century, the research of mathematician John Dee (1527–1608) was focused on advancements in both algebra and divine magic. In the seventeenth century, the pioneering experiments in chemistry of scientist Robert Boyle (1627–1691) recorded details not only about the materials and techniques he was using, but also about astronomical events that occurred during his experiments. The research of Dee and Boyle, both of whom worked *after* the scientific revolution, could now be interpreted as following a partially reactive (scientific) and partially regressive (superstitious) trajectory.

These three ways of thinking about the history of ideas—the revolutionary, the reactionary and the revivalist—are useful for conceptualizing patterns in the development and application of knowledge. They are, of necessity, only available to us in hindsight and are only ever true in a limited sense. Furthermore, this categorization of movements or trends is also, as Adorno suggests, more relevant to the history of ideas and aesthetics than to, for example, the history of technology. However, when examining connections and relationships between architecture and mathematics—fields that are integral to both the history of ideas and aesthetics—these categories can be useful. In particular, from the start of the sixteenth century, the field of mathematics could be regarded as pursuing a more consistently progressive path, while architecture's trajectory over the same period is marked by both forward and backward tendencies. This divergence between the paths taken by architecture and mathematics begins to explain the separation between these disciplines, which became apparent in the 1500s and widened in the eighteenth and nineteenth centuries, before showing more recent signs of realignment.

In the present chapter these three tendencies—the revolutionary, the reactionary and the revivalist—are used to frame changing relationships between architecture and mathematics. However, rather than explicitly adopting the Frankfurt school's or Kuhn's criticisms of regressive tendencies, here we take a more reconciliatory stance. This is because, throughout history, certain instances of revivalism provide moments of insight into the relationship between architecture and mathematics. Furthermore, while revolutionaries, by their very nature, outline brave or radical propositions, this does not necessarily make them more interesting than their reactionary counterparts. Thus, while we shall use these three concepts to frame an overview of the changing relationship between architecture and mathematics since the 1500s, for the most part they are simply used to suggest patterns of behaviour or attitudes towards knowledge.

Another reason for introducing these ways of conceptualizing the history of ideas and aesthetics is because they help to explain why, despite hundreds of scholarly works that identify connections between architecture and mathematics, there are historic moments and movements that are barely considered. This is because many of the most fertile connections between architecture and mathematics have occurred when architects have sought inspiration for their work through appeals to science or geometry (Evans 1995; Ostwald 1999; Di Cristina 2001). These moments are more likely to occur when architecture is pursued in a revolutionary or reactionary way, or when both architecture and mathematics have parallel, progressive trajectories. Examples of this type are especially prevalent in Renaissance, Baroque and Modern architecture. This is also potentially why few examples exist of innovative connections between architecture and mathematics during the Gothic Revival, Egyptian Revival or Post Modern Historicist movements. At such times architecture has sought inspiration primarily from within, and the only—if any—mathematics being used in the design process is adapted from a previous era.

Tracing Trajectories

The choice of the year 1500 as the starting point for the present volume was suggested more by changes in science, technology and mathematics than by specific developments in architecture. In the closing years of the fifteenth century the invention of the printing press by Johannes Gutenberg (1395–1468) gradually led to many changes in society, not the least of which was finally achieving a widespread acceptance of Arabic, rather than Roman or Indian, numeric systems. Furthermore, from the early years of the sixteenth century a growing number of people would view the world in an empirical manner rather than accepting explanations of faith or magic. Research was also undertaken from both an introspective (inquiry-based) and a more overtly global (discovery-based) frame of mind. In particular, the Age of Discovery, the beginning of which is conventionally dated by the fall of Constantinople in 1453, led to a remapping of the world based on exploration, with Ferdinand Magellan (1480–1521) circumnavigating the globe in 1522. Nicholas Copernicus (1473–1543) first considered the prospect that the earth might revolve around the sun in 1514, and expressed this in his *De revolutionibus orbium* in 1543. That same year saw the publication of *De Humani Corporis Fabrica* by Andreas Vesalius (1514–1564). This work provided the first detailed, illustrated and comprehensive description of anatomical dissection, but failed to identify the location of the human soul. The works of Copernicus and Vesalius, along with those of Tycho Brahe (1546–1601), Gerardus Mercator (1512–1594) and Galileo Galilei (1564–1642), are considered as having triggered the scientific revolution. While these events did not precipitate the type of social upheaval set in motion by the Protestant or English Reformations, they nevertheless signalled the first stage of a substantial philosophical paradigm shift. In architecture, awareness of these discoveries manifested themselves in a vocabulary of form that grew from being based on circles and squares to embrace the ellipse, although the language of that form remained classical. To be sure, mysticism and symbolism continued to shape architecture's formal and stylistic language, just as they continued to interest some mathematicians. But in the aftermath of the scientific revolution the influence of mysticism gradually waned (Shapin 1998; Henry 2008).

Technical advances that aided empiricism included the invention of optical microscopes (Hans Lippershey 1590), refracting telescopes (Hans Lippershey 1608, and Galileo 1609) and thermometers (Roger Fludd 1638). The sixteenth and seventeenth centuries also saw many important advances in mathematics which helped people to understand what the new instruments made it possible to observe. The age abounded in genius: John Napier (1550–1617), Henry Briggs (1561–1630), Johann Bernoulli (1667–1748), René Descartes (1596–1650), Pierre de Fermat (1601?–1665), and Blaise Pascal (1623–1662) all developed major insights in mathematics. Isaac Newton (1642–1727) and Christopher Wren (1632–1723) each proposed new applications of geometry and considered their use in architecture. Notwithstanding continued and often clandestine interests in alchemy (Isaac Newton is a case in point), the cabala and other hermetic traditions, mathematicians and scientists of the era were typically committed to the advancement of ideas through rigorous and empirical testing, rather than through the more symbolic or metaphysical properties of numbers. Thus, it is not surprising that the first scientific journals began to be published in the seventeenth century (the English *Philosophical Transactions of the Royal Society* was established in 1665) and the first practices of scholarly and scientific review can also be traced to this era, affirming the progressive (reactionary and revolutionary) agenda of science.

Architecture was slower to advance. Because of the time taken to design and construct buildings, architecture in the sixteenth and seventeenth centuries continued to use materials and methods similar to those employed in previous epochs. The works of Guarino Guarini (1624–1683), such as the church of San Lorenzo and the Chapel of the Holy Shroud, both in Torino, feature a Baroque mantel of architectural forms cloaking the Gothic masonry structure that holds it up. Moreover, many of the major works of architectural theory remained dominated by ideas and principles derived from classical Rome that had been rediscovered and refined two centuries earlier during the Renaissance (Kruft 1994). Because architecture serves both a cultural and social function, and as buildings can last for many generations, developments in architecture were also less dramatic than those in science or mathematics. For example, in Italy Mannerism, the name given to the style adopted by Michelangelo and others from about 1520 to about 1580, whose emphasis on tension and instability contrasted with the composed, static equilibrium of the Early Renaissance, could be regarded as a variation of Renaissance architecture, just as in England Jacobean architecture grew from the Elizabethan style. Both of these developments were reactionary in their trajectory and, for the layperson, neither of these had an especially immediate or obvious impact.

The most important stylistic movement of the seventeenth century in Europe, the Baroque, a development of late Renaissance architecture, started in central Europe (Italy, Germany and France) and spread to England, Spain and Portugal (and through them, to their colonies) and Russia. Baroque architecture was characterized by new explorations of form, light and shadow and dramatic intensity to produce new spatial effects such as dilatation and contraction. Baroque architecture was designed to provide extravagant evidence of the predominance and centrality of the Catholic Church during the Counter-Reformation, and was thus, by intent at least, not progressive. Another movement that flourished in the seventeenth century is what is today known as Palladian architecture, another reactionary descendent of the late Renaissance, characterised by symmetrical facades and plans and an appreciation of the impact of proportion and perspective. However, while Palladian architecture took advantage of some new ways of thinking about space, this was tempered with a desire to reference the seemingly timeless or transcendent forms of ancient Greek and Roman temples.

Such approaches underline the tension that existed between the progressive trajectory of the scientific and mathematical communities, and the more resistant views of the clergy, philosophers and the wider community. Architecture, which was informed by mathematics, but served a more general communal and cultural function by expressing social values, was frequently at the centre of such strained relations and sought ways to reconcile the two. This tension was to worsen across Europe in the eighteenth century, an era which is now seen as being responsible for reviving or revising various older styles, theories and beliefs. Neo-classical architecture and Gothic Revival architecture were amongst the largest stylistic trends in the eighteenth and nineteenth centuries. The neo-classicists included Karl Friedrich Schinkel (1781–1841) and John Soane (1753–1837) while the works of French neo-classical architects Claude Nicolas Ledoux (1736–1806) and Étienne-Louis Boullée (1728–1799) displayed a classical vocabulary of elements draped over their otherwise more Rationalist works. Conversely, in the hands of John Ruskin (1819–1900) and Augustus Pugin (1812–1852), the stark, Platonic forms of neo-classicism were rejected in favour of the allegedly more spiritual and phenomenally significant impact of a revived Gothic tradition. By the mid-nineteenth century there were relatively short-lived revivals of Romanesque, Egyptian, Greek and Renaissance (Italianate) styles. Despite some notable if isolated successes, architecture at this time must have looked increasingly moribund and self-referential to the younger generation of graduates. It is not surprising then that, from amongst the young architects of the late nineteenth century, and especially those who had been trained in these various revivalist traditions, the roots of the Modernist rebellion can be traced.

In the early years of the twentieth century the Futurist and Constructivist manifestos, each with their overt revolutionary agendas, strongly rejected the revivalist and retrograde tendencies of the previous generation. Even the more reactionary Arts and Crafts movement demonstrated an interest in themes which prefigured Modernist preoccupations with labour, materiality and emancipation. Organic and Functionalist architecture sought different inspirations, respectively nature and industrialization, for producing a contemporary architectural style (Le Corbusier 1927; Wright 1995). Both of these variations of Modernism, despite later criticism (Blake 1974; Brolin 1976), reunited the progressive trajectories of architecture and mathematics, which had been separated for 200 years. In the 1940s and 1950s theories of non-Euclidian geometry were swiftly adopted by architects and artists (Henderson 1983) and advances in complex surfaces and shells soon found their way into architectural design and then returned to mathematics by way of textbooks.

In the later years of the twentieth century (and following a different type of historical revivalism in the 1980s) a major revolution in geometric thinking—fractal geometry—was almost immediately appropriated by architects, leading to a series of works, unbuilt and built, which tried to capture the essence of fractal form (Ostwald 2001; Harris 2012). After that time, architects began to use topographic tiling, including aperiodic tiles, often less than a decade after particular sets (like Penrose or Conroy tiles) had been published in mathematical journals (Di Cristina 2001). By the late 1990s the disciplines of architecture and mathematics had become so specialized that few people could truly be considered to make advances in both fields, while the progressive, revolutionary and reactive trajectories of each had returned to a relatively close alignment, with new computer-aided design tools guaranteeing their continued interaction.

This Present Work in Context

The shifting relationships between the revolutionary, reactionary and revivalist trajectories of architecture and mathematics after 1500 are traced in different ways in the present work. Volume II continues the pattern developed in Volume I, of interweaving the theory and history of architecture and mathematics. It is divided into five parts, two of which are structured chronologically to cover the years between 1500 and 1800 (Part VIII) and from 1800 to the present day (Part IX). The 24 chapters in these 2 sections provide a broad coverage of the works of iconic architects from these eras along with analysis of major built and unbuilt works. Part IX is also the only section in these two volumes that is dominated by developments in North America; prior sections have tended to emphasise people, concepts and buildings in Europe, Asia, South America and the Middle East. The remaining three parts in Volume II are all about theories or practices which connect architecture and mathematics. The first of these, Part VII, is concerned with the modes of representation that played a significant role in shaping European architecture in the sixteenth and seventeenth centuries. The second of these, Part X, examines contemporary approaches to the use of mathematics in design and analysis. The final section, Part XI, concludes with eight chapters about computational, parametric and algorithmic approaches to architecture. Parts IX and X also feature examples of architecture from the Oceania region.

These final two parts of Volume II are the only ones which break from the alternating structure adopted across both volumes, which interweaves a historical chronology of buildings and architects with theories that were of relevance to the era. These last two parts are both classified as primarily concerned with theory, but they also feature research into specific buildings, designs or ideas. This reflects the fact that the closer we come to the present day, the less historical the tone of the research. Thus, even if a chapter is about a specific building or design, it is not generally considered through a historiographical framing, but rather as an example of the development or testing of a theory. The fact that several of the chapters in the last section are about computational design, an approach that often directly extrapolates or evolves a theory into the visualisation of a design, further reinforces this difference. Finally, the chapters in the last section include both forward-looking or projective research, and philosophical musings about the entire relationship between architecture and mathematics enabled by the computer. In a sense, Part XI, is neither purely historical nor purely theoretical, but rather comprises an extended conclusion where the two come together. In the sections that follow we will describe the content and themes in each of the five parts of this second volume.

Part VII: Theories of Representation

Part VII is about theories and practices of representation, focussing on perspective projections in art and architecture. The content of several of these chapters is based on works or advances which occurred prior to 1500 but which were to remain influential for several centuries. Prior to the Renaissance, artists used various techniques and practices to suggest depth in a representation. One of the most common involved the layering of objects in an image—so-called register perspective—such that those deeper in a scene (or further away from the viewer) were represented as being positioned partially behind those which were closer to the front of the scene. Despite such practices, the origins of perspectival approaches to constructing representational depth were only formalised in the fourteenth century (Andersen 2007). For example, historians commonly trace the origins of mathematical models of linear perspective to Filippo Brunelleschi's (1377–1446) paintings and panels (including the famous 'peep-hole' and mirror demonstration) of the Baptistery of Florence (Argan 1946; Grave 2010). *De Pictura*, the mid-fifteenth century treatise on painting by Leon Battista Alberti (1404–1472), further defined the geometric construction techniques used by Brunelleschi, making them more accessible. The contribution of Piero della Francesca (1415–1492), which followed within two decades of Alberti's work, was to refine the method used for depicting Phileban solids; a crucial step for architectural representation (Damisch 1995). Despite such progress, perspectival projections remained a major point of both fascination and contention throughout the Renaissance. During this era many different alternative construction techniques were proposed and tested, with some being closely protected secrets of particular artists or schools, while others were more widely disseminated by academicians (Damisch 1995).

The reason for the Renaissance world's fascination with perspective is itself a complex topic that has been extensively researched in the past. Perspective representations appeared to offer a geometry-based system that defined the relationship between the human body and the world. During the Renaissance this relationship was still regarded as being the province of theology, and thus perspectival constructions were of both practical and symbolic importance (Panofsky 1996). Indeed, as has often been the case throughout history, developments in mathematics and representation were frequently used to sustain arguments about metaphysics, social reform and political power. However, in a more subtle way, the rise of perspective techniques also brought into question the purpose of representation. These themes are considered in the four chapters in Part VII, all of which are about theories and techniques of representation, their geometric construction and significance.

In 'Architecture, Mathematics and Theology in Raphael's Paintings' (Chap. 49), David Speiser examines the geometric construction of perspective in two famous works by Raphael, *L'incoronazione della Madonna* (The Incoronation of the Madonna) and *Lo Sposalizio* (The Wedding of the Virgin), both dated 1504. Speiser argues that these present an early instance of multi-point perspective construction. Prior to this time, artists and architects used a type of artificial one-point perspective involving a single vanishing point to which all lines, which are not parallel to the viewing plane, converge. Through an analysis of the two paintings Speiser demonstrates the existence of a non-frontally-constructed perspective, which requires more than one vanishing point. Thereafter, he offers an explanation for how Raphael constructed these representations and why they were so revolutionary at the time. Kristina Luce's chapter 'Raphael and the Pantheon's Interior: A Pivotal Moment in Architectural Representation' (Chap. 50) is also about Raphael, but this time specifically concerning architecture. The subject is a much-copied sketch by Raphael of the interior of the Pantheon in which the artist attempted to represent the Pantheon simultaneously in both section and perspective. Luce argues that this signals a critical attempt to reconcile the conflicting geometries of architecture (the orthographic section) and of representation (the interior perspective). While such techniques are commonplace today, Raphael's simple sketch suggests a major conceptual advance in Renaissance representation.

One of the landmark images in the history of perspective is a simple line drawing of a chalice. Completed in the mid-fifteenth century, and inconclusively attributed to both Paolo Uccello and Piero Della Francesca, this drawing depicts an intricate, geometrically-faceted footed goblet. What is so striking about the image is its rigorous and transparent construction; in much of the drawing, lines that would otherwise be hidden are displayed. Richard Talbot's chapter, 'Design and perspective construction: Why is the Chalice the shape it is?' (Chap. 51) suggests that the elevation of the chalice is critical to the construction of its representation. This is important because, unlike many perspective drawings of this era, whose construction relied on the human eye and a rudimentary knowledge of vanishing points, there is evidence in the chalice drawing to suggest that it was directly projected from a designed object. Thus, this drawing is a true visualisation of intent, rather than a mediated representation of experience. While being completed several decades prior to Raphael's sketch of the Pantheon interior, both of these examples dramatize the subtle difference between representations of an object's physical characteristics (its dimensionality) and its visual ones (its phenomenological properties).

The final chapter in this section is about António Rodrigues's late sixteenth century treatise on architectural perspective. Rodrigues, a Portuguese architect and educator, was well-versed in geometry and like many of his contemporaries, worked on the design of fortresses and other military structures. In 'Perspective in António Rodrigues's *Tratado de Arquitectura*' (Chap. 52) João Pedro Xavier examines the book entitled *Liuro de Perspectiva*, considering both geometric and political properties of the work. As mentioned earlier, methods of perspectival representation were often regarded as having important theological or symbolic significance. Equally, they could be viewed as representing a rejection of previous practices. For both of these reasons, treatises like that of Rodrigues are important for understanding the techniques of architectural depiction, as well as the politics of representation.

Part VIII: Architecture from 1500 AD to 1800 AD

The chapters in Part VIII span 300 years of architectural history. The first group is focussed on the sixteenth century and includes research into the works of Mimar Sinan (1489?–1588), Andrea Palladio (1508–1580), Michelangelo Buonarroti (1475–1564) and Francesco Borromini (1599–1667). Thereafter two chapters consider the architecture of António Rodrigues (1520–1590) and Juan Bautista Villalpando (1552–1608) along with the latter's impact in the seventeenth and eighteenth centuries. The remaining chapters offer readings of the designs and theories of Inigo Jones (1573–1652), Christopher Wren (1632–1723), Robert Hooke (1635–1703) and Claude Perrault (1613–1688). Throughout Part VIII some of the chapters question the canonical interpretation of various architects, while others reinforce the significance of their buildings and writings. However, it is the influence of Vitruvius's *De Architectura*—re-discovered and disseminated across Europe at this time—that is the thread that binds many of these works together (Kruft 1994).

The first chapter in Part VIII is about the great Ottoman architect Mimar Sinan who was born in the late fifteenth century in Turkey. Serving three Sultans (including Süleyman the Magnificent), Sinan completed several hundred buildings in his career, the most famous of which are the Selimiye Mosque in Edirne and the Süleymaniye Mosque in Istanbul (Kuban 1987). A contemporary of Michelangelo, Sinan's influence in the sixteenth and seventeenth centuries was extensive, with several of his former students completing major works in Europe, the Middle East and Asia (Necipoğlu 2005). Indebted to his training as a military engineer, and inspired by the structure of the Hagia Sophia, Sinan's architecture has been repeatedly praised for both its geometric and structural properties and its careful layering of form and ornament. Zafer Sağdiç's chapter, 'Ottoman Architecture: Relationships Between Architectural Design and Mathematics in Architect Sinan's Work' (Chap. 53) stresses the importance of Euclidean geometric forms in Sinan's architecture, recognising a range of recurring proportional systems. Sagdic also describes the role Sinan's knowledge of geometry and construction played in his later works.

The second and third chapters in this section are dedicated to Andrea Palladio. Born in Padua (Italy), Palladio's work acknowledges a clear conceptual lineage to the ideas of the Roman architect Vitruvius (Ackerman 1974). Supported by wealthy patrons, Palladio was able to design and construct an influential series of houses and public buildings during his lifetime (Giaconi and Williams 2003). Palladio's classically-inspired forms and ideals were promulgated through his writings, notably including his *I Quattro Libri dell'Architettura* (The Four Books of Architecture). This work, illustrated with woodcuts of Palladio's own designs, was published in Venice in 1570 and sets out a series of detailed rules for design and construction (Tavernor 1991). Palladio's architecture and writings have long fascinated architectural historians and mathematicians for the way in which they depict the use of harmonic proportions in plan, elevation and section (Wittkower

1971; Rowe 1982). This interpretation of Palladio's work, made famous by German art historian Rudolph Wittkower in the early years of the twentieth century, is still accepted today, though not unconditionally (Mitrović 1990); we now know that it is not the only possible mathematical influence on Palladio's work.

In 'The Mathematics of Palladio's Villas' (Chap. 54), American mathematician Stephen Wassell argues that Wittkower's (1971) canonical reading of Palladio's villas tends to privilege just one of the major mathematical trends of the era, harmonic proportion. Wassell suggests that the significance of other, equally important theories of proportions, symmetry and geometry, has tended to be understated in Wittkower's work. For example, Wassell notes the significance of several geometric constructions in Palladio's designs including ratios derived from $\sqrt{2}$, $\sqrt{3}$ and $\sqrt{5}$. The following chapter, 'Golden Proportions in a Great House: Palladio's Villa Emo' (Chap. 55) by Rachel Fletcher, has a similar focus on alternative proportional systems. Constructed in the 1550s, the plan of Villa Emo was published almost unchanged in *I Quattro Libri dell'Architettura*, whereas many of the other plans in Palladio's treatise were presented in a more idealised manner. For this reason the Villa Emo is regarded as one of Palladio's most successful attempts to reconcile the practicalities of construction with the desire to reflect a perfect mathematical premise. However, as Fletcher makes clear, several features of the villa are proportioned in response to the golden section, even though it is the harmonic proportions of the plan that are typically praised. Fletcher, through a detailed review of the Villa, uncovers the appearance of hitherto unrecognised golden proportions in both the elevation and plan, as well as in the placement of individual doors and windows. Like Wassell, Fletcher's purpose is not to disprove Wittkower, but rather to uncover the existence of other significant, geometric systems in Palladio's oeuvre.

Trained in Florence and steeped in the humanist tradition, Michelangelo Buonarroti undertook early commissions in both Venice and Bologna. Over the next 50 years he created works of art for wealthy patrons from the Medici rulers to popes, cardinals and ambassadors in both Florence and Rome (Condivi 2007). Today Michelangelo's name is synonymous with the High Renaissance of the sixteenth century and he is revered as an artist, sculptor, poet and architect. Michelangelo is credited as initiating the Mannerist tradition in architecture; an intellectual, artificial (neither animistic nor naturalistic) and highly coded extension of the Renaissance tradition (Ackerman 1986). Two chapters here are about his architecture.

Constructed within the cloister of the Basilica di San Lorenzo, the Laurentian Library has long been regarded as the site of the genesis of Mannerism (Shearman 1991). While the Library itself has been the subject of many interpretations, it is the design of its red and white terracotta pavement that is the focus of 'The Hidden Pavement Designs of the Laurentian Library' (Chap. 56) by Ben Nicholson, Jay Kappraff and Saori Hisano. Concealed for much of the last two centuries by the wooden library desks placed on top of it, this elaborate pavement consists of 15 pairs of panels, each comprising a different geometric design. In this chapter, which is written in such a way as to reflect the differing views of its authors, a series of interpretations of the many unusual patterns and measures present in the pavement is offered. One such explanation is that the designs reflect the content of the works which were catalogued on shelves above them, thereby using geometry and mathematics as a type of ordering device or commentary. The authors admit that the original purpose of the panels will remain a mystery, but they underline the significance of intellectual and geometric tropes in Mannerist design.

In 'Measuring up to Michelangelo: A Methodology' (Chap. 57), Paul Calter and Kim Williams (also co-editor of the present volume) describe the method used to survey the Medici Chapel (the New Sacristy of the Basilica of San Lorenzo) in Florence. Using a theodolite to record specific points and trigonometry to triangulate their position in space, Calter and Williams outline a procedure which can be applied to the measurement of other historic structures, while also providing evidence that the Medici Chapel possesses a set of recurring proportional relations derived from a $\sqrt{2}$ rectangle. Calter and Williams emphasise the importance of accurate measured structures for the validation of theories both historic and new that seek to relate mathematics to architecture.

The next pair of chapters directs the reader's attention away from Italy and towards Portugal and Spain. In 'António Rodrigues, a Portuguese Architect with a Scientific Inclination' (Chap. 58), João Pedro Xavier returns to the topic of Rodrigues, featured previously in this volume for his work on perspective representation. The Onze Mil Virgens Chapel at Alcácer do Sal is the starting point for Xavier's second chapter and through a review of this building and Rodrigues's Santa Maria da Graça Church at Setúbal, Xavier identifies a particular way of working with geometry. Tracing the presence of a range of geometric construction and proportion systems—from *ad quadratum* geometry to the use of 5:4 and 6:7 proportions—Xavier offers an explanation of the role of mathematics in Rodrigues's work.

Born in Córdoba, Spain three decades after Rodrigues, Juan Bautista Villalpando was a renowned architect, mathematician and theologian. Villalpando designed several significant buildings for the Jesuit order, including the San Hermenegildo Church in Seville. However, his most enduring contribution to architecture is found in his elaborate reconstruction of the Temple of Solomon (Kravtsov 2005). Published in 1604 in *Ezechielem Explanationes*, Villalpando's Temple of Solomon, was designed around the principles of Platonic harmonies and ancient measurements. Tessa Morrison, who has also produced the first extensive English translation of part of *Ezechielem Explanationes* (Morrison 2009), provides the next chapter in this section. In 'Villalpando's Sacred Architecture in the Light of Isaac Newton's Commentary' (Chap. 59) Morrison notes that Villalpando imagined the temple as a perfect demonstration of the formal grammar of classical architecture, and sought to reconcile theological and architectural arguments in a single work. From this beginning Morrison considers the impact of Villalpando's design on later scholars and architects, in particular Isaac Newton, who undertook a detailed analysis of Villalpando's Temple of Solomon, arguing that it was one of the most important attempts to imagine an ideal architecture supported by perfect mathematical principles.

British architect Inigo Jones was influenced by both the classical tradition (by way of Vitruvius) and Palladio's interpretation of these principles. Jones, Palladio and Villalpando, each in different ways, sought to develop Vitruvian values through the practical and symbolic application of geometry in architecture. Completing his better-known works in the seventeenth century, Jones also held the post of the Royal Surveyor; it was while he was engaged in this role that he was asked by King James I to examine the origins of Stonehenge. In 'Coelum Brittanicum: Inigo Jones and Symbolic Geometry' (Chap. 60) Rumiko Handa describes how Jones interpreted Stonehenge, through the geometry of its plan, as a Roman temple of the Tuscan order. Handa demonstrates how Jones's reading of Stonehenge was part of his larger vision of *Coelum Brittanicum*, a poetic conceit popularised by writer Thomas Carew to reimagine Great Britain and its monarchy through a classical Roman allegorical lens (Sharpe 1987). Handa draws on an analysis of Jones's architecture and masques, informed by the mathematics of Robert Recorde and John Dee, to develop this thesis, which is as much about politics and symbolism as it is architecture and mathematics.

Typical characteristics of the seventeenth-century Baroque architectural tradition include the presence of complex curvilinear forms, elaborate spatial interpenetrations and a fascination with surface and naturalistic decoration (Wölfflin 1964; Norberg-Schulz 1971). The Italian Baroque is commonly regarded as arising from the works of Gian Lorenzo Bernini (1598–1680) and Francesco Borromini in central Italy and from the architecture of Guarino Guarini and Filippo Juvarra (1638–1736) in northern Italy. In Germany and across the Austro-Hungarian Empire, Johann Bernhard Fischer von Erlach (1656–1723), Johann Balthasar Neumann (1687–1753) and Christoph Dientzenhofer (1655–1722) all developed particular, geometrically-rich variations of the Baroque (Wölfflin 1964; Hubala 1989). Thereafter it spread across Europe, retaining some of its Central European characteristics, but also being modified to accommodate local traditions and materials in Spain and Portugal to the west, Poland and Russia to the east, and England and the Netherlands to the north (Hempel 1965).

Born near Lugano in Switzerland in 1599, Francesco Borromini was responsible for several great works of the Baroque style (Millon 1961). Educated in Milan, Borromini's first serious commissions were in Rome and it was there that he produced his most revered designs including the exquisite San Carlo alle Quattro Fontane, the church of Sant'Agnese in Agone and Sant' Ivo alla Sapienza (Wölfflin 1964). Unlike many architects of the preceding era, who published detailed treatises promoting their ideas, Borromini's sources are poorly documented. In 'The Science Behind Francesco Borromini's Divine Geometry' (Chap. 61), John Hatch suggests that one reason for Borromini's seemingly hermetic approach to design is that he was as much influenced by science as he was by architecture. From this premise Hatch traces a connection between the work of astronomer Johannes Kepler and Borromini's architecture, suggesting that features of S. Carlo alle Quattro Fontane and Sant' Ivo alla Sapienza are indebted to Kepler's interpretation of the cosmos. John Clagett's chapter, 'Transformational Geometry and the Central European Baroque Church' (Chap. 62) is also set against a backdrop of developments in science and mathematics although Clagett focuses on the work of Desargues, Newton, Leibniz and Descartes. From this foundation Clagett demonstrates the impact of mathematics on the design of space and form, transforming the ideally proportioned but often static architecture of the Renaissance into a more dynamic, transformative and challenging Baroque.

The next pair of chapters, both by American mathematician Maria Zack, are about English Baroque architecture and its mathematical and scientific properties and aspirations. In the aftermath of the Great Fire of London in 1666, Christopher Wren was appointed King's Surveyor of Works and Robert Hooke was named Surveyor to the City of London. They were both responsible for rebuilding large parts of the city as well as for the design of many major structures. Wren, trained as an astronomer, geometer and physicist, had at that time just returned from Paris, where he had viewed the plans of Bernini (Jardine 2003). Like Inigo Jones and Villalpando, Wren had studied Vitruvius and was interested in the classical Roman and Greek traditions, but the new Baroque buildings of Central Europe also shaped his architecture. Although English Baroque was rarely as extravagant as its Italian relative, it retained many of the latter's proclivities, although tempered by a more classical, geometric *parti* in plan and section (Downes 1966). In 'Are There Connections Between the Mathematical Thought and Architecture of Sir Christopher Wren' (Chap. 63) Zack investigates possible connections between Wren's building designs and his earlier mathematical theories, while in 'Robert Hooke's Fire Monument: Architecture as a Scientific Instrument' (Chap. 64), this time about Wren's colleague, scientist and architect Robert Hooke, she analyses the latter's Fire Monument (Stevenson 2005). This structure, consisting of a column on an orthogonal prism base, contained a zenith telescope, an instrument which could be used to measure gravitational affects or the position of stars.

The final chapter in Part VIII is about Claude Perrault's Observatoire de Paris. Designed around the same time as the Great Fire of London, this building was originally intended to house the Paris Academy of Sciences. In 'Practical and Theoretical Applications of Geometry in Perrault's Observatoire' (Chap. 65) Randy Swanson provides an overview of the building, focussing on the conceptual themes that shaped the design. Swanson stresses two particular achievements in the Observatoire, the first being in stereotomy—and associated with the design of a cantilevered, elliptical-vaulted semi-helical stairwell—and the second a series of previously unexplained dimensional eccentricities in the building's form. Swanson maintains that the Observatoire de Paris, presented Perrault with a unique opportunity to test his own theories of proportionality and geometry.

Perrault's work is also a fitting conclusion to Part VIII because not only was his architecture influenced by the theories of Vitruvius, but he also embraced a revived classicism more strongly than many of his contemporaries (Herrmann 1973). This is significant because, starting in the mid-eighteenth century, a series of stylistic revivals—led by neo-classicism but also including the Gothic revival—began to dominate the design of major public buildings in Europe and America.

Part IX: Architecture from 1800 to 2000

Many of the architects who are featured in Part IX were born in the closing years of the nineteenth century and came to prominence in the twentieth century. Trained in classical, traditional or arts-and-crafts-style architecture, these designers—including Frank Lloyd Wright, Le Corbusier and Louis Kahn—went on to define Organic, Functionalist and Brutalist variations of Modernism, as well as twentieth-century Rationalism (Jencks 1973; Frampton 2007). Whereas in previous centuries the symbolic potential of geometry was often tied to religion, in the twentieth century it was more commonly used to suggest a connection to nature (Portoghesi 2000; Steadman 2008). Moreover, in this era various structural and technical advances allowed architects to experiment with new formal and spatial compositions which were not easily available to previous generations. This same era is also marked by a gradual shift in focus away from a consideration of churches and palaces as key purveyors of architectural style and towards more profane, domestic and commercial structures (Walden 2011).

The first chapter in Part IX is about Claude Fayette Bragdon (1866–1946), an American architect who worked in architectural practice until the 1920s before becoming a stage designer. While Bragdon's early architectural designs were undertaken in a revivalist Italianate tradition, he is best known today as a proponent of a particular variant of the Arts and Crafts movement. Like his contemporaries Louis Sullivan and Frank Lloyd Wright, Bragdon showed a proclivity for ornamental and handcrafted methods, but was also informed by new industrial materials and production techniques. Indeed, the architecture of these three Americans varied considerably from that of the English Arts and Crafts movement, characterised by the early designs of Edwin Lutyens or the country houses of Charles Voysey. Instead, Bragdon is more correctly associated with the Prairie Style, not only because of the way this approach mediated between progressive and revivalist traditions, but also because of its rigorous approach to geometry. In 'Geomantic (Re)Creation: Magic Squares and Claude Bragdon's Theosophic Architecture' (Chap. 66) Eugenia Victoria Ellis examines Bragdon's fascination with theosophy, number symbolism and divine proportions and projections. Ellis traces the presence of the 'magic square' construction—a mathematical procedure with both geometric and geomantic significance—in Bragdon's First Universalist Church in Rochester, New York.

The next two chapters, both by eminent Wright scholar Leonard K. Eaton, are concerned with the Prairie Style and Usonian houses of Frank Lloyd Wright (1867–1959). Wright was born less than a year after Claude Bragdon, and he is typically regarded as having defined the basic principles of the Prairie Style. This style is characterized by horizontally exaggerated buildings, with low, sloping roofs, wide eaves, extensive terraces and clearly defined exterior spaces (Hess 2006). Organised internally around a central fireplace, Wright's prairie houses were less cellular than most housing of their era, characterised by a free flow of space that prefigured developments in European modern planning by over a decade (Lind

1994). Wright's prairie architecture has been celebrated for its rigorous geometric planning and geometric detailing and ornamentation (Koning and Eizenberg 1981). Both of Eaton's chapters are concerned with these geometric properties, which evoke connections to music and nature. In 'Mathematics and Music in the Art Glass Windows of Frank Lloyd Wright' (Chap. 67), Eaton examines Wright's predilection for embedding the geometric language of a house into the design of its windows. Within the context of patterns, both visual and musical, Eaton analyses five windows from Wright's Meyer May house in Grand Rapids, Michigan. Eaton especially notes in these designs the way in which Wright seemingly anticipates developments in non-figurative art, using a series of geometric ratios, which were largely intuitively defined, to create a system of recurring patterns. Eaton observes that, despite Wright's desire to emulate the structure or timbre of Beethoven, the geometric language of the windows more clearly prefigures the Modernist compositions of Bartok.

Recurring geometric patterns are also at the centre of Eaton's second chapter, 'Fractal Geometry in the Late Work of Frank Lloyd Wright' (Chap. 68), a reading of the plan geometry of the Palmer House. Like Louis Sullivan, Wright had a deep interest in geometry and patterns found in nature (Kubala 1990). This has led to the observation that the geometric patterns developed by these architects are potentially precursors to fractal geometry, a branch of geometry that was only formalised in the late 1970s. In this chapter Eaton notes the presence of triangular geometry, repeated across multiple scales, from siting elements through to decisions regarding design details. Eaton's argument is that Wright possessed an intuitive grasp of the way geometric systems, when repeated across multiple scales, have an affinity to natural systems. Thus, Eaton's argument could be defined as one tying Wright's architecture to fractal geometry, by way of analogy.

In the following chapter, 'Characteristic Visual Complexity: Fractal dimensions in the architecture of Frank Lloyd Wright and Le Corbusier' (Chap. 69), Michael Ostwald, Josephine Vaughan and Chris Tucker develop a very different connection between Wright's architecture and fractal theory. Ostwald (a co-editor of the present work) and his colleagues calculate the fractal dimensions of five of Wright's houses and then compare these results with the fractal dimensions of five of Le Corbusier's houses. Ostwald's and Eaton's chapters clarify the difference between 'fractal geometry' and 'fractal dimensions'. The former is an algorithmically defined process which generates a complex, deep and ordered figure from a recursive rule (Mandelbrot 1977; Ostwald 2001, 2003). In contrast, any object can have a fractal dimension, a measure of its spread of information across all scales of observation. There are several mathematical methods for determining the fractal dimension of an object and Ostwald, Vaughan and Tucker demonstrate one of the first computational applications of this approach. Thus, where previous chapters have sought to uncover the mathematical principles inherent in a design, this one instead describes the application of mathematics to architectural analysis.

The subject of the next chapter is the work of Walter Burley Griffin (1876–1937) and Marion Mahony Griffin (1871–1961), architects who, in their early careers, worked in Wright's Oak Park Studio in Chicago. In 1911 the Griffins submitted a master-plan concept for an international competition to design a new Australian capital city, Canberra (Reid 2002). The Griffins, like many of their Modernist contemporaries, were fascinated by the relationship between geometry and nature. In their case, this passion was furthered through their reading of Rudolph Steiner's spiritual and psychological theory, anthroposophy, as well as through various detailed investigations of the symbolic potential of geometry (Navaretti and Turnbull 1998). These proclivities are the subject of Graham Pont's and Peter Proudfoot's chapter, 'From Cosmic City to Esoteric Cinema' (Chap. 70), on the Griffins's attempts to evoke cosmic or spiritual ideals in architecture and urban design.

A different type of geometric application, for the purpose of constructing complex surfaces, is the subject of John Poros's chapter, 'The Ruled Geometries of Marcel Breuer' (Chap. 71). In the years following the Second World War, engineers and architects including Pier Luigi Nervi (1891–1979) and Felix Candela (1910–1997) had designed a series of structurally efficient shell surfaces using a technique that relied on ruled lines (Ching 2007). Marcel Breuer (1902–1981), a Hungarian-born Modernist architect who later designed the famous Whitney Museum of American Art in New York, not only adopted such ruled geometries to achieve structural solutions, but he also used them to shape space and form to achieve unexpected and sometimes contradictory results (Hyman 2001).

Another famous example of the curved surface in twentieth-century architecture is the subject of Alessandra Capanna's chapter, 'Conoids and Hyperbolic Paraboloids in Le Corbusier's Philips Pavilion' (Chap. 72). Designed for the Brussels World Fair and produced in collaboration with Iannis Xenakis (1922–2001), the Philips Pavilion uses conoids and hyperbolic paraboloids to create an enclosure for experiencing recorded music. Le Corbusier (1887–1965) designed the pavilion at a time when he was interested in double-ruled quadric surfaces, a geometric approach that is related to the one described in Poros's chapter about Breuer. A related, Modernist, curvilinear or parabolic approach to form is seen in the work of Brazilian architect Oscar Niemeyer (1907–2012). More than any other architect of the mid-twentieth century Niemeyer rejected simple orthogonal geometry, embracing sensuous, serpentine curves, shallow domes and partial conic sections. Niemeyer's language of curved form is the subject of Benamy Turkienicz and Rosirene Mayer's chapter, 'Oscar Niemeyer Curved Lines: Few Words, Many Sentences' (Chap. 73). Through a graphic comparison of the forms present in Niemeyer's canonical works, Turkienicz and Mayer demonstrate a particular set of recurring geometric tropes.

Dom Hans van der Laan (1994–1991) was born in the Netherlands. Prior to becoming a monk, he was trained as an architect, and his few completed buildings, like the Monastery church in Tomelilla, Sweden, were designed for the Benedictine order. Richard Padovan's chapter, 'Dom Hans Van Der Laan and the Plastic Number' (Chap. 74) is about one of van der Laan's major contributions to architecture and mathematics: a proportional system based on the 'plastic number'. Like the golden ratio and similar irrationals, plastic numbers are mathematical constants. However, unlike the so-called 'metallic ratios', plastic numbers are not derived from quadratic equations. Instead, they form a proportional system which is derived from cubic equations. The term 'plastic' refers to the extent to which a form is perceptible, in and of itself, while a plastic number is the ratio that describes the lower and upper limits of the human ability to perceive differences of size amongst three-dimensional objects. Padovan describes the basic ratios of the plastic number, 3:4 and 1:7, and how they have been used in architecture.

The final two chapters in Part IX are about the architecture of Louis Kahn (1901–1974). Kahn famously cultivated a metaphysical approach to materiality and form which inspired many attempts to interpret his architecture as being emphatically geometric in its proportions. Despite this, his buildings rarely reveal simple ratios or systems. Instead, Kahn described his geometric intentions as being mediated by the practicalities of siting, construction technology and budget. Like the French Rationalists, Kahn's use of Phileban forms was largely for their timeless or Platonic appeal, and not necessarily for their specific geometric properties. Nevertheless, spurred on by Kahn's association with Anne Tyng (who had completed research on complex geometries), various scholars have sought to find deeper mathematical patterns in Kahn's architecture. In 'Louis Kahn's Platonic Approach to Number and Geometry' (Chap. 75), Steven Fleming analyses several claims regarding such hidden geometric systems in Kahn's architecture. Fleming uses measured drawings and a detailed knowledge of the construction techniques applied in these buildings to show that Kahn's work rarely, if ever, displays the precise type of geometric order which he has since become famous for. Indeed, on many occasions it is apparent that Kahn avoided creating forms which would have been perfect squares, or provided ideal golden sections. In 'The Salk: A Geometrical Analysis Supported by Historical Evidence' (Chap. 76), Steven Fleming undertakes an analysis of the historical evidence surrounding the planning and geometry of Kahn's Salk Institute in La Jolla, California, while in parallel with this, co-author Mark Reynolds pursues a graphical analysis of geometry in the same building. Through a combination of Fleming's scholarly and Reynolds's intuitive geometric readings of the building, they uncover several large-scale patterns, reminiscent of those identified by previous scholars but not as clearly supported by a closer analysis of the working drawings and dimensions chosen by Kahn.

Part X: Contemporary Approaches to Design and Analysis

There are eight chapters in Part X, the first three of which concern surfaces, including minimal surfaces (soup bubbles) and two forms of topographic tiling (aperiodic and quasi-periodic). The architectural potential of each of these three systems is demonstrated using designs from historic and modern eras. The next three chapters describe different methods for using geometry to generate architecture. These include classical proportional systems, linear algebra and perspective hypercube constructions. The final pair of chapters contains accounts of the use of computational techniques for the analysis of geometry in historic structures.

Soap bubbles and soap films are the minimal surfaces required to define a three-dimensional space. Michele Emmer, in 'Architecture and Mathematics: Soap Bubbles and Soap Films' (Chap. 77), describes the history of the development of theories of soap bubble geometry in the late nineteenth century. Emmer then provides an overview of examples of historic applications of such isoperimetric forms, including the spherical designs of Ledoux and Boullée and the more literal application of minimal surfaces in Frei Otto's tensile structures.

Late twentieth-century applications of topographic tiling are the subject of the next pair of chapters. In a mathematical sense, tiles are a system of geometric patternation that fills a surface using a finite set of shapes without gaps or overlaps. Tiles are one of the oldest forms of decoration in architecture and square, rectangular, triangular and hexagonal sets have been used throughout history. Such periodic tiling systems (so-called because they repeat a core pattern at fixed intervals) are well known in architecture and, since the seventeenth century, have been considered of only limited interest to mathematicians. However, in the mid-twentieth century Hao Wang set out to discover if an aperiodic tile set exists, that is, one that would perfectly fill a plane, but would never repeat the same pattern. In 'Aperiodic Tiling, Penrose Tiling and the Generation of Architectural Forms' (Chap. 78), Michael Ostwald provides a background to the history of periodic tiling and then describes the search for an aperiodic tile set. Thereafter he analyses the application of one of the most efficient aperiodic sets—the Penrose tiling created by mathematician Roger Penrose (b. 1931)—in the refurbishment of Storey Hall in Melbourne, Australia. Ostwald's chapter finishes with a discussion of various tiling properties that are yet to be fully examined by architects. This interest in tiling surfaces is continued in 'Paving the Alexanderplatz Efficiently with a Quasi-Periodic Tiling' (Chap. 79), by Ulrich Kortenkamp, which describes a method for tiling a large, non-rectangular space. Using a refined version of the Penrose tiling system, Kortenkamp creates a polygonal set of four tiles to produce a quasi-periodic surface for a public plaza in Berlin.

The next three chapters examine mathematical systems for generating architectural form, but the approaches they take are from very different traditions and follow divergent historic trajectories. In 'Generation of Architectural Forms Through Linear Algebra' (Chap. 80), Franca Caliò and Elena Marchetti provide a mathematical taxonomy reliant on linear algebra which they use to classify architectural forms from different eras. They acknowledge that architects largely derive such forms intuitively, but demonstrate that these forms also have underlying mathematical rules. Through this process they establish a core set of forms which they use to demonstrate how variations in the underlying mathematics can produce alternative geometric constructions. In 'The Praxis of Roman Geometrical Ordering in the Design of a New American Prairie House' (Chap. 81), Donald J. Watts describes an equally rigorous, but innately revivalist approach to design using proportional and geometric systems. Watts demonstrates the application of classic Roman geometrical ordering systems in the design of a 1980s prairie style house in Kansas. Possibly the only Postmodern work in this volume, Watts's design is a homage to, and analysis of, two historic styles and their associated geometric themes. 'Exploring Architectural

Form in Perspective: A Fractal Hypercube-Building' (Chap. 82) by Tomás García-Salgado commences with a discussion of relatively recent applications of perspective, including references to twentieth-century architecture and cinema, to bring into focus the way perspective can be used to challenge representation. From this foundation García-Salgado demonstrates a perspective construction of a hypercube which he then "fractalises" to embed a smaller hypercube variant in one corner. Through this operation—geometry mediated through perspectival representation—García-Salgado generates a conceptual method for building design.

The final chapters in Part X examine computationally-based systems of measurement and analysis. In 'The Compass, the Ruler and the Computer: An Analysis of the Design of the Amphitheatre of Pompeii' (Chap. 83), Sylvie Duvernoy and Paul Rosin describe the application of two methods of measurement for historic structures. The first of these uses geometry to replicate or reconstruct the method used in ancient times for generating the form of a building; the example presented is a Roman amphitheatre. Duvernoy and Rosin next use modern digital tools to undertake an arithmetical analysis of the same structure. Finally, the two sets of results are compared, and in the case of the amphitheatre in Pompeii, the subject of their analysis, they independently arrive at the same conclusion. The final chapter in Part X, 'Correlation of Laser-Scan Surveys of Irish Classical Architecture with Historic Documentation from Architectural Pattern Books' (Chap. 84) by Maurice Murphy, Sara Pavia and Eugene McGovern, describes the application of three-dimensional surface modelling, derived from a laser scan, of historic architecture. This is another technically based and computationally intensive method for first recording, and then supporting the analysis of, the form of a historic building.

Part XI: Theories and Applications of Computing in Architecture

Four of the chapters in Part XI describe applications of digital technology in architectural design. These chapters typically celebrate the creative potential of the computer, identifying ways in which software allows architects to apply new mathematical knowledge in design. In contrast, three chapters consider computational design issues in a different light. One of these, the opening chapter in this section, offers an overview of the rise of computers in architecture, while the other two, are less sanguine in their assessment of the way architects have adopted and used computers for design, identifying ethical and philosophical dilemmas faced by architects in the digital design process. Taken together these seven chapters provide a balanced view of the challenges and opportunities of computational design.

Lionel March was one of the earliest innovators in the use of computational and contemporary mathematical approaches (like graph theory) in architectural design and analysis (March 1976). Starting in the 1960s, and working for over four decades at the forefront of this field, March completed several books on both historic and contemporary connections between architecture and mathematics (March and Steadman 1974; March 1998). In 'Architecture and Mathematics Since 1960' (Chap. 85), he looks back on this period of great change and provides an account of his personal experience in the field. That account commences with a consideration of the way computers (and computationally informed ways of thinking) have altered the relationship between architecture and mathematics, potentially returning architecture to its former, revolutionary, trajectory. March borrows Karl Friedrich Froebel's three categories of mathematical thought—the quantitative, qualitative and relational—to structure his review of the developments since the 1960s.

The next three chapters address closely related topics concerning the way in which digital modelling and animation software has provided architects with a means of generating form using parametric or algorithmic rules. This approach has been widely if somewhat controversially praised in architecture, triggering a plethora of publications and a growing body of research which tends to describe these advances as scientific or evolutionary, thereby seemingly providing authority for architecture through appeals to nature (Szalapaj 2005; Steadman 2008). In 'Bio-Organic Design. A New Method for Architecture and the City' (Chap. 86), Alessandra Capanna provides a synopsis of these emerging design paradigms which use software to generate complex, naturalistic forms. Citing advances in complexity theory and non-linear dynamics, since the 1980s architects have used the so-called "new sciences", coupled with advances in hardware and software, to visualize striking new forms. Andrejz Zarzycki develops this idea in 'Formal Mutations: Variation, Constraint, Selection' (Chap. 87), to consider the role of formal permutations in design decision-making. This type of outcome occurs because some of the parametric modelling and animation software used by architects does not generate a single design solution, rather it produces a myriad of alternative forms, each of which fulfil the starting parameters to a greater or lesser extent. Such variations often evocatively called 'mutations' by architects, are presented as striking and original alternatives for an architect to consider. In 'The Role of Mathematics in the Design Process under the Influence of Computational and Information Technologies' (Chap. 88), Arzu Gönenç Sorguç takes a larger scale view of these developments when she notes that the new focus of design is shifting towards process, rather than product, although digital manufacturing and printing techniques are also automating production at a similarly rapid rate. Sorguç also asks whether the emphasis of architecture has now shifted away from design and more towards presentation; that is, from ideation to representation. Focusing on the role of mathematics in the design process, Sorguç analyses this question in the context of contemporary software-driven design methods.

One of the first computational theories of generative design was developed by George Stiny in the 1970s and is called a "shape grammar" (Stiny 1980, 2008). Shape grammars could be regarded as early versions of more recent parametric approaches to design. Both of these methods use rules to generate or refine a design. In 'A Grammar for Dynamic and Autonomous Design in 3D Virtual Environments' (Chap. 89), Ning Gu demonstrates an auto-generative variation of the shape grammar for use in virtual environments. Gu's work shows that such a design grammar can use agent-based reasoning to begin to optimize the process of generating design variations. Whereas the earlier chapters by Capanna and Zarzycki highlighted the evocative and creative potential of computationally evolved forms, Gu's chapter is more in the spirit of March's research, which calls for the rigorous application of methods to make such processes useful.

Two of the final three chapters in Part XI are more critical of the use of computer technology in design. In 'Geometric Transformations and the Ethics of the Curved Surface in Architecture' (Chap. 90) Michael Ostwald reminds the reader that throughout history systems of geometry have typically been used with a knowledge and transparent demonstration of the basic mathematical principles underpinning that geometry. These properties are often associated with beliefs in correct, right or ideal applications of knowledge in architecture (Watkin 1977; Evans 1997). Grouped under the general heading of ethical considerations, these values have shaped some of the greatest architectural works of each generation (Harries 1997). However, in the late twentieth century, and despite the growing number of examples of applications of mathematics in architecture, the average architect's knowledge of mathematics has probably never been lower. This occurs because the computer allows a designer to create forms without necessitating even the most basic awareness of where these forms are coming from or what rules (geometric or numeric) are shaping them. Architecture's often shallow and opportunistic appropriation of geometry is something that this chapter warns against (Evans 1995; Ostwald 2010).

The future of architecture is sure to continue to include mathematics as a core source of inspiration and validation. Even today numbers surprise and fascinate. The field of number theory, aimed at discovering special qualities of numbers and their combinations, might still prove to be fertile ground for the architect's imagination. In 'Equiangular Numbers' (Chap. 91) mathematicians Henry Crapo and Claude Le Conte de Poly Barbut describe a class of numbers with peculiar properties that have yet to find an application in architectural design, but which future architects might find intriguing enough to pursue.

Finally, in his argument against the potential for digital architecture Alberto Pérez-Gómez follows a historical trajectory coupled with an ethical or humanist foundation. While Ostwald's ethical dispute is against the complacency and obfuscation which is common in digital design processes, Pérez-Gómez has repeatedly argued against such architecture, on the grounds that it lacks both phenomenal and spiritual depth. Pérez-Gómez (1992) was famously critical of both the loss of sense of purpose and of a more transcendent aspiration in much contemporary architecture and he sees these problems exacerbated in computational design practices (Pelletier and Perez-Gomez 1994). In particular, in 'Architecture as Verb and the Ethics of Making' (Chap. 92), Pérez-Gómez reiterates the trend identified earlier by Arzu Gönenç Sorguç that digital designers valorise the process rather than its product, but argues that this is a fundamentally flawed approach. Furthermore, this practice relies too much on the appearance of being scientifically based, rather than actually understanding the science on which its premise rests.

Through a review of the theories of Luca Pacioli and Le Corbusier, Pérez-Gómez suggests that the relationship between architecture and mathematics is at its most productive (both in ethical terms and in terms of appropriately aspirational or transcendent architecture) when designers demonstrate a deeper awareness of the challenges and opportunities of the geometric systems they are using. In a sense, Pérez-Gómez's work reinforces the observation expressed at the start of the present chapter, that the most productive and meaningful exchanges between architecture and mathematics have tended to occur when each are on a converging, but forward-looking trajectory.

Conclusion

In our opening chapter to Volume I, 'Relationships Between Architecture and Mathematics' (Vol. I, Chap. 1) we position the content of that volume against the backdrop of changing relationships between a field of knowledge—mathematics—and a discipline of practice—architecture. While the chapters in Volume I examined the works of particular architects and mathematicians, or provided descriptions of famous buildings or theories of geometry and design, taken collectively they also constructed a narrative thread through an era where architecture and mathematics were both respected, and often closely related, pursuits. However, impelled by increasingly specialised knowledge, supported by differing educational approaches, and triggered by the growing separation of the arts from the sciences, the years after 1500 begin to tell a different story. Nevertheless, while all of these forces have worked to divide the architectural profession from the discipline of mathematics, the two remain connected, as the chapters in Volume II demonstrate.

In order to begin to explain how, with so many forces separating the two, architecture and mathematics have continued to work productively together, here we have suggested a different framework. Drawn from critical theory and the history and philosophy of science, it has adopted three ways of viewing history as a trajectory though time and space. These three ways of conceptualising trends in aesthetics and knowledge—the revolutionary, the reactionary and the revivalist—allow us to see that the majority of the positive, productive or fruitful connections which have been proposed between architecture and mathematics appear to have occurred when both have followed a more rigorous forward trajectory, driven by revolutionary or reactionary agendas. At such moments, when the two disciplines have an equally progressive outlook, the gap between them has been at least partially bridged. Across 45 chapters and the work of 50 architectural historians, and designers, mathematicians, engineers, philosophers and computer scientists, this volume not only traces over 500 years of the history of the relationship between architecture and mathematics, but it is also drawn to the future, offering technical and philosophical views which will remain of relevance for many years, and will continue to shape new, creative opportunities for architecture and mathematics.

Biography Michael J. Ostwald is Professor and Dean of Architecture at the University of Newcastle (Australia) and a visiting Professor at RMIT University. He has previously been a Professorial Research Fellow at Victoria University Wellington, an Australian Research Council (ARC) Future Fellow at Newcastle and a visiting fellow at UCLA and MIT. He has a PhD in architectural history and theory and a DSc in design mathematics and computing. He completed post-doctoral research on baroque geometry at the CCA (Montreal) and at the Loeb Archives (Harvard). He is Co-Editor-in-Chief of the *Nexus Network Journal* and on the editorial boards of *ARQ* and *Architectural Theory Review*. He has authored more than 300 scholarly publications including 20 books and his architectural designs have been published and exhibited internationally.

Kim Williams was a practicing architect before moving to Italy and dedicating her attention to studies in architecture and mathematics. She is the founder of the conference series "Nexus: Relationships between Architecture and Mathematics" and the founder and Co-Editor-in-Chief of the *Nexus Network Journal*. She has written extensively on architecture and mathematics for the past 20 years. Her latest publication, with Stephen Wassell and Lionel March, is *The Mathematical Works of Leon Battista Alberti* (Basel: Birkhäuser, 2011).

References

Ackerman, James S. 1974. *Palladio*. London: Penguin.

———. 1986. *The Architecture of Michelangelo*. Chicago: University of Chicago Press.

———. 1981. *Prisms*. Cambridge Massachusetts: MIT Press.

———. 1998. *Critical Models, Interventions and Catchwords*. Henry W. Pickford, trans. New York: Columbia University Press.

———. 2003. *The Philosophy of Modern Music*. Anne G. Mitchell and Wesley V. Blomster, trans. New York: Continuum.

Andersen, Kirsti. 2007. *The Geometry of an Art: The History of the Mathematical Theory of Perspective from Alberti to Monge*. New York: Springer.

Argan, Giulio Carlo. 1946. The Architecture of Brunelleschi and the Origins of Perspective Theory in the Fifteenth Century. *Journal of the Warburg and Courtauld Institutes* **9** (1946): 96–121.

Benzer, Matthias. 2011. *The Sociology of Theodor Adorno*. Cambridge: Cambridge University Press

Blake, Peter. 1974. *Form Follows Fiasco: Why Modern Architecture Hasn't Worked*. Boston: Little Brown and Co.

Brolin, Brent C. 1976. *The Failure of Modern Architecture*. London: Studio Vista.

Ching, Francis D. K. 2007. *Architecture: Form, Space, and Order*. New York: John Wiley and Sons.

Condivi, Ascanio. 2007. *The Life of Michelangelo*. Charles Holroyd, trans. London: Pallas Athene.

Damisch, Hubert. 1995. *The Origin of Perspective*. John Goodman, trans. Cambridge, Massachusetts.: The MIT Press.

Di Cristina, Giuseppa (ed.). 2001. *Architecture and Science*. London: Academy Press.

Downes, Kerry. 1966. *English Baroque Architecture*. London: A. Zwemmer.

Evans, Robin. 1995. *The Projective Cast: Architecture and its Three Geometries*. Cambridge, Massachusetts: MIT Press.

———. 1997. Translations from Drawing to Building and other Essays. Cambridge, Massachusetts: MIT Press.

Frampton, Kenneth. 2007. *Modern Architecture: A Critical History*. London: Thames and Hudson.

Foucault, Michel. 1972. *The Archeology of Knowledge*. A. M. Sheridan Smith, trans. New York: Pantheon Books.

Giaconi, Giovanni and Kim Williams. 2003. *The Villas of Palladio*. New York: Princeton Architectural Press.

Gibbon, Edward. 1996. *The History of the Decline and Fall of the Roman Empire*, 3 vols. London: Penguin Classics.

Grave, Johannes. 2010. Brunelleschi's Perspective Panels: Rupture and Continuity in the History of the Image. Pp. 161–180 in *Renaissance Perceptions of Continuity and Discontinuity in Europe, c.1300 – c.1550*. Alexander Lee, Pit Péporté and Harry Schnitker eds. Leiden: Brill.

Harries, Kaarsten. 1997. *The Ethical Function of Architecture*. Cambridge Massachusetts: MIT Press.

Harris, James. 2012. *Fractal Architecture: Organic Design Philosophy in Theory and Practice*. Albuquerque: University of New Mexico Press.

Hempel, Eberhard. 1965. *Baroque Art and Architecture in Central Europe: Germany, Austria, Switzerland, Hungary, Czechoslovakia, Poland*. London: Penguin.

Henderson, Linda Dalrymple. 1983. *The Fourth Dimension and Non-Euclidean Geometry in Modern Art*. Princeton: Princeton University Press.

Henry, John. 2008. *The Scientific Revolution and the Origins of Modern Science*. New York: Palgrave Macmillan.

Herrmann, Wolfgang. 1973. *The Theory of Claude Perrault*. London: Anton Zwemmer.

Hess, Alan. 2006. *Frank Lloyd Wright Prairie Houses*. New York: Rizzoli.

Hubala, Erich. 1989. *Baroque and Rococo*. London: Herbert Press.

Hyman, Isabelle. 2001. *Marcel Breuer, Architect: The Career and the Buildings*. New York: Harry N. Abrams.

Jardine, Lisa. 2003. *On a Grander Scale: The Outstanding Life of Sir Christopher Wren*. New York: Harper Collins.

Jencks, Charles. 1973. *Modern Movements in Architecture*. New York: Anchor Press.

Kravtsov, Sergey R. 2005. Juan Bautista Villalpando and Sacred Architecture in the Seventeenth Century. *Journal of the Society of Architectural Historians* **3** (2005): 312–339.

Koning, H. and J. Eizenberg. 1981. The Language of the Prairie: Frank Lloyd Wright's Prairie Houses. *Environment and Planning B: Planning and Design* **8** (1981): 295–323.

Kruft, Hanno-Walter. 1994. *A History of Architectural Theory, from Vitruvius to the Present*. New York: Princeton Architectural Press.

Kubala, Thomas. 1990. Finding Sullivan's Thread. *Progressive Architecture* 71, 10 (October 1990): 102–104.

Kuban, Doğan. 1987. The Style of Sinan's Domed Structures. *Muqarnas* 4 (1987), 72–97.

Kuhn, Thomas. 1962. *The Structure of Scientific Revolutions*. Chicago: University of Chicago Press.

Latour, Bruno. 1987. *Science in Action: How to Follow Scientists and Engineers through Society*. Cambridge Massachusetts: Harvard University Press.

Le Corbusier. 1927. Towards A New Architecture. London: Architectural Press Limited.

Lind, C. 1994. *Frank Lloyd Wright's Prairie Houses*. California: Archetype Press.

Mandelbrot, Benoit B. 1977. *Fractals: Form, Chance, and Dimension*. San Francisco: W. H. Freeman and Company.

March, Lionel. 1976. *The Architecture of Form*. Cambridge: Cambridge University Press

———. 1998. *Architectonics of Humanism: Essays on Number in Architecture*. London: Wiley.

March, Lionel and Philip Steadman. 1974. *The Geometry of Environment: An Introduction to Spatial Organization in Design*. Cambridge, Massachusetts: MIT Press.

Millon, Henry A. 1961. *Baroque and Rococo Architecture*. Studio Vista: London.

Mitrović, Branko. 1990. Palladio's Theory of Proportions and the Second Book of *I Quattro Libri dell'Architettura*", *Journal of the Society of Architectural Historians*, 49 (1990), 279–292.

Morrison, Tessa. 2009. *Juan Bautista Villalpando's Ezechielem Explanationes: A Sixteenth Century Architectural Text*. Lewiston: Edwin Mellen Press.

Navaretti, Jeff and Jeff Turnbull (eds). 1998. *The Griffins in Australia and India: The Complete Works and Projects of Walter Burley Griffin and Marion Mahony Griffin*. Melbourne: Melbourne University Press.

Necipoğlu, Gülru. 2005. *The Age of Sinan: Architectural Culture in the Ottoman Empire*. London: Reaktion Books.

Norberg-Schulz, Christian. 1971. *Baroque Architecture*. New York: Harry N. Abrams.

Ostwald, Michael J. 1999. Architectural Theory Formation Through Appropriation. *Architectural Theory Review* 4, 2 (1999): 52–70.

———. 2001. 'Fractal Architecture': Late Twentieth Century Connections Between Architecture and Fractal Geometry. *Nexus Network Journal, Architecture and Mathematics* 3, 1: 73–84.

———. 2003. Fractal Architecture: The Philosophical Implications of an Iterative Design Process. *Communication and Cognition* 36, 3 & 4 (2003): 263–295.

———. 2010. Ethics and the Auto-Generative Design Process. *Building Research and Information*. 38, 4 (2010): 390–400

Panofsky, Erwin. 1996. *Perspective as Symbolic Form*. Christopher S. Wood, trans. New York: Zone Books.

Pelletier, Louise. and Alberto Pérez-Gómez (eds). 1994. *Architecture, Ethics and Technology*. Montreal: Institut de Recherche En Histoire de L'Archit.

Pérez-Gómez, Alberto. 1992. *Architecture and the Crisis of Modern Science*. Cambridge, Massachusetts: MIT Press.

Popper, Karl. 2002. *The Logic of Scientific Discovery*. London: Routledge.

Portoghesi, Paolo. 2000. *Nature and Architecture*. New York: Skira.

Reid, Paul. 2002. *Canberra Following Griffin: A Design History of Australia's National Capital*. Canberra: National Archives of Australia.

Rowe, Colin. 1982. *The Mathematics of the Ideal Villa and Other Essays*. Cambridge Massachusetts: MIT Press.

Shapin, Steven. 1998. *The Scientific Revolution*. Chicago: University of Chicago Press.

Sharpe, Kevin. 1987. *Criticism and Compliment: The Politics of Literature in the England of Charles I*. Cambridge: Cambridge University Press.

Shearman, John. 1991. *Mannerism*. London: Penguin.

Steadman, Philip. 2008. *The Evolution of Designs: Biological Analogy in Architecture and the Applied Arts*. New York: Routledge.

Stevenson, Christine. 2005. Robert Hooke, Monuments And Memory. *Art History* 28, 1 (2005): 43–73

Stiny, George. 1980. *Pictorial and Formal Aspects of Shape and Shape Grammars*. Basel: Birkhäuser.

———. 2008. *Shape: Talking about Seeing and Doing*. Cambridge Massachusetts: MIT Press.

Szalapaj, Peter. 2005. *Contemporary Architecture and the Digital Design Process*. Oxford: Architectural Press.

Tavernor, Robert. 1991. *Palladio and Palladianism*. London: Thames and Hudson.

Turchin, Peter. 2003. *Historical Dynamics: Why States Rise and Fall*. New Jersey: Princeton University Press.

Walden, Russell. 2011. *Triumphs of Change*. Bern: Peter Lang.

Watkin, David. 1977. *Morality and Architecture*. Oxford. Clarendon Press.

Wittkower, Rudolph. 1971. *Architectural Principles in the Age of Humanism*. New York: W. W. Norton and Company.

Wölfflin, Heinrich. 1964. *Renaissance and Baroque*. Kathryn Simon trans. Ithaca: Cornell University Press.

Womersley, David. (ed.). 1994. *Edward Gibbon's The History of the Decline and Fall of the Roman Empire. 3 vols.* (First published 1779). London: Allen Lane.
Wright, Frank Lloyd. 1995. The Language of an Organic Architecture. Pp. 60–63 in *Frank Lloyd Wright Collected Writings. Vol. 5, 1949–1949.* (First Published 1953). Bruce Brooks Pfeiffer, ed. New York: Rizzoli.

Part VII
Theories of Representation

Chapter 49
Architecture, Mathematics and Theology in Raphael's Paintings

David Speiser

Introduction

The subject I am going to talk about belongs, one may say, to the prehistory of descriptive geometry: it is part of our modern discovery of space. Three times a civilisation has made such an investigation: in ancient Egypt, in Antiquity, and in modern times, where perhaps we should speak of space-time. And each time, not only science, but also the arts participated in this endeavour. It is always extremely interesting to compare the progress of the sciences with the evolution of the arts, as well as their histories, their results, and also their methods. But it is fair to say that in spite of many valiant pioneering efforts, so far this has not been done systematically enough: think for instance of medieval architecture and its importance for the progress of technology and science.

This small contribution is devoted to two mathematical, that is, geometric, discoveries made in 1503 and 1504, and presented in two famous paintings by Raphael: *Lo Sposalizio (The Wedding of the Virgin)* and *L'incoronazione della Madonna (The Incoronation of the Madonna)*. It is especially in the second one that we find architecture, mathematics and theology closely intertwined in a way that is deeply characteristic for this artist, whom we can see here also as a great scientist.

First published as: David Speiser, "Architecture, Mathematics and Theology in Raphael's Paintings", pp. 147–156 in *Nexus III: Architecture and Mathematics,* ed. Kim Williams, Ospedaletto (Pisa): Pacini Editore, 2000.

D. Speiser (✉)
Université catholique de Louvain, Louvain-La-Neuve, Belgium
e-mail: rspeiser@intergga.ch

K. Williams and M.J. Ostwald (eds.), *Architecture and Mathematics from Antiquity to the Future*, DOI 10.1007/978-3-319-00143-2_2,

La Dolce Prospettiva

Attempts to represent buildings in perspective go back at least to Giotto and his school. But it seems that around 1400 Masaccio was the first to discover the law of the vanishing point; I remind you here simply of the Christ on the cross in Sta. Maria Novella and of his Frescoes in Sta. Maria del Carmine.

North of the Alps, the early Flemish painters approached this law step by step, by trial and error. This process is described in an essay by Erwin Panofsky (1962); it seems that the first correct painting is Dirc Bouts' *Last Supper* in St. Peter's in Leuven. However, in all these paintings we find only the use of what is sometimes called, a bit misleadingly, "central perspective". This means that all buildings are presented to us frontally, and the horizontal edges are either orthogonal to our view, in the line of our view, or converging with it. Thus, there is always only one "vanishing point", the point towards which the parallels converge. A typical example is the *Giving of the Keys to St. Peter* by Perugino in the Sistine Chapel (Canuti 1931) (Fig. 49.1). Please note that this restriction forced the painter to place all buildings parallel to each other and frontally with respect to the observer: a severe restriction indeed! So we may ask: who was the first painter who succeeded in representing correctly a building in other than the frontal position?

Perugino's fresco dates from 1480/1481, and in a moment you will see a second, very similar one. But in 1503 his pupil, Raphael Sanzio, was invited to paint for the church of the Franciscans in Perugia an *Incoronation of the Madonna*, which is today in the Pinacoteca del Vaticano (Fig. 49.2). I think that this is the first painting where a structure in a non-frontal position, the sarcophagus of the Madonna, is constructed rigorously. At least I have never seen an earlier one myself. So the question arises: how did Raphael do it? How did he achieve what so many others, presumably, had tried to do in vain?

But first: can we be sure that the sarcophagus of the Madonna is constructed correctly? It is fairly easy to convince yourself that the long edges do indeed converge to a vanishing point. For the short edges, this is obviously a bit more difficult; I convinced myself that they do so, but it seemed that the vanishing point to the left lies a tiny bit higher. But this may be due to my clumsiness together with the fact that I had to work with a comparatively small reproduction, or it may be due to the fact that according to Jones and Penny (1983) the painting was transported from wood on linen.

So how did Raphael do it? You can see the answer in Fig. 49.3: draw the crossing of the extended shorter edge at right with the horizontal that passes through the summit at left, and then descend from the upper summit to this horizontal line and extend it beyond. Now you see that this extension covers two lines in the painting: one that lies in the horizontal plane, and a second one that descends vertically from the central crossing point through the centre of the right side of the sarcophagus! This means that the central crossing point is the centre of the two squares. In the next figure you can now see how Raphael proceeded (Fig. 49.4).

Fig. 49.1 Perugino, Christ giving the keys to Saint Peter, Sistine Chapel, Vatican City. Image: Reproduced by permission, Musei Vaticani

Fig. 49.2 Raphael, *The Incoronation of the Madonna* (Oddi Altarpiece), Pinacoteca Vaticana, Vatican City. Image: Reproduced by permission, Musei Vaticani

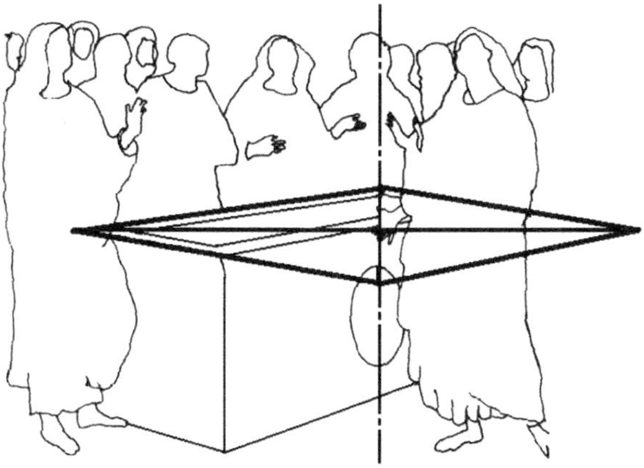

Fig. 49.3 The perspective scheme of the sarcophagus, *The Incoronation of the Madonna*. Drawing: Kim Williams

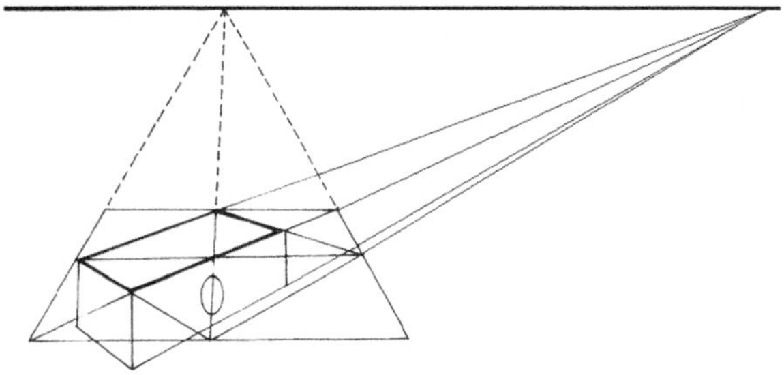

Fig. 49.4 Perspective construction of the sarcophagus. Drawing: Kim Williams

Raphael first drew a square, then he drew the diagonals that, finally, permitted him to draw the square whose corner points are the centres of the four sides. As often happens, it is all very simple, but one has to think about it! And this construction, while special indeed, is not only the very special case it might seem to be: namely a rectangle composed of two squares. You can, for example, extend the sidelines beyond the half-square and construct any rectangle with a line towards the vanishing point.

Having found the law of the two new vanishing points, Raphael has solved the problem of correctly constructing the right angle of any rectangle that is bisected by a line towards the central vanishing point. And my colleague M. Howald showed me a trick that allows one to draw correctly a building seen from an arbitrary angle. As he told me, it is contained in the writings of Leon Battista Alberti. That such a solution was looked for can be seen from paintings that try to give the impression of a building not placed frontally with respect to the spectator, although it is really placed this way. An example is Titian's famous *Madonna with Members of the Pesaro Family* in Venice: there this effect is achieved simply by placing the vanishing point far to the left, outside the painting. But buildings at which one looks obliquely remained rare for a long time. And in many of them a not-frontally-placed building is painted so that one cannot easily check its construction.

Now: theology. The painting makes indeed a theological statement, but one that is in accordance with the instruction of the Franciscans, who commissioned the painting: the Virgin is placed on the same level as the Christ, not lower, which is unusual. If Raphael himself made a theological statement here, it might be as I am about to explain, but I would not press the point. The axis of the geometric construction does not coincide with the axis of the painting itself. But the former concerns an earthly matter only, while the axis of the painting is determined by the heavenly order. Always remember, especially when we now go to *Lo Sposalizio*, that an altarpiece is a symbolic construction, and not only *un coin de la nature vu par un tempérament*! Indeed, at the time, some may have found Raphael's innovation too naturalistic.

Lo Sposalizio

Lo Sposalizio, originally painted for Città di Castello, is now in the Brera in Milan (Fig. 49.5). The High Priest celebrates the wedding of the elected Joseph with Mary. Only Joseph's stick bursts into flower, and thus his companions break their own sticks, which remained barren. But what catches the eye more than everything else is the building designed by Raphael, its grace, its lightness and, indeed its elegance: one must look far ahead, deep into the eighteenth century to find such a graceful building. Many things contribute to it: the cupola; the colouring; the elegant arcs (the same that Michelangelo will use in an inverted position on the sarcophaghi in the Medici Chapel). But if you look more closely, you see that the lightness is especially due to one accomplishment, with which the pupil beat his master, Perugino. Rather than the master's slightly heavy octagon, he constructed (and he was the first to accomplish this) a hexadecagon; the building has 16 sides! And this, as I shall now show, was no mean achievement.

It is well known that a regular polygon whose order is a power of two can be constructed simply by bisecting a number of times successively an angle with a ruler. Starting from this result, the construction of such a polygon in central perspective must be obtained in two separate steps. Recall that on the horizontal line you can always assume Euclidean geometry to be valid, and thus I have indicated for the octagon the relevant lengths, which you can transfer directly onto the frontal line. But while this Euclidean construction is almost trivial for the octagon (Fig. 49.6a, b), for the hexadecagon it is more cumbersome: you must construct the equivalent of the extraction of a square root of a term which contains a square root (Fig. 49.7a, b). And then the perspective construction proper must follow; there are two ways to do it. You choose freely the angle from which you see the square (for example, you may choose the rear edge) and with either method you then draw the diagonals. The first way to construct the perspective is to extend the diagonals to the vanishing points found by Raphael, then draw the other, lower edges, then the lines to the central vanishing point, and finally the second horizontal, which yields the last two edges. Mathematically, this is the

Fig. 49.5 Raphael, *Lo Sposalizio*. Pinacoteca di Brera, Milan. Image: Reproduced by permission, Ministero per i Beni e le Attività Culturali

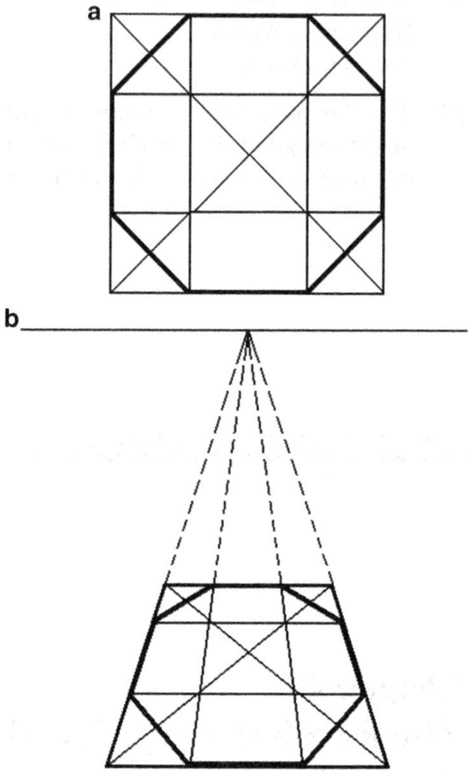

Fig. 49.6 (a, b) Perspective construction of an octagon. Drawing: Kim Williams

more transparent procedure, but you need to know the exact location of the new vanishing points, which, as was seen, are often way out of the painting. This difficulty can be avoided by drawing first the lines from all points marked on the frontal edge to the central vanishing point, and then directly the horizontals that yield the other edges. This method has the advantage that the painter need not work outside the painting. Which way Raphael actually used remains a guess. Indeed, both steps are fairly complicated for this figure, but Raphael worked rigorously: all parallels on any side of the hexadecagon meet at the same vanishing point; this is a grand accomplishment.

Theology in Lo Sposalizio

The investigation of theological points made in works of art was introduced in this century by Aby Warburg and Erwin Panofsky under the broader concept "Iconology" (Panofsky 1962).

How does theology come into this painting? I will begin with what you may call the lowest level of it (Fig. 49.5). It is, I think, recognizable that the heads of the

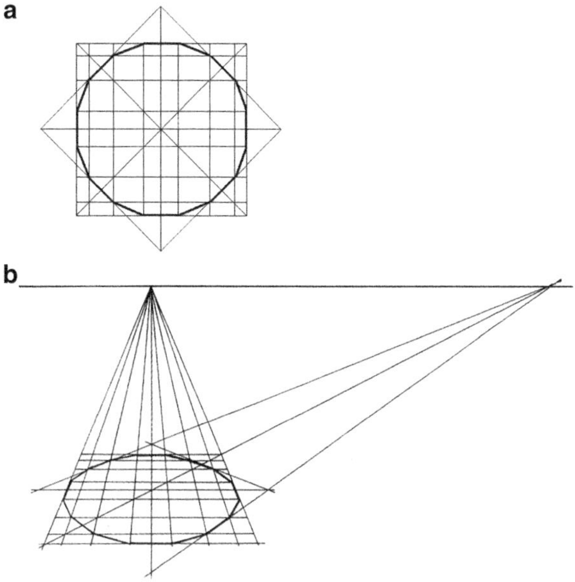

Fig. 49.7 (a, b) Perspective construction of a hexadecagon. Drawing: Kim Williams

Virgin's five bridesmaids form a regular pentagon, and those of the four companions behind Joseph a square. Of course, this may be due to nothing more than Raphael's well known desire to use geometric figures in his paintings for arranging the people in the painting, especially in depth, as well as to his unrivalled virtuosity of drawing in perspective. But he who draws (or, as here, forms) a pentagon, also draws a pentagram. Thus, where on so many churches, usually on the western front, we find a pentagram, even where it is artistically not especially wanted, there must be a special reason for it. As examples I mention Hannover, Strasbourg, Breisach, Basel and the largest cathedral in Italy at Raphael's time, the Duomo di Milano. No doubt, these pentagrams had a function: they served to keep out the devil, the *demonio*. Thus I suspect that the bridesmaids here are performing a pious service to the Virgin: they protect her, just as the four friends form a tower of strength for Joseph. But again, I will not press the point, and I pass right on to the higher level, that is, to the real theology, where we stay on safe ground.

Before doing so, I must say a word about the western tradition of the representation of the temple. There are two aspects to this tradition: one aspect follows the Biblical descriptions of Solomon's Temple, and Ezekiel's vision. The other aspect, which Raphael follows, goes back mainly to the crusaders and is based on what they saw, namely the octagonal mosque by Abd al-Malik, which they imagined to have been built in the tradition of Solomon's Temple: the seat of Wisdom. As such, it became a symbol, not only of the church and of wisdom itself, but also of Mary, the seat of true Wisdom. Still today in Louvain-la-Neuve the academic year is inaugurated, *Au nom de Notre Dame Siège de la Sagesse*. This tradition found expression also in numerous altar chapels, tabernacles etc., which always stand for the Temple, the Wisdom and the Virgin in one, as you can see in the interesting book by P. von Naredi-Rainer (1994).

But while the real Temple in Jerusalem was directed towards the East, since Yahweh enters into his house from the East, Raphael orients the front towards the West, as you can see from the shadows cast by the figures; thus the spectator looks towards the East.

I come now to the second step which Raphael took beyond his master, who also painted a picture with an octagonal temple, but the subject of which is a *Sposalizio* (Fig. 49.8), now in Caen. Both paintings were finished in the same year, and there is no doubt that the master had begun his considerably larger painting much before. Thus priority in choosing the subject belongs to him, as well as the idea of arranging the couple and the high priest frontally before the temple. In both paintings, contrary to Perugino's fresco in the Sistine Chapel, the door of the Temple is open, and there, in the opening, lies the vanishing point, so that you can follow the lines of view in the painting towards it. Is this innovation with respect to Perugino's first fresco due to the master or to the disciple? I do not know, and the question seems to be a difficult one to decide. In any case it was Raphael who saw the artistic possibilities that this seemingly small step permitted (Fig. 49.5). Perugino's painting is arranged in two frontally oriented layers: the group of the people, more or less on one line before the Temple, and the Temple itself, which serves mainly as a historical indicator for the narration.

Raphael's painting shows, as the German historian Hiller von Gaertringen (1999) says, *Tiefensog*, that is, a pull towards the pictorial depth. Besides breaking the one line of people up into geometrically composed groups, he accomplishes this by constructing these many squares, forcing the spectator to follow their edges with his eyes. And these edges meet where? Why, at infinity! And who has his seat at the infinitely distant? Of course: HE, God! Looking towards the vanishing point, you look towards infinity: you look towards God.

It is worthwhile to pause here. We often say that parallels meet at infinity, but this is not so according to Euclidean geometry. The figures can extend to the infinite, but Euclidean geometry makes only asymptotical statements about it, and this holds even more for mechanics, which it underlies and upon which its constructions depend. Bodies can go to and come from the infinite but we compute their behaviour in finite parts of space and only asymptotically with respect to the infinite.

It is different in projective geometry. There parallels do meet at infinity, and projective geometry underlies the perspective design and the corresponding theories of our view. Today we can formulate conceptually the difference between the two geometries; at that time this was not possible, there was only an idea, and it is natural that an artist could grasp it before and better than anyone else. This is why Raphael underlined with all means at his disposal this convergence towards the vanishing point in the open door: everything leads you to the infinite. Projective geometry and perspective serve here as a symbolic construction!

This is Raphael's first theological statement, and it is expressed through the mathematics of perspective. I doubt that any theologian could have expressed this

Fig. 49.8 Perugino, *Lo Sposalizio*. Image: Reproduced by permission, Musée des Beaux-Arts de Caen. photo: M. Seyve

idea; what was needed here was a mathematician, since mathematics is the very science of the infinite! But this statement was in line with the Florentine Platonism of the time, one of the roots of which was the "Docta Ignorantia" of the German philosopher and Cardinal, Nicolaus Cusanus.

That I am not imagining these things is attested by Raphael's second theological statement, and this time it is the architect who makes it by designing this magnificent edifice. Namely, you look towards God, if you look towards him through the church.

Altarpieces of this time are loaded with theological implications, but most often clad only in traditional, historic and sometimes accidental symbols. A construction like the one we find here, where architecture, mathematics and theology are so closely knit and intertwined, is surely extremely rare, if indeed not unique. For producing it, an architect, mathematician and theologian in one person was needed.

Acknowledgements This chapter was first presented as a lecture at the Symposium in Honour of Edoardo Benvenuto, organized by Profs. Massimo Corradi, Orietta Pedemonte and Patricia Radelet-de Grave.

In the 1950s, as a student I could discuss regularly the secrets of the use of perspective by the artists with my friend L. Burckhardt, and later, in the 1960s, I had the chance to have conversations on these questions with Erwin Panofsky. I could discuss these questions with Dr. Mario Howald, Dr. and Mrs. Th. Beck helped me with the literature and Mrs. M. Messmer pulled me out of laptop difficulties. I am greatly indebted to Kim Williams for translating my English and for drawing the figures! Last but not least I am indebted to my wife for linguistic advice and for proof reading.

Biography David Speiser is Professor Emeritus at the Catholic University of Louvain, where he taught physics and mathematics from 1963 to 1990. His research concerned elementary particles and physical mathematics. From 1990 to 2004, he gave lectures and seminars regularly at the Scuola Normale di Pisa. He was the general editor of the complete works of the mathematicians and physicists of the Bernoulli family from 1980 to 2004. His essays on the relationships between the history of art and the history of science have been collected in *Crossroads: History of Science, History of Art. Essays by David Speiser, Vol. II,* Kim Williams, ed. (Basel: Birkhäuser, 2011).

References

CANUTI. Fiorenzo. 1931 *Il Perugino.* 2 vols. Reprinted 1983. Siena: Editrice d'arte.

HILLER VON GAERTRINGEN. R. 1999. Raphaels Lernerfahrungen in der Werkstatt Peruginos. Munich: Deutscher Kunstverlag.

JONES, R. and N. PENNY. 1983. *Raphael* (German edition). München: Verlag C.H. Beck.

NAREDI-RAINER, P. von. 1994. *SalomonsTempel und das Abendland.* Köln: Dumont Verlag.

PANOFSKY, Erwin. 1962. Studies in Iconology. New York and Evanston: Harper & Row.

Chapter 50
Raphael and the Pantheon's Interior: A Pivotal Moment in Architectural Representation

Kristina Luce

In the first decade of the sixteenth century, several artists from Bramante's circle created a series of six related drawings of the Pantheon's interior.[1] Although each drawing exhibits unique traits, the set resulted from the copying of a single master or model drawing. I believe that the drawing catalogued as *Uffizi 164 A.r.*, attributed to Raphael, was that master image (Fig. 50.1).[2]

Although the debate over the primacy of *Uffizi 164 A.r.* may never be fully put to rest, I would like to leave aside these issues in favour of a return to the question originally asked by Hermann Egger. Why did this model drawing become so famous and so frequently copied? (Lotz 1977: 25).

First published as: Kristina Luce, "Raphael and the Pantheon's Interior: A Pivotal Moment in Architectural Representation", pp. 49–62 in *Nexus VII: Architecture and Mathematics*, Kim Williams, ed. Turin: Kim Williams Books, 2008.

[1] Three of these are now housed in the Uffizi, (U 1950 A r, U 4333 A r, and U 164 A r); one lies at the Universitätsbibliothek in Salzburg (Salzburg H 193/2 r), and another is folio 30 r of the Codex Escurialensis housed at the Biblioteca, El Escorial (Cod. Inv. 28.II.2). The sixth drawing, folio 33 r from the *Mellon Codex* is held at the Pierpont Morgan Library in New York (1978.44). This last drawing is clearly related to the others, sharing the same general point of view and compositional strategy. However, the *Mellon Codex* drawing is executed at a much smaller scale and was subsequently used to record what appears to be field measurements of the Pantheon, a particularly interesting use considering the drawing's deviation from that building's architecture.

[2] Subsequent discourse has offered alternative theories allowing the possibility of a lost model drawing or the suggestion that the version within the *Codex Escurialensis* was the master. However, none of these alternatives fully synthesize the various discoveries about the set. Although the intricacy involved in resolving the work of scholars such as Hermann Egger, Wolfgang Lotz, Gustina Scaglia and John Shearman is beyond the scope of this chapter, my work with the drawings, in tandem with the rich scholarship of these other authors, has made it possible for me to conclude that *Uffizi 164 A.r.* was the most likely master drawing. Their arguments and my attempt at resolution, along with my own observations are provided as Appendix to this chapter.

K. Luce (✉)
Department of Art, Western Washington University, Bellingham, WA 98225-9068, USA
e-mail: Kristina.Luce@wwu.edu

K. Williams and M.J. Ostwald (eds.), *Architecture and Mathematics from Antiquity to the Future*, DOI 10.1007/978-3-319-00143-2_3,

Fig. 50.1 Interior view of the Pantheon, attributed to Raphael. *Uffizi 164 A.r.*, Florence, Italy

This question is difficult to answer since *Uffizi 164 A.r.* defies our expectations of what constitutes a master drawing on a number of levels. The image presents an edited fiction of the Pantheon's architectural composition using an extremely idiosyncratic graphic structure. Generally, the drawing resists taxonomic categorization, and yet, in our recognition of the drawing's importance we have often elided these difficulties. For example, while the drawing's angle of view alone is enough to subvert the idea that *Uffizi 164 A.r.* is a perspective in any Albertian sense, this simple fact has not prevented the drawing from being "considered a masterpiece of applied perspective" (Lotz 1977: 25). Wolfgang Lotz writes that it may have been used "as an example in the teaching of perspective drawing," even as he writes just three paragraphs later that the drawing stands "in utter contradiction to Alberti's definition of a perspective view of an interior" (Lotz 1977: 26). Clearly *Uffizi 164 A.r.* is engaged in the creation of illusions, and yet just as clearly its perspectival structure, if indeed the drawing's structure is perspective-based, is highly irregular.

Similar arguments might be made about the drawing's portrayal of the vast space of the Pantheon's interior. By presenting the Pantheon from niche to vestibule, it seems logical that the drawing would capture a sense of the Pantheon's spatial feel. Further, in light of the building's compositional symmetry, this particularly wide-angle view allows the entire building structure to be inferred. All the information needed to understand the totality of the Pantheon seems to be presented. However, none of these proposals turns out to be accurate.

Rather than capturing the Pantheon's grand and centralized space, the architecture appears flattened in Raphael's image. The shallow sweeping curve at the base of the wall is more suggestive of a wide ellipse rather than the circular plan of the Pantheon's ideal architecture.

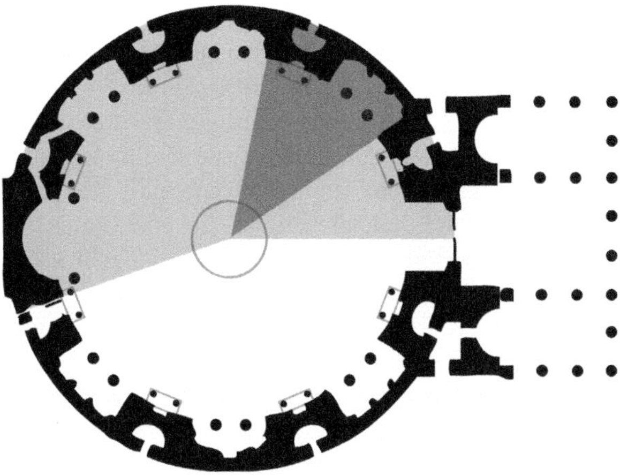

Fig. 50.2 Plan view of the Pantheon showing the extents of the building depicted by *Uffizi 164 A.r.* (*light grey*) and the area of the building that the drawing omits (*dark grey*). Image: courtesy George Weinberg, The Getty Research Institute

The dome above appears similarly compacted onto the drawing's surface. *Uffizi 164 A.r.* seems to offer little fidelity to the spatial experience of the Pantheon. If on the other hand, the drawing was meant to capture the total form of the Pantheon, the image still presents us with problems. Although the drawing has a naturalizing tendency, presenting itself as a faithful transcription of the Pantheon, as John Shearman has carefully demonstrated the image contains some rather glaring solecisms (Shearman 1977).[3] In its approximate 200° sweep, the drawing captures the Pantheon from the altar-niche on the left to the vestibule on the right, and between these are depicted two recesses and three aediculae. However, in the actual Pantheon, as Fig. 50.2 shows, there are three recesses and consequently, four aediculae.[4]

The arrangement of these in the building constitutes a carefully woven pattern of hierarchically received axes. While the omissions of Raphael's drawing might still reflect the idea of this rhythm, they destroy the actual composition's careful structure. Considering the attentive study of antique architecture underway during this stage of the Renaissance, the perturbation of the relationships the Pantheon exemplified seems curious.

In other words, *Uffizi 164 A. r.* presents us with an image that was clearly significant, having been the subject of study and replication by some of the most important architects of the Renaissance, and yet, as modern viewers we have very little ability to understand what it was our Renaissance counterparts saw as remarkable. Further, Raphael's association with the image becomes particularly

[3] However, Shearman is not the only scholar to have seen these documentary inaccuracies. Lotz mentions them as well (1977: 26).

[4] Lotz explains that these errors were a result of the author's desire to capture the opposing vestibule and niche, a goal that was impossible in terms of perspective given the "point of view" for the drawing. Certainly, I agree that this approximate 200° sweep was a motivating factor, but I disagree with Lotz's assumption that a single graphic structure, and therefore singular point of view, is reigning over the image.

important in light of his letter to Pope Leo X, in which he advocates for a certain methodology in the documentation of architecture: the combined use of plan, section and elevation. It is possible that if we can recapture an explanation of why the drawing served as such a significant model, this image may tell us something about the development of that triadic system of architectural representation, its eventual establishment as architectural convention and the hurdles that initially prevented its acceptance. Such an explanation is the goal of this brief chapter.

Thus far we have established many things that *Uffizi 164 A.r.* is not. It is not a traditional perspective. It is also not a reflection of the spatial feel of the Pantheon, nor is it an accurate portrayal of the Pantheon's architectural composition. Let us now turn to what the drawing is. It is of interest that in spite of the ways in which the drawing defies our expectations, it is still highly illusionistic. The space presented by the image is coherent, even if that space is a fiction. Lotz resolved this effect and his own contradictory impressions by concluding that the "Uffizi drawing consequently represents the sum total of many glances" (Lotz 1977: 26).

Interestingly, this description mirrors that of a drawing from two to three decades earlier. Although Juan Guas's presentation drawing for San Juan de los Reyes (Fig. 50.3), clearly captures a different architectural tradition, it engages in very similar imaging practices as that of *Uffizi 164 A.r.* (Sanabria 1992: 163).

It is further the case with both drawings that our own modern viewing habits mask many of the complexities of their compositions. The tendency is to read the perspectival qualities of the drawings as evidence of a resolved form of picturing space, but something much more transitional is taking place in both images. While the Prado drawing may resemble a perspective, or even a cavalier perspective, on closer examination the structure decomposes into an assemblage of elevations that have been perceptually pleated into place, surfaces unfolding whenever possible. "The space becomes like a folding polyptych, opened partially to reveal all sides" (Sanabria 1992: 168). The vaults appear to have been tipped backwards to reveal more of their surface. The rear transept walls likewise angle backwards in the space, allowing the front wall of the transept to be seen. The same is true of the clerestory windows, where the front-most jambs and their sculpture are visible. Sergio Sanabria, described this drawing as having been treated as a "fish-eye photograph.... The total space does not read as a unit; rather, there is a succession and articulation of parts, connected by the viewer, who processes nearsightedly through them" (Sanabria 1992: 168).

The Uffizi image functions similarly, but unlike the Gothic image where nearly unabridged detail is offered at the expense of spatial coherency, the space depicted in the Pantheon drawing functions as a complete unit. The drawing offers a strong and seemingly consistent spatial depiction that, without immediate recourse to other images or to direct experiences of the Pantheon, appears to be complete and highly illusionistic. *Uffizi 164 A.r.* creates this spatial fiction by marshalling together elements of multiple projective structures (perspective, cartography and orthography) in its attempt to resolve and portray the Pantheon. The resulting naturalistic impression of space makes it clear that perspective is playing a role in the composition of the Uffizi drawing, but further explanation is necessary to

Fig. 50.3 The presentation drawing for the altar piece of San Juan de los Reyes in Toledo, attributed to Juan Guas, c.1479–1480 (Prado D/5526). Image: ©Museo Nacional del Prado, Madrid, Spain, reproduced by permission

demonstrate how the drawing also exhibits affinities with orthographic and cartographic systems of projection.

The distinction between orthographic and perspective projections was voiced in Alberti's Book Two of *The Art of Building in Ten Books*, when he wrote that unlike the painter who engages the relief of objects, the architect "takes his projections from the ground plan and, without altering the lines and by maintaining the true angles, reveals the extent and shape of each elevation and side" (Alberti 1988: 34). In addition to describing the mechanics of creating an architectural drawing, the quality that Alberti is emphasizing here is that of preserved shape, or, to put a finer point on it, Alberti stresses the importance of formal commensurability between the drawing and the building. In such a system, objects that are similar will appear to be similar in the drawing because their true "shape and extent will be preserved." Perspective allows for no such preservation since identical objects depicted in perspective can have vastly different shapes and sizes depending on their relationship to the viewer/picture plane.[5] And, while it is true that the orthographic procedure Alberti outlines usually maintains the extent and shape of objects, this is only the case when the object's geometry is in harmony with geometry of orthographic projection. Such a harmony requires that the object reinforce the rectilinear projectors and 90° angles of orthographic projection with its own parallel lines and 90° angles. When planes occur at oblique angles, or worse, when they are round like the Pantheon, Alberti's imperative for commensurability within architectural drawings becomes impossible.

Uffizi 164 A.r., however, stands between the painterly and architectural models for drawing that Alberti described. It attempts to depict both the relief and the extents and shapes of objects. As resulted from the Prado image's spatial manipulations, by flattening the round and centralized space of the Pantheon, similar objects could be portrayed at nearly the same scale. In Raphael's drawing, the identical columns of the recesses are similarly sized, as they would be in an elevation. Likewise, the depicted sizes of aediculae are nearly identical even though the perspective should dictate that the outer two be larger. The artifice of the image and the alterations to the Pantheon's actual architecture are working to blend the two systems together. Perceptually, it is as if the Pantheon had been unrolled before it was depicted in perspective, or alternatively as if it was actually a drawing of an elevation partially bent into semi-cylindrical form.[6] The result is, as Lotz wrote, to make the drawing seem to occupy a place "halfway between the perspective image of the interior and the orthogonal projection of the inner wall" (Lotz 1977).

However, Lotz's explanation doesn't fully articulate what is happening within *Uffizi 164 A.r.* In addition to occupying the ground between perspective and section, Raphael's technique of unrolling or bending the Pantheon's architecture shares

[5] While we may perceive the shape and extents of these objects as identical, their presentation on the actual picture plane is not.

[6] This description is a reference to James Ackerman's analysis and description of Villard de Honnecourt's drawings of the choir at Reims cathedral (2002: 34).

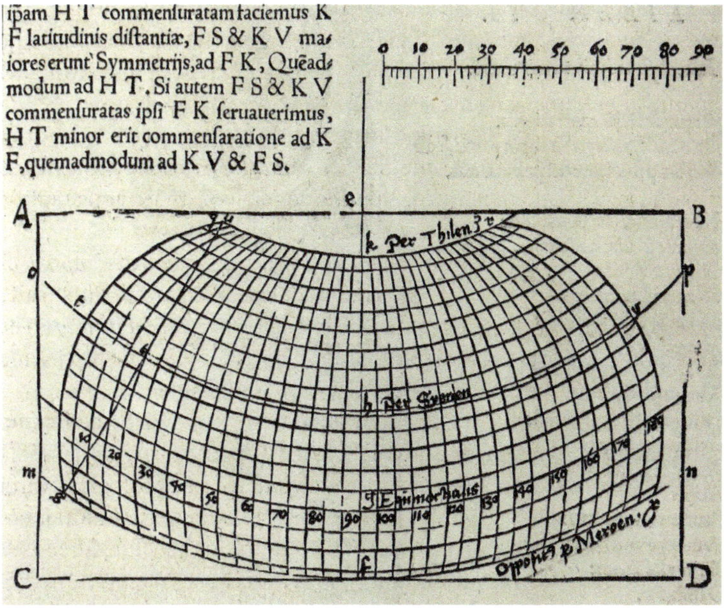

Fig. 50.4 Framework for the Ptolemaic projection of the globe, from *Geographicae enarrationis libri octo* (Ptolemy 1541: Book I, Chap. 24, p. 23)

similarities with another form of picturing during the Renaissance. The cosmographers, cartographers and choreographers of the late fifteenth and early sixteenth centuries also attempted to resolve curved forms into flat representations.[7] Just as the Pantheon problematized the projection of a sphere onto paper, so mathematical geography required the globe to be similarly flattened onto a surface. Raphael's strikingly flat drawing of the Pantheon seems to be informed by these contemplations of the globe. Although specific parallels might be drawn to several of these early global pictures, Ptolemy's generic framework projecting the world onto paper is enough to demonstrate the links between the cartographic images and Raphael's depiction of the Pantheon (Fig. 50.4).

Perez-Gomez describes the form of Ptolemy's global projection:

> Ptolemy's map itself is not a circle as would be formed by a section through the globe, nor an ellipse as argued later by Edgerton, but an elongated and curved stretch of land—the *oikumēnē*—whose center of curvature lies at the north pole (Pérez Gómez and Pelletier 1997: 95).

---

[7] "Ptolemy's *Geographia* was not included in the Ptolemaic *opera* introduced into the West in the twelfth century. It was only rediscovered in the West c.1406, when it was translated into Latin by Jacobus Angelus in Florence. In addition to numerous manuscript copies, it appeared in six printed editions in the fifteenth century: Bologna 1462 (1482?); Vicenza, 1475; Rome 1478; Ulm, 1482; Ulm, 1486; and Rome, 1490. It appeared in numerous editions in the sixteenth century in both folio and quarto; twenty in Latin, six in Italian and two in Greek" (Cormack 1991: note 17).

This account, and particularly that portion that describes the image as an elongated and curved stretch of land, could apply equally well to Raphael's depiction of the Pantheon. Both drawings demonstrate considerable flattening of their curvatures on the macro-level in order to portray more accurately the relative sizes and shapes of the objects within their projective frameworks. One need only imagine the interior of the Pantheon as the interior of a globe or as the celestial sphere, not a large ontological leap given the sensibilities of the sixteenth century, and even the curvature of the world map would then correspond to Raphael's image of the Pantheon.[8] Raphael's drawing, it would seem, actually stands between orthography, perspective and cartography.

Unfortunately, using Ptolemy's projection as a spatial framework, like the spatial schema used by Guas, still required compromise and undermined the Albertian goals for architectural drawing. By combining elements of orthography, perspective and cartography artists could attempt to go beyond what each system could capture on its own. However, this representational creativity undermined the geometric accuracy of each system of projection. Though these drawings struck a compromise in that they preserved a sense of commensurability, no actual measurements could be taken from them.

Other, slightly later, drawings partially resolved this problem. An image taken from the manuscript by the Giacomo Andrea da Ferrara (ca. 1490; Sgarbi 2004), one of the earliest illustrated versions of Vitruvius, today conserved in the Biblioteca Ariostea in Ferrara, is a drawing that pulls the tensions we have seen in *Uffizi 164 A.r.* into projective clarity (Fig. 50.5).

In this drawing we find the Pantheon with its interior surface fully unrolled; its dome broken into recognizable cartographically-influenced interrupted surface of four lobes, or gores. Waldseemüller is known to have used interrupted surfaces for cartographic images as early as 1507. While *Uffizi 164 A.r.* predates such images, these drawings do indicate that a graphic discourse was taking place, that a form of picturing and projection was being sought that could cope with centralized forms like the Pantheon while maintaining the aims Alberti laid out for architectural drawing.

Clearly, Raphael's drawing gains some of its significance because of the importance of this debate, but what is really at stake here? Alberti's call for formal commensurability is decades old, and yet for Raphael the Pantheon still presented a problem for architectural representation. The image of the Pantheon in the *Vitruvio ferrarese* suggestively gestures towards an increased acceptance of orthography, but neither image follows Alberti's description of how the architect draws. Even though measurements could theoretically be taken from the *Vitruvio ferrarese*, the image is not related to plan or section images of the Pantheon in a

[8] Although it changes the status of what we assume was Villard's knowledge of geometry, it is difficult not to see that the projection in *Geographia* also provides an explanation for Villard's visually ambiguous spatial contrivance for the Reims chapel drawings. Those structural features that cause its curves to create an equivocating spatial illusion, first projecting inward and then outward from the drawing's top to bottom, seem consistent with Ptolemaic projection.

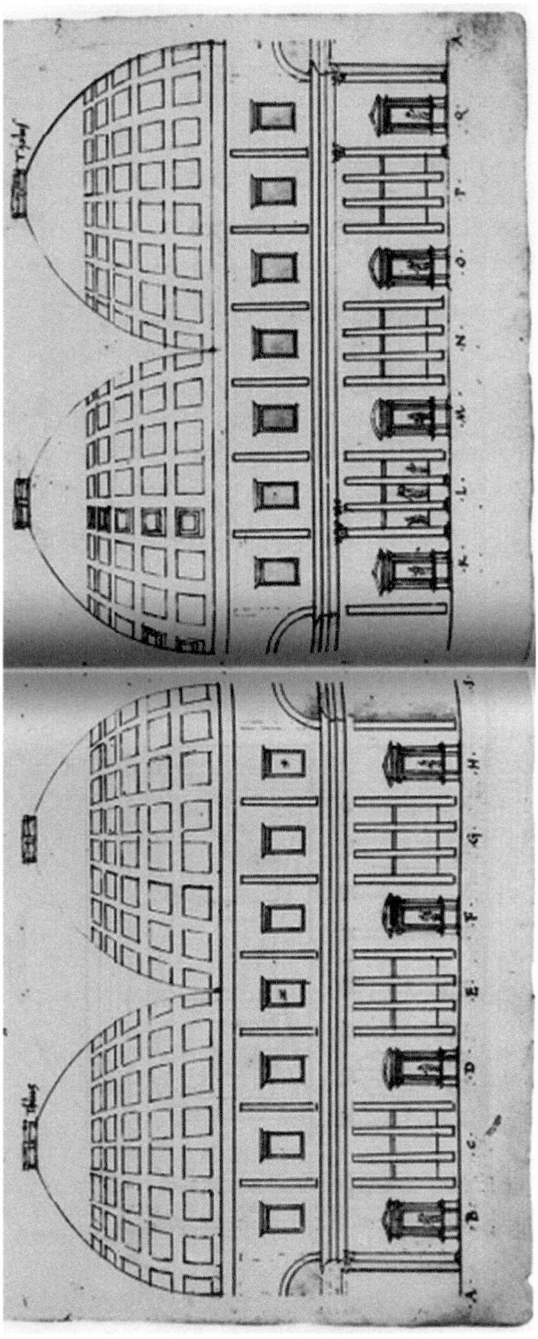

Fig. 50.5 Interior image of the Pantheon by Giacomo Andrea da Ferrara (attributed). Image: Vitruvius (ca. 1490), ms. cart., (1490–1518), Classe II 176, cc. 65v-66r. Reproduced by permission, Biblioteca Ariostea, Ferrara

linear manner. The projective links between the drawings were broken when the wall was unrolled.

In other words, one thing that was at stake was the representation of the particular type of building embodied by the Pantheon. Unlike the built form of San Juan de los Reyes, the structure of the Pantheon is not an assemblage. It is an interlocked totality of sphere, cylinder, axes and cross-axes.[9] The round, centralized form provides no opportunity to easily dis/re-aggregate building parts and drawn elevations as did the additive structure of a Gothic building. Unlike those Renaissance buildings that mirrored Gothic composition with a series of linearly arranged repeating bays (buildings like Saint Peter's or San Lorenzo), the Pantheon resisted fracture. Where those other buildings could be drawn in a manner that systematically portrayed some elements in one projective system and some in another, the importance of the Pantheon was that it challenged the apparent transparency of these hybrid techniques. Further, the building challenged Alberti's definition of the architectural drawing as an image that maintains commensurability by being projected from the plan. In the case of the Pantheon, such a projection would not preserve extent or shape. As a centralized space, the curves of the Pantheon distorted the true shape and width of every element along the wall when depicted in elevation. Nothing depicted in such a drawing would be commensurable with the building. *Uffizi 164 A.r.* illuminates these tensions. It demonstrates how the geometry of certain representational priorities may be at odds with certain buildings. By 1519, Raphael would write of this problem in his letter to Pope Leo X, when he identified domes and other inherently oblique geometries as those special cases where the ground plan, elevation, and section were ambiguous in themselves. In these cases, all three drawings were necessary, and only through a comparison of all three could a correct understanding occur.[10]

When Raphael penned this statement, he was not just talking about the craft of imaging. By allowing that no one representation could capture a building, he effectively advocated for a relocation of the realm in which architectural images could be verified. Where Brunelleschi's experiments functioned to link the image rhetorically to a reality against which that image could be measured and validated, Raphael acknowledged that architectural images should not be corroborated to vision, but instead to the mental constructs they created. He extended the primacy of the architectural quality of shape and allied it with ideal geometries and constructs. Where Alberti admonished the architect who tried to incorporate relief, Raphael's system dictated that such drawings should avoid perspective not

[9] The spatial composition of the Pantheon is sometimes referred to as an "ideal dome." This arrangement perfectly nests a complete sphere into a cylinder whose height matches its radius, allowing the base of the sphere to be exactly tangent to the base of the cylinder.

[10] There is some dispute as to whether this description of perspective, which is only found in the Munich copy of the letter, was actually authored by Raphael or was a later addendum by another author. I tend to think that the Pantheon drawing, which seems to problematize this very issue, makes a strong case for this thought being Raphael's even if it only made it into one copy of the letter. See Lotz (1977: 29 and n. 77).

just because measurements were necessary, but because perspective belonged outside the syntax of architecture. Through the projective grammar of plan, section and elevation, architectural representation became an internally coherent and verifiable system of architectural conception. Drawings that mimicked vision were insufficient for the expression of architecture, and multiplying and assembling views or glances could not solve the problem. Through the interplay of plan, section and elevation architectural images broke free of their links to the realities of vision and instead inhered to the abstract conception of a total building.

This shift in architectural conception is more than just a representational trope. It is indicative of the transformed manner in which buildings were understood during the Renaissance. Instead of the additive structure of a Gothic building conceived, built and altered through time, the Renaissance conceived the building as a total and coherent object. Fields outside architecture similarly shared this new grasp of the subject as a totality and the desire to represent it as such. When analysing the late fifteenth-century image depicting Florence, *Map with a Chain*, Samuel Edgerton described a similar tension between unity and assembly. The image, wrote Edgerton:

> makes it possible to grasp instantly the overall plan of Florence and its relationship to the surrounding countryside, but forces the viewer to lose tactile contact with the individual details that so delight all the senses when he walks through the city. The unity of the Renaissance view has replaced the diversity of the mediaeval one (Edgerton 1974: 277).

Again, a similar mode of thought and expression is presented by both architectural and cartographic fields. For architecture, the centralized form and ideal dome of the Pantheon was the ultimate example of this new conception of the building as a totality, and the challenge of picturing the Pantheon offered an especially timely problem: namely, if one begins to understand and conceive of the building as a totality, how should it be represented? In 1519, Raphael suggested that it should be represented within a system; not as a single image, but as a dialogue between three images whose interplay created a larger unified concept of the building. In essence, he suggests that if buildings are to be understood as totalities, then they should be represented as abstractions.

In some ways, then, hidden in the quiet of Raphael's drawing of the Pantheon is a very large conflict. The silence of *Uffizi 164 A.r.* comes from its illusionism, from the convincing way it mimics perspective, and therefore, speaks of the Pantheon with all of perspective's authority. Amusingly, however, the vision it presents is of reality reflected in a fun-house mirror rather than Brunelleschi's. The drawing is a kind of trompe l'oeil. Delightfully, even when we know the image presented is false, we are still convinced. Its deception throws into relief the opposition between visual and conceptual frames of knowledge. More than just representational play, the issue is one of abstraction, and in particular how a conceptual totality challenges images that acquire their authority through recourse to vision. The perspective, as explained by Damisch, was meant to be reality's mirror, but in the case of the Pantheon the building's conceptual totality competes with what the mirror can show (Damisch 1994). The impulse documented by *Uffizi 164 A.r.*, and its

200° + sweep, is to reach towards a total image of the Pantheon, but this mental schema resists graphic representation.

These are the conflicts Raphael captures with *Uffizi 164 A.r.* The image became so famous and so copied because it mounted so many questions about vision, formal totalities and representation. Together these questions mount one more: what is it that defines architecture? Is architecture of the world to be perceived and ordered by vision, or is it conceptual and abstract? Such matters had a particular valence during the Renaissance because the very discipline and definition of architecture was being reworked. *Uffizi 164 A.r.* holds us paused in that moment right before the decisions get made. The drawing captures neither the ideal nor the real, but is caught between the two. It mounts a mimetic masquerade which, once uncovered, highlights architecture's ineffability and the gap between vision and conception without picturing either.

Biography Kristina Luce is Assistant Professor of Art History at Western Washington University. She received her M. Arch in 1996. After working in preservation and historic rehabilitation for several years in Cincinnati, Ohio, Luce returned to academia to study drawing's role within the process of design. Her work defines drawing as a conceptual medium for architectural design, and goes on to explore how this medium encodes a certain definition of architecture and circumscribes the architects understanding of the design problem. Her dissertation is entitled "Revolutions in Parallel: The Rise and Fall of Drawing within Architectural Design." In 2006–2007 she was a fellow at Michigan's Institute for the Humanities, and in 2007–2008 she was a Pre-doctoral fellow at the Getty Research Institute. She was awarded a Ph.D. in the History and Theory of Architecture at the University of Michigan, Taubman College of Architecture and Urban Planning in 2010.

Appendix: Was *Uffizi 164 A.r.* the Primary Drawing? A Summary of the Arguments and Another Suggested Resolution

The question of dates and, by extension, the establishment of a model for this series of drawings was raised in 1956 by Wolfgang Lotz in his article "Das Raumbild in der Architekturzeichnung der italienischen Renaissance" (1956). Lotz proposed Raphael as the designer of the drawing, but a conflict exists with the date of the arrival of the *Codex Escurialensis* in Spain, which makes it nearly impossible for Raphael to have constructed the model given our current understanding of his arrival date in Rome.

John Shearman (1977) offers one resolution to this conflict, hypothesizing that *Uffizi 164 A.r.*, as it stands now, is a second, extended state of Raphael's original drawing, which was not in error. By identifying differences in the quality of ink and line, Shearman argues that the right most tabernacle and vestibule portion of the drawing, among other features, are later additions that sought to transform Raphael's working drawing into something that resembled a *veduta*. The reduced angle of view of Shearman's proposed original state for *Uffizi 164 A.r.* eliminates many of the perspectival irregularities that are apparent in the final drawing. Shearman believes that the artist who was later responsible for extending Raphael's drawing had not seen the Pantheon, but had in his possession another view that captured the right-most two recesses and the vestibule and that overlapped with Raphael's drawing. When fusing the two drawings together, the artist assumed that the drawings presented the same two recesses, rather than there being only one recess in common. This assumption resulted in an interior view of the Pantheon with only two recesses between the altar-niche and vestibule. Shearman's theory also explains the incorrect rhythm of the tabernacle pediments depicted in the final drawing. If the artist did indeed work with two overlapping drawings as Shearman thinks, the belief that only two recesses existed would consequently eliminate one of the segmental pediments, thus producing the incorrect alternating rhythm that *Uffizi 164 A.r.* demonstrates.

I find Shearman's theory intriguing, particularly because, through logical extension, it establishes that *Uffizi 164 A.r.* was the model copied by the other drawings, since those demonstrate only one state, not the two that Shearman sees. However, Gustina Scaglia (1995) argues that the opposite is the case. She believes that the Escurialensis drawing served as model to a now lost drawing, possibly by the artist of the Chinnery Album, which was subsequently copied by the others. Her argument is based on the use of abbreviated fluting seen in *Uffizi 164 A.r.* and all other copies. She sees these abbreviations as derivative of the complete fluting depicted in the Escurialensis version. However, Shearman points out that Raphael's abbreviated fluting also indicated the cabling at the bottom of the columns, an observation more accurate than the consistent fluting shown by the Escurialensis artist. Additionally, because the Escurialensis drawing maintains the segmental, triangular rhythm of the tabernacle pediments, I believe it must be a copy of *Uffizi 164 A.r.*

When considering these arguments together, it becomes significant that Shearman fails to acknowledge the copy of *Uffizi 164 A.r.* found in Salzberg. Scaglia quite convincingly argues that this drawing was also authored by Raphael, and the attribution complicates Shearman's theory. Even if a later artist altered Raphael's original version of the Pantheon interior, Raphael saw fit to make a copy of these alterations. There must have been something compelling about the new construction that made it worth recording, even in light of its documentary errors. If Raphael did not author *Uffizi 164 A.r.* in its entirety, his replication of it in the Salzberg drawing certainly legitimates his engagement with the unique features of the altered composition.

Further, after examining the actual drawings in the Uffizi, and high quality facsimiles of the drawings in Austria and Spain, I tend to believe that *Uffizi 164 A.r.* was the model for the other drawings. If the entire composition is not original to Raphael, I believe that Raphael drew his version in tandem with visits to the Pantheon. His drawing alone seems to engage in a process of sketching and correction as he matches the image to his conceptions. Other drawings seem to replicate his pentimenti and even attempt to resolve the ambiguities. The left-most aedicule is one area where these features are apparent. The other versions, including Raphael's copy of his own work in Salzberg, appear as simplifications of *Uffizi 164 A.r.*, and given this observation, I believe logic dictates that *Uffizi 164 A.r.* should be considered the model. As Shearman suggests, it may be our understanding of Raphael's travels that need some slight alteration, perhaps allowing for a visit to Rome on his journey to Florence.

References

ACKERMAN, James S. 2002. *Origins, Imitation, Conventions: Representation in the Visual Arts.* Cambridge, Massachusetts: MIT Press.

ALBERTI, Leon Battista. 1988. *On the Art of Building in Ten Books.* Joseph Rykwert, Neil Leach, and Robert Tavernor, trans. Cambridge, Massachusetts: MIT Press.

CORMACK, Lesley B. 1991. Good Fences Make Good Neighbours: Geography as Self-Definition in Early Modern England. *Isis* **82**, 4: 639–661.

DAMISCH, Hubert. 1994. *The Origin of Perspective.* Cambridge, Massachusetts: MIT Press.

GIACOMO ANDREA DA FERRARA. 1490 ca. *De architettura.* Biblioteca Ariostea, Ferrara.

EDGERTON, Samuel Y. 1974. Florentine Interest in Ptolemaic Cartography as Background for Renaissance Painting, Architecture and the Discovery of America. *The Journal for the Society of Architectural Historians* **33**, 4: 275–292.

LOTZ, Wolfgang. 1956. Das Raumbild in der Architekturzeichnung der italienischen Renaissance. *Mitteilungen des Kunsthistorischen Institutes in Florenz* **7**: 193–226.

___. 1977. *Studies in Italian Renaissance Architecture.* Cambridge, Massachusetts: MIT Press.

PÉREZ GÓMEZ, Alberto and Louise PELLETIER. 1997. *Architectural Representation and the Perspective Hinge* Cambridge, Massachusetts: MIT Press.

PTOLEMY. 1541. *Claudii Ptolemaei Geographicae enarrationis libri octo.* Vienna and Lyons: Gaspar Trechsel for Hugues de La Porte.

SANABRIA, Sergio L. 1992. A Late Gothic Drawing of San Juan De Los Reyes in Toledo at the Prado Museum in Madrid. *Journal of the Society of Architectural Historians* **51**, 2: 161–173.

SCAGLIA, Gustina. 1995. 11 Facsimile Drawings of the Pantheon's Vestibule and the Interior in Relation to the Codex-Escurialensis and Giuliano da Sangallo's Libro Drawings. *Architectura-Zeitschrift Fur Geschichte Der Baukunst* **25**, 1: 9–28.

SGARBI, Claudio, ed. 2004. *Vitruvio Ferrarese. De Architettura. La Prima Versione Illustrata.* Modena: Franco Cosimo Panini.

SHEARMAN, John. 1977. Raphael, Rome, and the Codex Escurialensis. *Master Drawings* **15**, 2: 107–146.

Chapter 51
Design and Perspective Construction: Why Is the *Chalice* the Shape It Is?

Richard Talbot

Introduction

The basis for this chapter is a study and examination of one of the most complex and well-known examples of early Renaissance perspective drawings, the *Chalice*. This drawing has become almost iconic within the history of perspective, although neither the author nor the exact date of its execution is certain. It is most likely to have been executed prior to 1460 and has been attributed variously to both Paolo Uccello and to Piero Della Francesca, based largely on references in Vasari's *Lives of the Artists*.[1] It is a drawing that I have long been fascinated and intrigued by because of its complexity, clarity and technical accomplishment.

I first attempted a reconstruction of the drawing over 20 years ago, primarily out of curiosity about the method of its perspective projection. I made a relatively simple version of the drawing, using only 16 facets (see Fig. 51.7 below), rather than the 32 in the original drawing. I quickly became aware, however, of the technical difficulties in making a similar drawing with such apparent accuracy and consistency at this scale the drawing is only 34 cm $\times$ 24 cm. I also realised that although the method behind the perspective projection is, in fact, very straightforward and is something that a good student could grasp in a few hours, many questions about the drawing remained.

First published as: Richard Talbot, "Design and perspective construction: Why is the Chalice the shape it is?", pp. 121–134 in *Nexus VI: Architecture and Mathematics,* Sylvie Duvernoy and Orietta Pedemonte, eds. Turin: Kim Williams Books, 2006.

[1] Vasari mentions *mazzocchi* and a faceted stellated sphere as being forms that Uccello attempted; he also mentions a faceted vase specifically in relation to Piero (Vasari 1996). Piero in *De Prospectiva Pingendi* constructs a *mazzocchio* (Fasola 1942: Tav. XXVII, Fig. LI).

R. Talbot (✉)
Fine Art, The Quadrangle, Newcastle University, Newcastle Upon Tyne NE1 7RU, UK
e-mail: richard.talbot@ncl.ac.uk

K. Williams and M.J. Ostwald (eds.), *Architecture and Mathematics from Antiquity to the Future*, DOI 10.1007/978-3-319-00143-2_4,

These questions partly concerned the practical aspects of making the drawing but primarily I felt that if someone had painstakingly constructed a drawing as complex as this then presumably the design of the *Chalice* was more than just something to hang a particularly clever and impressive technique on. The main question that arose, and which is the subject of this chapter, is 'why is the *Chalice* the shape it is?' The additional question of its attribution was not one that I had intended to engage during the course of this study, but I believe that my research has revealed aspects of the drawing that may have a bearing on it.[2]

The Subject Matter of the Drawing

The *Chalice* is a drawing of a completely self-contained object and does not appear to be a preparatory study for a painting or for an object that was intended to be set within a painting or some other kind of representation. It seems to fit into a category of drawings that could be considered autonomous, speculative or investigative. It is a 'wire frame' drawing a representation of an idealised three-dimensional form and as such does not imply any particular material or scale. However, the basic shape and specific forms and structures in the *Chalice* do reflect those of contemporary chalices and other ecclesiastic and secular objects; chalices often had hexagonal and octagonal elements in the base, a stem with a faceted widening and cut precious stones mounted around the surfaces. These all appear to be repeated, in some way, within the drawing, including the mixture of solid and transparent forms and the hexagonal and stellated octagonal sectioned *mazzocchi*. The *mazzocchio* is the semi-rigid frame within a particular fifteenth century head adornment but it has a form and structure that lends itself to being adapted, idealised and increased in complexity.[3] Whatever the source of the forms in the *Chalice*, they mostly appear to have simple geometric definitions. The forms themselves may have developed out of the process of geometric construction, rather than being developed from or referring to something already seen.

The object, therefore, could have been conceived as a pure crystalline structure and constructed on a purely theoretical and geometrical basis. This also means that, in theory, unravelling the procedures leading to the drawing should be more straightforward, as we are dealing, presumably, with the geometry of the square, circle and regular polygons and not with units, modules or possible wall thicknesses.[4] In addition, because of the way the *Chalice* is drawn, a vertical cross section parallel

[2] It forms part of a broader investigation on my part as an artist looking at how the physical layout of a drawing on the paper, the method and approach of the perspective construction, and the thinking processes relate to and influence each other.

[3] Kim Veltman (1986: 128–137) discusses the mazzocchio in relation to Leonardo and, in particular, his elaborate and playful transformations of it into spirals, wheels and cogs, etc.

[4] Given the size of the drawing and the degree of error inherent in the method of construction, I would expect high degrees of accuracy in the drawing. It seems to me the largest error that might legitimately arise is possibly 1.5 mm at the most.

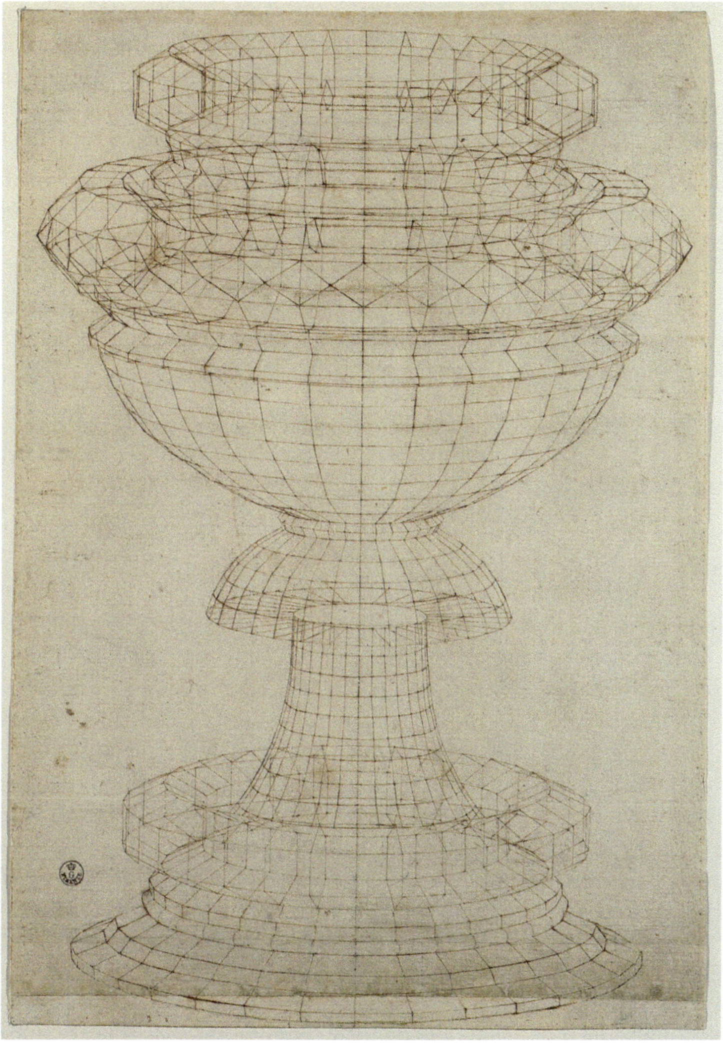

Fig. 51.1 Paolo Uccello (1397–1475), perspective study of the *Chalice*, 1430–1440 (pen and ink on paper). Gabinetto dei Disegni e Stampe, Uffizi, Florence, Italy. Image: ©Foto Scala Firenze. Reproduced by permission

to the picture-plane reveals its true elevation. Proportionally correct measurements relating to its elevation can therefore be taken directly from the drawing. A more direct relationship between the artist's intentions, the underlying construction and the final drawing should be apparent; however, in practise, many aspects remain very uncertain and any analysis will be qualified.

Physical Aspects of the Drawing

The visible lines that define the image are drawn in ink, but the overall impression of the actual drawing is of something more delicate than one would expect from seeing the various reproductions of it (Fig. 51.1).

The underlying construction lines are inscribed into the surface of the paper using a stylus and, where a specific point is plotted, the paper has often been pricked.[5] These various marks are almost invisible under normal lighting conditions and it is only in an early twentieth century photograph taken in a raking light that they are particularly evident. For this study, I have used a high-resolution scan and a much enlarged photographic print taken from this original photograph; I have also examined the actual drawing.

Building the Drawing

It is quite clear that the *Chalice* has been 'built' each point of every 32-sided polygon being located in space using information from a ground plan and an elevation. It is all constructed within a square-based, transparent rectangular block split into horizontal layers, each layer containing an individually drawn polygon. These make up separate but related interlinking geometric volumes, some of which are drawn as if transparent, others as if solid, and still others as a combination of both. The volumes that are drawn in their entirety are the self-contained forms, such as the mazzocchi. As everything that has been constructed has then been inked in, the decision as to what was to be visible was, I believe, taken while the drawing was being made.

The same procedures, and as much effort, would have been required to construct the smallest detail as would have been required to construct a larger or more visible form. In general, the method used throughout has been consistent, but there are exceptions. Two elements of the drawing appear to have been added by eye, without the aid of any construction lines. These are a tiny curved profile slightly below the lowest mazzocchio, and the vertical lip at the very base of the drawing, added, possibly, to confer a feeling of solidity. In addition, there is an inconsistency in the lowest constructed polygon, which could be the result of starting to use the coordinates from one circle and then inadvertently using those from another. It is here, at the bottom of the drawing that approximately 1 cm of extra paper has been attached and, although it is impossible to know at exactly what stage it was added, the scored lines on the surface appear to run seamlessly over the joint. However, even if this error in the lowest polygon had not been made, the drawing would still have fallen outside the bottom edge of the paper. As it is, the drawing appears to

[5] I do not believe that they are the result of the drawing being 'pricked through' as part of a copying process to enable the image to be used, possibly, in an intarsia design; see Kemp (1991: 241–242).

have been altered in order to fit into the available space. The topmost polygon, which is part of the upper hexagonal ring, is positioned as high as it could have been on paper of this size and it is therefore impossible to know how much the size of the paper ultimately dictated decisions regarding the design.

The top two mazzocchi are very accurately drawn, that is, they appear to contain true hexagonal and octagonal sections, but the lowest one, which at first sight appears to contain another regular hexagon, in fact, does not. Whether this was intentional or not we will never know, but as all the points of polygons of equal diameter should align vertically, it becomes apparent that this mazzocchio has been constructed using circles that are used elsewhere in the drawing. It transpires that several circles have been used more than once within the drawing and it appears that it is the circles from the upper regular forms that are used again lower down. This leads me to think that the drawing was, for the most part, constructed from the top down.

The section containing the uppermost mazzocchio apparently defies gravity it is not joined physically to the section below; this too supports the idea that it is an autonomous drawing and that it is not, and never was, intended to represent a 'real' chalice. It also suggests to me that the design was not fully resolved before the drawing was begun and that the progress of the drawing may have been relatively organic. The elevation, which in the case of the *Chalice* is what determines its design, originally may have been very simple and has grown in complexity, with decisions made and elements added during the process of its making.

The Perspective Construction: the Layout

The method used for the perspective projection is straightforward.[6] The preparatory stages of the drawing would have involved establishing a plan and elevation, although these could, to a certain extent, have evolved along with the drawing. One approach to the design would have been to draw the elevation first and then derive the plan from that elevation but, as I will demonstrate, I believe that the elevation has, in fact, been derived from geometry involved in drawing the plan.

There are good practical reasons for having a plan and an elevation laid out on the same sheet of paper as the main drawing, preferably positioned at the edges, as in Fig. 51.2.

Alternatively, they could occupy the same space, lying underneath the final drawing. However, examination of the *Chalice* suggests that the plan and elevation were not positioned at the edges, nor can they be found in the myriad of underlying

[6] Martin Kemp (1991: 241–242) suggests that the artist used some kind of orthographic projection, rather than perspective. Robin Evans also suggests that Piero's 'other method' was used for this drawing, but if this had been the case, different construction marks would have been present on the drawing; see Evans (1995: 173).

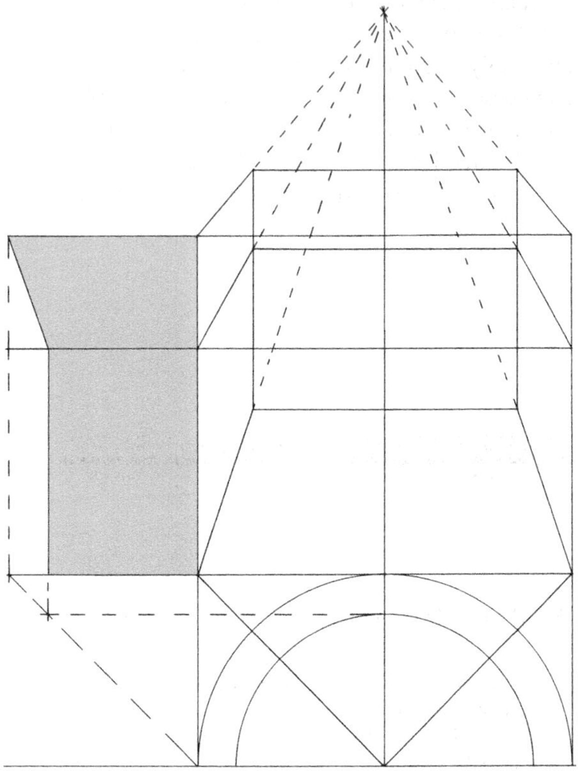

Fig. 51.2 The general principle of the perspective construction. Drawing: author

construction lines. This means that there must have been another drawing, or drawings, presumably now lost, which contained the plan and elevation from which the measurements required for the perspective construction were taken.

I propose that there were not two drawings, one each for the plan and the elevation, but a single drawing, containing a geometric construction that fulfilled both roles. This would certainly make sense on a practical level and, possibly, also on an aesthetic level. A simple example in support of this idea would be the preparatory drawings required to construct a faceted sphere; the elevation is constructed first and the plan is developed from it. The plan now effectively contains all the same information that is in the elevation, so the separate elevation can be discarded. The geometry and constructions in the plan and elevation become interchangeable the radiating lines required for establishing the elevation become the same radiating lines needed for the plan (Fig. 51.3).

The auxiliary drawing for the *Chalice* would have initially required the same construction procedures as those needed to create a 32-sided, faceted sphere a circle, square and octagon, sub-divided repeatedly in order to produce the 32 divisions around its surface. I therefore suggest that the elevation of the *Chalice* derived directly from these same constructions. The key to this is the construction of the octagon, which is clearly a prerequisite of drawing the largest

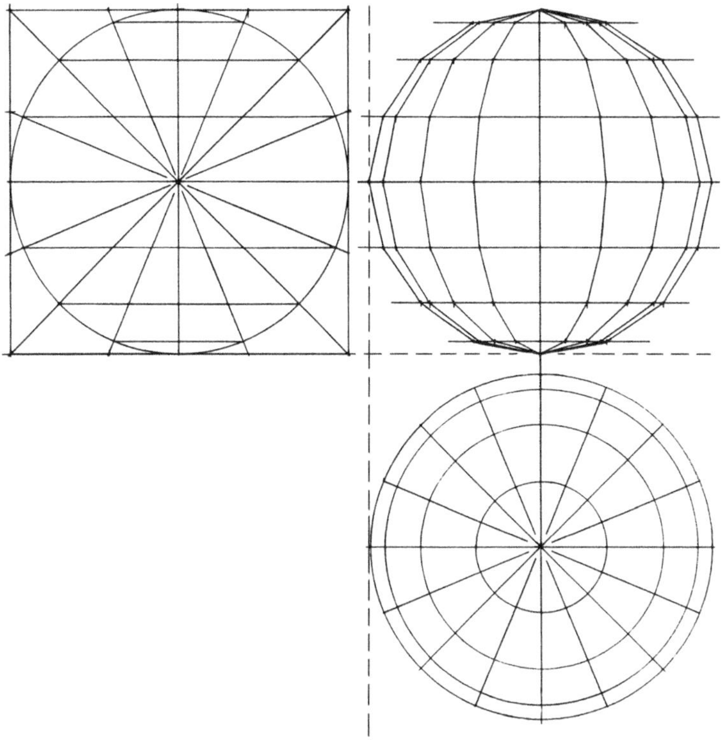

Fig. 51.3 Constructions in the plan and elevation are interchangeable. Drawing: author

mazzocchio but is also part of the construction of the radiating pattern that defines the points around the 32-sided polygon. The octagon is, therefore, required for the construction of elements within both the elevation and the plan (Fig. 51.4).

The Perspective Construction in Detail

The plan and elevation determined the dimensions of a square-based, transparent, rectangular block, which would be subsequently split into horizontal layers. Whatever the shape of the elevation, the plan would always be a set of concentric circles, split into 32 equal sections, equating to the 32 facets describing the surfaces around the object. The vanishing point of the *Chalice* has been placed centrally, at a level that is roughly twice the height of the drawing. The rear corners of the degraded square have been placed at a relatively low level, which means we are looking at the object from a distance of roughly nine times the height of the drawing. These measurements in themselves are not significant, but the relatively large viewing distance does reduce the possibility of uncomfortable distortions

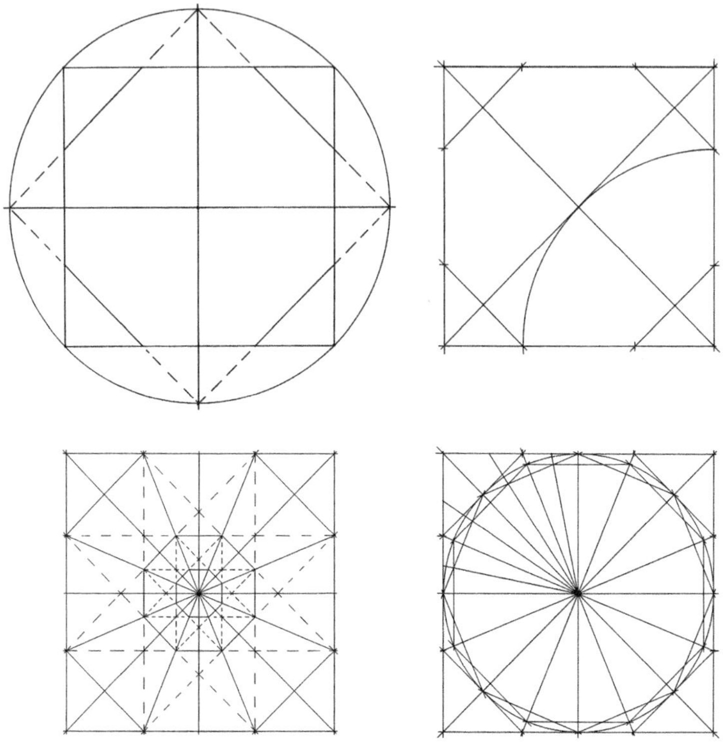

Fig. 51.4 Construction of the octagon and 32 divisions. Drawing: author

caused by being too close to the object. The construction has then taken place within this block, a separate polygon being constructed in each layer.

Each layer in the *Chalice* has been constructed within a few millimetres of the previous one, but it is only in those layers where a complete 32-sided polygon has been drawn, for instance in the mazzocchi, that a full set of radiating lines has been constructed. On the drawing the pricked marks indicating the ends of the radiating lines are visible. The radiating pattern may have been located by constructing it first in the base and then noting the points where the radiating lines touch the four edges of the base (Fig. 51.5).

The horizontal position of these points would then be the same for all the layers and so could be transferred using compasses, repeating their positions onto the four edges of each layer. Because of the very acute angle of the receding sides of each layer there is potential for inaccuracy if only the above method is used. Alternative simple geometric constructional routes, the kind described by Piero Della Francesca in *De Prospectiva Pingendi* and as demonstrated in Fig. 51.5, may have been used to locate, or at least double check, the positions of the points on the sides (Fasola 1942: Tav. VI, Fig. XIX).

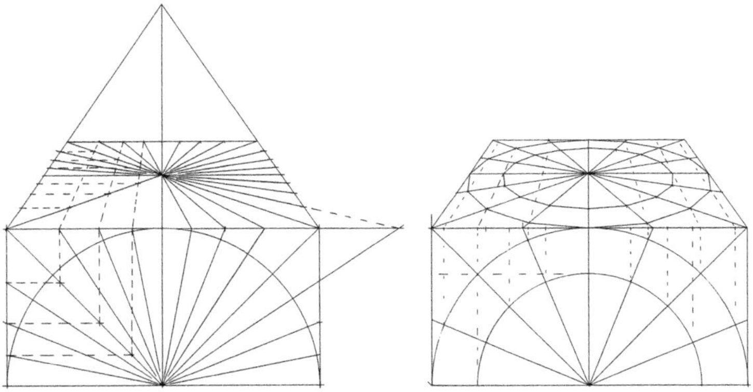

Fig. 51.5 Constructing the radiating lines and the polygons. Drawing: author

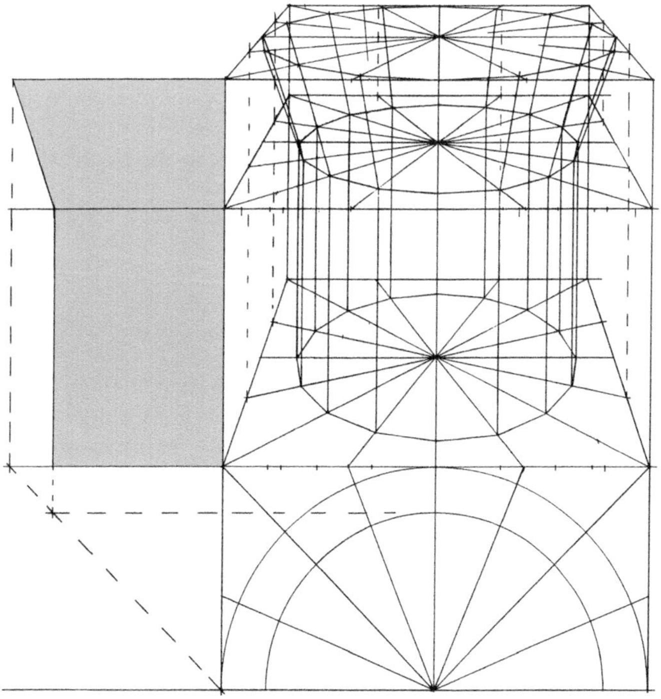

Fig. 51.6 The construction of polygons in each layer. Drawing: author

The points from the plan were then mapped onto the radiating pattern. The horizontal positions of the points of each 32-sided polygon have been measured using compasses and then transferred onto the front edge of the layer at which that polygon is drawn (Figs. 51.6 and 51.7).

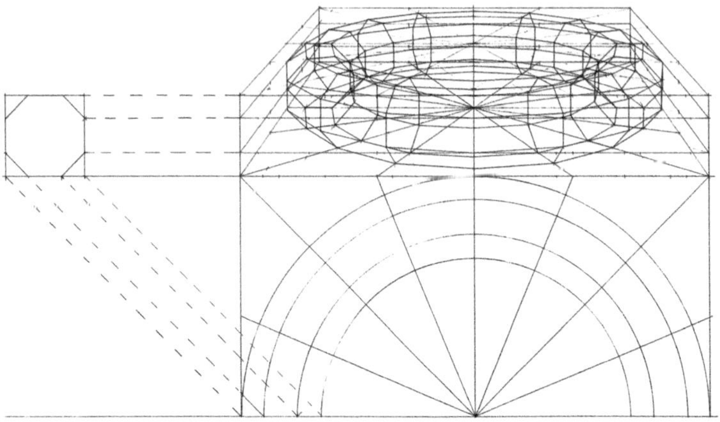

Fig. 51.7 The construction of a 16-sided octagonal sectioned mazzocchio. Drawing: author

In practise, this meant marking the same eight points on either side of the central vertical axis. These points in turn, were aligned with the vanishing point and their intersection with the appropriate radiating line located the points that form the polygon. Again, all the marks from this procedure are visible, either as scored lines from compasses on the leading edge, or as pricks in the paper where a point is established.

The Design: the Plan and the Elevation

Other evidence in support of my idea about the interrelation between the plan and the elevation comes directly from the drawing. For example, the object is symmetrical about a vertical axis but also appears to be symmetrical about a horizontal axis through the two concentric ellipses formed on the stem of the object. This suggests that the design was formed about a central point (Fig. 51.8).

Measurements from the drawing yield up simple relationships that, again, suggest that both the plan and the elevation were in some way, formed from the same geometric construction.[7] For instance, the radius **R** of the circle in Fig. 51.8, which appears to describe the curve of the stem in the vertical plane, is the same as the circle that defines the apexes of the pyramidal structures projecting from the uppermost mazzocchio in a horizontal plane. The size of this curve and its position may be defined by a construction that is itself part of the construction of an octagon. This circle also defines the dimensions of the block in which the whole construction has been made (Fig. 51.9).

[7] There is always a 'chicken and egg' situation in a geometric construction—what came first and what was derived from what?

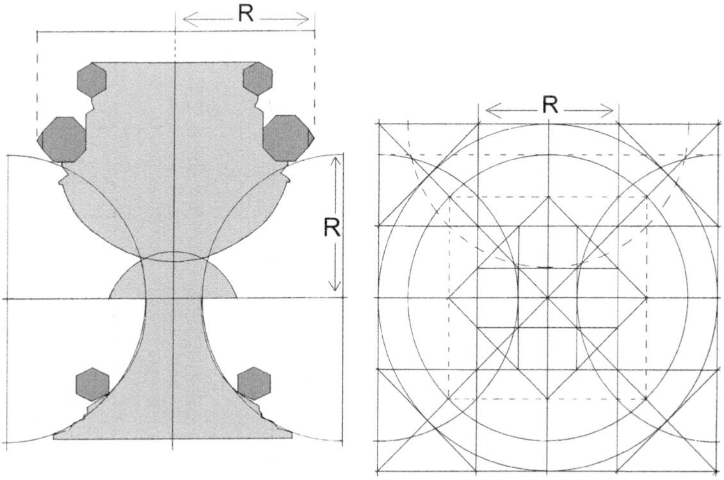

Fig. 51.8 Drawing: author

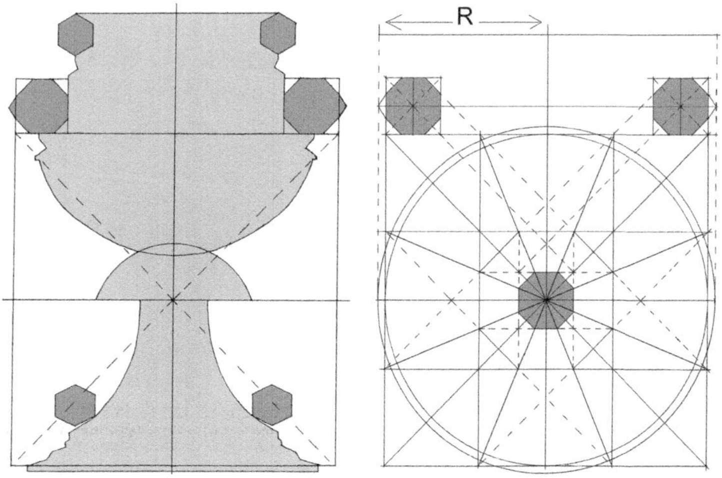

Fig. 51.9 Drawing: author

In addition, the height of the base of the octagonal mazzocchio above the central horizontal axis of the object corresponds with the outer radius of the mazzocchio. If the square is completed, as in Fig. 51.9, its base corresponds with the original level of the base of the object before the hand-drawn lip was added.

The dimensions of the octagon in the mazzocchio are very close to those of the small central octagon that is created as a consequence of repeatedly constructing octagons within the square, as shown in Fig. 51.9. The addition of the pyramidal structures on the outside of the mazzocchio then gives the radius **R**. The relationship between the outer radius of the mazzocchio and its octagonal section,

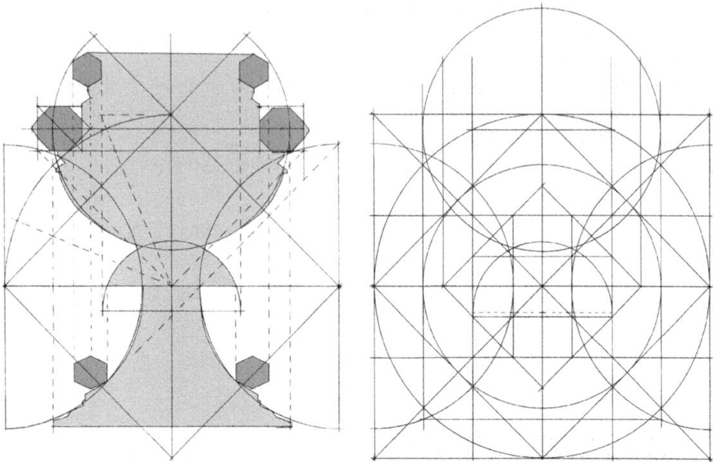

Fig. 51.10 Drawing: author

could of course, be coincidental but, as I will explain later, there are other examples of this. The main volumes of the object appear to be derived from, or are closely related to the construction in Fig. 51.10, although the curve of the bowl and its centre are particularly difficult to define.[8]

The Practicalities of Creating the Drawing

In the early stages of laying out the drawing the most crucial operations would have been to establish true perpendiculars and a method of maintaining horizontals. The accuracy and the logic of the drawing system rely on these and, as a consequence, any errors become more obvious. A draughtsman today would experience the same practical difficulties and use the same instruments as 600 years ago. While now they might not think of using a stylus to inscribe lines, overtly visible construction lines would quickly overwhelm the drawing process on an image of this scale and this complexity. To have successfully accomplished the drawing of the *Chalice* the use of a stylus was an absolute necessity. It is a much more accurate tool than a pencil, and I believe that there are other reasons why its use would have been important.

As there are no annotations on the drawing, it may seem that the main difficulty would be keeping track of all the information involved within a relatively small space. My experience, however, is that once the main "scaffolding" of the space is in place and visible, the logic of the process becomes clear and, in fact, annotations

[8] I am assuming that profiles of the curves are defined by circles, but I would not rule out them being parts of ellipses. Their construction would then have involved two concentric circles and a set of radiating lines.

would be a hindrance. In addition, the inscribed underlying structure, subtly emerging out of the paper by the action of light, would have played an essential role in helping the artist to follow the construction process. It would itself have been quite magical, but would also have been an aid to the imagination, enabling images and visual solutions to be found that otherwise might be inconceivable a process perhaps similar to Leonardo's use of indistinct images found within accidental marks on walls.

A separate sheet on which both the plan and elevation exist together may well have become an autonomous drawing a place for playful and creative thought, rather than simply being a means to an end. This drawing, which could have been relatively fluid and organic, would have provided measurements that were to be used in the other drawing, the one which was governed by the rational system of perspective. There would be a constant shift between the two drawings and elements in the drawings: between the plan, the elevation, the diagrammatic and the pictorial a quality that Martin Kemp has commented on specifically in relation to Piero Della Francesca's paintings (Kemp 1992: 20–40).

Thoughts on Its Attribution

There seems to be no conclusive proof of the *Chalice*'s attribution but its construction would have needed a good knowledge and understanding of geometry, including the concept of plan and elevation and the ability to rationalise a three-dimensional form, whether real or imaginary. Both Piero Della Francesca and Uccello are known to have drawn mazzocchi; Uccello uses them repeatedly as elements in his paintings, most notably in *The Deluge* and *The Battle of San Romano*. Their depiction was, however, to become a problem also tackled by many other artists, particularly those involved in intarsia.

There are drawings of a stellated sphere and a solid mazzocchio in the Paris Louvre, both attributed to Uccello, while catalogued alongside the *Chalice* in the Uffizi are two drawings of hollow mazzocchi, one hexagonal and one octagonal in section, (drawing 1756A), which are also usually attributed to him. These two drawings are on relatively rough paper, whereas that on which the *Chalice* is drawn is quite smooth. All three drawings in the Uffizi show evidence of the use of similar drawing instruments compasses, a straight edge, a fine sharp point and a stylus and they display, essentially, the same method of perspective construction.

In *De Prospectiva Pingendi* Piero demonstrates this same method, but he uses it only for relatively simple geometric objects. He also demonstrates the construction of a simple octagonal-sectioned mazzocchio, but using another more sophisticated method appropriate for any kind of object, in any orientation a method that is distinctly different from that displayed in all three drawings in the Uffizi. Nonetheless, it is clear that, as far as the method is concerned, Piero could also be the author of the drawing.

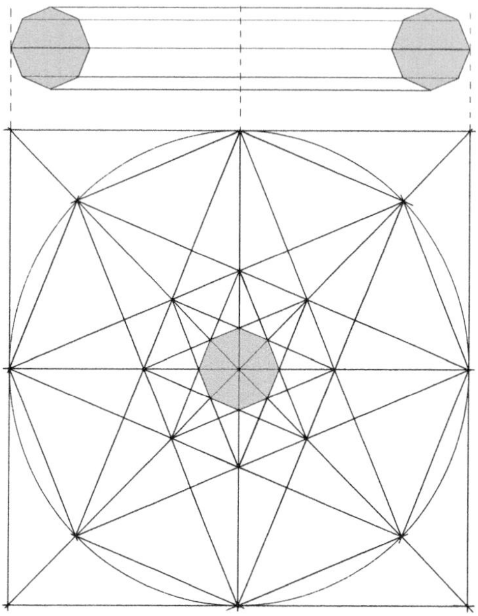

Fig. 51.11 The proportions of the octagonal mazzocchio (Uffizi 1756A). Drawing: author

The drawing in the Uffizi that has an octagonal section, drawing 1756A, is visibly of a much higher quality and accuracy than either of the other two mazzocchi drawings, which leads me to think that this drawing, in particular, relates directly to the *Chalice*. I base this partly on visual appearances, but also on the fact that the relationship between its octagonal section and its outer diameter and, therefore, its possible derivation, appears to match that of the octagonal mazzocchio in the *Chalice* (Fig. 51.11).

This relationship also appears to be the same as that of the octagonal-sectioned mazzocchio that Piero constructs in *De Prospectiva Pingendi*. He does not show how he arrived at the design of the particular mazzocchio that he uses, but he does show its plan and elevation, in which the same proportional relationship seems to be present. None of the mazzocchi in Uccello's paintings appear to show similar proportions. This evidence does not in itself prove that the *Chalice* is from Piero's hand, but it does favour him more than Uccello.

The Drawing in Context

Although the *Chalice* is, ostensibly, a drawing of a recognisable object, the methods and geometry used in its creation make it part, if not a forerunner, of a large genre of drawings of geometric objects. Some of Leonardo's drawings of churches show the same concern with the geometry of the octagon and its successive proportional division that seem to be demonstrated in the *Chalice*,[9] while a drawing by Nicéron, made at least 150 years later, shows an elevation being derived from a plan.[10] Unlike many methods of perspective projection, the method used in the *Chalice* positively lends itself to drawing individual, regular geometric forms. It is, in fact, akin to that of carving thinking and locating points and forms within the confines of a rectangular block. The concept of the transparent block, with its visible, internal three-dimensional logic, would facilitate thinking in three dimensions and, I believe, would have been a major factor in Piero's and, later, Leonardo's understanding and manipulation of the geometry of solid figures.

Biography Richard Talbot gained his BA in Fine Art at Goldsmiths' College, and his MA Sculpture at Chelsea School of Art. In 1980, he was awarded the Rome Scholarship in Sculpture and spent 2 years at the British School at Rome, travelling widely throughout Italy and Egypt. His drawings have been widely exhibited and he has recently completed a commission at Westminster Abbey. He is currently Head of Research and Head of Postgraduate Studies in Fine Art at the School of Arts and Cultures at Newcastle University.

References

Evans, R. 1995. *The Projective Cast: Architecture and Its Three Geometries*. Cambridge, MA: MIT Press.

Fasola, N., ed. 1942. *Piero della Francesca. De Prospectiva Pingendi*. Florence: Sansoni.

Kemp, M. 1991. *Circa 1492: Art in the Age of Exploration*. J. A. Levenson ed. Washington, D.C.: National Gallery of Art.

___. 1992. *The Science of Art: Optical themes in western art from Brunelleschi to Seurat*. New Haven, CT: Yale University Press. (1th. ed. 1990).

Nicéron, J.-F. 1663. *La perspective curieuse*. Paris: J. D. Puis.

Vasari, G. 1996. *Lives of the Painters, Sculptors, and Architects*. G. De Vere, trans. New York: Everyman's Library

Veltman, K. 1986. *Linear Perspective and the Visual Dimension of Science and Art*. Munich: Deutscher Kunstverlag.

[9] See, for instance, his plan for a centralised temple conserved in the Bibliothèque Nationale, Institut de France (Ms 2184; BN2037, fol. 5v).

[10] See the drawing in Nicéron (1663), British Library, shelfmark L.35/40. (1).

Chapter 52
Perspective in António Rodrigues's *Tratado de Arquitectura*

João Pedro Xavier

Introduction

The aim of this chapter is to discuss the *Liuro de Perspectiva* (*Book of Perspective*), found in the extended manuscript *Tratado de Arquitectura* of 1576, which has been attributed to António Rodrigues (c. 1520–1590) (Moreira 1982).

Considering the historical context of this work, its relevance lies mainly in the introduction of an innovative perspective rule that was designed to solve the questions raised by the propagation of inaccuracies and insufficiencies of previous methods, particularly evident in Serlio's second book, *Di Prospettiva*. However, the author's scientific limitations prevented him from fully understanding the enormous potential of his geometrically accurate construction. He employed traditional techniques of measuring distances, commonly used in maritime Portugal, related to the basic principle of similar triangles, shapes that Alberti could assemble as a *piramide visiva*, promoting its *intersegazione* with a surface (*finestra*), and thus obtaining a section that represents the exact perspective of the object. Decoding and verifying the validity of this peculiar perspective rule leads to the centre of the debate surrounding the origins of the *perspectiva artificialis*, which is still a matter of intense dispute in spite of new contributions, reinforcing the theory that considers practical geometry the mathematical basis of this representational system.

Rodrigues's work demonstrates a striking fidelity to central perspective, likely evidence of the Italian school Rodrigues belonged to. This is particularly obvious in some of his perspective representations of architectural objects, especially one

First published as: João Pedro Xavier, "The Book of Perspective of António Rodrigues's Architectural Treatise from 1576", pp. 63–78 in *Nexus VII: Architecture and Mathematics*, Kim Williams, ed. Turin: Kim Williams Books, 2008

J.P. Xavier (✉)
Faculdade de Arquitectura da Universidade do Porto, Rua do Gólgota 215, 4150-755 Porto, Portugal
e-mail: jpx@arq.up.pt

K. Williams and M.J. Ostwald (eds.), *Architecture and Mathematics from Antiquity to the Future*, DOI 10.1007/978-3-319-00143-2_5,

centrally-planned composition called *edeficio quadrado* (squared building) with *ad quadratum* and *ad circulum* geometry, traditionally discussed in the treatises on perspective and architecture at the time. In terms of architecture, we are dealing with one of the most perfect realizations of an anthropocentric vision of the world, the very core of the spatial research undertaken in the Renaissance, although by the time that this *Tratado* was written, this was already being questioned by the counter-reform movement.

Indeed, the inventory and analysis of architectural structures built following this typological unit, consecrated in the representation of this *edeficio quadrado* in the *Liuro de Perspectiva*, is one of the most important chapters in the history of modern architecture, one in which Rodrigues inscribed his name with the construction of Onze Mil Virgens Chapel (Xavier 2015), built some time before 1565 in Alcácer do Sal.

The *Tratado de Arquitecture* and Perspective in Sixteenth-Century Portugal

The *Tratado de Arquitectura* is a treatise related to the origins of architectural teaching and theorization in Portugal. It was used as a textbook at the Lição de Arquitectura Militar, a course in military architecture that began in 1573 under Rodrigues's direction in Paço da Ribeira School,[1] which had been founded by Pedro Nunes in 1559. According to Rafael Moreira,

> in addition to Pedro Nunes's lessons on Mathematics and Cosmography, António Rodrigues taught there the young nobles the elementary notions of Geometry applied to architectural drawing and perspective, the theoretical principles of engineering and fortification and methods and secrets of the art of building well and cheap in order to serve the best interests of the king (Moreira 1982: 75).

This was a Vitruvian curriculum, where one could not conceive of an architect's training that did not have a strong scientific foundation in mathematics, in arithmetic and especially geometry, but which included astronomy and music as well, completing the *Quadrivium*.

The treatise shows its inherent pedagogical inclination, especially obvious in our *Liuro de Perspectiva*[2]; it is clearly meant to be a textbook. The approach to this science, "which can be better learned by demonstration than by trial" (Rodrigues 1576: fol. 44v), begins with the foreshortening of surfaces, evolves to solid bodies, and culminates with the perspective representation of objects and architectural

[1] This school was shut down and transferred to Madrid by Filipe II, giving rise to the "*Academia Real Mathematica*" directed by Juan de Herrera. Later, this sovereign, by then Filipe I of Portugal, ordered its reinstitution in Lisbon, in 1594, with Filippo Terzi (c. 1520–1597) and João Baptista Lavanha (1550–1624) as its directors.

[2] An extended analysis of the *Book of Perspective* can be found in Xavier (2006).

spaces, revealing its purpose and the reason for its being part of an architectural treatise. Serlio did the same with his *Secondo Libro, Di Prospettiva* in 1545, and Pietro Cataneo reinforced this with *Libro Ottavo*, when he added, in 1567, four new books to his *Quattro primi libri di Architettura* of 1554.

The question was the definition of a new instrument for representation, one which was capable of contrasting the need for models, and which, in association with the orthogonal projections in use—plan, elevation and section—would considerably enrich the resources available to the architect for describing, and especially for visualizing, space.

Perspective was at that time regarded as essential among the elements of the Vitruvian *dispositio*, as António Rodrigues says,

> one of the parts that architects should master, [because] it was convenient for someone who wanted to practice architecture to understand perspective so that he could show the outside and the inside of the sketched building in order to avoid expenses with wood, wax or clay models (1576: fol. 11r).

Although Rodrigues, and Cataneo as well, praised perspective pragmatically for its economic value, I must emphasize that the claim that drawings can replace models is based in the consolidation of a technique of representation, perspective, that enables one to approach the three-dimensional nature of the architectonic object. This claim is strengthened if we add axonometry to perspective, which was already well known at the time.

As Gelabert Lino Cabezas states,

> one of the consequences of perspective will be to allow drawing architecture according to new spatial rules, both for representing pre-existing architecture, the ancient one, and for visualizing and projecting new works. The verisimilitude attained with this new perspective representation will allow new architects to control from the drawing (the *disegno italiano*) a new concept of architecture, even coming to replace models . . . in the presentation of works to be built. . . [Cabezas (1989): 167 (my translation)].

And so as the *disegno science* was growing, its first steps in Portugal were closely connected to the establishment of the Lição de Arquitectura Militar.

Considering its context, the importance of Rodrigues's *Liuro de Perspectiva* (*Book of Perspective*) lies mainly in the presentation of an original perspective rule intended to break a deadlock caused by the diffusion of inaccuracies—by Gaurico (1504), Dürer (1525 and 1538) and Serlio (1545)—although the author's scientific capacity wasn't sufficient for him to understand fully the potential of the geometrically accurate construction he produced. Most probably, Rodrigues tried to overcome the errors of Serlio's first rule (Fig. 52.1), his main reference, through the definition of a new non-canonical but flawless rule of his own.

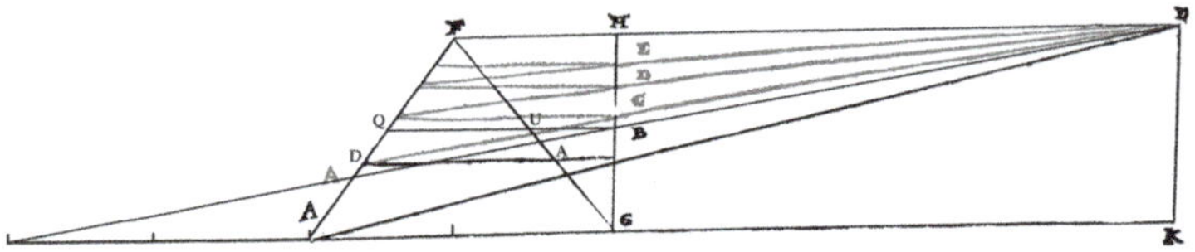

Fig. 52.1 Serlio's first perspective rule with the superimposition of the correct construction of the second square QUAD. Drawing: author, after (Serlio 1600: Libro secondo, fol. 19r)

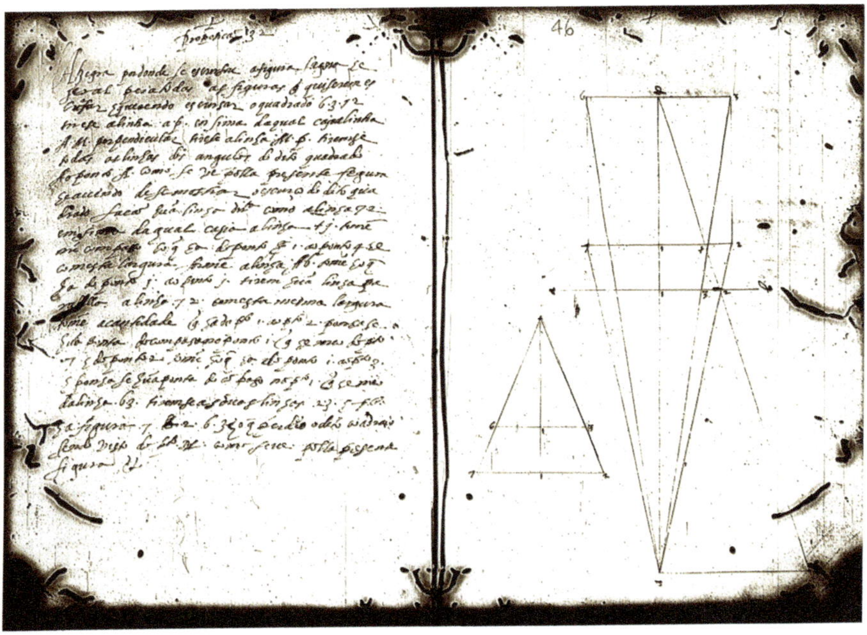

Fig. 52.2 Proposition 32, from Rodrigues (1576): fol. 45v/46r

Rodrigues's First Rule of Perspective

Rodrigues's first rule opens his *Liuro de Perspectiva* and is applied to the foreshortening of a hexagon.[3] As traditionally the square was the preferred figure for representing a plane, let us use instead the following example, found in proposition 32, where the ensemble of procedures to obtain its perspective is explained (Rodrigues 1576: fol. 46v) (Fig. 52.2).

[3] The second rule appears in proposition 34 and it is said to be the same rule used by Serlio: "Sebastianus Serlio bolognese in his "Book of Perspective" has foreshortened all the figures with this rule" (Rodrigues 1576: fol. 47v).

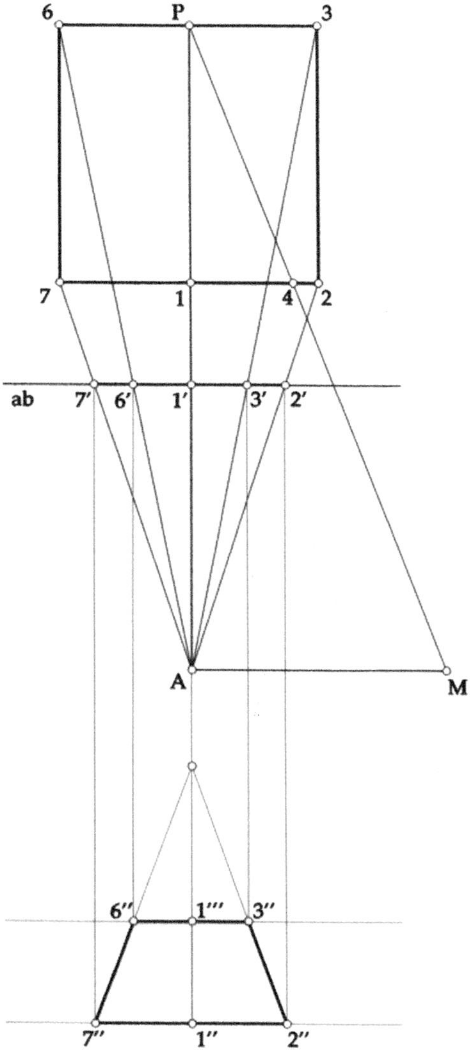

Fig. 52.3 Rodrigues's first perspective rule (Proposition 32). Drawing: author

Here is a modern translation of those steps, numbered in order to facilitate their identification (Fig. 52.3):

Proposition 32. The rule used to foreshorten the hexagon figure is general for all the figures we want to foreshorten. And if we want to foreshorten the square **6.3.7.2**:

1. draw the line **A.P**;
2. draw the line **A.M** perpendicular to it;
3. draw the line **M.P**;
4. join all the vertexes of the square to point **A**;
5. and if we want to show the foreshortening of this square, construct a straight line like line **7″.2″** and draw over it the line **1″.1‴**;

6. with the compass take the distance from point **1** to point **4**;
7. and with this length construct line **ab**;
8. take the distance from point **1** to point **1′**, and draw a parallel line to line **7″.2″**, with the same length
9. take the distance from pt **1′** to pt **2′**, put the point of the compass in pt **1″** which is the midpoint between pt **7″** and pt **2″**;
10. take the distance from pt **1′** to pt **3′** and put the point of the compass in pt **1‴** which is the midpoint of line **6″.3″**;
11. draw two lines **2″.3″** and **7″.6″**.

The figure **6″.3″.7″.2″** shows what is lost by the square when it is seen from point **A** as can be observed in the illustration.

Is it? Could figure **6″.3″. 7″.2″** be the exact perspective of the square **6.3.7.2**?

The first thing to note is that we have a plan and an elevation superimposed where line **A.P** is simultaneously:

a. the horizontal projection of the central visual ray, **A** being the Foot of the Observer and **6.3.7.2** the horizontal projection of the square;
b. the side projection of the ground plane, **M** being the Eye of the Observer and **1.P** the side projection of the square; here **M.P** is the side projection of visual ray **A. P** as it is of visual rays **A.6** and **A.3**.

In the original drawing the perspective construction is shown to the side of this system. I have aligned it with the horizontal projection in order to clarify the correspondence of widths.

We must then point to the striking lack of correspondence between the representation of the Picture Plane in plan and side elevation. Actually, line **ab** is its horizontal projection, while its side projection is a line coincident with points **1.2**.

Contrary to what might be expected, there is no relationship with *costruzione legittima* where the intersection of visual rays with the Picture Plane is achieved with the aid of a systematized double orthographic projection.

However, the foreshortening of the square is obtained by combining the widths taken from line **a.b** and the heights from the line passing through points **1.2**, as is the case of length **1.4** (equal to length **1.1′**) used to graduate the depth of the transversal side **6″.3″**.

We wonder if this can be possible!? … And surprisingly, the answer is yes!

Using an up-to-date drawing of proposition 32 (Fig. 52.4), with the lateral elevation placed to the side, I verified the exactness of Rodrigues's first perspective rule using distance point construction. I checked the relationship between the Observer and the Picture Plane and, as distance **PD.PS** [4] is equal to distance **PF**, I could be sure that line **ab** in plan indicates the correct position of the Picture Plane. So, the square represented in perspective is not in the square $\mathbf{A_1B_1C_1D_1}$ shown in plan, but the homothetic square **ABCD**, with its side **CD** coincident with the

[4] Rodrigues extends the orthogonal sides of the square to the central vanishing point (PS) but he doesn't take advantage of it in the construction.

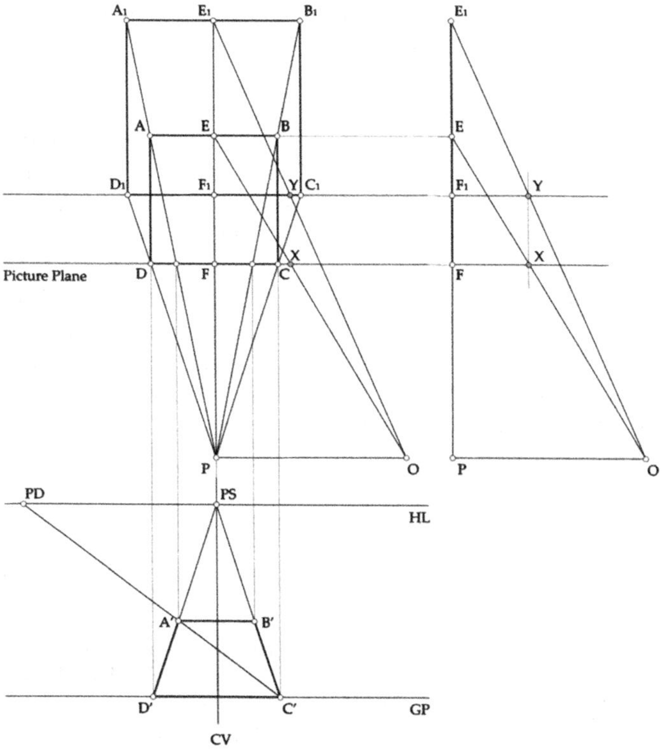

Fig. 52.4 Testing the validity of Rodrigues's first perspective rule. Drawing: author

Ground Line, although Rodrigues doesn't draw it there. But if this is true we have to check, using the side elevation, if height $\mathbf{F_1Y}$ used to graduate the depth of transversal side $\mathbf{A_1B_1}$ is equal to height **FX** which determines the depth of side **AB**.

More than this, we should prove that for any position of the square $\mathbf{A_1B_1C_1D_1}$, similar to square **ABCD** in a homothety of centre **P**, the segment **FX** is always equal to segment $\mathbf{F_1Y}$.

With the aid of the famous theorem attributed to Thales we can assert that the triangle **PEO** is similar to triangle **FEX**, and so:

$$\frac{\mathbf{PE}}{\mathbf{PO}} = \frac{\mathbf{FE}}{\mathbf{FX}}. \tag{52.1}$$

On the other hand, the triangle $\mathbf{PE_1O}$ is similar to triangle $\mathbf{F_1E_1Y}$, so:

$$\frac{\mathbf{PE_1}}{\mathbf{PO}} = \frac{\mathbf{F_1E_1}}{\mathbf{F_1Y}}. \tag{52.2}$$

From these two proportions we can deduce the following equality:

$$\mathbf{PO} = \frac{\mathbf{PE}}{\mathbf{FE}} \cdot \mathbf{FX} = \frac{\mathbf{PE_1}}{\mathbf{F_1E_1}} \cdot \mathbf{F_1Y}. \qquad (52.3)$$

As there is a homothetic relationship of centre **P**,

$$\frac{\mathbf{PE}}{\mathbf{FE}} = \frac{\mathbf{PE_1}}{\mathbf{F_1E_1}}, \qquad (52.4)$$

one may conclude from expression (52.3) that

$$\mathbf{FX} = \mathbf{F_1Y}.$$

QED!

Finally, one might be puzzled with the indication of step 7 to place the line **ab**, the Picture Plane, taking the height **1.4**. There is no geometric reason for that, although it might be a way to control the dimensions of the perspective result. I tried to overlap the orthogonal projections with the perspective drawing and at least it is possible to recognize a good adjustment of the whole (Fig. 52.5). But, unfortunately, this is more a sign of the author's incapacity to understand the implications of his own perspective rule fully, namely the perfect control of viewing distance in relation to the object and the Picture Plane locations.

Rodrigues's *Liuro de Perspectiva* in Context

If we look at Rodrigues's *Liuro de Perspectiva* within an international frame of reference we have to admit that it never achieved a position of great importance. Although Vignola's *Le due regole della prospettiva pratica* (ca. 1545) was only printed in 1583 by Egnatio Danti, the truth is that these perspective rules were already known to a select few, those who went down in history.

However, I believe that Rodrigues's *Liuro de Perspectiva* is much more interesting than it might appear on the surface if we look at the reasons that led the author to search for a thoroughly successful solution where Serlio and his predecessors failed.

We must remember that Serlio's first, erroneous, rule of perspective is one more in a series of attempts to draw the geometric construction described by Alberti in *De Pictura* (1435). So, by criticizing Serlio's work, Rodrigues ended up close to the methodological assumptions underlying Alberti's construction. When I realized this unexpected similarity in methodology it became necessary for me to re-examine the different interpretations of the perspective representation described in *De Pictura*. The absence of graphic illustrations in Alberti's work has given rise

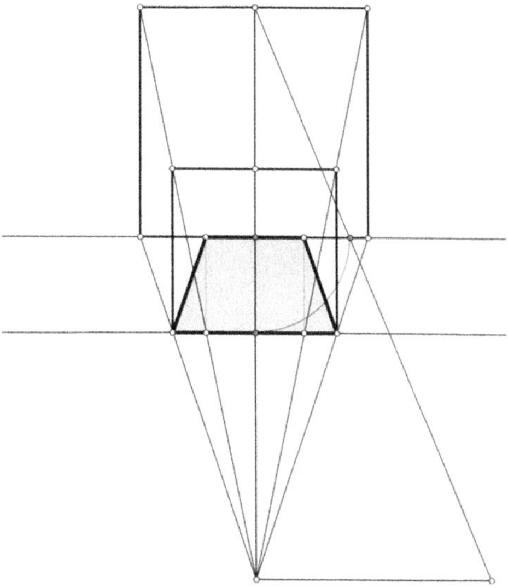

Fig. 52.5 Rodrigues's first perspective rule with the foreshortening of the square overlapping the plan and side elevation. Drawing: author

to the ongoing proliferation of several hypotheses that intertwine with the discussion of the origins of the *perspectiva artificialis* and take us back to Brunelleschi, to whom Alberti dedicated his work. I undertook this journey back to the origins starting with a thorough review of Alberti's original work, and I verified that, within the time span from its appearance until the edition of Rodrigues's book, several authors translated its *modo ottimo* with only a graphic representation—such as Filarete (1461–64, Fig. 52.6), Francesco di Giorgio (c. 1485), Leonardo da Vinci (c. 1492) and Jehan Cousin (1560)—or with some theoretical contents as well—as was the case with Piero della Francesca (c. 1460, Fig. 52.7), Giacomo Vignola (c. 1545), Federico Commandino (1558) and Daniel Barbaro (1568), although this last made exclusive use of Piero della Francesca's contribution.

The problem was, in fact, the noise coming from works produced in the first half of the 1500s, which determined a state of disturbing uncertainty due to the ignorance of reliable sources (or to their inadequate decoding) and the prevalence of practical recipes, not always correct, and without a theoretical basis.

Viator's handbook, *De Artificiali Perspectiva* (1505), was an exception, as it was irreproachable regarding operational matters, but lacked an indispensable conceptual foundation. Viator took it upon himself to give legitimacy to the idea of perspective as a graphic representation of natural as well as panoramic vision, remaining within the sensory concept of a virtual pyramid (Fig. 52.8), in opposition to Alberti's attitude, of a rational nature, expressed in the concept of *piramide visiva* and *intersegazione* (Mesa Gisbert 1994: 112). In the beginning of the following century the first picture in *Perspective* (1604) by Hans Vredeman de Vries (Ioannes Frisius) (1526–1609) (Fig. 52.9) will express eloquently the visual theory already present in Viator's work, placing the observer in the centre of the Horizon and the pyramid vertexes in the circular line that defines it (Alpers 1983: 57).

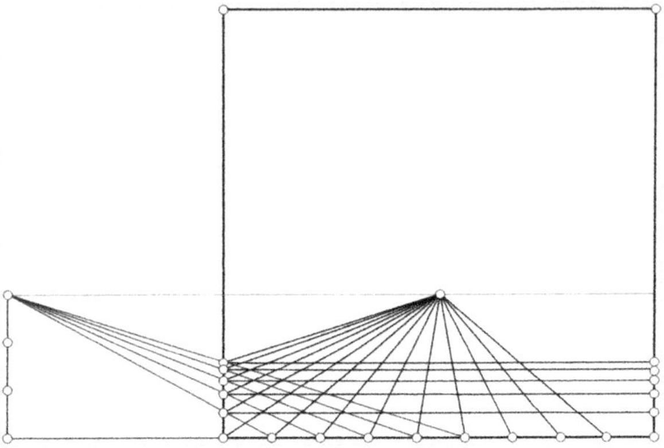

Fig. 52.6 Alberti's *modo ottimo* by Filarete. Drawing: author, after *Tratatto di Architettura* (Bk. XXIII, fol. 177v)

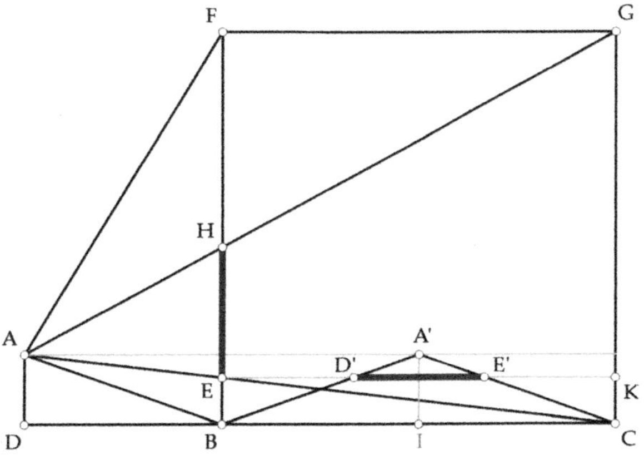

Fig. 52.7 Piero's perspective proof. Drawing: author, after *De Prospectiva Pingendi* (Bk. I, prop. XIII)

It was against this destabilized background that Rodrigues worked to legitimise his book and its original perspective rule. That rule rests in the methodology and instruments concerning the evaluation of inaccessible distances, which the discipline called "practical geometry", whose main driving force was nautical science, along with astronomy and cosmography, in Portugal as well as in other countries. In consequence, it is the actual application of the principle of similar triangles—in which the observer's eye is at one of the vertexes, and at the others there may be a tree, or a tower top, or a mountain summit, or else the sight of the coast line from a caravel or a twinkling star in the sky—that drove Rodrigues to the discovery of a functional perspective rule, in spite of its lack of the conceptual purity already known at the time but which he hadn't heard of.

Even so, the fact that it was supported by the proportionality principle applied with remarkable flexibility, fully mastering Thales theorem, usually employed in practical measuring tasks, is sufficient in itself to make it worthy of note.

Indeed, this principle is the essence of the perspective representation system, although its simple application occurs at a relatively basic stage of development,

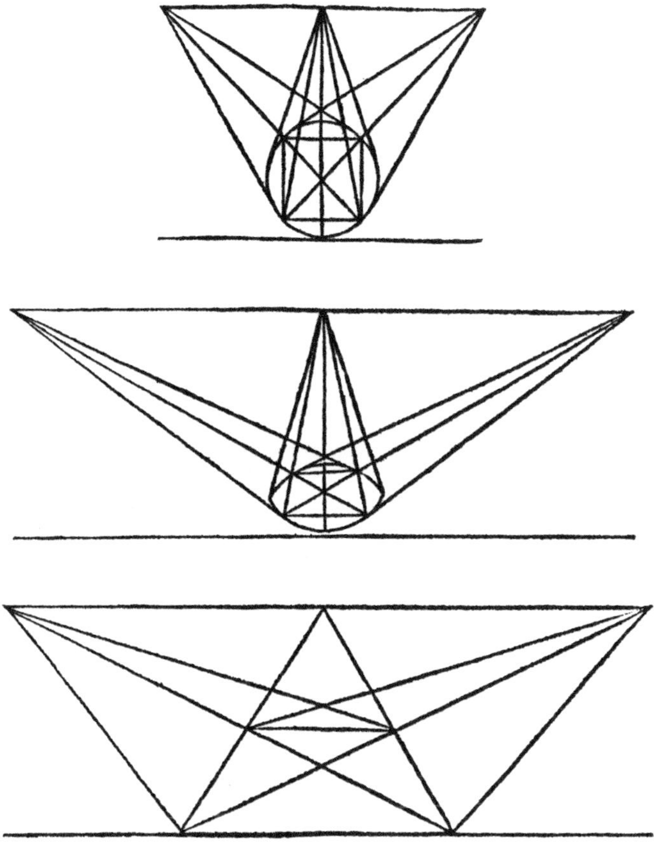

Fig. 52.8 The visual pyramids and the 'tiers-points' perspective construction. Image: Viator (1505)

taking us back to the origins of *perspectiva artificialis* and to the issue about the mathematic principles underlying its genesis.

I suspect that the capacity to deal easily with proportionality, based in similarity and homothetic relationships, contributed to Alberti's definition of his *modo ottimo*. The precedence of a practical activity of measuring for the definition of his rule can be felt in his work, especially in the *Ludi Matematici*. The illustrations of Francesco di Giorgio that appear in his *Trattati* (Fig. 52.10) in the context of a thorough inventory of typical problems of practical geometry unequivocally warrant that connection.

Following Alberti's requirement, Giorgio isolated the side elevation and physically materialized geometric entities, drawing planes with rods and visual rays with threads. In addition to the requirement of drawing the side elevation separately, a decisive step for achieving the accurate definition of depth grading, Alberti gave us the enigmatic indication that a small space (*picciolo spazio*) would

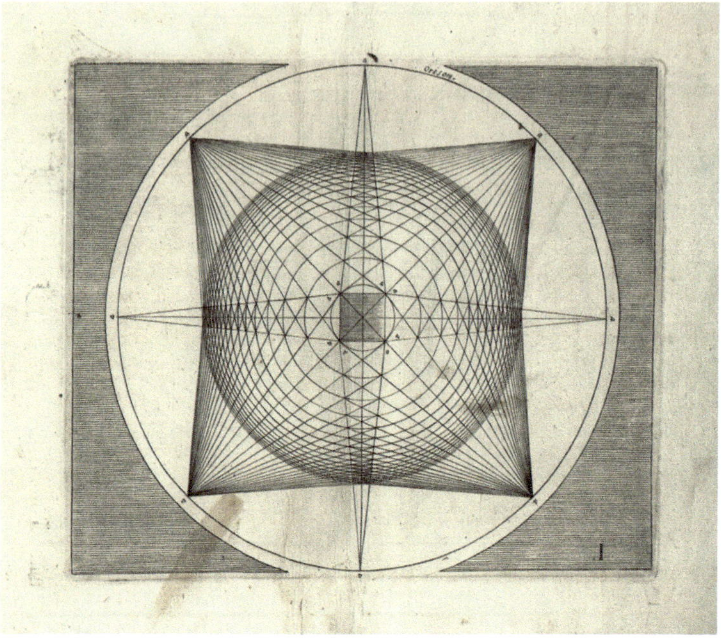

Fig. 52.9 Panoramic natural vision. Image: De Vries (1604)

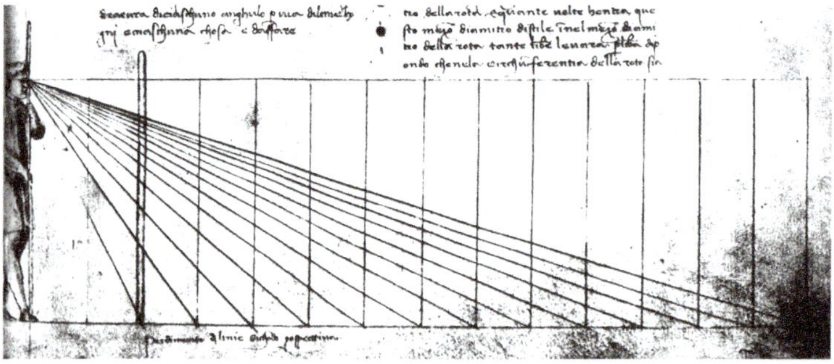

Fig. 52.10 Alberti's modo ottimo by Francesco di Giorgio Martini (ca. 1490: fol. 33)

be enough for its execution. This compression of the drawing frame allows us to think of the possibility of working with dimensions smaller than the *braccio*, the unit that divides the ground line (Fig. 52.11), as suggested by Pietro Roccasecca (2001). And then their mutual dependence could be supported by a proportional relationship, without compromising the transverse lines, which together with the orthogonal lines already drawn would define the ground plane or *pavimento*.

In this particular matter, Rodrigues's approach to perspective through a rule based in homothetic principles was similar in its essence to methodology suggested by Alberti, even though it appeared later.

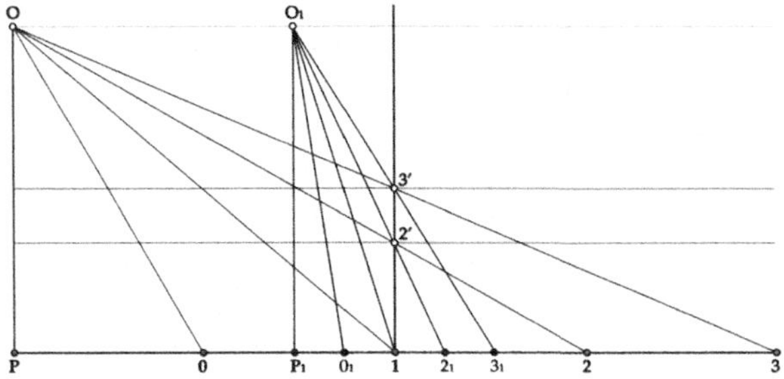

Fig. 52.11 Alberti's *modo ottimo*. Interpretative drawing of Roccasecca's proposal (2001: 66). Drawing: author

Rodrigues's Rule of Perspective and Centrally-Planned Architecture

The mathematical systematization of the representational space conceived by Alberti is primarily a project for space itself, which runs parallel to the need to map the earth and sky for exploration.

If the general order of the universe was mathematical, a Pythagorean belief that rested on Plato's *Timaeus*, and geometry was the connection to subjects that dealt with spatial problems, then the issue of the representation of visual space would be dealt with by perspective, while cosmography would be charged with representing the earth and sky. So there was a need for improved knowledge about the sphere, with the contributions of geography and astronomy, facilitated by the discoveries of the epoch. Eventually it was acknowledged that the first accurate maps of the terrestrial globe and the celestial sphere were based on conic projections that had been known since Antiquity: gnomonic in Mercator maps, in that case the surface of projection being a cylinder tangent to the equator, and stereographic in Ptolemy's Planisphere. And, as in this last situation the surface of projection was a plane that cut the sphere along the celestial equator, Federico Commandino realized (at last!) the intimate relationship between this cartographic projection and perspective (Commandino 1558).

The core idea of space codified by Alberti, which translated into a very particular type of perspective—central perspective—corresponded to a concept of centralized space, with its inherent fidelity to central (or at least bilateral) symmetry and, naturally, the strict obedience to a measurement system, which implied a system of proportional relationships of an arithmetic and geometric nature and which, in terms of perception, would change into a system of harmonic proportions that, through the correct rule, could be transferred to paper in order to become perspective.

This could be a description of San Sebastiano, an exceptional centralized space, but can also be applied to S. Andrea, both in Mantua. It is possible to evoke Brunelleschi's works, San Lorenzo or Santo Spirito, on which Wittkower founded his attribution of the discovery of *perspectiva artificialis*. It is Leonardo's thorough investigation of *prospettiva liniale*, and his studies of projects of buildings with central plans (Xavier 2008) and, from there, the effective materializations by Bramante and others that disclose the first evidence of this connection. Even so, I think it is sufficient to recall the panels produced by the circle of Urbino regarding the depiction of an ideal city in order to verify the convergence of a global plan of space centralization, which is inevitably anthropocentric, and because of that, avails itself of central perspective to affirm its value, but also to persuade.

In architectural treatises in general, and in the books on perspective that became part of them, as well as in specialized textbooks—even in the first of them, *De Prospectiva Pingendi*—centralized spaces from cities to architectural objects, closely associated with central perspective, are always given a place of privilege. In parallel, a specific representational system that will end up in axonometry was being developed mainly for cities, with an intuitive approach by Francesco di Giorgio, already with its own rules by Maggi and Castriotto, and would soon be preferred for military architecture and engineering.

Because perspective apprenticeship always started with a plane represented by a square, which could be squared according to the most convenient measurements conveying the Albertian spatial core, buildings or small spaces with square plans have always been given preferential treatment. One can see them in Filarete's early sketches, in the treatises of Piero della Francesca, Francesco di Giorgio, Serlio and Hernan Ruiz, and we should not forget that this was one of Leonardo's favourite subjects, although his spatial research was linked to his own particular representational system (Xavier 2008).

Generally speaking, the Observer, standing, perceives space according to his plane of axial symmetry—*sia sempre la sua distantia all'entrare di esse*, as Serlio stated (1600: fol. 18r)—emphasizing that formal characteristic and amplifying the centrality that the space already possessed.

It was in this context that one *edeficio quadrado* appeared in *Proposição* 42 of Rodrigues's *Liuro de Perspectiva*, foreshortened according to the first rule. It was crowned with a dome, embodying the *ad quadratum* and *ad circulum* geometric composition that was synthesized so well in Caesare Cesariano's representation of the Vitruvian man (Fig. 52.12).

I tested the reconstruction of this ideal building (Fig. 52.13) and compared it with the sepulchral space of Onze Mil Virgens Chapel (Figs. 52.14, 52.15, and 52.16) and I found that this space could be considered as typological variation of that ideal building. As the Onze Mil Virgens Chapel, according to Rafael Moreira, was designed by the author of the sixteenth-century architectural treatise and its book of perspective, António Rodrigues, I think that this attempt to represent an archetype in perspective is a proof of the decisive role this unit played in the formalization of centrally-planned churches, from its appearance in Byzantine through the Mannerist period.

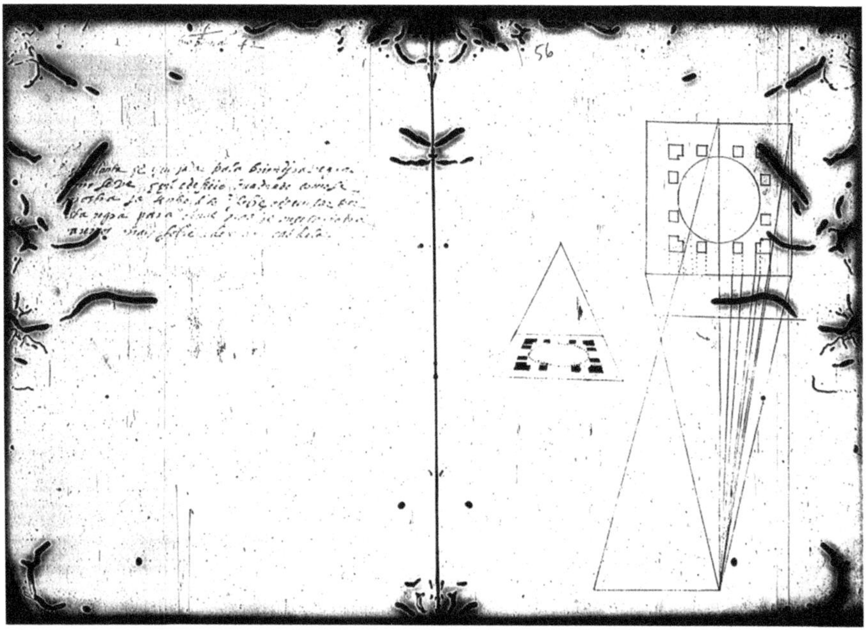

Fig. 52.12 Proposition 42 from Rodrigues (1576): fol. 55v/56r

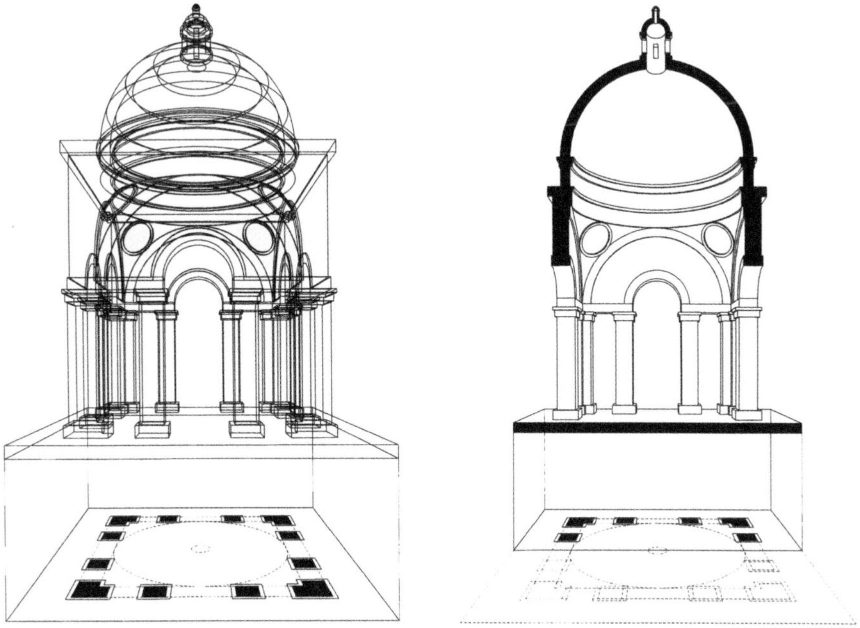

Fig. 52.13 3D models of the proposal for the 'squared building' reconstituted from the foreshortened plan shown in Rodrigues (1576: Prop. 42, fol. 56r). Drawing: author

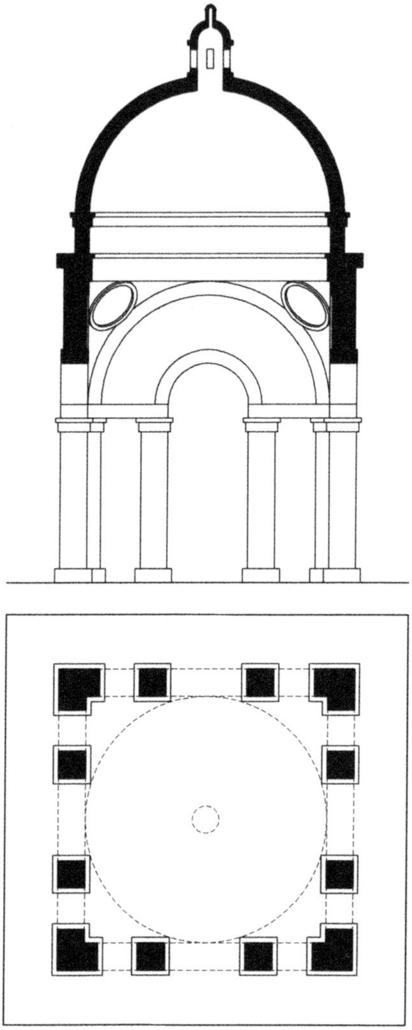

Fig. 52.14 The 'squared building' of Proposition 42. Drawing: author

Based on the extensive analysis done from the Onze Mil Virgens Chapel (Xavier 2015), in which one can recognize mainly rectangular shapes that express consonances from the Pythagorean tetrachord, as well as other secondary consonances, as well as other relationships that are specifically geometric, I must emphasize that the utilization of the proportional ratios built as homothetic situations evolving from centres, carefully located according to the modular structure of the temple (Xavier 2006: 378–444), shows much the same methodology that underlies the definition of Rodrigues's perspective rule.

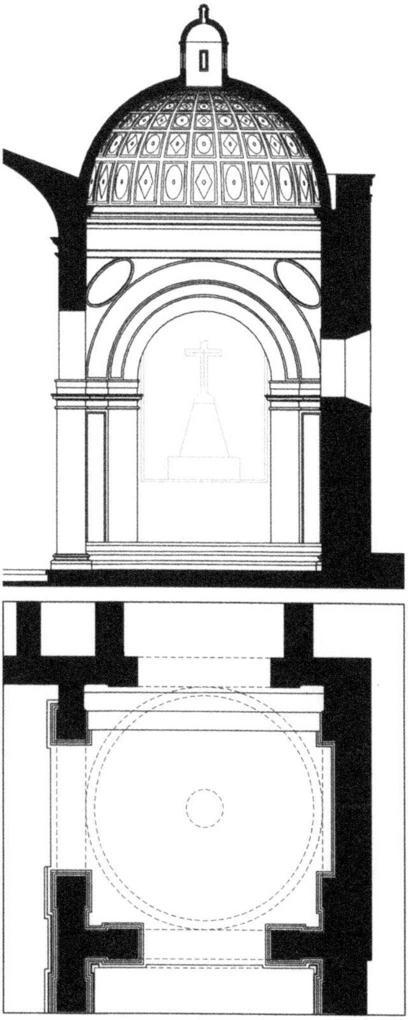

Fig. 52.15 The sepulchral space of the Onze Mil Virgens Chapel. Drawing: author

Biography João Pedro Xavier is an architect and a Professor of Architecture in the Faculty of Architecture of the University of Porto (FAUP), where he received his degree and Ph.D. in Architecture. He worked in Álvaro Siza's office from 1986 to 1999. At the same time, he established his own practice as an architect. He is a member of the research group for Theory, Design and History of architecture at the Centro de Estudos de Arquitectura e Urbanismo at FAUP. His research focus is on architecture and mathematics, and in particular on perspective. He is the author of *Perspectiva, perspectiva acelerada e contraperspectiva* (FAUP Publicações, 1997) and *Sobre as origens da perspectiva em Portugal* (FAUP Publicações, 2006). He has participated in conferences, lectures and exhibitions, and has published a number of papers on the subject. He is on the Editorial Board of the *Nexus Network Journal* and is a member of the executive council for the journal *Resdomus*.

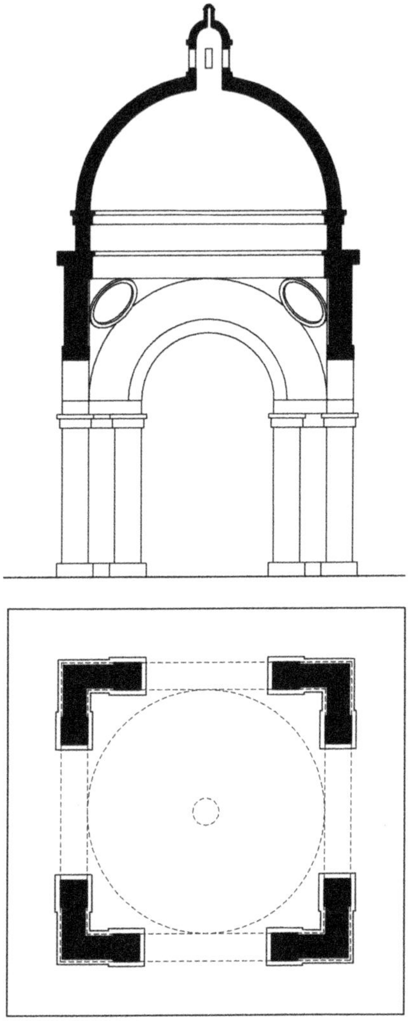

Fig. 52.16 Typological variant of the 'squared building'. Drawing: author

References

ALPERS, Svetlana. 1983. *The Art of Describing*. Chicago: University of Chicago Press.

BARBARO, Daniele. 1568. *La pratica della perspettiva...*Venice: C. & R. Borgominieri.

COMMANDINO, Federico. 1558. *Ptolemaei Planisphaerivm. Iordani Planisphaerivm. Federici Commandini Vrbinatis in Ptolemaei Planisphaerivm commentarivs. In quo uniuersa Scenographices ratio quambreuissime traditur, ac demonstrationibus confirmatur*. Venice.

DE VRIES, Hans Vredeman (Ioannes Frisius). 1604. *Perspective*. Leiden.

FRANCESCO DI GIORGIO MARTINI. ca. 1490. *Tratatti di Architettura, Ingegneria e Arte Militare*. Florence, Biblioteca Nazionale Centrale ms. Magliabecchiana 11.1.141. Published as: *Tratatti di Architettura, Ingegneria e Arte Militare*, 2 vols. Corrado Maltese, ed. Milan: Cassa di Risparmio di Jesi, 1967.

LINO CABEZAS, G. 1989. La "Perspectiva angular" y la Introducción de la Perspectiva artística en la España del siglo XVI. *D' Art* **15:** 167–180.

MESA GISBERT, Andrés de. 1994. Entre la Práctica Artesanal y la Teoría de la Visión. El Concepto de Pirámide Visual en el Tratado de Perspectiva de Jean Pélerin (Viator). *D'Art*, 1 (20): 59–113.

MOREIRA, Rafael. 1982. Um Tratado Português de Arquitectura do séc. XVI (1576–1579). Master's thesis, Universidade Nova de Lisboa, Faculdade de Ciências Sociais e Humanas.

ROCCASECCA, Pietro. 2001. La Finestra Albertiana. Pp. 65–78 in *Nel Segno di Masaccio: l'invenzione della Prospettiva*. Filippo Camerota, ed. Florence: Giunti.

RODRIGUES, António. 1576. *Tratado de Arquitectura*. Lisbon: Biblioteca Nacional de Lisboa, Codex 3675.

SERLIO, Sebastiano. 1600. *Tutte l'Opere d'Architettura, et Prospettiva di Sebastiano Serlio Bolognese. Diviso in sette Libri*. Dam. Gio. Domenico Scamozzi Vicentino. Venice: Presso Heredi F. de' Franceschi.

VIATOR (Jean Pélerin Le Viateur). 1505. *De Artificiali Perspective*. Toul. (2nd ed. 1509.)

XAVIER, João Pedro. 2006. *Sobre as Origens da Perspectiva em Portugal. O "Liuro de Prespectiua" do Códice 3675 da Biblioteca Nacional, um Tratado de Arquitectura do século XVI*. Porto: FAUP Publicações.

———. 2008. Leonardo's Representational Technique for Centrally-Planned Temples. *Nexus Network Journal* **10**, 1 (Spring 2008): 77–100.

———. 2015. António Rodrigues, a Portuguese Architect with a Scientific Inclination. Pp 165–181 in Kim Williams and Michael J. Ostwald eds. *Architecture and Mathematics from Antiquity to the Future: Volume II the 1500s to the Future*. Cham: Springer International Publishing.

Part VIII
1500 A.D.–1900 A.D.

Chapter 53
Ottoman Architecture: Relationships between Architectural Design and Mathematics in Sinan's Works

Zafer Sağdiç

Introduction

To know and understand the relationship between mathematics and architecture better, it is needed to make searches not only on west architectural styles, but also on east architectural styles, as well. Thus, it is useful to study one of the unique and magnificent architectural style of the world, the Ottoman Empire architecture as its military and political achievements on three continents. Within Ottoman architecture, the works of the great architect Sinan (1496–1588) occupy a particular place, especially with regards to aesthetic qualities and highly advanced structural characteristics. The relationship between design and mathematics is also significant in Sinan's works. In order to understand why, a general review of the relationship between architectural design and mathematics is helpful.

Mathematics can be roughly divided into two main parts: one is "practical mathematics", such as the operations we use in daily life; the other is "pure mathematics", which is used in establishing the complex mathematical relationships in the positive sciences. The study of the relationship between architectural design and mathematics depends on "pure mathematics".

We must first accept that all the objects in nature and the relationships between these objects are governed by rules of geometry. As Galileo stated, nature has a certain mathematical design: "the book of nature can only be read by those who

First published as: Zafer Sagdic, "Ottoman Architecture: Relationships between Architectural Design and Mathematics in Architect Sinan's Works", pp. 123–132 in *Nexus III: Architecture and Mathematics,* ed. Kim Williams, Ospedaletto (Pisa): Pacini Editore, 2000.

Z. Sağdiç (✉)
Faculty of Architecture, Department of Architecture, Branch of History of Architecture, Yildiz Technical University, Istanbul, Besiktas D-217, Turkey
e-mail: zafersagdic@hotmail.com

K. Williams and M.J. Ostwald (eds.), *Architecture and Mathematics from Antiquity to the Future*, DOI 10.1007/978-3-319-00143-2_6,

know its language, which is mathematics" (King 1997). Thus, if we accept "pure mathematics" as a kind of game, it can be said that nature is the medium in which this game is played, and the constituent parts of nature are the symbols used in mathematics; in other words, they are the pawns used in the game. Throughout the game we can assume that the single symbols, by further developing the metaphor, will form groups of symbols. As a result of these formations, we can see the geometrical rules governing the symbols, that is, of the groups of objects existing in nature. To put it more clearly, the rule(s) governing the relationship between architectural design and mathematics can be found among the rules of forming/ establishing the relationships in nature.

Architects, while analysing the designs of various buildings, almost always turn to nature, observing it carefully and transforming their observations into design elements. Consequently, it can be seen that architects create their designs by studying the geometrical rules that establish the various natural collaborations with a concern for architectural style. In that sense, architects can be defined as the organizers of the relationships between the forms and functions of buildings. As Monroe Beardsley said, "the form of an aesthetic object is the total web of relations among its parts" (Sağdiç 1998: 383). In this way, the geometrical rules in nature can be taken as the rules of "Beauty". It is known that, throughout the history of design, architects have always created the designs for "the artificial environment" within the framework of the geometrical rules (Beardsley 1958).

In the light of this preface, the question of how the mathematical rules (symmetry, proportion, geometry, etc.) affected Sinan's style can be answered more clearly. But in order to consider the question in a certain framework, a few more basic questions should be answered. What are the structural and aesthetic rules that were used in Sinan's works? Why is it so important even today to take these rules into consideration? Why is Sinan's period still being defined as the "Classical Period" of Ottoman architecture?

The Classical Period of Ottoman Architecture

The development of Ottoman Architecture to its highest level, called "The Classical Period", during which it achieved its finest features, was due to the development of building technology as well as its aesthetic value. Both of these were the results of the political achievements and the economical growth of the Ottoman Empire. The Classical Period of Ottoman architecture is sometimes referred to as the "period of Architect Sinan" because of the fact that the most original and monumental religious and public buildings in this period were created by this great Master-builder. The fundamental principles underlying the dialectics of building developed by Sinan provide the solution(s) for the system of its structure. In this context, the most important point is that there exists a concept of a certain form and aesthetics originating from the type of construction. In Sinan's architecture, the most important element is the dome, which forms the central point of the whole structure; therefore the overall structure of the building is oriented towards carrying and supporting the dome. Considering the elements of this monumental architecture, it can be seen that the system of relationships between the elements establishing the dimensions and constituting the buildings has been established on the basis of the dimension of the human body. These dimensions were observed in storey layers related to the façade, in the choice of the proportions of building elements such as windows and doors, and in calculating the proportional relations of such elements/details. Sinan, who was called "the Euclid of his age" (Kuban 1988: 620), developed a system of forms through an understanding of architecture based on geometry, and through the consideration of functions. According to Freely, "Euclidean geometry had a certain influence on Sinan's spatial compositions as well as the rules of construction" (Freely and Burelli 1992: 124). It can be said that the key that allowed intensive building activities to spread throughout the large geographical area ruled by the Ottoman Empire is the fact that Sinan had a common system of design in all his constructions.

In European architecture, the ideal form of the system of structure was attained in Gothic architecture. It can be claimed that a similar development was brought to a peak in Ottoman architecture by Sinan in the late fifteenth and the sixteenth centuries. This means that in studying Sinan's period, we will also be examining the main lines and the general rules of Ottoman architecture.

Proportional Relationships in the Classical Period of Ottoman Architecture

When we look into the relationship between mathematics and the architectural design found in the monumental architecture of the Classical Ottoman period, it will be observed that certain proportional relationships were used, the existence of which cannot be denied. According to Arpat (1984: 40), these proportional relationships can be divided into two groups: those that use modules or some religious-symbolic figures as in-principle arrangements, and those that use proportions.

It can be seen that in three of his mosques—Pasha, Shehzade and Mihrimah Sultan (Fig. 53.1)—Sinan used a modular network obtained by placing an octagon within a circle with a radius of 68 cm (3 *arshin*) or multiples of this unit-measure.

It has also been established that the main module commonly used in all the mosques mentioned above are multiplied by 3 (that is $3 \times 3 = 9$ arshin). This measure, 204 cm (9 arshin), was also employed in establishing the design principles of such elements as the levels in buildings, the overhangs of the eaves of domes, etc. (Arpat 1981: 33; Schemmel 1998: 78). Another fact that is important to emphasize here is that, in the said period, the design principles governing monumental architecture, both European and Ottoman, depended on some symbolic values along with functionality. In Christian Architecture, the number 3 and its multiples have been used symbolically in organizing space because of the Holy Trinity. In

Fig. 53.1 The Mihrimah mosque. Photo: author

Islam, it is believed that there are three spirits: the good spirit, the evil spirit, and a third spirit that tempts people to "evil deeds" (Egli 1954: 79). Moreover, in the monumental religious buildings of both Christianity and Islam, a centralized plan, believed to represent the monotheistic belief in the organization of space, is commonly used. Here it also should not be forgotten that in Islamic religion, *namaz*[1] is the most important ritual of the daily life. On an Islamic point of view, it has done for five times during a day. It also should be done on a mosque which has a square like space in plan because of the tradition of to keep/be on the line of the brothers of the religion which is called as "*safh tutma*", while in Christianity according to the ritual of the religion of taking bread and wine from the priest on a corridor like rectangular space in shape, even the apsid part of the church is centralized.

In Classical Ottoman Architecture the best example of the centralized plan is the Süleymaniye Mosque with its huge dome placed at the epicentre of the complex design (Fig. 53.2).

According to Ernest Egli, the proportion between the width and the length of the big courtyard, in which the mosque and the enclosed cemetery (*hazire*) are located, is 2:3, which is equal to the proportion between the axes. Again, measuring along the vertical axis, Egli established a regular increase in the spaces covered by the outer courtyard, inner courtyard, the area covered by the mosque building itself, and the enclosed cemetery, which corresponds to 4, 5, 6 and 7. On the horizontal axis,

[1] *Namaz* is originally from Persian; it is *salat* in Arabic. In a general view it means praying and giving greetings to Allah, the one and the only creator, the God.

Fig. 53.2 The Süleymaniye mosque. Photo: author

the measurements of the widths of the outer courtyard on the right, the space covered by the mosque building itself, and the outer courtyard on the left correspond to 4, 7 and 4. As for the measurements of the specific spaces, the inner courtyard is 5:7, the space covered by the, mosque building is 6:7, and the space of the enclosed courtyard is 7:7 (Haidar and Yazar 1986: 33).

In Shehzade Mosque (Fig. 53.3), another religious monumental building designed by Sinan, the *naos* or cella (*sahin* in Turkish, an inner sanctuary) consists of nine modules, and the inner space of the mosque has 25 modules, as does the courtyard (Fig. 53.4). On the vertical plane, the relationships between the measurements seem to have proportions of $1:\sqrt{2}$, 2:3, 3:5, and 5:8 (Tuncer 1997: 30).

Another version of the same plan scheme applied by Sinan can be found in the Beyazid Mosque, where the square plans consisting of 16 modules each for the mosque and the courtyard are equal. The cella in this mosque has the size of six modules, four covered by the main dome and two covered by the semi-domes at the sides (de Launay et al. 1873: 15).

Generally, it can be said that in the mosques of the Classical Ottoman period, the width of the capital on a column inside the mosque is taken as the module. The height of the columns used in monumental buildings is usually 10–18 times the radius of the capital. In all the orders used for columns, the radius at the bottom is the size of six modules and the radius of the top is a five-and-a-half module. In this connection, it might be useful to remember the orders used in columns: the conical order (*tarz-i mimari-i mahruti*); the multiple-plane order (*tarz-i mimari-i müstevi*); the stepped/crystal order (*tarz-i mimari-i mücevheri*). To define these orders briefly, it can be noted that the conical order has columns whose maximum height is six modules. In the multiple-plane order, the height of the whole column, including the base and the capital, is ten modules. In the stepped/crystal order, which is both spectacular and sophisticated, the maximum height of the column together with the base and the capital is 18 modules (Soylemezoglu 1986: 69; Tuncer 1997: 132).

Fig. 53.3 The Shehazade mosque. Photo: author

The relationship between architectural design and mathematics will be discussed with regards to the drawings of two major mosques designed by Sinan, the Rustem Pasha Mosque and the Sulemaniye Mosque.

The Rustem Pasha Mosque

Soylemezoglu (1986) studied the geometrical arrangements by using the quadrated system using the available existing measured drawings of the Rustem Pasha Mosque (Fig. 53.5).

In this study, the projection of the dome was drawn in a square and then another square was superimposed on the first one at a 45° angle. An octagon was drawn by using those squares. The axes of construction were created by the lines drawn on the corners of this octagon, shown as *SA* in Fig. 53.5. The axes of the columns, by which the dome was supported, were found from the intersection of axes *SA* and two superimposed squares. The second octagon was drawn by connecting the centres of columns, shown from *1* to *8*. The inner borders of the mosque building were found by the intersection of the line *ZE1* and line *CD*, and then points *K, L, P, and H*. In the second octagon, points *G2, G3* and *V* are given to the outer borders of the mosque building. Through similar methods, all the necessary points of the design could be found, such as the outer and inner borders of the spaces, centres and locations of the columns, etc. (Tuncer 1997: 111). The relationships of the third dimension on the Rustem Pasha Mosque can be seen in the section in Fig. 53.6. By using the intersections of the two big circles, which are superimposed on each other, and six squares, all the necessary points for the creation of the design can be found.

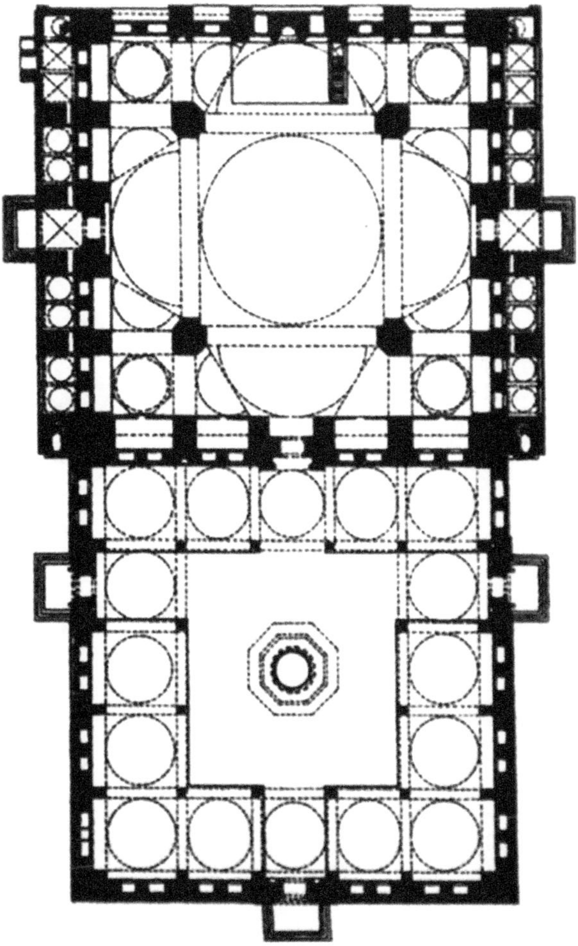

Fig. 53.4 The plan of the Shehzade mosque. Drawing: author, after Goodwin (1997)

The Süleymaniye Mosque

The Süleymaniye Mosque, designed for Süleyman the Magnificant (Kanuni Sultan Süleyman) between the years of 1550 and 1557, has a centralized plan. The plan scheme of the mosque (Fig. 53.7) is a perfect square; the main cella was covered by a huge dome, which was supported by two semi-domes.

Side cellae were covered by five small domes. The mosque has a nice courtyard, which had aisles all around and four minarets. According to Tuncer, the golden rectangles *BDGH*, *ACEF*, *CDIJ* and *ABMN*, which were derived from the main rectangle *ABCD*, were drawn. By using the connections of those rectangles, all of the creative possibilities could be found, such as the inner and outer borders of the mosque building, the borders of the mosque's courtyard, etc.

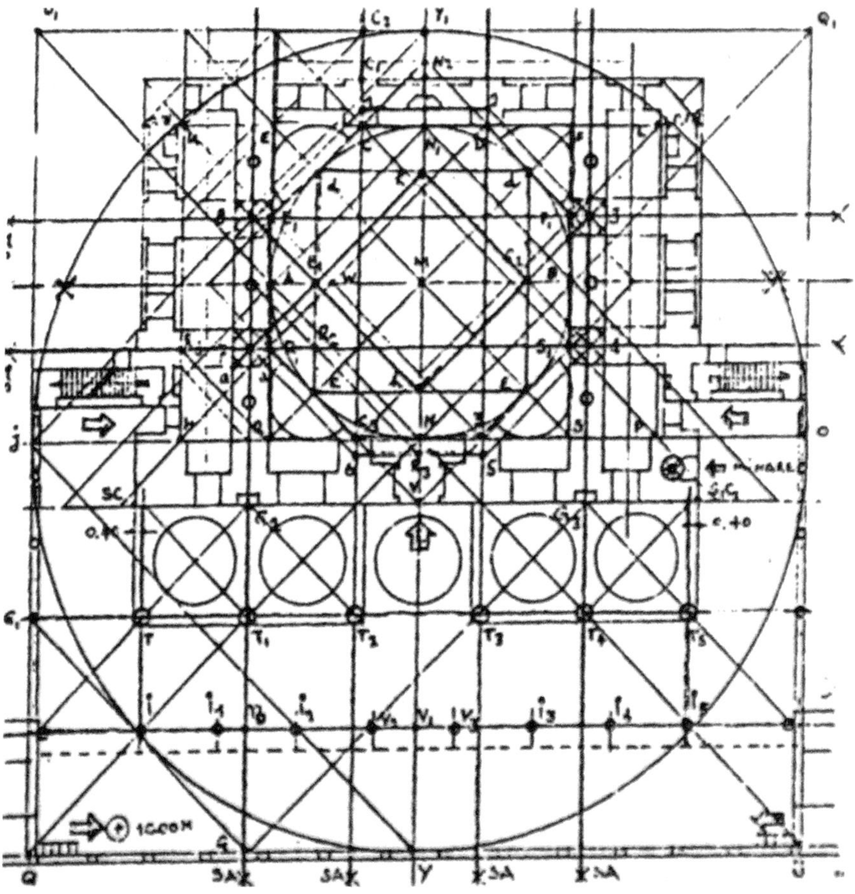

Fig. 53.5 The plan of Rustem Pasha mosque. Drawing: author, after Tuncer (1997)

The quadrated system, which was applied to the mosque's measured drawings, can be seen in Fig. 53.8.

The side wall axes are drawn on the eleventh and twelfth squares; the axes of the columns are drawn on the seventh and eighth squares; the axes between the *mihrap* and the columns are drawn on the ninth and tenth squares; the axes of the huge dome are drawn on the third and the fourth squares; the borders of the portal are drawn on the fifth and sixth squares; finally, the locations of the flying buttresses are drawn on the first and second squares.

It can be said that similar methods were used in the design of the façades of the Süleymaniye Mosque. One-quarter of the main module used in the plan was used as the main module for the creation of the façades. The golden rectangles that were applied to the design of the façades can be seen in Fig. 53.9 (Tuncer 1997: 111).

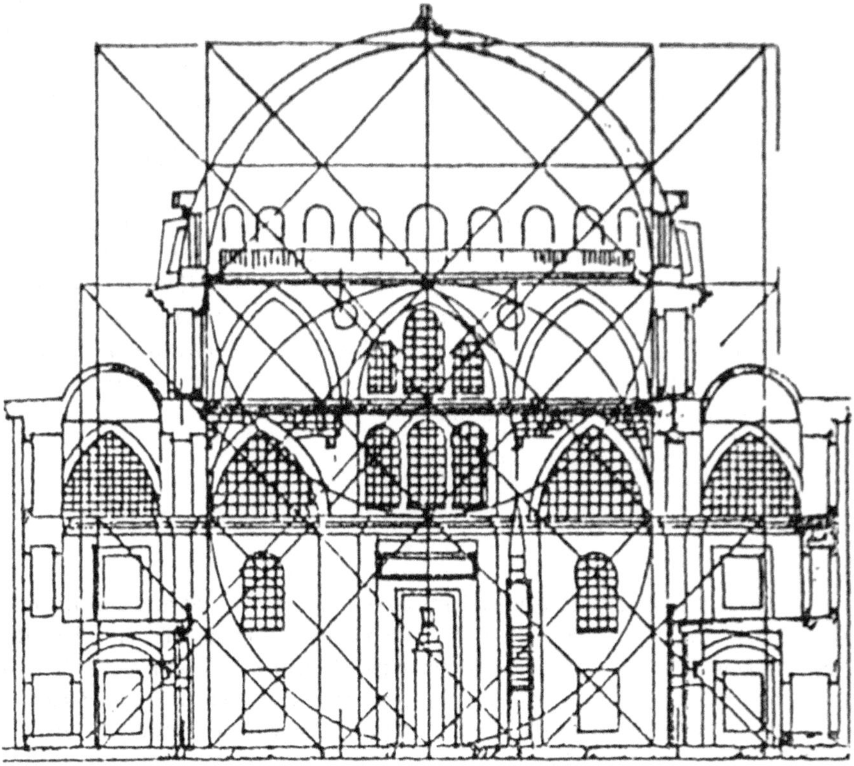

Fig. 53.6 The section of Rustem Pasha mosque. Drawing: author, after Tuncer (1997)

Conclusion

It can be easily understood from the two examples given above that Sinan's architecture defines the Classical period of Ottoman Architecture because of the sophisticated relationship between the architectural design and mathematics. Sinan always used a specific space organization that was created in the light of the structural rules and mathematical rules. By reflecting the idea of functional design onto the creation of monumental buildings in the pre-modern world, he became one of the most important and interesting master-builders in history. It may be said that, even today in our modern age, Sinan's compositions should be observed because their rules of functionality reflect the relationships between architectural design and mathematics.

Fig. 53.7 The Süleymaniye mosque. (*top*) Plan, after (Tuncer 1997); (*bottom*) Design phases 1–4. Drawings: author

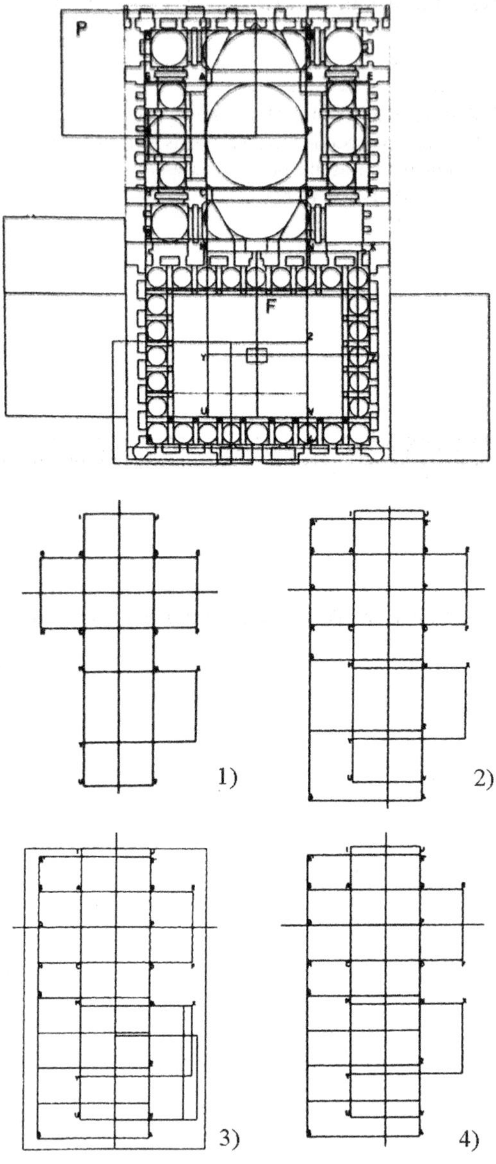

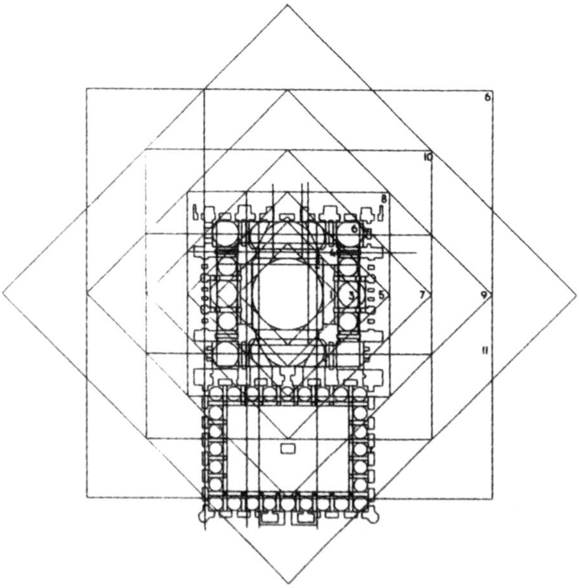

Fig. 53.8 The plan of Süleymaniye mosque with geometrical scheme. Drawing: author, after Tuncer (1997)

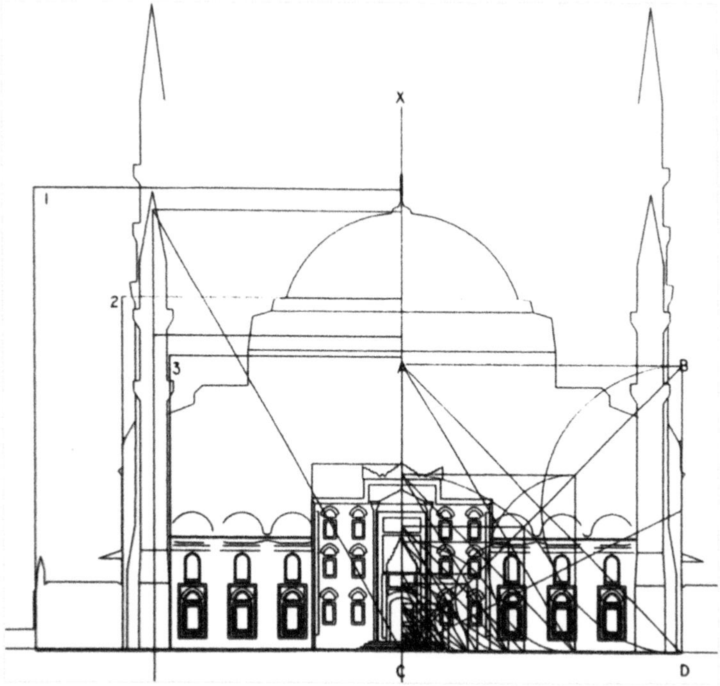

Fig. 53.9 The section of Süleymaniye mosque. Drawing: author, after Tuncer (1997)

Biography Zafer Sagdic is a Turkish architect living and working in Istanbul. She has taught "History of Art and Architecture", and "Architectural Thought" in the Department of Architecture of the Faculty of Architecture in Yildiz Technical University since 1996. She received her degree from the Faculty of Architecture of Yildiz Technical University in Istanbul, and her master's degree from the Master Program of the Architecture History of Istanbul Technical University in 1999, and her Ph.D. from the Faculty of Architecture of Yildiz Technical University in 2006. Her research is focussed on the relationships between mathematics and architectural design, especially in Seljuck Architecture, Ottoman Architecture and Islamic Architecture, as well as modern theories of design in art and architecture in light of philosophers such as Kant, Baudrillard and Derida.

References

Arpat, Atilla. 1981. Osmanli Dini Mimarisi'nde Modul ve Duzenleyici Geometri. *MTRE Bulteni* **13-14** (1981): 29–35.

———. 1984. Osmanli Camileri'nde Modüler Düzen ve Boyutsal Sembolizm. *Yapi* **54,** 2 (1984): 40–42.

Beardsley, Monroe. 1958. *Aesthetics: Problems in the Philosophy of Criticism*. Michigan: Harcourt College Publishers.

De Launay, Marie, Boghoz Chachian, Em Maillard, P. Montani, and Pascal Sebah. 1873. *L'architecture Ottomane/Die Ottomanische Baukunst*. Constantinople.

Egli, E. 1954. *Sinan der Baumeister Ossmanischer Glanzzeit*. Stuttgart: Verlag für Architektur Erlenbach.

Freely, J. and A. Burelli. 1992. *Sinan*. New York and London: Thames and Hudson.

Goodwin, G. 1997. Ottoman Architecture. London: Thames and Hudson.

Haidar S and H. Yazar. 1986. Implict Intentions and Explicit Order in Sinan's Work, II. Pp. 29–42 in *Uluslararasi Turk ve Islam Tarihi Kongresi*, Cilt no. 2. Istanbul: Istanbul Teknik Üniversitesi.

King, Jerry. 1997. *Matematik Sanati*. 2nd edn. Ankara: Tubitak.

Kuban, D. 1988. Sinan'in Dünya Mimarisindeki Yeri. Pp 581–624 in *Mimarbagi Koca Sinan*. Istanbul: Vgmy.

Sağdiç, Z. 1998. Mathematics and Design Relationship in Architecture. Pp. 383–391 in *Mathematics & Design 98. Proceedings of the Second International Conference*. Javier Barrallo ed. San Sebastian, Spain: The University of the Basque Country.

Schemmel, A. 1998. *Saylarin Gizemi*. Istanbul: Kabalci.

Soylemezoglu, K. 1986. Istanbul Rustem Pasa Camii. Pp. 105-114 in *Uluslararasi Turk ve Islam Tarihi Kongresi, Cilt no.2*. Istanbul: Istanbul Teknik Üniversitesi.

Tuncer, N. 1997. Klasik Dönem Osmanli Mimarisi'nde Iç mekan ve Cephelerde Oran. Ph.D. thesis. Yildiz Teknik Üniversitesi, Istanbul.

Chapter 54
The Mathematics of Palladio's Villas

Stephen R. Wassell

> *By . . . showing to what extent [Palladio] was a natural geometer, we do not make him less the great architect; on the contrary, we show, in a way that gives more than mere lip service to the proposition, how great architecture may flow from geometry*
>
> (Hersey and Freedman 1992: 12).

Introduction

Much has been written about the mathematical qualities of Andrea Palladio's architecture, including his own *I quattro libri dell'architettura.* Often this has been analysed within the context of a larger collection of architectural treatises, including Vitruvius' *De architectura* and Alberti's De re *aedificatoria*, as well as works by contemporaries of Palladio, such as Daniele Barbaro, Cesare Cesariano, Sebastiano Serlio, and Giacomo Barozzi da Vignola. These Cinquecento writings underscore the importance of proportion, symmetry and geometry in Renaissance Italy; for example, Barbaro maintains that "some arts have more of science and others less," and the "more worthy (are) those wherein the art of numeracy, geometry, and mathematics is required" (Puppi 1975: 18). Lionello Puppi concludes, "Architecture obviously came into this category. . . . Palladio . . . bring [s] to the concrete stage of his planning operation a single-minded scientific approach, arrived at through 'lofty speculation' into number and proportion" (Puppi 1975: 18). Rudolph Wittkower asserts, "[t]he conviction that architecture is a science, and that each part of a building, inside as well as outside, has to be

First published as: Stephen R. Wassell, "The Mathematics of Palladio"s Villas", pp. 173–186 in *Nexus II: Architecture and Mathematics,* eds. Kim Williams, Fucecchio (Florence): Edizioni dell'Erba.

S.R. Wassell (✉)
Department of Mathematics and Computer Science, Sweet Briar College, Sweet Briar, VA 24595, USA
e-mail: wassell@sbc.edu

K. Williams and M.J. Ostwald (eds.), *Architecture and Mathematics from Antiquity to the Future*, DOI 10.1007/978-3-319-00143-2_7,

integrated into one and the same system of mathematical ratios, may be called the basic axiom of Renaissance architects" (Wittkower 1952: 89). Many modern authors have analysed Wittkower's thesis that harmonic proportions derived from musical scales played a central role in the minds and designs of Renaissance theorists and architects. Central to this debate is Palladio's oeuvre—his architecture and his *Quattro Libri*.

This chapter provides a review of the mathematical aspects of Palladio's work as it has been discussed in the literature and offers a novel perspective on his mathematical approach to architectural design. The argument is made that, whether or not harmonic proportions played a major role in the beauty of Palladio's architecture, it is now time to search further for other mathematical facets of his design philosophy. For convenience the body of analysis is arranged in three sections, based on the categories of geometry, proportion and symmetry.

Geometry

One of Palladio's great gifts was his ability to analyse the ancient and contemporary architecture of Rome, visualize the key elements of plan, section and elevation, and extract the forms that were appropriate to his own design needs. His interest in geometrical form and the process of the extraction of that form from classical elements of architecture developed throughout his career and may be traced in the evolution of his villas.

The Villa Godi at Lonedo di Lugo (1540) has a façade devoid of classical orders.[1] "The central spine ... is simply inserted into a rectangular block rather than being integrated into it by interlocking parts or by the proportions of the plan or elevation" (Ackerman 1966: 164). The "ornamentation" of the facade is simply the pure form of the geometry. The Villa Valmanara at Vigardolo (1541) takes this a step further. The simple façade is articulated at the entrance with a Serlian arch, accented by two oculi flanking the arch and square window and a third oculus placed well above it. Two drawings for possible villas from this period also demonstrate Palladio's early recognition of the natural elegance of simple geometric forms. The first is a plan and elevation study for a villa featuring a square perimeter, a biapsidal loggia, and a cruciform, cross-vaulted *salone* (Fig. 54.1). This study anticipates the Villa Malcontenta and the Villa Rotonda. The second drawing is the penultimate plan and elevation for the Villa Pisani, Bagnolo (Fig. 54.2).

The major difference between paper and building is the hemicyclical portico of the former. The Villa Poiana at Poiana Maggiore (c. 1548) is an illustration of Palladio's geometric interpretation of Roman elements. "The familiar Roman columns and tabernacles were transformed into cubic blocks in a composition that depends wholly on geometric form for its effect" (Ackerman 1967).

[1] For a detailed discussion of Palladio's architecture, see Boucher (1994).

Fig. 54.1 Study of a ground plan and elevation for Villa Valmarana, Vigardolo. Image: RIBA31775, by permission of the RIBA Library Drawings & Archives Collections

Although his mature villas are not abstracted to the point where architectural forms give way to purely geometric ones, their treatment is governed by a notable rigour often mathematical in nature. In the Villa Pisani at Montagnana (1552), the

Fig. 54.2 Study of a ground plan and elevation for the Villa Pisani, Bagnolo. Image: RIBA31796, by permission of the RIBA Library Drawings & Archives Collections

salone is square in plan, four free-standing columns marking the central area, with eight engaged columns and four corner pilasters flanking apses in the corners. The vaulting system is rich in geometrical intricacy. Several techniques are used by Palladio to help integrate the entire design in the Villa Cornaro at Piombino Dese (1552): the squarish *salone* has a flat beamed ceiling supported by four free-standing Ionic columns which align with the second and fifth columns of the

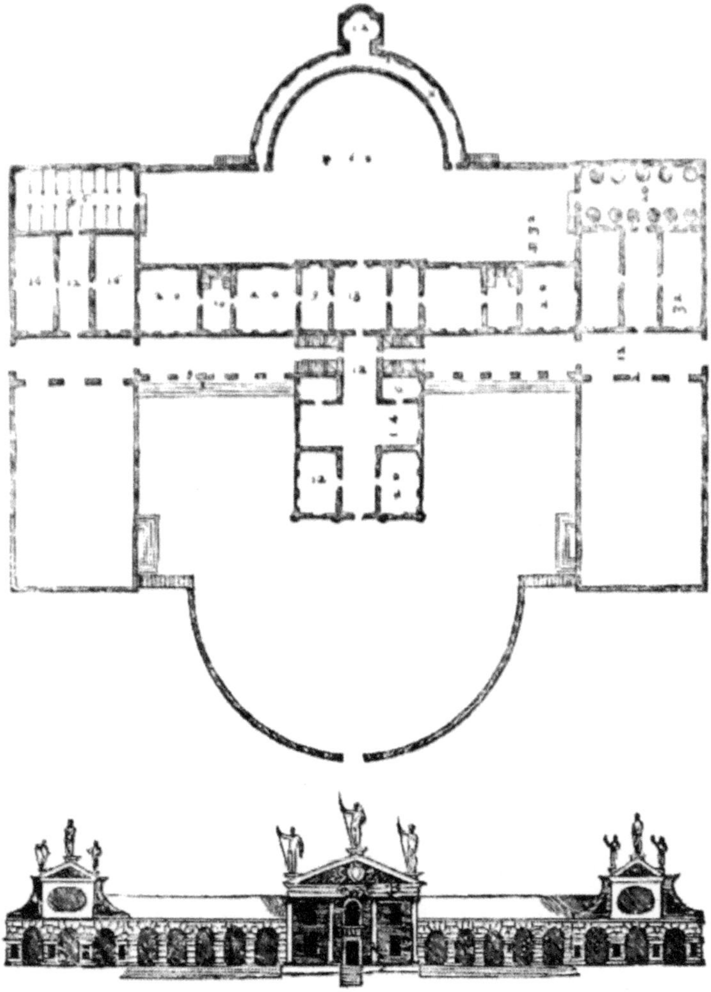

Fig. 54.3 Plan and Elevation of the Villa Barbaro from *I Quattro Libri*. Image: Palladio (1570: II, xiii, 51)

loggias; the Corinthian columns in the upper loggia are 1/5 thinner than the Ionic columns below, lending them a strong verticality; as with the Villa Pisani, the entablature of the first-storey loggia is continued around the entire elevation. At the Villa Badoer at Fratta Polesine (1556) the most purely geometrical of the experiments with the classical elements is the use of colonnades in the form of *quadrants* (quarter circles) to integrate the agricultural outbuildings within the design of the villa. The rear of the complex of the Villa Barbaro at Maser (late 1550's) opens onto a hemicyclical *nymphaeum*, answered by a sweeping semicircular exedra facing the road at the front of the villa (Fig. 54.3).

The plan of the Villa Malcontenta (also known as Villa Foscari) at Gambarare di Mira (1560) is dominated by a Latin cross *salone* with a "semicircular cross vault," the impost of which is "as high above the ground as the hall is broad" (Palladio 1997: II, xiv, 128). Of the rear facade, Rowe writes, "it is by vertical extension into arch and vault, diagonal of roof line and pediment that Palladio modifies the geometrical asperities of his cube; and this use of the circular and pyramidal elements with the square seems both to conceal and to amplify the intrinsic severity of the volumes" (Rowe 1976: 11).

In summary, the main focus of the analysis of plan is the *salone*, since this is often the most geometrically powerful room: it is given the form of the Latin cross in the Villa Malcontenta and the Villa Pisani at Bagnolo, the Greek cross in one of his theoretical villas (Fig. 54.1), the highly articulated square in the Villa Pisani at Montagnana, and the pure circle in the Villa Rotonda and the Villa Trissino, which will be discussed further below.

In Palladio's mature architectural vocabulary, the elements he chose to extract, taken from classical Roman architecture, were often geometric in form, in plan and section but also largely in elevation. In a sense, Palladio developed a geometric toolkit that included linear, planar and spatial tools, from mouldings and rustication to arches and oculi to vaults and pediments. As we shall see in the next sections, Palladio treated most of his classical elements in mathematical ways through the measured use of proportion and symmetry.

Proportion

> Beauty will derive from a graceful shape and the relationship of the whole to the parts, and of the parts among themselves and to the whole, because buildings must appear to be like complete and well-defined bodies, of which one member matches another and all the members are necessary for what is required (Palladio 1997: I, I, 7; similar statements are found in II, i-ii, 77–78).

These words of Palladio essentially restate principles that Vitruvius and Alberti had embraced (Vitruvius 1960: I, ii, 13–14 and VI, ii, 174; (Alberti 1986): I, I, 1; VI, ii, 113; and IX, v, 195). They suggest (at least) two criteria: (1) the parts of a building must relate among themselves, and (2) the parts of a building must relate to the whole (and vice versa). However, depending on the interpretation of the first criterion, the two criteria may be in conflict! Thus it is necessary to consider what the first criterion meant for Renaissance architects, Palladio in particular. "Parts" may be defined in different ways. First, consider "parts" as the components of the orders. These were governed by precise mathematical rules, but different theories were developed to satisfy different requirements. Palladio had two sources for the rules he considered legitimate: the authority of Rome, and mathematics.[2] He omits

[2] For a complete discussion of Palladio's extensive rules governing the proportion of the orders, Palladio (1997: xiii–xix, 18–55).

the anthropomorphic origins of the column types discussed by Vitruvius and Alberti; he also sides with the direct teachings of antiquity over Vitruvius when they differ. The aspect of Palladio's ornamentation that at first seems the least mathematical is in fact rich in proportional content.

Architectural historians have focused principally on two related aspects of "parts": on individual rooms and on their dimensions. The relationships between the length, width, and height of a room were highly important to Renaissance theorists. The underlying reason for this is the focus of the debate over harmonic proportions mentioned in the Introduction. The idea that harmonic proportions are beautiful to the ear because they are part of a higher universal design and thus should be equally beautiful to the eye is traced by Wittkower to the Pythagoreans via Plato, who explained in *Timaeus* that "cosmic order and harmony are contained in certain numbers" (Wittkower 1952: 91). Alberti draws on Pythagoras when he "conclude [s] that the same Numbers, by means of which the Agreement of Sounds affects our Ears with Delight, are the very same which please our Eyes and our Mind" (Alberti 1986: IX, v, 196–197). Palladio seems somewhat ambivalent on the subject:

> He subscribed to the ancient topoi that the macrocosm of the world was reflected in the microcosm of man and that the rules of architecture refer to the rules of nature, but there is very little evidence that Palladio treated such concepts as more than metaphors. Indeed he once remarked appositely that "just as the proportions of voices are harmony to the ears, so those of measurement are to the eyes, which according to their habit delights [in them] to a great degree, without it being known why, save by those who study to know the reasons of things" (Boucher 1994: 239).

Although Palladio makes no specific mention of analogies to music in his *Quattro Libri*, there is substantial use of harmonic proportions in that treatise. One conclusion reached by Deborah Howard and Malcolm Longair (1982:121ff) in their study of all 44 plans of Book II in order to measure statistically Palladio's use of "harmonic numbers" is that about 2/3 of the dimensions followed harmonic proportions, whereas only 45 % would be harmonic had Palladio picked dimensions at random. Branko Mitrović contrived an explanation of how $\sqrt{2}:1$ can be viewed as a musical ratio using the augmented fourth of a tempered scale. After taking into account heights as well as lengths and widths, he concludes that Wittkower's thesis was more consistent than it seemed (Mitrović 1990: 281–285).

If harmonic proportions really are at work in Palladio's architecture, does it not imply that some mathematical proportions are inherently more beautiful than others? If so, does this not admit the possibility of additional mathematical components of beauty? But if harmonic proportions are not at work, the search for the operative factor must need to be expanded! One such operative factor may lie in pure mathematics.

An examination of the plans of Book II for Palladio's seven preferred room shapes found significant evidence that room shapes were more important to him than harmonic ratios. Howard and Longair suggest that either "Palladio used a system of musical harmonies ...; or ... that he adhered to his own simpler recommendations concerning room shapes; or ... that he recognized the practical

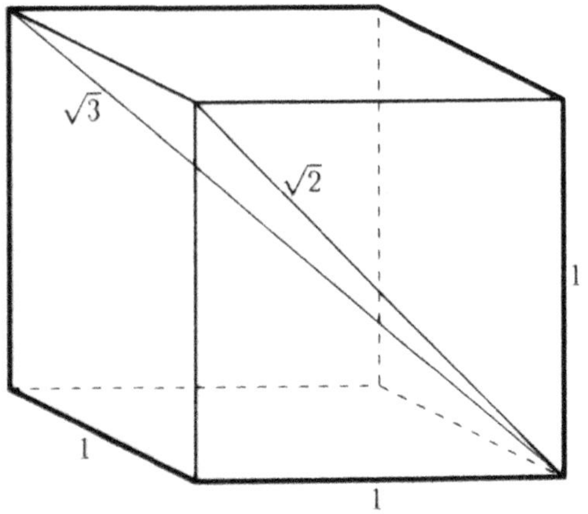

Fig. 54.4 Illustration of Alberti's construction of a cube exhibiting the proportions √3 and √2: "...we may consider the Line drawn from one Angle of the Cube to that which is directly opposite to it, so as to divide the Area of the Square into two equal Parts, and this is called the Diagonal. What this amounts to in Numbers is not know: Only it appears to be the Root of an Area, which is as Eight on every Side; besides which it is the Diagonal of a Cube which is on every Side, as twelve" (Alberti 1986: IX, vi, 199). Image: author, after (Alberti 1986: Pl. 64)

advantages of using simple, easily divisible numbers."[3] Mitrović is more provocative. He finds that six unexplained ratios are close to √3:1, with the closest being the four large corner rooms of the Villa Rotonda. Each corner room has dimensions 26 by 15, which differs from √3:1 by only 0.07 %, a deviation smaller than the allowable error used in many rigorous scientific experiments! (Mitrović 1990: 285–286).

The ratio √3:1 is referred to as *triangulature* since it can be derived as the ratio of the height of an equilateral triangle to half of the base. Mitrović informs us that this method was well known in Renaissance times. In fact, Alberti (1955: IX, vi, 199) describes precisely this construction, but immediately before this, he describes a construction based on "some other natural Proportions for the Use of Structures, which are not borrowed from Numbers, but from the Roots and Powers of Squares" (Fig. 54.4). This construction simultaneously produces the ratios √3:1 and √2:1 by way of a perfect cube, and this may have appealed greatly to Palladio.

Regarding room proportions, many writers have argued that the difference between a 30 × 30 room and a 29-1/2 × 30-1/2 room might be imperceptible and that the concept of a proportional system (harmonic or otherwise) is thus meaningless. This is not the point, however, with Palladio, who governed himself

[3] Howard and Longair (1982: 136). For the seven preferred room shapes, see Palladio (1997: I, xxi, 57). He recommends circles, squares and rectangles of proportions √2:1, 4:3, 3:2, 5:3 and 2:1. The last four are harmonic proportions; all are consistent with Vitruvius and/or Alberti, though circles are discussed only in terms of temples; see Vitruvius (1960: IV, viii, 122–124 and VI, iii, 177–179); Alberti (1955: VII, iv 138–139 and IX, v–vi, 197–199).

by principles, many of which were mathematical in nature. While it was impossible for him to capture the ratio $\sqrt{3}:1$ with integers, it was still possible to capture the principle of the perfect cube, in a sense, by using an extremely close approximation to this ratio.

Proceeding now with the other interpretation of the first criterion, we may take *whole rooms* as the 'parts' of a building and consider ways in which multiple rooms can relate among themselves. This was of explicit concern to Palladio:

> But the large rooms should be distributed with the medium-sized, and the latter with the small rooms in such a way that . . . one part of the building corresponds to the other so that the whole body of the building would have an inherently suitable distribution of its members, making the whole beautiful and graceful (Palladio 1997: II, ii, 78).

Wittkower maintains Palladio's

> systematic linking of one room to the other by harmonic proportions was the fundamental novelty. . . . Those proportional relationships which other architects had harnessed for the two dimensions of a façade or the three dimensions of a single room were employed by him to integrate a whole structure (Wittkower 1952: 113).

What, then, was Palladio's method? A simple answer lay in the restriction of the dimensions of individual rooms to the "harmonic numbers," thus the rooms would relate to each other via harmonic proportions. Another, more ingenious approach is found in Palladio's rules for determining the heights of rooms. For flat ceilings, the height is taken to be equal to the width, $h = w$. For vaulted ceilings in square rooms, Palladio's rule is simply $h_s = (4/3)w = (4/3)l$. For vaulted ceilings in rectangular rooms, the height is determined in three possible ways, corresponding to the arithmetic, geometric, and harmonic means: using the arithmetic mean, $h_a = (w + l)/2$; using the geometric mean, $h_g = \sqrt{wl}$; using the harmonic mean, $h_h = 2wl/(w + l)$. Of course, Palladio uses neither these names nor this modern notation; his definitions are purely numerical, and he supplies examples with numbers aligned in particular ways for ease of comprehension. (Alberti defines the three means in his treatise, naming them *arithmetical, geometrical* and *musical* (Alberti 1986: IX, vi, 199–200). Why Palladio does not use these names is a good question.) More importantly, Palladio supplements the numerical methods with illustrations of geometric constructions for each mean, employing a geometric approach to ensure correct proportional relationships, both within rooms and between rooms (Fig. 54.5).

Of the use of the three means for the heights of vaulted rooms, Palladio writes,

> [W]e should make use of each of these heights depending on which one will turn out well to ensure that most of the rooms of different sizes have vaults of an equal height and those vaults will still be in proportion to them, so that they turn out to be beautiful to the eye and practical for the floor or pavement which will go above them (Palladio 1997: I, xxiii, 59).

This can be done with a (limited) number of Palladio's preferred proportions in such a way that the height/width ratios are also among the Mitrović smaller rectangular rooms of the Villa Rotonda are related to the large corner rooms. Recall that each corner room has length/width ratio of 26:15, approximately equal

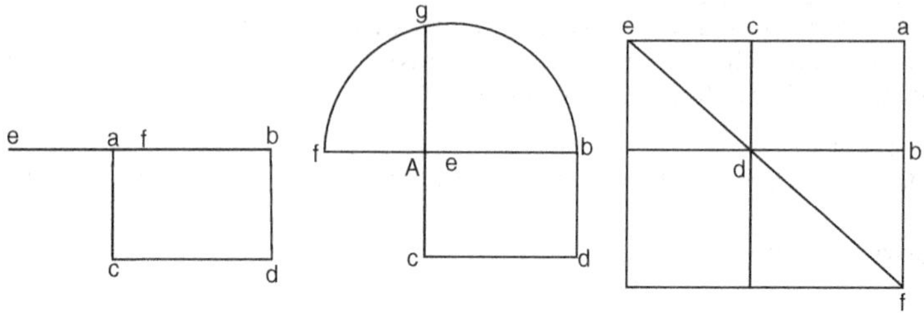

Fig. 54.5 Palladio's constructions for the three means (equations 3–5) from the *Quattro Libri*. For those wishing to verify Palladio's definitions, use the Pythagorean theorem on the second and similar triangles on the third. Image: Palladio (1570: I, xxiii, 53–54)

to √3. The height of the corner room is determined using the arithmetic mean, $h_a = (26 + 15)/2 = 20\ 1/2$, so that the height/width ratio is 20 1/2: 15 = 1.3666. Each smaller room has dimension 15 × 11 (they share a dimension with the large room), a length/width ratio of 15/11 = 1.3636![4] The closeness of 1.3666 and 1.3636 suggests that Palladio was very careful about the proportional relationships of his most celebrated villa, and it appears that he relied on pure mathematics as opposed to harmonic proportions.

Finally, let us turn to the criterion that the parts of the building must relate to the whole, where we still consider "parts" to be rooms and their dimensions. The "additive problem," choosing rooms from a small set of ratios such that they add to produce another one of these ratios, is not easily solved, especially if the ratios must be commensurate.[5] Therein lies the inherent conflict between the two criteria. Order may be introduced through the use of a square grid, as, for example, a 3 × 3 square grid containing only the ratios 1:1, 3:2, 2:1, and 3:1 (as opposed to a generic 3 × 3 grid, which can have as many as 36 different ratios embedded in it[6]) but this may not always be either practical or aesthetically pleasing. Often the most interesting solutions involve incommensurate ratios based either on √2 or √5. Though solutions may be found with commensurate ratios, Scholfield notes, "Palladio omits the overall dimensions from his plans, and so avoids the problem of adding the separate dimensions together ... His system of proportion integrates the whole structure in the sense that it links the parts, or separate rooms, to each other, but it still fails to relate them to the whole" (Scholfield 1958: 64). Howard and Longair addressed this while trying to discover whether or not Palladio used wall thicknesses to help develop additive solutions, (Howard and Longair 1982:

[4] Mitrović (1990: 289–291). Both decimal figures are close approximations of (1 + √3)/2; for those interested in pure trigonometry, this equals $\sin 30° + \cos 30° = \sqrt{1 + \cos 30°} = \sqrt{\frac{1+\sqrt{3}}{2}}$.

[5] Although Palladio did allow himself the use of approximations of the incommensurate ratio √2:1, he did not use it very often; see Howard and Longair (1982: Appendix, Table A4, 141–143), where this ratio is found only four times out of over one hundred entries.

[6] For more on the additive problem, see Scholfield (1958: 132–134).

128–129) concluding that in some cases he did use wall thicknesses, while in others he missed opportunities to use them.

In fact, Palladio recognized the occasional need to bend the rules, as his instructions on correct vault heights indicate: "There are other heights for vaults which do not come under any rule, and the architect will make use of these according to his judgment and practical circumstances" (Palladio 1997: I, xxiii, 59). Many of his rules for dimensions of doors and windows are practical rather than being based on abstract mathematics or harmonics. "Palladio's intelligence and experience would not have allowed him to suggest that a single proportional theory alone would enable one to design a beautiful building, any more than a musician could compose a great symphony merely with a knowledge of harmony and counterpoint" (Howard and Longair 1982: 137). Indeed, Palladio's toolkit contained many tools, including one that was especially effective in solving the problem of relating the parts to the whole.

Symmetry

The discussion of proportion often focuses on the rooms flanking the *salone*, especially with consideration to the relationships of sequences of rooms to each other. Palladio's commitment to symmetry, simply yet forcefully expressed, ties these elements together into a cohesive whole. His main paradigm is reflective symmetry, the type of symmetry found in the bodies of so many of Earth's creatures. Palladio usually employs "a triadic composition with a central block built around the axis of the entranceway, and two symmetrical flanking blocks. . . . The design was thus tightly knit as an organism" (Ackerman 1966: 160–161).[7] The major events of his designs occur on axis, both in plan and in elevation, relating the two to achieve a more integrated whole. Further, no walls are aligned on top of the axis, and there are almost invariably doors at the perpendicular intersection of walls with the axis, so that one has the pleasure of experiencing the design from anywhere on the axis.

Palladio uses an especially rich symmetry in the Villa Rotonda and the Villa Trissino at Meledo di Sarego.[8] Both are sites on hilltops with excellent views in all directions. Palladio uses the sacred circle for the shape of the *salone* and provides loggias on all four sides, creating two perpendicular axes of symmetry that result in 180° rotational symmetry.[9] The clarity of geometry and depth of symmetry make these villas two of Palladio's most influential designs.

---

[7] See also Ackerman and James (1967: 11–12).

[8] The latter was designed ca. 1567 but never completed, see Palladio (1997: II, iii, 94–95 and II, xv, 138); Puppi (1975: 384–388).

[9] To be precise, the rotational symmetry is broken in the Villa Trissino by the forecourt, the arcades of which project from the central block in quadrants as with the Villa Badoer. The Villa Rotonda,

The symmetry concept underwent a rapid evolution during the Renaissance (Hersey and Freedman 1992: 15–37). Symmetry's original meaning was closer to our concept of commensuration or correspondence in measure, and related more to proportion than to our modern concept of symmetry. Vitruvius often employed the term along with the concept of proportion, as in the phrase "symmetrical proportions" (Vitruvius 1960: VI, ii–iii, 174–180). Hermann Weyl writes:

> [i]n the one sense symmetric means something like well-proportioned, well-balanced, and symmetry denotes that sort of concordance of several parts by which they integrate into a whole. *Beauty* is bound up with symmetry. ... [T]he second sense in which the word symmetry is used in modern times [is] bilateral symmetry" (Weyl 1952: 3–4).

When symmetry took on its current meaning "the word's ancient association with 'beautiful' probably strengthened the idea that a design with two identical halves was more beautiful than one without" (Hersey and Freedman 1992: 16). In addition to bilateral or reflective symmetry, translatory and rotational symmetry were also regarded as common denotations of symmetry. Wittkower points out that Alberti, Leonardo, Francesco di Giorgio, and Serlio were quite intrigued by central plans for churches (Wittkower 1952: 1–28, plates 1–13), their drawings showing an explicit interest in rotational symmetries.

Because Vitruvius prescribed symmetry only for public buildings, the use of symmetry for house plans in the Quattrocento "required vigorous reinterpretation" (Hersey and Freedman 1992: 31). To this end, Cesare Cesariano was more than willing to "clarify and extend" Vitruvius' notion of symmetry so that it applied to domestic as well as public architecture (Hersey and Freedman 1992: 33). Daniele Barbaro insisted that private houses should be equipped with all the refinements of public buildings, including the rigours of proportion and symmetry. Palladio's illustrations for Barbara's edition of Vitruvius include a plan, section and elevation exhibiting the hybrid design.

For his part, Palladio justified symmetry on structural grounds:

> Rooms must be distributed at either side of the entrance hall, and one must ensure that those on the right correspond and are equal to those on the left so that the building will be the same on one side as on the other and the walls will take the weight of the roof equally [. . .] if the rooms on one side are made large and those on the other side small, the former will be more capable of resisting the load because of the thickness of their walls, while the latter will be weaker, causing grave problems. (Palladio Palladio, Andrea 1997: I, xxi, 57).

Renaissance theorists, including Palladio, had thoroughly convinced themselves that symmetry, at very least reflective symmetry, was the only correct design choice. Though he did not give an anthropomorphic rationale for symmetry, he did use the analogy of the human body in explaining the proper placement of rooms. There are, in fact, many rationales for and interpretations of symmetry besides the anthropomorphic and economic. The kinaesthetic rationale is related to the experience of architectural space:

---

on the other hand, has essentially 90° rotational symmetry, except that the rectangular rooms do not quite align in 90° rotation.

> [a] single axis of symmetry in a space impels the spectator smoothly along it, whereas two precisely balancing cross-axes ... as in ... the Villa Rotonda, impart a sense of static serenity (Tabor 1982: 21).

For Palladio, despite his structural claims, a combination of the political and aesthetic arguments seems to be at work. On the one hand, the link through Vitruvius to Rome provided legitimacy; on the other, the use of symmetry went a long way towards solving the aesthetic problem of relating the parts to the whole.

Conclusion

Palladio exhibits a strong interest in geometry, both in the crafting of architectural spaces and in geometric constructions for the correct design of architectural elements and their interrelationships. We have explored Palladio's concern with the proportional relationships of parts among themselves and to the whole, as in the elements of the orders and in the dimensions within and between rooms. Concerning the latter, the whole numbers and occasional simple fractions used are chosen for a number of reasons: to employ proportions suggested by Vitruvius and Alberti, possibly informed by analogies to musical theory; to reference proportions derived from simple, pure geometry; and to provide practical solutions to the problems of a particular design. Finally, Palladio's consistent use of symmetry was an aesthetically pleasing and seemingly correct way to link plan and elevation into a cohesive whole.

Acknowledgment I would like to acknowledge Dr. Carroll William Westfall in gratitude for the fruitful discussions and invaluable suggestions that were crucial to the preparation of this manuscript.

Biography Stephen R. Wassell received a B.S. in architecture in 1984, a Ph.D. in mathematics (mathematical physics) in 1990, and an M.C.S. in computer science in 1999, all from the University of Virginia. He is a Professor of Mathematical Sciences at Sweet Briar College, where he joined the faculty in 1990. Steve's primary research focus is on the relationships between architecture and mathematics. He has co-authored various books, one with Kim Williams entitled *On Ratio and Proportion* (a translation and commentary of Silvio Belli, *Della proportione et proportionalità);* one with Branko Mitrović entitled *Andrea Palladio: Villa Cornaro in Piombino Dese,* and another with Kim Williams and Lionel March entitled *The Mathematical Treatises of Leon Battista Alberi*. Steve's overall aim is to explore and extol the mathematics of beauty and the beauty of mathematics.

References

ACKERMAN, James S. 1966. *Palladio*. Baltimore: Penguin Books.

———. 1967. *Palladio's Villas*. Locust Valley, New York: J. J. Augustin.

ALBERTI, Leone Battista. 1986. *The Ten Books on Architecture* (1755), Giacomo [James] Leoni, trans. New York: Dover Publications.

BOUCHER, Bruce. 1994. *Andrea Palladio: The Architect in His Time*. New York: Abbeville Press.

HERSEY, George and Richard FREEDMAN. 1992. *Possible Palladian Villas (Plus a Few Instructively Impossible Ones)*. Cambridge, Massachusetts: MIT Press.

HOWARD, Deborah and Malcolm LONGAIR. 1982. Harmonic Proportion and Palladio's Quattro Libri. *Journal of the Society of Architectural Historians* **41**, 2 (May 1982).

MITROVIĆ, Branko. 1990. Palladio's Theory of Proportions and the Second Book of the Quattro Libri dell'Architettura. *Journal of the Society of Architectural Historians* **49**, 3 (Sept. 1990).

PALLADIO, Andrea. 1570. *I Quattro Libri dell'Architettura*. Venice: Appresso Dominico de' Franceschi.

PALLADIO, Andrea. 1997. *I Quattro Libri dell'Architettura (The Four Books on Architecture)*. Robert Tavernor and Richard Schofield, trans. Cambridge, MA: MIT Press.

PUPPI, Leonello. 1975. *Andrea Palladio*. Pearl Sanders trans. Boston: New York Graphic Society.

ROWE, Colin. 1976. *The Mathematics of the Ideal Villa and Other Essays*. Cambridge, Massachusetts: MIT Press.

SCHOLFIELD, P.H. 1958. *The Theory of Proportion in Architecture*. Cambridge: Cambridge University Press.

TABOR, Philip. 1982. Fearful Symmetry: A Reassessment of Symmetry in Architectural Compositions. *Architectural Review* **171**, 1023 (May 1982).

VITRUVIUS, Marcus. 1960. *De Architectura (The Ten Books on Architecture)*. Morris Hicky Morgan trans. New York: Dover.

WEYL, Hermann. 1952. *Symmetry*. Princeton, New Jersey: Princeton University Press.

WITTKOWER, Rudolf. 1952. *Architectural Principles in the Age of Humanism*. London: Alec Tiranti.

Chapter 55
Golden Proportions in a Great House: Palladio's Villa Emo

Rachel Fletcher

Build them...with such proportions that together all the parts convey...a sweet harmony.
–Palladio, I Quattro libri, Book IV, Foreword (1997: 213)

Introduction

Among the great houses of the Renaissance is Palladio's Villa Emo at Fanzolo in Northern Italy, built in the late 1550s by Leonardo Emo to realize a family seat in the country. Created originally to support a farming economy in grain, spinning mills and, eventually, maize, or Indian corn, from the New World, the villa until recently belonged to the Emo family for 18 continuous generations and was family farmed. The estate, which consisted of some 300 ha at the time of Palladio, lies on an open plain in Italy's Veneto region, the terrafirma countryside where Venetian Renaissance patricians developed farm retreats to escape the city heat, diversify family holdings, and develop food-producing economies.[1]

Villa Emo is the collaboration of its founding owner Leonardo Emo, architect Andrea Palladio, and fresco painter Giovanni Battista Zelotti. Leonardo's scheme united a diverse community of nobles, landowners, farmers, peasants, craftsmen, and artists about a single site. To support Leonardo's social vision, Palladio

An earlier version of this chapter was published as: Rachel Fletcher, "Golden Proportions in a Great House: Palladio's Villa Emo", pp. 73–85 in *Nexus III: Architecture and Mathematics*, ed. Kim Williams, Ospedaletto (Pisa): Pacini Editore, 2000.

[1] According to recent scholarship, the patron who commissioned Palladio to build Villa Emo was Leonardo Emo, who died in 1586. He was the grandson of the Leonardo Emo who owned and made improvements to the land in Fanzolo, likely had plans to build on it, and who died in 1539. The date of Villa Emo is uncertain, but is believed to have been built c.1558 (Beltramini 2008: 11; communication with Caroline Emo: October, 2001; May, 2012).

R. Fletcher (✉)
New York School of Interior Design, New York, NY, USA
e-mail: rfletch@bcn.net

K. Williams and M.J. Ostwald (eds.), *Architecture and Mathematics from Antiquity to the Future*, DOI 10.1007/978-3-319-00143-2_8,

connected luxurious family quarters to practical farm buildings, then situated the villa complex to face nearby peasant dwellings of the villa's borgo, or hamlet.

Natural Harmony

At Villa Emo, Palladio blended Vitruvian rules of measure with local customs by combining the grace and elegance of classical mathematical proportions with sustainable methods of farm planning and landscape siting. The long arcaded wings flanking the central block derive from vernacular farm dwellings, or *barchesse*, and, prior to modern-day renovations, were fitted with granaries on the first floor, storage rooms on the ground floor, and dovecotes on the outer ends. The villa is situated to promote energy and health and utilizes sun and wind for heating and drying. The entire complex faces approximately 10° east of south, receiving direct solar gain earlier in the winter day. The orientation aligns with the ancient Via Postumia and a landscaped grid of cultivated fields, dating to late Roman times, and with an owner-improved canal and irrigation system introduced prior to Leonardo's occupancy. At one time, water drawn from the Brentella, an artificial canal diverted from the Piave River, serviced the villa's fields on an 11-day cycle and powered mills for grinding corn.[2]

Villa Emo is sited with respect to the cardinal points. Taking into account the 10° solar adjustment, a hill of the distant Asolo range to the rear and an open avenue of poplars planted in front mark north and south, respectively. The spacious colonnades project into the countryside and express east and west. The axes converge at the central block, where lofty rooms on the main floor (*piano nobile*) are decorated with frescoes by Zelotti that depict mythological subjects and scenes of agrarian life. An open corridor extends through the portico and central hall (*sala*), lightening the overall mass and producing an interior to exterior flow.

Palladio's methods for harmonizing buildings with their physical surroundings borrow from rural traditions and local farm customs. The villa's south-facing ramped entry in front may have served as a threshing floor, where grain could be spread in the sun to air and dry.[3] Palladio adopted a number of classical rules for building, originally recorded in *De architectura libri decem* (*Ten Books on Architecture*), the master treatise of ancient Roman author and architect Marcus Vitruvius Pollio. Vitruvius is well known for applying mathematical proportions to a building's measures, but was equally adept at achieving harmony in more pragmatic and natural ways.

[2] Water rights were awarded by 1536 to the senior Leonardo who made improvements to the land (Beltramini 2008: 11; communication with Leonardo Marco Emo Capodilista: December, 1993; Cook (n.d.): 4; Favero 1972: 13–14).

[3] The date of the ramp and its attribution to Palladio are questioned, but its presumed function as a threshing area is in keeping with the owner's agricultural vision. Structural components underneath suggest it is contemporary with the original project (Fondazione Villa Emo Onlus).

Vitruvius recommends running water and moving air for cooling and purification, and orienting different room types to specific light exposures. Winter dining rooms and baths ought to face southwest to receive the early evening light and warmth of the winter's setting sun. An eastern orientation permits bedrooms to receive morning light, prevents the contents of libraries from discoloring and decay, and moderates the temperature of spring and autumn dining rooms as the sun travels westward through the day. Meanwhile, summer dining rooms, studios, and picture galleries face north to avoid the summer heat and take advantage of indirect light protection (Vitruvius 1960: 180–181; Vitruvius 1999: 80, 260).

Inspired by Vitruvius, Palladio promoted similar rules in his own architectural masterwork *I quattro libri dell'architettura* (*The Four Books on Architecture*). The owner's house should connect to farm buildings by long colonnades that provide undercover passage, shade, and fuel wood protection. Haylofts should open to the south or west to prevent fire and fermentation. Wine cellars exposed to indirect eastern or northern light prevent the wine from weakening. Storerooms and granaries should be elevated, well vented, and face north to stay wind cooled and dry. Threshing floors should be spacious and face south for full sun exposure (Palladio 1997: 123). At Villa Emo, storerooms along the *barchesse* open to the south, but the loggias are sufficiently wide to accommodate wind cooling and shade protection.

Harmony in Number

In addition to these vernacular methods for integrating built forms and the natural landscape, harmony appears mathematically in the villa's dimensions. In fact, there are two distinct versions of Villa Emo—the plan that Palladio published in *I quattro libri* and the villa he actually built and that survives today. The discrepancy between the two was known as early as the 1770s when Ottavio Bertotti Scamozzi published *Le fabbriche e i disegni di Andrea Palladio* and attempted to reconcile the built and published versions of Palladio's works (Bertotti Scamozzi 1976: 75–76).

Palladio's published plan of Villa Emo is measured in Vicentine feet, or *piedi*, and describes a central block that consists of a 27 × 27 central hall, or *sala*, and additional rooms of 12 × 16, 12 × 27, 16 × 16, and 16 × 27 framing the hall and portico (16 × 27) (Fig. 55.1).

Various chambers in the wings measure 12, 24, and 48 *piedi* across. Rudolf Wittkower, interpreting Italian humanist and architect Leon Battista Alberti, and classical scholar Francis M. Cornford, commenting on Plato's *Timaeus*, have identified these measures in a mathematical system of harmony that is consistent with the whole number ratios of musical consonance. The numbers of music emerge when harmonic and arithmetic means are applied to the geometric sequences 1, 2, 4, 8 and 1, 3, 9, 27 that comprise the Platonic *lambda*.[4]

[4]The numbers in each series are multiplied by the number six (Plato 1948: 66–72; Wittkower 1971: 110–111).

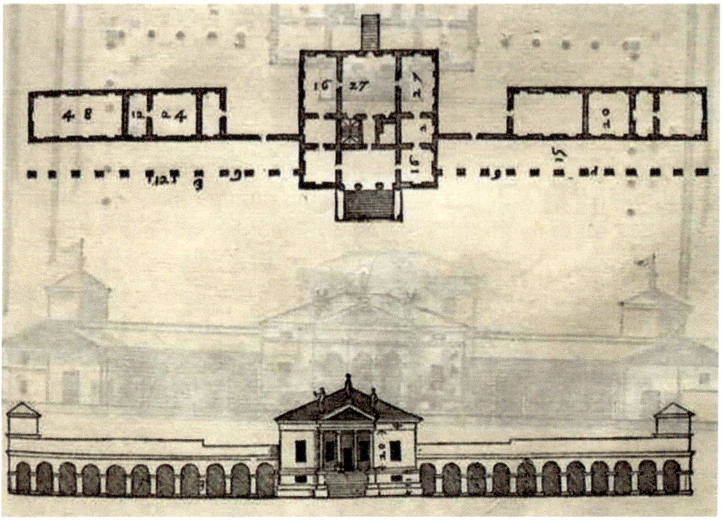

Fig. 55.1 Plan and façade elevation of Palladio's Villa Emo by Andrea Palladio, 1570. Image: Palladio (1570: II, xiiii, 55)

Alberti, whose *De re aedificatoria* (*On the Art of Building in Ten Books*) was the first architectural treatise of the Renaissance, revived the ancient theories of Vitruvius, Plato, and the Pythagoreans, and translated these numbers of musical consonance into spatial ratios and proportions: "The very same numbers that cause sounds to have that *concinnitas*, pleasing to the ears, can also fill the eyes and mind with wondrous delight" (Alberti 1988: 305). Thus evolved architectural rules for orchestrating a building's individual measures and harmonizing the parts with the whole.

At Villa Emo, Wittkower explains that the chambers in ratio 16:27 would be viewed as a compound ratio generated from 16:24:27 and comprised of 16:24 and 24:27. In music, this translates to a fifth (16:24 or 2:3) and a major tone (24:27 or 8:9). The 12 × 27 room would read 12:24:27, or an octave (12:24 or 1:2) and a major tone (24:27 or 8:9). Meanwhile, the measures of 12, 24, and 48 in the wings express two musical octaves.[5] Lionel March observes still more connections to the *lambda*, and to the Pythagorean 3:4:5 right triangle (March 2001: 96–100).[6]

[5] As Wittkower puts it, the measures would read (24:12:48) or (2:1:4), with 1:4 expressing two octaves (Wittkower 1971: 131).

[6] March's article was written as part of a debate with (Fletcher 2001).

Symmetry and Proportion

The appearance of musical ratios in the measures of Villa Emo reflects the Renaissance quest for order in structure and harmony in measure. From classical times, a quadrivium of mathematical arts codified theoretical mathematics into four distinct studies and provided a basis for order and harmony in numeric terms. One reference to the quadrivium appears in Plato's *Republic*. Besides music, or harmonics, which addresses the laws of audible motion; arithmetic is the theory of pure number; geometry, both plane and solid, conveys the eternal nature of mathematical objects; and astronomy codifies the pure laws of bodies in motion (Plato 1945: 235–250). This fourfold curriculum had far-reaching implications in the classical world and was a cornerstone of Plato's view of creation and doctrine of unity, which he defined as harmonized diversity.

ΑΓΕΩΜΕΤΡΗΤΟΣ ΜΗΔΕΙΣ ΕΙΣΙΤΩ

Inscribed at a later time over the door of Plato's Academy, the school founded by the Greek philosopher in Athens, this phrase from Euclid translates to: "Let no one ignorant of geometry enter." To Plato, mathematical symmetry demonstrated that the universe is alive and endowed with intelligence, purpose, and order, with number essential to its creation and geometric proportion a means of unifying its endless variety. The *Timaeus* (32d) presents a universe "in the fullest measure a living being whole and complete, of complete parts" (Plato 1948: 52). Thus, it is called *kosmos*, the Greek word for "order," "good order" and "form." Before Plato, Pythagoras proclaimed that, "All things accord in number" (Iamblichus 1986: 87).

To most people, "symmetry" is the division of space into identical fragments, as in the bilateral organization of elements in anatomy, or the axial arrangement of crystals where the whole is divided into identical parts and uniformly distributed around a point, line, or plane. But "symmetry" can be synonymous with the quality of harmony that relates unique and individual differences. The Greek for "symmetry" is *symmetria*, which means "suitable relation" and "due proportion." In physiology, it refers to the harmonious working of the bodily functions, producing a healthy temperament or condition. In this context, "symmetry" is synonymous with "proportion," which means "the harmonious relation of parts to each other or to the whole."

Incommensurable Proportions

In addition to the whole number ratios of musical harmony, the *Timaeus* introduces a set of incommensurable ratios that characterize geometry's elementary shapes and achieve proportion in yet another way. Here, Plato's Demiurge, a divine creator and craftsman, organizes the elemental world by endowing fire, air, water, and earth with certain mathematical properties of regular solid bodies. Specifically, the planar faces of different volumes divide into constituent triangles. Two 30°–60°–90° triangles comprise each equilateral triangular face of the tetrahedron, octahedron, and icosahedron, which Plato associates with fire, air, and water, respectively. Two 45°–45°–90° triangles comprise each square face of the hexahedron or cube, which Plato assigns to earth (Plato 1948: 210–218).[7] Briefly mentioned is a fifth construction used "for the whole, making a pattern of animal figures thereon." It is assumed Plato means the dodecahedron, whose 12 pentagonal faces represent the signs of the zodiac and are comprised of 36°–72°–72° isosceles triangles (Euclid 1956: II, 98–99; Plato 1948: 218–219).

The specific connection between these geometric shapes and elementary material particles is not made clear. As Cornford explains, Plato's cosmology is not an exact account of the physical laws of modern science, but rather a poetic statement about the imposition of exact mathematical principles on the sensible world to bring order out of chaos and produce "an intelligent and intelligible design" "made after the likeness of an eternal original" as near as possible (Plato 1948: 30–33, 36, 39).

Together, the tetrahedron, octahedron, icosahedron, cube, and dodecahedron constitute the regular or "Platonic" solids.[8] By describing their constituent triangles, Plato introduces a set of incommensurable ratios that cannot be expressed in whole numbers, but that inhabit the elementary shapes of geometry precisely. The half-side and altitude of any equilateral triangle (or the two sides of a 30°–60°–90° triangle) are in ratio 1:1.7320... or $1{:}\sqrt{3}$. The side and diagonal of any square (or the side and hypotenuse of a 45°–45°–90° triangle) are in ratio 1:1.4142... or $1{:}\sqrt{2}$. And the side and diagonal of any regular pentagon (or the short and long sides of a 36°–72°–72° triangle) are in ratio 1:1.6180... or $(1{:}\sqrt{5}/2 + 1/2)$. This ratio is commonly known as the "golden mean," "golden section," or "extreme and mean" ratio, and is written $1 : \phi$ (Fig. 55.2).

The incommensurable ratios that characterize these regular geometric shapes have the potential to divide space proportionally, as they replicate through endless divisions and relate one level of subdivision to the next. Twentieth-century artist and scholar Jay Hambidge calls this method of spatial composition "dynamic symmetry," which he observes in the way that root rectangles divide into reciprocals, or smaller rectangles of the same proportion (Hambidge 1967). A $1{:}\sqrt{2}$ rectangle divides continuously into two reciprocals in root-two ratio. The area of each reciprocal is one-half the area of the whole (Fig. 55.3).

[7] Each triangular face divides further into six constituent 30°–60°–90° triangles and each square face divides into four 45°–45°–90° triangles. Thus, the four triangular faces of the tetrahedron contain 24 right triangles. The eight triangular faces of the octahedron contain 48 right triangles. And the 24 triangular faces of the icosahedron contain 120 right triangles. The six square faces of the cube contain 24 right triangles. In this way, Plato is able to pose mathematical formulas that describe how elements transform into one another.

[8] Each regular solid is a convex polyhedron in which: all faces are the same; all faces are regular polygons (squares, triangles, or pentagons); the same number of edges meet at each vertex; and all vertices touch the surface of a circumscribing sphere.

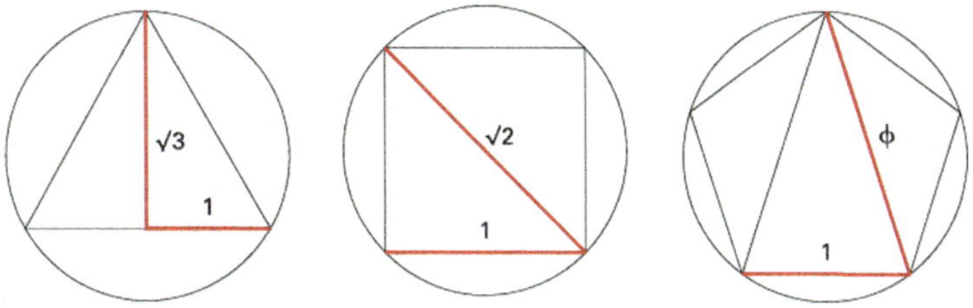

Fig. 55.2 Incommensurable ratios in basic geometric shapes

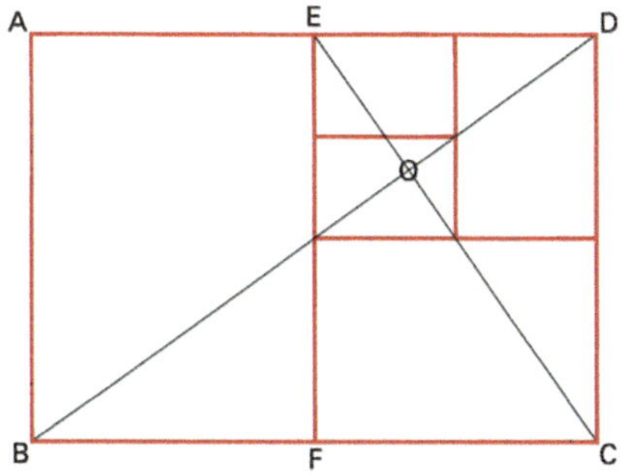

Fig. 55.3 Root-two rectangle and reciprocals

A 1:√3 rectangle divides continuously into three reciprocals in root-three ratio. The area of each reciprocal is one-third the area of the whole (Fig. 55.4).

The rectangle in ratio 1:1 : ϕ is called a whirling square rectangle because it divides continuously into a reciprocal and a square (Fig. 55.5). Harmony is sustained each time the governing incommensurable ratio repeats at a new spatial level.

The Golden Section

The golden mean, or golden section, provides a remarkably efficient way to achieve unity among a diversity of elements, for it increases simultaneously by geometric progression and simple addition. This unique characteristic may account for its appearance in natural form and human anatomy.

The ratio is identified by the Greek letter *phi* ($\phi = \frac{\sqrt{5}+1}{2}$ or 1.618034…), after the Greek sculptor Phidias. It is found when a line divides into two unequal lengths such that the shorter length relates to the longer in the same way as the longer length relates to the whole. If the whole equals 1, the proportion translates to $\frac{1}{\phi^2} : \frac{1}{\phi} :: \frac{1}{\phi}$

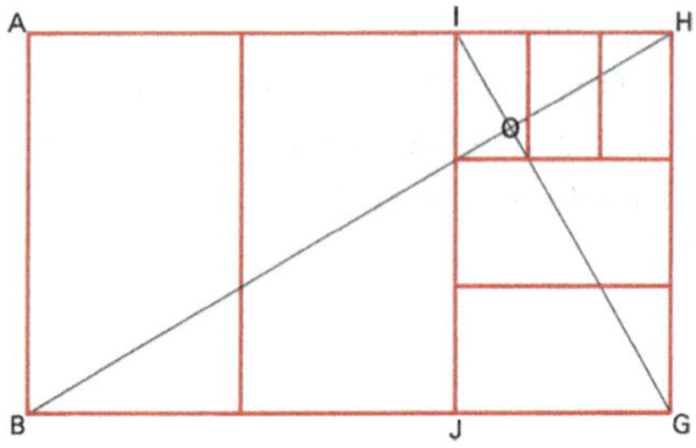

Fig. 55.4 Root-three rectangle and reciprocals

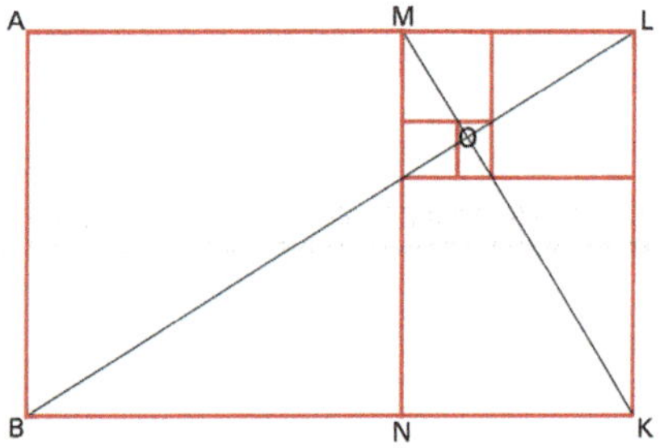

Fig. 55.5 Rectangle of whirling squares

: $\left(\frac{1}{\phi^2}+\frac{1}{\phi}\right)$ or 1. The reciprocal of *phi* ($\frac{1}{\phi}$) equals ($\sqrt{5}/2 - 1/2$) or 0.6180…. The ϕ number sequence increases by geometric proportion ($\frac{1}{\phi}$: 1 :: 1 : ϕ) and by simple addition ($\frac{1}{\phi} + 1 = \phi$), since each new term is the sum of the previous two.

$\frac{1}{\phi^3}$	$\frac{1}{\phi^2}$	$\frac{1}{\phi}$	1	ϕ	ϕ^2	ϕ^3…
0.236…	0.382…	0.618…	1.0	1.618…	2.618…	4.236…

Although some do not agree, the extreme and mean proportion frequently appears in nature, from sunflowers, apple blossoms, and daisies in the plant world to spiral shells beneath the seas. Allowing for individual differences, the proportions of the human body demonstrate golden mean geometry. Fingers divide at the joints in golden mean progression. The face generally conforms to a golden mean rectangle. The tradition of rendering the human body divided at the navel in golden ratio dates to ancient Egyptian times and continued in the twentieth century with Le Corbusier's Modulor.[9]

The spiral-like triton shell approaches golden mean proportions, when viewed in cross section. Its central pillar or stem divides continuously in golden ratio, similar to the way that human fingers divide at the joints (Fig. 55.6a, b). Each curved whorl

[9] A condition of the Egyptian system is that a small portion of the crown is subtracted from the total height (Le Corbusier 1980: I: 50–51; Schwaller de Lubicz 1998: 325).

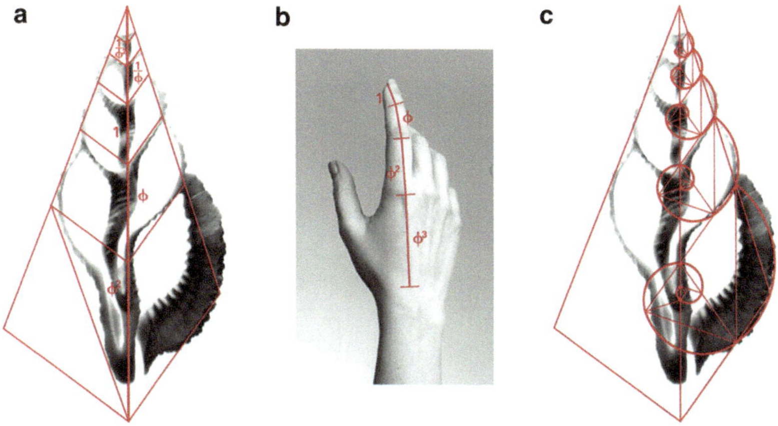

Fig. 55.6 (**a**, *left*) The central pillar of a triton shell divides in golden ratio; (**b**, *centre*) human fingers divide at the joints in golden progression. Photo: George Leisey for Brattleboro Museum & Art Center, Brattleboro, VT; (**c**, *right*) the curved whorls of the shell approximate a portion of the same spiral-like figure

of the triton approximates a portion of the same golden mean spiral-like figure (Fig. 55.6c).[10]

As a mathematical principle, the extreme and mean ratio appears as early as Euclid, if not before (Heath 1981: I, 168). But the origin and history of its use in art and architecture are rigorously debated. Opponents are careful to distinguish *phi* as a mathematical principle from its design application. Marcus Frings argues that the golden ratio does not appear in the canon of proportion attributed to Vitruvius and, therefore, architects and artists of the Renaissance who rediscovered Vitruvian principles are unlikely to have adopted it (Frings 2002: 9–20).

Golden Proportions at Villa Emo

And yet, golden mean proportions appear in the constructed Villa Emo. In 1967, architects Mario Zocconi and Andrzej Pereswet-Soltan completed a definitive survey for the Centro Internazionale di Studi di Architettura Andrea Palladio (C.I. S.A.). The plan and elevation from that survey (Figs. 55.7 and 55.8) provided the

[10] In a true logarithmic or equiangular spiral, a radius vector "makes a constant angle with the tangent to the spiral." For a given angle of rotation, the distance from the pole is multiplied or divided by a specific amount. Each point on the spiral develops from the same center or pole, and the radius vector changes constantly (Sharp 2002: 59–62). The spiral in Fig. 55.6c resembles a logarithmic spiral in ϕ ratio, but its individual arcs develop from different points and the radius vector remains constant until a new arc is taken.

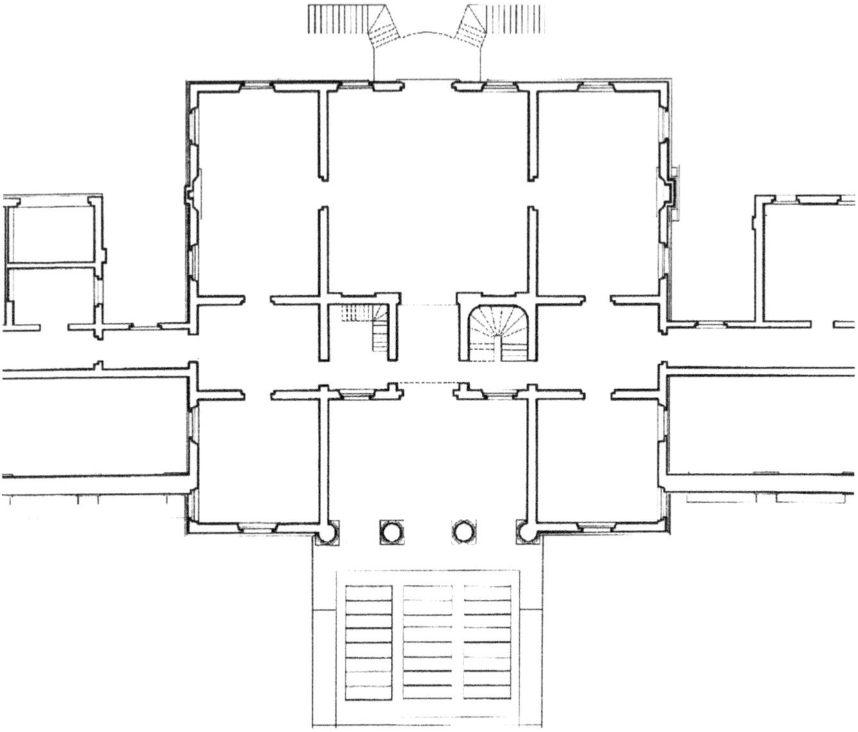

Fig 55.7 First-floor plan of Palladio's Villa Emo, Fanzolo. Survey drawing by Mario Zocconi and Andrzej Pereswet-Soltan (1977: Pl. III). Reproduced by permission

basis for the geometric analysis that follows. The discrepancy between the built plan and Palladio's published version is subtle, but sufficient to allow for a different proportional system (Favero 1972: 31–32). In the built version, the golden ratio prevails throughout the elevation and plan, appearing repeatedly in the building's measures, from the overall proportions of the central block to the placement of individual doors and fireplaces.

The plan of the central block is not perfectly square, but proportioned to a circle inscribed by two smaller squares (Fig. 55.9). Room dimensions on the first floor are in golden mean ratio. The system of dividing a golden rectangle into a square and reciprocal golden rectangle is evident (Fig. 55.10). The central passage is in golden proportion (Fig. 55.11). The placement of doors and fireplaces derives from a regular pentagon whose base is drawn on the front edge of the portico (Fig. 55.12). Approximate golden mean spirals flow through the scheme (Fig. 55.13).

Fig 55.8 Façade elevation of Palladio's Villa Emo, Fanzolo. Survey drawing by Mario Zocconi and Andrzej Pereswet-Soltan (1977: Pl. VI). Reproduced by permission

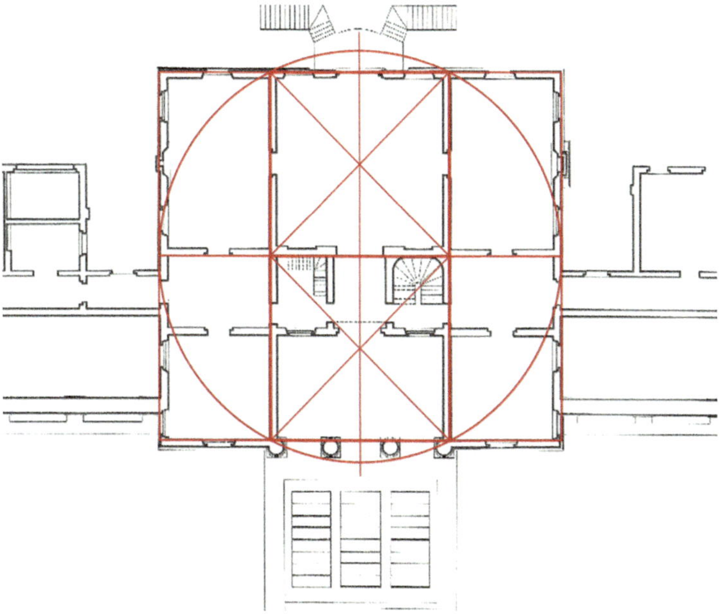

Fig 55.9 The plan of the central block is not perfectly square, but proportioned to a circle inscribed by two smaller squares. Overlay: author

On the front façade of Villa Emo, double squares delineate the flanking walls. The diagonals of the double squares locate the endpoints of the hip of the roof, in golden mean proportion (Fig. 55.14). Proportions derived from other geometric shapes are present. A rectangle proportioned to the half-side and altitude of an equilateral triangle, in ratio 1:√3, defines the height and width of the upper two

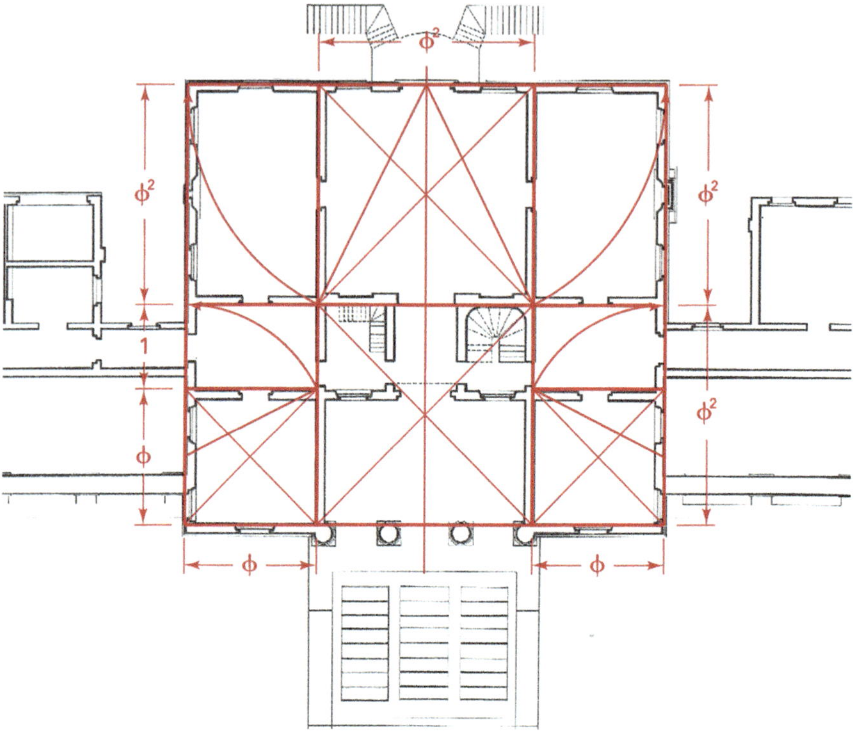

Fig 55.10 Room dimensions are in golden mean ratio. The system of dividing a golden rectangle into a square and reciprocal is evident. Overlay: author

stories (Fig. 55.15). But golden mean proportions dominate the façade overall (Fig. 55.16).

Total Harmony

Palladio's design reflects an integrated system of proportion in which a simple geometric theme binds plan and elevation, interior and exterior, and whole and part to achieve dynamic unity in three dimensions. Harmony is preserved at Villa Emo not merely in the style and proportions of the architect's classical building, but in numerous aspects of villa life. The villa sustained a nurturing agricultural relationship with the land for more than four and a half centuries.

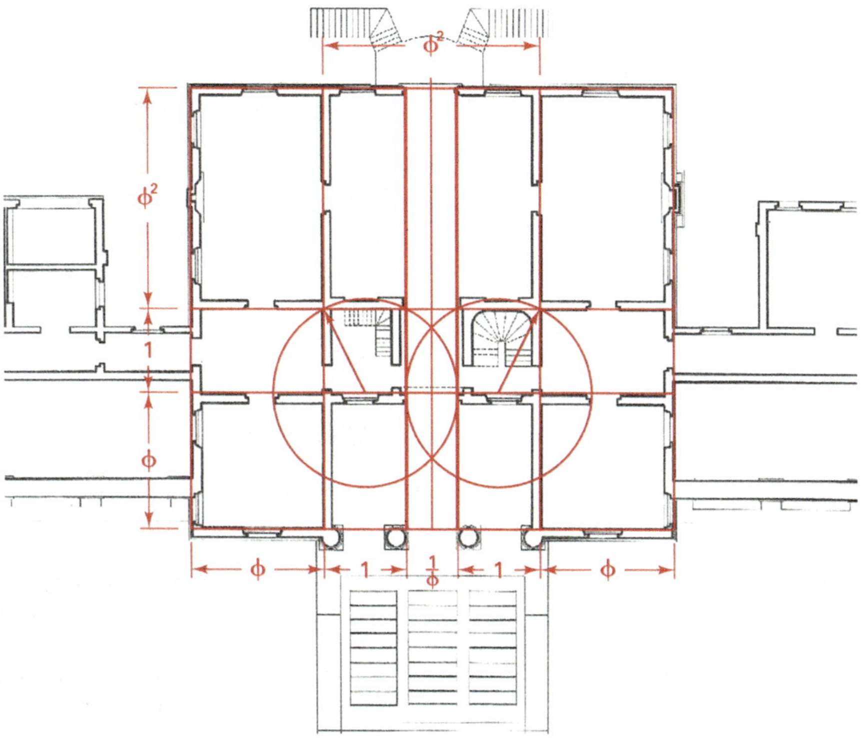

Fig 55.11 The central passage is in golden proportion. Overlay: author

In *I Quattro libri*, Palladio recommends that houses and temples be built "with such proportions that together all the parts convey to the eyes of onlookers a sweet harmony..." (Palladio 1997: 213). Villa Emo achieves beauty and integrity not merely in the mathematical design of spatial elements, but in the blending of mathematical proportions with organic methods of land-use planning and siting. Ultimately, Palladio's vision of harmony evokes the heavens as a pattern for architecture. As celestial revolutions produce the seasons, each in its proper place and all in agreement, architecture should orchestrate the elements of building, landscape, and the human community so that beauty will arise from "the relationship of the whole to the parts, and of the parts among themselves and to the whole" (Palladio 1997: 7, 213). To this Villa Emo aspires in each particular.

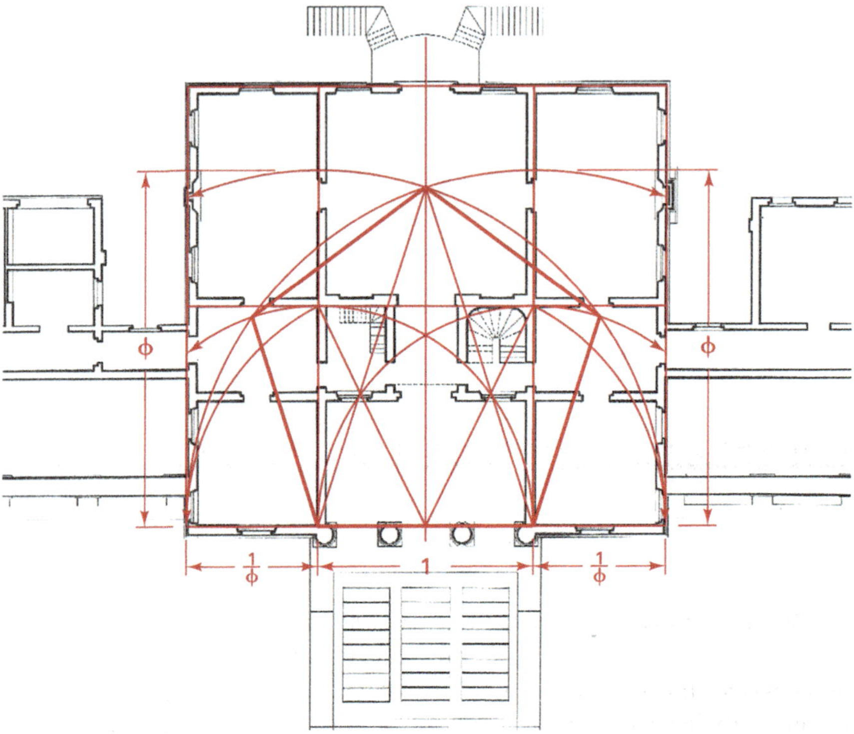

Fig 55.12 The placement of doors and fireplaces derives from a regular pentagon whose base is drawn on the front edge of the portico. Overlay: author

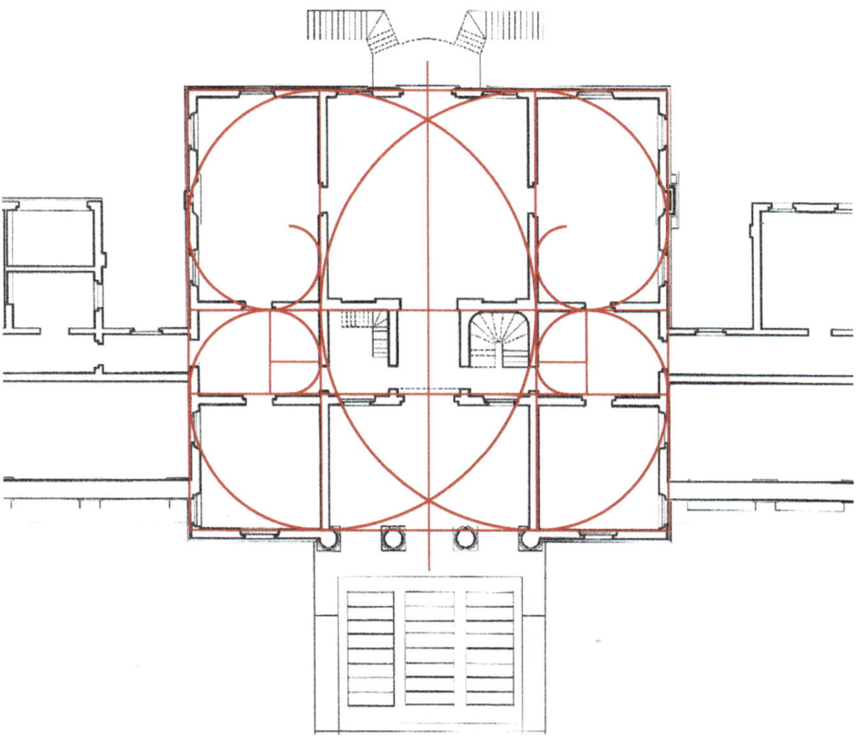

Fig 55.13 Approximate golden mean spirals flow through the scheme. Overlay: author

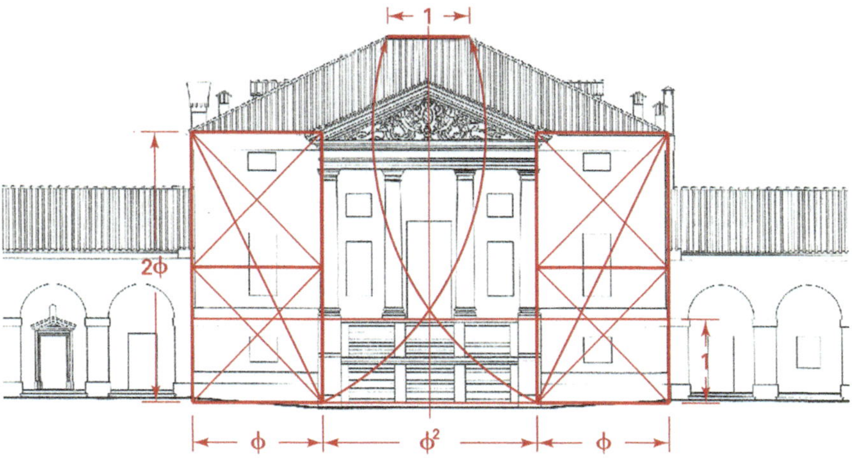

Fig. 55.14 Double squares delineate the flanking walls. Their diagonals locate the endpoints of the hip of the roof, in golden proportion. Overlay: author

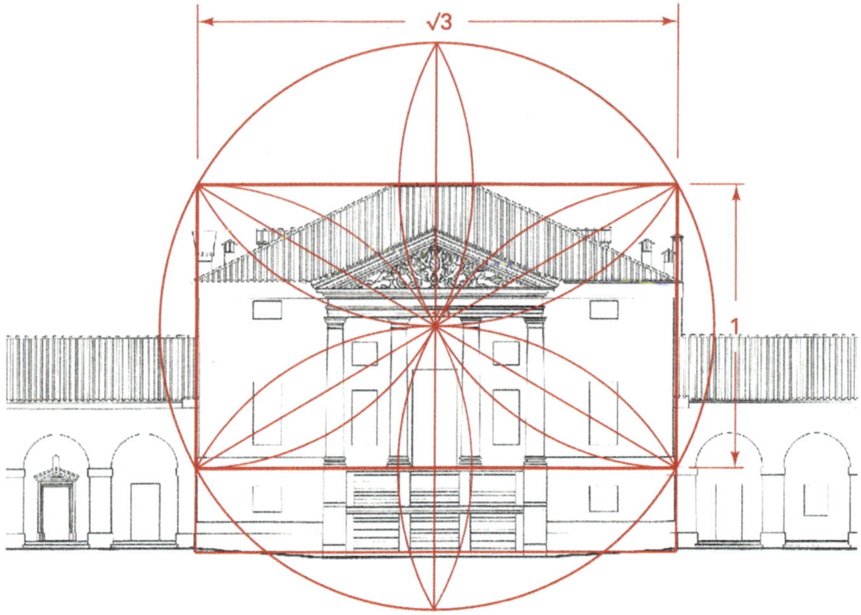

Fig. 55.15 A rectangle proportioned to the half-side and altitude of an equilateral triangle locates the upper two stories. Overlay: author

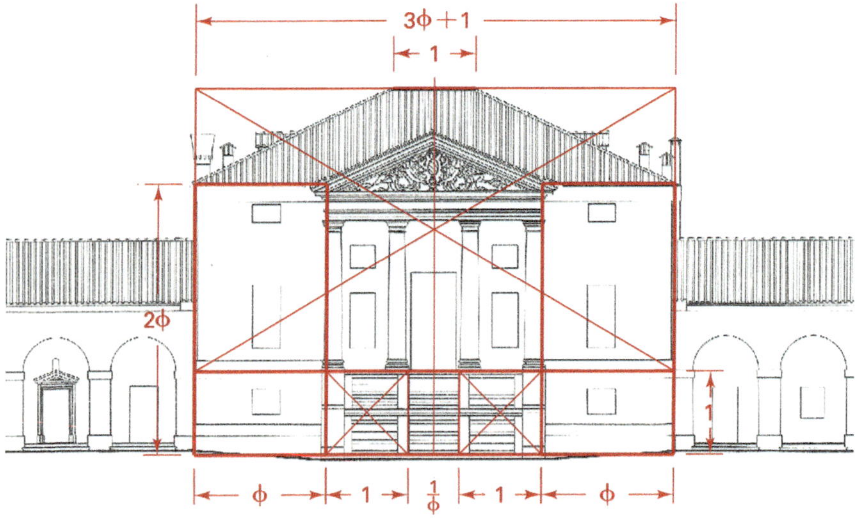

Fig. 55.16 Golden mean proportions dominate the façade. Overlay: author

Acknowledgment The author wishes to thank Count Leonardo Marco Emo Capodilista and Caroline Emo who generously opened the doors of Villa Emo and shared their family history. The original article has been revised to include new information about the villa and family genealogy and to recognize and address challenges to the author's premise made by Lionel March in "Palladio's Villa Emo: The Golden Proportion Hypothesis Rebutted" (March 2001). Alexander Skillman, assisted by Samuel B. Seigle, reviewed and contributed to the etymologies of mathematical terms.

Biography Rachel Fletcher is a geometer and teacher of geometry and proportion to design practitioners. With degrees from Hofstra University, SUNY Albany and Humboldt State University, she is the author of *Infinite Measure: Learning to Design in Geometric Harmony with Art, Architecture, and Nature* (GFT Publishing, 2013). She was the creator/curator of the museum exhibits "Infinite Measure," "Design by Nature" and "Harmony by Design: The Golden Mean." She is founding director of the Housatonic River Walk in Great Barrington, Massachusetts, and founding co-director of the Upper Housatonic Valley African American Heritage Trail. She has been a faculty member of the New York School of Interior Design since 1996 and a contributing editor to the *Nexus Network Journal* since 2005. She resides in Great Barrington, Massachusetts, and her website is http://www.rachelfletcher.org.

References

ALBERTI, Leon Battista. 1988. *On the Art of Building in Ten Books*. Joseph Rykwert, Neil Leach, and Robert Tavernor, trans. Cambridge, MA: The MIT Press.

BELTRAMINI, Guido. 2008. Andrea Palladio 1508-1580. Pp. 2-15 in *Palladio*, Guido Beltrammi and Howard Burns, eds. London: Royal Academy of Arts.

BERTOTTI SCAMOZZI, Ottavio. 1976. *The Buildings and the Designs of Andrea Palladio*. Howard Burns, trans. Trento, Italy: Editrice La Roccia.

COOK, Jeffrey. n.d. Orientation in Palladio's Villas. Unpublished article. Tempe: Arizona State University, Department of Architecture.

EUCLID. 1956. *The Thirteen Books of Euclid's Elements*. 3 vols. Sir Thomas L. Heath and Johan Ludvig Heiberg, trans. New York: Dover.

FAVERO, Giampaolo Bordignon. 1972. *The Villa Emo at Fanzolo*. Douglas Lewis, trans. University Park, PA: Pennsylvania State University Press.

FLETCHER, Rachel. 2001. Palladio's Villa Emo: The Golden Proportion Hypothesis Defended. *Nexus Network Journal* 3(2): 105–112.

———. 2013. *Infinite Measure: Learning to Design in Geometric Harmony with Art, Architecture, and Nature*. Staunton, VA: George F. Thompson Publishing.

Fondazione Villa Emo Onlus. n.d. The Villa: Architecture. http://www.villaemo.org/index.asp?menu = 0203&lingua = EN (accessed 26 November 2013).

FRINGS, Marcus. 2002. The Golden Section in Architectural Theory. *Nexus Network Journal* **4**, 1 (February 2002): 9-32.

HAMBIDGE, Jay. 1967. *The Elements of Dynamic Symmetry* (1926). New York: Dover.

HEATH. 1981. *A History of Greek Mathematics* (1921). 2 vols. New York: Dover.

IAMBLICHUS. 1986. *Iamblichus' Life of Pythagoras, or Pythagoric Life* (1818). Thomas Taylor, trans. Rochester, VT: Inner Traditions.

Le Corbusier (Charles-Edouard Jeanneret). 1980. *Modulor I and II* (1948, 1955). Peter de Francia and Anna Bostock, trans. Cambridge, MA: Harvard University Press.

March, Lionel. 2001. Palladio's Villa Emo: The Golden Proportion Hypothesis Rebutted. *Nexus Network Journal* **3**, 2: 85-104.

Palladio, Andrea. 1570. *I Quattro Libri dell'Architettura*. Venice: Appresso Dominico de' Franceschi.

———. 1997. *The Four Books on Architecture*. Robert Tavernor and Richard Schofield, trans. Cambridge, MA: MIT Press.

Plato. 1945. *The Republic of Plato*. Frances Macdonald Cornford, trans. London, UK: Oxford University Press.

———. 1948. *Plato's Cosmolology: The* Timaeus *of Plato*. Frances Macdonald Cornford, trans., with running commentary. London, UK: Routledge & Kegan Paul.

Schwaller de Lubicz, R. A. 1998. *The Temple of Man: Apet of the South at Luxor*. Vol. 1. Deborah Lawlor and Robert Lawlor, trans. Rochester, VT: Inner Traditions..

Sharp, John. 2002. Spirals and the Golden Section. *Nexus Network Journal* **4**, 1: 59-82.

Vitruvius. 1960. *The Ten Books on Architecture*. Morris Hickey Morgan, trans. New York: Dover Publications.

———. 1999. *Ten Books on Architecture*. Ingrid D. Rowland and Thomas Noble Howe, eds. Cambridge, UK: Cambridge University Press.

Wittkower, Rudolf, *Architectural Principles in the Age of Humanism* (1949). New York: W. W. Norton, 1971.

Zocconi. Mario and Andrzej Pereswet-Soltan. 1977. *Villa Emo di Fanzolo*. Vol. 1. of *Rilievi delle fabbriche di Andrea Palladio*. Vicenza: Centro Internazionale di studi di Architettura Andrea Palladio.

Chapter 56
The Hidden Pavement Designs of the Laurentian Library

Ben Nicholson, Jay Kappraff, and Saori Hisano

Introduction

Although standard measure may have been used in ancient civilizations to measure fields or for other practical considerations, it is fair to assume that the architectural masterpieces of antiquity were created by using the elements of pure geometry. It is our conjecture that the architects of antiquity used various geometrical tools of the trade: the Sacred Cut, based on √2 geometry; a geometrical construction based on the geometry of the half-square, discovered by Danish engineer Tons Brunés (Brunés 1967) and referred to by us as the *Brunés star*; the square-within-a-square (*ad quadratum*); circle grids; a system of proportions originating in ancient Rome; the law of repetition of ratios (Kappraff 1998).

There is scant evidence to support the use of this sacred geometer's tool kit since there are few records of the architects' plans for these ancient structures, but several scholarly investigations have been made of ancient Roman rains that support our point of view (Watts and Watts 1986: 132–139, 1987: 265–276). However, even in

First published as: Ben Nicholson, Jay Kappraff and Saori Hisano, "The Hidden Pavement Designs of the Laurentian Library", pp. 87–98 in *Nexus II: Architecture and Mathematics,* ed. Kim Williams, Fucecchio (Florence): Edizioni dell Erba, 1998.

B. Nicholson (✉)
School of the Art Institute of Chicago, 36 South Wabash, Chicago, IL 60603, USA
e-mail: bnicholson@saic.edu

J. Kappraff
Department of Mathematics, New Jersey Institute of Technology, University Heights, Newark, NJ 07102, USA
e-mail: Kappraff@njit.edu

S. Hisano
2-4-16-503 Nishikicho, Aobaku-ku, Miyagi, Sendai-shi 980-0012, Japan
e-mail: saorikojin@gmail.com

K. Williams and M.J. Ostwald (eds.), *Architecture and Mathematics from Antiquity to the Future*, DOI 10.1007/978-3-319-00143-2_9,

these instances there is a great deal of speculation. Therefore, Ben Nicholson's detailed study of a series of geometric pavement designs executed in terracotta that lie beneath the floorboards of the Laurentian Library in Florence and his investigation of their geometry is of great interest. The results of his work have given additional evidence of the use of the sacred geometer's tool kit.

The Laurentian Library

In 1774, a portentous accident occurred in the Reading Room of the Laurentian Library, designed by Michelangelo (Nicholson 1997). The shelf of desk 74, overladen with books, gave way and broke. During the course of its repair, workmen found a red and white terracotta pavement hidden for nearly 200 years beneath the floorboards. The librarian had trapdoors, still operable today, built into the floor so future generations could view these unusual pavements. In 1928 another mishap resulted in the exposure of the entire pavement, which allowed photographs to be made of the 15 panels on the West side of the library before the wooden floor was replaced.

Overall the pavement consists of two side aisles and a figurative centre aisle. Each side aisle is composed of a series of 15 panels, each measuring about $8'6'' \times 8'6''$, and is of a different design. The 15 panels mirror each other's form but differ by a very small degree and in subtle ways. When juxtaposed in a series, the 15 pairs of panels appear to tell a story about the essentials of geometry and numbers. Each panel settles upon a theme: the *tetractys* (Panel 5); Brunés's star, $\sqrt{2}$ and the Sacred Cut (Panels 7 and 11); Plato's *lambda* (Panel 14); the Golden Mean (Panel 13). When assembled together they form an encyclopedia of the essential principles handed down from ancient geometers.

Although they are hidden from view today, Nicholson believes that the panels were laid according to a plan for a furniture layout that would have exposed them, but that this plan was changed after the panels had been made. He suggests that the original intention was to infuse the spectator with the foundations of ancient geometry as he walked through the Reading Room of the Laurentian Library, the geometry being a perfect complement for the 3,000 classical texts chosen to reveal the body of ancient and modern learning.

Interpreting the Pavement

Ben Nicholson has worked with students over a 10-year period to reconstruct the system that the team of geometers and theologians, which may have included Michelangelo, used to create the original panels. He has recently collaborated with artist Blake Summers and architecture graduate student Saori Hisano to replicate all 15 panels at full scale. Working largely with straightedge and

Fig. 56.1 Laurentian Library, Florence, panel based on Plato's *lambda*

compass, Nicholson and his team have stumbled upon ancient tenets of geometry which have been infused with an ingenious Mannerist twist to promote ideas of irrationality and void, criss-crossed with the witty numerical conundrums so admired in the late Renaissance.

At first glance, the panels appear to be square. However, curious irregularities guide the dimensioning of each panel. Each panel is set into a rectangular frame that measures approximately 4 *braccia* (233 cm) by 4 1/4 *braccia* (248 cm), but the size of each panel is slightly different. Nicholson proposes that the geometric grids and numerology in the pavement respond to the essential theological and scholastic questions posed in the sixteenth century. For example, Plato's *lambda* orders panel 14, which is aligned in its general appearance with descriptions in the *Timaeus* setting the *lambda* within four interconnecting circles (Fig. 56.1).

Nicholson's proposition that each pair of panels differs very slightly from East to West now becomes relevant to the discussion. For example, there is evidence to suggest that Panel 14 East is laid out on a grid of 81 parts, and that Panel 14 West is laid out on a grid of 80 parts. The ratio 80:81 is a measure of the comma's difference between the Pythagorean and Just musical scales (Kappraff 1998). Can these matters have been intentional? In the remainder of this chapter, we shall present some details of two panel reconstructions.

Fig. 56.2 Reconstruction of the "Medici panel" based on the rosette

The Medici Panel

Panel 2 is called the "Medici panel". The painted reconstruction by Nicholson and Summers of the original terracotta design is shown in Fig. 56.2.

It seems to be wholly symmetrical and it has the same appearance as the antique rosettes, of which there are many examples in Renaissance design. At the centre of the rosette lies the *emblema* of Cosimo I from the house of the Medici, advertising the threefold symmetry of the pattern. However, at second glance the panel exhibits the mannerist tell-tale irregularities that are common to all the Laurentian pavement designs: the panel is not square, but a rectangle with sides in the ratio of 12:13; curving white bands radiate from the centre, a graphic treatment never present in the antique form; ovals are inserted into the residual spaces between these bands.

The following steps show that Panel 2 is created by superimposing 96 circles on a 12:13 rectangle that is composed upon a 13×13 square:

1. Form a 12:13 rectangle by dividing one axis of a square into 26 parts and extend the left and right edges of the rectangle to form a 13:13 square as shown in Fig. 56.3a. The diagonals of the 12:13 rectangle and 13:13 square are shown (Fig. 56.3a).
2. Draw a second 12:12 square within the 13:13 square and place the x- and y-axes at the centre of the squares.
3. Draw an equilateral triangle with side equal to the base of the 12:12 square. The distance from the apex of this triangle to the centre of the square determines the

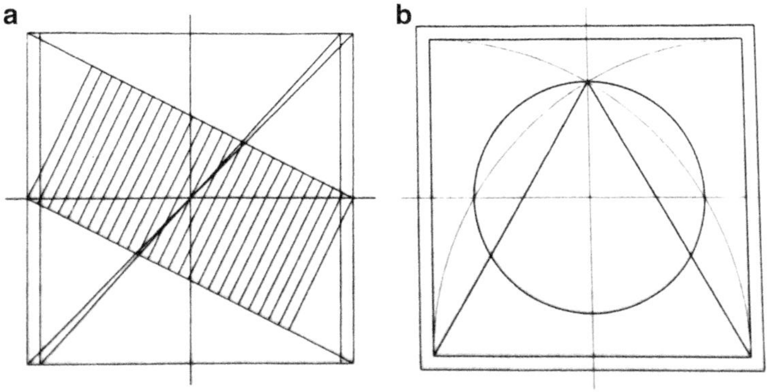

Fig. 56.3

radius of a circle (Fig. 56.3b). This circle is called the *pitch circle*. The radius of the pitch circle differs from 1/4 the diameter of the 12:13 rectangle by less than 1 %. Either value can be used for this construction. However we use the first because of its elegance.[1]

4. Place x- and y-axes at the centre of the rectangle and draw a rosette pattern with 24 circles by the following procedure:

   (a) Draw six circles whose radii are the same as the radius of the pitch circle. The first circle has its centre point at the intersection of the pitch circle and the upper y-axis; each of the other five circles' centre points are intersection points between the pitch circle and the preceding circle (Fig. 56.4a). The centres of these circles define two sets of three axes through the centre of the rectangle corresponding to the threefold axes of the Cosimo symbol in the centre of the design. Notice that four of the circles intersect the vertices of the 12:12 square. It is also worth noting that the rosette produces a series of intersections between adjacent circles, known as the *Vesica Pisces*, a key figure of sacred geometry (Kappraff 1991: 54, 87). It was in this region that images of Christ were placed in many sacred designs.

   (b) Draw six more circles using the same method as step 4a, beginning this time with the intersection point of the pitch circle and the right hand x-axis as the first centre point.

   (c) Repeat steps 4a and b by using intersection points between the pitch circle and diagonal lines of the 13:13 square as centre points to create 12 additional circles forming the 24-circle rosette pattern (Fig. 56.4b).

5. The rosette is composed of a grid of curvaceous diamonds formed by the intersection of the first 12 circles and the second 12 circles. Using as centres the midpoints of the arcs on the pitch circle connecting adjacent circles of the

[1] This construction was also described by Paul Marchant, a member of Keith Critchlow's London-based group studying traditional geometry; see Marchand (1997).

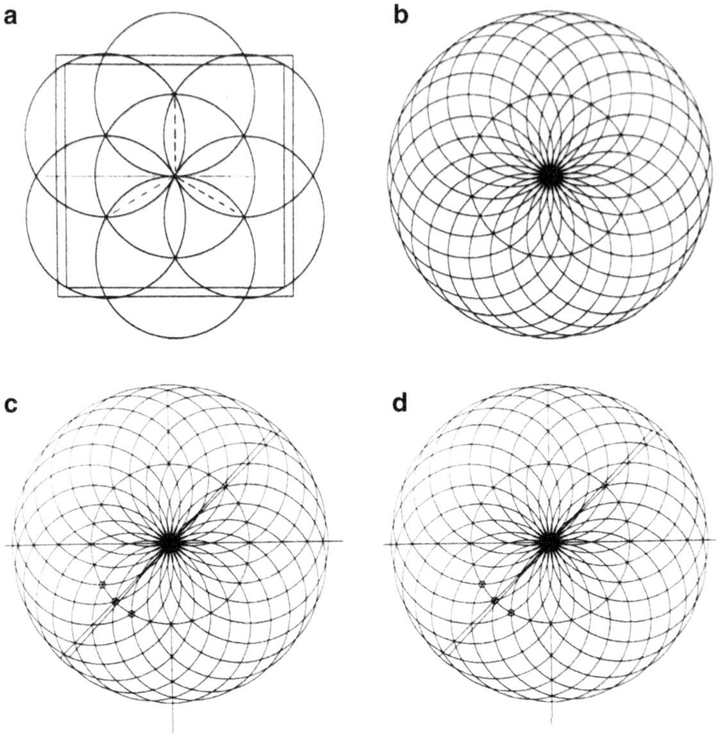

Fig. 56.4

rosette pattern, draw 24 additional circles using the same method as in step 5 (Fig. 56.4c).

6. The small mismatch between the diagonals of the square and rectangle leaves space to construct a *reference circle*. Draw 24 circles, identical to the reference circle, with centres at the intersection points of the pitch circle and the latest 24 circles. Two of these 24 reference circles are shown on opposite sides from the original reference circle.
7. Draw 48 circles with the same radius as the pitch circle and the intersection points of the pitch circle and the 24 reference circles from the previous step as centres (Fig. 56.4d). These circles are to become the white bands of the Medici Panel. This step demonstrates how the panel makes "Mannerist space" out of the difference between the series of circles generated by the 12:13 and 13:13 diagonals.
8. In the final step ovals of eight different types are created, to fill the diamond shapes. The detail of this step is beyond the scope of this chapter.

Nicholson hypothesizes that this design represents an interplay between the circle, representing the cosmic realm, and the square and rectangle, representing the earthly domain. The ratio of 12:13 represents the solar and lunar cycles since the sun goes through the 12 signs of the zodiac in approximately the time that the Moon undergoes 13 revolutions about the Earth. The 96 circles that make up the pattern and the original pitch circle are grouped in the series: 1, 3 + 3, 6, 12, 24, 48. We recognize this series as the series that was used by Bode to determine the relative distances to the sun of the planets up to Saturn (all the planets known in the year 1550) (Kappraff 1998). Therefore, the designer of this pavement was able, either consciously or unconsciously, to compress a great deal of cosmic information into a geometrical setting. The panel on the other side of the library is identical except that it fits into an 11:12 rectangle. Nicholson conjectures that the numbers 12 and 11 refer to the number of Christ's disciples with and without Judas.

The Mask Panel

Nicholson has named Panel 13 the "Mask" panel (Fig. 56.5). When looked at either directly or from the side it appears like the classical masks of the theatre with either a happy or sad face popular at the time. Michelangelo made a number of carvings of the "mask of night." Saori Hisano and Hingan Wibisono were able to use a combination of the Golden Mean and the Brunés star to reconstruct this panel.

To construct a rectangle with Golden Mean proportions, begin with a square and add the semilength of a side to the length from a vertex to the midpoint of the opposite side (Fig. 56.6).

To construct a Brunés star, begin with a square divided into four half squares, and place a pair of diagonals into each half-square (Fig. 56.7a). The Brunés star provides a means for dividing a line segment into from 1 to 12 equal subdivisions. Figure 56.7b shows positions for subdividing a line segment into 1, 2, 3, . . ., 8 equal segments (7 is missing but can be approximated by another construction) (Kappraff 1998).

Rather than go into detailed explanations, we present in Fig. 56.8 one of the diagrams in which Nicholson has defined a central unit square and two symmetrically placed golden rectangles (proportions $1{:}\phi$).

The construction lines to create the golden rectangle are shown in the figure. The central square is trisected by the methods of Brunés into first a 3×3 grid, then each third is divided again by the Brunés star into a 4×4 grid. As a result the original square is divided into a 12×12 grid. Notice that the star and the golden rectangle share construction lines. Once again, the panel is just off from being a square, with the difference between the length and width being equal to 1/6, i.e., a width of 1/12 placed on either side of the short side.

In Fig. 56.9 four circles of radius $1/\phi$ are drawn about each vertex of the central square as centres. The width of the four oblong regions of intersection of these circles equals the diameters of the four black circles of the mask pattern. They are equal to $1/\phi^2$ units. These oblong regions are somewhat reminiscent of the *Vesica Pisces* regions that formed the basis of panel 2. From this construction, Nicholson was able to deduce that the widths of the white annuli around the black circles and

Fig. 56.5 The "Mask" panel

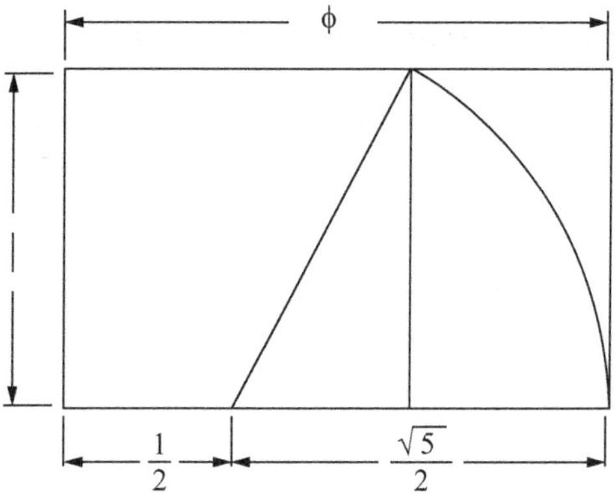

Fig. 56.6

the narrow white annuli in the left and right sections of the pavement were related to the Golden Mean and summed to 1/12, the width derived from the Brunés star.

The reconstructed terracotta Mask panel was created by Nicholson, Hisano, and Wibisono (see Fig. 56.5). The geometer, who may have been Michelangelo himself, appears to have found ingenious ways of wedding two geometrically different worlds, the one of the Golden Mean and the other of the Brunés star.

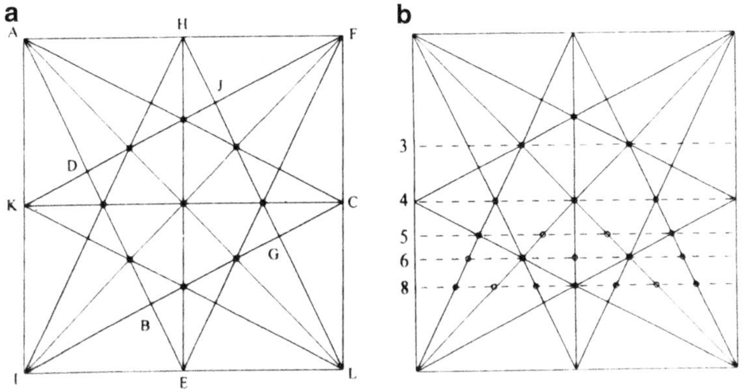

Fig. 56.7

Fig. 56.8

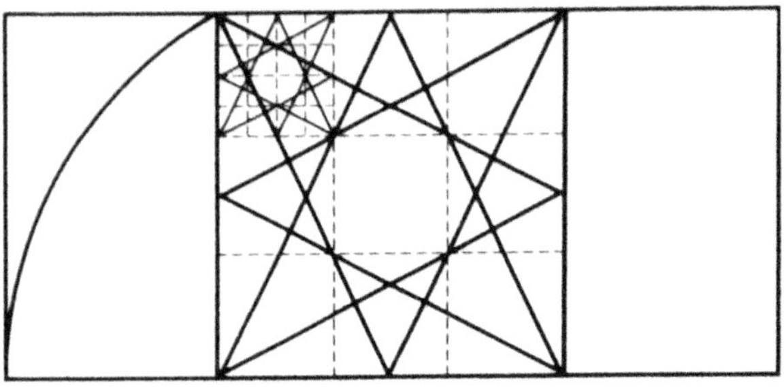

Fig. 56.9

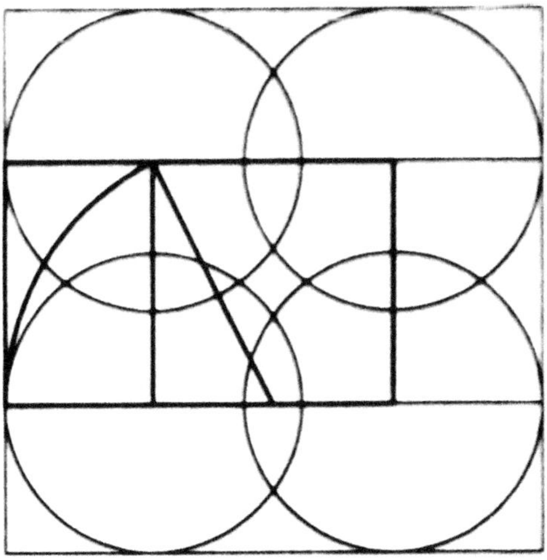

Conclusion

The pavement panels of the Laurentian library have presented us with a set of geometrical puzzles and are remarkable because they represent, in a single building, an almost complete set of the predominant forms of ancient geometry. It is Nicholson's hope that his work will generate sufficient interest among the officials at the Laurentian Library to open these beautiful pavements to public view, as well as affording the possibility of making a conclusive set of measurements against which his thesis can be checked. Nicholson believes that the panels are an unambiguous expression of Mannerism, which has long been associated with a somewhat wilful and uncontrolled adjustment of Renaissance principles of perceived wholesomeness. He considers the pavement as constituting a 'document' of Mannerist number theory, albeit expressed in the difficult language of geometry, that spells out, in a wholly reasoned way, issues of paradox, void and a confrontation of the status quo for which Mannerist art is so famous. It is also possible that this pavement forms the treatise on proportion that Michelangelo had in mind to write and that was alluded to by Condivi in his biography of Michelangelo (1553).

The panels are certainly in accordance with the ancient tenets of sacred geometry, but new procedures are added to the construction of each panel inferring a considered development of the extant body of knowledge. Panel 2 does spring forth from a single point that leads to a circle representing the undifferentiated cosmos, but the geometer introduces a radius based upon a distillation of the numbers 12 and 13, which throws a controlled, but irrational, haze over the whole construction. Elsewhere, the appearance of a square shows up in several panels: their well-defined vertical and horizontal axes classically imply the immutable polarity of opposites—made more succinct by the crossing at its centre, representing the essential generative qualities of male and female, necessary for the sustenance of earth and nature. The crossing in Panel 14 follows this tradition, but one of the limbs is measured with a string of whole number integers and the second limb is formed by a $\sqrt{2}$ evolution of the first limb: thus the crossing is caught in the tension of two systems that pit rational against irrational—and the observer is left to gaze upon the whole and wonder why.

Acknowledgements Author Ben Nicholson thanks the Laurentian Library for permission to make rubbings of the panels to assemble a set of accurate measurements. All figures herein are by the authors.

Biography Ben Nicholson, British born, and educated at the AA, Cooper Union and Cranbrook, is now Associate Professor at SAIC in Chicago. He was guest professor at the Bartlett, SCI-Arc, Royal Danish Academy, Universities of Cornell, Edinburgh, Michigan & Houston and a fellow of Chicago Institute for Architecture and Urbanism. His publications include: *The Appliance House, Thinking the Unthinkable House,* and a satire *The World Who Wants It?* He has exhibited at the Canadian Center of Architecture, Cartier Foundation, Whitney Museum, and three times at the Venice Biennale of Architecture. Currently living in New Harmony, Indiana where he sorts through issues of rural & urban life in America—from urban agriculture to gun culture. His work on the hyper-minimalist architectural plan resulted in 200 hand-drawn labyrinths exhibited at the *2008 Venice Biennale of Architecture*. He contributed to the book *Ineffable Architecture (2009), and to AD* Architectures of the Near Future (2009).

Jay Kappraff holds a Ph. D. from the Courant Institute of Mathematical Science at New York University. He was associate professor of mathematics at the New Jersey Institute of Technology where he has developed a course in the mathematics of design for architects and computer scientists. Prior to that, he taught at the Cooper Union College in New York City and held the position of aerospace engineer at NASA. He has published numerous articles on such diverse subjects as fractals, phyllotaxy, design science, plasma physics, passive solar heating and aerospace engineering. He has also lectured widely on the relationships between art and science. He is the author of *Connections: The Geometric Bridge Between Art and Science* (1st ed, 1991; 2nd ed., 2001). His book, *Mathematics Beyond Measure: A Random Walk Through Nature, Myth and Number,* was published in 2003.

Saori Hisano obtained an M. Arch at Illinois Institute of Technology after she graduated from Yokohama national University at Japan. She works for A&I Architectural office at Yokohama, Japan. She has been engaged in the projects to improve the quality of the houses provided by a company of housing manufacture. As part of it, she conducted a study and training program for the stuff of the company, and also wrote articles to a magazine intended for the people who are going to build a house. As a design work, she worked on renovation of residence and apartment house.

References

BRUNÉS, Tons. 1967. *The Secrets of Ancient Geometry – and Its Use*. Copenhagen: Rhodos.

CONDIVI, Ascanio. 1553. *Vita di Michelagnolo Buonarroti raccolta per Ascanio Condivi da la Ripa Transone*. Rome: Appresso Antonio Blado Stampatori.

KAPPRAFF, Jay. 1998. *Mathematics Beyond Measure: A Random Walk through Nature, Myth, and Number*. New York: Plenum Press.

———. 1991. *Connections: The Geometric Bridge Between Art and Science*. New York: McGraw-Hill.

MARCHAND, Paul. 1997. *Unity in Pattern: A Study Guide in Traditional Geometry*. London: The Prince of Wales Institute of Architecture.

NICHOLSON, Ben. 1997. *Thinking the Unthinkable House*. CD ROM. Chicago: Renaissance Society.

WATTS, Carol Martin and Donald J. WATTS. 1986. A Roman apartment complex. *Scientific American* **255** 6 (December 1986): 132–139.

———. 1987. Geometrical Ordering of the Garden Houses at Ostia. *Journal of the Society of Architectural Historians* **46,** 3 (September 1987): 265–276.

Chapter 57
Measuring up to Michelangelo: A Methodology

Paul A. Calter and Kim Williams

Introduction

In the fall of 1998, we began a new survey of Michelangelo's earliest built work of architecture, the New Sacristy of the Basilica of San Lorenzo in Florence, also known as the Medici Chapel. Kim Williams had surveyed the ground plan of the Sacristy in 1993 (Williams 1997: 105–112), and found a recurring series of proportional relationships related to the $\sqrt{2}$ rectangle. In order to establish a systematic use of $\sqrt{2}$ proportions in the three-dimensional space of the Sacristy, the dimensions of the interior elevations of the space were necessary. The Sacristy had been surveyed by hand in 1939 by a group of students from the University of Florence under the supervision of Armando Schiavi.[1] But because the 1939 survey contained dimensions that differed widely from Williams's 1993 surveyed dimensions, we had decided it was necessary to resurvey the entire interior using a theodolite and a trigonometric method devised by Paul Calter (2014).

At the same time we were in Florence to survey the New Sacristy, Ben Nicholson, who has spent many years studying Michelangelo's Laurentian Library (Nicholson et al. 2015), also in the San Lorenzo complex, approached us about surveying a doorway in the library. We were happy to comply, although due

First published as: Paul Calter and Kim Williams, "Measuring up to Michelangelo: A Methodology", pp. 23–34 in *Nexus III: Architecture and Mathematics*, ed. Kim Williams, Ospedaletto (Pisa): Pacini Editore, 2000.

[1] This survey appears in (Real Accademia d'Italia 1934; Schiavo 1949, Fig. 6; Schiavo 1990, Fig. 118).

P.A. Calter (✉)
Vermont Technical College, Randolph, VT, USA
e-mail: pcalter@sover.net

K. Williams
Kim Williams Books, Corso Regina Margherita, 72, 10153 Turin (Torino), Italy
e-mail: kwb@kimwilliamsbooks.com

K. Williams and M.J. Ostwald (eds.), *Architecture and Mathematics from Antiquity to the Future*, DOI 10.1007/978-3-319-00143-2_10,

Fig. 57.1 The Laurentian Library doorway designed by Michelangelo. Photo: Kim Williams

to restoration work in the vestibule of the library, we were unable to survey the doorway that Nicholson had requested. Instead, we surveyed key dimensions of the portal on the opposite side of the wall, that is, the portal that exits the reading room and leads into the vestibule (Fig. 57.1). What follows is a description of our findings. Significant as they may be in terms of what they reveal about Michelangelo's use of a proportional system, we have organized the present chapter in order to concentrate on a methodology of obtaining data, organizing it, and estimating its uncertainty. The intent is to begin to provide information that may help establish standards relating to these tasks.

Description of the Surveying Method

Here we briefly summarize the trigonometric method for measuring façades using surveying instruments.

We start by marking two theodolite setup locations A and B on the pavement. They can be at different heights and at different distances from the façade. We

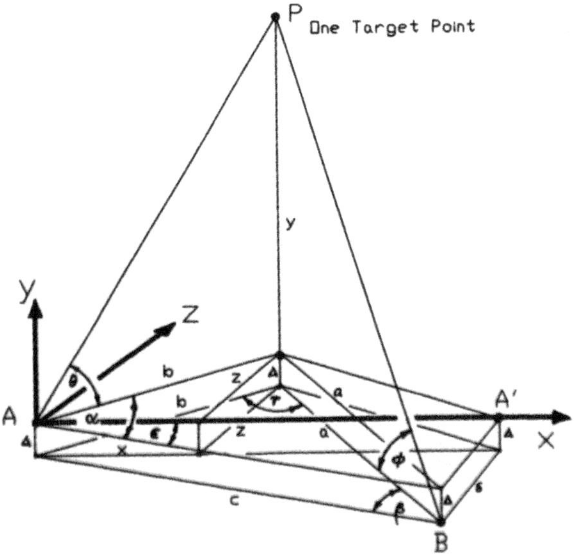

Fig. 57.2 An oblique view of the theodolite setup for the trigonometric method for measuring façades using survey equipment. Drawing: Paul Calter

record their horizontal distance c apart and the horizontal distance d from A to the façade.

Next we set up and level the theodolite at location A. With the telescope horizontal, we sight a vertical scale at the façade and record the height of the instrument above the pavement (Fig. 57.2). We then sight setup location B at which we zero the horizontal scale of the theodolite. We are now ready to take a series of measurements. We sight each target point on the façade, recording the horizontal angle α and the vertical angle θ for each.

After each target has been sighted, we move the theodolite to setup location B and level it. As we did from position A, we record the height of the instrument, and zero the horizontal scale of the theodolite while sighting location A. Again we sight each target point on the facade, recording the horizontal angle b and the vertical angle j for each.

Finally we enter all measurements into the computer spreadsheet, which is programmed with the following equations, and print out the x, y, and z coordinate of each target point.

Façade Equations

For each target point P we first calculate three intermediate values, γ, a and b (shown in Fig. 57.2):

$$\gamma = 180 - \alpha - \beta$$
$$a = c(\sin\alpha/\sin\gamma)$$
$$b = c(\sin\beta/\sin\gamma)$$

Where

$c =$ horizontal distance between theodolite locations
$\alpha =$ horizontal angle at A from B to the target
$\beta =$ horizontal angle at B from A to the target.

The x and z coordinates of point P are then

$$x = b \cos\alpha$$
$$z = b \sin\alpha.$$

We get two values of the y coordinate, from

$$y = b \tan\theta$$

and

$$y = a \tan f$$

where

$\theta =$ vertical angle at A from the horizontal to the target
$\phi =$ vertical angle at B from the horizontal to the target.

The values of y found from each setup position are independent, and their comparison can be used to indicate the accuracy of the measurements.

Our coordinate axes will be as shown in Fig. 57.2, with the origin at A, with the x-axis in the direction of B, the y-axis vertical and directed upwards, and the z axis perpendicular to the x- and y-axes, directed towards the façade. A simple translation of axes will later place the origin at any selected point, such as a corner of the building. We have used this method to survey interior and exterior façades of the Torre Bernarda in Fucecchio, both the interior and exterior, and two interior façades of the New Sacristy. Kim Williams used it in order to assist Mark Reynolds in studying the façade of the Pazzi Chapel (Reynolds 2015).

The Setup in the Library

The floor plan of the library placed some constraints on our instrument setup positions. The benches lining the aisle prevented us from laying out a base line anywhere but in the centre of the library, at a distance of about 19 m from the doorway. There, we could locate the two theodolite setup positions about 5.3 m apart. For our calculations, these dimensions were each measured three times to the nearest millimetre, and the readings averaged. Instead of marking setup points on the floor, we selected joints between paving tiles for this purpose.

We used a wooden triangle to protect the library floor from the steel tips of the tripod. To overcome the dim lighting in the library, Williams, standing near the doorway, located each target point with a tripod-mounted laser pointer. Calter then used the laser spot to approximately aim the theodolite telescope, and then refined the adjustment to the actual target feature (corner, edge, crack, etc.), which we illuminated by flashlight. The internal angle scales of the theodolite were illuminated by lamps powered by a battery pack. The data were reduced using Microsoft Excel and Quattro Pro computer spreadsheets.

The x and y coordinates of the target points, shown in Fig. 57.3, are given in Table 57.1. Depth, or z coordinates, were also computed for reference, but due to the great distance from the base line to the façade, they are of a lower order of accuracy than the x and y coordinates. As they are not needed for the proportional analysis that follows later, we omit them here.

Uncertainty in Coordinates

This trigonometric method is exact, in that it contains no approximations. However it does rely on measured quantities, and no measurement is exact. In this method, the location of each point is found from four measurements, which are then used to calculate the coordinates of that point. We want to determine how inevitable random errors in the four measurements, propagated through the computation, affect the accuracy of the final coordinates. After that we can estimate the uncertainty for *dimensions,* or the distances between points, and for the *ratio* of two dimensions. To estimate the uncertainty in our computed x and y coordinates we ran two computer simulations. Starting with the four measurements for a typical point on the doorway (Point 1 in particular) we set each measurement in turn, to its worst possible value, both low and high, and observed how this affected the computed coordinates.

We had measured the base line with a tape graduated in millimetres, and estimated each reading to the fraction of a millimetre. However, considering the uncertainty in setting the end of the tape over one station, uncertainty about reading the tape at the other station, and uncertainty about the exact location of the station (that is, having to estimate the centreline of a mortar joint between tiles), we took

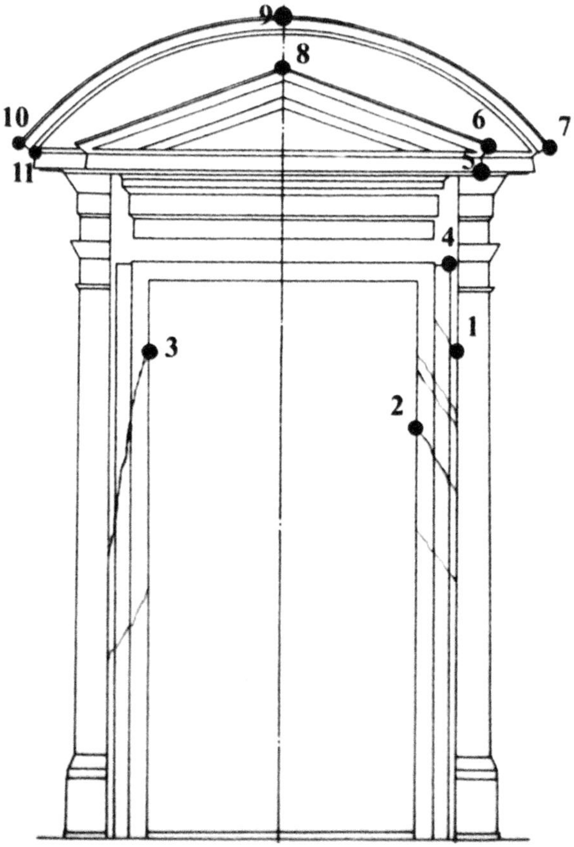

Fig. 57.3 Schematic scale drawing showing the location of survey target points. Drawing: Kim Williams

Table 57.1 Target Point Coordinates, with x measured left or right of the doorway centreline and y measured up from the pavement

Point	x	y
1	125.7	337.1
2	96.7	305.3
3	−95.6	316.2
4	120.7	400.8
5	142.6	467.7
6	150.4	485.4
7	192.9	484.9
8	−0.3	537.3
9	0.3	575.5
10	−192.4	484.5
11	−181.3	480.3

the uncertainty of the baseline measurement to ± 2 mm. Our theodolite could read angles to the nearest second of arc. Again, taking into account uncertainties of levelling the instrument and uncertainties about the location of a target point (perhaps a chipped corner of a ledge, or the centre of a mortar joint, etc.) we took the overall uncertainty of angular measurements as ±1/2 min of arc.

For four measurements, each set to two extreme values, we have

$$2^4 = 16 \ \textit{possible combinations}.$$

Of these 16, certain combinations of extreme values "lined up" in a way that gave the poorest results. We found that:

$$\text{Maximum uncertainty in x or y coordinate} \approx \pm 6\,\text{mm}.$$

This is the worst possible range of values attainable using the method, in this situation.

But random errors in measurement seldom line up in the worst possible way; in fact, they often cancel out (Briker and Taylor 1966). To get a better idea of what errors to expect, we next altered each of the four measurements not to their two extreme values, but to *randomly selected* values uniformly distributed between those same two extremes. Again using the four measurements for target point 1, with values from the spreadsheet's random number generator, we repeated the computation 100 times. We thus obtained 100 random values for both x and y. A frequency distribution of these values reveals that most were clustered near the mean value, with relatively few towards the tails. None of the random combinations gave errors as large as the ±6 mm obtained using the extreme values.

This means, for example, that 50 of the 100 computations gave the coordinate within ±1 mm of the mean. In other words, any one computed coordinate has a 50 % chance of being within ±1 mm of the mean.

It was noted earlier that the trigonometric method gives two independent measures of the height, or y coordinate, of each point. Another measure of the uncertainty in coordinates is the difference between each pair of y values. For this survey we found that 91 % of the target points had y differences within ±4 mm of their mean value, a good agreement with the values obtained in the above test.

In surveying parlance, the *probable error* of a measurement is the range of values within which 50 % of the random errors are expected to lie, and the *standard error* is that range within which 68 % of the errors lie. Thus we can say, approximately, that

$$\text{probable error in a } \textit{coordinate} = \pm 1\,\text{mm}$$
$$\text{standard error in a } \textit{coordinate} = \pm 1.6\,\text{mm}$$

with the standard error found by interpolation in Table 57.2.

Table 57.2 Frequency distribution of points about the mean, for coordinates

Distance from the mean (mm)	Percent of points
±1	50
±2	75
±3	92
±4	99

Uncertainty in Dimensions

Up to now we have been calculating the error of each *coordinate*. But we use coordinates to find *dimensions*, by subtracting one coordinate from another, and we now want to estimate the uncertainty of such a dimension.

When adding or subtracting two values, each with its own uncertainty, the theory of errors as applied to surveying states that it is not correct to simply add the uncertainties of the two values to get the uncertainty of the difference. Because random errors partly cancel each other out, it can be shown that it is more reasonable to give the final uncertainty as the square root of the sum of the squares of the individual errors, a quantity called the *mean-square error* (Kissam 1966).

If E is the mean-square error of a difference, and e is the error in each of the two values being subtracted, then

$$E = \pm\sqrt{e^2 + e^2} = \pm\sqrt{2e^2} = \pm\sqrt{2}e.$$

Thus, if $e = \pm 4$ mm, for example, then $E = \pm 4\sqrt{2} = \pm 5.7$ mm. Rewriting our table of frequency distributions for *dimensions*, we get (Table 57.3):

For example, a horizontal dimension obtained by subtracting two coordinates has a 50 % chance of being within 1.4 mm of the mean. Our probable and standard errors for dimensions become,

$$\text{probable error in a } \textit{dimension} = \pm 1.4\,\text{mm}$$

$$\text{standard error in a } \textit{dimension} = \pm 2.4\,\text{mm}, \text{ by interpolation}$$

It should be mentioned that these figures will vary depending on the particular set of random numbers chosen by the computer and the particular target point. We are also using the same values for both vertical and horizontal measurements, a further simplification.

Table 57.3 Frequency distribution of points about the mean, for dimensions

Distance from the mean (mm)	Percent of points
1.4	50
2.8	75
4.2	92
5.7	99

Key Dimensions

In order to derive key dimensions, we compared the dimensions found in the survey with taped dimensions by Kim Williams and dimensions recorded on the drawing of the upper part of the doorway by Carl von Stegmann and Heinrich von Geymueller.[2] Table 57.4 gives the key dimensions and their derivation.

We follow accepted practice of including the rightmost digit for which there is some uncertainty, but no digits to the right of that one. Thus we give survey dimensions to the nearest tenth of a centimetre (that is, to the nearest millimetre) with it being understood that there is an uncertainty in the last digit of a few millimetres.

The digits in each number, including the tenths of a centimetre, are called *significant digits*. Thus 575.5 has four significant digits, while 90.6 has three significant digits. We will need this distinction when computing ratios.

Uncertainty in Ratios

Usually we want not just dimensions, but *ratios* of dimensions. When we divide one dimension by another, each with its own uncertainty, what is the uncertainly in the ratio?

The usual practice in technical computation is to keep as many significant digits in the quotient as contained in the divisor or dividend having the fewest significant digits (Calter 1999). Thus, 385.4 has four significant digits, as does 575.5, indicating that there is some uncertainty in the rightmost digit of each. When we divide one by the other, ($385.4 \div 575.5 = 0.6696$) we keep only as many digits (four) in our quotient as contained in the original numbers, but no more. This indicates that the rightmost digit (6) in our quotient has some uncertainty. We have followed this procedure in Table 57.5.

[2] Carl von Stegmann and Heinrich von Geymueller's measurements of the upper part of the Laurentian Library doorway appeared in *Die Architektur der Renaissance in Toscana*, published serially in 11 issues from 1885–1908. For the measurements referred to here, see (Stegmann and Guymueller 1920: vol. 2, 42).

Table 57.4 Key doorway dimensions

Feature	Dimension	Percent error[a]
1. Overall height of door (floor to survey point 9 vertical)		
From the Calter-Williams survey	575.5 cm	±0.04 %
From Geymüeller + Williams's taped	574.4 cm	
2. Overall width of arched pediment (pt. 7 to pt. 10 horizontal)		
From the Calter-Williams survey	385.4 cm	±0.06 %
3. Overall height of arched pediment (pt. 7 to pt. 9 vertical)		
From the Calter-Williams survey	90.6 cm	±0.26 %
4. Overall width of triangular pediment (obtained by symmetry) 2 × (pt. 6 to pt. 8 horizontal)		
From the Calter-Williams survey	150.7 × 2 = 301.4	±0.08 %
5. Overall height of triangular pediment (pt. 6 to pt. 8 vertical)		
From the Calter-Williams survey	51.9 cm	±0.046 %
6. Clear width of door opening survey (inside to inside of jamb, horizontally; pt. 2 to pt. 3 horizontal)		
From the Calter-Williams survey	192.4 cm	±0.13 %
From Williams's taped dimensions	191.5 cm	
7. Clear height of door opening (underside of jamb to sill)		
From Williams's taped dimensions	383.5 cm	–
8. Door jamb width		
From the Calter-Williams survey	29.0 cm	±0.83 %
From Williams's taped dimensions	28.6 cm	
From Geymueller	28.7 cm	
9. Overhang of arched pediment to inside of door jamb (pt. 7 to pt. 2 horizontal)		
From the Calter-Williams survey (left side)	96.8 cm	±0.25 %
From the Calter-Williams survey (right side)	96.2 cm	±0.25 %

[a]Based on an estimated error of ±2.4 mm for survey dimensions

Setting an Acceptable Margin of Error

An indication of a range of acceptable percentage deviations from ideal dimensions was given by Howard Saalman, who once called a 2.43 % deviation "too large to be accepted on its face", while sustaining that a deviation of 0.8 % "cannot be dismissed ... lightly" (Saalman 1979: 1–5). In order to determine the range of our margin of error for the doorway survey, we examined the dimensions for the width of the doorway jamb. We chose this dimension because it is small enough to readily indicate discrepancies, because it may be readily measured by hand, and because three individual measurements existed for comparison. Again, the three sets of dimensions derived from the Stegmann-Geymueller survey, Kim Williams's

Table 57.5 Significant proportional relationships in the doorway

Features	Ideal ratio	Actual ratio[a]	Difference between actual and ideal
Overall width to overall height	2:3	385.4 ÷ 575.5 = 0.6696	0.30 %
Clear opening width to doorway clear opening	1:2	191.4 ÷ 383.5 = 0.4991	0.09 %
Doorway clear opening height to overall door width	1:1	383.5 ÷ 383.4 = 1.000[b]	0
Overhang of arched pediment to doorway clear opening width	1:2	96.8 ÷ 192.4 = 0.5031	0.31 %
Clear width of doorway to overall doorway height	1:3	192.4 ÷ 575.5 = 0.3343	0.10 %
Overhang of arched pediment to inside of jamb to overall door width	1:4	96.8 ÷ 385.4 = 0.2512	0.12 %

[a]Using survey dimensions, where available

[b]We get 1.000 (not 1.0003), because computationally that final 3 is meaningless

Table 57.6 Comparison of a survey dimension with taped dimensions

Source	Dimension in centimetres	% Deviation from Williams's taped	% Deviation from Stegmann-Geymueller	% Deviation from Calter-Williams survey	% Deviation from Average
Williams	28.6	–	0.35	1.5	0.63
Geymüeller	28.7	0.35	–	1.1	0.28
Survey	29.03	1.5	1.1	–	0.87
Average	28.78	0.63	0.28	0.87	–

taped measurements on site at the time of the survey, and the calculated dimension derived from the Calter-Williams survey (Table 57.6).

Our margin of error ranges from 0.28 to 1.5 %, well within the range indicated by Saalman. The largest percentage deviation, 1.5 %, is equal to a value of .43 mm. We generally assume that in taped measurements it is possible to be accurate to a millimetre.

In larger dimensions, the margin of error remains similarly small. The largest dimension taken was the overall height of the entire doorway, calculated from the surveyed points as 575.49 cm. Deriving a composite dimension from Williams' taped dimensions for the height of the clear doorway opening (383.5) plus the height of the sill above the floor (4.9) plus Stegmann-Geymueller's dimensions for the upper moulding details (186.0), the resulting overall dimension is 574.4. The actual difference between the dimensions is 1.09 cm; the percentage deviation is 0.19 % (Table 57.7).

Table 57.7 Matrix of significant dimensions of the doorway and their relationships

	575.0	385.4	90.6	301.4	51.9	191.9	383.5	28.8	96.2
575.0	—	2/3			1/11	1/3	2/3	1/20	1/6
385.4	2/3	—				1/2	1/1		1/4
90.6			—	6/20					
301.4			6/20	—					
51.9					—				
191.9	1/3	1/2				—	1/2	3/20	1/2
383.5	2/6	1/1				1/2	—		1/4
28.8	1/6					3/20		—	6/20
96.2	1/6	1/4				1/2	1/4	6/20	—

Conclusions

Out of 72 possible relationships between dimensions, only 29 reveal themselves to be ratios of whole numbers. This means that about 40 % of the relationships are significant, but that 60 % are not. This does not seem to statistically support a thesis that Michelangelo relied on a proportional system for his design of the doorway. On the other hand, when the six most significant relationships listed above in Table 57.5 are represented graphically, as in Fig. 57.4, we are allowed to see how one governing rectangle with proportions 2:3 is subdivided into smaller rectangles to establish key relationships in the doorway.

Of course, these are only the first steps in what could be a very complex geometric arrangement. In order to do a complete geometric analysis, many more points would have to be surveyed. However, this simple geometric analysis of an isolated element designed by Michelangelo affords us an exercise in organizing the data and establishing a method of analysis that can be used in a much larger project. In terms of the implications of this small study for the New Sacristy project, finding rational proportions in the Laurentian Library doorway that are similar to those previously noted in the ground plan of the Sacristy reinforces our theory of Michelangelo's use of proportion and helps focus our search for significant relationships in the volume of the Sacristy.

What is significant for the present chapter, however, is that we feel we have met our criteria of accurately surveying the portal, arranging our data so that our accuracy is ascertained, and applying our data to the actual portal so that it may be interpreted. Providing this kind of transparent approach in the analysis of measurement may mean that multiple surveys of the same space by different scholars can be avoided, because scholars may use the data provided by others with the reasonable certainty that the data is valid. The interpretations of the data may differ widely but of course, that is where science leaves off and imagination takes over.

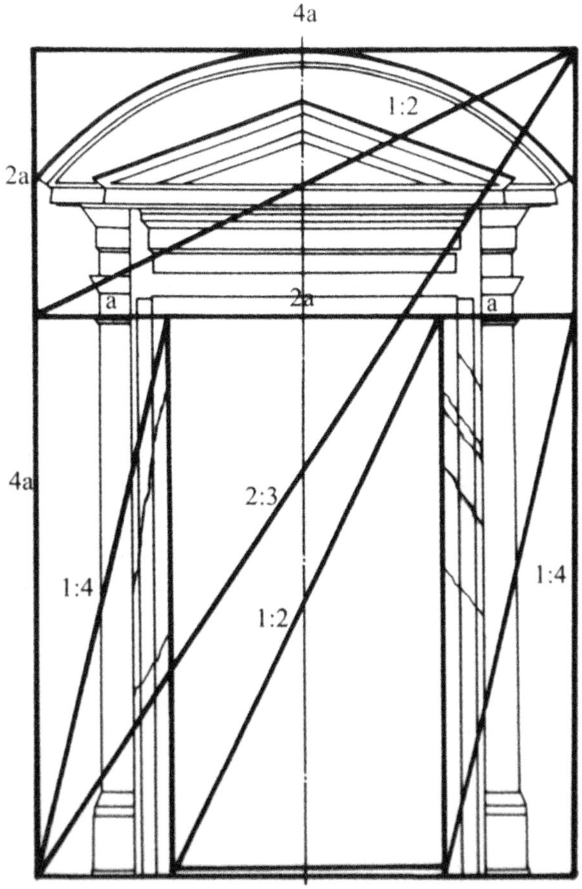

Fig. 57.4 The proportional rectangles showing relationships between elements of the doorway. Drawing: Kim Williams

Acknowledgment The authors most heartily thank Susan Knight for recording data on-site, inserting the survey data into the spreadsheet program, and providing moral support during the project. We thank also the Laurentian Library for permission to measure the doorway.

Biography Sculptor Paul A. Calter is Professor Emeritus of Mathematics at Vermont Technical College and was a Visiting Scholar at Dartmouth. He has interests in both the fields of mathematics and art. He received his B.S. from Cooper Union and his M.S. from Columbia University, both in engineering, and his Masters of Fine Arts Degree at Vermont College of Norwich University. Calter taught mathematics for over 25 years and is the author of ten mathematics textbooks and a mystery novel. He has been an active painter and sculptor since 1968, has participated in dozens of art shows, and has permanent outdoor" sculptures at a number of locations in Vermont. For the "Mathematics Across The Curriculum program, Calter developed the course "Geometry in Art & Architecture" and has taught it at Dartmouth and Vermont Technical College, as well as giving workshops and lectures on the subject.

Kim Williams was a practicing architect before moving to Italy and dedicating her attention to studies in architecture and mathematics. She is the founder of the conference series "Nexus: Relationships between Architecture and Mathematics" and the founder and editor-in-chief of the *Nexus Network Journal*. She has written extensively on architecture and mathematics for the past 20 years. Her latest publication, with Stephen Wassell and Lionel March, is *The Mathematical Works of Leon Battista Alberti* (Basel: Birkhäuser, 2011).

References

BRIKER, R. and W. TAYLOR. 1966. *Elementary Surveying*. 4th ed. Scranton: International.

CALTER. P. 2014. Façade Measurement by Trigonometry. Springer, Heidelberg, Vol. I, pp. 261–269.

———. 1999. *Technical Mathematics with Calculus*. Fourth edn. New York: Wiley.

KISSAM, R. 1966. *Surveying Practice*. New York: McGraw-Hill.

REAL ACCADEMIA D'ITALIA (ed.). 1934. *Opere architettoniche di Michelangelo a Firenze, Roma*. I Monumenti Italiani, fasc. II. Rome: Istituto Poligrafico e Zecca dello Stato.

NICHOLSON, Ben, Jay KAPPRAFF and Saori HISANO. 2015. The Hidden Pavement Designs of the Laurentian Library. Pp 139–149 in Kim Williams and Michael J. Ostwald eds. *Architecture and Mathematics from Antiquity to the Future: Volume II the 1500s to the Future*. Cham: Springer International Publishing.

REYNOLDS, Mark. 2015. A New Geometric Analysis of the Pazzi Chapel in Santa Croce, Florence. Chapter 46 in this present volume.

SAALMAN, Howard. 1979. Designing the Pazzi Chapel: The Problem of Metrical Analysis. *Architectura* **9**, 1 (1979): 1–5.

SCHIAVO, Armando. 1949. *Michelangelo Architetto*, Rome: Libreria dello Stato.

———. 1990. *Michelangelo Nel Complesso Delle Sue Opere*. Rome: Istituto Poligrafico e Zecca dello Stato, Libreria dello Stato.

STEGMANN, Carl von and Heinrich von GEYMUELLER. 1920. *The Architecture of the Renaissance in Tuscany*. 2 vols. New York: The Architectural Book Publishing Company.

WILLIAMS, K. 1997. Michelangelo's Medici Chapel: The Cube, the Square and the Root-2 Rectangle. *Leonardo* **30**, 2 (1997): 105–112.

Chapter 58
António Rodrigues, a Portuguese Architect with a Scientific Inclination

João Pedro Xavier

Introduction

Mestre António Rodrigues (ca. 1520–1590) has been only recently acknowledged as an important personality of Portuguese architecture, in spite of his promotion to the post of First Architect of the Realm by D. Sebastião, in 1565, after Miguel de Arruda, and to Master of Fortifications, in 1575, after the death of Afonso Álvares. He performed both tasks for 15 years, which constitutes a unique event in our country (Moreira 1993: 148).

We owe to Rafael Moreira this re-evaluation; he calls Rodrigues "a Portuguese architect with a scientific inclination" (Moreira 1982: 56). Moreira attributed to Rodrigues the authorship of a *Tratado de Arquitectura* (Treatise on Architecture),[1] which was the textbook for the course in military architecture (Lição de Arquitectura Militar) for which Rodrigues was responsible in the School of Moços Fidalgos do Paço da Ribeira, as well as the architectural design of the church of Santa Maria da Graça in Setúbal (presently the Cathedral), for which Moreira found documentary evidence. On the basis of stylistic evidence, he also attributed to Rodrigues the Onze Mil Virgens Chapel at Alcácer do Sal, an addition

First published as: João Pedro Xavier, "António Rodrigues, a Portuguese architect with a scientific inclination", pp. 253–268 in *Nexus IV: Architecture and Mathematics,* Kim Williams and Jose Francisco Rodrigues, eds. Fucecchio (Florence): Kim Williams Books, 2002.

[1] Two versions of this *Treatise* exist: a preliminary version (Biblioteca Nacional de Lisboa, Cod. 3675 (Rodrigues 1576)) and another, revised for print ca. 1579 [Biblioteca Pública Municipal do Porto Ms. 95 Rodrigues (1579)]; both are incomplete.

J.P. Xavier (✉)
Faculdade de Arquitectura, Universidade do Porto, Rua do Gólgota 215, 4150-755 Porto, Portugal
e-mail: jpx@arq.up.pt

K. Williams and M.J. Ostwald (eds.), *Architecture and Mathematics from Antiquity to the Future*, DOI 10.1007/978-3-319-00143-2_11,

to the church of the convent of Santo António built as a mausoleum for Dom Pedro de Mascaranhas.[2]

In fact, these attributions confirm:

- A solid theoretical formation, classically based, built on treatises, mainly those written by Vitruvius, Serlio and Pietro Cataneo, and most probably nurtured by direct contact with the Renaissance homeland. King João III used to encourage learning in Italy; besides it is hard to believe that the architect of the Onze Mil Virgens Chapel did not have firsthand knowledge of his sources.
- A pedagogical viewpoint in conformance with a teaching model in which, according to Vitruvius, an architect's formation could not be conceived without a strong scientific basis of mathematics, especially geometry, but where astronomy, music and the disciplines of the *trivium* were also taught.

  This was indeed the method of teaching that took place at Paço da Ribeira's School founded by Pedro Nunes, who was the director of the courses of mathematics and cosmography (the humanist philosopher João Pedro Lavanha was also part of its teaching staff). It was later closed by Filipe II, the first king of the Spanish dynasty in Portugal, and transferred to Madrid where it gave rise to the Academia de Matematicas y Arquitectura directed by Juan de Herrera.
- The capacity to translate this knowledge into built work and to achieve results of undeniable architectonic quality, the greatest being, in fact, the Onze Mil Virgens Chapel (c. 1565), built of pink marble from Estremoz (Fig. 58.1). The Cathedral of Setúbal (c.1570), with specific typological and programmatic constraints, is nevertheless a well-executed work.

In fact, both the Onze Mil Virgens Chapel and the Cathedral of Setúbal show a rigorous geometric structure, as expressed in the proportions and the purity of the stereometric shapes that were used as well as in the clarity of their spatial articulation. These characteristics are reinforced by the assuredness of a strong, clean drawing technique, which is a fundamental vehicle for its poetry.

The Onze Mil Virgens Chapel

The founder of the Onze Mil Virgens Chapel, Dom Pedro de Mascarenhas, was the well-known ambassador for King João III at the papal court in Rome and at the Emperor Charles V's Austrian House in Brussels. This has most certainly influenced the exceptional character of his chapel.

[2] Other works by António Rodrigues are the Igreja de São Pedro de Palmela and the Igreja da Anunciada de Setúbal, no longer in existence, as well as the Chapter Room and Sacristy (demolished) from the Convento de Jesus de Setúbal, according to documentary evidence found by João Custódio Vieira da Silva.

Fig. 58.1 The Onze Mil Virgens Chapel. Interior view looking east along the longitudinal axis towards the presbytery and reliquary. Photo: author

This highly cultivated and cosmopolitan man, who was familiar with the artistic ways of his time as well as to some of its key personalities (he met Michelangelo during his stay in Rome, according to Francisco d'Ollanda), died in 1556 while a viceroy in Portuguese Indian colonies. He had trusted António Rodrigues with the mission of erecting at his homeland a Roman-style work worthy of his prestige—a sepulchral "temple" for himself and his family (where the relics of the virgins, among others, would be kept) near the church of the convent of Santo António commissioned by his mother, Dona Violante Henriques, the wife of Dom Fernando de Mascarenhas, Governor of Alcácer.

This determination to tie the new sepulchral chapel to the church, permitting their interconnection through an opening in the wall that becomes the link between the two spaces symbolically celebrating the union of mother and son, was bound to determine the spatial design of the new chapel, as well as its dimensions and proportional relationships.

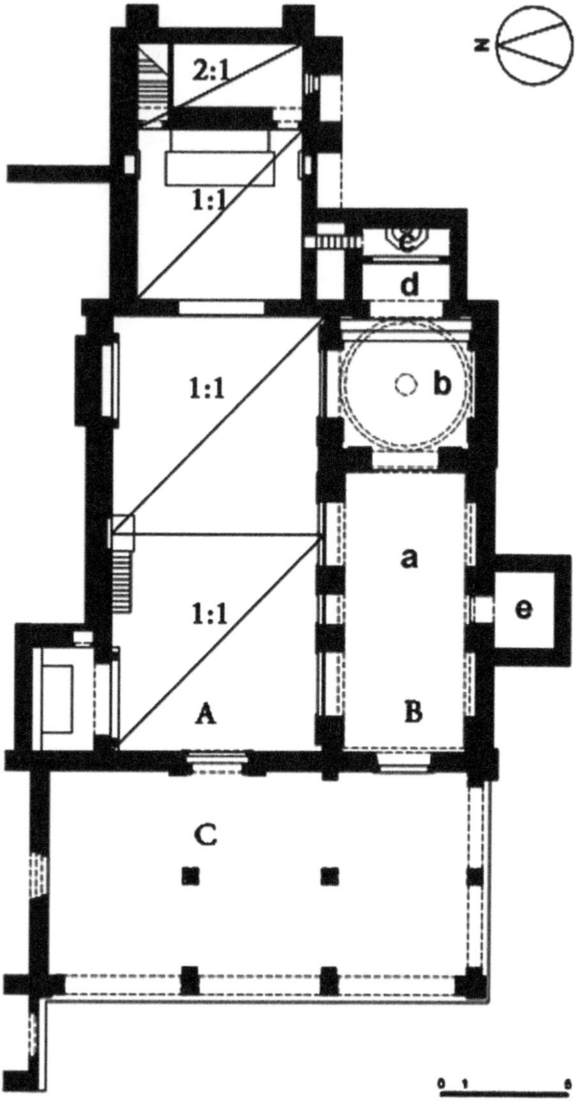

Fig. 58.2 Plans of (A) the Santo António Church (*left*, 1528) and (B) the Onze Mil Virgens Chapel (*right*, 1565 ca.) [(a) nave; (b) sepulchral chapel; (c) reliquary; (d) altar; (e) sacristy]; (C) galilee, constructed after 1565. Drawing: author

The Franciscan church commissioned by Dona Violante between 1524 and 1528 (Fig. 58.2) is formed by a double-square nave, and a presbytery corresponding to a square plus a half to include the sacristy and a stairway, the whole being a rectangle of the proportions 3:2 Belonging typologically to the family of small single-aisled Romanesque and Gothic churches, it differs in the scale and in the nearly-square proportion of its transversal section, as well as in the presence of Renaissance architectural elements. Its main porch, indisputably Lombard, testifies to Dom Pedro's mother's appreciation of new architectonic styles.

The Onze Mil Virgens Chapel (the Portuguese name means "chapel of the eleven thousand virgins") was designed to sit side by side with the convent church nave, sharing its delimiting south wall, and adopting the same length (16.60 m or 77 palms)[3] through a complex construction process, which entailed the demolition and subsequent reconstruction of this shared wall.

If this dimension, 16.60 m, is divided into three parts, and an allowance made for twice the thickness of the pilasters that support the triumphal arch (62 cm, nearly two palms and 7 in.), the result obtained is the side of the square sepulchral chapel where Dom Pedro's remains were buried, the nave of which consists of a double square plus the above-mentioned dimension (Fig. 58.3).

The square piers of the *serliana* that divides the nave of the chapel and the church have a square section with sides measuring 62 cm. The theme of the repeated use of sides and diagonals of squares, reminiscent of *ad quadratum* geometry, begins here: the sides of the piers, 62 cm, and their diagonals, 87.7 cm, form the modules, combining to produce the length of the sides of the three squares that form the chapel, 5.11 m (4 sides × 62 cm + 3 diagonals × 87.7 = 5.11 m, a deviation of only −0016 %) (Fig. 58.3a).[4] The fact that the width of the pillars is not a round number (for instance, three palms) may be due to the fact that the dimensions of the pre-existing church conditioned those of the new chapel. The chosen dimension could be more convenient for establishing the connection between the general dimensions and the modular ones. Under these circumstances, it seems that the definition of the module was not done a priori.

As the sepulchral chapel is the main purpose of the new "temple", this space becomes the prevailing element of the composition. On axis with the nave, in the position of a non-existent crossing, it precedes the presbytery in the east, which is subdivided into two spaces, the altar area and the relics' enclosure. Presently separated by a railed opening, once the enclosure for the relics must have been closed by a double portal (the rabbet and some traces of former hinges are visible), which probably was a diptych. A difference in the floor level accompanies this spatial sequence: the chapel is 15 cm above the nave, the altar 72.7 and the reliquary 183.5.

The role of the sepulchral chapel is of such importance that it is tempting to consider the nave as a large antechamber, as happens in Alberti's Tempio Malatestiano in Rimini. From another perspective, considering the sequence of the reconstructed wall, a *serliana* with an elegant alternation of arches and pilasters, the only open arch of which corresponds to the nave area, the fluidity between the two naves contributes to the preponderance of the sepulchre.

[3] King Manuel I's measurement system included the palm, 21.56 cm, which is divided in 8 in. One foot corresponds to 1½ palms; one ell to five palms; and one fathom to ten palms.

[4] Surveying drawings were based on manual measurements taken by the author. Concerning the 5.11 m dimension, I verified the measurements of 12 sides of all 3 squares and the greatest deviation found was less than 0.4 %.

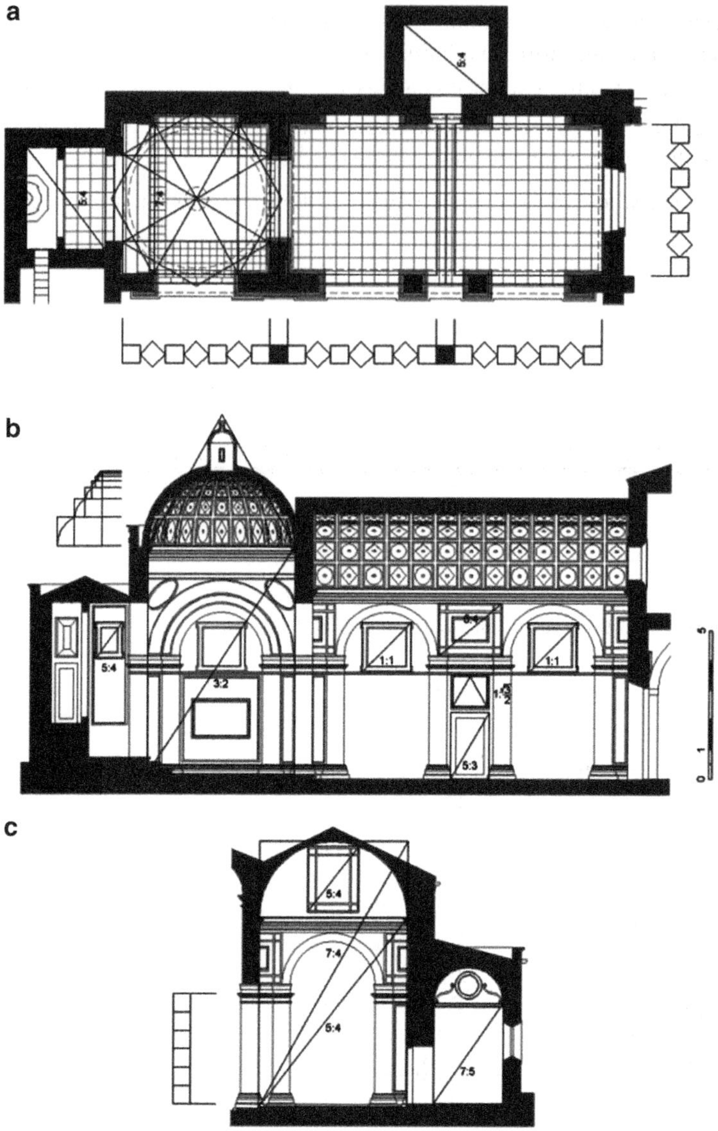

Fig. 58.3 Geometrical analysis of the Onze Mil Virgens Chapel: (**a**) plan; (**b**) longitudinal section; (**c**) transversal section. Drawing: author

Other factors contribute to the importance of this central-plan space, namely, its resemblance to its more obvious models, the New and Old Sacristies of San Lorenzo, by Michelangelo and Brunelleschi, respectively.

António Rodrigues makes masterly use of this mighty spatial *macchina*, redefining it with marble and jasper, exploring the translucency and reflections of those materials while proudly asserting its geometry. And he was the first one to do that in Portugal.

It should be emphasised that this spatial construction has an underlying *ad quadratum* geometry as well, as seen in the relationship between the sepulchral chapel and the dome that crowns it (Fig. 58.4).

The quadrangular coffers with inscribed circles that are present in the dome (and in the nave vault) allude to the geometry, and the alternate ones, with inscribed squares with vertexes in the middle position of the sides, reinforce this idea. In fact, the circle that might circumscribe the square of the dome coffers, the area of which would be the double of the inscribed circle, cannot be seen but is implicitly there. It would be the horizontal projection of the semi-sphere cut by vertical planes that correspond to the walls and that define the inscribed square. This square, in turn, circumscribes the circle resulting from the intersection of the same semi-sphere by a horizontal plane, where a small cylindrical wall sits. Over it stands the hemispherical dome crowned by a lantern. The part resulting from the first semi-sphere is apparent. The pendentives are spherical triangles that circumscribe blind circular oculi. I don't think it would be otherwise even if Donatello were around (Fig. 58.5).

The transition between a cubical space bounded by four walls and the hemisphere of the dome was made with the insertion of a small drum that is not present in either of the sacresties of San Lorenzo but which can be found in the crossings of Santo Spirito in Florence or Sant'Andrea in Mantua. By means of this small elevation the springpoint of the dome is made to correspond to a side and a half of the square of the plan, obtaining in this way a 3:2 ratio.

The arches reveal the recurring influence of Serlio's treatise both by its stereotomy and by the coffered interior vaults. They sit on Doric pilasters and form a portal 2.92 m wide. As the square space of the sepulchral chapel has a 5.11 m side, the dimension of the arches was calculated so that the vertexes of its opposing pilasters would form a 7:4 rectangle, a rational convergent of $\sqrt{3}$, which determines the arrangement of the tombs. As the lesser side of that rectangle can be defined by opposing sides of a hexagon the presence of that figure can be indirectly acknowledged. This should not surprise us, as the hexagon's connotation to lifelessness suits the chapel programme (see Fig. 58.3).

The dome deserves some special attention because of its remarkable formal expression, stereotomy, accuracy and material. It is divided in 24 semi-meridians and 7 parallels (the distance between them varies according to a gnomonic growth pattern), numbers of self-evident cosmic significance, aiming to symbolize, through expressly declared geometry, the mathematical order of the universe. But it is the translucency of the stone that increases its charm.[5] Thanks to this singularity it becomes more than a metaphor to the celestial dome as the sun actually projects

[5] It was not possible to determine the thickness of the stones, but they must be thin, underlining the quality and rigour of the construction. It is possible that its thickness diminishes approaching the lantern, but this does not prevent the dome from revealing itself on the exterior as a perfect semi-sphere. No covering was used, obviously, which is remarkable!

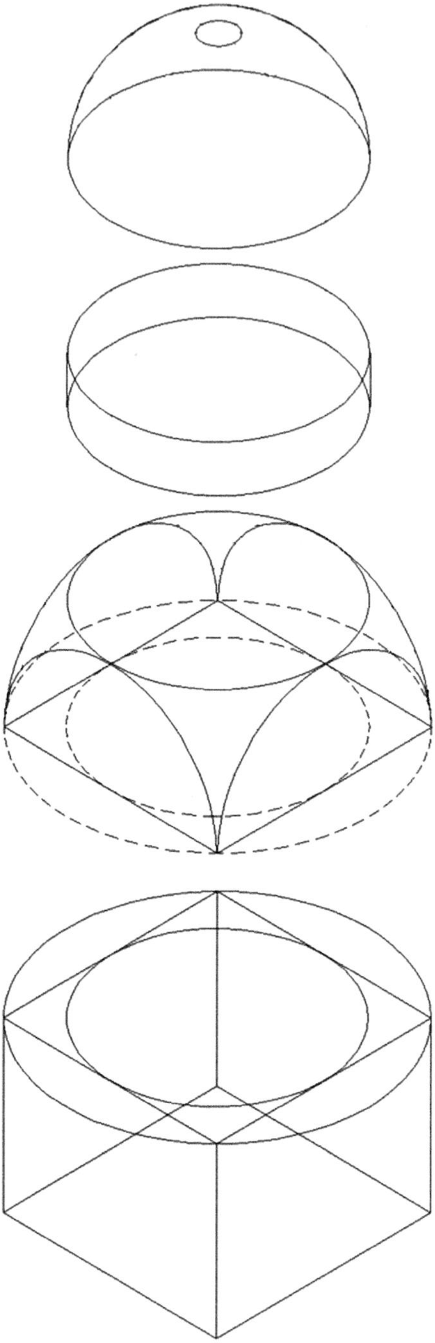

Fig. 58.4 Sepulchral chapel elements and its *ad quadratum* geometry, from bottom: chapel walls; arches and pendentives; drum; dome. Drawing: author

Fig. 58.5 Corner of the sepulchral chapel showing the pendentives and blind oculi. Photo: author

through it: a savoury Aristotelian touch within a neo-Platonic conception of the cosmos.

I believe that this cosmic dimension, well expressed in the first version of Mestre António's *Tratado*, is strongly corroborated by the frequent utilisation of the 5:4 rectangle—*hua sexquiquarta proposição de hum quadrado e hu quarto*[6]—in the making of different spaces (main chapel, sacristy, transversal section of the nave) and different architectural elements (the altar window, blind niches) of the Onze Mil Virgens Chapel.

If a gnomon is dimensioned according to the larger side of that rectangle, the length of its noonday equinoctial shadow will be its lesser side, or as Vitruvius

[6] "A sesquiquartal proposition of a square and a quarter", proposition 2 from António Rodrigues, *Proposições Matemáticas* (BPMP, Ms 95).

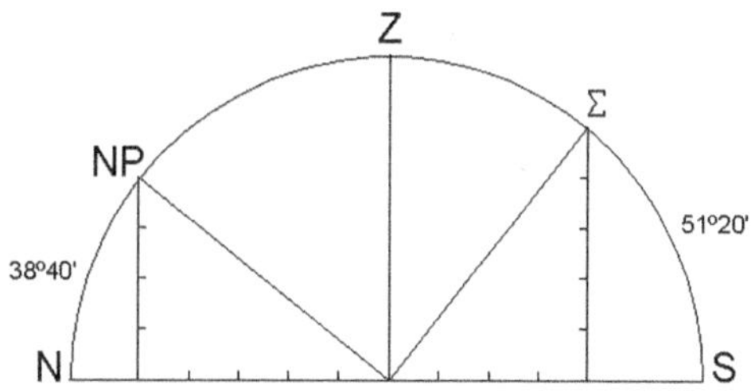

Fig. 58.6 Section of celestial hemisphere visible at Alcácer with the position of the pole. Drawing: author

would say,[7] if the gnomon length is divided in five parts, the shadow will be four at Alcácer pole altitude (Fig. 58.6).[8]

I think there is a strong possibility that this relationship is intentional because of the references concerning these matters in his *Tratado* and also because of this connection with the Great Cosmographer of the Realm, Pedro Nunes (born at Alcácer) who lent him his own translation of the Vitruvian text, although the Cesariano edition was already known among the Portuguese.[9] In the Codex 3675 of the Biblioteca Nacional de Lisboa, there is a general description of the celestial sphere and the apparent movement of the sun. There is even an explanatory drawing for proposition 29, indicating his extensive knowledge of position astronomy (Fig. 58.7).

The 5:4 rectangle is present in several building spaces: the nave, the main chapel and the sacristy. I found that the springpoint of the vault is located at 6.42 m above the ground level, so that the transversal section is made of a rectangle of 6.42 × 5.11 m, very close to 5:4 (a deviation of +0.995 %). Considering the semicircle of the dome, the radius of which is half of the nave width, the section can be inscribed in a 7:4 rectangle, which was taken into consideration when discussing the width of the arches. The plan of the main chapel measures 3.47 × 2.78. The sacristy, measuring 2.965 × 2.42 m, is not exactly a 5:4 rectangle, but even so the deviation, −0.98 %, is acceptable.

With respect to the architectural elements, we must mention the blind niches placed between the arches over the entablature of the *serliana* measuring 2.205 × 1.764 m, or 5:4 (see Fig. 58.3b). When the *serliana* turns the corner, the blind niche is subdivided by the right dihedron angle this wall forms with the side ones. This division gives origin to two 8:5 rectangles, approximating the golden

[7] *Car le soleil étant au temps de l' Equinoxe dans les Beliers aux dans les Balances, si la longuer du Gnomon est divisée en neuf parties, l'ombre en a huit à l'élevation du Pole de Rome* (Perrault 1979: IX, viii, 283).

[8] Alcácer do Sal latitude is 38º 30′ and the smallest angle of the right triangle of 5:4 cathetus measures 38º40′, a rather small difference. I thank Prof. Fernanda Alcântara (Geometry Course supervisor at FAUP) for the suggestion made by about the relationship between the 5:4 rectangle and the latitude or the celestial pole at Alcácer.

[9] The whereabouts of this translation is not exactly known. It may have been sent to the Madrid Academia of Juan de Herrera.

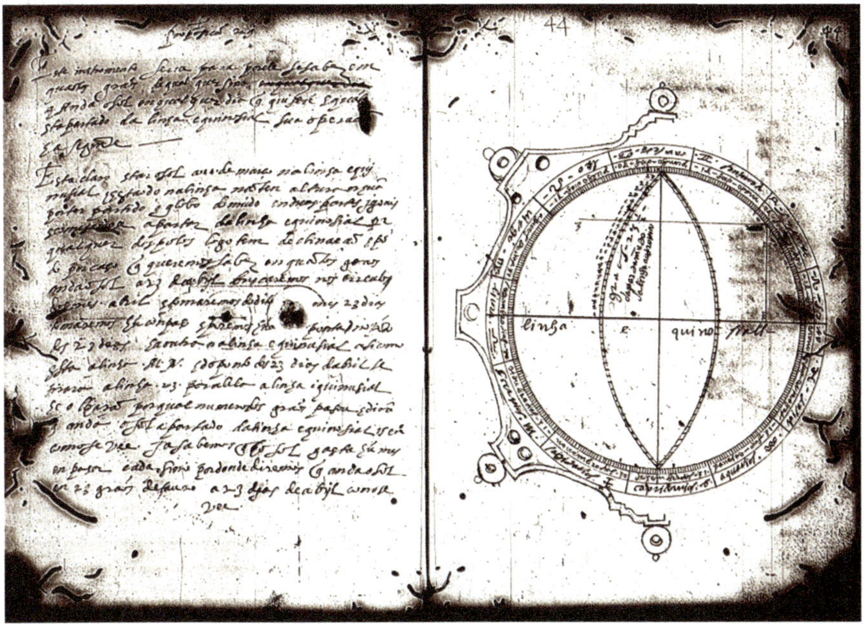

Fig. 58.7 The page from António Rodrigues's *Tratado* showing the explanatory drawing for proposition 29 (Rodrigues 1576: 43v–44r)

section. The small window from the altar, measuring 120.8 × 95.9 cm, is also an element of 5:4 family.

In addition to the similitude, I looked for a proportional relationship between these figures. I found that, taking as a reference the longer side of the rectangle formed by the nave, 6.42 m, and successively dividing it by 7/6, it was possible to find the longer side of the rectangle represented by the great chapel plan. Following through with this operation values were obtained that were quite close to those for the other elements, which allows them to be considered as part of a geometrical progression (Fig. 58.8).

I found the 7:6 ratio of the geometrical progression in the rectangle formed by the back wall of the reliquary under the statue of Christ.

It must be added that, as the chapel plan is based on three squares, that of the sepulchral chapel having a special importance, the visitor entering the space can sense it as a well-proportioned harmonious whole. He has only to place himself at the intersection of the axis and the west side of the first square and look towards the altar. The concern with this work's visual impact, or the way it would be seen and perceived, should not be undervalued, especially at a time in which perspective was being rediscovered. That António Rodrigues was concerned with perspective and visual impact is shown by the chapter on perspective (*Livro de Perspectiva*), included in his *Tratado*.

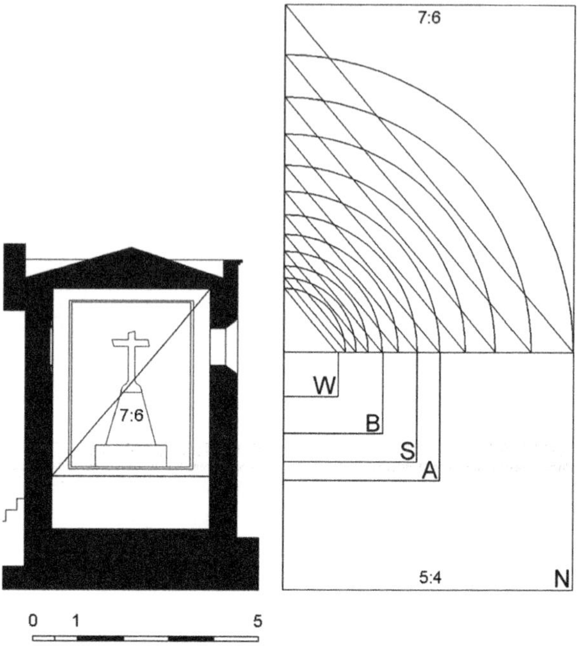

Fig. 58.8 Diagram of a 7:6 geometrical progression found in 5:4 rectangles corresponding to architectonic elements: N—nave, A—main altar, S—sacristy, B—blind niche, W—window. The reliquary transversal section is a 7:6 rectangle. Drawing: author

One of the most puzzling aspects of this temple concerns the main façade, which has been adulterated, or maybe never built according to the original design, due to the addition of a two-storied body that sits transversally to both churches. Both churches acquired a *galilee* (small porch, in Portuguese *galilé*) and the church of Santo António was given a choir, but the space over the chapel porch was obliterated (Fig. 58.9).

However, in spite of the running over of the mighty Chapel pilaster on the southwest corner, which, together with the body of the sacristy, sustains the stresses produced by the nave vault, the new building still manages to maintain some stylistic affinity with the Onze Mil Virgens. However, its disrespect towards the church of Santo António is complete: one of the new pilasters stands inexplicably in front of the Renaissance porch. It is a "plain",[10] unpretentious work,[11] somewhat uneven, and that is why it stands closer to the chapel architecture, one of the first

[10] Georges Kubler emplyed the expression "plain" to name works made in our country from the second half of the sixteenth to the mid-seventeenth century. According to Horta Correia, "by the time ideological superstructures of counter-reformist nature grabbed the power in Portugal . . . the tendency to decorative simplicity, the adoption of a certain classicism based on treatises and an austerity with religious and military features converged to define a new architectonic era dominated by what Kubler called plain style" (Horta Correia 1991: 48).

[11] In the west elevation I found that there were originally three more rectangular openings identical to the one still visible. Two of the openings were located where the oculi are presently. The third one was to the right of the existing one, over the central arch.

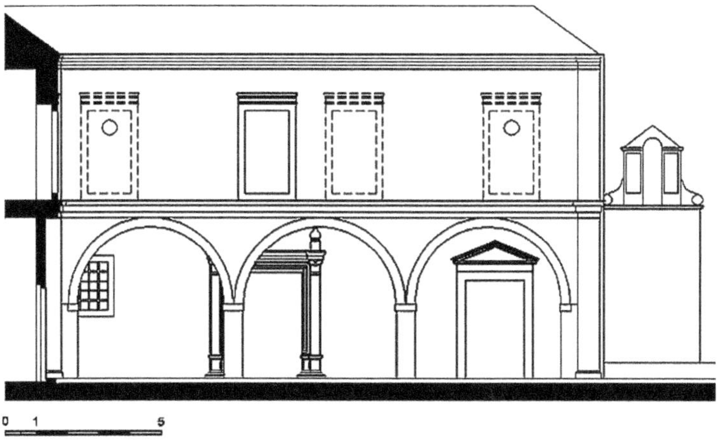

Fig. 58.9 West Façade. The porticos of Santo António Church and Onze Mil Virgens Chapel stand below the galilee. Drawing: author

works in our country to show that style, which preceded the austere "Herreran" style.

One of the surprising elements is the interior elevation of the façade. The tile wall abruptly cuts the lateral *serliana*s that are unfinished, in what seems to be a provisional solution. Only the cornice maintains the connection. I would prefer to see the *serliana* arching as it does on the elevation opposite, together with the cornice, the main entrance remaining confined to the inner arch. Figure 58.10 shows my hypothesis of a possible reconstruction that would make a lot more sense.

From the exterior, based on some existing evidence, I decided to try out one hypothesis for the façade drawing (I am somewhat less confident, but even so, it is a possibility to be considered). The galilee pavement is at the same level as the thresholds of the porticoes of both churches, higher than both naves, which is not logical. The chapel threshold is still there, untouched, and it seems to have been a step. The Tuscan pilaster of the southwest corner is unfinished at ground level; the base is missing, probably buried. Its reconstitution makes me believe that three steps would be needed to get to the chapel. As the pilaster can be measured on the south side, its size can be inferred in the main façade. Symmetry would require one pilaster on the other side of the porch and, besides, structural reasons would justify its presence. The former chapel could resist lateral forces produced by its vault, but, without the additional galilee, there would be no support to forces acting on the façade. Further, it seems obvious that the façade should be finished at the top. Observing the crowning of the existing pilaster, one can see that its capital is different from the cornice that goes around the church and where the roof sits. It is a detail of a portico, waiting for its pediment. In the present case, it is interrupted by a central window, as happens in San Sebastiano in Mantova, for instance. Finally, it must be remarked that if the façade were to be hypothetically rebuilt, the surface between the pilasters and the base of the pediment would form a square

Fig. 58.10 Inner main façade at *left* and its possible reconstitution at the center. At *right*, a hypothetic reconstitution of the main façade. Drawing: author

and a half, the same figure as that of the sepulchral chapel. The level where the dome springs is one and a half times the length of the side of its plan (see Fig. 58.10).

All these possibilities seem more credible in the context of the reconstitution of the design. Even so, some doubts remain, as were pointed out. It is hard to believe that there was once a façade more or less similar to the one I have tried to design and that the building of the galilee body had entailed its almost complete destruction. That is why it appears that the decision of building that new area must have been taken before the completion of the chapel, the project being adapted to the new situation.

In the façade, the portico, which is still present, shortened because at least one step is missing, deserves comment. It is possible to compare its design to the picture included in the second version of the author's treatise (the only one representing an architectural element), which differs from the one in Serlio's treatise only by the suppression of the pinnacles over the pediment (Fig. 58.11).

Surprisingly, the trisection method cannot be applied to the portico of the Onze Mil Virgens, but even so, the upper corners of the door fall on the diagonals of the circumscribing square. Even more interesting is the fact that the subdivision of that square in 7:7 results in squares with sides of 62 cm, that is the dimension of the pillar that corresponds to the temple module, and that the portico's opening forms a 5:3 rectangle (3.10 $\times$ 1.856 m), approximating the golden section.

As for the south elevation, distinguished by a strong and minimal design, it is impossible to hypothesize about its geometrical structure at present, as it there is no certainty about the ground level and the way the building was attached to it. However, it is possible to detect the important role of the square windows, opened in the limed white walls, contributing to impression of massiveness

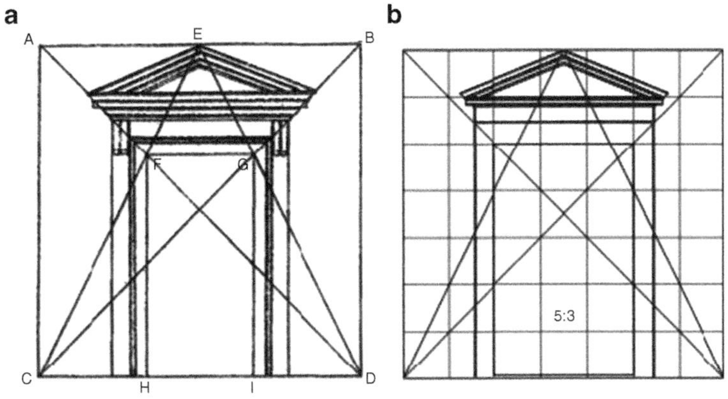

Fig. 58.11 The portico: (**a**, left) an illustration from *Proposições Matemáticas* (BPMP, Ms. 95) attributed to António Rodrigues; (**b**, right) drawing of the portico of the Onze Mil Virgens. Drawing: author

conveyed by the building. Paradoxically, its marble chamfered mouldings are extremely delicate, a subtle announcement of the preciousness of its interior.

Conclusion

Although the study of the Onze Mil Virgens Chapel is not yet completed, it has been demonstrated that this work discloses in several ways the omnipresence of geometry, ever in association with numbers and calculus, essential to the definition of an idea of architecture. The same can be said concerning the church of Santa Maria da Graça in Setúbal, the analysis of which is underway at the moment.

This realization is not a surprise, but I did not expect the relationship to be so close. It should be noted that there is still a lot to be discovered and that these results are preliminary.

In the chapters on geometry (*Livro de Geometria*) and perspective (*Livro de Perspectiva*) in the treatise by António Rodrigues are found a myriad of shapes and constructions that were also present in the treatises that inspired his. But even without a written sequence, lost or never completed, it is nevertheless possible to find them in the configurations of the architectural spaces he created.

On the other hand, the cosmic sense of this close connection is undeniable. In accordance to the neoplatonic ideal, every work is made as a microcosms, becoming the representation of a mathematically arranged macrocosms.

Although António Rodrigues considered that "he who will be an architect should also be a musician to understand the proportions of sounds, because their proportions will make him understand the proportions of buildings" [Rodrigues 1576: fol. 10v (my translation)], I don't think that the proportional relationships he employed can be interpreted only in the light of musical theory. They have their own mathematical meaning, which can be sometimes related to musical intervals, but essentially follows the intrinsic logic of the building. And we have seen, in the present work, that its structure was conditioned from the start by the need to conform to the pre-existing building.

Like Vitruvius, Rodrigues warns his readers that it is "*nesessario ser esperto na Giometria*" (Rodrigues 1576: fol. 10v). Geometry is the instrument he uses to build bridges between reality and transcendence.

In the introduction to the chapter entitled, "What is geometry?" he explains, "Geometry is no more than figures, that can not be done without lines, angles and points... By the same Geometry we will see that nothing can be done without it, and Mathematical Art cannot be understood without one being a Geometry expert...."

He concluded with some worthy advice: "The one who is curious about this art should study Euclid, and will find there something to wonder about".[12]

Acknowledgment The author wishes to thank Alexandre Alves Costa, History of Portuguese Architecture Lecturer and Professor at FAUP, for his advice on this chapter, and architect Kim Williams who so carefully revised the text.

Biography João Pedro Xavier is an architect and a Professor of Architecture in the Faculty of Architecture of the University of Porto (FAUP), where he received his degree and Ph. D. in Architecture. He worked in Álvaro Siza's office from 1986 to 1999. At the same time, he established his own practice as an architect. He is a member of the research group for Theory, Design and History of architecture at the Centro de Estudos de Arquitectura e Urbanismo at FAUP. His research focus is on architecture and mathematics, and in particular on perspective. He is the author of *Perspectiva, perspectiva acelerada e contraperspectiva (*FAUP Publicações, 1997) and *Sobre as origens da perspectiva em Portugal* (FAUP Publicações, 2006). He has participated in conferences, lectures and exhibitions, and has published a number of papers on the subject. He is on the Editorial Board of the *Nexus Network Journal* and is a member of the executive council for the journal *Resdomus*.

References

Horta Correia, J. E. 1991. *Arquitectura Portuguesa: renascimento, maneirismo, estilo chão.* Lisboa: Editorial Presença.

Moreira, R. 1982. *Um tratado português de arquitectura do séc. XVI (1576-1579).* Master Diss., Faculdade de Ciências Sociais e Humanas, Universidade Nova de Lisboa.

[12] (Rodrigues 1576: fol. 25v). The original text is as follows: *Quem for coriozo desta harte estude Hoclides, e nele achará bem couza em que se desemfade.* It is part of the introduction to the chapter "What is Geometry".

———. 1993. A Arquitectura Militar. P. 148 in V. Serrão, ed. *História da Arte em Portugal, O maneirismo*. Vol. VII. Lisboa: Publicações Alfa.

Perrault, c. 1979. *Les dix livres d'Architecture de Vitruve corrigés et traduits en 1684 par Claude Perrault* (1684). Bruxelles: Pierre Mardaga Éditeur.

Rodrigues, A. 1576. *Tratado de Arquitectura*. Lisbon, Biblioteca Nacional de Lisboa, Codex 3675.

———. 1579. *Proposições Matemáticas*. Biblioteca Pública Municipal do Porto, Ms 95.

Chapter 59
Villalpando's Sacred Architecture in the Light of Isaac Newton's Commentary

Tessa Morrison

Introduction

The three-volume commentary on the Book of Ezekiel was to be a collaborative project by two Spanish Jesuits priests, Hieronymus Prado and Juan Bautisa Villalpando. Originally the project was led by Prado, and although it was to be collaboration, Villalpando's main contribution was to have been on chapters 40–42, which consist of Ezekiel's vision of the Temple of Jerusalem. The first of the three volumes was published in 1596 as *Ezechielem Explanationes et Apparatus Vrbis Templi Hierosolymitani,*[1] and deals with the first 26 chapters of Ezekiel and was mainly written by Prado. However, Prado died before the publication of this volume and Villalpando was left to complete the project alone. Volumes II and III were subsequently published in 1604. Volume II, *De Postrema Ezechielis Prophetae Visione*, contains Villalpando's famous reconstruction of the Temple along with his justification for it. Volume III, *Apparatus Vrbis ac Templi Hiersolymitani,* consists of explanatory notes for the first two volumes. The overall project is a massive body of extraordinary and detailed scholarship. Villalpando was a highly skilled architect and draftsman and his reconstruction of the Temple is illustrated by a portfolio of exceptionally detailed architectural drawings. The project was an expensive one and it was only made possible through the financial support of Philip II of Spain.

First published as: Tessa Morrison, "Villalpando's Sacred Architecture in the Light of Isaac Newton's Commentary". Pp. 79–91in *Nexus VII: Architecture and Mathematics,* Kim Williams, ed. Turin: Kim Williams Books, 2008

[1] The translations of Villalpando's *Ezechielem Explanationes* and Newton's Babson MS 0434 and Yahuda MS 14 from Latin are by the author.

T. Morrison (✉)
The School of Architecture and Built Environment, The University of Newcastle, Callaghan, NSW 2308, Australia
e-mail: Tessa.Morrison@newcastle.edu.au

K. Williams and M.J. Ostwald (eds.), *Architecture and Mathematics from Antiquity to the Future*, DOI 10.1007/978-3-319-00143-2_12,

Villalpando studied mathematics under the royal architect, Juan de Herrera, who at that time was involved with the construction of the Escorial. Herrera had an extensive library of books on the occult; these books indicated a strong interest in Hermetism, which is also supported by Herrera's treatise *Sobre la figura cúbica* (1935) on the Hermetic philosopher Ramón Lull. Fundamentally, Renaissance Hermetism promulgated a belief in an astrologically ordered cosmology where a geo-centric universe was divided into three worlds: the world of man, the celestial world of the planets and the fixed stars, and the super-celestial world of God (Taylor 1972: 63–64). The Christian Hermetism that was practiced in the Renaissance was a combination of Christianity and *prisca theologia* (ancient Knowledge). Ancient mystical mathematics of music, geometry and arithmetic became prominent in Renaissance Hermetism. This atmosphere of Hermetic learning pervaded the Spanish Court, affecting even Philip II himself, and Villalpando's *In Ezechielem Explanationes* was a product of this atmosphere.

Villalpando's "Flawless System"

In Volume II, Villalpando laid out a reconstruction of Solomon's Temple based on the vision of Ezekiel. Rudolf Wittkower described the rationale used in Villalpando's reconstruction as an "absolutely flawless system" (Wittkower 1988: 122). This flawless system is a combination of: the three hermetic worlds of microcosm-macrocosm; the Pythagorean-Platonic musical harmonies; a cosmic-astrologic plan which determined the plan of the temple precinct; Vitruvian anthropomorphism; and the module that governs the buildings. All of this was supported and justified by a deep knowledge of both Christian and Hebrew Sacred Scripture.

Villalpando clearly distinguished sacred architecture from the profane architecture of Vitruvius. He claimed that "Sacred architecture constitutes the origin of architecture, and the profane one is like a copy, or better still, like a shadow of sacred architecture" (Villalpando and Prado 1604: 414). The purpose of Vitruvius, who Villalpando described as "the pioneer of our architects," was to equip the architect with the norms of architecture.

But Villalpando's purpose was to examine the origins of architecture and to extract the norms that were derived from God's plan and promulgated by the sacred scriptures; this natural order was followed by Vitruvius in his *Ten Books on Architecture*. Villalpando's reconstruction envisaged the Temple to be a building that encapsulated the entire formal grammar of classical architecture, which begins with the harmonic ratios.

Villalpando carefully defined all the measurements of the Temple as being derived from the sacred texts. He provides all the measurements of the three main floors of the buildings of Solomon; the measurements in column one are for the house of the Lord, in column two for the atrium, and in column three for the house of the king (Fig. 59.1). These are grouped under the headings: the diameter of the columns; the height of the columns; the height of the entablature; the height of the

Fig. 59.1 Symmetry of sacred architecture. Image: author, after Villalpando and Prado (1604)

TOMI II· EXPLANATIONVM PARS II·

SYMMETRIAE ARCHITECTVRAE SACRAE.

DIAMETRI COLVMNARVM.

	DOMVS DOMINI. Cubiti.	Digiti.	ATRIORVM. Cubiti.	Digiti.	DOMVS REGIAE Cubiti.	Digiti.
Primi ordinis	$4\frac{1}{6}$	100	$2\frac{1}{12}$	50	$1\frac{1}{24}$	25
Secundi.	$3\frac{1}{8}$	75	$1\frac{9}{16}$	$37\frac{1}{2}$	$\frac{25}{32}$	$18\frac{3}{4}$
Tertij.	$2\frac{1}{2}$	60	$1\frac{1}{4}$	30	$\frac{5}{8}$	15

ALTITVDO COLVMNARVM.

Primi ordinis	40	960	20	480	10	240
Secundi.	30	720	15	360	$7\frac{1}{2}$	180
Tertij.	24	576	12	288	6	144

TRABEATIONVM ALTITVDO.

Primi ordinis	10	240	5	120	$2\frac{1}{2}$	60
Secundi.	$7\frac{1}{2}$	180	$3\frac{3}{4}$	90	$1\frac{7}{8}$	45
Tertij.	6	144	3	72	$1\frac{1}{2}$	36

ORDINVM CELSITVDO.

Primi ordinis	50	1200	25	600	$12\frac{1}{2}$	300
Secundi.	$37\frac{1}{2}$	900	$18\frac{3}{4}$	450	$9\frac{3}{8}$	225
Tertij.	30	720	15	360	$7\frac{1}{2}$	180

PODIOLI ALTITVDO.

Secundi ord.	$1\frac{1}{4}$	30	$\frac{5}{8}$	15	$\frac{5}{16}$	$7\frac{1}{2}$
Tertij.	$1\frac{1}{4}$	30	$\frac{5}{8}$	15	$\frac{5}{16}$	$7\frac{1}{2}$

VNIVERSA ALTITVDO.

Domus Domini		Atriorum		Domus Regiae.	
120	2880	60	1440	30	720

floors; the height of the balcony and the overall height of the buildings. All of the measurements of the atrium are double that of the house of the King, and the measurements of the house of the Lord are double that of the atrium. In all of the columns the numbers reveal that the second floor is a quarter part smaller than the first floor, or a third part of its own measure smaller that the first floor; the third floor is a fifth part less than the second floor, or the fourth part of its own measure smaller than the second floor; proceeding in the same way it is possible to find out the other measurements of the other floors, i.e., the fourth floor will be a fifth part smaller than the third, or a sixth part of its own measure smaller than the third, and so on.

Villalpando also examined the heights of the columns and the entablatures (Fig. 59.1). Considering the columns of the atrium, the overall height of the atrium is 60 cubits: the height of the first floor columns is 20 cubits, a third of the overall height, the height of the second floor columns is 15 cubits, a quarter of the overall height and the height of the third floor columns is 12 cubits, a fifth of the overall height. Thus the above ratios are reflected in the height of the columns in relation to the overall height. The height of the entablature of the first floor is 1/12 of the overall height; for the second floor it is 1/16 the height and for the third floor, 1/20 of the overall height. The ratio between the first and the second floor is 4:3 and between the second and the third is 5:4 (Villalpando and Prado 1604: 441–443).

To deduce the heights of the entablature's elements, divide the height of the entablature by eight; two parts will be the height of the architraves and three parts each for the heights of the frieze and crown. The width of the triglyph and the metopes, are calculated from the distance between the centres of the columns. In the first floor the height of the architrave was equal to the width of the triglyph, and that of the metopes equal to the frieze (Fig. 59.2). The width of the triglyph of the second floor to that of the first is in the ratio 3:2, and the third to the second 4:3 (Villalpando and Prado 1604: 449). For Villalpando the proportions of the columns are the foundation of all other measurements and proportions of the entire temple (Villalpando and Prado 1604: 423).

Villalpando used the monochord to demonstrate the relationships between the widths of the triglyph and the metopes which resulted in intervals of the Pythagorean musical scale. Figure 59.2 shows the inter-relationships between the three buildings of Solomon. Although Vitruvius outlined six harmonic ratios—the quarter (diatessaron), the fifth one (diapente), the eighth (diapason), the quarter of the eighth (diapason with diatessaron), the fifth of the eighth (diapason with diapente) and the double of the eighth one (disdiapason) (Vitruvius 1960: Bk. 5, ch. 4)—Villalpando rejected the quarter of the eighth. Throughout his commentary Villalpando followed Daniel Barbaro's commentary on Vitruvius's *De Architectura*; Barbaro opposed the musical theory of Vitruvius. For Villalpando "the quarter of the eighth, called '*superpatiens*', is truly a dissonant chord, and consequently, the chords are simply five: three simple and two composed" (Villalpando and Prado 1604: 458).

Villalpando claimed that these harmonic proportions are most apt for a building of divine origins and he implied the existence of a link between the harmonic proportions and the celestial bodies. For Villalpando the Temple reflected the

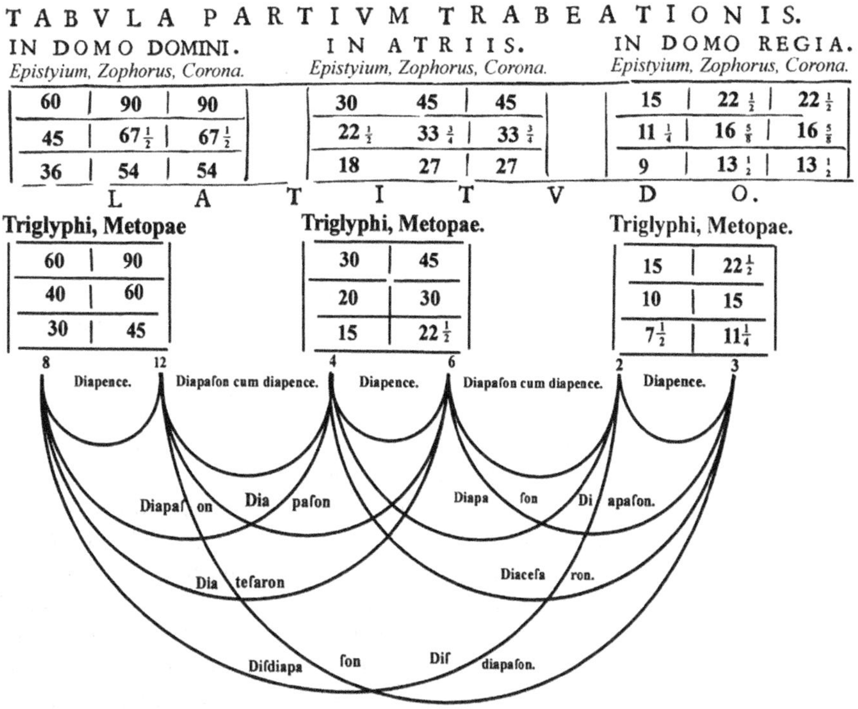

Fig. 59.2 Table of the parts of the entablature. Image: author, after Villalpando and Prado (1604)

creation of God and thus had to incorporate itself into the universal harmony according to the movements of the planets and the fixed stars. To this end he examined the Tabernacle of Moses, since it prefigured the plan of the Temple. The camp of the tribes of Israel that surrounded the Tabernacle is a primitive plan of the Temple precinct (Villalpando and Prado 1604: 466). Villalpando first established that the proportion of the atrium that surrounds the immediate temple and the altar is a double square; he then considered the configuration of the camp of the tribes of Israel. The configuration of the camp was highly structured, with the Tabernacle placed in the centre, fortified by the four camps of the Levites (Moses and Aaron; Caathi, Gerson and Merari). Surrounding them were the 12 tribes of Israel, each tribe camped under its banner that declared its ancient lineage.

The distribution and placement of the tribes in the camp was determined by a perfect plan with nothing left to chance, since it reproduced the plan of the Temple and thus was the microcosm of the universe (Fig. 59.3). The four tents of the Levites in the centre that fortified the Tabernacle corresponded to the four simple elements of the sub-lunar world, and represented the world of man. These were encircled by the celestial orbits made up of the seven atriums. The orbits are positioned on the plan as Ptolemy assigned them in *Almagest*: "Thus Saturn is situated between Capricorn and Aquarius; Jupiter in Pisces; Mars in Aries; Venus in Libra;

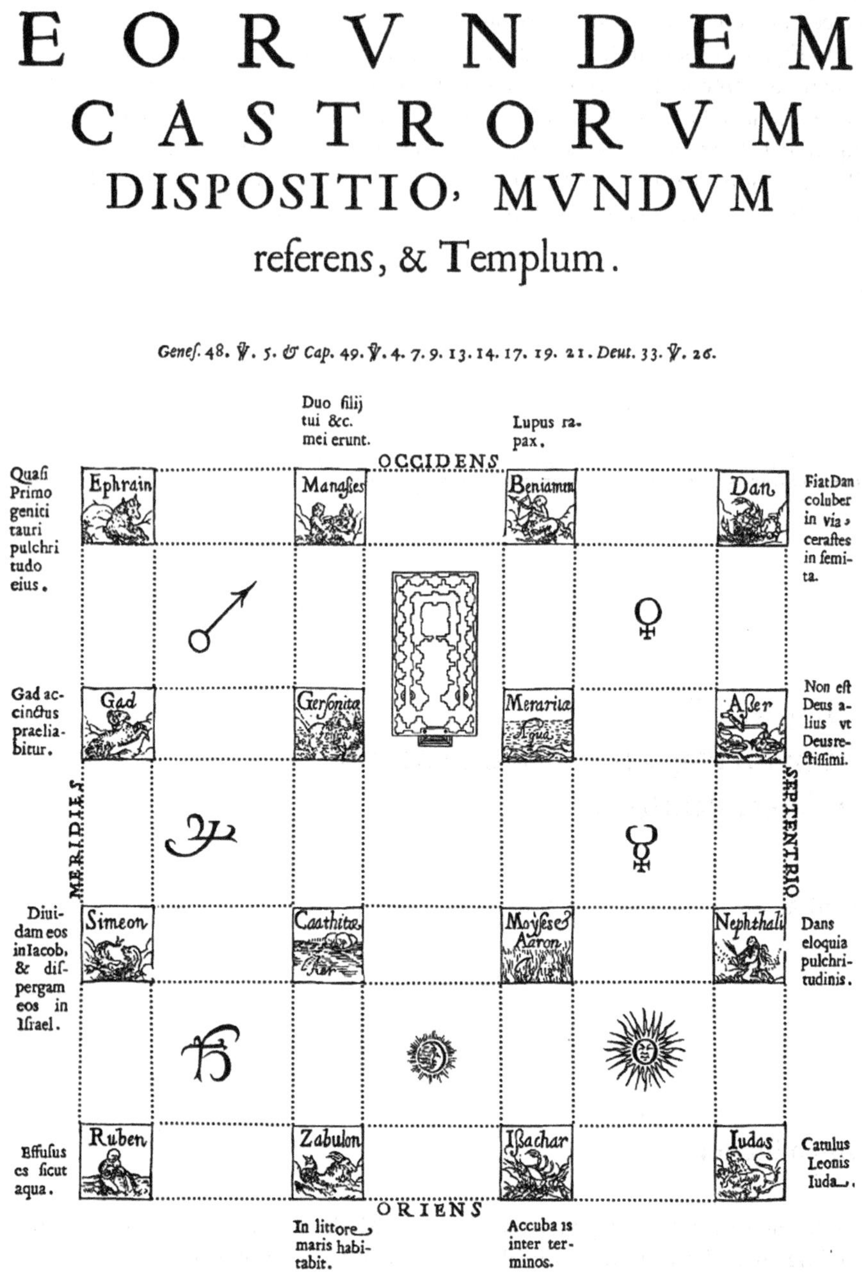

Fig. 59.3 The arrangement of the heavenly fortress. Drawing: author, after Villalpando and Prado (1604: vol. 2, 470)

Mercury in Virgo; the Sun in Leo and the Moon in Cancer" (Villalpando and Prado 1604: 469). Surrounding the 7 courts or celestial orbits were the 12 fortifications or bastions of the Temple precinct perimeter, which corresponded to the 12 tents of the tribes of Israel. Judah was represented by the lion, Ruben by the water-bearer, Ephraim by the bull, Dan by the Scorpion, and so on, so that the tribe's banners represented the signs of the zodiac. In the centre was the Temple, 'dedicated to the profit of man,' that represented the 'true Sun' of super-celestial world of the Church. This true Sun is Christ, the 'Sun of Justice' whose light is salvation. This light illuminates the 7 planets and the 12 constellations, and the centralized Earth is illuminated by the Planet Sun that is located in Leo. This perfect plan represented the three worlds of the microcosm and macrocosm: in the centre was the super-celestial world of God; this is surrounded by the world of man; and this in turn is surrounded by the celestial world of the seven planets and the fixed stars encircling the Earth—a perfect hermetic vision of a geo-concentric universe.

Villalpando fully endorsed the anthropomorphic theories of Vitruvius. He perceived that the humanity assumed by God is reflected in the measurements and geometry of the Temple, which prefigured the perfection of the mystical body of the Church. Man has a height of 6 ft, and this measurement agrees with that of his arms extended; but if the arms are doubled in front of the chest, so that the end of the longest finger of the right hand touches the end of the middle finger of the left hand, then the width of man will be one and a half cubits, or 3 ft. The colonnades of the Temple have eight intercolumniations and are divided into three promenades or galleries that correspond to the barrel of the chest with the arms (Fig. 59.4). These colonnades correspond to the proportion of 1:2, not only a double square but also the harmonic ratio of an eighth, an octave. Here Villalpando portrayed Christ taking the appearance of man as the cosmological man, which emphasizes the microcosm-macrocosm analogy.

The gridded floor plan of Villalpando's reconstruction (Fig. 59.5), corresponding to the plan that represented the three worlds of the microcosm and macrocosm, was crowded with colonnades and incorporated 1,500 columns. The Temple precinct was 500 × 500 cubits and the exterior boundary 800 × 800 cubits. Its height, including the foundation, was a massive 420 cubits. Every part or element was in a harmonious ratio to the entire building. For Villalpando this was the greatest building ever built and no building could ever surpass it. His was the first full-scale reconstruction of the divine archetype and this reconstruction inspired not only other commentaries and other reconstructions of Solomon's Temple but it also stimulated discussion on the very nature of the origins of architecture.

However, his reconstruction was not without its critics. In the seventeenth and eighteenth centuries, critics included Louis Cappel, Samuel Lee, Louis Compiègne de Veil, Nicolaus Goldmann and others who produced alternative reconstructions.[2]

[2] Taylor (1972) is an excellent summary of the support and criticism of the Villalpando's reconstruction.

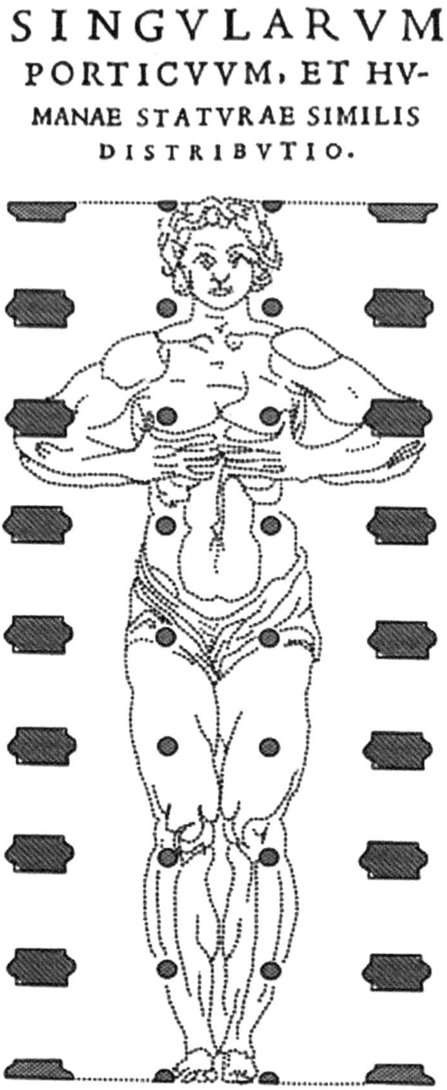

Fig. 59.4 A single colonnade and the resemblances to the division of the human stature. Drawing: author, after Villalpando and Prado (1604: vol. 2, 472)

There were many points of disagreement between the critics, including whether Ezekiel's vision was the same as the Temple of Solomon, whether the architecture of the Temple could ever be surpassed, whether the Temple set the norms of architecture and thus was the origin of architecture. There was also a vast array of different interpretations of the sacred text, which resulted in many different reconstructions. Cappel wrote a commentary on Villalpando's reconstruction in Brian Walton's *Biblia Sacra Polgotta* (Cappel 1657), and a revision of this is in *Critici Sacri* (Cappel 1660). Both contain large paraphrased sections of Villalpando's work, which continued to stimulate interest in Villalpando's work

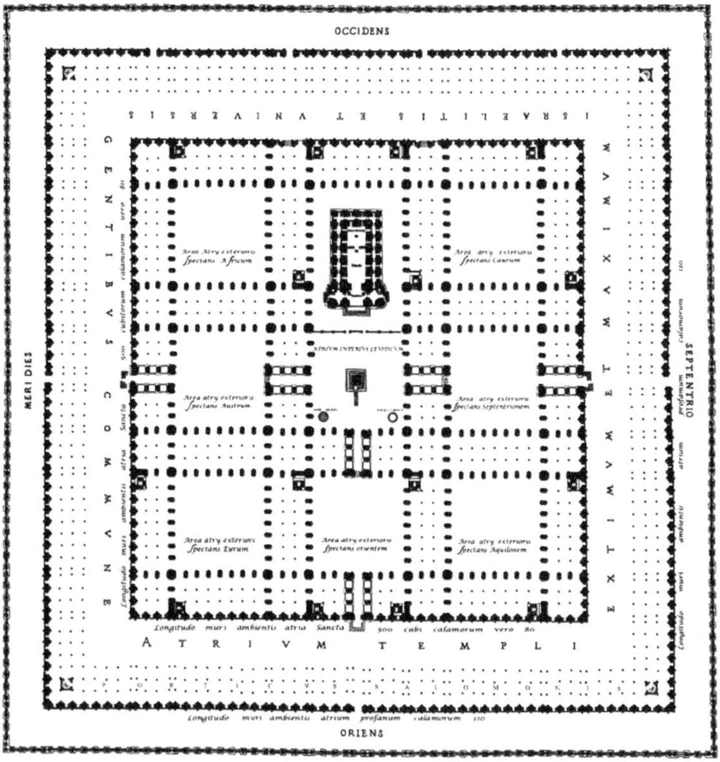

Fig. 59.5 Villalpando's floor plan of Solomon's temple. Drawing: author, after Villalpando and Prado (1604: vol. 2, not numbered)

and Solomon's Temple; many of the later critics often quoted from the paraphrased version rather than from the original.

Newton's Commentary

In the twentieth century, criticism from what appears to be an unusual source was uncovered. Newton's unpublished manuscripts reveal that he had a long-running interest in Solomon's Temple, yet the only published work of his on it was a very brief summary of the floor plan in *The Chronology of Ancient Kingdoms Amended*, which was printed posthumously in 1728 (Newton 1988). Two of Newton's manuscripts that contain a commentary on Villalpando are *Prolegomena ad Lexici Propretici partem Secundam: De Forma Sanctuary Judaici*) known by its call number Babson Ms. 0434 (Newton c. 1680) and *Miscellaneous Notes and*

Extracts on the Temple, the Fathers, Prophecy, Church History, Doctrinal Issues, known as Yahuda Ms. 14 (Newton n.d.).[3]

Newton's comments are a mixture of both support and criticism. He says, for instance, "Villalpando, although the best (and) the most eminent commentator on Ezekiel's Temple: yet (he is) out in many things" (Yahuda Ms. 14: 32r). Like Villalpando, Newton strongly believed that Ezekiel's vision of the Temple was the plan of Solomon's temple. In Babson MS 0434 he studied Ezekiel's vision of the Temple and confirmed this vision with readings of the ancient Hebrew and Greek texts. From this study he reconstructed the structure of the temple, revealing it to be mathematically perfect. However, his floor plan and description of the Temple are remarkably different from Villalpando's (compare Figs. 59.5 and 59.6). He believed that Villalpando's errors in his design were primarily derived from his failure to take advantage of Jewish sources and his misinterpretation of the Latin texts (Yahuda Ms. 14: 32r). Newton pointed out that the Latin text that Villalpando used sometimes differed in its translation to the Hebrew texts, for instance in the Latin version in Ezekiel 42:3 Villalpando translated 'colonnades united' to be a triple colonnade but in the Hebrew text it translated to 'colonnade against colonnade three times' indicating three storeys (Babson Ms. 0434: 12).

According to Newton, Villalpando created his grid plan of the Temple precinct from an "incorrect translation' and his plan "has no support and is lacking in reason" (Babson Ms. 0434: 46). Villalpando interpreted Ezekiel 40:19–20 as meaning that the length of the atrium from the south to the north is the distance between the gates, a 100 cubits, and this divided the area of the precinct into nine small atriums or anterooms, two of which formed the temple atrium and seven exterior to it (see Fig. 59.5). These anterooms are divided from each other by triple colonnades 50 cubits wide. Newton pointed out that these anterooms not mentioned in Ezekiel. Regarding the 30 chambers that flank the sides of the gate, which are expressly mentioned by Ezekiel, it is impossible to arrive at the number 30 for these chambers if the spaces of the gates are not counted. However, this goes against the text of Ezekiel. In addition, Newton also claimed that Villalpando's grid plan cannot be accepted "unless we want to move away from the proportion of Moses' atrium that surrounds the immediate temple and the altar, which was established by Villalpando himself as being a length over double its width" (Babson Ms. 0434: 46).

These criticisms based on Villalpando's interpretation of the Biblical texts challenge the basis of his reconstruction. The triple colonnades that Newton claimed was a mistranslation were important to Villalpando's plan. First, they portrayed man/Christ as the cosmological man, emphasizing the microcosm-macrocosm analogy. Second, they divided the gridded plan into the seven small ante rooms and the temple atrium, which Newton considers to be 'lacking in reason' and their creation goes against the proportions of the Temple atrium that

[3] For a full account of Newton's comments on Villalpanda's reconstruction and his own attempt, see Morrison (2011).

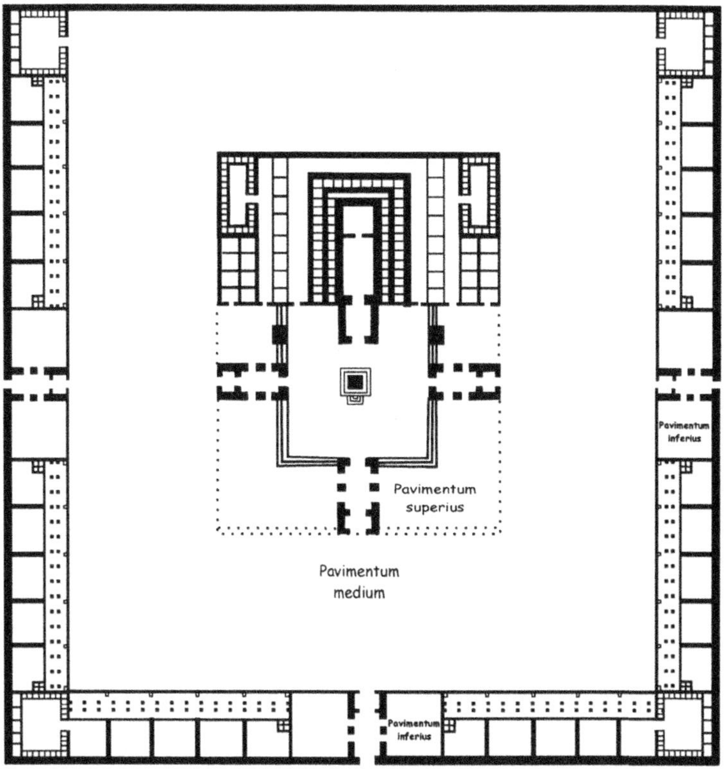

Fig. 59.6 Newton's floor plan of Solomon's temple. Drawing: author, after Babson (Ms. 0434, fol. 9r)

Villalpando had himself established. These triple-colonnaded atriums not only formed a considerable part of Villalpando's reconstruction they are also significant for the plan of the three worlds of microcosm-macrocosm (Fig. 59.3). Their removal from his plan changes his reconstruction so that it becomes unrecognizable. Furthermore, Newton referred to Villalpando's reconstruction as a 'fantasy' (Babson Ms. 0434: 56). All of these criticisms beg the question of why Newton considered him "the best (and) the most eminent commentator on Ezekiel's Temple?"

In Yahuda MS 14 (32r–33v) Newton agreed with Villalpando's symmetrical layout of the camp around the Tabernacle and with the heraldry of the tribe's standards. He accepted that this plan prefigures the plan and the proportions of the temple, which were double that of the Tabernacle as proven in detail by Villalpando. In addition, Newton is in agreement that the perfect architectural harmony of the Temple represents a microcosm of the perfect harmony of the macrocosm. However, Newton misread Villalpando's geo-centric plan of the microcosm-macrocosm and considered it to be a heliocentric system.[4] Newton claimed that "Temples were anciently contrived to represent the frame of the Universe as the true Temple of the great God" (quoted in (Castillejo 1981: 33)). Newton established that Solomon's Temple was the model of all temples (Yahuda Ms. 14: 1r, 6r). Thus this was the model microcosm of the universe and revealed the mind of the Supreme Architect, that is, the mind of God. This precise concentric model of the heliocentric universe is particularly strange given that Newton established the orbital paths of the planet to be elliptical before his making his comments on the Temple as the model microcosm of the universe.

A final point of agreement is that the Temple of Solomon was a masterpiece of architecture and was not only the model for all future temples but also established the norms of architecture as practiced by the later Greek and Roman architects and codified by Vitruvius (Yahuda Ms. 14: 32r). In Newton's reconstruction the "capitals were carved in the Corinthian style of a beauty that was a miracle" (Babson Ms. 0434: 15) and there were bronze Corinthian columns that were "covered with a great deal of silver and adorned with gold" (Babson Ms. 0434: 19). There were also massive Doric columns (Babson Ms. 0434: 35, 38); like Villalpando, Newton's reconstruction was the perfect classical building.

In 1752, William Stukeley wrote his *Memoirs of Sir Isaac Newton's Life*. He recalled a conversation with Newton in 1725 on Solomon's Temple, where Newton claimed that the architecture of the temple was Doric and "the Greeks advanced it into Ionic and the Corinthian, as the Latins into Composite" (Stukeley 1936: 19). But the architecture of Newton's reconstruction of the 1680s is both Corinthian and Doric, and there is nothing in his papers to demonstrate this development, mentioned by Stukeley, of the architectural orders, or that he had ever changed his mind. Stukeley's reminiscences appear to support his own concept of architectural development rather that Newton's. For although Newton does not directly state that Solomon's temple was the pinnacle of architecture that could never be surpassed, he does point to proportions where "the columns will be less numerous than in the proportion of the eustyli of Vitruvius, and far more beautiful" (Babson Ms. 0434: 37). Architecture had not been improved by the Romans, but had in fact declined with the loss of a most beautiful proportion.

---

[4] In Chapter XXX of *Ezechielem Explanationes* Villalpando clearly describes and illustrates a geo-centric system, where the Sun completes a circuit around the Earth every 24 h. There is some ambiguity in the light of the centralised "Sun of Justice" Christ reflecting back to the Earth through the illumination of the planet Sun. However the planet Sun is circling the centralised earth. Villalpando clearly held a hermetic view of geo-centricism. Since he also lived and worked in Rome in the early seventeenth century, he would have been aware of, if he had not personally witnessed, the burning of Giordano Bruno in 1600 in Rome for promoting a heliocentric view of the universe. Furthermore, *Ezechielem Explanationes* was published in Rome and would have been censored if there had been any hint of the promotion of a heliocentric system. It is possible that Newton did not see the original volumes of Villalpando, and that his knowledge of Villalpando came from the paraphrased section of Cappel (1660), which is not as detailed as the original. It is known that Newton owned a copy of Cappel; see Harrison (1978).

Conclusion

Newton's criticisms not only highlight fundamental errors in Villalpando's reconstruction based on his interpretation of the scriptures but also point to inaccuracies in the rationale behind his "flawless" system. The anterooms and their divisions and the triple colonnades are features that are fundamental to both Villalpando's architectural reconstruction and his concept of the microcosm/macrocosm. Although Newton concurred with the temple as the microcosm of the universe, he perceived this microcosm as a heliocentric system with the Temple as the hearth—the sun—of the universe, not the complex geo-centric universe of Villalpando. But despite In Ezechielem Explanationes being a book of the Renaissance and Newton's manuscripts being works of the Enlightenment, both Villalpando and Newton strongly believed in the medieval concept of the Divine Architect. In their architectural reconstructions they both attempted to find mathematical and geometrical formulations of divine truths. For Villalpando architects were the first Apostles, since they continued the work of the Divine Architect (Villalpando and Prado 1604: 464). The image depicted by both is not far away from the thirteenth-century illustration of the Divine Architect wielding his compass in the Bodleian Bible Moralisé (Friedman 1974: pl. VII). Yet their works remain exemplary of their respective periods, for the origins of architecture is to be found in the Temple of Solomon, which was the perfect model for all sacred architecture, and all profane architecture will ever be as a "shadow of sacred architecture."

Biography Tessa Morrison is a senior lecturer at the School of Architecture and Built Environment at the University of Newcastle Australia. Her academic background is in art history mathematics and philosophy. She has published extensively on geometric spatial symbolism and has an interest in examining and reconstructing the plans and structures of architecture in medieval poetry through poems such as the eighth-century Gaelic poem *Saltair na Rann* and the fourteenth-century *Pearl*. She has translating Book V of Villalpando's *De Postrema Ezechielis Prophetae Visione* and Newton's work on the Temple of Solomon from Latin into English. Her research on this culminated in the publication of *Isaac Newton's Temple of Solomon and his Reconstruction of Sacred Architecture* (Basel: Birkhäuser 2011).

References

Cappel, Louis. 1657. Trisagion Sive Templi Hierosolymitani Triplex Delineatio. In: *Biblia Sacra Polgotta*, Brian Walton, ed. London.

———. 1660. Excerpta Ex Villalpando Ad Ezechielem. In: *Critici Sacri*, John Pearson, ed. London.

CASTILLEJO, David. 1981. *The Expanding Force in Newton's Cosmos*. Madrid: Ediciones De Arte Y Bibliofilia.

FRIEDMAN, John Block. 1974. The Architect's Compass in Creation Miniatures of the Later Middle Ages. *Traditio* **XXX**: 419–429.

HARRISON, John R. 1978. *The Library of Isaac Newton*. New York: Cambridge University Press.

HERRERA, Juan de. 1935. *Sobre la Figura Cúbica*. Santander: Universidad de Cantabria.

MORRISON, Tessa. 2011. *Isaac Newton's Temple of Solomon and his Reconstruction of Sacred Architecture*. Basel: Birkhäuser.

NEWTON, Isaac. (n.d.). *Miscellaneous Notes and Extracts on the Temple, the Fathers, Prophecy, Church History, Doctrinal Issues, ets* (Yahuda Ms. 14). Unpublished manuscript, Jewish National and University Library.

———.1680. *Prolegomena ad Lexici Prophtici partem secundam in quibus agitur De forma Sanctuarii Judaici* (Babson Ms. 0434). Unpublished manuscript, Babson College, Massachusetts.

———. 1988. *The Chronology of Ancient Kingdoms Amended*. London: Histories and Mysteries of Man.

STUKELEY, William. 1936. *Memoirs of Sir Isaac Newton*. London: Taylor and Francis.

TAYLOR, Rene. 1972. Hermetism and Mystical Architecture in the Society of Jesus. Pp. 63–97 in Rudolf Wittkower and Irma B Jaffe eds., *Baroque Art: The Jesuit Contribution*. New York: Fordham University Press, 1972.

VILLALPANDO, Juan Bautisa and Hieronymus PRADO. 1604. *Ezechielem Explanationes Et Apparatus Urbis Hierolymitani Commentariis Et Imaginibus Illustratus*. Rome.

VITRUVIUS. 1960. *The Ten Books on Architecture*. Trans. Morris Hickey Morgan. New York: Dover.

WITTKOWER, Rudolf. 1988. *Architectural Principles in the Age of Humanism*. London: Academy Editions.

Chapter 60
Coelum Britannicum: Inigo Jones and Symbolic Geometry

Rumiko Handa

> *Many of these monuments remain in the British islands, curious for their antiquity, or astonishing for the greatness of the work: enormous masses of rock, so poised as to be set in motion with the slightest touch, yet not to be pushed from their place by a very great power; . . . displaying a wild industry, and a strange mixture of ingenuity and rudeness. But they are all worthy of attention not only as such monuments often clear up the darkness and supply the defects of history, but as they lay open a noble field of speculation for those who study the changes which have happened in the manners, opinions, and sciences of men. . .*
>
> Edmund Burke (1887: 188)

Introduction

Inigo Jones's interpretation that Stonehenge was a Roman temple of Coelum, the god of the heavens, was published in 1655, 3 years after his death, in The most notable Antiquity of Great Britain, vulgarly called Stone-Heng, on Salisbury Plain, Restored.[1] King James I demanded an interpretation in 1620. The task most reasonably fell in the realm of Surveyor of the King's Works, which Jones had been for the preceding 5 years. According to John Webb, Jones's assistant since 1628 and executor of Jones's will, it was Webb who wrote the book based on

First published as: Rumiko Handa, "Coelum Brittanicum: Inigo Jones and Symbolic Geometry". Pp. 109–126 in *Nexus IV: Architecture and Mathematics*, Kim Williams and Jose Francisco Rodrigues, eds. Fucecchio (Florence): Kim Williams Books, 2002.

[1] There are two modern facsimile reproductions of Jones's *The most notable Antiquity of Great Britain, vulgarly called Stone-Heng, on Salisbury Plain, Restored*: of the 1655 edition (Jones 1972) and of the 1725 edition (Jones 1971). Although the 1655 edition was narrowly distributed, and the 1666 Great Fire destroyed the unsold copies, it was re-issued in 1725, together with Walter Charleton's refuting account of 1662 and John Webb's rebuttal of 1665. The page numbers referred to in this article refer to Jones (1971).

R. Handa (✉)
College of Architecture, University of Nebraska-Lincoln, Room 237 Arch Hall West, Lincoln, NE 68588-0107, USA
e-mail: rhanda1@unl.edu

K. Williams and M.J. Ostwald (eds.), *Architecture and Mathematics from Antiquity to the Future*, DOI 10.1007/978-3-319-00143-2_13,

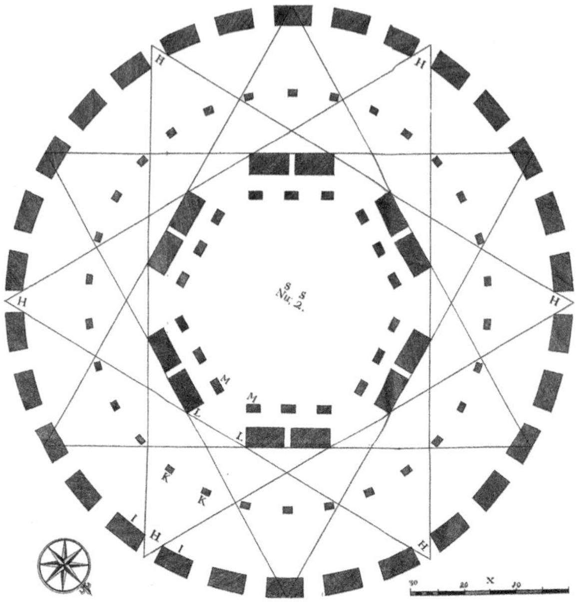

Fig. 60.1 Inigo Jones, "Groundplot" of Stonehenge. Image: Jones (1655: Pl. 40)

Jones's "few indigested" notes, on the recommendation of William Harvey, physician to James and to Charles I, and John Selden, antiquarian.[2] The treatise included a plan of the megalith restored (Fig. 60.1).

On the outer circle were 30 columns, to which a concentric circle of 30 smaller columns corresponded, the radius of the latter tracing the outermost intersections of the four equilateral triangles within the first circle. On the hexagon resulting from two of the four triangles were six sets of two stones each. A side of this hexagon was as wide as that of the dodecagon.

John Aubrey, seventeenth-century antiquarian and Royal Society member, characterized Jones's theory by a "Lesbian rule", a soft lead ruler that fits curbs of stones: Jones "had not dealt fairly, but had made a Lesbian's rule, which is conformed to the stone; that is, he framed the monument to his own hypothesis which is much differing from the thing it self" (Aubrey 1862: 315; Hunter 1975: 179–180). Since then, scientific archaeology has advanced our knowledge of the monument. Thirty stones make up the outer circle, as Jones depicted. However, no hexagon exists, but rather a U-shape of ten stones. No indication of Tuscan order is found in the crude cuts of the stones. Isotopic method has proven several construction stages between 2000 and 1600 B.C., ruling out the Romans, who reached the British isles in 43 A.D. Some present-day scholars have suspected that Webb published the theory of which Jones was not convinced, or simply borrowed the master's name to publish his own idea. However, the idea, if not the writing, should be attributed to Jones, and reveals the architect's sense of the past and imagination. The symbolism of Coelum are also found in other works associated with Jones.

[2] These two individuals both have connections to Robert Fludd; see Yates (1969): 64, Rykwert (1980).

Jones's Stonehenge interpretation reveals an important difference between his world and ours, as Edmund Burke's statement above suggests. Jones demonstrated the ideal through architecture, no matter if, as was in fact the case, the ideal was far from the real. Mathematics, and geometry in particular, enabled him to do so. Stone-Heng was not so much related to the original as to its ideal. It not only idealized the megalith but also the nation and monarch. It further idealized Jones's own realm, that is, architecture, the architect, and his own being. To compare, today's advanced technology makes almost any construction possible but at the same time allows us to be oblivious to what ought to be built. Professionals might ask what is timely, but often fail to question whether being timely is always good. Positivistic clarity in the matters of economy and efficiency makes it difficult for us to see ethical values. In order to fully appreciate Jones's world, we need to get at the provenance of his knowledge.

Jones's Intellectual World and His Mathematics Education

Called by John Summerson "England's first classical architect" (Summerson 2000: 1), Jones himself listed the Vitruvian qualifications for the architect:

> An Architect; who, (as Vitruvius saith) should be . . . perfect in Design, expert in Geometry, well seen in the Opticks, skilful in Arithmetick, a good Historian, a diligent hearer of Philosophers, well experience'd in Physick, Musick, Law and Astrology (Jones 1971: 3).

Artisans of Elizabethan England could have known of Vitruvian qualifications through English authors as well, including John Shute in *The First and Chief Ground of Architecture* (1563) and John Dee in his preface to Euclid's *Elements of Geometry* (1570).

Jones himself must have made efforts in acquiring these qualities. *Altro diletto che Imparar non trovo* ("I find no other pleasure than learning"), Jones wrote decoratively for frontispiece of his sketchbook in January 1614. Jones owned a copy of Daniel Barbaro's translation and commentary of Vitrivius's *Dieci Libri dell'Architettura* in Italian of 1567, now in the Devonshire Collection. When and how Jones acquired this book published before his birth, or any others, would be an interesting but difficult topic to pursue. Out of about 50 volumes of Jones's extant library, five bear the dates of purchase, the earliest of which is 1601, in which the 28-year-old paid two gold coins for a copy of Palladio in Venice.[3] Only a portion of Jones's library has survived, which range from architecture to history, geography, mathematics and philosophy.[4] Most of these books are in Italian, some with Jones's

[3] The particular dated inscription has been considered by generations of scholars "not by Jones" or "by an Italian bookseller". For the present author's argument to affix the authorship to Jones, see Handa (2006).

[4] For a list of Jones's extant books, see Johnson (1997, Anderson 1993). For the provenance of the George Clark (1661–1736) collection which makes up the Worcester books, see Wilkinson (1926). See also Sayce (1970).

annotation, while the only two are in Latin without notes. Jones must have been proficient in Italian but not in Latin. Jones cited about 60 authors in his *Stong-heng*, out of which only about 10 are among his extant books. The fact that quotations are in Latin with an English translation following should not refute the assumption about his language capabilities. Jones could well have drawn quotations from Italian, while Webb could have searched equivalent passages in Latin editions. An observation can be made: While much was quoted from the 1567 Latin edition of Barbaro's Vitruvius, one statement was in Italian with a reference to the 1584 edition. The statement being in Italian is natural, for the 1567 Latin edition did not include that precise passage by Barbaro. The citation is unusual, however, as the only one that specifies a particular edition, and puzzling for not referring to the 1567 Italian, which Jones owned and which included the quoted passage. What should be deduced is that Jones's 1567 copy had left Webb's hands before 1655. This might begin to explain why the book was separated from others, which stayed with the assistant until his death, and most of which are at Worcester College, Oxford.

Although we do not know where Jones learned mathematics, he grew up in a time when mathematics was valued in practical trades. A number of individuals, including Robert Recorde and John Dee, had spread the mathematics discipline.[5] Robert Recorde, whose life ended before Jones was born, taught mathematics in London (Taylor 1954). His books, *Grounde of Artes*, 1542, on arithmetic, Pathway to Knowledge, 1551, on geometry, and *Castle of Knowledge* (1556) on astronomy, written in vernacular English, were meant for tradesmen and artisans (Johnson and Larkey 1935). Geometrical operations needed for Jones's diagrams were in book I of *Pathway*, such as drawing an equilateral triangle within a given circle, a circle within a given triangle, and a hexagon within a given circle.

Inigo Jones's Stonehenge Interpretation

Jones refuted preceding theories of Stonehenge and presented a new interpretation:

> Wherefore leaving these, Stoneheng was dedicated, as I conceive, to the God Coelus, by some Authors called Coelum, by others Uranus, from whom the Ancients imagined all things took their beginning (Jones 1971: 101).

In the last portion of the book Jones gave principal reasons for his interpretation. He first listed the surrounding environment: "My reasons are, first, in respect of the situation thereof; for it stands in a Plain, remote from any Town or Village, in a free and open air, without any groves or woods about it" (Jones 1971: 67). Jones had rejected the popular belief of Stonehenge as a Druids' temple for the reason that Druids, who according to Julius Caesar lived in groves and woods, would not have been involved in complex building such as Stonehenge. Jones quoted Vitruvius:

---

[5] Taylor (1954) includes 582 individuals whose lives ranged from 1486 to 1768, and 628 printed books and manuscripts on mathematics and related subjects.

> In the first Age of the World (saith he) Men lived in Woods, Caves, and Forests, but after they had found out the Use of Fire, and by the Benefit thereof were invited to enter into a certain kind of Society, . . . Some of them began to make themselves Habitations of Boughs, some to dig Dens in Mountains; other some, imitating the Nests of Birds, made themselves places of Lome and Twigs, and such like Materials, to creep into, an shroud themselves in (Jones 1971: 8).

Jones's second reason came from observing the roofless nature of Stonehenge: "...in regard of the Aspect; for Stonehenge was never covered, but built without a roof" (Jones 1971: 67). Jones had learned the term *hypaethros* from Vitruvius, noting in his copy, "7/hipteros the open or uncovered".[6] Jones listed the suitable deities, quoting Vitruvius: "To Jove the Lightner, and to Coelus, and to the Sun, and to the Moon, they erected buildings in the open air and uncovered" (Vitruvius 1567: III, 2). These deities should be presented "in a clear and open view", which required the edifice not to be enclosed by walls but instead to be surrounded by columns. Jones had earlier observed in Vitruvius: "Temples open to the Air, and without Roofs, have columns on the Inside, distant from the Walls, as Courts Porticoes about them" (Vitruvius 1567: III, 1; Jones 1971: 46). Additionally, Jones considered it a "hainous matter to see those Gods confined under a roof, whose doing good consisted in being abroad".

Jones's third reason concerned the circular plan (Jones 1971: 67). Pierio Valeriano, Leon Battista Alberti, and Philander on Vitruvius were his sources. Quoting from Philander (1549: 137–138), Jones observed:

> Although (saith he) the Ancients made some Temples square, some of six Sides, others of many Angles, they were especially delighted with making of them round, as representing thereby the Form or Figure of Coelum, Heaven (Jones 1971: 67).

In Philander (1549: 138) Jones found a reference to circular temples with double columns:

> Varro de re rustica (as I find him cited by Philander) tells us, that they had in use amongst them a round Building without any Walls, having a double Order of Columns round about, this he calls by the name of Tholus, . . . A round Edifice (saith he) environed about with a double Order of Columns (Philander 1549: 138).

Jones was ready to compare two types of circular temples. Referring to Vitruvius, Jones stated, "there were amongst others two Forms of round Temples, commonly in Use amongst them, the one called Monopteros (Fig. 60.2), the other Peripteros" (Fig. 60.3) (Vitruvius 1567: IV, 7; Jones 1971: 51).

Earlier Jones had mentioned that *peripteros* "has the Cell enclosed about with a continued Wall, and at a proportionate Distance from it, the Columns place which made a Portico round about it, clean different from Stone-Heng", while monopteros was "made open, and instead of a Wall encompassed with a Row of Pillars only, having no enclosed Cell within it at all, as much conducing to our Purpose in Hand". Jones had noted on the illustration of a *monopteros* in his copy of Vitruvius,

[6] According to John Newman (1992: 33), the spelling mistake, came from the fact Jones did not know the Greek language. The annotation is in Vitruvius (1511: 121).

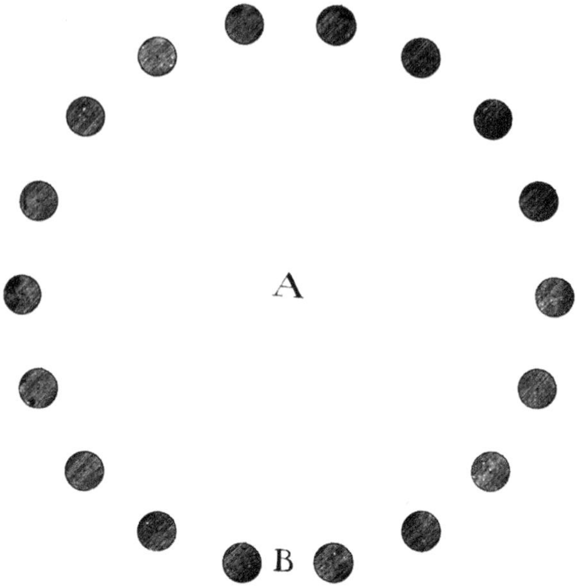

Fig. 60.2 The monopteros. Image: Jones (1655: 55)

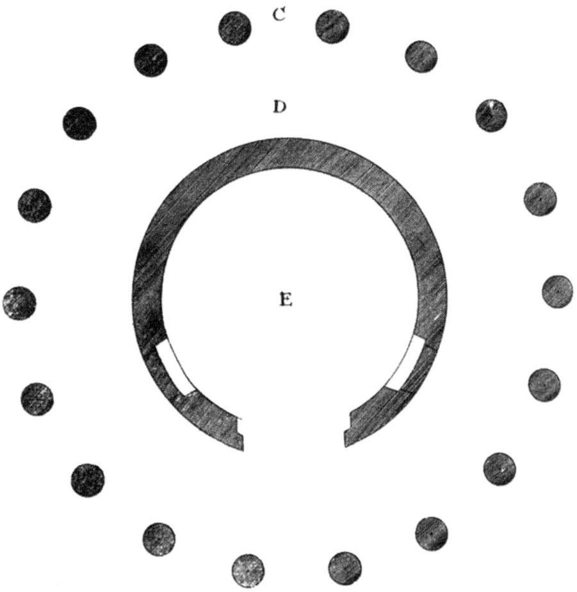

Fig. 60.3 The peripteros. Image: Jones (1655: 56)

"the one without sell and only with Colloms/the other winged about" (Newman 1992: 48). Jones stressed the relevance of *monopteros* to Stonehenge, quoting Barbaro in Italian: "I believe that Temple without Walls (speaking of the Monopteros aforesaid) had a Relation to Coelum (Heaven) because the Effects thereof are openly displayed to the full View of all Men" (Jones 1971: 71). A sort of evolution can therefore be observed about the circular temple, from monopteros, peripteros, and to Stonehenge. The changes had taken place, for

> Architect disdaining usual and common Forms, of both the aforesaid Forms [monopteros and peripteros] composed one. For, taking the outward Circle from the Monopteros, he made it open also as in that, but instead of the continued Wall circularly enclosing the Cell of the Peripteros, at Stone-Heng he made only an Hexagon about the Cell, leaving the same open in like Manner (Jones 1971: 51).

Architectural order of Stonehenge was the next topic. The megalith must have a specific order, for "it was the Custom of the Ancients (as in Part I remembered before) to appropriate the several Orders of Architecture, according to the particular Qualifications of those they deified" (Jones 1971: 67–68). It must have been the severest and most grave order: "[I]t is in mine Opinion," Jones stated, "Respecting therefore this Decorum used by the Ancients in building their Temples, and that this Work Stone-Heng is principally composed of a most grave Tuscan Manner, by just Proportions of an agreeable Form." Jones might also have read Shute's account:

> Then the Tuscanes, beginning to builde, having knowlaige of the pillor, whiche was firste invented by the Ionians, upon the Symetrie, of a strong manne invented to buylde stronglye after the maner aforsayde, yea, and to garnishe also theyr cyties and townes beautifullye with a pillour of their owne devise whyche yet at this present time, remayneth wholle in the citie of Forence and in the countreis there about they fourmed and fashioned that pillor, whyche to thys daye is named after the sayde countrey Tuscana.... This pillor is the strongest and most able to beare the greatest of burteofal the others. And that same his stregthe cometh by his shortenes, ... (Shute 1563).

This severest and the most grave order was appropriate for Coelum, the "ancientist" and "father of Saturn". His understanding of Coelum came from both classical and contemporary authors, Apollodorus, Boccaccio, Diodorus Siculus, Plutarch, Johannes Rosinus,[7] Thomas Godwin[8] and Valeriano. Book 1 of *Bibliotheca* of Apollodorus (1997), a grammarian of Athens of the second century B.C., was a common guide to Greek mythology, which drew from older sources like Hesiod's *Theogony*, of the eighth century B.C., in which Ouranos, Greek equivalent to Coelum, was described as the first deity:

> Ouranos was the first ruler of the universe. He married Ge, and fathered as his first children the beings known as the Hunred-Handers, Briareus, Cottos, and Gyes, who were unsurpassable in size and strength, for each had a 100 hands and 50 heads. After these, Ge bore him the Cyclopes, namely, Arges, Steropes, and Brontes, each of whom had a single eye on his forehead (Hesiod 1968).

---

[7] Johannes Rosinus (1551–1626) was Jones's contemporary. Jones cited Book 2, Chap. 5 of Rosinus's *Antiquitatum Romanarum* (1645).

[8] Thomas Godwin (1587–1642) was another contemporary of Jones. His *Romanae Historiae Anthologia. An English Exposition of the Roman Antiquities, wherein many Roman and English Offices are parallelled, and diverse obscure Phrases explained* of 1614 was intended for the use in Abingdon school in Berkshire, where he was the schoolmaster, and was revised and reprinted a number of times, till the sixteenth edition in 1696. Jones cited its Book 1, Chap. 20.

Diodorus of Sicily, who in turn drew from Apollodorus, Greek historian of the first century B.C., stated in his Bibliotheca historica:

> As for the Muses, since we have referred to them in connection with the deeds of Dionysus, it may be appropriate to give the facts about them in summary. For the majority of the writers of myths and those who enjoy the greatest reputation say that they were daughters of Zeus and Mnemosyne; but a few poets, among whose number is Alcman, state that they were daughters of Uranus and Ge... (Diodorus 1935: IV, Chap. 7).

Diodorus, earlier, told the stories of Uranus, their first king, who improved his subjects' ways of living, and introduced the year, months and seasons based on the observations of the stars (Diodorus 1935: III, Chap. 56).

In time, according to Diodorus, the people accorded Uranus with immortal honours and made him the king of the universe.

In his copy of Vitruvius, Jones had made this note: "in musicke the(re) must be a proportionatt distance between the low and heaygh/the same symphathy is in the stares/the ruels of arethematicke that unite musick wth astrologiy" (Newman 1992: 27; the annotation is on Vitruvius 1567: 24). Near the end of the *Stone-Heng* book, Jones discussed the correlation of architecture, astrology and music, made possible through mathematics:

> Lastly, that Stone-Heng was anciently dedicated to Coelus I collect from the Conformation of the Work. For the Conformation of the Cell and Porticus in the Plan, was designed with four equilateral Triangles, inscribed in a Circle, such as the Astrologers use in describing the twelve celestial Signs in musical Proportions (Fig. 60.4).

He quoted Vitruvius:

> In the Conformation thereof, let four Triangles be inscribed of equal Sides and Intervals, which may touch the extreme Part of the Circumference: ... by which Figures also, Astrologers from the musical Harmony of the Stars ground their Reasonings, as concerning the Description of the twelve celestial Signs (Jones 1971: 70; Vitruvius 1567: V).

Jones added that the hexagon, which made Stonehenge's inner cell, was also a tool of astrologers. He quoted Philander: "The Astrologers make use of three Sorts of Figures, the Triangle, Tetragon, and Hexagon" (Jones 1971: 70). The four equilateral triangles determined not only the hexagon, but also the openings, or "comparting", of the outer columns. According to Jones, "the three Entrances leading into the Temple from the Plain, were comparted by an equilateral Triangle; which was the Figure whereby the Ancients expressed what appertained to Heaven, and Divine Mysteris also." This must have stemmed from Jones's careful observation of an illustration of a theater in Barbaro's translation and commentary of Vitruvius (Fig. 60.5).

Jones reinforced the symbolism of the equilateral triangle, referring to Valeriano: "The Magi add, that a Triangle of equal Sides is a Symbol of Divinity, or Sign of celestial Matters" (Valeriano 1556: Bk. 39; Jones 1971: 70). Finally, Jones related the stars' circular movements in the heavens to the plan of Stonehenge: "those several Stars which appearing to us in the Heavens in Form of a Circle," or "the celestial Crown", was not "improbable" for the Stonehenge

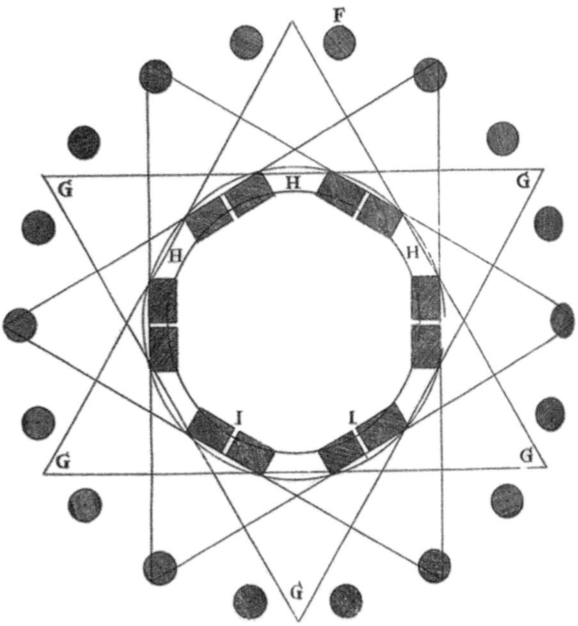

Fig. 60.4 The four equilateral triangles that govern Stonehenge. Image: Jones (1655: 58)

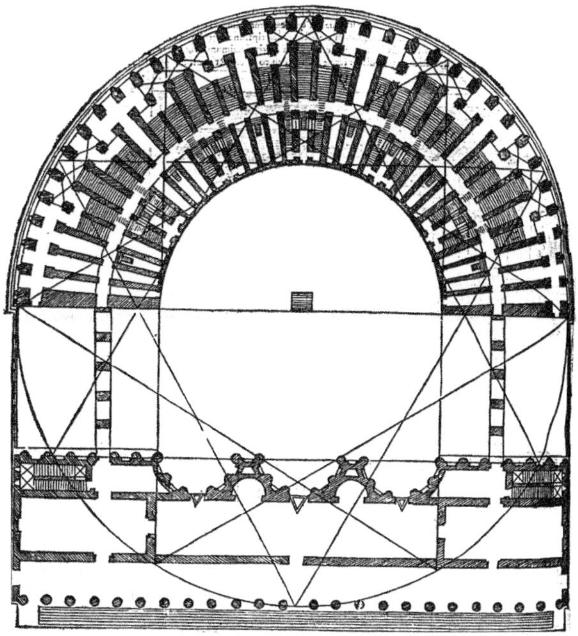

Fig. 60.5 The illustration of a theater from Daniele Barbaro's Latin edition of Vitruvius. Image: Vitruvius (1567: 188)

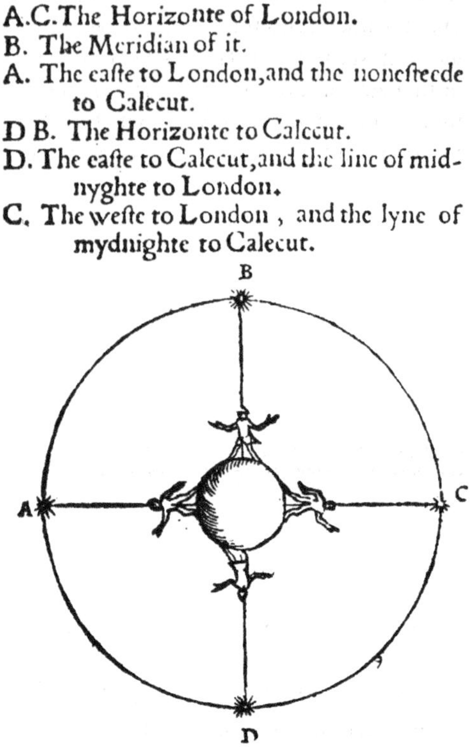

Fig. 60.6 Circular movements of celestial bodies. Image: Recorde (1556)

composition, for "after Ages might apprehend, it was anciently consecrated to Coelus or Coelum Heaven" (Jones 1971: 70).

Although Jones did not identify the "Astrologers", Recorde and Dee, and also Robert Fludd, are possibilities. Although Fludd's books were in Latin, Jones's personal acquaintance on medical matters has been found by Joseph Rykwert. With Recorde's Castle alone Jones would have known circular movements of celestial bodies (Fig. 60.6).

Jones's Ideal Vision of Britain and the British Monarch

Jones's interpretation of Stonehenge as a Roman monument was "profoundly informed by Jones's vision of Britain as the true heir of Roman culture" (Strong 1973: 82). Justifying the present by the virtues of the past had been in practice before Jones. Brutus the Trojan and King Arthur represented English chivalry. The poem Faerie Queen by Edmund Spenser (d. 1599), for example, had deliberately linked Queen Elizabeth to Prince Arthur, and to Brutus, in order "to fashion a gentleman or noble person . . . to be of good birth and to be aware of your past and of the obligations imposed upon you by your past was an urgent first rule" (Kendrick 1950: 130).

Early Stuarts also likened themselves to historical figures, including James I who styled himself as King Solomon for uniting Scotland and England. In reality he was never Elizabeth's match officially or personally. The schisms between the monarch and his subjects would continue with Charles I, eventually culminating, of course, in beheading the monarch in the Civil Wars. And yet the sovereigns had an extremely high vision, as James wrote in Basilikon Doron: "A King is as one set on a stage, whose smallest actions and gestures, all the people gazingly doe behold." The monarch must therefore exemplify good laws

> with his vertuous life in his owne person, an the person of his court and company; by good example alluring his subjects to the love of virtue, and hatred of vice . . . Let your owne life be a law-booke and a mirrour to your people, that therein they may read the practise of their owne Lawes; and therein they may see, by your image, what life they should leade (Orgel 1975: 42–43; see also McIlwain 1918).

Such symbolism extended even to equate the king to the sun and to the god.

Why was the Roman origin advantageous? Other possibilities included, as John Speed listed, Britons, Saxons and Danes. In emphasising the Roman past Jones was not alone, however. Emerging historiography tended to discredit old chroniclers like Geoffrey and instead to rely on artefacts and vocabularies found at the site. According to William Camden, the word Britannia had nothing to do with Brutus, but was a Celtic and Greek compound meaning "land of the painted people" (Woolf 1990). Others who rejected Brutus included John Clapham (1602, 1606), John Selden and Richard Rowland (1605), and eventually Oxford University Almanac in 1675 (Levine 1987; Smuts 1987).

Coelum was the oldest in Roman theogony, and yet it was not necessarily a perfect representation, for Coelum was an archaic, and therefore less popular deity, and even in the Roman pantheon

> had a rather shadowy existence . . ., for he was more a personification of the heavens than a god who was worshipped in the ancient world, and although he would have been credible as a figure in a Renaissance masque, he was less so as the centre of a Roman cult (Parry 1981: 157).

The choice of Coelum must have been architectural: it could easily be tied to a specific geometry, thus making the architect the supplier of symbolism, providing him with an advantage over theologians or poests. Jones's famous quarrel with poet Ben Jonson, long-term collaborator of court masques since 1605, stemmed from the desire of each to be superior to the other (Gordon 1975). A symbolism that was geometrical and therefore architectural must have made the architect the idea generator, while the poet was only the executor.

Where else did the notion of Coelum Britannicum appear?

If Jones considered Coelum important, then it should be natural for the same symbolism to appear in other works among his opus. The first of such instances is the design for James's catafalque of 1625 (Fig. 60.7) (Peacock 1982: 1–5; Harris and Higgott 1989). Its design sources included Domenico Fontana's Catafalque for

Fig. 60.7 Jones's design for the catafalque of James I, 1625. The Provost and Fellow of Worcester College, Oxford

Pope Sixtus V (Fig. 60.8) and Bramante's Tempietto; however, differences between Jones's and Fontana's designs are important here. Fontana's looks Corinthian in order, while Jones's was likely Tuscan. While Fontana used six sets of double columns on a circumference with an arched opening in-between, Jones's sets of two columns appear to line up in the radius, with a complete opening below the entablature. Jones's design is therefore closer to Vitruvius's description of monopteros. Another difference is in the dome, Fontana's being pointed and Jones's semi-spherical. All these characteristics correspond to Jones's symbolism of the heavens.

Jones's masque designs included Coelum Britannicum written by the young poet Thomas Carew, performed on Shrove Tuesday in 1633/1634 at Banqueting House, Whitehall. The allegory originated in Giordano Bruno's 1584 *Spaccio de la bestia trionfante* (The Explusion of the Triumphant Beast (1964)). Here the central figure was Jove, ageing father of the gods, who was to bring forth much-needed reform both of microcosm and macrocosm, or society as well as man, both disturbed by religious, philosophical, and scientific crises (Giordano Bruno 1964: 27).

The Devonshire Collection includes a scenery design that generations of scholars left unidentified (Fig. 60.9). Knowing the 60-year-old Jones had much control over author, story line and allegory, one cannot help but notice a small yet distinct depiction of a ring of stones in the center of this drawing. Additionally, the opening scene matches the features of this drawing, making it highly probable the

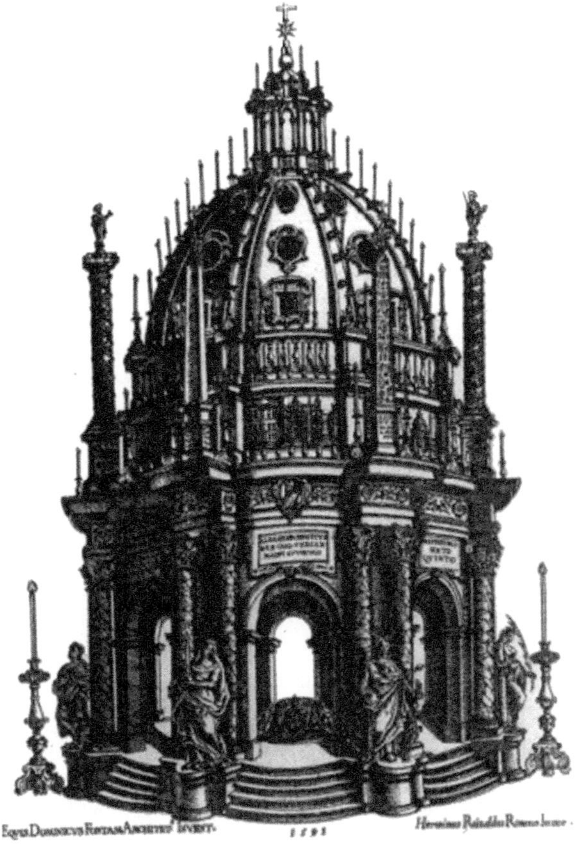

Fig. 60.8 Domenico Fontana's catafalque for Pope Sixtus V. Image: Catani (1591: Pl. 24)

Fig. 60.9 Sketch for a scenery design by Inigo Jones, with a small yet distinct ring of stones in the center. Image: Devonshire Collection, Chatsworth. Reproduction by permission of the Duke of Devonshire and the Chatsworth Settlement Trustees

Fig. 60.10 Depictions of Atlas holding the spherical cosmos on his shoulders: (*left*) from Valeriano's *Hieroglyphihca* (1602); by Inigo Jones. Image: Devonshire Collection, Chatsworth. Reproduction by permission of the Duke of Devonshire and the Chatsworth Settlement Trustees

drawing was for this masque: "the scene, representing old arches, old palaces, decayed walls, parts of temples, theatres, basilicas and Thermae, with confused heaps of broken columns, bases, cornices and statues, lying as underground, and altogether resembling the ruins of some great city of the ancient Romans or civilised Britons" (Orgel and Strong 1973: vol. 2, 571). John Peacock has traced many elements of this scenery to Willem van Nieulandt's (Peacock 1995: 315–320). What is important, however, is the ring of stones appears only in Jones's scenery.

Jones's costume design for Atlas, a character in this masque, holding the spherical cosmos on the shoulders, resembles an illustration from Valeriano's Hieroglyphica, one of Jones's sources for Coelum (Figs. 60.10 and 60.11).[9] Atlas's characteristics matched the Tuscan order:

> As namely Tuscana, is applied unto Atlas, the kynge of Mauritania . . . This [Tuscan] pillor is the strongest and most able to beare the greatest of burteofal the others. And that same his stregthe cometh by his shortenes, therefore he is linked unto Atlas, kynge of Maurytania, and the piller is named Tuscana (Shute 1563).

In the same year as this masque production, Peter Paul Rubens was working in Antwerp on what would become the ceiling paintings of Banqueting House (Strong 1980: 13). A panel depicted James I as King Solomon in a circular edifice of Tuscan order under a semi-spherical dome (Fig. 60.11). Provided that Jones supplied the

[9] Frances A. Yates (1969: 180) made a passing remark that Jones used the 1602 edition of Valeriano's *Hieroglyphica* in Italian.

Fig. 60.11 Rubens's depiction of James I as King Solomon (Crown copyright)

allegory, we see that Jones made an explicit association of the deceased British monarch as Coelum Britannicum. This then constitutes the third instance of the symbolism.

The fourth possible instance is Charles I's portrait by Anthony Van Dyck (1638), who had come to London 6 years earlier on a royal invitation. The monarch, clad in Roman armour, is passing through a triumphal arch of Tuscan order (Fig. 60.12). Equestrian positions induced chivalry, endowing the monarch with much needed powers and virtues. The second sitter, carrying Charles's helmet and wearing a medal, stands slightly ahead of the horse, and looks up and back at the monarch. According to Oliver Millar and recent findings at Royal Collection,[10] the figure is Antoine Bourdin, French equestrian teacher to Charles I. A teacher in an authoritarian portrait seems contradictory, however. Is it possible that the standing figure was our Jones himself? Enough resemblance points to Van Dyck's depiction of Jones (Fig. 60.13), including facial features, hair and scull cap, and plain but wide collars and shirt with many front buttons. While drawing the equestrian teacher officially, could Van Dyck have secretly depicted another individual? A concrete instance of such is *Emperor Theodosius Refused Entry into Milan Cathedral* by St

[10] Letter from the Royal Collection to the present author.

Fig. 60.12 Anthony van Dyke, Charles I with M. de St. Antoine. Image: The Royal Collection ©2002, Her Majesty Queen Elizabeth II

Ambrose, Archbishop of Milan, in which Van Dyck copied Rubens's painting,[11] but cast his contemporaries so that the allegory made sense (Gritsai 1996: 28). Just as Jones was the mastermind of court masques and of Rubens's court paintings, the architect could also have advised Van Dyck, a fairly new arrival in British court.

The painting then would reveal Jones's ideal image of the architect. To see it, we must go back to James I's coronation. A royal procession took place in London in March 1604, with a performance devised by poets Ben Jonson and Thomas Dekker and seven triumphal arches designed by Stephen Harrison.

Among them was Fenchurch arch (Fig. 60.14) with a London cityscape for the pediment, the British monarch immediately below, and a figure further below who

[11] Van Dyck's copy is in the National Gallery, London.

Fig. 60.13 Van Dyke's portrait of Inigo Jones. Image: Devonshire Collection, Chatsworth. Reproduction by permission of the Duke of Devonshire and the Chatsworth Settlement Trustees

looked up the rest. Graham Parry identified this figure as Theosophia, or divine wisdom. Jonson characterized her as, all in white, a blue mantle seeded with starres, a crowne of starres on her head . . . Shee was alwayes looking up; in her one hand shee sustayned a dove, in the other a serpent: . . . Intimating, how by her, all kings doe governe, and that she is the foundation and strength of kingdomes, to which end, shee was here placed, upon a cube, at the foot of the Monarchie, as her base and stay.

The inscription in the entablature, "Par Domus Haec Coelo Sed minor est domino," predicated the city the monarch resided as Coelo, the heaven (Hart 1994). Now looking back in Van Dyck's composition, we see the architect, the source of wisdom to Charles I, who shone under a triumphal arch. And back in the masque Coelum Britannicum, Jones might have portrayed himself as Atlas.

Fig. 60.14 Fenchurch arch by Stephen Harrison. Image: British Library shelfmark G10866. By permission of the British Library

Conclusion

Jones's theory of Stonehenge is not a singular instance of erroneous interpretation, but an important piece of the grand ideal vision. We might interpret it as a political maneuver, but that would describe nothing but our present conditions. Jones believed in architectural symbolism if not for present, then for future, and if not for future, then for utopia. Geometry collaborated in the construction of the ideal.

Acknowledgments The author gratefully acknowledges supports provided by the Graham Foundation for Advanced Studies in the Fine Arts, and the University of Nebraska-Lincoln Research Council. Research assistants have been made available through the University Undergraduate Creative Activities and Research program.

Biography Rumiko Handa has a Bachelor of Architecture from the University of Tokyo, and a Master of Architecture, a Master of Science in Architecture, and a Doctor of Philosophy degrees from the University of Pennsylvania. She is a licensed architect (Japan), and taught architecture at the University of Michigan, Ann Arbor, and at Texas Tech University. She is currently a Professor at the College of Architecture at the University of Nebraska-Lincoln. Her research has been supported by the Graham Foundation for Advanced Studies in the Fines Arts and the University of Nebraska Research Council. She has won numerous teaching awards. Her latest publication is *Conjuring the Real: The Role of Architecture in Eighteenth- and Nineteenth-Century Fiction* (co-edited with James Potter, University of Nebraska Press, 2011).

References

ANDERSON, C. J. 1993. *Inigo Jones's Library and the Language of Architectural Classicism in England*, 1580-1640. Cambridge: MIT Press.

APOLLODORUS. *The Library of Greek Mythology*. R. Hard, trans. 1997. Oxford: Oxford University Press.

AUBREY, J. 1862. *Wiltshire. The Topographical Collections of John Aubrey*. J. E. Jackson, ed. Devizes: Wiltshire Archaeological and Natural History Society.

BURKE, E. 1887. An Essay towards an Abridgment of the English History in Three Books (1757). Pp. 159-488 in *The Works of the Right Honourable Edmund Burke*, vol. VII. London: John C. Nimmo.

CATANI, B. 1591. *La pompa funerale fatta dall'Ill.mo & Rev.mo S.R. Cardinale Montalto nella trasposizione dell'ossa di papa Sisto il quinto scritta & dichiarata da Baldo Catani*. Rome: Stamperia Vaticana.

CLAPHAM, J. 1602. *The Historie of England*. London: Printed by Valentine Simmes, for Iohn Barnes, dwelling in Fleete-streete, at the signe of the Great Turke.

———. 1606. *The Historie of Great Britainnie*. London: Printed by Valentine Simmes.

DIODORUS OF SICILY. *Bibliotheca historica*. C. H. Oldfather, trans. 1935. Cambridge: Harvard University Press.

GIORDANO BRUNO. 1964. *The Expulsion of the Triumphant Beast*. A. D. Imerti, trans. and ed. Lincoln and London: University of Nebraska.

GORDON, D. J. 1975. *The Renaissance Imagination*. Berkeley: University of California Press.

GRITSAI, N. 1996. *Van Dyck*. Kunsthistorisches Museum, Vienna.

HANDA, R. 2006. Authorship of The Most Notable Antiquity (1655): Inigo Jones and Early Printed Books. *The Papers of the Bibliographical Society of America* **100**, 3: 357-378.

HARRIS, J. and HIGGOTT, G. 1989. *Inigo Jones: Complete Architectural Drawings*. New York: Drawing Center.

HART, V. 1994. Art And Magic in the Court of the Stuarts. London and New York: Routledge.

HESIOD. The Works and Days, Theogony, the Shield of Heracles. R. Lattimore, trans. 1968. Ann Arbor: The University of Michigan Press.

HUNTER, M. 1975. John Aubrey and the Realm of Learning. London: Duckworth.

JOHNSON, A. W. 1997. Three Volumes Annotated by Inigo Jones: Vasari's Lives (1568), Plutarch's Moralia (1614), Plato's Republic (1554). Abo: Abo Akademi University Press.

JOHNSON, F. R. and LARKEY, S. V. 1935. Robert Recorde's Mathematical Teaching and the Anti-Aristotelian Movement. Huntington Library Bulletin, 7: pp. 59-85. Cambridge: Harvard University Press.

JONES, I. 1655. *The most notable Antiquity of Great Britain, vulgarly called Stone-Heng, on Salisbury Plain*. London:.

———. 1972. *The most notable Antiquity of Great Britain, vulgarly called Stone-Heng, on Salisbury Plain* (1655). Restored ed. London: The Scholar Press.

———. 1971. *The most notable Antiquity of Great Britain, vulgarly called Stone-Heng, on Salisbury Plain* (1725). Restored ed. London: Gregg International Publishers Limited.

KENDRICK, T. D. 1950. British Antiquity. London: Methuen.

LEVINE, J. M. 1987. Humanism and History: Origins of Modern English Historiography. Ithaca and London: Cornell University Press.

MCILWAIN, C. H., ed. 1918. Political Works of James I. Cambridge: Harvard University Press.

NEWMAN, J. 1992. Inigo Jones's Architectural Education before 1614. *Architectural History* **35**: 18-50.
ORGEL, S. 1975. The Illusion of Power: Political Theater in the English Renaissance. Berkeley: University of California Press.
ORGEL, S. and STRONG, R. 1973. Inigo Jones: The Theatre of the Stuart Court. Berkeley and Los Angeles: University of California Press.
PARRY, G. 1981. The Golden Age Restor'd. New York: St. Martin's Press.
PEACOCK, J. 1982. Inigo Jones's catafalque for James I. Architectural History, 25: pp. 1-5. London : The Society of Architectural Historians of Great Britain.
———. 1995. The Stage Designs of Inigo Jones The European Context. Cambridge: Cambridge University Press.
PHILANDER, G. 1549. Gulielmi Philandri castilionii galli civis ro. in decem libros M. Vitruvij Pollionis de Architectura Annotationes, ad Franciscum Valesium Regum Christianisimum, cum Indicibus Graeco & Latino locupletissimis. Parisiis: Ex officiano Michaelis Fezandat, in domo Albretica, e regione divi Hilarij.
RECORDE, R. 1556. *Castle of Knowledge*. London.
ROSINUS, J. 1645. Antiquitatum Romanarum. Köln: Kalcovius.
ROWLAND, R. 1605. Restitution of Decayed Intelligence in Antiquities concerning the most Noble and Renowned English Nation. Anversa: Bruney.
RYKWERT, J. 1980. *The First Moderns*. Cambridge: MIT Press.
SAYCE, R. A. 1970. Preface. In: I. Jones, *Inigo Jones on Palladio being the notes by Inigo Jones in the copy of I Quattro Libri Dell'Architettura Di Andrea Palladio*. London: Oriel Press.
SHUTE, J. 1563. The First and Chief Groundes of Architecture. London: Country Life Limited.
SMUTS, R. M. 1987. Court Culture and the Origins of a Royalist Tradition in Early Stuart England. Philadelphia: University of Pennsylvania Press.
STRONG, R. 1973. Jones and Stonehenge. In: J. Harris, S. Orgel and Roy Strong, *The King's Arcadia: Inigo Jones and the Stuart Court*. London: Arts Council of Great Britain.
———. 1980. Britannia Triumphant: Inigo Jones Rubens and Whitehall Palace. London: Thames and Hudson.
SUMMERSON, J. 2000. Inigo Jones. New Haven and London: Yale University Press.
TAYLOR, E. G. R. 1954. Mathematical Practitioners of Tudor and Stuart England. Cambridge: Cambridge University Press.
VALERIANO, G. P. 1556. Hieroglyphica, sive de sacris Aegyptiorum aliarumque gentium litteris commentariorum libri LVIII. Basel.
VITRUVIUS. 1511. *M. Vitruvius per iocundum solito castigatior cactus, cum figuri et tabula, ut iam legi et intelligi possit*. Venezia: Joannes Iucundus da Verona [Giovanni Taccuino].
———. 1567. *M. Vitruvii de architectura libri decem....* Daniele Barbaro, trans. Venice: Rev. ed. Venezia: Francesco de' Franceschi.
WILKINSON, C. H. 1926. Worcester College Library. Transactions of the Oxford Bibliographical Society, I, iv: pp. 261-320. Oxford: Oxford University Press.
WOOLF, D. R. 1990. The Idea of History in Early Stuart England. Toronto: University of Toronto Press.
YATES, F. A. 1969. Theatre of the World. Chicago: University of Chicago Press.

Chapter 61
The Science Behind Francesco Borromini's Divine Geometry

John G. Hatch

Introduction

The popular notion of religion and science being at opposite poles within the intellectual currents of the seventeenth century is challenged by the designs and architectural iconography underlying the churches of Francesco Borromini (1599–1667). Described as something of a licentious eccentric by Gian Lorenzo Bernini and his contemporaries (Blunt 1979: 212–213; Steinberg 1977: 15–16), Borromini nonetheless relied upon a complex geometric system in his architectural designs, which ruled both the layout and elevation of his buildings. In turn, this use of geometry also seems to have had an important theological justification, namely that of stressing the underlying divine order of the universe whose existence or revelation can only be perceived by the faithful. In essence, this is simply a restatement of the Medieval idea of "God as Divine Geometer" except that, for Borromini, God is no longer depicted in the garb of scholastic rationalism, but rather, as I will show, of seventeenth-century scientific rationalism: a rationalism that embraces the notion of divine revelation. Although Borromini's interest in ancient mathematics and geometry, as well as his interest in the work of Galileo, are commonly known, one potential source for the architect's interest in divine geometry and cosmology has yet to be acknowledged fully. It is specifically in the writings of Johannes Kepler that one finds the most consistent explanation for Borromini's use of geometry in architecture, as well as a source for the unusual

First published as: John G. Hatch, "The Science Behind Francesco Borromini's Divine Geometry". Pp. 127–139 in *Nexus IV: Architecture and Mathematics*, Kim Williams and Jose Francisco Rodrigues, eds. Fucecchio (Florence): Kim Williams Books, 2002.

J.G. Hatch (✉)
Department of Visual Arts, The University of Western Ontario, London, ON, Canada N6A 5B7
e-mail: jhatch@uwo.ca

K. Williams and M.J. Ostwald (eds.), *Architecture and Mathematics from Antiquity to the Future*, DOI 10.1007/978-3-319-00143-2_14,

cosmological Trinitarian references found at the churches of S. Carlo alle Quattro Fontane (Rome: 1638–1641) and S. Ivo della Sapienza (Rome: begun 1642).[1]

S. Carlo alle Quattro Fontane

Upon entering the church of S. Carlo, it would appear preposterous even to suggest the existence of a geometrical scheme (Figs. 61.1, 61.2). Our first impression of the interior is of a flowing, almost dizzying, sense of movement. The walls are composed of shallow and deep curved bays all linked by straight horizontal elements. The whole interior surface is articulated by columns set into walls, while the surfaces themselves are pierced by a series of niches of varying sizes, adding yet another rhythmic dimension. All these elements combine to create a sense of spatial plasticity, serving to dematerialize any notion of architectural solidity. This interior has been described by various architectural historians as "willfully complex", producing a "rolling, rocking effect", "a delightful confusion", etc., all epitomizing Heinrich Wölfflin's characterization of the Baroque's "wild desire for movement" (Steinberg 1977: 15–16, 169; Wölfflin 1984: 59). In fact, this interior is so complex that countless interpretations, many contradictory, have been set forth to explain the design logic of the interior, in spite of the existence of Borromini's drawings (Steinberg 1977: 18–41).

The dizzying character of the interior and, potentially, the diversity of interpretations offered by scholars, simply reflects the intentions of the architect. The main body of the interior is meant to confuse and destabilize one's sense of spatial orientation. There is no ideal viewpoint, but rather a multiplicity of viewpoints which, as a totality, are not meant to exhibit a coherent whole. However, despite this apparent ambiguity and "willful complexity," the ground plan is nonetheless based on a series of rather simple geometric manipulations. Borromini's preliminary drawing 173 in the Albertina Collection, Vienna, reveals that the church interior is structured upon two equilateral triangles sharing a common side, with two circles inscribed within them. The two circles are combined to form an oval, describing the area of the dome. The longitudinal chapels are defined by the end points of the two triangles, while the lateral chapels are marked out by the shared corners of the triangles. Finally, the four piers of the church are defined by crossing diagonals originating from the shared corners of the two triangles, cutting the centre of the circles. In this manner, the layout of the building is based on two triangles and two circles circumscribed by an oval.

What then are we to make of this paradoxical relationship between the actual appearance of the interior and its underlying geometrical skeleton? The answer to

[1] Blunt (1979: 47). Leo Steinberg does hint at a possible connection between Kepler and Borromini, but more in terms of a shared *zeitgeist* than any direct links; see Steinberg (1977: 240–241).

Fig. 61.1 Altar wall, S. Carlo alle Quattro Fontane, Rome, 1638–1641. Photo: Kim Williams

Fig. 61.2 Dome, S. Carlo alle Quattro Fontane, Rome. Photo: author

this query is partially revealed in the upper half of the church and, specifically, the oval dome (Fig. 61.3). The shape of the dome represents the essential underlying scheme of the lower part of the church, complemented by the decorative motifs found in the coffering; in essence, the upper half of the church represents a synopsis of the lower half. At this level of the church, one reaches the point of divine revelation, where the complexity of the forms below are clarified through a process of religious enlightenment. The dome space is clearly meant to be understood as a metaphorical representation of the heavenly realm, which is

Fig. 61.3 Dome, S. Carlo alle Quattro Fontane, Rome. Photo: Kim Williams

explicitly shown in the lantern where we find the symbol of the Holy Spirit and behind it a series of rays depicting the spiritual light of revelation.[2] This light is complemented by the natural light that emanates from the windows at the base of the dome, making this area the most brightly lit space in the whole church and, consequently, reinforcing the notion of a progression from temporal reality towards the perfection of the heavenly sphere, through the process of spiritual revelation.

A similar reading emerges with regards to the church of S. Ivo della Sapienza, which Rudolf Wittkower describes as symbolizing a "movement downward from the chastity of forms in the heavenly zone to the increasing complexity of the earthly zone" (Wittkower 1973: 138). Thus at S. Ivo we should find a similar symbolic program as we saw at S. Carlo, but with certain modifications more ideally suited to the church of the future University of Rome.

[2] This idea of the dome or vault of heaven may have been inspired by Byzantine architecture since we known that Borromini was collecting information on the Hagia Sophia and San Vitale in the early 1640s; see Connors (1996: 50–51).

Fig. 61.4 Interior, S. Ivo della Sapienza, Rome. Photo: Kim Williams

S. Ivo della Sapienza

The interior of S. Ivo is as confused as that at S. Carlo (Fig. 61.4). Borromini presents another complex fusion of architectural forms serving to create a sort of amorphic structure. Though the central space is essentially circular, it is composed of six bays, three of which are semi-circular, and three others of an irregular shape. The wall surfaces are articulated by a series of niches and a string course serving to divide the wall into two sections. Instead of columns, we find at S. Ivo the use of pilasters, distributed in a complex rhythm, which are in turn combined with a series of broken pilasters. On the ground level then, this interior serves to create a similar sense of spatial disorientation as that found at S. Carlo. And like S. Carlo, we also find at S. Ivo a controlling, geometrical skeleton underlying this plan (Fig. 61.5).

As was the case with S. Carlo, the key to understanding the geometry of design at S. Ivo lies in the dome area (Fig. 61.6). Standing in the centre of S. Ivo, our eye is gradually led upwards towards the dome.

Again we find expressed a process of design clarification where, at one level, the entablature partially reveals the geometric structure, which in turn is further clarified as our eye moves along the vertical ribbing of the dome towards its apex. The whole underlying order is beautifully unfolded in a very gradual process, finally culminating in the figure of the Holy Spirit bathed in a symbolic representation of divine light. As is the case with S. Carlo, we are in the presence of the vault of heaven at S. Ivo, punctuated by the stars decorating the dome and the

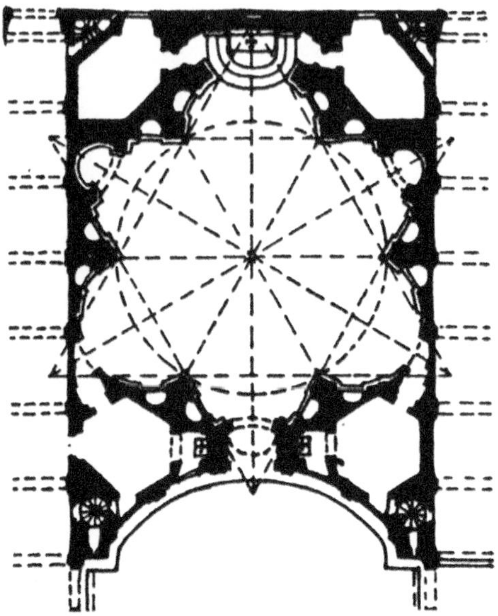

Fig. 61.5 Plan, S. Ivo della Sapienza, Rome. Drawing: Author

Fig. 61.6 Interior of the Dome, S. Ivo della Sapienza, Rome. Photo: Kim Williams

two superimposed circles formed by the Cherubim and Seraphim (Portoghesi 1968: 149–158).

Borromini's use of geometry has always been puzzling. Comparing his architecture with that of Palladio, we observe that where Palladio's use of geometry is explicitly demonstrated in the final structure, Borromini's still remains implicit to a large extent. Even in the buildings of such contemporaries as Bernini, the geometry of design is still evident in controlling the unfolding of his structures. In Borromini's case, geometry is not necessarily meant to only generate form, but also remains an implicit justification of the building as a whole. It becomes a sort of hidden dimension, disguised by a profusion of structural and decorative manipulations.[3]

Borromini and Kepler's Geometric Universe

To my mind, the parallels are difficult to avoid between Borromini's use of geometry and the role geometry plays in the work of Johannes Kepler (1571–1630). Kepler's primary goal was to discover the invisible skeleton of the universe; a skeleton that Kepler believed to be geometric (Fig. 61.7).

In fact, for Kepler, the eternal and ultimate truths of the universe are based on a "divine geometry". More importantly, geometry becomes a principle link between the human and the divine. According to Kepler, God endowed "man" with an understanding of geometry, providing him with the tools to deduce a priori the whole blueprint of the universe and, through this process, come in contact with the mind of God (Koestler 1979: 263–264; Martens 2000: 32–38, 48–51). In Borromini's case a similar interpretation emerges. The churches of S. Carlo and S. Ivo represent, in a sense, a microcosm of the universe; the lower storeys reflecting nature in its apparent haphazard and accidental form, the upper storeys representing the divine in its simplicity and perfection. But the lower storeys of S. Ivo, and particularly S. Carlo, as accidental and haphazard as they may appear, are, as pointed out above, nonetheless conceived and ruled by a strictly organized geometric system, echoing the inherent skeleton of the universe as a whole. In this manner, Borromini recreates nature in its complexity, while in his designs he supplies by analogy, obviously, its underlying order: a divine order ruling an apparent chaos, as Kepler expressed it, and as Galileo spoke of when he wrote that, "...the great book of nature ... is written in the language of mathematics, and its characters are triangles, circles and other geometrical figures without which it is humanly impossible to understand a single word of it" (Blunt 1979: 46–47). This idea of a hidden order may explain why Borromini guarded his plans so adamantly (Blunt 1978: 48; Connors 1995: 590). It is most likely that Borromini did not want to reveal the geometric underpinnings of his buildings largely because he expected the viewer to undertake, unassisted, the process of decoding the substructure of his buildings in a manner analogous to that outlined by Kepler; to become, as Kepler puts it, "the priests of God, called to interpret the Book of Nature" (Koestler 1979: 264; Kozhamthadam 1994: 41–42).

The contrast of geometric complexity dictating the lower zones of Borromini's churches and the geometric simplicity of the upper zones is meant to reflect the apparent imperfections of the terrestrial versus the perfection of the heavenly realm. Such a contrast does not exist in the churches of Borromini's contemporaries, where

[3] Wittkower argues that Borromini's use of geometry represents a move away from "the classical principle of planning in terms of modules" (Wittkower 1973: 132).

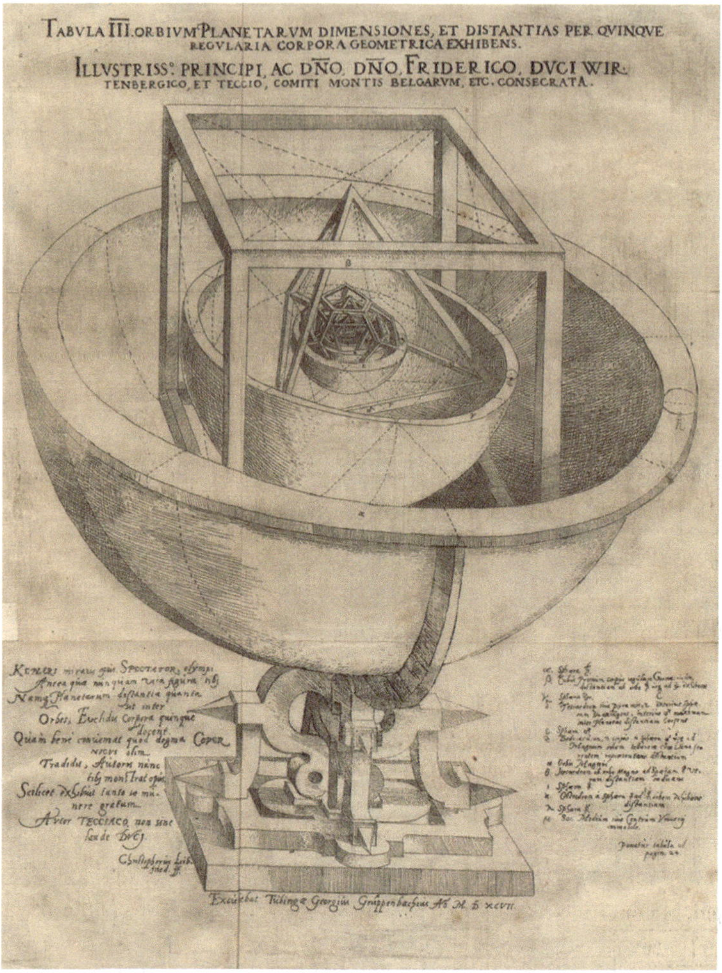

Fig. 61.7 Johannes Kepler's nested sphere model of the solar system. Image: Kepler (1596: Chapter III, tab. II [inserted between pp. 24 and 25)]

geometric clarification rules throughout despite the proliferation of decoration. In Borromini's case, we are witnessing a strong debt to Neoplatonic philosophy; a philosophy which serves as a foundation for modern science, and which influenced extensively the work of Michelangelo whom Borromini admired profoundly. It is Neoplatonism which rules the scientific inquiries of, for example, Nicholas Cusa, Giordano Bruno and Kepler. In fact, the reading of the dialogue between the earthly and the heavenly Borromini establishes at S. Carlo and S. Ivo, is succinctly echoed in Marcellus Palingenius's statement in the *Zodiacus vitae* (1534) where he posits, "the opposition between the terrestrial and the celestial regions, where the former is imperfect and nothing more than a shadowy reflection of the perfection of the latter" (Koyré 1957: 25). In sum, Borromini's debt to Neoplatonism plays, in my opinion, a substantial part in his vision of architecture as a microcosmic version (or vision) of the universe, taking on the form of an architectural hieroglyphics dictated by the creator whose interpretation must be pried by the power of reason and observation in order to discover and understand its order, a process whose interpretation can only occur through the contemplation of the divine. It is this same Neoplatonism which might have made Borromini receptive to Kepler's ideas to begin with.

Borromini's Cosmology

With Borromini's religious architecture acting as a microcosm of the universe, it is interesting to note that where the universe of Copernicus (and subsequently Galileo and Kepler) removed the earth from its centre, Borromini shifts away substantially from the classical conception of anthropomorphic architecture. Bernini himself observed that architecture, in its design, depended on the proportions of the human body; what distinguishes Borromini from his contemporaries was the break with this tradition (Wittkower 1973: 130). Thus as science dislocated man from the centre of the universe, Borromini abandoned the use of the human body as a model for architecture.[4] But the parallel does not end there. The sun quickly assumed the heralded position at the centre of the universe in science, and with it occurred a theological revision. Where God had previously been situated at the outer realm of the universe, the Baroque saw him introduced, with some well-documented difficulties, into the centre with the sun as his attribute. This change is reflected in the words of Monsignor Giovanni Battista Agucchi who wrote

> God himself may reasonably be designated and recognized as the middle because the created things are outside Him yet always return to Him as do the rivers to the sea (Panofsky 1954: 39).

Kepler echoes these words in the *Mysterium Cosmographicum* (1596) where, in attempting to justify the placement of the sun at the centre of the universe, he writes,

> that the sun must be in the centre of the world because he is the symbol of God the father, the source of light and heat, the generator of the force which drives the planets in their orbits... (Koestler 1979: 263).

The sun is significantly the central symbol of the churches of S. Carlo and S. Ivo. Both are crowned by lanterns through which sunlight enters and illuminates the interiors. In the case of S. Carlo light acts as a symbol of divine revelation, where

[4] It should be noted that Borromini does not sever completely his ties with the anthropomorphic in architecture. One does still find occasional references to parallels between the human body and architecture in Borromini's notes and letters. In turn, when Borromini does reference the human body, he adopts the more organic relationship between the body and architecture defined by Michelangelo; see Ackerman (1966).

the process of design clarification between the lower and upper storeys seems rather sudden, while at S. Ivo light represents divine wisdom, where the movement from the lower to upper story is more gradual.

But both S. Carlo and S. Ivo appear to carry this religious cosmology further. For Kepler, the visible universe as a whole is a symbol of the Trinity. Not only is God represented by and as the sun, but the sphere of the fixed stars represents Christ and the invisible forces emanating from God represent the Holy Spirit. As Kepler himself put it in the *Mysterium Cosmographicum*:

> the Sun in the center, which was the image of the Father, the Sphere of the Fixed Stars, or the Mosaic waters, at the circumference, which was the image of the Son, and the heavenly air which fills all parts, or the space and firmament, which was the image of the Spirit. (Martens 2000: 40; Kozhamthadam 1994: 29–34.)

Significantly, S. Carlo is co-dedicated to the Trinity, while S. Ivo contains numerous symbolic references to the Trinity, and both contain symbols of the Holy Spirit at the top of their respective domes. S. Ivo also has stars decorating its dome, while at S. Carlo we find the presence of an oval dome. The latter is a particularly curious feature since Kepler himself discovered in the *Astronomia Nova* (1609) that the planets revolve around the sun elliptically rather than in a circular manner.[5] All of these factors seem to further reveal that Borromini was aware of Kepler's cosmology and applied some of its basic ideas to his architecture.

As suggestive as these parallels between Kepler's scientific speculations and Borromini's architectural iconography are, how Borromini became familiar with Kepler's work is difficult to establish in any concrete way. Unfortunately, we know little of Borromini's formal education or what books were collected in his extensive library, nor do we possess any first-hand documents explicitly outlining Borromini's position *vis-a-vis* science. But we do know this, that Kepler's popularity at the time had definitely reached Italy. He was protected often by the Jesuits (despite the fact that Kepler was Lutheran), was offered the Chair of Mathematics at Bologna, the highest position for an astronomer in Italy, and was nominated to the *Academia dei Linceie* (Koestler 1979: 282, 353–354, 387.) We also find instances recorded of Borromini's contacts with Galileo's students, who could easily have made the architect aware of Kepler's work (Connors 1996: 50–52) In turn, one cannot ignore the possibility that the patrons of S. Carlo, the Discalced Trinitarians, an order with which Borromini formed a lifelong relationship, may have been attracted to Kepler's rather unique Trinitarian explanation of the cosmos. By whatever means Borromini came to know of Kepler's work, once exposed to it, he would certainly have been attracted to

[5] It will, obviously, be pointed out that the dome at S. Carlo is oval rather than elliptical. I suspect that Borromini may have decided that designing an elliptical dome was far too difficult; an oval is much simpler to handle geometrically and does have architectural precedents. It should be noted, however, that Kepler himself had toyed with the idea of planetary motion describing an oval around the sun, but the observational data Kepler inherited from Tycho Brahe simply did not support this; see Martens (2000: 87–90).

Kepler's unusual geometric and aesthetic approach to astronomy, which was rather unique (Martens 2000: 12–13).

Conclusion

There can be little doubt that both S. Carlo and S. Ivo are microcosmic models of the universe, devised along general Neoplatonic lines, and use geometry to represent the underlying order of that universe. This is further supported by Borromini's belief that most ancient architecture was concerned largely with two things, astronomy and mathematics (Steinberg 1977: 355). The specific connection to Kepler is, admittedly, more difficult to establish with any certainty. However, in closing, I would like to propose two more links that seem to point to such an intellectual interchange. Kepler believed that the sun was the prime mover of the planets. He was uncertain as to how this actually occurred but speculated that it may be related to magnetism (Martens 2000: 83–84). Although he was uncertain about the actual material forces at play, in terms of his Trinitarian model, the sun/God acts through the Holy Spirit in moving the planets. Significantly, at S. Carlo, one finds inscribed above the entrance that "the august Trinity is revealed as ruler of the world's circuit, the *orbis* terrarum", an inscription that is underscored by the fact that the Holy Spirit crowns the dome before a symbolic representation of the sun. At S. Ivo, the lantern at the top of the dome is similarly crowned and the church itself is dedicated to the *Divina Sapienze*—"the planet which leads men aright along every path" (Steinberg 1977: 316, 386, n. 49).

Lastly, one finds a consistent ABA rhythm along the wall surfaces of both S. Carlo and S. Ivo, often suggesting the shape of a triangle. This rhythm is most definitely an inference to the Trinity, while also referencing the triangles used in the designs of both S. Carlo and S. Ivo, yet it bears a possible connection to one more aspect of Keplerian cosmology. Kepler held fast to the Pythagorean notion of the music of the spheres which he reformulated in his book, *Harmonice Mundi* (1619) (Kepler 1995: 199–200). The earth "sings", as Kepler puts it, MI FA MI, which in musical notation (Fig. 61.8) forms a visual analogue to the type of rhythms Borromini produces along the walls of the lower parts of both S. Carlo and S. Ivo. If both churches are dedicated in some form to planetary motion, and echo in their lower storeys the terrestrial realm, as I have argued above, it may not be so far fetched to assume that Borromini has included a musical reference as well.

Obviously, this chapter has not exhausted all the links that might exist between Borromini and Kepler. It is hoped that the above speculations have at least pointed to a new and fruitful direction that ultimately complements many of the existing interpretations of S. Carlo and S. Ivo, by adding yet another fascinating wrinkle to the church architecture of Francesco Borromini.

Fig. 61.8 Musical notation attributed to the planets by Johannes Kepler. Image: Kepler (1619: 207)

Acknowledgments I must thank Ann Marie Carroll, J. Gerard Curtis, Karen Hatch (and the boys), Laurier Lacroix, Jody McNabb, and Irene Mordas, for their invaluable assistance and encouragement during the various stages of research and writing.

Biography John G. Hatch is Associate Professor of Art History at Western University (London, Canada) where he also serves as an Associate Dean for the Faculty of Arts and Humanities. He received his PhD from the University of Essex and his area of research is twentieth-century European and American art and theory, with a special focus on the influence of the physical sciences on modern art. Recent publications include the articles "Nature, Entropy, and Robert Smithson's Utopian Vision of a Culture of Decay" (2012), "Modern Earthworks and Their Cosmic Embrace" (2011), and "Some Adaptations of Relativity in the 1920s and the Birth of Abstract Architecture" (2010).

References

Ackerman, J. S. 1966. *The Architecture of Michelangelo*. London: A. Zwemmer.

Blunt, A. 1978. The Architecture of Rome, Borromini. P. 48 in A. Blunt, ed. *Baroque and Rococo: Architecture and Decoration*. New York: Harper and Row.

———. 1979. *Borromini*. London: Penguin Books.

Connors, J. 1995. A Copy of Borromini's S. Carlo alle Quattro Fontane in Gubbio. *Burlington Magazine* **137**, 1110 (Sept.): 588-599.

———. 1996. S. Ivo alla Sapienza: The First Three Minutes. *Journal of the Society of Architectural Historians* **55**: pp. 38-57.

Kepler, J. 1596. *Mysterium Cosmographicum*. Tübingen: Georgius Gruppenbachius.

———. 1619. *Harmonices mundi*. Linz: J. Plancus.

———. 1995. *Epitome of Copernican Astronomy and Harmonies of the World*. Trans. C. G. Wallis. Amherst, New York: Prometheus Books.

Koestler, A. 1979. *The Sleepwalkers*. London: Penguin Books.

Koyré, A. 1957. *From Closed World to the Infinite Universe*. Baltimore and London: The Johns Hopkins University Press.

Kozhamthadam, J. 1994. *The Discovery of Kepler's Laws: The Interaction of Science, Philosophy, and Religion*. Notre Dame and London: University of Notre Dame Press.

MARTENS, R. 2000. *Kepler's Philosophy and the New Astronomy*. Princeton and Oxford: Princeton University Press.

PANOFSKY, E. 1954. *Galileo as Critic of the Arts: Aesthetic Attitude and Scientific Thought*. The Hague: Martin Nijhoff.

PORTOGHESI, P. 1968. *The Rome of Borromini: Architecture as Language*. New York: George Braziller.

STEINBERG, L. 1977. *Borromini's S. Carlo alle Quattro Fontane: A Study in Multiple Forms and Architectural Symbolism*. New York: Garland Publishing.

WITTKOWER, R. 1973. *Art and Architecture in Italy, 1600-1750*. Harmondsworth: Penguin Books.

WÖLFFLIN, H. 1984. *Renaissance and Baroque*. London: William Collins Sons & Co.

Chapter 62
Transformational Geometry and the Central European Baroque Church

John Clagett

> *The parts of a continuum, like a line, can only be individuated if distinguished by an indivisible boundary between them. Without such an indivisible divisor between the parts, the parts themselves would compenetrate one another; more than one part would be in the same place; and parts would no longer be individuated by their situs. They would in fact melt into the confusion of the indeterminate*
> (Smith 1954: 53–54).

The Central European Baroque church (CEBc) appears to be in endless conflict with itself: it is both unified and chaotic, continuous and fragmented. Although the plasticity of its composition augments a sense of singularity, its separateness is heightened through incompleteness and irregularity of placement. Its walls are a pliant, impermanent membrane; its vaulting an elastic, undefined surface, out of synchronization with the ground plan. This is not a style of simple black and white opposites, but rather one with infinite variables, as if it were created to be an insoluble problem. Nonetheless, all of these observations are paradoxically correct—to a degree—for it was the Baroque architect's intention to hold a multiplicity of ideas in harmonious suspension. Its theoreticians and practitioners sought to achieve a sense of both *Gesamtkunstwerk,* where architecture and the plastic arts merged into an interconnected, symphonic whole, and *Zweischaligkeit,* in which contrasting tectonic systems coexisted as a composite envelope, serving to dissolve sharply defined boundaries (Feulner 1923: 52–55).

Thus the arts acted to harmonize, while the structural systems acted to disseminate. This effect of oscillating proximity resulted from an intention to create a spectrum of elements eternally approaching singularity; to establish a dynamic continuum.

First published as: John Clagett, "Transformational Geometry and the Central European Baroque Church", pp. 37–51 in *Nexus I: Architecture and Mathematics,* ed. Kim Williams, Fucecchio (Florence): Edizioni dell'Erba, 1996.

J. Clagett (✉)
165 Jane Street, Engelwood, NJ 07631, USA
e-mail: jclagett1@verizon.net

K. Williams and M.J. Ostwald (eds.), *Architecture and Mathematics from Antiquity to the Future*, DOI 10.1007/978-3-319-00143-2_15,

Table 62.1 Mathematical and architectural developments: a comparative chronology, 1639–1827

Desarques: *Brouillon projet*	1639	
Descartes: *La Géométrie*	1673	
Leibniz's first paper on the calculus	1684	G. Dientzenhofer: Pilgrimage church at Kappel begins construction
Newton: *Principia*	1687	Approximate start of CEBc
	1694	Fischer v. Erlach: Project for Dreifaltigkeitskirche
	1713	C. Dientzenhofer: Smirice, chapel
Taylor: *Methodus incrementorum*	1715	
	1726	J. Dientzenhofer: Holzkirchen
Sacheri: *Euclid Vindicated*	1733	Neumann: Wüzburg, Hofkirche
	1737	Guarini: *Architettura civile*
D'Alembert: *Traité de dynamique*	1743	
	1744	J.M. Fischer: St. Michaelskirche
Euler: *Introductio*	1748	
	1763	Neumann: Vierzehnheiligen
John Love: *Geodaesia*	1768	
Hyperbolic trigonometry	1770	Approximate close of CEBc
Lobachevskian geometry	1827	

Compiled from the following sources: Boyer (1991: 686–687), Norberg-Schulz (1974: 369–379), Hugh of St. Victor (1991: 7)

The scientific/mathematical developments of the seventeenth and eighteenth centuries, as Table 62.1 illustrates, place the CEBc in context. During this time, reasoning shifted from the isolated to the integrated. In Desargues' *Brouillon projet* (1639), a treatise on projective geometry, he postulates that all lines converge, including those that are parallel, which intersect at infinity (Field and Gray 1987: 47–59). In Newton's *Principia* (1687) the phenomena of dynamics was so lucidly formulated that the physical world was redefined through motion. Newton and Leibniz's syntheses of integral and differential calculus were important, first, in that they created a composite mathematics from unrelated studies, and second, in that they achieved this synthesis via the concept of the limit, where a progressively diminishing ratio was applied to problem solving. Descartes' *La Géométrie* (1673) also created a mathematical synthesis, in this case of algebra and synthetic geometry, by formulating analytic or "arithmetized" geometry (Körner 1960: 67–75). Descartes defined a process by which "any problem in geometry can easily be reduced to such terms that a knowledge of the lengths of certain lines is sufficient for its construction" (Descartes 1925: 2–3).

Given these advances, one could conclude that the prevailing mathematical/theological order was thoroughly supplanted; that an opposing, superior system had superseded it. Yet, the situation can be interpreted otherwise; that the pre-existing structure was not obviated, but gradually transformed. This can be compared to a structure that is progressively transformed by the act of building, or to geometry's subject matter that has been refined throughout history—many of its classical assumptions persist, but every facet is continually redressed (Hugh of St Victor

1991: 17–23). This notion of the continually evolving sets the compositional principles of the CEBc in a clearer light. Its architects sought a dynamic continuum in which all subject matter was interrelated, in motion, and progressing toward a spiritually ideal point. Their methodology involved applying a set of operations which transformed space from the pure, static, and isolated into the composite, dynamic, and interpenetrating. This strategy is reflected in the vaulting of the CEBc:

> I describe it as "unexpected" when domes rest upon lunettes (as in the church of the Fraternity of St. Michael in Berg-am-Laim near Munich, 1737), that is, when there are openings in the pendentives that are not the usual horizontal barrel vaults but vertical cylinders covered by domes. In the parish church at Elwashausen, 1741, vaults that are customarily found above circular plans are used above rectangular ones, and vice versa (Frankl 1968: 68).

By examining such examples—in which the horizontal is rotated to the vertical, or the circle is thrust into the square—the relationships between pre-figurative intentions, a formal system of composition, and the built-form of the CEBc can be interrelated. To this end, geometry has been referenced for specific reasons; first, it is immediately relevant—the architects of the CEBc generated their plans through strict geometric methods (Otto 1979: 37–38). Second, the process of geometry involves identifying individual rules and grouping these into sets and subsets of rules until a generalized, functioning system of logic is assembled, such as Euclid achieved in his *Elements*. Third, architecture and geometry both involve problems of figure-construction; that is to say these two fields make use of figure-construction to demonstrate the cohesiveness and consistency of their system of logic.

The specific set of geometric operations known as transformation is applicable to the CEBc and its property of dynamic continuum, for transformational operations define the formal operations of repositioning and reconfiguration which characterize the vigorous nature of the CEBc. These procedures allowed the distinct elements of Euclidean geometry—plane, square, cube, etc.—to be interpreted; these transformations in effect set them in motion, agitating the elements so that they visually interpenetrated as they approached an in-motion singularity.

A transformation is a type of mapping, and so is analogous to and, in fact, influenced by the methods of cartographers such as Gerhardus Mercator. Mapping can be illustrated by placing a globe onto a sheet of paper, and projecting the points of the globe onto that sheet; in other words, projecting a three-dimensional form onto two-dimensions. Every point on the globe except the uppermost pole can be mapped. During this transformation certain relationships are distorted; for example, distances between points become increasingly elongated the further those points on the globe are from the paper, while other relationships remain constant-circles on the globe that are parallel to the sheet of paper remain circular. This cartographic projection is one specific case of transformation (Gray 1979: 46). In addition to projection, contemporary geometricians identify a number of these operations, including: area, rotation, reflection, translation, dilatation, and coordinate transformation.

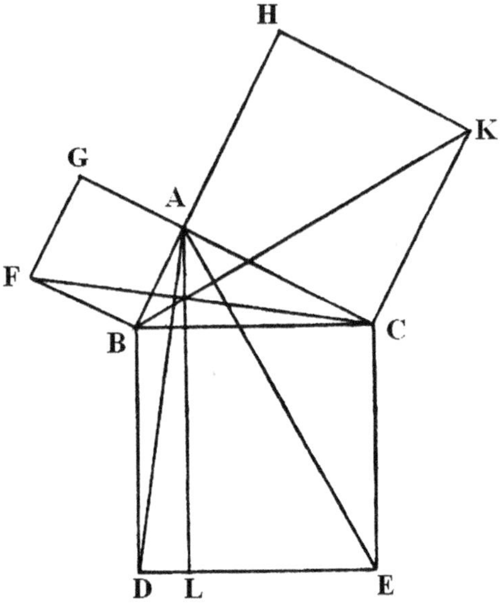

Fig. 62.1 The Pythagorean theorem, from Euclid's *Thirteen Books of the Elements* (I Book, Proposition 47). Image: author

An area transformation alters the perimeter form of a closed figure while its area remains invariant. Area transformations were the first of this class of operations to be explicitly recognized as such. Pythagoras is attributed with having developed a system of area transformation by which polygons can be reconfigured through a manipulation of parallel lines and area formulae (Eves 1990: 90–91). Euclid made use of area transformation when proving geometry's most well-known theorem; that for right-triangles, the square of the hypotenuse is equal to the sum of the squares of the remaining face (Fig. 62.1). Beginning with the right triangle *ABC*, each face of the triangle is 'squared'—that is, three squares, *ABFG, ACKH,* and *BDEC* (which I call *a, b*, and *c*) are constructed. Then the line *AL* is constructed parallel to *BD* and *CE* and passes through *c*, which abuts the hypotenuse of the right triangle *ABC*. The line *AL* passes through *c* so as to divide it into two rectangles which I term *a'* and *b'* ; *a'* is equal to *a, b'* to *b* (Euclid 1956: Vol. 1, 349–360).

One of three constructive geometry problems posited but never solved by classical geometricians, the quadrature of the circle, was seem as an area transformation problem. Although it was finally proven insoluble twenty-one centuries after *Elements* (Boyer 1991: 573), an equivalent architectural idea of transforming the rectilinear into the curvilinear was an important concern for the artists of the CEBc. The term "Borrominian transformation" describes the mutation of a rectilinear spatial organization into an equivalent curvilinear structure, while maintaining the starting rectilinear diagram; as is the case with Christoph Dietzenhofer's St Miklas on the Kleinseite, Prague (1711), a vertical hall with side chapels (Franz 1987: 34–36), and Kilian Dientzenhofer's St. Margaret, Brevnov (1715), a *Wandpfeilerhalle* "wall-pillar-hall." These Bohemian examples of Borrominian transformation closely model the geometric notion of

Fig. 62.2 Würzburg, Hofkirche. Photograph: author

transformation, for certain aspects of the form are transfigured, while others remain invariant. The Borrominian transformation from the rectilinear to the curvilinear has a number of variations, each with significance to other geometric transformational qualities. The strategy predates the CEBc; Vignola in his Sant' Andrea in via Flamina, Rome (1554) fused the line and circle into the form of an oval (Murray 1978: 117), yet the CEBc substantially reinterpreted the idea. In Johann Dientzenhofer's plans for Holzkirchen (before 1726), the oval is substituted with a bi-axially symmetrical, hemiolic curvilinear grouping—a dominant central element flanked by symmetrical ancillary elements. The equivocal, semi-longitudinal/semi-centralized organization of the oval's singular space was replaced by an increased indeterminacy; Johann Dientzenhofer's schemes both oscillate between a three-space interlocking composition and a two-space overlapping composition consisting of a radially organized central structure and an orthogonally organized longitudinal structure.

Balthasar Neumann's Würzburg Residenz Hofkirche (1733, Figs. 62.2 and 62.3), designed in collaboration with Lucas von Hildebrandt, shares many of the properties of Borrominian transformation expressed by Dientzenhofer. The Würzburg palace's planning is governed by a series of orthogonally subdivided wings intersecting to produce four interior courts and a central *cour d'honneur*. The Hofkirche, by occupying one of the south wing's orthogonal compartments, explicitly maintains the rectilinear organization at the exterior, and implicitly on the interior. The area transformation from the rectilinear to the curvilinear expresses *Zweischaligkeit;* from the interior, while the curvilinear order dominates, the planar, orthogonal superstructure can be perceived through the curved form.

The linear transformations of *rotation, reflection, translation,* and *dilatation* are each detectable in the CEBc. Rotation is the displacement of a form about a point. A reflection is a mirror-image transformation in which the initial and resultant images, when viewed together, are symmetrical about a line. A translation is a shift, analogous to pulling a tablecloth set with dishes across a table: the relative positions of the dishes themselves remain the same, but their positions with respect to the table changes. A dilatation uniformly enlarges or reduces a figure without changing the form or focus of dilatation.

Rotation is present throughout the underlying planning geometry of St. Michaelskirche, Berg am Laim (1744). Its central rotunda has 12 circles ringing the space. Because these circles are equal in radius, and are positioned at equal increments, a strong sense of rotation is present. As well, a 12-pointed star is present which is constructed by rotating an equilateral triangle at 90° increments about the rotunda's centre.

Unlike plans which clearly indicate reflection by maintaining a single axis of symmetry only, in the Hofkirche the ground-plan diagram maintains bi-axial symmetry. By doing so, rotation is expressed—the north-west and south-east quadrants are rotational transformations of one another, as are the south-west and north-east. Yet the ground plan's north half is also a reflection of the south, as the east side is of the west. While the mezzanine plan is more clear—it has a single line of symmetry—it nonetheless adds to the transformational dynamism of the Hofkirche in other ways. The ceiling plan, though, once again displays the duality of rotation and reflection. Its dome-like structure, like the ground plan, appears rotational, reflective, and translational. It can be read as comprising five intersecting quasi-ellipsoidal domes, or three tangentially intersecting domes, with the voids between these taken up by an additional pair of dome-fragments, or as the section suggests, three interpenetrating domes. While the realized building does deviate from Neumann's plan and section, the sense of interpenetrating structure is evident. When looking directly upward at the first full-height space after entering—thereby simulating the Euclidean space of an architectural plan—the line of the mezzanine produces a nearly complete oval, which visually frames the vaulting above (Fig. 62.2). It also appears to be cast from the same mould that produced the coupled transverse arches—all are ovals. Yet, like the crossing of Neumann's Pilgrimage Church of Vierzehnheiligen (1763), the centre of the mezzanine oval is split by the transverse arches—the two plans are not in synchronization. This phenomenological shift is the result of translation; the coupled arches appear to have been repositioned by one-quarter period of the rhythmic cycle. Translation is also

Fig. 62.3 Würzburg, Hofkirche. Photograph: author

detected in the Hofkirche's fenestration. The masonry vault- and wall-construction of the chapel suggests that its fenestration be vertically stacked so that the resultant structure acts as a series of inverted U-shaped frames, letting the building's loads be transferred through the walls directly as compressive forces. This is common sense—masonry is strong in compression, but weak in shear, bending, and tension. Yet Neumann's section and the interior view discussed above show the transverse arches seem to rest above glazed openings, as if Neumann once again deliberately shifted the layers of the chapel's plans. A similar shift is apparent in the relationship between the interior cornice line and fenestration. The cornice line winds along the interior walls, creating a series of alternating solid convex projections and concave voids. Yet at the mid-point of the longitudinal walls—where a nearly undetectable transept occurs—the cornice line cavity is aligned not with a glazed opening, but a wall-pier separating two openings. Again translation identifies Neumann's recurring motif: solid and void alternate vertically through shifted plan layers (Frankl 1968: 62–65) (Fig. 62.3).

Dilatation can be detected in the CEBc in conjunction with its sense of near-symmetry, elasticity, and loss of centre-dominance. The plan of Neumann's Benedictine Church of the Holy Cross at Neresheim (commissioned 1748) appears at first to be bi-axially symmetrical; a longitudinal nave interfused with a centralized dome. Curiously, the choir/altar half of the nave is slightly narrower than that of its near-mirror-image. The difference is not because these are intended to be perceived as different building parts; in fact, their repeating organization affirms they are two halves of a whole. The scale change is much more plausible as a dilatation; a reduction in scale. The elastic nature of Neresheim is also found in the Hofkirche's section at its vaulting. The court chapel presents a rigid rectilinear system of fenestration running along the space's perimeter. Contrasting this are the two scales of curvilinear form passing in front of the mezzanine-level fenestration. The smaller scale exists as curved arches tightly surrounding this grid-work of glazing, sometimes overlapping and obscuring the fenestration. Several arches, unlike the glazing behind, vary slightly in width and height, as if their surfaces were not only malleable—demonstrating plasticity—but capable of being stretched. The larger scale exists as the grouping of shallow domes described earlier. Again a sense of elasticity is present. The domes are not rigidly defined as simple geometric figures such as a cylinder and half-sphere, but appear as continuously dilating forms, whose scale is dynamically fluctuating according to their importance of position.

During the development of the CEBc, several characteristics evolved in concert to fully transform its centre. Initially, as with Theatinerkirche at Munich (begun 1663), the Baroque church's centre was defined vertically by a dome and lantern resting on a drum, and horizontally by a fully extended transept. Together these emphasized the hierarchical dominance of the centre. A gradual process of dilatation occurred in which the drum was reduced in scale and finally eliminated, allowing it to merge with the main aisle. The dome itself eventually was transformed from a hemispherical to a flat dome. Similarly, the transept was progressively decreased in scale (Feulner 1923: 5–9). Norberg-Schulz attributes this loss of the centre's dominance to the rise of democratic spirit in the eighteenth century, but that period's geometric developments are also a factor. By the start of the CEBc period, geometricians had absorbed and began surpassing the work of classical mathematicians. In Descartes' *Géométrie,* number and geometric form were fused. This ability to fuse or merge separate ideas is comparable to the CEBc's transformation of centre and transept, reinforcing a progressively increasing continuity of spatial form. Neumann's Hofkirche is an appropriate example of this architectural tendency. Its transept exists as a shallow altar, basically absorbed into the surrounding space (Fig. 62.3), while the domes' flatness and elasticity also increase the holistic unity of the chapel. In the case of Christoph Dientzenhofer's Chapel of the Castle at Smirice (1713), the transept is not merely aligned with the nave, but compressed to the point that it forms a convexity of form at the chapel's centre.

The architect Guarino Guarini shortly predates the CEBc. Guarini is important for disseminating the Italian Baroque throughout Europe, through his travels and posthumously-published *Architettura civile.* In this treatise are included a number of coordinate transformations, including a study transforming a polar coordinate system to a Cartesian system (Fig. 62.4). A coordinate transformation allows for a system's scalar or angular units to be re-configured, and can also involve Guarini's transformation of coordinate systems. In a sense, this is a variation of an area transformation. Yet, the quadrature of the circle was never resolved and finally proven insoluble. Also, Guarini's drawing illustrates that not only the perimeter, but the full extent of the space is acted upon. The idea of an interrelated duality appears prominently in the CEBc, and can be associated with two of its fundamental principles: *Zweischaligkeit* and the interpenetration of polar and Cartesian coordinate systems; in other words, centralized and orthogonal building-types. The idea of *Zweischaligkeit* was raised earlier, yet the planning grid per se is not a structural component. Instead, the perimeter wall-diaphragm which surrounds the glazed openings is, as are the wall-pillars which abut the wall-diaphragm; together they form a *Wandpfeilerhalle* (Norberg-Schulz 1974: 65–76). It is in this sense that the double-delimitation of space is typically referenced, for they are clearly two

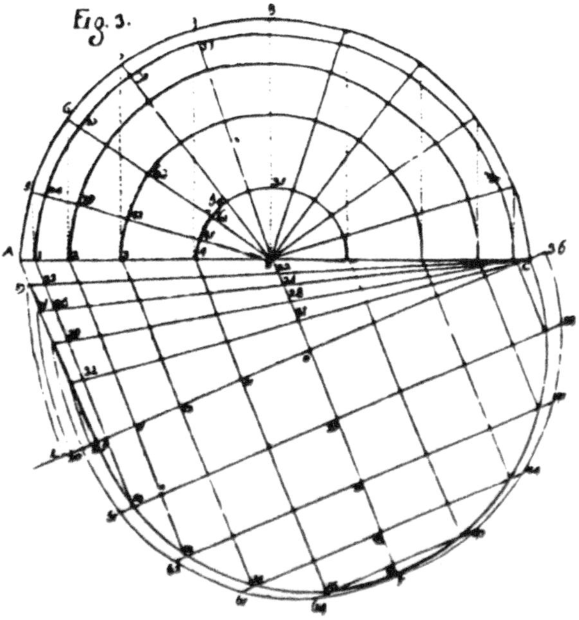

Fig. 62.4 Guarini's study transformaing a polar coordinate system to a Cartesian system. Image: Guarini (1737: Lastra XIV, Fig. 3)

distinct structural systems—one a continuous-diaphragm, the other a point-load system—which are smelted into a composite system of an interrupted diaphragm welded to pillars rotated at right angles. Although a twice-defined structural system might seem redundant, with the complex architecture of the CEBc it is inspired, for the rectilinear wall-envelope is efficacious in defining the spatial boundary (both literally as weather-proofing, and figuratively as a conceptual boundary) while the curvilinear intermittent pillars effectively transform the stolid, physical *Schlichtkeit* "modesty" of the exterior into a dynamic, curvilinear three-dimensional space.

A projective transformation is related to the cartographic analogy made earlier, in which a form is projected, for instance, from a three-to two-dimensional surface. In a correct geometrical sense, these projections originate not from a pencil of parallel lines, but a pencil of diverging lines, as the Hofkirche's plan shows. One sees in Neumann's plan a series of irregular voids which run the length of each of the longitudinal walls. These segment similarly scaled, non-parallel wall-pillars. Seen as a group, these voids and solids appear to radiate from a mirror-image pair of points, external to the chapel, creating fan-shaped compositions. Geometrically related to this is the correlation between the lower and upper limits of the chapel's space—the ground floor and vaulting. The ground floor is geometrically a two-dimensional space; the vaulting three-dimensional. Interestingly, the fan-shaped voids and ground floor/vaulting relationships can both be connected to the geometry of projective transformation, a correspondence noted by Wittkower in his description of Guarini's *Architettura civile*:

> More than one-third of the text is concerned with a new kind of geometry, namely the plane projection of spherical surfaces and the transformation of plane surfaces of a given shape into corresponding surfaces of a different shape. Guarini was perhaps the only Italian architect who had studied Desargue's *Projective Geometry*, first published in Paris in 1639, which was informed by the modern conception of infinity (Wittkower 1958: 27).

Desargues' study of perspective, and the intersection of the cone and plane, preceded his treatise on projective geometry. Both subjects involve a sectional plane (a conic section or picture plane), a point or points at infinity, and deformations which are continuous and definable geometrically (Field and Gray 1987: 14–30). Bernini's Scala Regia (1666) and Palladio's Teatro Olimpico (1580) employ projective geometry in the form of *trompe l'oeil*. However, in the vaulting of the CEBc a more far-reaching potential for transformational geometry to act as a generative operation for form may be recognized: hyperbolic parabloid and other complex, geometrically-indeterminant forms may be generated through the projection of two-dimensional, Euclidean ground-plan forms onto negative- and positive-curvature, three-dimensional surfaces. Neumann's projects in particular grew from a planar to a volumetric state (Otto 1979: 39), substantiating that the vaulting's form was based on ground-plan projection. Thus this "new geometry" was directly generated from figures based on constructive geometry. The Dientzenhofers' Convent Church at Oboriste (1712), Chapel of the Castle, and St. Margaret at Brevnov (1715) as well as Neumann's churches at Neresheim, Vierzehnheiligen, and Kitzingen-Etwashausen (1745) all exhibit a developed awareness of projective geometry and its possibilities, as do, to a lesser degree, J.M. Fischer's St. Michaelskirche and the Abbey Church at Ottobeuren (1766).

In assessing the importance of transformational geometry and the CEBc, two major conclusions can be drawn. One is that the architectural principle of the dynamic continuum which can be applied to the individual built work is also a characteristic of the ongoing creative process in architecture. Both the building and the process can be enunciated through geometry, for their common subject matter is continually evolving. As Hanfried Lenz has shown, a contemporary projective geometric premise conjectures that that which originates as a point can be transformed to an ellipse, and then to an indeterminate curve (Fig. 62.5). At the very least the architects who composed the unexpected vaulting of the CEBc grasped this as a precept. The vaulting cannot be derived simply as an extrusion of the ground plan; a more complex, projective process was involved. But a train of thought can be followed in the opposite direction as well. In *The Elements*, its first definitions—of point, line, and plane—are based on the intellectually-genetic, transformational nature of geometry, and of architecture.

The second conclusion is that transformational operations were employed to make perceptible the experience of approaching singularity. These geometric operations were intended to stir the emotions, engaging the participants in an

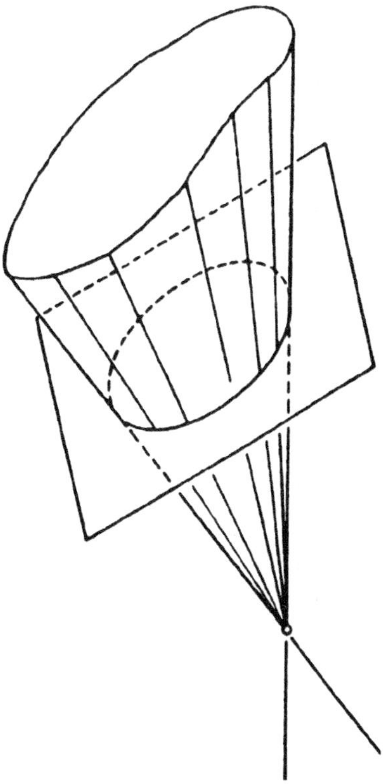

Fig. 62.5 Transformation of a point to an ellipse to an indeterminate curve. Image: author, after Lenz (1965)

active state of interpenetration, flexuosity, and transfiguration, drawing them close to a point of convergence at infinity.

Acknowledgement This project was supported by a grant from the Graham Foundation for Advanced Studies in the Fine Arts.

Biography The architectural work of John Clagett has followed two streams: research and practice. On the former, he has investigated Central European Baroque ecclesiastical architecture with the intention of formulating a general theory unifying the mathematical, philosopical and architectural thinking of the seventeenth and eighteenth centuries.

References

Boyer, Carl B. 1991. *A History of Mathematics*, 2nd edn. Revised by Uta C. Merzbach. New York: John Wiley & Sons.

Descartes, René. *The Geometry of René Descartes*. 1925. David E. Smith and Marcia L. Latham, trans. Reprinted 1954. New York: Dover Publications.

EUCLID. 1925. *The Thirteen Books of the Elements*. 2nd edn. Thomas L. Heath, trans. Reprinted 1956. New York: Dover Publications.

EVES, Howard. 1990. *An Introduction to the History of Mathematics, with Cultural Connections*. Philadelphia: Saunders College Publishing.

FEULNER, Adolph. 1923. *Bayerisches Rokoko*. Munich: K. Wolff.

FIELD, Judith and Jeremy GRAY. 1987. *The Geometrical Work of Girard Desargues*. New York: Springer Verlag.

FRANKL, Paul. 1968. *Principles of Architectural History: The Four Phases of Architectural Style, 1420–1900*. James F. O'Gorman trans. Cambridge: MIT Press.

FRANZ, Heinrich Gerhard. 1987. Balthasar Neumanns kurvierte Räume und ihre Vorstufen bei Borromini, Guarini und in Böhmen. In *Balthasar Neumann*, Thomas Korth and Joachim Poeschke, eds. Munich: Hirmer Verlag.

GRAY, Jeremy. 1979. *Ideas of Space: Euclidean, Non-Euclidean, and Relativistic*. Oxford: Clarenden Press.

GUARINI, GUARINO. 1737. *Architettura civile*. Torino: Gianfrancesco Mairesse all'insegna di Santa Teresa di Gesù.

HUGH OF ST. VICTOR. 1991. *Practica Geometriae* (Practical geometry). Frederick A. Homann, trans. Mediaeval Philosophical Texts in Translation, no. 29. Milwaukee: Marquette University Press.

KÖMER, Stephan. 1960. *The Philosophy of Mathematics: An Introductory Essay*. Reprinted 1986. New York: Dover Publications.

LENZ, H. 1965. *Vorlesungen Über Projektive Geometrie*. Leipzig: Akademische Verlagsgesellschaft Geest & Portig.

MURRAY, Peter. 1978. *Renaissance Architecture*. Milan: Electa Editrice; New York: Rizzoli International Publications, 1985.

NORBERG-SCHULZ, Christian. 1974. *Late Baroque and Rococo Architecture*. New York: Harry N. Abrams.

OTTO, Christian F. 1979. *Space Into Light: The Churches of Balthasar Neumann*. Architectural History Foundation, Cambridge: MIT Press.

SMITH, Vincent Edward. 1954. *St. Thomas on the Object of Geometry*. Aristotelian Society of Marquette University, Aquinas Lecture. Milwaukee: Marquette University Press.

WITTKOWER, Rudolf. 1958. *Art and Architecture in Italy 1600 to 1700*. London: Penguin Books.

Chapter 63
Are There Connections Between the Mathematical Thought and Architecture of Sir Christopher Wren?

Maria Zack

Introduction

After the Great London Fire of 1666, Sir Christopher Wren was appointed as a member of the group that was charged with rebuilding the City of London. Over the next several decades this massive reconstruction effort was lead by Wren and Robert Hooke. Both Wren and Hooke were founding members of the Royal Society and until the Great Fire both were best known for their work in mathematics, physics and astronomy.

Wren has been credited with designing St. Paul's Cathedral as well as more than fifty parish churches that were rebuilt in the City of London. Current scholarship indicates that the parish churches were most likely designed by Wren, Hooke and others that worked in their office. Using parish church architecture as illustrations, this chapter explores evidence for and against the notion that Wren's mathematical knowledge directly influenced his architectural designs.

First published as: Maria Zack, "Are There Connections Between the Mathematical Thought and Architecture of Sir Christopher Wren?", pp. 171–180 in *Nexus VI: Architecture and Mathematics*, Sylvie Duvernoy and Orietta Pedemonte, eds. Turin: Kim Williams Books, 2006.

M. Zack (✉)
Department of Mathematical, Information and Computer Sciences, Point Loma Nazarene University, 3900 Lomaland Drive, San Diego, CA 92106, USA
e-mail: MariaZack@pointloma.edu

K. Williams and M.J. Ostwald (eds.), *Architecture and Mathematics from Antiquity to the Future*, DOI 10.1007/978-3-319-00143-2_16,

Wren the Mathematical Scientist

In *On a Grander Scale*, Lisa Jardine observes "Wren lived on the cusp of the modern era, between a world saturated in received wisdom and one whose fundamental scientific beliefs were grounded in observation and meticulously recorded data" (Jardine 2002: xv). She goes on to cite this illustrative entry from Hooke's diary for Thursday August 16, 1677:

> At the Crown [public house], Sir Christopher told me of killing worms with burnt oyle. . .and of curing his lady of thrush by hanging a bag of live boglice about her neck. Discoursed about theory of the Moon which I explaind. Sir Christopher told his way of solving Keplers problem by the Cycloeid.

Though Newton called Wren one of "the foremost geometers of the previous generation" (Newton 1999: 424), very little of Wren's mathematical and astronomical work was ever published. It appears that, like Newton, Wren was interested in a wide variety of fields and often left projects incomplete and without thorough documentation.

Thomas Sprat, in his 1667 *History of the Royal Society*, says of Wren "For in turning over the Registers of the Society, I perceiv'd that many excellent things, whose first invention out to be ascribed to him, were casually omitted." He then goes on to list work on refraction, the theory of motion, accurate optical measurement, work on the rings of Saturn, the lunar globe, the weather, magnetism and navigation, microscopy, injections and anatomy. Sprat says "I know very well, that some of them he did only start and design, that they have been since carried out to perfection, by the Industry of other hands." Yet he argues that Wren deserves credit for his work in each of these areas (Tinniswood 2001: 81–82).

Sir John Summerson in *Sir Christopher Wren* suggests that Wren wrote very little down because so many branches of science were easy for him (Summerson 1965: 59). Tinniswood says:

> [T]here is a restlessness about him which can't simply be attributed to the conventionally catholic interests of the seventeenth-century virtuoso a hunger, almost, as he pursued first one thing, then another and another, switching between disciplines, hunting for answers, chewing over problems and spitting out solutions (Tinniswood 2001: 83).

Perhaps it was this restlessness that kept him from publishing much of his work. We do know that he did some significant mathematical work on cycloids in response to a Kepler-related question posed by Pascal (under the name of Jean de Montfert) and published that result. Perhaps because Wren gave well-attended public lectures as the Gresham Professor of Astronomy and later as the Savilian Professor of Astronomy at Oxford, he felt that it was sufficient to make his ideas known via lectures. In a letter to Sir Paul Neile over his unpublished work on the rings of Saturn (*De Corpore Saturni*), Wren says that in a Gresham lecture he gave "fuller discourses on the same subject, which he thought was publication enough" (Tinniswood 2001: 64). Much of his other mathematical or scientific work is incorporated into larger documents written by others, among them John Wallis and Robert Hooke (Huxley 1960: 204, 207), or is documented in the correspondence of Henry Oldenburg, the secretary for the Royal Society.[1]

Wren was firmly rooted in the English tradition of practical mathematics that began in the mid-sixteenth century. At that time there was a demand for improved techniques related to navigation, surveying, cartography and fortification and thus there was a close relationship between mathematicians and instrument makers. Bennett says "It can be summarized in their two fundamental tenets: that mathematics is both certain and useful" (Bennett 1982: 7). Oldenburg reported that Wren's theory *Lex Naturae de Collisioine Corporum* (The Theory of the Collision of Bodies) is a "synthesis of experimental results, constrained in its expression by certain unspoken regulative principles of simplicity and symmetry" (Bennett 1982: 72). Wren simply offered experimental results as proof. Neile reported to Oldenburg that Wren "says that the appearances carrie reason enough in themselves as being the law of nature" (Bennett 1982: 119).

This very practical attitude toward rules can be seen in Wren's architecture as well. He used the five orders of classical architecture in many of his buildings but with some of the freedom usually associate with Baroque design. For him it appears that theory was justified by utility. In a 1662 address to the Royal Society Wren emphasized that the role of the members of the Society was "approving themselves Benefactors to Mankind, and of perfecting something, for which Posterity may be really obliged to us" (Wren 1750: 221). Clearly Wren himself would go on to benefit mankind in a practical way and gain recognition for his architectural work.

Wren the Architect

There are a great many myths surrounding the architectural legacy of Wren, in part due to personal attacks that were launched at him near the end of his life. In an attempt to defend his father's memory, Wren's son Christopher (and later his grandson Stephen), worked to compile and then publish *Parentalia: or Memoirs of the Family of the Wrens*. This volume provides a family history and contains detailed accounts of Wren's astronomy, mathematics and buildings. Among the claims made in this volume is that Wren designed all 51 of the post-fire City of London churches. Current scholarship suggests that Wren and Hooke had an active architectural practice and that the designs for the city churches were a mixture of Wren's work, Hooke's work and the work of others in their office. In *The City Churches of Sir Christopher Wren*, Paul Jeffery has done a meticulous job of researching individual churches and attempting to attribute their design.

In considering the city churches, even those attributed to Wren, it is wise to remember that much of the finishing detail was the work of craftsmen and may not have been specified by Wren (with the notable exception of St. Paul's Cathedral).

[1] For further details about Wren's mathematical work, see Bennett (1982).

The available records (architectural drawings, building accounts, Hooke's diaries and parish vestry minutes) contain little about that level of detail in the church designs. It makes sense, given the number of building projects that Wren and Hooke were managing at any given time, that the details were left to others. So to look for connections between Wren's mathematics and his architectural work, it is reasonable to focus on the general designs and not the decorative details.

In Wren's architectural writing labeled Tract I in *Parentalia* he says:

> There are two Causes of Beauty, natural and customary. Natural is from Geometry, consisting in Uniformity (that is Equality) and Proportion. Customary Beauty is begotten by the Use of our Senses to those Objects which are usually pleasing to us for other Causes, as Familiarity or particular Inclination breeds a Love to Things not in themselves lovely. Here lies the great Occasion of Errors; here is tried the Architect's Judgment: but always the true Test is natural or geometric Beauty (Wren 1750: 351).

Clearly Wren thought that true beauty came from using geometric forms, but does that really mean that his mathematical ability influenced his architectural work?

Arguments for and Against a Relationship Between Wren's Architecture and Mathematics

In his work on the Banqueting Hall of Whitehall Palace (1622) and the new portico for the pre-fire St. Paul's (1640), Inigo Jones introduced continental design in the classical tradition to England. Jones's ideas can be traced to his travels through continental Europe and his contact with both the great architecture of the time and a canon of architectural texts that included Vitruvius, Alberti and Palladio. Sir John Summerson states that in the minds of seventeenth-century scholars, Vitruvius provided the same basis for architecture that Euclid did for geometry, that to them "the natural equivalent of scientific thought in architecture was classical design" (Summerson 1960: 102).

We do know that Wren had access to this same canon of classical texts and all of them advocate geometry, proportion and symmetry as the basis of architectural design.[2] Wren himself states in Tract I:

> Geometrical Figures are naturally more beautiful than other irregular; in this all consent as to a Law of Nature. Of Geometrical Figures, the Square and the Circle are most beautiful; next, the Parallelogram and the Oval. Straight Lines are more beautiful than curve (Wren 1750: 351).

Like many gentlemen of his time, Wren went to the continent to tour great buildings and meet fellow scientists. He spent from July 1665 to February 1666 in France. During that visit he formed some clear opinions about French architecture

[2] Some further details can be found in Falter (2015) and Wassell (2015).

that would later influence the classical nature of his work. Of the Palace of Versailles he wrote "Works of Filgrand, and little Knacks are in great Vogue: but Building certainly out to have the Attribute of eternal, and therefore the only thing uncapable of new Fashions" (Wren 1750: 261). He would later write in Tract I:

> An architect, out to be jealous of Novelties... and to think his judges those who will live five Centuries after him,...That which is commendable now for Novelty, will not be a new Invention to Posterity... The Glory of that which is good of itself is eternal (Wren 1750: 352).

While in Paris, Wren spent time with Adrien Auzout, a physicist, mathematician and astronomer who, like Wren, had an interest in architecture (Tinniswood 2001: 121–123). Through friends in England, Wren also had a limited encounter with Gian Bernini, who was considered the best living architect and was in Paris, having been summoned by King Louis XIV to complete the east wing of the Louvre (Tinniswood 2001: 126–129). By the time Wren began his post-fire designs, he was conversant in both the philosophy and practice of architectural design in the classical style and had seen many examples of Baroque buildings in Paris.

In 1981, in a tantalizing *Scientific American* article, Harold Dorn and Robert Mark analyzed the structural aspects of Wren's Sheldonian Theater in Oxford and St. Paul's Cathedral. They paid particular attention to Wren's triple dome design for St. Paul's. Wren probably saw double domes during his 1665–1666 visit to Paris; these were built based on a tradition that can be traced back to the thirteenth century (Dorn and Mark 1981: 164). A careful analysis of the dome, piers and buttresses indicates that the dome was over-engineered and that the buttresses (and hence the second level façade that conceals them) are not necessary (Dorn and Mark 1981: 167). Certainly Wren was aware of Leonardo da Vinci's analysis of the cracks in domes presented in *Codex Arundel*, and in particular the cracks in the dome of St. Peter's in Rome, which resulted from what today we would call "hoop stress" (Falter 2015). Wren's clever triple dome design, with the central cone bearing the weight of the lantern for St. Paul's, was an elegant solution to the problems observed in St. Peter's. But there is very little documentary evidence that any of the engineering was based on more than the rules of thumb and intuition used by craftsmen for centuries.

Dorn and Mark do point out an intriguing bit of information. Hooke had noted that a catenary arch acts only in compression and the shape of the conic dome in St. Paul's is approximately a catenary. It is unclear if this was intentional or just coincidental (Dorn and Mark 1981: 173). In Tract II, written near the end of his life, Wren gave a mathematical and mechanical analysis of the arch and of vaulting (Wren 1750: 353–358). So it is clear that he was thinking about architecture in some "mathematical ways" by that time.

When seeking connections between the mathematics and architecture of Wren, many authors point to the strong use of geometry in his buildings. Derek Whiteside says of Wren, "Perhaps his greatest mathematical gift was his visual sensitivity and feeling for form which are obvious in his architectural designs and scientific sketches" (Whiteside 1960: 107). He goes on to say:

> But at a deeper level a sense of precise mathematical form and feeling of order and clarity suffused many of his architectural designs indeed, when it has tended to the monotonous use of the simpler geometric shapes in uncomplicated juxtaposition, it has been seen by some as a fault (Whiteside 1960: 111).

In Tract I Wren is very specific about beauty coming from regular geometric shapes. These ideas can be traced to Vitruvius, Palladio and Alberti. However, Wren "criticized architects who tried to reduce the proportions of the orders to rules 'to strict and pedantick'" (Bennett 1982: 122).

J. A. Bennet in *The Mathematical Science of Christopher Wren* summarizes these ideas by saying that Wren designed his buildings to satisfy the three criteria of Vitruvius, "beauty," "firmness" and "convenience." He used geometric proportions to create "beauty" and symmetry to make buildings "firm." Wren's churches were "convenient" because he used the existing land and foundations (where possible) and because his designs took into account the worship functions conducted in the church (Bennett 1982: 92–93).

Wren used his mathematical mind in a further way to design the post-fire churches in the city of London. Wren resolved over-determined parametric systems in designing these small churches. I do not mean that he wrote down long strings of linear equations and calculated a numerical solution, but rather that he balanced a collection of needs, desires and constraints to find a design for each parish with which he worked. To illustrate this, let us consider the elegant designs for St. Stephen Walbrook and St. James Garlickhythe, both of which provide good evidence supporting the belief that Wren designed these buildings himself (Jeffery 1996).

The Parameters of Rebuilding

Let us begin by considering the general parameters:

- *The need to keep costs down and construct the churches as quickly as possible.* The reconstruction of London put enormous strain on the financial resources of the nation. In 1670 the coal tax was raised to help rebuild the city and construction was begun (Tinniswood 2001: 187). In addition, many parishes were partially funding the reconstruction and had limited resources. Because of the sheer number of buildings that needed to be rebuilt, Wren and Hooke were constantly involved in a juggling act to set priorities. Many congregations were worshipping in hastily repaired parts of their unsound buildings or in temporarily erected structures. In the cold London winters, this was a less-than-ideal solution. To try to speed up the process, the old foundations of the destroyed buildings were used wherever possible in the construction of the post-fire churches. Many of these foundations were for buildings of irregular shape on land-locked pieces of property. In some cases, the post-fire widening of some roads took land away from the churches (for further details see (Jeffery 1996)).

- *The need for a church to have certain architectural features that by tradition have specific theological meaning.* Tradition holds that Christian churches should face east, though Wren says "Nor are we, I think, too nicely to observe East or West in the Position, unless it falls out properly" (Wren 1750: 319). Each of the narthex, nave, aisles, altar and apse have very specific theological meaning and were essential ingredients in the shape of a church. For further details, see Margaret Visser's excellent book *The Geometry of Love*.[3]
- *The new demands put on worship spaces by the Reformation.* The increased emphasis on preaching influenced church architecture. Wren specifically stated that "in our reformed Religion, it should seem vain to make a Parish-church larger, than that all who are present can both hear and see" (Wren 1750: 320). In the same document, he also gives some rough calculations to aid in the placement of the pulpits (Wren 1750: 320–321).

St. Stephen Walbrook and St. James Garlickhythe as Examples

The Vestry minutes of St. Stephen Walbrook (1648–1699, Guildhall Library Manuscript MS 594, vol. 2, 128) state that "Dr. Christopher Wren in consideration of his great care and extraordnary pains taken in the contriving the designe of the Church and assisting in the rebuilding the same" should be given a gift and invited to dinner.

The good people of St. Stephen's had lost part of their property in a street widening and were left with an irregular shaped piece of land. There is good evidence that the pre-fire foundations were used for part of the building (Jeffery 1996: 338). In this church, Wren skillfully uses geometry to impose a centrally-planned church on a longitudinal space, turning a very plain box into the most elegant of his parish churches. Lawrence Weaver says

> Wren contrived to give the effect of nave, aisles and crossing to a plain room by his ingenuity in carrying a circular dome on eight arches which rest on an entablature supported by 12 columns. East of the dome is one groined bay, and west of it two groined bays divided by four more columns: the side aisles having flat ceilings. The plan is thus an oblong room with 16 free columns but so cleverly disposed as to produce the variety of effect described above (Weaver 1923: 88–89).

Among the ideas Wren employs:

- A central plan in a cruciform shape was defined by the use of columns (Fig. 63.1);

[3] For further details, see Margaret Visser's excellent book *The Geometry of Love* Visser (2000).

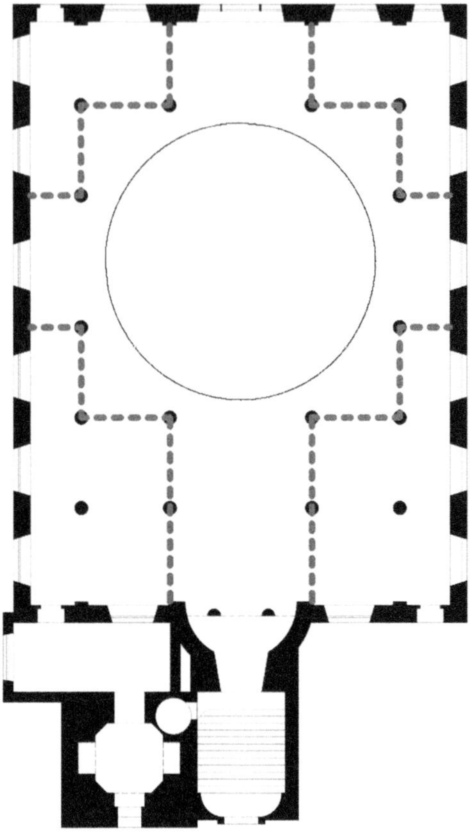

Fig. 63.1 St. Stephen Walbrook, plan. Image: courtesy Dan Schanaker using dimensions in Jeffery (1996)

- A nave and aisles that were created using those same columns to support an entablature that formed an unusually-shaped clerestory resulting in a lowered ceiling over the aisles (Fig. 63.2);
- The dome is supported on an octagon sitting inside of a square rather than a circle. This particular configuration allows the aisles to be narrower than the nave. This is an elegantly simple "mathematician's solution" to the problem (Fig. 63.3);
- Groin vaulting and tunnel vaulting are employed in various locations so that the geometric shapes fit together like pieces of a graceful puzzle.

St. James Garlickhythe is known as "Wren's Lantern" because of the very large number of windows. In this building Wren also created a cruciform shape within a rectangular box. The nave is the tallest of Wren's parish churches with a lovely clerestory and the four-column barrel vault.

Among the ideas Wren employs in this church:

- A cruciform shape was defined by a clever use of the columns supporting the clearstory and an entablature (Figs. 63.4 and 63.5);

Fig. 63.2 St. Stephen Walbrook. Photo: author

- Barrel vaulting is used to give the crossing the feel of a "central plan" church. This was much more evident in the church in the past. Over time some of the doors and windows have been bricked closed (Fig. 63.6).

The one discordant feature is that the church has a separate chancel, a rarity among city churches and somewhat in conflict with a church that has the feel of a central plan. It is believed this was retrained at the insistence of the parish vestry (Jeffery 1996: 248). Vestry minutes from the period (*Vestry Minutes* of St. James Garlickhythe, 1615–1693, Guildhall Library Manuscript MS 4813, vol. 1, overleaf of 161) indicate that the parish vestry took a very active role in the rebuilding, including paying two of Wren's clerks to hasten completion of the building before a planned dedication ceremony.

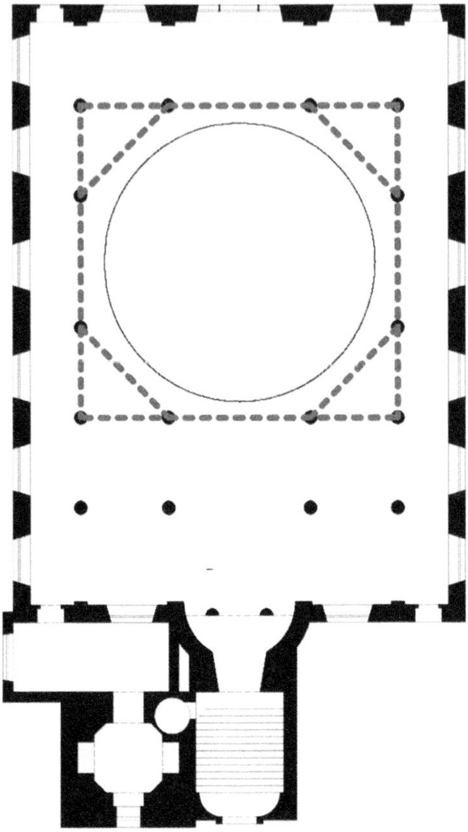

Fig. 63.3 St. Stephen Walbrook. Image: courtesy Dan Schanaker using dimensions in Jeffery (1996)

Conclusion

Aldoux Huxley says of Wren:

> a great gentleman: one who valued dignity and restraint... one who desired that men and women should live with dignity, even grandeur, befitting their proud human title; one who despised meanness and oddity as much as vulgar ostentation; one who admired reason and order, who distrusted all extravagance or excess (Tinniswood 2001: 377).

Wren's clear love of order and his belief that lasting beauty is derived from regular geometric shapes is evident in his architectural work and philosophical writings about architecture. There are no clear and documented links between his work in pure mathematics or astronomy and his building designs. However it is reasonable to say that he used his mathematically-trained mind to find elegant solutions to many of the complex design problems that he encountered.

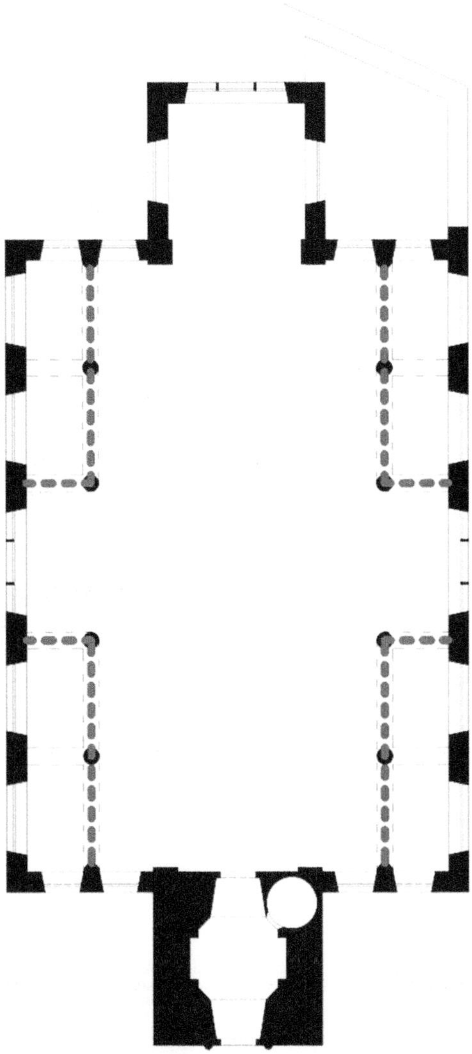

Fig. 63.4 St. James Garlickhythe. Image: courtesy Dan Schanaker using dimensions in Jeffery (1996)

Biography Maria Zack received her BA (1984) and PhD (1989) in Mathematics from the University of California at San Diego. She has held posts at Texas Tech University, The Center for Communications Research and Point Loma Nazarene University, where she is currently a Professor as well as the Chair of the Department of Mathematical, Information and Computer Sciences. Her research interests include the history of mathematics in seventeenth- and eighteenth-century England.

Fig. 63.5 St. James Garlickhythe, interior. Photo: author

Fig. 63.6 St. James Garlickhythe, interior. Photo: author

References

Bennett, J. A. 1982. *The Mathematical Science of Christopher Wren.* Cambridge: Cambridge University Press.

Dorn, H. & Mark, R. 1981. The Architecture of Christopher Wren. *Scientific American*, **245**, 1: pp. 160-173. London: Nature Publishing Group.

Falter, Holger. 2015. The Influence of Mathematics on the Development of Structural Form. Pp 81–93. in Kim Williams and Michael J. Ostwald eds. *Architecture and Mathematics from Antiquity to the Future: Volume I Antiquity to the 1500s*. Cham: Springer International Publishing.

Huxley, G. L. 1960. The Geometrical Work of Christopher Wren. *Scripta Mathematica*, **XXV**: pp. 201-208. New York: Yeshiva University.

Jardine, L. 2002. *On a Grander Scale: The Outstanding Life of Sir Christopher Wren.* New York: Harper Collins.

Jeffery, P. 1996. *The City Churches of Sir Christopher Wren.* London: Hambeldon Press.

Newton, I. 1999. *The Principia.* I. B. Cohen and A. Whitman, trans. Los Angeles: University of California Press.

Summerson, J. 1960. Sir Christopher Wren P.R.S. (1632-1723). *Notes and Records of the Royal Society*, **XV**: pp. 99-105. London: The Royal Society.

Summerson, J. 1965. *Sir Christopher Wren.* London: Collins.

Tinniswood, A. 2001. *His Invention So Fertile.* London: Jonathan Cape.

Visser, M. 2000. *The Geometry of Love.* New York: North Point Press.

Wassell, Stephen R. 2015. The Mathematics of Palladio's Villas. Pp 107–120. in Kim Williams and Michael J. Ostwald eds. *Architecture and Mathematics from Antiquity to the Future: Volume I Antiquity to the 1500s*. Cham: Springer International Publishing.

Weaver, L. 1923. *Sir Christopher Wren, Scientist, Scholar and Architect.* London: Country Life Ltd.

Whiteside, D. 1960. Wren the Mathematician. *Notes and Records of the Royal Society*, **XV**: pp. 107-111. London: The Royal Society.

Wren, C. 1750. *Parentalia: or Memoirs of the Family of the Wrens.* Facsimile ed. Farnborough: Gregg Press Limited, 1965.

Chapter 64
Robert Hooke's Fire Monument: Architecture as a Scientific Instrument

Maria Zack

Introduction

After the Great London Fire of 1666, Robert Hooke was appointed to work in the office of the City Surveyor of London. With that appointment, a scientist best known as the Curator of Experiments for the Royal Society, whose research encompassed both the microscopic (*Micrographia*) and the astronomical, embarked on a second career as an architect and surveyor. For the next several decades the massive effort to reconstruct London was led by Hooke and his long-time friend, fellow scientist and co-founder of the Royal Society Christopher Wren.

Hooke was involved extensively in all aspects of the rebuilding of London, both the mundane (widening streets and establishing property boundaries) and the creative (designing churches and civic buildings). Very little of Hooke's architectural work has survived the passage of time. However, one shining example of his creativity remains in London: the Monument to the Great Fire.

At the time of the monument's design, Hooke was conducting experiments on both the motion of the earth and the effects of gravity. The monument is an elegant column that was constructed to contain a zenith telescope to further Hooke's research. This ingenious building is an excellent example of the intersection between Hooke's architectural and scientific work.

First published as: Maria Zack, "Robert Hooke's Fire Monument: Architecture as a Scientific Instrument". Pp. 117–126 in *Nexus VII: Architecture and Mathematics*, Kim Williams, ed. Turin: Kim Williams Books, 2008.

M. Zack (✉)
Department of Mathematical, Information and Computer Sciences, Point Loma Nazarene University, 3900 Lomaland Drive, San Diego, CA 92106, USA
e-mail: MariaZack@pointloma.edu

K. Williams and M.J. Ostwald (eds.), *Architecture and Mathematics from Antiquity to the Future*, DOI 10.1007/978-3-319-00143-2_17,

Hooke the Scientist

Robert Hooke was born in Freshwater on the Isle of Wight in 1635. Hooke was the youngest of four children of John Hooke, an Anglican priest with deeply Royalist leanings. In 1648, Hooke's father died "by suspending himself" (Aubrey 1957: 165) leaving him "forty pounds of lawful English money, the great and best joined chest, and all my books" (Cooper 2003: 12). John Aubrey, the diarist and friend of Robert Hooke tells us:

> When his father died, his Son Robert was but 13 years old, to whom he left one Hundred pounds, which was sent to London with him with the intention to have bound him Apprentice to Mr. Lilly the Paynter, with whom he was a little while upon tryall; who liked him very well, but Mr. Hooke quickly perceived what was to be done, so, thought he, why cannot I do this by myself and keep my hundred pounds? (Aubrey 1957: 164).

By 1649 Hooke had left Mr. Lilly (Lely) and was a student of Dr. Richard Busby at the Westminster School. At Westminster, Hooke's talent for mathematics and mechanical devices became apparent.

> H[ooke] fell seriously upon the study of Mathematicks, the Dr. encouraging him therein, and allowing him particular times for that purpose. In this he took the most regular Method, and first made himself Master of Euclide's Elements, and thence proceeded orderly from that sure Basis to the other part of the Mathematicks, and thereafter to the application thereof to Mechanicks, his first and last Mistress (Hooke 1705: iii).

Through connections at the Westminster School, Hooke entered Oxford in 1653 as a "servitor" student with a choral scholarship and was awarded a Master of Arts degree in 1662 or 1663 (Cooper 2003: 19). At Oxford Hooke developed the scientific relationships that would define the rest of his professional life. Hooke and Christopher Wren began a life-long friendship while they were both students and in 1656 Hooke became Robert Boyles's experimental assistant. By the early 1660s Hooke and many other virtuosi had moved to London, and this active group of scientists formed the Royal Society. In 1662 Boyle nominated Hooke for the position of Curator of Experiments for the Royal Society. Hooke's duty was "to furnish them every day, on which they met, with three or four considerable experiments" (Cooper 2003: 28). Hooke became a Fellow of the Royal Society in 1663 and would take an active role in the affairs of the Society until his death in 1703.

The Royal Society saw itself as a "Baconian" scientific institution. The members knew Francis Bacon's writings intimately and reflected them in their own philosophy and work.

> [I]t was Bacon's general statements about the aims and methods of modern science that early English experimentalists based themselves on. In Bacon's view the business of modern science was to amass a great corpus of precise information based on experiment and observation, to generalize from this by the process of induction, and to do all for useful ends (Espinasse 1956: 19).

In his role as the Curator of Experiments, Hooke himself performed many experiments ranging from the microscopic to the astronomical. He used his mechanical genius to design a wide variety of instruments and to manufacture them in partnership with many of London's most skilled craftsmen (Cooper 2003: 45). Hooke believed that instruments were an essential part of experimental science. He said:

> It is the great prerogative of Mankind above other Creatures, that we are not only able to behold the works of Nature, or barely to sustain our lives by them, but we have also the power of considering, comparing, altering, assisting, and improving them to various uses. And as this is the peculiar privilege of human Nature in general, so is it capable of being so far advanced by the helps of Art, and Experience, as to make some Men excel others in their Observations, and Deductions, almost as much as they do Beasts. By the addition of such artificial Instruments and methods, there may be, in some manner, a reparation made for the mischiefs, and imperfection, mankind has drawn upon itself (Hooke 1665: Preface).

From the very beginning Hooke also made use of buildings as instruments. As early as 1662 he conducted experiments on gravitational attraction by dropping items from the top of Westminster Abbey (Cooper 2003: 46) and he conducted additional experiments from the top of the pre-fire St. Paul's Cathedral, leading to his 1666 paper *On Gravity* (Cooper 2003: 54–65). Certainly the monument provided another great height for Hooke's gravitational experiments, but its unique role was as a telescope. The telescope was designed to gather data to measure the parallax and thus resolve one of the great scientific questions of the time, the motion of the earth.

The Parallax

Hold a finger at arm's length and look at a distant object beyond your finger. Close each eye in turn. The position of your finger relative to the distant object appears to change because of the alteration in your viewing angle. This apparent shift is known as a parallax. This simple phenomenon was at the heart of 2,000 years of debate about the nature of our universe.

Aristarcus (c. 310–230 B.C.) calculated that the sun was significantly larger than the earth, and thus much more likely to be the centre of the solar system. Aristarcus believed that the earth revolved around the sun and rotated on an axis.

> [H]e realized that his model gave him a method for measuring the distance to the stars because the motion of the Earth from one point in its orbit to the extreme opposite point would cause the stars to show a parallax, that is, they would appear slightly shifted in the sky (Wilson 1997: 32).

However, because his instruments were too crude for the distances involved, Aristarcus was unable to detect a parallax. This lead Aristarcus not to abandon his theory, but rather to conclude that the universe was very large (Wilson 1997: 33).

Ptolemy's (c. 100–170 A.D.) elegant *Almagest* was used for roughly fourteenth centuries to calculate with a high degree of accuracy the motion of the sun, moon, planets and stars based on a geocentric model of the universe. In *De Revolutionibus Orbiun Coelestium* (1543) Copernicus proposed a heliocentric universe with the sun in the centre of a fixed sphere of stars and the planets rotating around the sun. Copernicus launched significant debate in the scientific community.

> There were no astronomical observations that specifically favoured the Copernican system over that of Aristotle and Ptolemy. The heliocentric model made the specific prediction of stellar parallax—the apparent wobble in the position of the star as the earth moved from one side of its orbit to the other side—whereas the geocentric model predicted none, and, indeed none had been detected. . ..[A]t the time, the only argument in favour of the heliocentric system was an aesthetic one—it had great simplicity and form—yet this was sufficient to convince the most important scientific minds who were to follow Copernicus in the Renaissance period (Wilson 1997: 54)

Those who gazed at the sky looking for signs of a parallax included Tycho Brahe (1546–1601), Galileo Galilei (1564–1642), John Flamsteed (1646–1719) and Robert Hooke. They knew that if they could find a stellar parallax this would prove that the earth moves through space.

Zenith Telescopes and the Motion of the Earth

Tycho Brahe held to a "modified Copernican" cosmology. He believed that the planets orbited the sun but still maintained that the Earth was stationary and fixed at the centre of the universe (Wilson 1997: 62). Galileo agreed with Copernicus and believed that the Earth joined the planets in revolving around the sun. In *Dialogue Concerning the Two Chief World Systems* (1632), Galileo suggested using two stars, a near star and a background star, as a way to compute the parallax (Hoskin 1997: 210) and thus prove that the heliocentric theory is correct.

In 1669, Hooke proposed a series of experiments to attempt to measure stellar parallax. His stated purpose was "to furnish the Learned with an experimentum crucis to determine between the Tychonick and Copernicuan Hypotheses" (Hooke 1674: 2). He chose the star Gamma Draconis primarily because it is bright and it daily passes directly overhead (near the zenith point) in London (Hooke 1674: 13).

Picking a star that passes near the zenith simplifies the experiment because gravity defines the zenith exactly, so the telescope could be aligned simply by using a plumb bob. Hooke said "by this way of observing I avoid all the difficulties that attend to the making, mounting and managing of great Instruments." In addition, because the star's light passes perpendicularly through the Earth's atmosphere, calculations did not need to be adjusted to account for refraction (Hooke 1674: 15).

Hooke described his experiments in *An Attempt to Prove the Motion of the Earth from Observations*. In this paper he gave details of the construction, installation and alignment of his 36 foot telescope (Hooke 1674: 17–23). To accommodate it, he needed to cut a hole through the floor and ceiling of his lodgings at Gresham College (Fig. 64.1).

In June of 1669, Hooke began his experiment; however, he made only four observations (July 6, July 9, August 6 and October 21, 1669) before declaring:

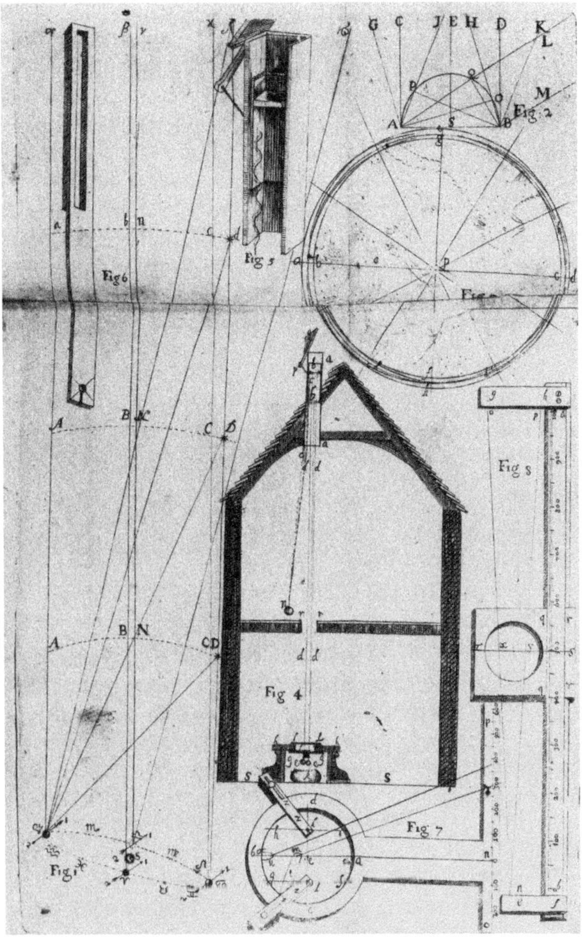

Fig. 64.1 Hooke's drawing of his zenith telescope. Image: Hooke (1674: Table III)

> Tis manifest then by observations ... that there is a sensible parallax of the Earths Orb to the fixt Star in the head of Draco, and consequently a confirmation of the Copernican System against the Ptolomaick and Tichonick (Hooke 1674: 25).

Hooke was however conscious of the possibility of experimental error particularly in setting the plumb bob and keeping the telescope in position (Hooke 1674: 24). By the 1660s astronomers realized that the distance between the two extremes of the earth's orbit around the sun was relatively small compared to the size of the universe, thus slight instrumental inaccuracies could invalidate an observation. Hooke's 1669 parallax angle of 27 arc sec seemed unexpectedly large (Chapman 2005: 93). Hooke believed that there was more to be learned using a longer telescope. In his paper he considered the benefits to be gained from building a 144 ft telescope and described the possibility of putting such an instrument in a well to provide greater stability (Hooke 1674: 22). Hooke knew that the longer the focal length of a telescope, the larger the image and hence the greater the ability to see a difference in the parallax angle. The monument was designed to provide his desired long (nearly 200 ft) focal length.

The Monument

On October 4, 1666, just a few weeks after the Great Fire of London, the City of London appointed Robert Hooke to the Rebuilding Commission and King Charles II appointed Christopher Wren to the same Commission (Jardine 2004: 144). There were four others appointed to the Commission, but less than 6 weeks after the fire Henry Oldenberg (the Secretary of the Royal Society) wrote to Robert Boyle stating that the rebuilding of London "is to be forthwith taken in hand, and that by the care and management of Dr Wren and Mr Hooke" (Jardine 2004: 147). For the next 37 years, Hooke and Wren's scientific collaboration was expanded to include architecture. The collaboration ended only with the death of Hooke in 1703.

The 1670 City Churches Rebuilding Act provided funds for a monument to "preserve the memory of this dreadful visitation" (Jardine 2002: 316). Not all of the destroyed parish churches were to be rebuilt. The parish of St. Margaret's Church on Old Fish Street was merged with St. Magnus the Martyr and the new church for the joint parish was constructed on the previous foundations of St. Magnus. Because of its proximity to Pudding Lane where the fire began, the location of the destroyed St. Margaret's was deemed ideal for the construction of the memorial pillar and a surrounding square (Cooper 2003: 198–199). The memorial was intended to be viewed from a great distance (Fig. 64.2) and that was the case until the relatively modern construction of high rise building around the square (Fig. 64.3).

Hooke's diaries from the 1670s show Hooke and Wren meeting almost daily for both professional and social reasons (Batten 1937: 83). Certainly Hooke and Wren collaborated on the Monument to the Great Fire, but it is now generally accepted that Hooke was the designer of the Monument. There is a single drawing of the Monument by Wren (at All Souls in Oxford), but it is not the design that was built. Hooke's drawings for the Monument and for the urn at the top are the ones that were executed. These drawings are in Hooke's hand and are part of a collection of "Dr Hooke's drawings" housed in the British Museum/Library. The confusion over attribution is most likely rooted in the fact that these drawings, along with several other designs of Hooke's, were published by the Wren Society in vol. V of the twenty-volume collected works of Wren (Robinson 1948: 51–52).

On January 26, 1671 the Court of Alderman considered "the draught now produced by Mr Hooke one of the Surveyors of the new buildings of the Pillar to be erected in memory of the Late dismall Fire." Approximately 2 weeks later the design was approved, the drawing signed by Wren and construction authorized to begin (Cooper 2003: 200).

By 1673, the Monument was under construction. Hooke's diaries indicate that he was involved in each step of the construction process. On October 19, 1673 he

Fig. 64.2 The Monument To The Great Fire as depicted by engraver Sutton Nicholls, ca. 1750

wrote "perfected module of Piller"; on June 1, 1674 "At the pillar on Fish Street Hill. It was above ground 210 steps"; on August 7, 1674 "At the Pillar in height 250 steps"; on September 21, 1675 "at fish-street-hill on ye top of ye column"; and on April 11, 1676 he was with Wren "at the top of ye Pillar" (Batten 1937: 84). Lisa

Fig. 64.3 Contemporary view of the Monument to the Great Fire. Photo: author

Jardine speculates that Hooke used the same attention to detail in the construction of the Monument that he employed when designing other scientific instruments:

> From the precision of the elements in the column as built (the accuracy of the height of each individual stair-riser, the breadth of the circular apertures) it appears that Hooke took very particular care with the construction of this single, vertical shaft, exercising close control over its execution, which increased the period of completion significantly (Jardine 2002: 317–318).

With a bit of political manoeuvring, Wren ensured that the Monument would be a scientific instrument. The City Lands Committee had thought it appropriate to place a statue of the King at the top of the column, which would have made it difficult to see straight though the column to the sky and would have ruled out its use as a zenith telescope (Cooper 2003: 202). In July of 1675 Wren sent the Committee a letter with some proposals for what might be placed at the top of the column. He offers several suggestions: a gilt ball, a statue, a copper ball with flames of gilt or a phoenix. He rules out the phoenix as being dangerous and emphasizes the usefulness of either of the spherical options, claiming that they would "give Ornament to the Town at a very great distance" and because "one may goe up into it; & upon occasion use it for fireworks" (Jardine 2002: 316–317).

Fig. 64.4 The sphere and urn at the top of the Monument to the Great Fire. Photo: author

Though Wren carefully keeps the option of a statue as a possibility, he discourages it because of the great expense. Fortunately Wren was persuasive and the Monument's use as a telescope was preserved.

The Monument was completed in 1677 and its use as a scientific instrument began. It has an underground chamber set in a bed of gravel, which was the location of the eyepiece of the telescope. The objective lens of the telescope was mounted 200 ft above, near the top of the pillar inside the ball but below the hinged doors to the flaming urn (Fig. 64.4). The accuracy of the observations made by this zenith telescope depended on maintaining the alignment of the eyepiece and objective lens. Unfortunately, the vibrations caused by air currents traveling down the core of the column and from the wheeled traffic passing by the pillar caused a misalignment in the lenses that was greater than the changes in parallax that Hooke was trying to measure (Cooper 2003: 201). In *An Attempt to Prove the Motion of the Earth from Observations* Hooke discussed alignment difficulties with his 36 foot telescope:

> I was forced to adjust the Instrument at every observation I made, both before and after it was made, which hath often made me wish that I were near some great and solid Tower, or some great Rock or deep well, that so I might fix all things at once, and not be troubled continually to adjust the parts of the said Instrument (Hooke 1674: 22).

Though the Monument was built on a foundation of gravel to help provide stability, his "solid Tower" was not solid enough. Perhaps he would have had better success with a well, though architect Michael Cooper says "as so often was the case with Hooke's ingenious instruments … the methods and materials available to him prevented him from making instruments accurate enough to do what he wanted" (Cooper 2003: 201). It would be another 165 years before technology advanced sufficiently to measure the parallax. In 1838 Friedrich Bessel computed the parallax for 61 Cygni, whose angle of change is much greater than that of Hooke's Gamma Draconis (Wilson 1997: 101). Once again, Hooke was ahead of his time.

Hooke did conduct some experiments on barometric pressure at the monument. On May 16, 1678 his diary says "At Fish Street pillar tried mercury barometer experiment. It descended at the top about 1/3 of an inch." The proceedings of the Royal Society for May 30, 1678, contain a report from Hooke about these barometric experiments. He also continued some of his gravitational experiments with pendulums at the Monument (Jardine 2001: 300–301).

Though the Monument was not as successful a scientific instrument as had been hoped, it has remained an enduring memorial to the Great Fire of 1666 as well as a symbol of Hooke and Wren's enduring partnership in both architecture and experimental science.

Biography Maria Zack received her BA (1984) and PhD (1989) in Mathematics from the University of California at San Diego. She has held posts at Texas Tech University The Center for Communications Research and Point Loma Nazarene University where she is currently a Professor as well as the Chair of the Department of Mathematical Information and Computer Sciences. Her research interests include the history of mathematics in seventeenth- and eighteenth-century England.

References

Aubrey, John, 1957. *Aubrey's Brief Lives*. Ann Arbor: University of Michigan Press.

Batten, M.I. 1937. The Architecture of Dr. Robert Hooke F.R.S. *Walpole Society* **25** (1936–37). London: Walpole Society.

Chapman, Allan. 2005. *England's Leonardo: Robert Hooke and the Seventeenth-Century Scientific Revolution*. Bristol: Institute of Physics Publishing.

Cooper, Michael. 2003. *A More Beautiful City: Robert Hooke and the Rebuilding of London After the Great Fire*. Stroud: Sutton Publishing.

Espinasse, Margaret. 1956. *Robert Hooke*. London: William Heineman Ltd.

Hooke, Robert. 1665. *Micrographia*. Rpt. 1961, New York: Dover Publications.

———. 1674. *An Attempt to Prove the Motion of the Earth from Observations*. London: Royal Society.

———. 1705. *Posthumous Works*. Richard Waller, ed. Rpt. 1968, New York: Johnson Reprint Corporation.

Hoskin, Michael. 1997. *The Cambridge Illustrated History of Astronomy*. Cambridge: Cambridge University Press.

Jardine, Lisa. 2001. Monuments and Microscopes: Scientific Thinking on a Grand Scale in the Early Royal Society. *Notes and Records of the Royal Society* **55**, 2 (May 2001): 289–308.

———. 2002. *On a Grander Scale: The Outstanding Life of Sir Christopher Wren*. New York: Harper Collins.

———. 2004. *The Curious Life of Robert Hooke: The Man who Measured London*. London: Harper Perennial.

Robinson, H.W. 1948. Robert Hooke as Surveyor and Architect. *Notes and Records of the Royal Society* **6**, 1 (Dec 1948): 48–55.

Wilson, Robert. 1997. *Astronomy through the Ages: The Story of the Human Attempt to Understand the Universe*. Princeton: Princeton University Press.

Chapter 65
Practical and Theoretical Applications of Geometry at Claude Perrault's Observatoire de Paris (1667–1672)

Randy S. Swanson

Introduction

Dr. Claude Perrault (1613–1688) was, among many things, a Vitruvian scholar who developed unsettling questions and theories concerning the relation between the perception of beauty and the proportions of form. He is credited with producing the striking design of the double column colonnade for the East Façade of the Louvre (1667–74), and yet he argued that architectural beauty did not rely upon the fixed or idealized proportions that were then thought to be embodied by nature, music, or ancient works of architecture. Perrault recognized the need for classical ornament, but unlike the architects of his era, he argued that the minor adjustments of proportions rampant throughout built works and architectural treatises made no discernible difference to the public. He proposed—to the outrage of many—that the proportions for architecture could be simplified and standardized without loss since ornamental forms were transitory and matters of fashion or taste. Lasting beauty, he proposed, was to be found in

> ...richness of materials, the size and magnificence of the building, the precision and cleanness of execution, and symmetry, which in French signifies the kind of proportion that produces an unmistakable and striking beauty (Perrault 1993: 50–51).[1]

First published as: Randy S. Swanson, "Geometry in Perrault's Observatoire", pp. 237–251 in *Nexus IV: Architecture and Mathematics*, Kim Williams and Jose Francisco Rodrigues, eds. Fucecchio (Florence): Kim Williams Books, 2002.

[1] The conventional notion of building proportion was rooted in the analogy of the human form with the cosmos where the relation of parts to the whole provided the highest expression of the act of creation and remained at the core of architectural efforts. In this approach, beauty resided principally in proportion. Perrault argued otherwise believing that the adjustment made by most architects were to small to be appreciated. He had less faith in the ability of the public to perceive the subtleties that most architects claimed were required for the perfection of beauty.

R.S. Swanson (✉)
Randy Swanson Architect, 300 2nd Ave. S.E. #39, St. Petersburg, FL 33701, USA
e-mail: rsswanson@earthlink.net

K. Williams and M.J. Ostwald (eds.), *Architecture and Mathematics from Antiquity to the Future*, DOI 10.1007/978-3-319-00143-2_18,

Fig. 65.1 Claude Perrault, Observatoire de Paris, 1667–1672. Photo: author

This position was first let loose upon the Academy of Architecture early in 1672 and was fully developed for the publication of his treatise, *Ordonnance For The Five Kinds Of Columns After The Method Of The Ancients* in 1683. In his day, his position was regarded as a direct contradiction to his own practice and the canons of Classicism. This view held that proportions were indivisible with beautiful forms; that proportions with all their complexities were the terrestrial embodiment of cosmic harmonies and were the basis through which an ideal beauty was sought. Many of Perrault's personal theories created in support of his position were also dismissed in that era, but in the last quarter of the twentieth century have been viewed as pivotal in the development of architectural theory.

Claude Perrault is less well remembered as the architect of the Observatoire de Paris. Its external appearance, simple and severe without classical ornament, contrasts thoroughly with the Louvre (Fig. 65.1).

The project is little known in standard architectural histories despite its importance as Perrault's only freestanding building design and as the first building contributing to the institutionalization of science.[2] The time frame for design and construction coincide directly with the formation of his personal ideas and his translation of Vitruvius's *Ten Books of Architecture* (1667–1673, first edition). When facing the prospect of forced changes to his plan after the building construction had begun (1669), it is reported that Perrault strongly resisted and appealed to Louis XIV, stating that the design was already perfect. Some members of the Academy of Sciences even suggested that in his resistance,

[2] The Observatoire has been touched upon in Herrmann (1973). The best scholarly treatment of this project is Petzet (1967). Since that publication, the Observatory has been reviewed in (Picon 1987: 197–219). Picon's work reinforces Petzet's scholarship.

he had placed the building form above its ability to serve the needs of science and the state.[3]

What was so important to Perrault about the form of the Observatoire? What can we learn from it concerning his position on beauty and proportion? This chapter will take a preliminary step in addressing these questions by examining several prominent features of the Observatoire design. The stairwell form and the general building form will be examined for their constructional, proportional and formal attributes. The results are the product of both archival and on-site field research. A brief overview of the original facility programme, the building siting and general construction will first be provided.

Background

The Observatoire was originally conceived to house the Academy of Sciences with the consent of Louis XIV, and was sanctioned by Jean Baptiste Colbert (1619–1683) after his first year in office as *Intendant* of France (controller general). The Academy had been founded in 1666; Colbert immediately proceeded to have Claude Perrault design the facility (1666–1669); and construction was undertaken (1669–1672). It was Colbert's intention that the Academy was to be comprised of the most notable scientists and philosophers of Europe whose efforts were to be directed in part, to the resolution of the problems of state. The facility was to provide a meeting space for the members of the academy; be the center for astronomical observations; provide laboratories for "chemistry"; provide space for the display of new inventions, mechanical models and machines; provide laboratories for anatomical dissections; as well as house the royal collections of natural history objects (Cassini 1810; see also Hahn 1971: 18–60). Those familiar with the work of Frances Bacon will no doubt recognize a resemblance to the centralized scientific endeavor described in *New Atlantis* (1626).

This endeavor represented a turning point for the development of astronomy and geography, and was appreciated as such from the outset. The new facility was expected to provide a permanent platform to insure the consistent collection of astronomical data over the next several centuries. It was intended to be a benchmark from which the most exacting measurements were to be made—of Paris, of France, of the earth, out to the planets, and ultimately beyond. A committee of the Academy was formed to select the site. They chose the highest point south of Paris to obtain the best views of the night sky. On 21 June 1667 (date of the summer solstice), the astronomical and mathematical members of the Academy ceremoniously

[3] The brevity of this chapter will not permit the relation this interesting incident. However, varying descriptions can be found in the following works: (Wolf 1902: 19–37; George 1938; Perrault 1909a: 42–50, b: 219–221).

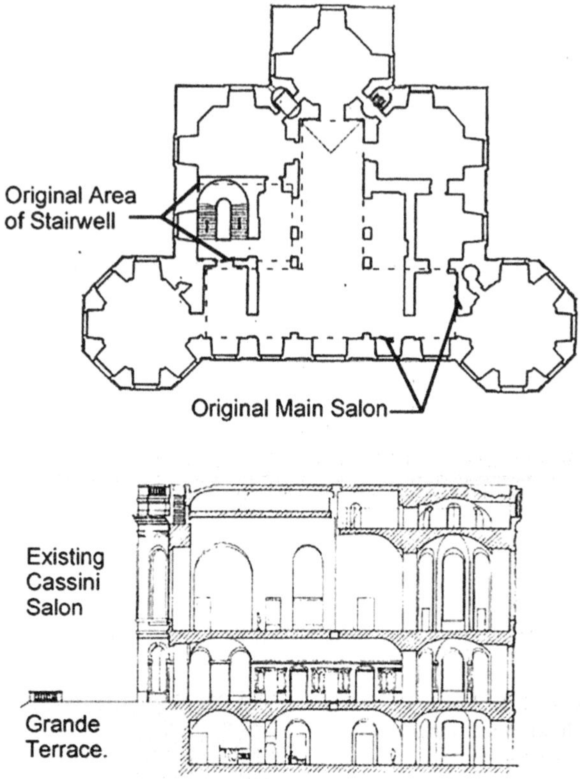

Fig. 65.2 (*above*) Main floor plan with Cassini Salon. Existing conditions: double vaults with piers, 1786 renovation; (*below*) South–north section. Base drawings (1987) provided by permission of M. Herve Baptiste, Architect en chef des Monuments Historiques, Paris

established a true north orientation at the site. This act established the prime meridian of France and the principal point of reference from that date forward for their geographic and astronomical measurements. The meridian was adopted by Perrault as the principal axis of the design and was signified within the building by an inlaid bronze floor strip on the main (uppermost) level.

Perrault proposed a generous three-story building placed into the northern face of a small hillside, which today lies just to the south of the heart of Paris (Fig. 65.2).

The stone is from the quarry of Montemarte, and the quality of finish and jointing throughout is superior. Several construction strategies are revealed from a careful examination of the construction and period figures. The walling indicates the construction proceeded course by course, layered evenly throughout.[4] An interlocking pattern of voussoir construction can be observed at all apertures, as demonstrated at the exterior window arch of the first floor salon, south facade (Fig. 65.3).

[4] A detail of "Louis XIV being shown round the Academy of Science by Colbert" (1671), from the frontispiece of Perrault's *Histoire des Animaux* prepared by Sebastien Leclerc, also suggests the construction proceeded layer by layer, evenly throughout construction.

Fig. 65.3 The interlocking pattern of voissoir construction in the exterior window arch of the first floor salon, south façade. Photo: author

This approach offered maximum stability for the arch and wall (Fitchen 1961; Heyman 1996). The method was more expensive than others since more surface area had to be cut, but doing so prevented the slippage of stones over time. The transition of the arched window opening to the interior vault is handled as an intersection between two cylindrical surfaces, resulting in a visibly skewed ellipse. The masonry jointing in the soffit of the window vault demonstrates that the jointing of exterior and the interior stonework flow directly from one to the other in a smooth pattern that can be found in most interior vaults (Fig. 65.4).

An examination of the earliest detailed sectional drawing (dating from ca. 1692) reflects the heavy masonry construction of sufficient wall thickness to restrain the vaults. What is not reflected is that each room varies modestly so that it's likely that each vault would have been laid out directly on the stone floor of each space to find the final dimensions. The result is a remarkably harmonious effect of construction and a fluidity of space. The fundamental construction strategies of simple stack construction, interlocking prisms, working in situ, and the continuity of construction geometry from interior to exterior constitute the basis of the observatory construction.

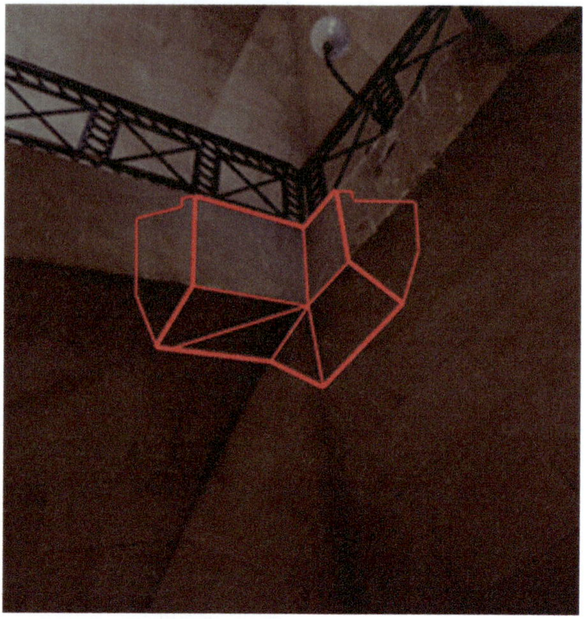

Fig. 65.4 The interlocking pattern of voissoir construction in the exterior window arch merging with the second floor cantilevered stair vault. Landing ascent block and window keystone outlined in *red*. Photo: author

Practical Application of Geometry

Connecting all levels of the design is a semi-helicoidal stone stair hall. The stair hall is a marvelous and arresting form (Fig. 65.5).

Immediately it prompts the questions, "How was this done?" and "Could Claude Perrault have been responsible?" Up to this point there is no reason to suspect that anyone other than a master mason might have been relied upon by Perrault in the production of this building. The stone vault prisms with complex warped surfaces present an apparently impossible geometrical task of dimensioning, cutting and shaping. A careful inspection of the stair demonstrates a structural and constructive logic that would be prized in any era.

Dimensional data gathered in the field determined that the stair vault is elliptical in section (Fig. 65.6).

The height and slope of the stairwell vaults made the act of taking consistent dimensions physically improbable. The use of an acoustic range finder was ineffective (June 1997) due to all surfaces being so highly curved and reflective. A Pulse Laser Range finder (PLRF) was found to provide the best solution for taking accurate dimensions in this environment. This type of instrument measures a distance to any surface that the laser strikes up to a 1,000 ft with exceptional accuracy and without the use of a reflector. It also permits measurements to be taken by a single individual.

All readings in the stairwell were taken on 9 June 1999 by the author. The instrument, a tripod mounted PLRF, was calibrated on site prior to beginning. The

Fig. 65.5 The semi-helicoidal stair hall of the Observatoire. Photo: author

check was made against a tape measure reading. The tape reading was 15′-2½″±. The PLRF reading was 15.22′, and was the average of three readings. 15′-2½″ = 15.21′, leaving a difference between tape and instrument readings of 0.01′ or roughly 3 mm. Readings for each angle of inclination were provided by the instrument in degrees. Ray length is from the instrument centerline to each joint between voussoirs, starting from the top of the stair stringer to the vault spring line joint. Each ray length dimension is the average of three separate readings. All readings were rounded off to the nearest hundredth of a foot or degree.

Readings began on the main floor (upper level) starting with the vault above the landing. The tripod was placed to permit the instrument the best opportunity to take readings from all the construction joints within the vault. When the readings of the vault above the landing were finished the instrument was placed at the center of the landing and turned to face the rear of the stairwell and readings of the vault in that location were taken. When completed, the tripod was moved down to the first floor level and the process repeated (although these dimensions are not shown here) (Tables 65.1 and 65.2).

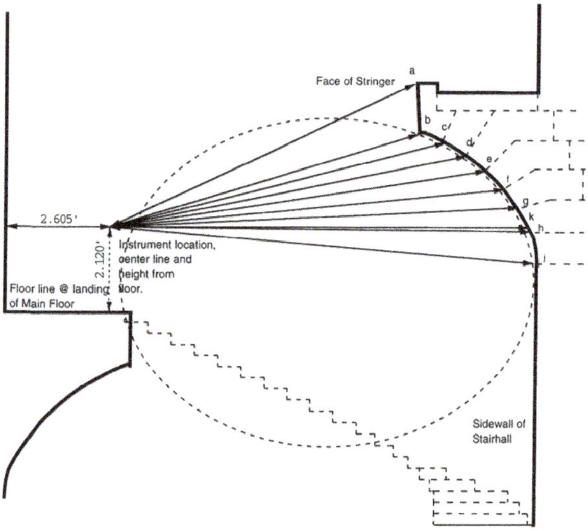

Fig. 65.6 (**a**, *Above*) Rear wall of semi-helicoidal stone stair hall and Pulse Laser Rangefinder; (**b**, *Below*) Vault dimensions, rear stairwell, main floor. Photo and drawing: author

No readings were taken of the ground floor vault. The arc of the vault creates an ellipse in section, shown as a dotted line that is consistent from landing to landing. The dimension of the ellipse falls outside the width of the stair hall landing, so that it seems doubtful that this form was a direct result of the medieval method of constructing an ellipse. It is more likely that the ellipse was generated by placing the arcs of two circles, a full stair hall width diameter and one-half diameter, into simple relation with one another. Shown as solid lines, these circles also equal the length of the stairwell dimension in plan when combined, as can be seen in Fig. 65.7. This approach is much simpler and allows the form to be resolved in situ.

Table 65.1 Rear stairwell vault dimensions, main floor (see Fig. 65.6a, b)

	Top of stringer								Spring line	Horiz. distance
Point	a	b	c	d	e	f	g	h	j	k
Avg. ray length	17.00′	16.42′	17.55′	18.48′	19.30′	20.15′	20.85′	21.45′	21.70′	21.35′
PLRF angle of inclination	25.05°	16.49°	13.91°	11.08°	8.31°	5.35°	2.70°	−0.52°	−4.80°	0.00°

Table 65.2 Stair hall vault dimensions, main floor landing (see Fig. 65.7a, b)

	Top of stringer								Spring line	Horiz. distance
Point	a	b	c	d	e	f	g	h	j	k
Avg. ray length	19.65′	18.06′	18.28′	18.40′	18.52′	18.42′	18.38′	18.20′	17.36′	15.16′
PLRF angle of inclination	63.35°	59.80°	55.73°	51.68°	47.36°	43.24°	39.07°	34.90°	30.04°	0.05°

Several details of the stair begin to reveal the concerns of the designer. As the stairs run down from each landing, the treads interlock with the sidewall throughout the semi-circular rear wall to insure that all the tread work would remain fixed in place. The lapping of the vault voussoirs suggest the stonework was completed incrementally, progressing from the base of the wall upward and outwards, in stair-step fashion, minimizing the formwork necessary in construction. The interlocking nature of the last and first voussoirs of the stair stringer fixes the stair form at each landing, and acts as a lead block for construction. The elliptical form provided a modest weight reduction beyond what a simple circular section would have provided.

A suggestion of how the stair vault may be constructed can be found by re-examining the interlocking voussoirs in the arched window of the south façade in Fig. 65.3. Visualizing this arch in section through the cantilever stairwell, would result in the production of interlocking prisms with centers of gravity that would fall behind the edge of the stone beneath it, thereby permitting each stone to be put in place without the workman's fear that it would fall from the vault (Rondelet 1817: Pl. LXIV). Of course, there is the added complexity: that of being inclined and curving in three dimensions.[5] The quality of construction in the stair and the building certainly attests to a societal shift toward increased precision that is also being experienced in science and particularly in surveying and astronomy at this time. Who might have been responsible for this: Perrault, members of his scientific circle, or perhaps the master mason?

[5] The traditional method for cutting stonework with warped surfaces and compound curvatures appears embarrassingly simple when described in nineteenth-century texts; see French and Ives (1902: 26–28).

No archival drawings for the stair or its design have been found, from that era or any other![6] The stair form was a resultant of the changes imposed after construction had begun and is quite radical in comparison with what was originally intended for the building—a set of stairs that were similar to those Perrault designed for the Louvre. As a minimum, Perrault would have been responsible for selecting the solution as it was implemented. While Perrault was quite skilled in drawing, nothing in his background suggests that he was prepared to solve graphically a construction problem of this complexity. More importantly, it is quite likely that no one at that time could have drawn the stone prisms accurately enough to direct the construction.

No record has been found that astronomical or surveying members of his scientific circle were employed to solve this problem. The only individual who might have had the skill at that time was M. Bosse (1602–1676), a student, illustrator and refiner of the work of Girard Desargues (1591–1662). Desargues's method of a descriptive geometry may have provided an indirect source for the helicoil stair form. The geometer, inventor and architect, developed a projective method with sufficient precision to credit him as the first to design an *escalier a jour ou vis suspendu* (a suspended day-lit elliptical stair), for the Hotel de Ville de Lyon, in 1646 (Chaboud 1996). It is reported to be a stone staircase that appeared as a seamless plastic form defying gravity.[7] Further evidence of Desargues's influence may be due to the iron panel geometry of the handrail, which employs vertical dividers, an issue of visual correctness argued for by Desargues. The influence however would seem limited to the form alone, since none of the figures by Desargues himself or by Bosse (1643, 1648, 1664) appear to demonstrate an interlocking masonry construction technique.[8] This observation would seem to substantiate the criticism that Desargues received through the 1640s, that his efforts did not fully correspond to the reality of masonry construction (Schneider 1983: II).[9] The suggestion being made here, given the conventional limitations of projective geometry and the absence of any other drawings at this time, is that the

[6] See Evans (1995: Chap. 5), in which he posits that accurate stone dimensions might have been produced as early as 1550 through drawing. However, on the advice of Mr. Arthur French, if the form is too difficult to be defined by calculation then the approximate solution must be arrived at by drawing alone—and the accuracy of drawing alone was not sufficient for good masonry construction; see French and Ives (1902: Chap. 2). No finish dimensions should be taken from the projections since the reduced scale of the drawings would not allow a sufficient accuracy nor account for the changes that creep into a project due to the necessities of construction. The final responsibility for accuracy of construction was the stonemason's, who was expected to make full-scale working drawings and then find the finish dimension of each prism in situ. This approach seems to be a reasonable explanation for the case of the Observatoire.

[7] Girard Desargues's (1591–1662) method of projective geometry was rejected by the professors at the Academie Royale and masons alike, for different reasons, despite the concerted efforts of his student, Abraham Bosse; see Schneider (1998). See also Schneider (1983).

[8] See Bosse (1643, 1648, 1664).

[9] Desargues countered that he was not a craftsman and that his intention was to improve the methods of geometric projection, not necessarily improve the craft of construction.

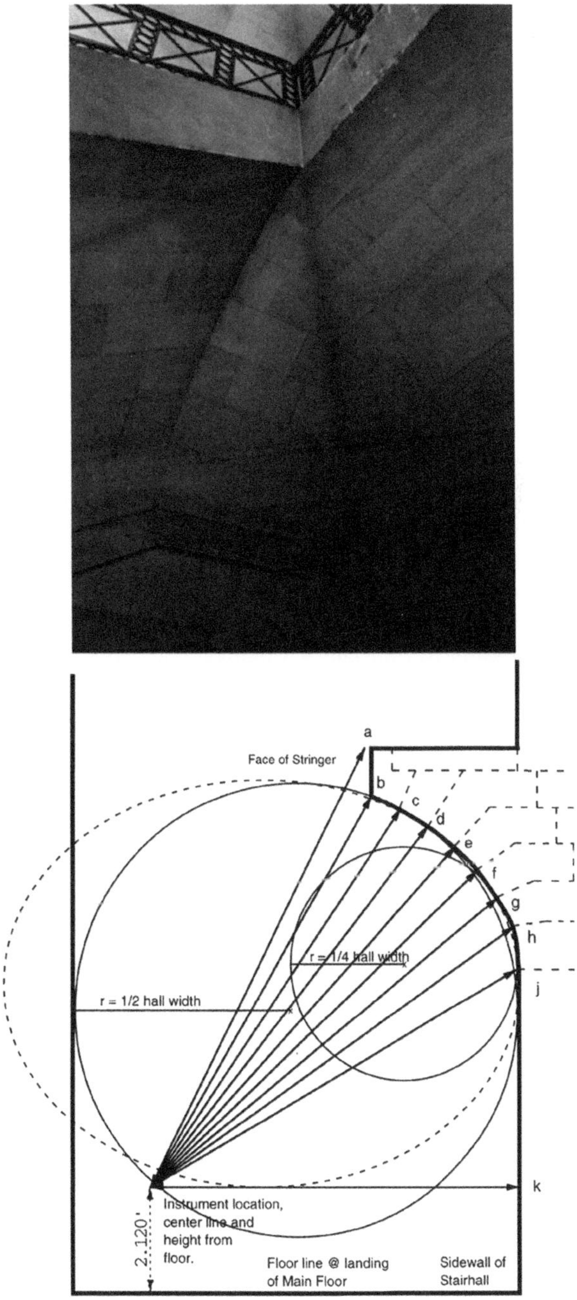

Fig. 65.7 (**a**, *Above*) Stair hall vault, main floor; (**b**, *below*) Vault dimensions, main floor landing. Photo and drawing: author

craft skill of the master mason was the only source sufficient for the resolution of this problem.

Theoretical Application of Geometry

The *Ordonnance* is best known for a proposed series of standardized proportions for the five orders, but it is less well known that Perrault's writings viewed the application of proportion as a necessity to the totality of a design. Perrault did not employ the orders at the Observatoire as it was a modern building without ancient precedent, nor was it connected to Royal ceremonies that he knew would require a classically mythic background. The present examination then will be limited to the broader question regarding Perrault's approach to proportion and beauty of the general form and the disposition of elements.

Is it reasonable to expect that the Observatoire (1667–1672) or Perrault's published drawings from 1673 might provide sufficient evidence about his point of view as expressed in the *Ordonnance* of 1684? The views presented in the *Ordonnance* first appeared as opinions in his Vitruvius (1673) and were first discussed before the Academy of Architecture in January and February 1672. The commission to translate Vitruvius coincided almost directly with the commission for the design and construction of the Observatoire. The argument over the serviceability of the design, between the Head Astronomer, Cassini, and Perrault, occurred in the fall of 1669, at which time he was reluctant to make substantive changes. It would not have been unlike Perrault to use the design of the Observatoire even in the late 1660s as an experiment to develop and test his views. In this context then, it might interesting to ask what evidence exists in the Observatoire and his published drawings concerning the use of a module of design and how that module might relate to the proportions of the general form? It might also be interesting to ask what proportions can be found that relate directly to the traditional tools of the architect?

In working towards an answer for these questions, numerous attempts were made to discover the module of the design. The scale shown in Perrault's drawing (Book I, Plate II, p 13) indicates *toises*, roughly 6½ feet per unit, but this does not appear to easily relate to the general disposition of the design (Fig. 65.8).

After an examination of the masonry coursing, the focus shifted to the main floor plan, and eventually upon the Cassini Salon. The salon was originally judged on site as not the best place to start an analysis, given the substantive changes that were completed there in 1734 by M. Soufflot. New roof vaults were put in place supported by the insertion of new piers but the plan otherwise remained unchanged. Only after having completed data collection on site did the diameter of the vertical viewing ocular appear to demonstrate a useful module. This module could be used with relative ease throughout the design in finding the length, width and height of fundamental elements, as well as indicate minor elements such as window openings. It became clear that a unit approach to the design was probably used and

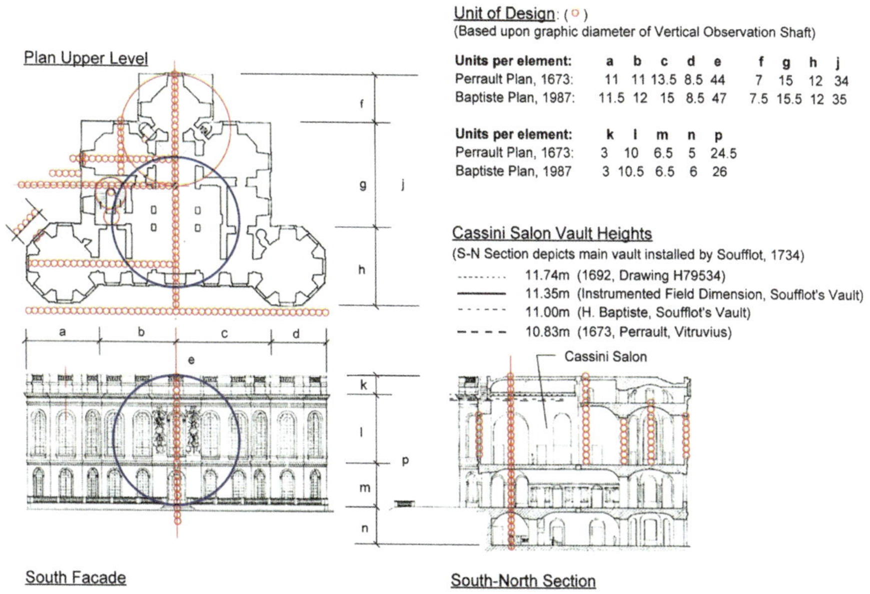

Fig. 65.8 Module/design analysis, Observatoire de Paris. Image: author's overlays on base drawings (1987) provided by permission of M. Herve Baptiste, Architect en chef des Monuments Historiques, Paris

guided construction (Fig. 65.5). With the module proving equally useful in elevation and section the sense developed that a proportional guide was also likely in design development. The result was that a module was found that could have been used to describe most volumes and masses in some combination of simple whole and half modules. The simple and straightforward application of a module for achieving form was an primary objective underlying Perrault's theoretical efforts.

To what extent does the module proposed here relate to the drawings and actually have meaning? A dimensional analysis was performed on the drawings of both Claude Perrault (prepared by S. le Clerc, 1673) and the office of M. Baptiste (del J. F. Gordon, 1987). Both sets of drawings were assessed in relation to instrumented field dimensions recorded by the author. The difference between the gross dimensions found in the Perrault drawings and those of M. Baptiste showed a consistent proportional difference of roughly 1 %. The dimensional difference between the module as drawn was 3.7 %, where the vertical viewing ocular diameter used by Perrault was roughly 1.12 m and that by Baptiste was 1.08 m at the roof. Selected checking of M. Baptiste's drawings against instrumented field dimensions recorded by the author revealed a consistent gross difference of roughly 3.4 % in both section and plan. The practical result is that the difference between the general building form and the drawings is negligible for the questions we're asking. These differences do not seem to significantly betray a good relation between either set of drawings and the physical building. It would be best however to return to the site to verify this module and its relation to the built form.

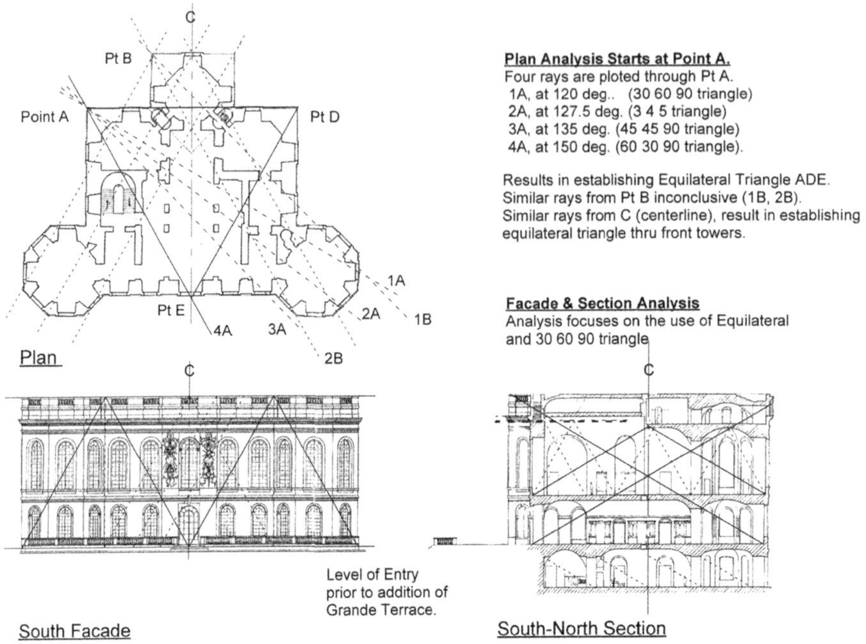

Fig. 65.9 Proportional analysis, Observatoire de Paris. Image: author's overlays on base drawings (1987) provided by permission of M. Herve Baptiste, Architect de Chef des Monuments Historiques, Paris

Like the module/design analysis, the proportional analysis also proceeded indirectly. Perhaps through habit, the analysis began on the main floor plan with the application of a 45° triangle that only led to frustration. Working through the gambit of traditional triangles, the use of a 30°–60°–90° triangle and an equilateral triangle proved fruitful. Since many principal points of the plan were located with equilateral triangles, an elevation and sectional analysis followed (Fig. 65.9).

The south elevation offers the strongest suggestion that Perrault may have directly applied the use of 30°–60°–90° and equilateral triangles for limiting the field (area) of the façade and for the disposition of elements as well. The south–north section seems less compelling, although is still susceptible to their application. With significant points of the design appearing to be located in plan, section and elevation with the use of an equilateral or a 30°–60°–90° triangle, the proportional analysis stopped. Again, the application of these proportions is straightforward and in retrospect, seems simple almost to the extreme. It was also surprising in this instance that the octagonal end pavilions of the south façade fell into the two-dimensional elevation scheme without significant manipulation.

Supposing that Perrault did use equilateral triangles to order his design, why should he have chosen to employ those specific proportions? The best answer appears to be provided by Vitruvius. In Book IX, Chap. 1, where Vitruvius discusses the mechanics of the cosmos, he links the equilateral triangle with a description of how the thermal rays of the sun are prevented from setting fire to those planets with the nearest orbits: "The sun's rays stretch out into the cosmos along lines that take the form of a triangle with equal sides," warming distant planets at certain angles (Vitruvius 1931: 221–227; Vitruvius 1999: 109–111 and 285). Immediately upon this, Vitruvius presents the power of the sun as likened to the power of a king with a line borrowed from Euripides, "the hot fire of the king rising over the earth: He burns the distant; what is near he keeps temperate" (Vitruvius 1931: 221–227, 1999: 109–111 and 285).

The ultimate client was Louis XIV and the classical allusion to his reign through this passage might have provided Perrault with added grace at court. It also helps to explain why Perrault initially chose to ornament the ceremonial south entry with astrological symbols, but at the Head Astronomer's objection, was induced to change those images to contemporary instruments of astronomy and surveying. Perhaps this is why Perrault was so vocal in maintaining the fundamental proportions of the design in the presence of Louis XIV. The analogy also provides a further glimpse into the larger social context of that era as still being rooted in the faithful application of ancient analogies to contemporary life, in this case, relating the forcefulness of Louis XIV to the power of the sun.

Conclusions

At this point the Observatoire de Paris seems to correspond quite closely to the objective conditions of beauty that Perrault championed. The building not only appears quite simple compared to its contemporaries, it also may have been the result of applying a module and proportional system with simplicity. In its day the building was only politely acknowledged (Herrmann 1973: 133–138.). It would not be a surprise to learn that his peers may have viewed it as lacking a certain *grace de la forme*, a condition ultimately causing his design for the triumphal arch at the Porte Saint Antoine to be terminated (Pérez-Gómez 1993: 8). However, Perrault's choice for a module or proportion does not seem simple.

The present day re-examination of Perrault's position is principally the result of scholarly efforts by Wolfgang Hermann, Antoine Picon and Alberto Pérez-Gómez (Herrmann 1973; Picon 1987: 197–219; Pérez-Gómez 1993). There are modest differences in the appraisal of his position. On one hand, the work has been viewed by Hermann as "less radical and far-seeing" than appears and was intended only to provoke rather than initiate a full transformation of the classical view (Herrmann 1973: 138–140 and 188–189). But, on the other hand, Perrault, like the best theorists before him, placed a high value upon innovation and invention, which in this building can be found in the stairwell, the original Cassini Salon vaults, and in the building form. In Hermann's appraisal (following Blondel's criticism), Perrault may have been insincere with his arguments aimed only at trying to make a practical simplification of the proportional divisions for each order acceptable to the architects of his era. But Perrault appears to have been faithful to many of his assertions at least in this review of the Observatoire.

On the other hand, his efforts have also been characterized by Pérez-Gómez as the first step toward the modern eradication of classical architectural values for those of a technological position (Pérez-Gómez 1993: 36–37). In this view, the creation of simplified proportions for the orders results in a prescriptive approach to design that is viewed as the first reductionist step toward emptying architectural form and practice of relevant meaning. The simplification of means however, whether in planning or construction, doesn't necessarily imply a paucity of idea or heritage. The resulting design seems neither simple nor necessarily empty.

There is no doubt that this building was meant to endure the ages and to provide an additional richness to French intellectual culture. It may also reach back to a remote past albeit in a manner less obvious than one might expect. Without the added richness of detail provided by classicism, an overt and public representation was absent. But then, this building was not intended for the public. For those for whom it was intended, its intricacies of meaning may have been more obvious. Perrault doesn't seem to have been insincere or entirely committed to overturning the classical heavens. The Observatoire for Perrault may have simply been an engaging way to explore a curious inconsistency of theory between buildings and human beings. This was the kind of experimental activity he consistently pursued in his chosen profession. The truth seems to turn upon both these views leading one to think that Perrault was himself somewhat of a paradox, able to pursue the design of the Observatoire with a bit of experimental objectivity and a measure of the passion of the faithful.

Acknowledgments This research was undertaken with support provided by the University of North Carolina Charlotte. I would like to thank Madame Danielle Michoud, le Service des Relations Exterieures of the Observatoire de Paris, and Madame Claudine Laurente, Astronomer, also of the Observatoire de Paris, as well as the Office of Herve Baptiste, Architecte en Chef des Monuments Historiques, Paris, for their very generous gift of time and interest.

Biography Randy Swanson is a registered architect and architectural scholar. He studied at the University of Illinois, Champaign-Urbana and the University of Pennsylvania, receiving the Ph.D. in Architecture in 1993. As an academic, he has lectured widely and held positions at The University of North Dakota (1987–1989), and The University of North Carolina (1989–2004) where he was Associate Professor of Architecture, History and Theory of Technology and Design. Licensed since 1984, he has practiced in Chicago, Illinois, Virginia, Washington, D.C., and Maryland. Randy Swanson is currently in private practice on the Gulf Coast of Florida (2005 to present) providing residential and commercial design services. Further information on his practice and publications on the history of technology and design can be found at: http://www.randyswansonarchitect.com. He can be contacted at http://rsswanson@earthlink.net.

References

Bosse, A. 1643. *La Pratique du Trait A Prevues, de Mr. Desargues pour la Copupe des Pierres en l'Architecture*. Paris: Imprimerie de Pierre Des-Hayes.

———. 1648. *Maniere Universelle de M. Desargues pour pratiquer la Perspective*. Paris: Imprimerie de Pierre Des-Hayes.

———. 1664. *Traite des Manieres de Dessiner les Orderes de l'architecture Antique en Toutes Leurs Parties*. Paris: Colbert.

Cassini IV, J. D. 1810. *Memoires pour servir a l'Histoire des Sciences et a celle de l'Observatoire Royal de Paris*. Paris: Observatoire de Paris.

Chaboud, M. 1996. L'Hotel de Ville de Lyon. Les correspondances scientifiques, 1643–1648. Chap. 6 in M. Chaboud. 1996. *Girard Desargues*. Lyon: Aleas Editeur.

Evans, R. 1995. *The Projective Cast*. Cambridge, MA: MIT Press.

Fitchen, J. 1961. *The Construction of Gothic Cathedrals*. Oxford: Oxford University Press.

French, A. & Ives, H. 1902. H. *Stereotomy*. New York: John Wiley & Sons.

George, A. J. 1938. The Genesis of the Academie des Sciences. *Annals of Science* **3**, 4: pp. 372–401.

Hahn, R. 1971. *The Anatomy of a Scientific Institution: The Paris Academy of Sciences, 1666–1803*. Berkeley, University of California Press.

Herrmann, W. 1973. *The Theory of Claude Perrault*. London: A. Zwemmer.

Heyman, J. 1996. The Stone Skeleton. *Arches, Vaults, and Buttresses*. Norfolk: Variorum.

Pérez-Gómez, A. 1993. Introduction. In: C. Perrault, *Ordonnance*. H F. Mallgrave, trans and ed. Santa Monica: The Getty Center for the History of Art and the Humanities.

Perrault, C. 1683. *Ordonnance des cinq especes de colonnes selon la methode des anciens*. Paris: Jean Baptiste Coignard.

———. 1909a. *Memoires de Ma Vie*. P. Bonnefon, ed. 1909. Paris: Librairie Renouard.

———. 1909b. *Voyage A Bordeaux*. P. Bonnefon, ed. Paris: Librairie Renouard.

———. 1993. *Ordonnance For The Five Kinds of Columns After The Method of The Ancients by Claude Perrault*. H. F. Mallgrave, trans and ed. Santa Monica: The Getty Center for the History of Art and the Humanities.

Petzet, M. 1967. Claude Perrault als Architekt des Pariser Observatoriums. *Zeitschrift Fur Kunstgeschichte*, **30**, I: pp. 1–54. Basel: Kunsthistorisches Seminar der Universität.

Picon, A. 1987. *Claude Perrault ou la Curiosite d'un Classique*. Paris: Picard, Caisse Nationale des Monuments Historiques et des Sites, and Delegation a l'action artistique de la ville de Paris.

Rondelet, J. B. 1817. *Traite Theorique et Pratique de l'Art de Batir*. Vol. VI, *Planches*. Paris: chez l'Auteur.

Schneider, M. 1983. Girard Desargues, the Architectural and Perspective Geometry: A Study in the Rationalization of Figure. Ph.D. Diss. Blacksburg: Virginia Politechnic Institute and State University.

———. 1998. Review of *Desargues en son Temps*. *Journal of the Society of Architectural Historians* **57**, 3 (Sept 1998): 333–335.

Vitruvius. 1931. *On Architecture* F. Granger, ed. 1931. Cambridge, MA: Harvard University Press.

———. 1999. *Ten Books On Architecture*. I. D. Rowland and T. N. Howe, eds. Cambridge: Cambridge University Press.

Wolf, C. 1902. *Histoire de L'Observatoire de Paris*. Paris: Gauthier-Villars.

Part IX
1800–2000

Chapter 66
Geomantic (Re)Creation: Magic Squares and Claude Bragdon's Theosophic Architecture

Eugenia Victoria Ellis

> *If it be true that the soul of the world is about to animate the materialism of modern life it will create for itself a new language of power and beauty, and architecture will again become a living art, for architecture deals in visible symbols, and visible symbols form the very language of mysticism*
> (Bragdon 1901).

Claude Bragdon's Theosophic Architectural Theory

At first glance there appears to be nothing unusual about Claude Bragdon's First Universalist Church in Rochester, New York (Fig. 66.1). In looking at its west façade, the trained architectural eye would guess correctly that the plan of the church was patterned after the traditional Byzantine Greek-cross arrangement: its forms pile up around a central dome-like structure, the square base of which projects through and is revealed to the exterior at the inter cardinal directions of the crossing, the limbs of the cross itself gesturing toward the cardinal directions.[1]

Universalism, the faith for which Bragdon designed his church, is built upon a belief in the universals common to all life and the unity that binds all life into one indivisible whole; for this reason, it embraces all the major world religions (Morrison-Reed et al. 1983: 4–5). In keeping with this faith, Bragdon looked toward the universal in the church's design. Bragdon would have described the

First published as: Eugenia Victoria Ellis, "Geomantic (Re)Creation: Magic Squares And Claude Bragdon's Theosophic Architecture", pp. 79–92 in *Nexus V: Architecture and Mathematics*, Kim Williams and Francisco Delgado Cepeda, eds. Fucecchio (Florence): Kim Williams Books, 2004.

[1] For the architectural symbolism of the Byzantine church, see Mathews (1995: 11–21). See also Mathews (1998).

E.V. Ellis (✉)
Department of Civil, Architectural and Environmental Engineering, College of Engineering, Drexel University, 3141 Chestnut Street, 71–134, Philadelphia, PA 19104, USA
e-mail: genaellis@drexel.edu

K. Williams and M.J. Ostwald (eds.), *Architecture and Mathematics from Antiquity to the Future*, DOI 10.1007/978-3-319-00143-2_19,

Fig. 66.1 First Universalist Church, 1907, west façade. Photo: author

design of the First Universalist Church as being in the "Gothic spirit" for the similar reasons he labeled its inspiration, the Byzantine Hagia Sofia in Istanbul, a "Gothic building." However, Bragdon's definition of "Gothic" was neither the popular conception at the time, nor its present-day, historical, usage meaning an architectural style distinguished by such characteristics as pointed arches, groined vaulting, and buttressed walls. Rather, Bragdon was referring to Gothic in the metaphysical sense, in terms of a design process that responds to prevailing contextual conditions: "...Gothic means organic, as opposed to arranged architecture,—spontaneous, as opposed to deliberate. It is a manner of building in which the form is everywhere determined by the function, changing naturally and inevitably as that changes..." (Bragdon 1908). When designed and built in the "Gothic" way, architecture becomes a "living art" that relates to universals and to life itself.

The First Universalist Church is a visible symbol of Universalism composed using Bragdon's invisible, mystical, architectural vocabulary that conflated Gothic mysticism with Eastern spirituality. Bragdon's architectural theory was influenced by his interest in Theosophy, a belief system similar to Universalism that also brought together Eastern and Western religious traditions. Four interrelated parts, distinct and yet indivisible due to their mutual correspondences—nature, the human body, number and geometry, and music—formed the basis of Bragdon's theosophic architectural theory, which emphasized a cosmological relationship between the body and the building through number, geometry, and harmonic proportions.

Bragdon's theosophic architectural theory influenced those aspects of architecture *that could not be seen*. This invisible aspect of design, which is consistent throughout his work, is a symbolic emphasis on the crossing and the cardinal directions, together with encircling—a duality of polar opposites united within one "divine androgyne" (Fig. 66.2). Historically, this symbol has both

Fig. 66.2 Bragdon's example of a *magic square*. The columns, rows and diagonals all sum to 34; the sum of the numbers of opposing corners equals the *square* of the *square's* base number plus unity ($13 + 4 = 17 = 4^2 + 1$; $1 + 16 = 17 = 4^2 + 1$). Image: Bragdon (1928)

Gothic and Eastern origins—it can be traced to the "rose" and the "cross" of the Rosicrucians; and to the Hindu "Golden Person" or *puruṣa*—and embodies a cosmological relationship uniting aspects of nature, the body, number and geometry, and music. The encircled crossing is a "circle of orientation" (Ghyka 1931: 142) that represents the primal act of architectural creation—a talismanic operation with divinatory roots in the geomantic act of mathematical creation (and recreation) known as the magic square.

Magic Squares and the Cardinal Directions: Divining the Constructed World

> Though the formation of magic squares is classed today among mathematical recreations... a certain religious or mystical significance has always attached to such arrangements of numbers—perhaps because they are so mysterious, and suggest the operation of some supernal intelligence. Be that as it may, they are conspicuous instances of the intrinsic harmony of number, and as such serve as an interpreter to man of that cosmic order which permeates all existence (Bragdon 1928: 164–165).

Magic squares came to be known in the Western world via the Islamic texts of the tenth century, most probably brought into Europe by Jewish traders. They were common to the Muslims, Hebrews and Hindus, but the most ancient documented magic square came from China. The magic square is a numerical acrostic disposed so that, when summed, each column, row, and diagonal equal the same number. Numeric squares become "magic" when the numbers in opposing corners sum to $n^2 + 1$, where n equals the base number of the square and the added unity is the universal symbol for the divine creator—the Tao, Allah, or the Brahman (see Fig. 66.2).

The square of three is the smallest that can be constructed with magical properties. Common to the Muslims, Hebrews, and Chinese was the belief that knowledge of magic squares was divine: not a human invention, but a revelation.

The magic square of three was first published in the Islamic world around the year 900 A.D. in an Arabic treatise traditionally ascribed to Jāsbir ibn Hayyān (known to Europeans as Geber). It was presented as a charm known as the *badūh* seal, which was written using the first nine numbers of an Islamic alphabet that equilibrates letters with corresponding numbers, called the *abjad* system.[2] The first set of Islamic magic squares were presented in an encyclopedia published in 989 A. D. called the *Rasā'il*, which was composed by the Ikhwān as-Safā, a Muslim brotherhood known as the Brothers of Purity. These squares were presented as illustrations and text, and were described as small models of a harmonious universe. Along with magic squares, the canon of the Brothers of Purity included numerical and musical correspondences that described a system of proportions based on the human figure (Panofsky 1955: 76–77).

In the Islamic world, each of the four archangels are associated with their own sigil derived from one of the first four magic squares; the *badūh* seal is considered to be the seal of the archangel Uzrā'īl. It also represents the planet Saturn when spelled in *abjad* letter-numerals, the sum equals 45, the magic sum of all the numbers in the square. Islamic magic squares often took the place of words themselves, and by interchanging numbers and letters to create sums and words that spelled the names of gods or planets, they became talismans to bring good fortune. The first seven Islamic magic squares, with from three to nine squares per side, came to be associated with the seven planets. In later European occult circles, the magic path of the square of three, which connects the numbers in sequence, became known as "the seal of Saturn." This Islamic practice of calculating with letters was considered to be a secret science, known only to "the authorities in divine learning" (Canaan 1936: 89–92).

The first evidence of magic squares in Europe was the square of three discussed by Abraham ibn Ezra of Toledo in twelfth-century Spain, written with Hebrew letter-numerals in a style that became popular with later Cabalists. The earliest set of European magic squares was most probably derived from the Islamic planetary squares that were translated by thirteenth-century Jewish and Christian scholars living in Spain. These are considered to be the source of the Islamic astrological and cosmological lore that helped to build the tradition of the Cabala (Cammann 1969: 204). The Islamic method of interpreting letter-numeral squares and associating

[2] The *abjad* system was a way of using Arabic letters with numerical values instead of numbers, as was done before the introduction of numerals from India. Since these letters were arranged in an archaic sequence, following the order of the Hebrew alphabet (the first four were *alef*, *bā*, *jīm* and *dāl* and corresponded to the numbers one through four), the initials of these names for the Arabic letters were taken to make the–otherwise meaningless—name *abjad* as a mnemonic device to be used in remembering them (Cammann 1969: 190).

mathematical sums with names for God may have been influential in the development of gematria, a cabalistic method of interpreting the Hebrew scriptures by interchanging words whose letters have the same numerical value when added.

In the Table of Jupiter derived from seventeenth-century alchemical and Rosicrucian sources (Fig. 66.3), we can see the "first Persian square of four" written in both Hebrew letter-numerals and Arabic numerals adjacent to a list of "divine names" and their magic sums. The Islamic origins of this square can be seen in its construction technique, which according to the Islamic method of writing begins with the upper right-hand corner: the diagonals are marked with a dot and the numbers are written horizontally in the dotted squares from right to left. The unmarked "houses" are filled in reverse by beginning with the lower left-hand corner. The square's construction emphasizes the diagonals and their crossing so that both the diagonals and the pairs of numbers encircling the crossing sum to 34. Due to the mathematical reasoning of the square's construction, the "intelligence of Jupiter" is indicated by a talismanic sigil that is a cross inscribed within a circle, which is also one "divine" source for the "rose" and the "cross" of the Rosicrucians as well as Bragdon's "divine androgyne."[3]

The Islamic magic square was worn as a necklace or ring to ensure one's well-being and to bring good fortune to one's immediate environment. To cast out evil spirits, magic circles, squares, and figures were sketched on the ground, aligned with the cardinal directions and the demoniac person seated in the center while an incantation was read (Shurreef 1863: 218, 3321–239). The encircling of a person in danger was a means of protection from evil: the disciple would be protected within and oriented to the earth below and the universe above by becoming "square with the world."

Divining the constructed world was a geomantic, magical procedure used by ancient cultures to orient their built world with the cardinal directions of the earth and with respect to the cosmos (Vitruvius 1931: I, vi, 6). The divine coordinating principle used was geomancy, which is derived from the Greek words *geo*, literally meaning the earth, and *manteia*, meaning divination or coming from above. Geomancy is the act of projecting lines onto the earth from the cosmos above through marking the ground and encircling. This talismanic operation projects the magic square's properties upon the ground at the human scale as regulating lines to provide auspicious conditions for the construction of the built environment and to protect the constructed world. This is a "divine" act with heavenly origins. The divine is embodied in an earthly construction that begins with the human body at its center and origin. The body marks the beginning and the first point of contact with the divine through its *axis mundi*. The divine resources for ancient geomantic

[3] Bragdon's square was created using the Indian method of writing, which would begin in the upper left-hand corner and move horizontally through the rows from left to right. Although we refer to our numbering system in the West as being "Arabic numerals," these numbers most probably came from India. The first recorded use of "Arabic" numerals is in Baghdad, where an Indian scholar appeared in 771 A.D. with a treatise on astronomy using the Indian numerical system that was later transmitted to the West by the Arabs by way of Moorish Spain.

Fig. 66.3 The table of Jupiter. Image: Kim Williams, after McLean (1989: 111)

ד 4	יד 14	טו 15	א 1
ט 9	ז 7	ו 6	יב 12
ה 5	יא 11	י 10	ח 8
יו 16	ב 2	ג 3	יג 13

procedures included the positions and the paths of the sun, the moon, the stars and the planets. The instrument used to take their measurements was the *gnomon*, literally, interpreter. The gnomon is a vertical element, often in the form of the human body, which is used to orient one's position on earth with respect to the greater universe of the cosmos by being encircled: the intersection of the gnomon's cast shadow and the circle in the morning and the evening at the summer solstice locate solar east and west from which north and south can be determined (Fig. 66.4).

In the Indian tradition, it was necessary to have "knowledge of the circle and the line" in order to square the earth (Boner et al. 1996).[4] Geomancy in India is called *Vāstu Śāstra*: *vāstu* means to dwell and *śāstra* means science.[5] *Vāstu Śāstra* is used to orient the built world to the cosmos through squaring the circle using a gnomon and the *vāstu puruşa mandala* (Fig. 66.5), a magic square in the anthropomorphic form of a sacrificial victim lying face down in a *yantra* that serves "as an interpreter to man of that cosmic order which permeates all existence." A *yantra* is literally a visual and graphic symbol in the form of a square with four gates at the cardinal

[4] In India there is no principle distinction made between the Fine Arts and the Practical Arts. A *Sthāpaka* is one who has knowledge of the circle and the line and can be either a sculptor or an architect. *Śilpa* refers to sculpture and *vāstu* refers to architecture. Although this *upanişad* is directed toward both the sculptor and the architect, it is called a *vāstusūtra* most probably because it does not have to do with "modelling" the three-dimensional qualities of a work or sculpture, but has to do with its layout and geometric composition, which in either discipline has to do with the organization of the work in plan and in elevation.

[5] *Vāstu Śāstra* is a geomantic art that was fully developed by the first century A.D. Its tenets were not written down in any one ancient manuscript but developed out of a corpus of over 20 texts referred to as *Vāstu Vidya* dating back to the sixth century A.D., most of the previous material having been lost; see Padam (1998), Chakrabarti (1999). See also Nathan (2014).

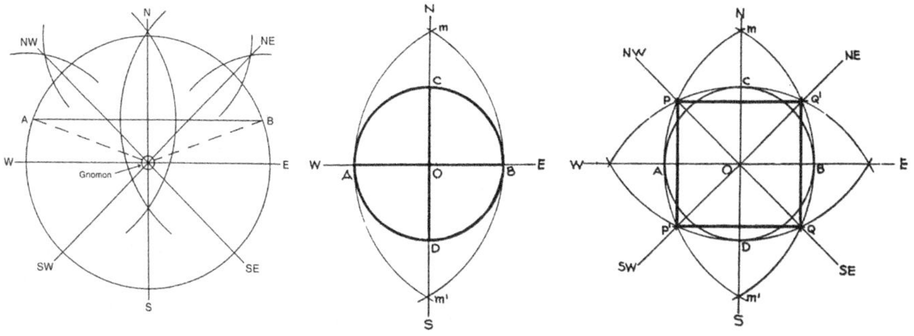

Fig. 66.4 Determining the cardinal points with a gnomon and using *circles* to define the original *square*. Image: author

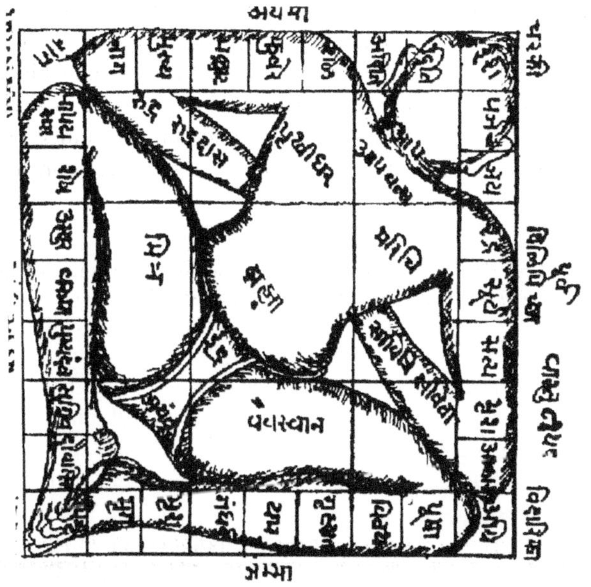

Fig. 66.5 *Vāstu Puruṣa Mandala*

directions that is used for meditation purposes (Zimmer 1946; Mookerjee and Khanna 1977).

The individual "houses" (*pāda*) of the square correspond with parts of the human anatomy; the navel of the *vāstu puruṣa* is located at the square's origin in the center. The *pāda* (literally, feet) are to human scale and are "regulated" by the human pace in moving through the houses. The individual *pāda* control the disposition of the whole and determine the function or use of the corresponding space to be constructed. In India, the human sacrifice is related to the sun, because "the

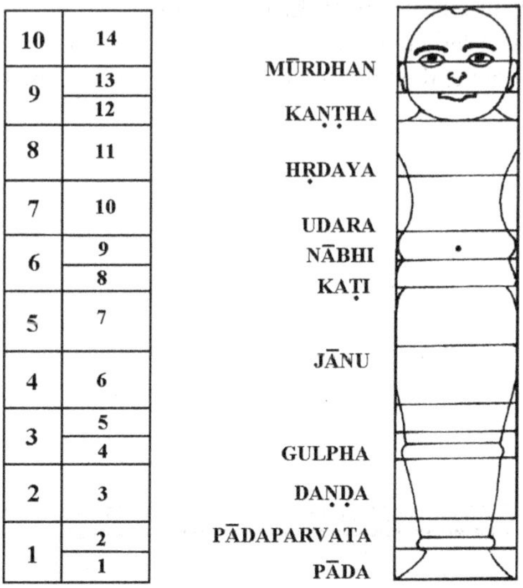

Fig. 66.6 *Yūpa*, or sacrificial post. Image: Kim Williams, after Boner et al. (1996)

Brahman in the Sun and the Brahman in Man are One."[6] The human frame, the constructed temple, and the whole of the universe are analogical equivalents; therefore, the parts of the temple correspond both to the parts of the human body and to the parts of the universe itself. The *yantra* represents an idealized sanctuary of the mind and is also characteristic of Hindu temple architecture.[7]

The temple's square-shaped central space, called the *brahmasthāna*, is open to the sky and acts as a conduit for the exchange of communication between the earth and the cosmos. The vertical central *axis mundi* of both the human body and the Hindu temple communicate with the universe, located in the spine of the body and the *brahmasthāna* of the temple. The *brahmasthāna* is set within another square, both oriented to be "square" with the path of the sun. The generic Hindu temple form is a nine-square grid, or magic square of three (Bunce 2002).

The magic square of the *puruṣa* is a *yantra* that becomes a *mandala* when it is encircled during the act of shadow casting to orient it with the sun. The geomantic art in India begins at the navel with the casting of a shadow using a gnomon, synonymous with the sacrificial post, or *yūpa* (Fig. 66.6), and symbolic of the upright sacrificial person.

The geomancer is a "sacrificer" and the casting of the initial shadow to orient the *vāstu puruṣa mandala* to be "square with the world" is a rite of initiation and

[6] This is a reference that Bragdon copied down from Max Müller's *Sacred Books of the East: The Upanishads* (Bragdon Family Papers, Department of Rare Books and Special Collections, University of Rochester Library, A.B81, Box 36, Folders 1–3, dated 04/20/1891).

[7] The Kandariya Temple in Bragdon's illustration is comprised of three sequential magic squares, or *yantras*.

symbolic death.[8] *Vāstu Śāstra* is a body-centered science that relates the body of the building to the human body through divine analogy. This is a procedure that "protects" the constructed world by encircling and by providing a "divine" framework to ensure that architecture will correspond to cosmic laws. This is an anthropomorphic procedure that through analogy guarantees alignment of the earthly with the divine.

Body Building: The Occult Anatomy of the First Universalist Church

> If the body is a temple, it is not less true that a temple, or any work of architectural art, is a larger body which man has created for his uses, just as the individual self is housed within its stronghold of bones and flesh. Architectural beauty, like human beauty, depends upon the proper subordination of parts to the whole, a harmonious inter-relation between and adjustment of these parts, the expressiveness of each of its functions, and when such functions are many and diverse, their reconcilement, one with another. In the ideally perfect human form are exemplified all those principles of natural beauty dispersed throughout nature (Bragdon 1901: 10–14).

The magic square is a universal symbol that was used to align the earthly with the cosmos. It is a mathematical creation that is an operation of the active intellect, which guides the architectural imagination in orienting the constructed world to be square with the universe. Although the magic square itself is not visually apparent in the First Universalist Church, it is symbolically present in its overall design; and the symbol of the encircled crossing, or the "rose" and the "cross," is evidenced throughout.

The organization of the First Universalist Church is clear from schematic plans and the section (Fig. 66.7). The tripartite scheme is composed of the large sanctuary and two arrangements of smaller support spaces. Its east–west axis, a "universal" element consistent with both Eastern and Western traditions, is emphasized through the hierarchical configuration lengthening the building along this line. Bragdon has selected symbolic attributes from several sources and combined them into one unique whole. His fusion of Eastern and Western traditions and the human body are evident in Fig. 66.8.

In the section of the First Universalist Church, the Byzantine Greek-cross plan of the sanctuary can be seen to be rising out of the whole. Characteristically, the movement into and through a traditional Byzantine church is axial: the participant would enter from the west and move in an easterly direction through the nave toward the altar; this is also true of the Western Christian church. In Bragdon's church, contrary to Eastern tradition, there is an axial shift at the entrance.

The western façade of the First Universalist Church is on Clinton Avenue, which in 1907 was a broad thoroughfare with little traffic but is today heavily trafficked.

[8] The symbolism of the temple and the body in India was found in several writings of A. K. Coomaraswamy: *Svayamātrnnā*: Janua Coeli; An Indian Temple: The Kandarya Mahade; The Symbolism of the Dome, all in Coomaraswamy (1997).

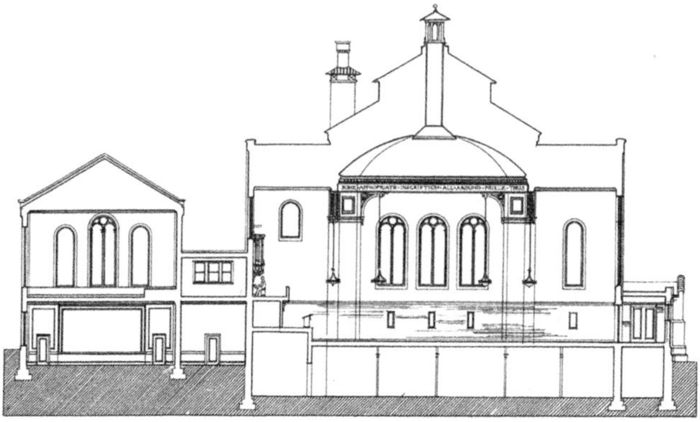

Section Looking South

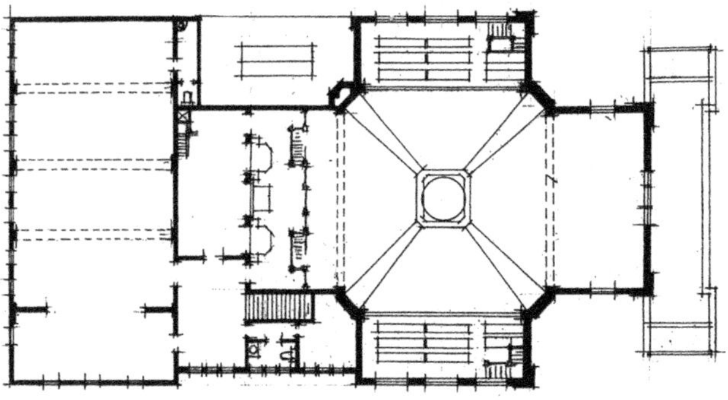

Second Floor Plan

EAST **WEST**

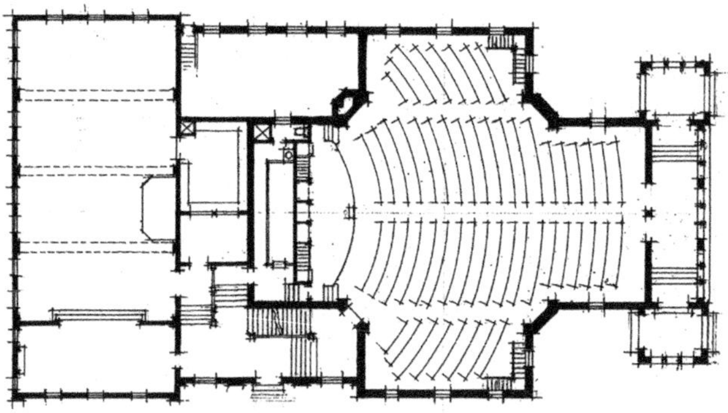

First Floor Plan

Fig. 66.7 First Universalist Church, 1907, schematic floor plans and section. Image: author

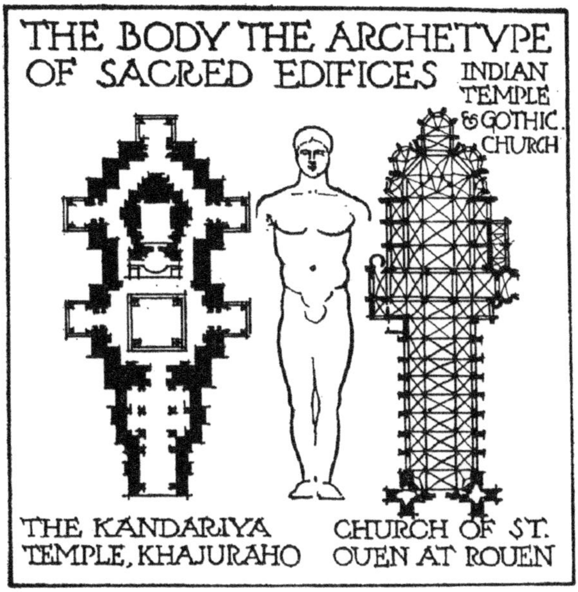

Fig. 66.8 "The body, the archetype of sacred edifices". Image: Bragdon (1910)

Nowadays, members of the congregation enter by the north elevation, but originally, churchgoers entered through the two symmetrical porticos on Clinton Avenue. These provided a transition space between the sanctuary and the outside world. The participant entered from the west, and turned to the north–south axis of the porticos prior to entering the nave itself. This axial shift is a very "modern" solution to a traditional design problem with medieval origins.

In the first floor plan, the two porticos and loggia appear almost as appendages to the Greek-cross, or as feet to an anthropomorphic plan. Bragdon's inspiration came from a diagram in Hargrave Jennings's *The Rosicrucians*, first published in 1870, the key points of which Bragdon reproduces for the readers of his *House and Garden* article of 1902 on "The Bodily Temple" (Fig. 66.9) (Bragdon 1902: 195).

It is apparent from Bragdon's drawing that he sees a relationship between the body and the building: symmetrical to the upper left-hand corner drawing of a Christian church is a representation in the upper right-hand corner of a crucified Christ. The male and female pillars of the central diagram form a gateway into the nave, or belly, of the building. In the First Universalist Church the two porticos were most probably modelled after these pillars; however, Bragdon's porticos are transitional gateways between the sacred and profane spaces.

The second floor plan shows simultaneously both the plan and the reflected ceiling plan of the sanctuary: to the north and south are the two choir lofts, and to the east is the organ loft; however, the reflected ceiling plan is represented with a "belly-button" at its center. In section, this center-point is connected directly to the lantern above.

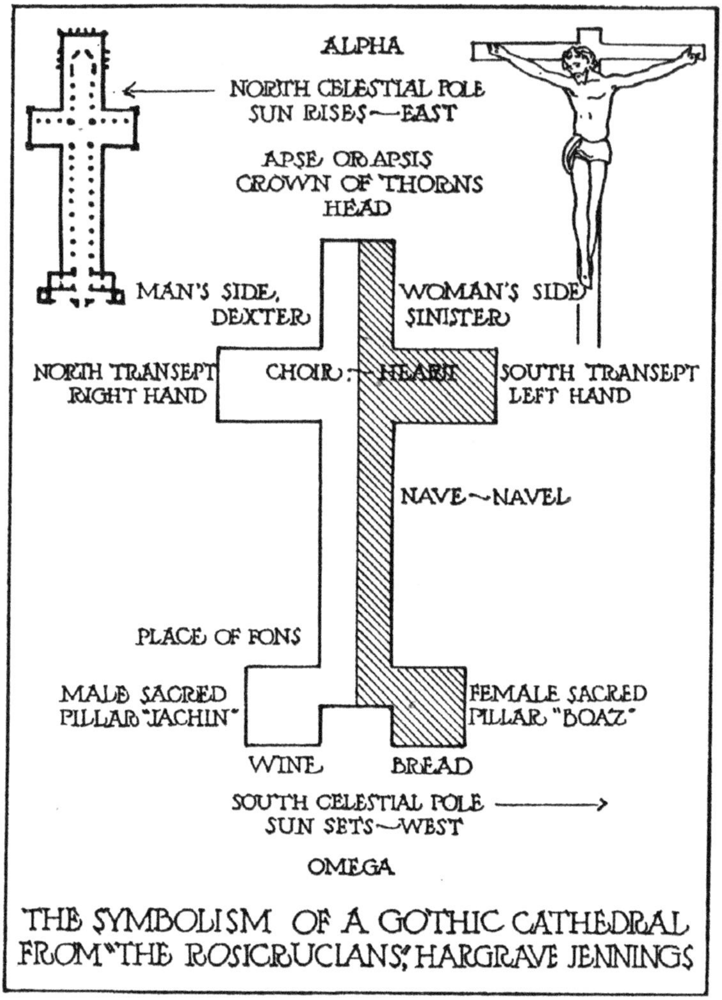

Fig. 66.9 "The symbolism of a Gothic cathedral". Image: Bragdon (1902)

The original First Universalist Church in Rochester was built in 1846 and was extensively renovated in 1901, at which time many magnificent stained glass windows were added. In 1907, the parishioners were made a generous offer for the church property by real estate promoters. Bragdon was given the commission for the design of a new building to be located a few blocks south of the original church, with the caveat that the stained glass windows be incorporated into the new church. An unusual feature of Bragdon's design was the relocation of the "rose window" to the ceiling at the center of the nave. Although the church's design is a conflation of Eastern and Western Christian traditions, Bragdon's intention was other than simply replacing a Byzantine Christ the *Pantokratōr* with a Christian rose window. In 1911 Bragdon wrote "that God works by sacrifice: that His universe is itself His broken body." At first, it would seem that Bragdon was referring to Christ; however, he prefaces this comment by referring to "the Golden Person, the Light of the World" (Bragdon 1911: 308). The "Golden Person" is the *puruṣa*, the cosmic person who is man and God alike. Unlike the Byzantine who identifies with Christ through enlightenment and communion, the Hindu body-image is inseparable from God, because Brahman is one in both man and the sun—hence, the Golden Person. The First Universalist Church is a temple modelled after God's broken body, and as such man's broken body, which "depends upon the proper subordination of the parts to the whole." Used in the design process, this symbolic sacrifice regulates the lines of architecture by reconciling functions part by part to be in harmony with the whole, which is what makes architecture a "living art."

In looking at the section in Fig. 66.7, it is apparent that Bragdon's intent was for the First Universalist Church to correspond to the bodily temple. This is a schematic drawing and not the final design, and more clearly illustrates Bragdon's imaginative design process does than the final built work. It is evident that the sanctuary was thought of as a *brahmasthāna* that could communicate directly with the cosmos through its central vertical *axis mundi*. The light from the lantern above theoretically could be channelled to the rose window below to become a sun-column for the sanctuary. In plan the rose window is represented as a navel; in section it defines the central spine of the sanctuary. Although the functional use of the rose window may have been to provide light central to the nave, it has far greater symbolic meaning as a Rosicrucian rose positioned at the intersection of a cross—a symbol that is repeated as ornament throughout the building.

The First Universalist Church is a visible symbol expressing Bragdon's invisible, theosophic architectural theory that conflated Gothic mysticism and Eastern spirituality. Embodied within this church is a symbolic emphasis on the crossing and the cardinal directions, together with encircling—a united duality of polar opposites within one "divine androgyne" that Bragdon symbolized through number in the magic square. The magic square is a mathematical creation and operation of the architectural imagination that demonstrates the cosmological relationship between the body and the building through number, geometry and harmonic proportions. According to Bragdon, number is the foundation of all creation, and because the ideally perfect human form exemplifies all the principles of nature, Man *is* the Magic Square (Fig. 66.10).

Biography Eugenia Victoria Ellis is a practicing architect and educator. She is a principal at BAU Architecture dedicated to sustaining, preserving and cultivating the natural and built environment. Her research interests include the visual and non-visual effects of light, (eco)logical building technology, architectural theory and wellbeing. Her aim is to create frameworks for the design of smart, sustainable buildings at the nexus of health, energy and technology. At Drexel University she teaches advanced seminars in theory and interdisciplinary experimental courses in

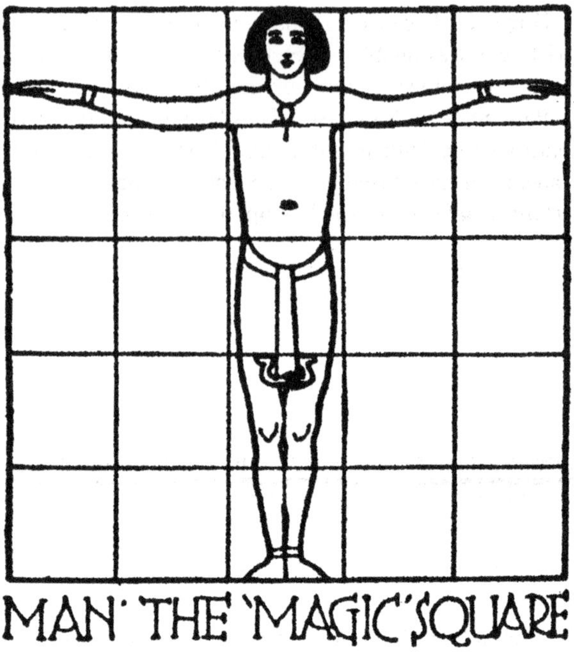

Fig. 66.10 "Man the 'magic' square". Image: Bragdon (1928)

sustainable and smart design. Dr. Ellis has contributed to *Writing Urbanism* (Routledge, 2008) and has journal articles published in *Intelligent Buildings International, Journal of Architectural Education, Architectural Theory Review* and *Dichotomy*. Her latest book is *Claude Bragdon and the Beautiful Necessity* (co-edited with Andrea Reithmayr, RIT Cary Graphic Arts Press, 2010).

References

Boner, A., Sarma, S. R. and Bäumer, B. 1996. *Vāstusūtra Upaniṣad: The Essence of Form in Sacred Art* (1982). Delhi: Motilal Banarsidass Publishers.

Bragdon, C. 1901. Mysticism and Architecture. *The Interstate Architect and Builder* (July 13 and 20): 10–14.

———. 1902. The Bodily Temple. *House and Garden* (May): 192–196.

———. 1908. The Gothic Spirit. *Christian Art* **II/IV** (January): 65–172.

———. 1910. *The Beautiful Necessity. Seven Essays on Theosophy and Architecture.* New York: Cosimo.

———. 1911. Symbols: A Fragment of Thought. *Orpheus* **15** (July): pp. 307–311.

———. 1928. *The New Image*. New York: Alfred A. Knopf.

Bunce, F. W. 2002. *The Iconography of Architectural Plans*. New Dehli: D. K. Printworld (P) Ltd.

Cammann, S. 1969. Islamic and Indian Magic Squares, Part I. *History of Religions* **8/3** (February): 181–209.

Canaan, T. 1936. Arabic Magic Bowls. *Journal of the Israel Oriental Society* **XVI**: 79–127.

Chakrabarti, V. 1999. *Indian Architectural Theory: Contemporary Uses of Vastu Vidya*. Delhi: Oxford University Press.

Coomaraswamy, A. K. 1997. *The Door in the Sky*. Princeton: Princeton University Press.

Ghyka, M. 1931. *Le Nombre d'Or. Tome I: Les Rythmes*. Paris: Gallimard.

Mathews, T. F. 1995. *Art and Architecture in Byzantium and Armenia*. London: Variorum.

———. 1998. *Byzantium: From Antiquity to the Renaissance*. New York: Abrams.

McLean, A. 1989. *A Treatise on Angel Magic: Being a Complete Transcription of Ms. Harley 6482 in the British Library*. Grand Rapids, MI: Phanes Press.

Mookerjee, A. and Khanna, M. 1977. *The Tantric Way*. London: Thames and Hudson.

Morrison-Reed, M. D., Stott H. K. and Bramlet, R. 1983. *Reaching for the Infinite: The First Universalist Church*. Rochester, NY: The First Universalist Society.

Nathan, V. 2014. Vastu Geometry: Beyond Building Codes. Vol. I, Chap. 26, pp. 375–388 in this present volume.

Padam, J. P. A. 1998. *Văstu: Reinventing the Architecture of Fulfillment*. Dehradun, India: Management Publishing Co.

Panofsky, E. 1955. *Meaning in the Visual Arts*. Chicago, University of Chicago Press.

Shurreef, J. 1863. *Qanoon-E-Islam, or Customs of the Mussalmans of India*. Madras: Higginbotham. Republished 1973. Lahore, Pakistan: Zulfiqar Ahmad.

Vitruvius, M. P. *De Architectura*, Book I, chapter VI.6 of trans. F. Granger. 1931. Cambridge MA: Harvard University Press.

Zimmer, H. 1946. *Myths and Symbols in Indian Art and Civilization*. Princeton: Princeton University Press.

Chapter 67
Mathematics and Music in the Art Glass Windows of Frank Lloyd Wright

Leonard K. Eaton

Introduction

> It's not poetic or particularly pleasing to hear but we humans are basically pattern recognition devices. Our eyes take in the world but what we really see are intricate patterns of lines and curves and colours and brightness. Our ears hear sound, but we only recognize music and language when we decode the signals into discrete patterns of tone and rhythm. We crave finding patterns in the world around us because it is the only way that we can give meaning to anything, including ourselves. Nature may loathe a vacuum but humans cannot stand a lack of patterns (Blatner 1997: 72).

As in most other matters, Frank Lloyd Wright was articulate about the role of leaded glass in his buildings. In 1908 he wrote:

> The windows usually are provided with characteristic straight line patterns, absolutely flat and usually severe. The nature of the glass is taken into account in these designs as is also the metal bar used in their construction, and most of them are treated as metal "grilles" with glass inserted forming a simple rhythmic arrangement of straight lines and squares made as cunning as possible so long as the result is quiet (Wright 1959: 59).

As his autobiography reveals, from his earliest days as a boy on the farm in Wisconsin he was sensitive to the patterns of nature. Many years later he recalled,

> I used to love to sit down at the drawing board with a T-square and triangle and concoct these patterns that you will see in the windows. I evolved a whole language of my own in connection with these things (Wright 1952).

What was this language? We believe that in some houses it was a mathematical language which was essentially musical. It was musical in the sense that it was

---

First published as: Leonard K. Eaton, "Mathematics and Music in the Art Glass Windows of Frank Lloyd Wright", pp. 57–71 in *Nexus III: Architecture and Mathematics,* ed. Kim Williams, Ospedaletto (Pisa): Pacini Editore, 2000.

Leonard K. Eaton (1922–2014)

K. Williams and M.J. Ostwald (eds.), *Architecture and Mathematics from Antiquity to the Future*, DOI 10.1007/978-3-319-00143-2_20,

related to the mathematics that one finds in an ordinary piano octave. It was one feature of the design system which he used to control the entire building.

In this chapter we will study closely five windows from the Meyer May house (Grand Rapids, 1908). One of the attractions of the May house is that practically all of the glass has survived intact. When it was restored by Steelcase Inc. in 1986–1987, the windows were taken down, numbered, and carefully inspected before reinstallation. Only a few minor repairs were necessary. The firm which made the windows is not known. Purely on the basis of quality, we think that they came from Chicago. There was a local art glass industry in Grand Rapids, but it is difficult to believe that it produced these windows. They are not mentioned in the daybooks of Niedecken-Walbridge, who supervised other aspects of the interior. And we know that Wright was in the habit of going back to craftsmen who had served him well. For the purposes of this chapter, then, we are assuming that the Meyer May windows were made from a missing cartoon produced in the studio after a design by Wright himself. This cartoon would have had the exact dimensions which are the key to the problem. It is these dimensions which we observe in the windows themselves.

Window Design and Construction

Because it affects our analysis, we will first say a word about the technology of window design and construction. None of the drawings from Wright's studio have dimensions between cames[1] or of individual pieces of glass. Many do not have any measurements at all other than the scale. Some windows were drawn freehand, possibly by Wright, and these are remarkable for their precision. Again, they are not dimensioned. In all this material one important type of drawing is missing: the full size shop drawing or cartoon. This would have been an absolute necessity for the glaziers at Linden Glass or Gianini and Hilgart, the two firms which did most of Wright's windows. The cartoon was laid down on the glazing table, or bench, and the window assembled on top of it. For multiple windows the pattern may have been drawn on the bench top. This was the procedure in the Middle Ages before large sheets of paper became available. If a cartoon was used, it would have been in tatters by the time the job was finished. In my view the only way to replicate these cartoons is by extremely careful measured drawings of the windows themselves.[2]

---

[1] A "came" is the metallic division and structural support for individual panes of art glass; in cross section the metal has the shape of an I-beam. The metal used must be malleable (or soft) to conform to the various curving shapes of the individual pieces of glass. In medieval times lead was most commonly used, hence the term "leaded glass windows." Wright's windows are predominantly made using zinc came, sometimes coated with bronze or copper for particular colour effects.

[2] Letter from Ms. Julie Sloan to the author, 23 June 1995. I am much indebted to Ms. Sloan for assistance with the technical side of this chapter. It is only fair to say that we differ in our views of

Wright owed much to his glaziers. They had to be precise in cutting the glass and came in order to build his windows. A geometric pattern requires much more precision than a floral window because if the lines do not match, it is very obvious and the window will quickly get out of square. It is likely that the pieces of Wright's windows were cut with jigs. Similarly the came would be jig cut; zinc was cut with saws and lead with knives, but it was basically a hand process. Frank Linden Jr. remembered Wright coming to his father's studio and actually sitting down with the manager to lay out sample windows. David Hanks notes that bins of glass were available for Wright's inspection. These visits must have been familiarizations. It should be stressed that unlike the furniture or fabrics, the windows were never turned over to Niedecken Walbridge, the Milwaukee interior design firm with whom Wright worked on several important projects. They were part of the architecture and were therefore the responsibility of the Studio.[3]

There is the additional problem of the glazier's shop. Ms. Julie Sloan states that there is no way that glaziers then or even now would be able to measure to dimensions of thousandths of an inch. And instruments are only a part of the problem. There is also a materials constraint. It is difficult enough to cut a piece of double thick glass (which is 1/8″ thick) to 1/4″ tolerances simply because of the way the glass handles. The 1/8″ tolerance which we found in the root figures of the May windows, was difficult but possible. And there is the problem of the cartoon. In the cartoon the lead lines are often drawn with a wide marker or a paint brush as wide as the lead is. But this is not a precise width. With the line on the pattern paper, the cut sizes of the glass are created by using three blade shears which remove a strip from between the pieces to allow for the came. It is not an exact process. Wright's windows might not have had patterns cut for them since they were geometric, meaning that measurements could be used instead of patterns, but we do not know the details of how his studio worked.

It is also important that the studios would not have bothered to cut to extremely tight tolerances in view of what they could charge for their work. Astonishing though it may seem, Wright's art glass windows were not particularly expensive. Wright advocated these windows as "art glass," not "fine art" i.e., the windows of Tiffany or La Farge. After all, "art glass" was a commodity sold through the catalogues of Sears and Roebuck and Montgomery Ward. Sears' windows and Wright's windows averaged around $1.00 or $1.50 per square foot. By contrast, between 1900 and 1920 John La Farge's windows averaged at least $100.00 per square foot. La Farge even quoted some at $1,000.00 per square foot. Wright's lower pricing held because there were so few individual designs. A good example is the Robie house. There were a dozen different patterns to be applied to 175 window

the importance of mathematics in these windows. She has set forth her views in Sloan (1997). I agree with her conclusions but believe that the windows which I am discussing in this chapter present a different set of problems.

[3] Relying on Duthe (1908), David Hanks has beautifully described the craftsmanship in Wright's windows Hanks (1979: 57). Mr. Jim Hofstra of Tilton and Lewis was in charge of windows during the Meyer May restoration and is the source of my information.

openings. Hence 14–15 windows would be made in each design. This is very cost effective. The only exception to this generalization is the Dana house where there were a great many unique patterns.[4]

Musical Theory

Let us also define our musical terms. Remembering that Wright was a good pianist let us pursue the musical analogy. We begin with the familiar diatonic scale as it appears on a piano keyboard. We select the tone of C which is in approximately the centre of the keyboard and play upward to the octave using only the white keys. The succession of sounds produces the scale, which at some point of our education, most of us have been taught to play or sing. This scale is really rather peculiar. The composer Douglas Moore wrote:

> Look at the keyboard and you will see that between the third and fourth note or E and F there is no black key providing for intermediate tone, nor is there any to be found between the seventh note B and the octave. The interval between E and F and between B and C is the same as the interval between C and the black note above it. By interval is meant the number of vibrations of the higher tone to that of the lower. If you were able to count the number of vibrations, you would find the ratio exactly the same. The piano is conveniently divided into 88 tones which are an equal interval apart counting both white and black notes. In our system we call this interval between tones a halftone. The interval between C and D which have a black note between them is therefore that of a wholetone and between E and F and B and C where the black key is missing, the interval is only that of a half tone (Moore 1937: 22–23).

The concept of interval is at the beginning of the idea of harmony. Every note of the scale, when combined with a tone, has a sound which varies greatly in consonance. Most Western musicians agree that the order from consonance to dissonance is octave, fifth, fourth,[5] third, sixth, second, and seventh. There are

---

[4] Professor Narciso Menocal was the first to demonstrate a system for determining the elegant geometric ratios which the architect employed to control his window designs. By careful measurement Menocal elucidated the modular system of squares and rectangles which Wright used to organize the design of a Tree of Life Window from the Martin house (Buffalo, 1904) at the Elvehjem Museum in Madison, Wisconsin. While acknowledging Professor Menocal's primacy in the geometric analysis of Wright's windows, we differ with him in certain respects. Menocal essentially argues that Wright used two modules. One was 0.472 of an inch across by 0.486 in the vertical dimension. The other module, for the area at the bottom of the window, is slightly different in measurement but very close to the first. We question Menocal's findings because we do not see how such fine tolerances could have been obtained with the measuring devices of 1904. Wright and all his staff and colleagues of that time would have done design and layout with a conventional architect's scale or a full-scale metal rule. Menocal's decimal divisions, 0.472 and 0.486, could not have been laid out with such devices. They do not relate to the halving subdivisions of an inch (1/2, 1/4, 1/8, etc.) and they are carried to the third decimal place. They differ from one another by only 0.014, a tolerance which cannot be measured with pre-1985 instruments.

[5] Professor Steve Larson of the Music School at the University of Washington points out to me that the fourth was at one time considered a consonance but for many years has been classified as a dissonance.

various reasons for our preferences among the consonances, but a common explanation emphasizes the simplicity of the ratios between the vibration numbers of the limiting sounds: Octave, 2:1; Fifth, 3:2; Fourth, 4:3; Major Third, 5:4; Minor Third, 6:5; Major Sixth, 5:3; Minor Sixth, 8:5.

Every pianist, whether a beginner or virtuoso, must observe these mathematical relationships. With even this minimal background it is easy to agree with the philosopher/mathematician Gottfried Leibniz that "music is the pleasure the human soul experiences without being aware that it is counting".[6] In his Norton lectures Leonard Bernstein remarked, "the fact is that music is not only a mysterious and metaphorical art; it is also a form of science. It is made of mathematically measurable elements: frequencies, durations, decibels, intervals."[7] More recently Edward Rothstein has claimed that music and mathematics have the same kind of "inner life" (1994). In an intricate argument he holds that musical and mathematical beauty are substantially alike. Architecture is only beginning to receive such rigorous treatment. Here we simply want to demonstrate that certain of the art glass windows of Frank Lloyd Wright have a musical quality which is dependent on mathematics. We believe that for these windows he used a set of dimensions which, in an abstract way, resemble a musical scale. It is no wonder that many critics have found a musical quality in Wright's art glass.

The Windows of the Meyer May House

In 1908 Frank Lloyd Wright built a house for Meyer May, who owned a department store in Grand Rapids. The glass in the library corner provides a good overview of the glass in the house (Fig. 67.1).

There are five different types of art glass designs for the May house: windows for the living room, dining room, bedroom, main door and kitchen door. All are linked in terms of dimension and proportion; the unit dimensions which are repeated throughout the windows are, to say the least, provocative. They are 2″, 1-3/4″, 1-1/2″, 1-1/4″, 1-1/8″, 1″, 7/8″ and 5/8″. The unit of organization or cell is 1/8″. The intervals are somewhat more understandable if we write them as: 16/8, 14/8, 12/8, 10/8, 9/8, 8/8, 7/8, and 5/8. At first glance this appears to resemble an octave with eight major tones and two types of interval. There are intervallic relationships analogous to whole tones and half tones in the diatonic scale. But the intervals do not fall as they would in a conventional octave. We conjecture that it is a mathematical series which Wright devised in order to secure conveniently the maximum number of consonances—that is, combinations of small whole numbers

[6] The quotation from Leibniz is at the head of Chapter XIX in Kline's *Mathematics in Western Culture* (1964: 287) and is followed by a brilliant discussion of the work of Joseph Fourier.

[7] Bernstein (1976: 9). Bernstein's Norton lectures, in which he attempts to apply Chomskyan linguistics to music, are extremely technical.

Fig. 67.1 Library corner in Meyer May House, Frank Lloyd Wright, 1908. Photo: author

pleasing to the eye as consonances are pleasing to the ear in music. We could even say that Wright generally avoids dissonant ratios such as the Major Seventh (15:8), the Minor Seventh (16:9), and the Tritone (45:32). The repetition of 1-1/8″ and 2″ is especially significant. The ratio of these two numbers is 9:16, equal to the double sesquialtera of 4:9, a significant musical proportion which Alberti describes as "twice the width plus a further double tonus".[8] We do not claim that Wright knew a lot of music theory—but he was knowledgeable enough to remark that in his own work he preferred consonances and would leave the dissonances to others. With this scale of his own devising he could work very much as a composer at the piano. The 1/8″ dimension is, of course a constraint, like the limited range of the flute or clarinet. In mathematical terms he could easily secure a variety of musical ratios.

8 Leon Battista Alberti, *De Re Aedificatoria*, IX.6 (Alberti 1999: 306 and 409 n, 94). I owe this last observation to Ms. Kim Williams in a letter of September 18th, 1998. References by Wright to the importance of music for his architecture are abundant in his writings; the most important is Wright (1946: 200–202).

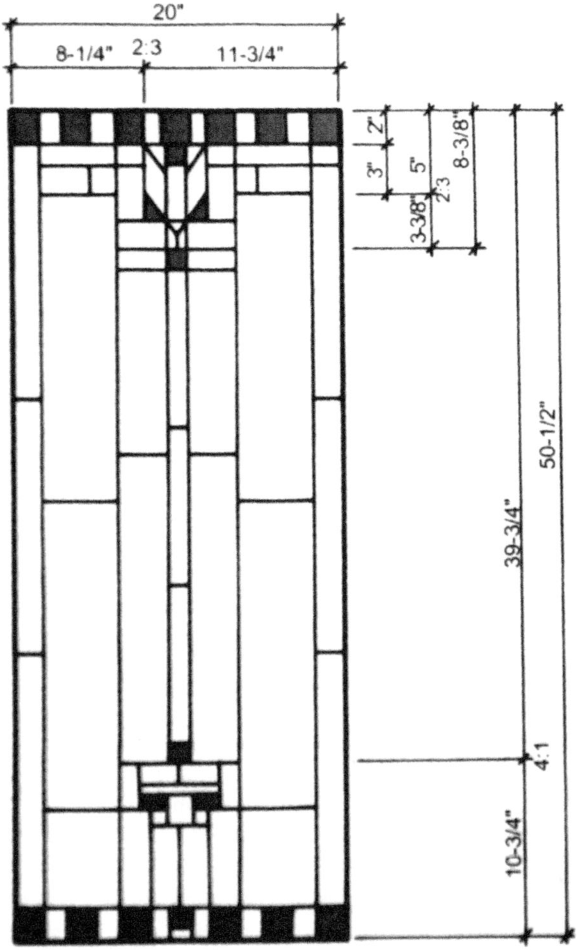

Fig. 67.2 Living room window. Drawing: Doug Smith

The units 1″, 7/8″, and 5/8″ are each in a 1:2 ratio to 2″, 1-3/4″, and 1-1/4″ respectively. All the glass except the front door is of a simple floral or tree motif which has three divisions: the base (or root), the trunk (or stem), and the branches (or crown). Our major tool for the examination of this glass is the analytical drawing. We begin with a window in the living room (Fig. 67.2).

The base appears in all but the bedroom window and the entry door. Let us examine it closely. It is like a small song within a larger composition (Fig. 67.3).

Centred in the lower portions of the design, the figure is contained within a 7-1/4″ by 2-3/4″ rectangle. A 1-1/4″ border runs on both sides. Two horizontal lines cross the centre of the design. One is 1-1/4″ below the top. The other is 5/8″ or half the upper width below the first for a 1:2 ratio. This arrangement leaves a 7/8″ band along the bottom. At the heart of the design is a 1-3/4″ square, twice the 7/8″ dimension on a side, with the top edge 7/8″ above the bottom line, leaving 7/8″

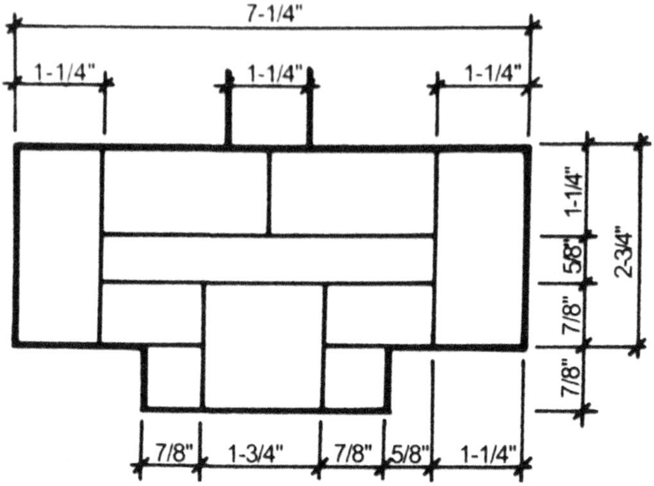

Fig. 67.3 Detail of base figure, living room window. Drawing: Doug Smith

below the rectangle. Adjacent to the 1-3/4″ square and below the main rectangle are two 7/8″ squares.

This root figure is a small composition based on 1-3/4″ and 1-1/4″ and their half dimensions, 7/8″ and 5/8″ (Fig. 67.4).

It reveals many 1:2 relationships. The height of this figure is a constant 2-3/4″ in all the windows where it appears, but the width is adjusted at the outside vertical band to fit the upper border unit widths to which it visually relates. The figure is slightly altered in the entry door composition. In studying it I am very much struck by the resemblance to the "line ideas" of Arthur Wesley Dow.[9] Kevin Nute has noted that Wright never acknowledged that he had met Dow, but that it was very likely he did so. Dow was a disciple of Fenollosa, the famous Boston orientalist. Here, then is another example of the syncretic quality of Wright's mind.

In these living room windows a 1-1/4″ vertical element centred in the composition connects the base figure to the floral crown. The crown shows a delicate system of triangles and parallelograms, which introduce a thematic variation within the design (Fig. 67.5). The central 1-1/4″ band is symbolically the stem or trunk linking the base or roots to the flower or crown. It terminates each central figure with a 1-1/4″ square of coloured glass. The entire colour scheme is restrained. It consists of clear glass, yellow and amber. The mid-section of the window is clear glass except for the crossing bars. Carla Lind remarks that the row of windows dissolves the wall and obliterates the corner. At the same time, it is a complicated exercise in geometry—and very musical.

[9] For a discussion of Dow's line-ideas, see Nute (1993: 86–96).

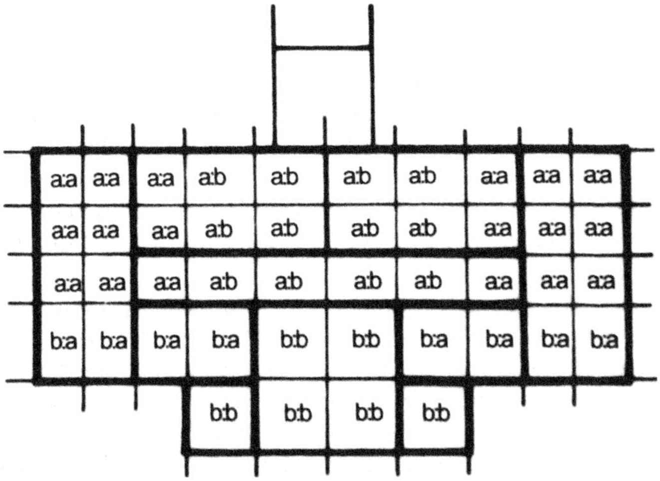

Fig. 67.4 Base figure with proportional relationships. Drawing: Doug Smith

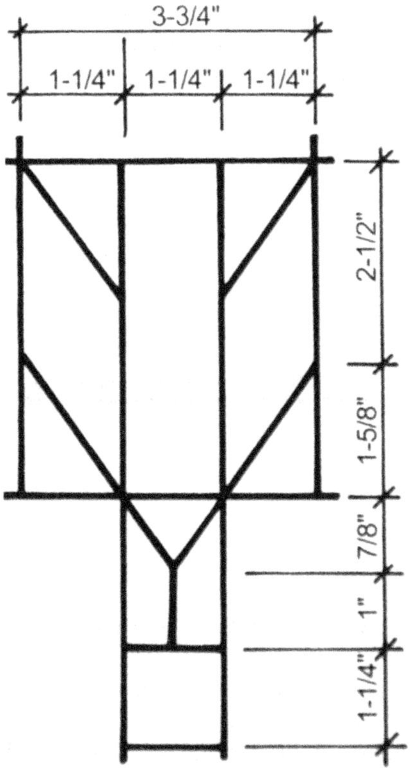

Fig. 67.5 Detail of upper floral figure, living room window. Drawing: Doug Smith

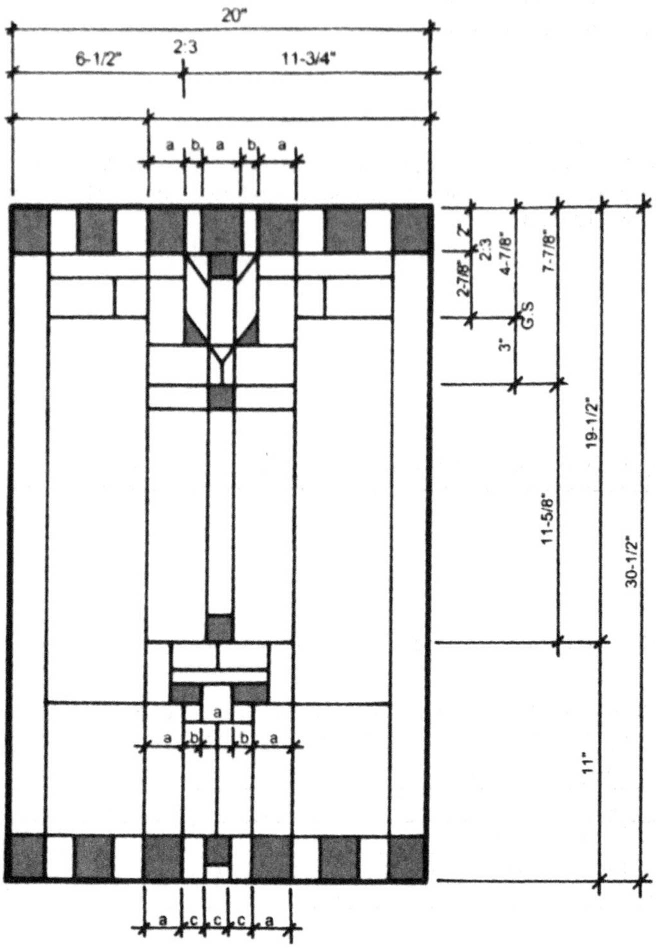

Fig. 67.6 Dining room window. A = 1-3/4″; B = 7/8″; C = 1-3/16″. Drawing: Doug Smith

The Dining Room Windows

In the dining room windows we again encounter Wright's mathematical series (Fig. 67.6).

Its most important property is that it generates ratios which approximate those of the piano scale. Thus we find a good many measurements in the approximate ratios of 5:3 (the Major Sixth) and 8:5 (the Minor Sixth). These ratios yield decimal values of 1.666 and 1.60. The values are close to that of the golden section (1.618). It would be tempting to say that Wright sought to base his windows on the golden section, but that statement would be misleading. Because he used a series which inevitably generated a number of approximate golden sections, some are present, but in the overall scheme they are less important than the other ratios which come close to those of the musical scale such as 2:1 (the Octave) and 3:2 (the Fifth). We are constantly bumping up against the fact that exact decimal values require a precision which is difficult, if not impossible, to achieve in leaded glass. Often we run into a deviance which is too great to be convincing if we hypothesize that Wright was designing on the basis of the golden section. Thus the design measures 30 1/2″ overall and is divided 19-1/2″ from the top so that we obtain a decimal value of 0.56—too great a deviance to be allowable. A truer approximation of the golden section would demand a division at 18-7/8″. Although the came does not reach the edges of the glass, it becomes the top of the stem and root portion of the design. The 19-1/2″ portion is divided at 11-5/8″ above this line. For a more proper golden section the division should be at 12.05″—and again we run into an unallowable deviance. The lower limit of the floral portion of the design is stopped at this point. In this small section of the window we have two more approximate golden sections. The remaining 7-7/8″ is divided at a point 4-7/8″ below the top of the design (in this case we are close to a proper decimal value [4.87 vs. 4.85]). The 4-7/8″ is again divided at a point 2″ below the edge of the window. The horizontal ordering of the window is linked through a series of approximate golden sections which develop into an approximate Fibonacci relationship (2, 3, 4-7/8, 7-7/8 as 2, 3, 5, 8).

Further, we can say that the vertical divisions of the window show a few golden sections. The importance of these should not be overestimated. We are immediately reminded of the wise words of Marco Frascari and Livio Ghirardini: "Approximate measures are tangible and tamable, whereas uncompromising measures are elusive. The Golden Mean is an untamable and intangible measure, since, in order for it to be real and efficient, it must be explicitly exact" (Frascari and Volpi Ghirardini 2015). They go on to say that for the architect who wants to employ the golden mean in an actual building, some real world factor such as the thickness of a mortar joint is always getting in the way. With Wright it was the dimension of a window came. However, we suspect that the approximations which produced the approximate golden sections are a consequence of the musical series which he had set up as an ordering device. We are, after all dealing with an architect who had a remarkable mathematical and musical intelligence. We cannot refrain from observing that the window might be interpreted as a musical composition in which the Minor Sixth and the Major Sixth are dominant chords.

The middle and lower sections of the design mix units from the above and introduce new ones as in the development of thematic material in a musical composition. The horizontal spacing of the elements which connect the 2″ base to the root figure mirrors that of the top band (a̲, b, a̲, b, a̲). As noted earlier, the spacing across the central section of the bottom band mirrors that of the main floral figure (a̲, c, c, c, a̲). A complex rhythmic pattern is thus established between the two

Fig. 67.7 Upper bedroom windows. Photo: author

sections as they exist in the upper and lower portions of the window. There is a strong analogy to the mirroring effect in Bartok. A good example is in his "Sonata for Two Pianos and Percussion" (1937).[10]

Upper Bedroom Windows

The typical upper bedroom window (Figs. 67.7 and 67.8) is a bilaterally symmetrical floral design with overall dimensions of 27″ wide by 44-1/2″ high. Close examination reveals the same kind of approximation of the golden mean that we encountered earlier. Again we think that it is adventitious and that the musical series are more important. These dimensions are close to a golden section relationship since the width to height ratio is 0.606 (golden section = 0.6181). A 2″ high band runs along the top and bottom of the design. The top band is divided into 1-3/4″ and 1-1/2″ wide units except at the point where the 1-7/8″ floral design joins the band. The bottom band differs from the top in that it utilizes a combination of 1-3/4″ and 2″ width divisions outside the vertical floral design zone and a 1-3/4″ and 1-5/8″ width divisions below the floral design. This variation with the top band spacing establishes a polyrhythm which is sensed through the interplay of alternating vertical lines from these lower spacings into the upper section of the design.

The window is divided horizontally 27″ from the bottom so there is again an inadmissible golden section (0.64) of the overall height (a truer golden section

[10] There is substantial literature on the employment of the Fibonacci Series and the Golden Mean in twentieth century music. This mathematics is especially important with Debussy and Bartok. A short bibliography should include Bachmann and Bachmann (1979), Howat (1983), Kramer (1973), Lendvai (1962), Lowman (1971). We will note that Bartok was fascinated with the occurrences of the Fibonacci series in nature. In his response to the mathematics of the natural world he closely resembles Wright.

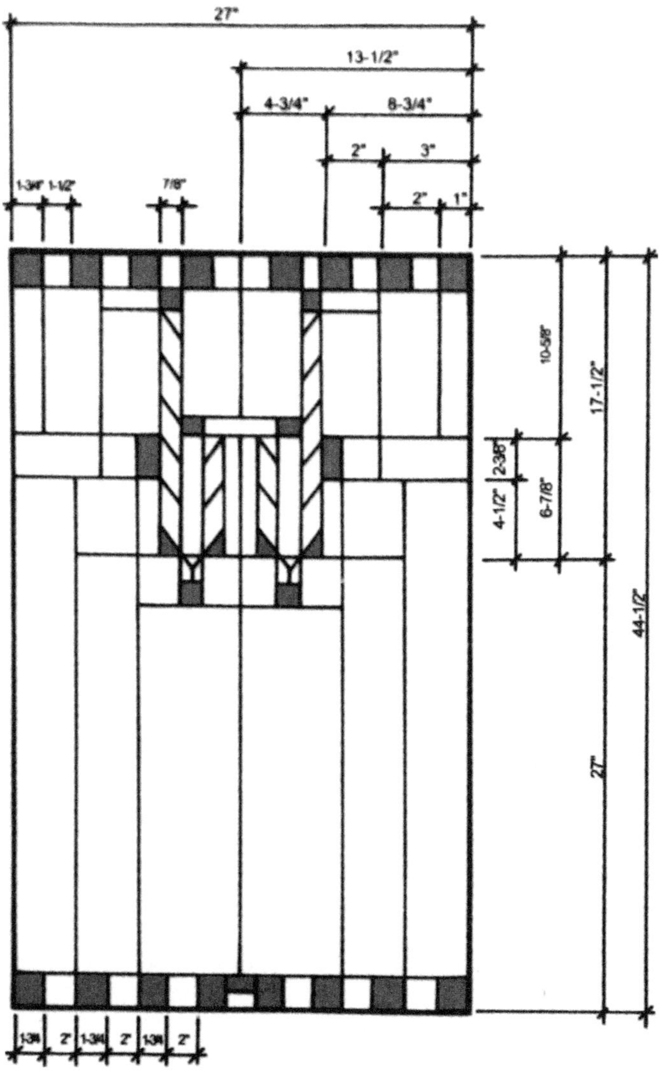

Fig. 67.8 Upper bedroom window. Drawing: Doug Smith

would fall at 27.5″). This line, which does not extend to the outside edges of the window, acts as a baseline for the major floral vertical elements. The remaining 17-1/2″ overall height is further divided at 10-5/8″ from the top which is almost an exact golden section (0.61). The remaining 6-7/8″ further divides at 4-1/2″ for an inconvenient ratio of 4.5–6.87. All of these divisions define significant horizontal structural elements within the design.

In addition to those relationships, there are other mathematical devices in these windows. The vertical line at 8-3/4″ establishes the outside edge of the diagonal elements in the floral pattern. The 8-3/4″ distance is comprised of five 1-3/4″ segments divided at the 2:3 ratio point so that we encounter a partial Fibonacci series. The 5-1/4″ distance is split into three 1-3/4″ segments which are again divided at the 1:2 ratio point for yet another series (1, 2, 3).

The central floral portion of the design is basically constructed within a 9-1/2″ by 17-1/2″ rectangle. This floral design consists of two 1-1/4″ vertical bands so arranged that there is a 1″ distance to the centreline. The central 1-1/4″ vertical band extends below the baseline established by the height into a 3″ high band below. The final 1-1/4″ forms a square which is filled in with white glass and acts as a lower terminus to the floral form.

Main Entry Door

Like the windows in the house, the doors are bilaterally symmetrical. The main entry door (Fig. 67.9) has overall dimensions of 32″ wide and 54-3/4″ high so that we are again confronted with an inadmissible golden section of 0.68. The design, which contains variations of the root and stem motif of the windows, is more purely geometric in character. As in the living room windows there is a 2″ border at the top and bottom of the design. A 1-1/2″ border frames the sides. A 1-1/2″ high band runs directly below the top 2″ border but does not extend to the outer edges.

The height of the design is divided by two horizontal lines into three sections. The bottom of the upper section is defined by a line which does not cross the entire composition but stops at the edge of the primary vertical elements. This line is 36-1/4″ above the bottom of the design at approximately two thirds of the 54-3/4″ height. It is in a 1:2 ratio with the 18-1/2″ remaining above. This 18-1/2″ piece is split into a 2:3 ratio at the base of the largest coloured rectangles. This point, 11-1/8″ below the top of the design, marks a distinct change in the character of the upper section of the design from repeated vertical squares to shifting rectangular blocks. The "root" figure in the upper section of the door rests on a line 5-3/4″ below the top of the glass. These two figures are variations on the "line ideas" used at the base of the living room windows. They divide the top 11-1/8″ into two parts having a 1:1 ratio. The lower section of the design is defined by a line 15-1/2″ above the bottom of the glass. It divides the 36-1/4″ into a 3:4 ratio. Wright projects a 1″ square form above and below the lines which define the lower and upper sections of the design.

The width of the window between the outside borders is divided by four vertical lines into five sections. The lines which frame the central panel are 19-1/2″ from each outside edge. Here we are close to 3:2, the ratio of the Major Fifth. Moving 5-7/8″ to the outside of the central lines are other parallel lines from top to bottom. The primary design elements are placed within these two 5-7/8″ bands. The widths of the vertical bands across the 8″ width of the central figure of the design are also related through small number ratios. The widths of *a* and *b* (1″ and 1-1/2″) are in a 2:3 ratio and widths *b* and *c* (1-1/2″ and 3″) are in a 1:2 ratio.

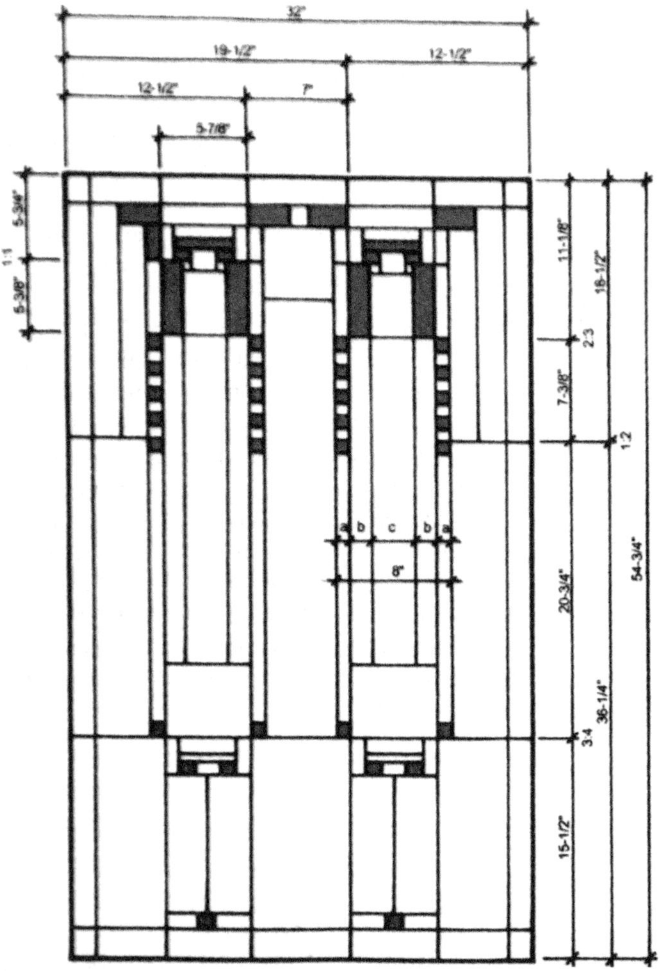

Fig. 67.9 Main entry door. a = 1″; b = 1-1/2″; c = 3″. Drawing: Doug Smith

Kitchen Door to Terrace

The door from the kitchen to the east terrace (Fig. 67.10) is a symmetrical floral design which incorporates elements from the living room windows and the main entry door.

The glass measures 22″ wide by 57″ high for a 1:2-1/2 ratio of width to height. A 2″ band crosses the bottom of the design. A top band is lacking. The divisions across the bottom vary greatly but establish a sequence (a, b, a, c, a, d, d, d, a, c, a, b, a) reminiscent of symmetrical arch forms in music (A, B, A, C, A, B, A). Bartok was fond of such forms, especially in his string quartets. Vertical bands 1-3/4″ in diameter enclose the sides of the glass. In the bottom band the dimensional units are

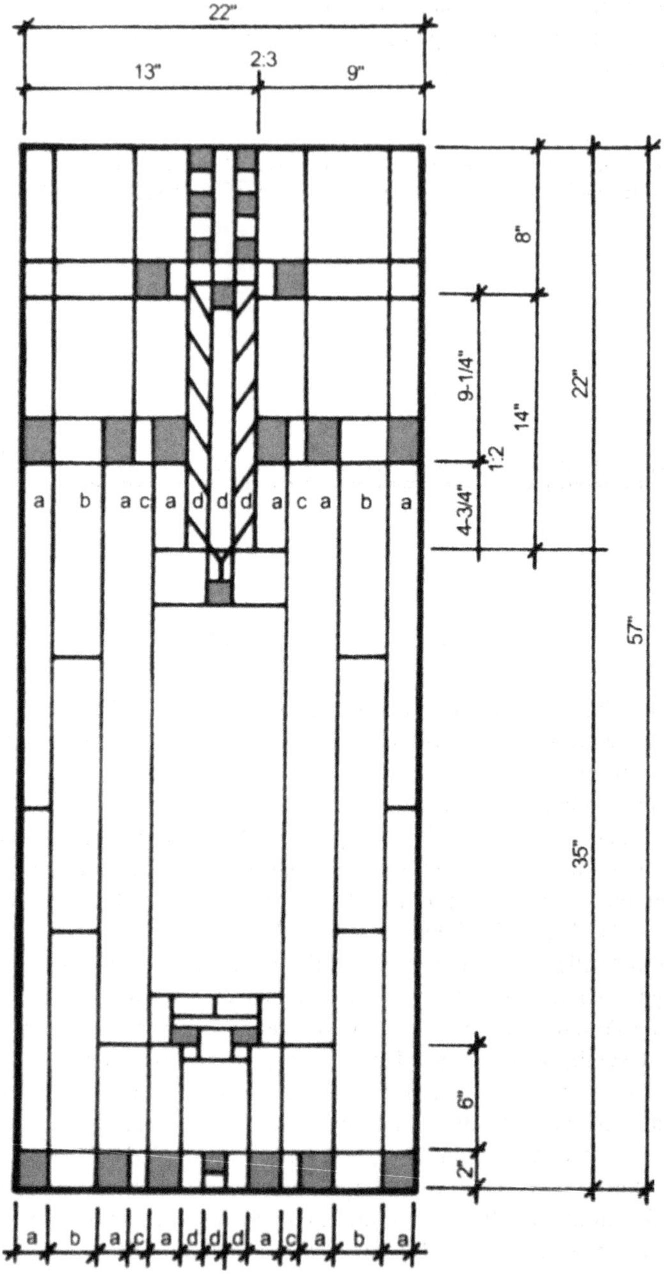

Fig. 67.10 Kitchen door. A = 1-3/4″; b = 2-3/4″; c = 1-1/8″; d = 1-1/4″. Drawing: Doug Smith

also of great interest. All are 2″ in height, but the widths vary: 1-3/4″, 2-3/4″, 1-1/8″, 1-1/4″. The rectangles thus generated measure 2″ × 1-3/4″, 2″ × 2-3/4″, 2″ × 1-1/8″, and 2″ × 1-1/4″. This is an extremely delicate series of relationships, much like the line idea which reoccurs 8″ above the bottom of the glass.

The design is divided by a short horizontal line 35″ from the bottom leaving 22″ above—an amount equal to the overall width of the design. This point approximates (0.628) the golden section of the height of the glass (a more exact golden section would fall at 21.63). Wright creates a square in the upper portion of the design within which the floral elements are developed Once again one has the feeling that Wright might have achieved a true golden section but was constrained by the size of the door and by came dimensions. The line, which does not extend to the outside edges of the design, is a baseline for the floral elements. Only the 1-1/4″ wide centre element projects 3″ below this line. This upper 22″ square is divided at a point 14″ above the baseline; again we have a barely admissible approximate golden section of.636 (a truer golden section would fall at 13.6). This important line establishes the bottom of a 2″ high horizontal band which crosses the composition and acts as a transition point. It marks a change. The outside bands of the central floral element change from diagonal lines to repeated squares. The 14″ below this band is split into 4-3/4″ and 9-1/4″ (1:2 ratio) segments above the baseline. This line establishes the bottom of the 2-1/2″ high band which crosses the door and is almost identical to the band at the base. If the design is viewed as a whole from this band downwards, the door glass contains nearly the exact design elements and position as the living room windows. The line which supports the lower root or line idea is 6″ above the top of the 2″ band—a 1:3 ratio.

The horizontal spacing of the two upper bands and the vertical lines linked to them are in a slightly varied pattern which creates an interesting rhythmic counterpoint between the upper and lower sections of the design. The location of the vertical elements in the upper section of the design is such that the outside edge of the triple banded floral element is located at a point having approximately a 2:3 ratio to the 22″ width of the glass.

The other terrace doors in the house are of a slightly different width and height, typically 28″ wide by 60″ high. The basic design elements are the same with the adjustments occurring in the width of the outside vertical panels and in the height of the central open glass area. These doors are, then, an excellent example of Wright's ability to take the basic arrangement of elements in a design, vary the relationship of these to the overall frame, and thereby secure a new and strikingly different composition.

In 1941 Henry-Russell Hitchcock observed that the windows of Wright's Coonley Playhouse of 1912 foreshadowed certain varieties of non-objective painting in Europe. Perhaps it is not too much to say that these windows at the May House anticipated the *process* which Piet Mondrian went through in the design of the paintings in his mature style of the 1920s. Concerning these works Robert Hughes remarks, "he did not calculate mathematical proportions. He had no special belief in the golden section or anything like it. His way of painting was wholly intuitive, a matter of inspired guesswork and adjustment" (Hughes 1995: 93).

We believe that much the same comment could be made about Wright's design process for many of these windows. Interestingly, Hughes argues that the high point of Mondrian's intuitiveness came with his "Boogie Woogie" paintings. These were a response to modernist African American music. With Wright the situation is paradoxical. In his architecture he *wanted* to emulate Beethoven. In these windows he is much closer to Bartok. And he was constrained by came width and by the fact that windows and doors had to fill openings dictated by architectural requirements. And there was his determination to employ a strictly limited palette and to create a "light screen". Mondrian could select the size of his canvas and had a full palette of colours at his disposal. But the design process must have been strikingly similar. If, as Goethe remarked, "Architecture is frozen music", Wright came as close as any architect in recorded history to the idea in these windows.

Acknowledgment The author is pleased to acknowledge a grant received from the Kittredge Fund, Cambridge, Massachusetts, and a grant from the Graham Foundation for Advanced Studies in the Fine Arts. Chicago. Figures 67.2, 67.3, 67.4, 67.5, 67.6, 67.8, 67.9, and 67.10 drawings by Doug Smith, reproduced by permission.

Biography Leonard K. Eaton was Emil Lorch Professor of Architecture Emeritus, the University of Michigan, where he taught architectural history from 1950 to 1988. He has also taught at Wayne State University, Michigan State University, and the University of Victoria (British Columbia). In 1985 he was Margan Professor at the University of Louisville. He took his B.A. with highest honors at Williams College in 1943, and after war service with the 10th Mountain Division, received an M.A. and Ph.D. from Harvard University. His publications include: *Landscape Artist in America: the Life and Work of Jens Jensen* (1964), *Two Chicago Architects and their Clients* (1969), *American Architecture Comes of Age* (1972) and *Gateway Cities and Other Essays* (1989). He is best known for his work on Frank Lloyd Wright. His most recent book is *Hardy Cross: American Engineer* (University of Illinois Press, 2006).

References

ALBERTI, Leon Battista. 1999. *On the Art of Building in Ten Books*. Joseph Rkywert, Neil Leach and Robert Tavernor, trans. Cambridge, MA: MIT Press.

BACHMANN, Tibor and Peter J. BACHMANN. 1979. An Analysis of Bela Bartok's Music Through Fibonaccian Numbers and The Golden Mean. *Musical Quarterly*, **LXV**, 1 (Jan., 1979): 72–82.

BERNSTEIN, Leonard. 1976. *The Unanswered Question: Six Talks at Harvard*. Cambridge, Massachusetts: Harvard University Press.

BLATNER, David. 1997. *The Joy of Pi*. New York: Walker and Co.

DUTHE, Arthur. 1908. *Decorative Glass Processes*, New York: Van Nostrand.

FRANK LLOYD WRIGHT to the Taliesin Fellowship, 27 August 1952, FLW Archives, AU.1014.045

FRASCARI, Marco and Livio VOLPI GHIRARDINI. 2015. Contra Divinam Proportionem. Pp 619–626. in Kim Williams and Michael J. Ostwald eds. *Architecture and Mathematics from Antiquity to the Future: Volume I Antiquity to the 1500s*. Cham: Springer International Publishing.

HANKS, David. 1979. *The Decorative Designs of Frank-Lloyd Wright*. New York.
HOWAT, Roy. 1983. Review-Article: of Bartok, Lendvai and the Principles of Proportional Analysis. *Musical Analysis*, **2**, 1 (1983): 69–95.
HUGHES, Robert. 1995. Purifying Nature. In *Time*, October 23, 1995.
KLINE, Morris. 1964. *Mathematics in Western Culture*. New York: Oxford University Press.
KRAMER, Jonathan. 1973. The Fibonacci Series in Twentieth-Century Music. *Journal of Music Theory*, **17** (1973): 111–149.
LENDVAI, Erno. 1962. Duality and Synthesis in the Music of Bela Bartok. *New Hungarian Quarterly* **3**, 7: 91-114. Reprinted in Gyorgy Kepes ed. *Module, Proportion, Symmetry, Rhythm*. New York: George Braziller, 1966.
LOWMAN, Edward A. 1971. An Example of Fibonacci Numbers Used to Generate Rhythmic Values in Modern Music. *Fibonacci Quarterly*, **9**, 4: 423-426.
MOORE, Douglas. 1937. *Listening to Music*. New York: W. W. Norton.
NUTE, Kevin. 1993. *Frank Lloyd Wright and Japan*. New York: Routledge.
ROTHSTEIN, Edward. 1994. *Emblems of Mind: the Inner Life of Music and Mathematics*. New York: Random House.
SLOAN, Julie. 1997. The Echo of the Eaves: The Elevation in Frank Lloyd Wright's Prairie Window Designs. *Nineteenth Century* **XVII**, 2 (Fall 1997): 44–49.
WRIGHT, Frank Lloyd. 1946. *An Autobiography*. London: Faber and Faber.
———. 1959. *In the Cause of Architecture, Essays for the Architectural Record 1908–1952*, Frederick Gutheim ed. New York: Architectural Record.

Chapter 68
Fractal Geometry in the Late Work of Frank Lloyd Wright: The Palmer House

Leonard K. Eaton

During the incredibly long and fruitful career of Frank Lloyd Wright there are two constants: Nature and Geometry. That Nature was Wright's deity is well known. He summed up his attitude in the following language: "I wish more life to creative rhythms of great Nature, Nature with a capital N as we spell God with a capital G. Why? Because Nature is all the body of God we mortals will ever see." Several scholars have commented on this attitude. Donald Hoffmann wrote that, "Nature was, of course, Wright's deity," and quoted various passages from the architect's 1908 and 1910 writings to support his contention. Hoffmann traced this vein of thinking to the organic analogy in the works of Viollet-Le-Duc and noted that it was present in a variety of other thinkers as well (Wright 1957; Hoffmann 1969; Hertz 1993; Creese 1985). Wright placed Pythagoras *first* in the list of 33 historical figures to whom he acknowledged indebtedness at the conclusion of *An Autobiography*. We submit that "Pythagoras" should not be understood as a specific individual who created the theorem which bears his name. Rather, in this context "Pythagoras" must mean something like "the spirit of geometry." Recent scholarship has, in fact, stressed the importance of Wright's feeling for geometry. Anthony Alofsin has pointed out the impact of Wright's contact with the geometric forms of the Vienna Secession and examined the Secession influence in the Midway Gardens (1914) and the Imperial Hotel (1916–1922). Referring to Wright's use of the rectilinear grid, Narciso Menocal writes that it "was contingent on his conception of the universe as a geometric entity that architecture mirrors." Menocal traces this concern with geometry to Wright's absorption in the thought of Viollet-Le-Duc, whose *Dictionnaire Raisonné* (Viollet-Le-Duc 1856) he probably encountered while in the offices of Adler and Sullivan. Wright himself admitted that Sullivan's ornament had a powerful impact on him, and Sullivan's *System of Architectural Ornament*

First published as: Leonard K. Eaton, "Fractal Geometry in the Late Work of Frank Lloyd Wright: the Palmer House", pp. 23–38 in *Nexus II: Architecture and Mathematics,* ed. Kim Williams, Fucecchio (Florence): Edizioni dell'Erba, 1998.

Leonard K. Eaton (1922–2014)

K. Williams and M.J. Ostwald (eds.), *Architecture and Mathematics from Antiquity to the Future*, DOI 10.1007/978-3-319-00143-2_21,

had a strong foundation in geometry. In short, the sources of Wright's fascination with geometry are manifold. Despite his assertion that only mathematics had meaning for him during his time at the University of Wisconsin, the evidence is that he did not learn much. He received a grade of "C" in descriptive geometry. A possibility not to be lightly dismissed is that he worked out a lot of plane geometry for himself without help from anyone. In any event, by 1904 he had sufficient command of the discipline to organize the windows of the Darwin D. Martin house in a complex series of ratios, which have a distinct musical quality. Whether or not he was aware of such concepts as the Golden Mean and the Fibonacci series is a moot point. Wright used nature as the basis of his geometrical abstraction. His objective was to conventionalize the geometry which he found in Nature, and his method was to adopt the abstract simplification which he found so well expressed in the Japanese print. Therefore, it is not too shocking perhaps that in this quest his work would foreshadow the new mathematics of nature: fractal geometry (Alofsin 1993; Menocal 1992).[1]

In the work of Wright's Prairie years, his attitude toward Nature and Geometry is perhaps most easily seen in the art glass of his windows. These windows, however, were placed in houses that, like those for hundreds of years past, were built with a Euclidean geometry of right angles, rectangles, and squares. Sometime in the early 1920s Wright became discontented with this conventional geometry and became interested in composing his plans in a radically new manner. There is nothing unusual about the plan of the Alice Millard house of 1923, but only a few months later he produced his amazing project for the Little Dipper Community Playhouse at Olive Hill, the estate of Aline Barnsdall [see the plan in Pfeiffer (2011: 190)]. Robert Sweeney has rightly noted that the plan of the Little Dipper features the intersection of an irregular square and a circle. It is symmetrical on two axes but it employs a non-conforming wing so that the balance of the composition is disturbed. Sweeney writes:

> The enclosed schoolroom, the square, is oriented on its diagonals; the stage, in one corner, is on axis with the outdoor seating area. The cross-axial space is defined in plan and section by a hexagonal 'overhead' lantern built of open trusses separated by windows, which extends the length of the building (1993: 45).[2]

---

[1] Professor Menocal very kindly sent me a copy of Wright's transcript at Madison in a letter of 24 May 1992. The suggestion that Wright worked out much of plane geometry for himself comes from Professor Grant Hildebrand. See also Wright (1967). Richard Joncas in his Ph.D. thesis Joncas (1991) argues that Wright's fascination with geometry stemmed from the Transcendentalist ideals instilled in him since childhood. Joncas sees the beginnings of a non-rectangular geometry in the early works of the 1890s, many of which featured polygonal and circular forms. I am unable to accept this contention but agree with Joncas that Buckminster Fuller may have been a powerful influence on Wright during the 1920s. Fuller's Dymaxion house dates from 1927, and the first project in which a triangular geometry was used as a module seems to have been one of the outbuildings for San Marcos in the Desert (1928).

[2] Sweeny (1993) is an excellent book on Wright's work of the 1920s. This kind of diagonal planning reoccurs in several of Wright's projects of that decade, notably the Lake Tahoe Summer Colony of 1923.

It must be emphasized that the geometry that governs this project is still Euclidean, though it is certainly non-rectangular. There is only a suggestion of the amazing developments that were to come.

With Wright, the experience of a new section of the country and the encounter with a new kind of landscape often coincided with a renewal of artistic inspiration. Such a renewal certainly occurred with his move to Arizona. Concerning the Ocatillo Desert Camp of 1927, he wrote:

> The one-two triangle we used in planning the camp is made by the mountain ranges around about the site. And the one-two triangle is the cross section of the talus at their bases. This triangle is reflected in the general forms of all the cabins as well as their general plan. We will paint the canvas one-two triangles in the eccentric gable scarlet. The one-two triangles of the ocatillo bloom itself are scarlet. This red triangular form in the treatment is why we called the camp "Ocatillo 'Candle flame'" (Wright 1946: 274).[3]

The ocatillo cactus does indeed flower a brilliant scarlet in the spring and its blossom can easily be interpreted as triangular. According to Webster, talus is "a slope formed (esp. by) an accumulation of rock debris." It usually occurs at the foot of a cliff, and is commonly the result of an earthquake but sometimes of glaciation.

The first executed building in which the triangle was used as a planning module is the Desert Camp of 1927 (Fig. 68.1).

Although this structure lasted only a season it was extremely important in Wright's career. Thereafter the triangular module reappeared in a large number of both unexecuted projects and finished buildings. At this point we should emphasize that the so-called "one-two" triangle to which Wright was alluding was the 30°–60°–90° triangle (Hoffmann 1998: 68), found with most drafting kits. This triangle has a short side of 1 and a long side of 1.7321 ($=\sqrt{3}$) with a hypotenuse of 2. Most importantly, it is exactly one-half of an equilateral triangle. The long side is an irrational number, the first of many which we find in Wright's later work.

The shift to equilateral triangles which serve as modules in themselves, or can serve as units in a parallelogram, occurred in the Carleton D. Wall house (Plymouth, Michigan, 1941).

Here Wright used a 30°–60° parallelogram grid with lines 2 ft on centre as the basis for the plan. As a gridded plan this house is fairly successful, with most of the wall elements falling near or on the grid. But some of the walls in the building fall across grid points as if the house were actually laid out on a triangular grid rather than a parallelogram. Further, the grid is dimensioned along the grid lines, making it difficult for the builder to lay out. Notwithstanding this problem, Wright suggested the direction that his architecture would take during the rest of his life with the name that he gave to the building: Snowflake. He would seek to emulate the beauty of one of Nature's most perfect crystalline forms. Perhaps the name foreshadows his later preoccupation with fractals.

---

[3] Wright used comparable language in a letter of June 1, 1928 to his son, Eric. In the first edition of his *Autobiography* (1932) he wrote of the one-two triangle as "magical". In later editions of the work he excised this word [but see (1977: 335) where it is restored]. The deletion might have been an indication of his desire to appear as an exponent of a rational and technological architecture rather than as an architect in tune with the mystical vibrations of the universe.

Fig. 68.1 Desert camp. Photo: author

In the Palmer house (Ann Arbor, 1950–1951), Wright evidently decided that the equilateral triangle was the most elemental shape in a grid based on 30°–60°–90° triangle and therefore was the most versatile shape with which to work. The key dimension is the 4-ft *altitude* of the triangle rather than the side. Wright's instructions on the working drawings for the house are clear: "The house is planned on a triangular shaped unit. All masonry walls have one face on the unit or half-unit lines. Wood interior partitions centre on unit or half-unit lines." The accompanying drawing, however, makes it clear that he was thinking in terms of his favourite one-two triangle. However, the hypotenuse of the triangle at 4′–7 9/16″ is now merely a fact of trigonometry, irrelevant to the basic design procedure:

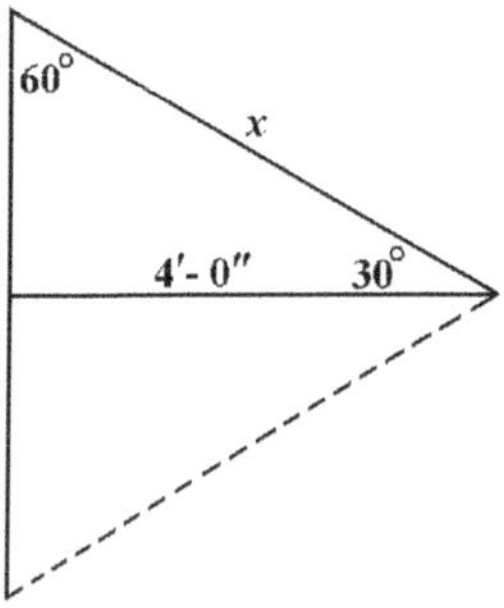

$$x = \frac{4'}{\sin 60^\circ} = 4.61888' \approx 4' - 7\,\frac{9}{16}''$$

(actually 4.630, a 0.2 % deviation).

The builder need not deal with the difficult measurement of the hypotenuse of Wright's beloved 30°–60°–90° triangle but can concentrate on the 4-ft altitude of the equilateral triangle. Testimony from the builder, Paul McDowell, is that he found the system complicated at first but simple when he became accustomed to it.

In the equilateral triangle Wright came closest to fulfilling his desire to create a geometry which would emulate Nature. In taking this direction Wright may have been unconsciously following D'Arcy Thompson, who studied the basic geometry of Nature in his great work of 1917, *On Growth and Form*. Thompson, whose reputation grows with the years, was probably the most prescient biological scientist of the twentieth century. His major effort was to put the biological sciences on the same mathematical footing as physics and chemistry. To do so he studied in depth the geometrical structure of a large number of natural forms, including beehives, snowflakes and rams' horns. Thompson was, for his time, a good mathematician, but today most scientists would say that he lacked the tools for the task which he had set himself.

Richard Voss correctly notes that the inability of Euclidean geometry to describe accurately the natural world has been overlooked until recently. "In retrospect," he remarks, "clouds are not spheres, mountains are not cones, coastlines are not circles, and bark is not smooth, nor does lightning travel in straight lines" (Voss 1985: 1). To describe these phenomena, a new geometry is needed.

"A fractal," says Hans Lauwerier, "is a geometrical figure in which an identical motif repeats itself on an ever diminishing scale" (Lauwerier 1991: 11).[4] These figures were at one time mathematical curiosities, but in the last two decades they have received a great deal of attention from both pure mathematicians and physical scientists. The father of fractal geometry is the Franco-American mathematician B. B. Mandelbrot. He states that a fractal must be self-similar, which means that each part is a linear geometric reduction of the whole, with the same reduction ratios in all directions. Our first example is a commonly cited pure linear fractal called a Sierpinski Gasket (Fig. 68.2).

It is simply a solid equilateral triangle which features a series of ordered reductions in the same shape. Theoretically these reductions could be continued to infinity, in which case we would have an object with a perimeter but no area. A similar reduction in three dimensions can be performed with an equilateral pyramid called a Sierpinski arrowhead.

The other most commonly shown example of a pure linear fractal is a Koch snowflake (Fig. 68.3).

To construct this fractal, begin with an equilateral triangle with sides of length 1 as in the diagram. This figure is the generator. We proceed to add a new triangle

[4] For the mathematically minded, I offer the following definition: "A fractal is a type of set produced by a rule called recursive—one keeps applying the same transformations to parts of a set that one applies to the whole. This means that any portion of a fractal curve contains the same types of movements as the whole; any portion, magnified, will reveal as much information as the whole. Thus, out of a simple set of proportions, the most complex curves and properties can be described" (Rothstein 1994: 162). For a discussion of fractals in architecture, see Ostwald (2001).

Fig. 68.2 Sierpinski gasket. Drawing: Kim Williams

Fig. 68.3 Koch snowflake. Drawing: Kim Williams

one third the size as in the intermediate stages of the diagram and continue the process indefinitely. How long is the perimeter? After n iterations it will have increased $(4/3)^n$ fold over the perimeter of the initial triangle. Hence, as we approach infinity, the perimeter becomes infinitely long. To describe the perimeter's size we can therefore no longer use its length and must use the new mathematics of fractal geometry.

We have been discussing pure linear fractals which are mathematical abstractions; the starting point of the discipline was Mandelbrot's *The Fractal*

Geometry of Nature (1982). The attraction of fractal geometry derives from the fact that it offers a method of describing and measuring all kinds of natural phenomena that have hitherto resisted analysis. Many of these phenomena turn out to display essentially the same kind of iteration and invariant scaling as the Sierpinski gasket and the Koch snowflake. The fractal approach has, for example, proved especially helpful to geophysicists whose task it is to measure and describe the extremely bumpy surface of the earth. Consider the problem of measuring and describing a classic geological formation, a talus slope on a mountainside, such as the one Wright mentioned in *An Autobiography*. From a distance the talus appears to be a Euclidean shape, perhaps a pyramid or a cone. Presumably one could walk across it, measure it, and calculate its dimensions. But, says James Gleick, as the geologist approaches, he finds that he is not so much walking *on* it as *through* it. The stuff is composed of jagged boulders in a variety of sizes. Its three-dimensional surfaces hook over and wrap around the human being. Crossing a field of talus can be a strenuous business. It is not at all like walking across a tennis court. The tennis court is a man-made bit of Euclidean geometry. The field of talus is a piece of fractal geometry built by nature. Notwithstanding its apparently random character, geologists have discovered that talus slopes are usually from 34° to 37°. Using fractal geometry we can determine from this slope that a man scrambling across the field will go about 2.7 times as far as he would if he were traversing a flat surface like a tennis court. This figure of 2.7 is considered the effective, or fractal dimension of the field of talus. It is analogous to the 4/3 which governs the iteration in the Koch snowflake (Gleick 1988: 103–107).[5] The field of talus is a homely example of the great class of random fractals which are so common in the natural world. Its analysis by fractal geometry confirms Eugene Wigner's famous remark that mathematics is "unreasonably effective" in describing the world of nature.

But an architect is by definition a master builder, one who is concerned with man-made objects, not with inert natural things. Is there any analogue among man-made constructions to the field of talus? James Gleick suggests that the structure of the Eiffel Tower is a good 3-dimensional approximation.[6] Its beams, trusses, and girders branch into a lattice of increasingly fine members so that one is lost in a fantastic network of increasingly fine detail. Eiffel could not carry the scheme to infinity, but the effect is striking. We think that the work of Frank Lloyd Wright's last decade is an even more striking anticipation of fractal geometry.

[5] The example of the fractal dimension in the field of talus is crude. For a proper description of the derivation of the fractal (or Hausdorff) dimension, see pp. 9–10 in Schroeder (1991: 9–10). For a provocative application of fractals to a scientific problem, see Outcalt and Melton (1992).

[6] Dr. Leonard Schlain, author of *Art and Physics* (1993), points out to me that the Eiffel Tower was a structure fully in tune with the scientific and mathematical developments of its period. It provided a platform for the yet to be invented transmission of radio waves. Also, it was the first major structure whose construction was fully documented in photographs. Conversation with Dr. Schlain, October 20, 1994.

We have concentrated on Wright's Palmer house in Ann Arbor, Michigan of 1950–1951. Its basis, as we have remarked earlier, is an equilateral triangle as module. Let us then see how the new geometry worked out in the Palmer house.

In analysing the house we must remember that it is neither a mathematical abstraction like a Koch snowflake nor a random fractal like a field of talus. By its very nature it cannot be a pure fractal because it is a structure designed by a real architect and built by real craftsmen for real people living in a real world. Nonetheless we will find the basic fractal elements of iteration and invariant scaling. We can see these qualities beginning to develop in Wright's earlier houses on a hexagonal module and in the Carl Wall house as well, but nowhere are they carried so far as in the Palmer house. Edgar Kaufmann Jr., a great admirer of the Palmer house, wrote that it was a "fugue." If repetition and variation of a single theme are the essence of a fugue, Kaufmann may have been right on target. William Gray Purcell observed that a fugue was essentially a development from the old round song and gave as an example:

Soprano—Scotland's burning, pour on water—Scotland's burning, pour…
Contralto—Scotland's burning, pour on water—Scotland's burning, pour…
Tenor—Scotland's burning, pour on water—Scot…
Bass—Scotland's burning, pour on…

By the time all four parts of Scotland are burning at the same time, said Purcell, we have the makings of a fugue. If we play it with an orchestra until the listeners are also burned up, we will have classical music.[7] This, of course, is a humorous over-simplification, but there is a kernel of truth in it. If we consider the triangular module of the Palmer house as the basic statement of the fugue (Scotland's burning), we will note that it has the musical (and mathematical) property of maintaining form regardless of magnification or diminution. Wright was here achieving spatially what he had been able to do with his windows in two dimensions. Once again the musical analogy is powerful. The invariant form, then, is the equilateral triangle. Aside from its presence in the brick piers, which compose into parallelograms, there are no less than 11 scales of equilateral triangles ascending and descending from the basic triangle, which is 4′ in altitude (Fig. 68.4).

In our view the fractal tendencies in the design reinforce this position that Wright was seeking a biologically sound architecture. The new geometry was simply another means to that end.

Consider the fractal qualities in the Palmer house. The entrance, one of the finest that Wright ever designed, is marked by a triangular lamp. As the visitor proceeds up the delicately scaled flight of steps, on his left are rows of clay blocks cut out in a complicated triangular pattern. Under foot the slabs of tinted concrete are canted slightly (Fig. 68.5). As he passes the threshold, his movement is temporarily blocked by a bookcase. Thus the great space of the living room is all the more

[7] For William Gray Purcell's acute but sometimes cranky notion of the fugue, see Purcell (1950). A more sophisticated analysis is in Aaron Copland (2011, Chap. 12).

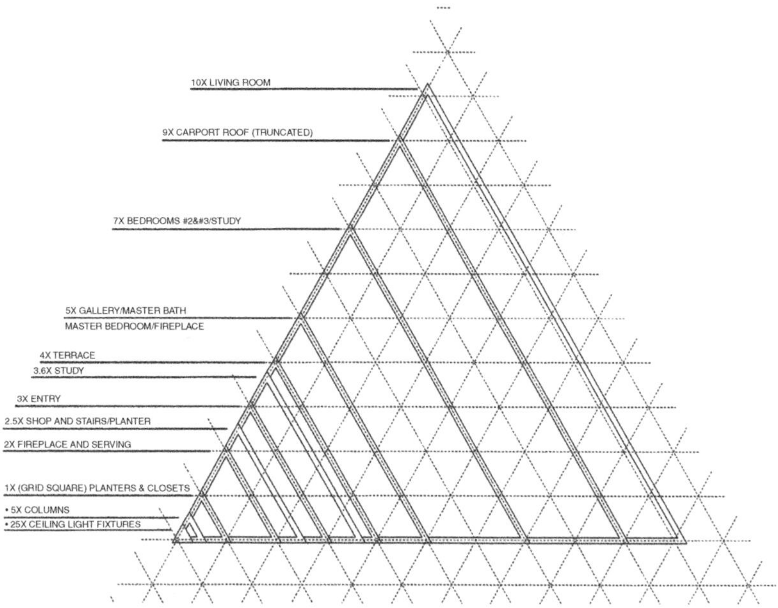

Fig. 68.4 Triangles and scale in the Palmer house. Drawing: courtesy Eric A. Murrell

Fig. 68.5 Entrance of Palmer house. Photo: author

effective when he turns the comer and it explodes out in front of him. The floor consists of slabs of concrete cast in the basic triangular module of the house (Fig. 68.6).

The slabs are tinted red, waxed, and over the years have acquired a wonderful glowing patina. Overhead are the great triangles of the ceiling. A view of the kitchen shows a 120° angle in the brick wall and the complex triangularity of the

Fig. 68.6 Interior of Palmer house. Photo: author

clay block. And at the low end of the scale are the small inset triangular lights on the soffit of the cove which runs around the room. A pair of drawings by architect Eric A. Murrell show the variety of dimension in the triangles which are visible in plan (Figs. 68.7 and 68.8). Remembering that we defined a fractal as "a geometrical figure in which an identical motif repeats itself on an ever diminishing scale," the Palmer house is an excellent illustration of the concept. An equal compatibility between definition and illustration would be found in a number of other possible diagrams. Even finer details, such as the perforations of the ceramic blocks or the entry lamp, are omitted from this diagram for reasons of size. Were the diagramming done on a larger scale, those details would add yet another micro-dimension of scaling. Nor is the recent garden wall included. It, too, would enrich the diagram. Furthermore, similar diagrams could be drawn in section; the fractal scaling of the Palmer house would then be seen to be eminently three-dimensional, as one would expect of the late work of the most three-dimensional of architects.

These observations have experiential value. Many of Wright's clients have observed that he makes little spaces seem large and offers new discoveries over a lifetime of contact. It is typical of fractals—the Koch snowflake is an instance—that they have long perimeters—even theoretically infinite perimeters—enclosing a finite area. So, too, Wright's architecture meanders without end, leading us along and through infinities of experience among its fractally bounded spaces, often within what, as here, is not at all a large building.

In addition, a kind of "nesting" of fractal forms can be observed at two points in the Palmer house: the entry way and the fireplace. At these places one encounters not only actual triangles but also implied (truncated) triangles. The result is a three-dimensional geometry of bewildering complexity. At the entrance there are not only the triangles composing the ceramic ornament, there is also a triangular light fixture atop of a triangular pier. There is a triangle jutting forward overhead, and triangles in the red-tinted concrete slabs underfoot. The fireplace hearth is a triangular cavity enclosed between triangular piers. The concrete slab on which the grate rests is a triangle. Overhead are the triangular soffit lights and the larger

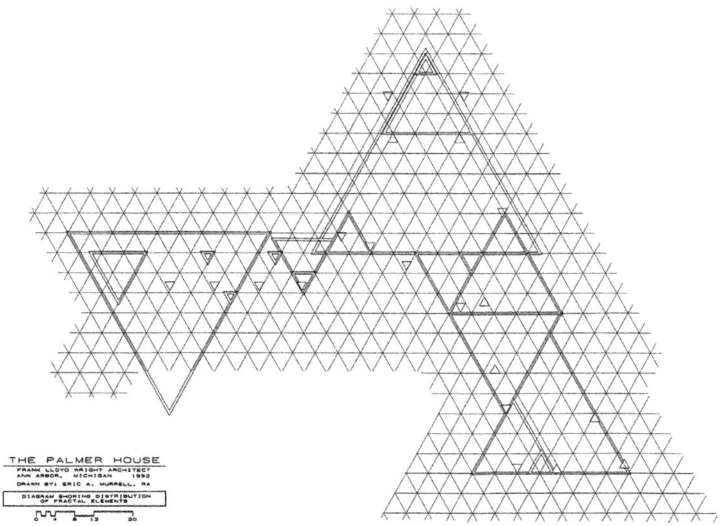

Fig. 68.7 Fractal elements in the Palmer house. Drawing: courtesy Eric A. Murrell

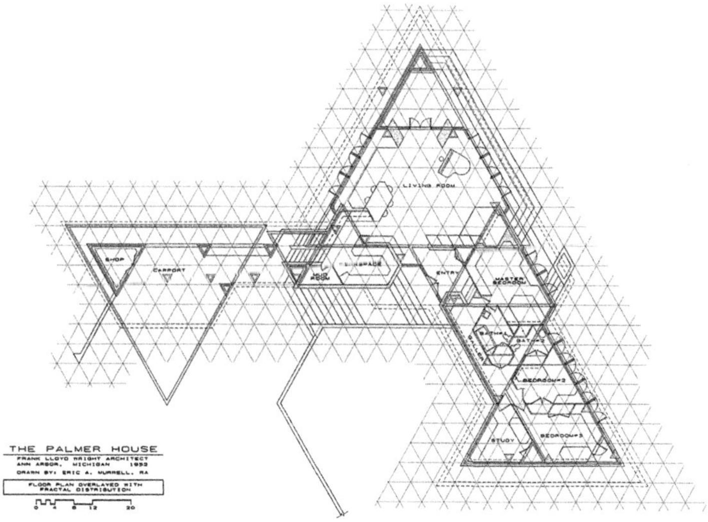

Fig. 68.8 Fractal elements of Fig. 68.7 overlaid on the plan of the Palmer House. Drawing: courtesy A. Eric Murrell

triangles of the ceiling. The hassocks are truncated triangles. Perhaps the most amazing detail in the entire house is the cast iron triangle on which the fireplace implements rest. This is almost a complete anticipation of Eric Murrell's drawing made to show the different scales of triangle in the house. It is as if Wright were hiding a clue at the hearth, or heart, of the structure.

It might be objected that diagrams such as these could be drawn for any building and that therefore the distinction claimed for this example is false. There is only a small measure of truth to this assertion. A typical floor plan of a rectilinear building by almost any architect would, it is true, offer the opportunity to pick out rectilinear patterns of floor tiles, appliances, bricks and the like, and thus could claim limited illustration of fractal approximations. But the Palmer case is entirely different. Firstly, the Palmer house presents iterations of precisely similar geometric units, not approximations of varying sizes. Secondly, none of the iterations is the serendipitous result of available manufactured materials; in the Palmer house the fractal quality is in every case the result of a *specific and conscious* act of design. We are not talking here about approximations or the odd chances of available catalogue choices. At this point, one might perhaps cast an eye backward to find similar conditions in much of Wright's earlier work, thus opening insights on his whole career. But it is the particular contribution of the Palmer house that Wright's manipulation of the triangular module reveals with special clarity, dramatically and beyond debate, his intuition of what we now recognize as fractal geometry—a discipline that was neither named nor recognized in his lifetime.

Acknowledgment A grant from the Graham Foundation supported the research in this chapter.

Biography Leonard K. Eaton was Emil Lorch Professor of Architecture Emeritus, the University of Michigan, where he taught architectural history from 1950 to 1988. He has also taught at Wayne State University, Michigan State University, and the University of Victoria (British Columbia). In 1985 he was Margan Professor at the University of Louisville. He took his B.A. with highest honors at Williams College in 1943, and after war service with the 10th Mountain Division, received an M.A. and Ph.D. from Harvard University. His publications include: *Landscape Artist in America: the Life and Work of Jens Jensen (*1964), *Two Chicago Architects and their Clients* (1969), *American Architecture Comes of Age* (1972) and *Gateway Cities and Other Essays (*1989). He is best known for his work on Frank Lloyd Wright. His most recent book is *Hardy Cross: American Engineer* (University of Illinois Press, 2006).

References

ALOFSIN, Anthony. 1993. *Frank Lloyd Wright—The Last Years: 1910–1922*. Chicago: University of Chicago Press.

COPLAND, Aaron. 2011. *What to Listen For in Music*. Reissue ed. New York: Signet Classics.

CREESE, Walter. 1985. *The Crowning of the American Landscape: Eight Great Spaces and Their Buildings*. Princeton: Princeton University Press.

GLEICK, James. 1988. *Chaos: Making a New Science*. New York: Penguin USA.

HERTZ, David Michael. 1993. *Angels of Reality: Emersonian Unfoldings in Wright, Stevens and Ives*. Carbondale, Illinois: Southern Illinois University Press.

HOFFMANN, Donald. 1969. Frank Lloyd Wright and Viollet-Le-Duc. *Journal of the Society of Architectural Historians*. **XXVIII**: 178.

———. 1998. *Frank Lloyd Wright, Louis Sullivan, and the Skyscraper*. New York: Dover Publications.

JONCAS, Richard. 1991. Pure Form: the Origins and Development of Frank Lloyd Wright's Non-Rectangular Geometry. Ph.D. thesis, Stanford University.

LAUWERIER, Hans. 1991. *Fractals: Endlessly Repeated Geometric Figures*. Princeton: Princeton University Press.

MANDELBROT, Benoit B. 1982. *The Fractal Geometry of Nature*. New York: W. H. Freeman.

MENOCAL, Narciso. 1992. Taliesin, The Gilmore House, and The Flower in the Crannied Wall. *Wright Studies*. Carbondale, Illinois: Southern Illinois University Press.

OSTWALD, Michael J., 2001. Fractal Architecture: Late Twentieth Century Connections Between Architecture and Fractal Geometry. *Nexus Network Journal* **3**, 1: 73-84.

OUTCALT, Samuel L. and Mark A. MELTON. 1992. Geomorphic Applications of the Hausdorff-Besicovitch Dimension. *Earth Surfaces Processes and Landforms* **17**: 775–787.

PFEIFFER, Bruce B. and the Frank Lloyd Wright Foundation. 2011. Frank Lloyd Wright Designs: The Sketches, Plans and Drawings. New York: Rizzoli.

PURCELL, William Gray. 1950. Noise as Music: How Historic Architecture Refurnished the Listening Ear. *Northwest Architect* **XIV**, 32: 8-10, 30-32.

ROTHSTEIN, Edward. 1994. *Emblems of Mind: The Inner Life of Music and Mathematics*. New York: Random House.

SCHLAIN, Leonard. 1993. *Art and Physics: Parallel Visions in Space, Time, and Light*. New York: William Morrow.

SCHROEDER, Manfred. 1991. *Fractals, Chaos, Power Laws*. New York: Freeman.

SWEENEY, Robert L. 1993. *Wright in Hollywood: Visions of a New Architecture*. Cambridge, Massachusetts: MIT Press.

VIOLLET-LE-DUC, Eugène. 1856. *Dictionnaire raisonné de l'architecture française du XIe au XVIe siècle*.

VOSS, Richard. 1985. Random Fractal Forgeries: From Mountains to Music. In *Science and Uncertainty*. Dana Nash ed. London: IBM United Kingdom.

WRIGHT, Frank Lloyd. 1957. Architecture and Music. *Saturday Review of Literature* **40** (September 28).

———. 1946. *An Autobiography*. London: Faber and Faber.

———. 1967. *The Japanese Print: An Interpretation*. New York: Horizon Press.

———. 1977. *An Autobiography*. New York: Horizon Press.

Chapter 69
Characteristic Visual Complexity: Fractal Dimensions in the Architecture of Frank Lloyd Wright and Le Corbusier

Michael J. Ostwald, Josephine Vaughan, and Chris Tucker

Introduction

In the late 1970s Benoit Mandelbrot proposed that natural systems frequently possess characteristic geometric complexity over multiple scales of observation (Mandelbrot 1977). In mathematics this realization lead to the formulation of fractal geometry and was central to the rise of the sciences of non-linearity and complexity (Mandelbrot 1982). While architectural designers adopted fractal geometry within a few years of Mandelbrot's initial formulation, more than a decade passed before fractal geometry began to be more widely used for the analysis of the built environment (Ostwald 2001). For example, Batty and Longley (1994) and Hillier (1996) have each developed methods for using fractal geometry to understand the visual qualities of urban space. Oku (1990) and Cooper (2003, 2005) have separately used fractal geometry to provide a comparative basis for the analysis of urban skylines. Yamagishi et al. (1988) have sought to determine geometric complexity in street vistas and various other groups have applied fractal geometry to the analysis of historic street plans (Kakei and Mizuno 1990; Rodin and Rodina 2000). While these projects rely on a range of methods, the majority of examples of the fractal analysis of architecture possess a more common lineage.

Carl Bovill's *Fractal Geometry in Architecture and Design* (1996) demonstrates how Mandelbrot's "box-counting" approach to determining approximate fractal dimension can be applied to the analysis of architectural elevations and plans.

First published as: Michael Ostwald, Josephine Vaughan and Chris Tucker, "Characteristic Visual Complexity: Fractal Dimensions in the Architecture of Frank Lloyd Wright and Le Corbusier". Pp. 217–231 in *Nexus VII: Architecture and Mathematics,* Kim Williams, ed. Turin: Kim Williams Books, 2008.

M.J. Ostwald (✉) • J. Vaughan • C. Tucker
School of Architecture and Built Environment, University of Newcastle, Callaghan, NSW 2308, Australia
e-mail: michael.ostwald@newcastle.edu.au; josephine.vaughan@newcastle.edu.au; chris.tucker@newcastle.edu.au

K. Williams and M.J. Ostwald (eds.), *Architecture and Mathematics from Antiquity to the Future*, DOI 10.1007/978-3-319-00143-2_22,

Bovill (1997) then offered an extrapolation of this method, and Bechhoefer and Appleby (1997) used this approach to examine the visual qualities of vernacular architecture. Bovill's method has also been repeated by Makhzoumi and Pungetti (1999) and Burkle-Elizondo et al. (2015). Importantly, in the original 1996 work, Bovill demonstrates how fractal dimension can be used to analyse two façades; one from Frank Lloyd Wright's Robie House and the other from Le Corbusier's Villa Savoye.[1] Bovill's analysis of the two façades has been used to support a wide range of arguments about architecture and, more specifically, a range of criticisms of modernist approaches to design, but it has rarely been tested and never expanded or developed (Lorenz 2003).

The present research undertakes a comprehensive analysis of the fractal dimension of five houses each from the early careers of Wright and Le Corbusier. The fractal dimensions of the elevations and plans of these houses are calculated using TruSoft's *Benoit* (vers. 1.3.1) program and *Archimage* (vers. 2.1), a program developed by the authors. The following section explains what is meant by fractal dimension and provides an overview of the box-counting method. Thereafter, the chapter describes how the present study was undertaken and why the particular houses were chosen. The chapter concludes with a review of the results of the study and any questions raised by these results.

Determining Fractal Dimension

Mandelbrot argues that Euclidean geometry, the traditional tool used in science to describe natural objects, is fundamentally unable to fulfil this purpose. To paraphrase Mandelbrot (1982), mountains are not conical in form, clouds are not spherical and rivers are not orthogonal. While science has historically considered roughness and irregularity an aberration disguising underlying ordered systems with fixed-state or finite values, Mandelbrot argues that the fragmentation of all naturally occurring phenomena cannot be so easily disregarded: a coastline is not straight and no Euclidean geometric construct can approximate the form of a coastline without serious abstraction or artificiality (Ostwald 2003). As a result of this natural fragmentation, mathematicians have shown that the length of a coastline cannot be determined at all (Feder 1988). However, Mandelbrot postulates that the degree of geometric irregularity or complexity that is visible in a coastline at one scale (from a satellite) may be similar to that when viewed from another scale (from a helicopter). If this is the case, then the coastline may possess a form of consistent complexity, or characteristic irregularity, that can be measured.

[1] While Bovill's Robie House façade is a relatively close approximation of Wright's original elevation, the elevation Bovill uses of the Villa Savoye lacks the same level of detail and it is less consistent in its relationship with the original.

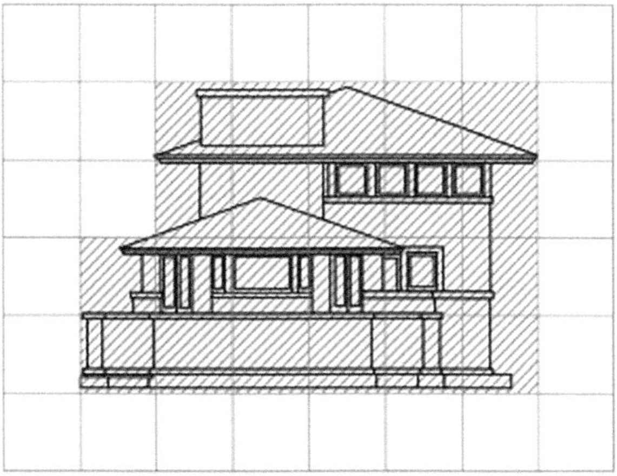

Fig. 69.1 Starting grid placed over the east elevation of the Tomek house showing box-counting. Image: authors

The characteristic irregularity of a coastline may be measured by imagining that the increasingly complicated and detailed path of the coastline is actually somewhere between a one-dimensional line and a two-dimensional surface (Schroeder 1991). The more complicated the line, the closer it comes to being a two-dimensional surface. Therefore coastlines, as well as many similar natural lines, can be viewed as being fractions of integers, or what Mandelbrot describes as "fractal geometric forms". Thus, fractal geometry describes irregular or complex lines, planes and volumes that exist between whole number integer dimensions. This implies that instead of having a dimension *D* of 1, 2, or 3, fractals might have a *D* of 1.51, 1.93 or 2.74. One way of determining the approximate fractal dimension of an irregular or complex object is to apply the box-counting method.

Consider a drawing of an elevation of a house. A large grid is placed over the drawing and each square in the grid is analysed to determine whether any lines from the façade are present in each square (Fig. 69.1). Those grid boxes that have some detail in them are recorded. Next, a grid of smaller scale is placed over the same façade and the same determination is made of whether detail is present in the boxes of the grid. A comparison is then constructed between the number of boxes with detail in the first grid and the number of boxes with detail in the second grid. Such a comparison is made by plotting a log-log diagram for each grid size (Bovill 1996). By repeating this process over multiple grids of different scales (Fig. 69.2), an estimate of the fractal dimension of the façade is produced (Fig. 69.3). This method is central to Bovill's explanation and is also critical to the operations carried out by *Benoit* and *Archimage*. However, there are many variations of how this method is used to calculate *D*. For example, Bovill halves the grid dimension for each comparison, whereas *Benoit* and *Archimage* use a range of scaling coefficients to gradually reduce the grid size and generate a more accurate result. Other factors that alter the way in which *D* is determined include the width of the lines in the elevation, the position of the elevation in the image, and the way in which statistical variations are handled.

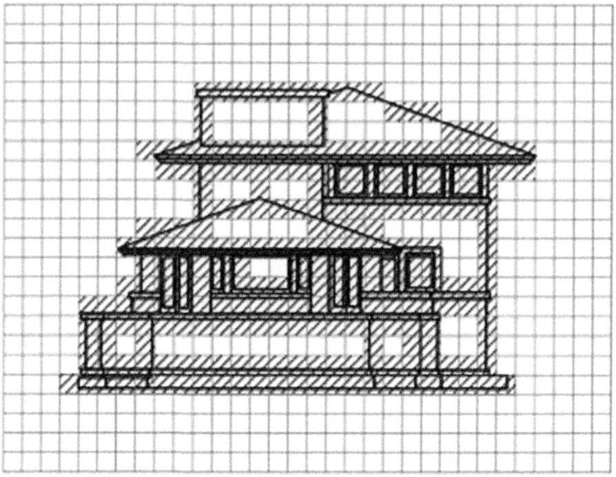

Fig. 69.2 Third stage grid placed over the east elevation of the Tomek house showing box-counting. Image: authors

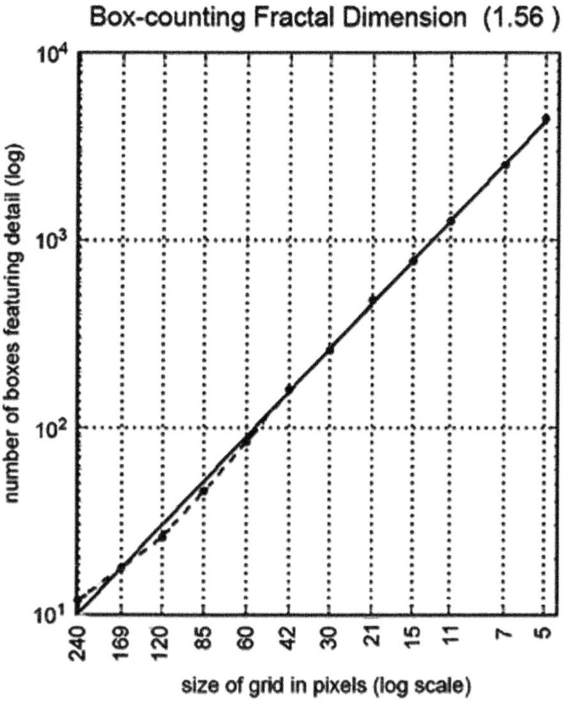

Fig. 69.3 Log-log diagram of the comparison between the number of boxes counted in a grid and the size of the grid. Image: authors

The wider the lines in the source image, the more chance they have of being counted twice when grid sizes become very small, leading to artificially increased *D* values. To counter this situation, *Archimage* software pre-processes images using a line-detection algorithm (Sobel gradient technique) that produces images for analysis that are one pixel wide. *Benoit* overcomes this problem by allowing the analytical grid to be rotated or resized to minimize the impact of line weight at each scale of observation. In addition to the line width problem, the volume and distribution of white or empty space around the source image can also alter the result. To solve this, Foroutan-Pour et al. (1999) offer an algorithm to optimize the way in which an image is positioned against its background and suggestions on how to derive an ideal analytical grid. A further, related issue is that the proportions of the image being analysed also influence the result. If the original image being analysed is not pre-sized to produce a clear starting grid, then an additional step must be added to ensure that a divisible starting grid is determined. *Benoit* solves this problem by cropping the image size to achieve a whole-number starting grid.[2] In contrast, *Archimage* enlarges the image by adding small amounts of empty space to the boundaries. While neither of these variations change the elevation in the source image, they produce subtle variations in the resultant *D*.

A final challenge for any application of the box-counting method is the problem of statistical divergence. The average slope of the log-log graph may be the approximate *D* value, but the points generating the line are not always consistent with it. The *D* value is only a reasonable approximation when most of the points in the chart correspond with the resultant average line. The question then becomes, how are divergent points handled? While there is no definitive answer to this question, divergent results tend to occur primarily at the extremes of the graph; with the largest and smallest grid sizes but not normally those in between. Bovill is aware of this problem and solves it by intuitively determining where the practical limits of scaling in an image can be found. *Benoit* similarly allows the operator to intuitively deactivate certain data points or use a range of algorithms to determine best-fit for the data. While both of these are possible solutions, neither of them are useful for producing a consistent analysis of almost 50 images.

For the present research, similar settings for starting grid proportion and size minimize the number of divergent results associated with the largest grid dimensions. However, for divergences associated with small scale grids different tactics are used. The parameters of *Benoit* may be set to limit the smallest scale grids uniformly, creating a consistent set of results. *Archimage* has no artificial limit on the small grid size, and any divergences it produces are averaged into the log-log graph. This means that, while the actual differences will be minor, *Archimage* is likely to produce higher results in general, as well as slightly more accurate results

[2] In its default mode, *Benoit* determines starting grid size by dividing the shortest input boundary by four. The grid is then extrapolated across the image until the last complete grid-line fits on the page. Any additional space in the source image is deleted. Thus, if the input image was 500 × 1020 pixels, then Benoit identifies a 4 × 8 grid of 125 × 125 pixels and it discards a 20 pixel wide slice of space at the boundary of the input image.

for objects which exhibit characteristic irregularity over a large range of scales. *Benoit*, as set up for consistent application, is likely to produce slightly lower results but may be marginally more accurate for objects which exhibit fractal characteristics over a smaller range of scales.

Method

The early house designs of Frank Lloyd Wright and Le Corbusier are the focus of the present study. Houses are ideal for this purpose because they possess similar scale, program and materiality. While a study of commercial, urban or religious projects by these architects is possible, none of these alternative sets have the same potential for producing consistent, verifiable and statistically valid results.

A total of five houses by each architect was selected. In each sequence there are four houses that prefigure the completion of an acknowledged masterwork of architecture, along with the masterwork itself. For Wright, the masterwork is the Robie House; for Le Corbusier it is the Villa Savoye. In both sequences of houses, no more than 10 years separates the earliest design from the last. While within these parameters a range of possible designs could be selected, preference was given for single houses (rather than pavilions or estates), completed works (ensuring a similar level of development) and houses with a relatively tight geographic distribution (to limit the impact of climate on the form of the house).

No clear precedents exist for determining the holistic D value of a house. The present research proposes that a holistic D may be produced either by averaging the results for each elevation of the house, or by averaging the results for all elevations and one or more plans of the house. The first of these methods produces a $D_{(\text{Elev})}$ result which is a reasonable approximation of the visual qualities of the building when viewed from the exterior. The second method is more controversial and results in a $D_{(\text{Plan + Elev})}$ outcome which might be seen as reflecting the two- and three- dimensional qualities of the building's form. The research focuses primarily on the former method for two reasons. First, while two-dimensional views of architecture can be combined together to describe a three-dimensional building, the D values derived from two-dimensional plans and elevations cannot be averaged to produce a D value for the complete three-dimensional object. Instead, the combined result, either $D_{(\text{Elev})}$ or $\text{D}_{(\text{Plan + Elev})}$, describes average visual complexity for two-dimensional images of the building. Because elevations are conventionally regarded as the primary determinants of visual character (plans are more experiental) the former method is more appropriate. The second reason for focusing on $D_{(\text{Elev})}$ results is that past research, including Bovill's, suggests that elevations will more tightly clustered in their results than plans, leading to more consistent results for $D_{(\text{Elev})}$.

To produce a holistic result for each of the ten houses being considered, between four and five data sources (elevations and plans) were developed for testing. In all but two cases, the chosen houses possess traditional orthogonal or right-angled plans with four elevations. However, two of Le Corbusier's villas are sited on urban streets and possess party-walls with neighbouring structures. These walls are excluded from the analysis, and average results for the house are only taken across the number of visible elevations. Where an almost blank wall is part of one of Le Corbusier's designs (because the wall could have become a future party-wall), it was still included in the analysis. For each house only a single plan was subjected to analysis. Typically, this plan was of the primary living level; for Wright's houses this was the ground floor; for Le Corbusier's houses it was often the *piano nobile*.

Following Bovill's convention, the primary lines considered in the analysis correspond to changes in form, not changes in surface or texture. Thus, major window reveals, thickened concrete edge beams, and steel railings are all considered, while brick coursing and control joints are not. Furthermore, the images of the ten houses were all redrawn using consistent graphic conventions and scales prior to analysis. The image source for Wright's work was his original working drawings reproduced by Storrer (1993) and Pfeiffer and Futagawa (1987). Where the particular house was altered by Wright during construction, or only an incomplete set of working drawings is available, the measured drawings of the *Historic American Buildings Survey* were used to supplement the originals. For all of Le Corbusier's houses the definitive set of drawings developed by the University of Tokyo was adopted (Ando 2001).

Twenty-five images of Wright's houses and 22 of Le Corbusier's houses were prepared, each with consistent line weight and positioning within their respective frames. These 47 images were analysed by placing, on average, 12 grids of decreasing size over each image and making a comparison of the presence of detail at different scales. The almost 600 data points were processed, leading to 94 separate $D_{(Archi/Benoit)}$ values being calculated.[3] For each house $D_{(\mathrm{Elev})}$ and $D_{(\mathrm{Plan\ +\ Elev})}$ outcomes, for both *Archimage* and *Benoit*, were then produced, leading to 20 holistic results for the houses. Composite results ($D_{(\mathrm{Elev},\ Archi)}+D_{(\mathrm{Elev},\ \mathrm{Benoit})} \div 2 = D_{(\mathrm{Comp})}$) were then produced for each of the ten houses. Finally, the composite results for the complete sequence of five designs were aggregated into a summary result for each architect, $D_{(\mathrm{Agg})}$. The end result of the research is a distillation of almost 600 operations into 126 sets of data to inform a comparative reading of the early houses of Wright and Le Corbusier.

The Ten Houses

The five houses by Wright were constructed between 1901 and 1910. Four of the five are in the state of Illinois and the fifth is in Kentucky. All of the houses are associated with Wright's Prairie Style, an approach characterized by strong

[3] The settings of *Benoit* were modified to match *Archimage*'s. Thus scaling coefficients, grid angles, and line widths are consistent between the two programs.

horizontal lines, overhanging eaves, low-pitched roofs, an open floor plan, and a central hearth. Importantly, the five houses span the period between the first publication of Wright's Prairie Style, in the *Ladies Home Journal* in 1901, and what is widely regarded as the ultimate example of this approach, the Robie house.

The first design is the F. B. Henderson House (1901) in Elmhurst, Illinois. The house is a wooden, two storey structure with plaster rendered elevations. A range of additions were made to the house in the years following its completion until, in 1975, the house was restored to its original form. The Tomek House (1904–1907) in Riverside, Illinois, is also a two storey house although it possesses a basement and is sited on a large city lot. This house is finished with pale, rendered brickwork, dark timber trim and a red tile roof. Storrer notes that, in response to the Tomek family's needs, Wright later allowed posts to be placed beneath the cantilevered roof to heighten the sense of support and enclosure (Storrer 1993: 128). As the posts were not required for structural reasons, and Wright found them personally unnecessary, they have been omitted from the analysis. The Robert W. Evans House[4] (1908) in Chicago, Illinois, features a formal diagram wherein the "basic square" found in earlier Prairie Style houses is "extended into a cruciform plan" (Thomson 1999: 100). The house is set on a sloping site and possesses a plan similar to one Wright proposed in 1907 for a "fireproof house for $5,000". The Evans house was later altered to enclose the porch area and the stucco finish on the façade was also cement rendered. The Zeigler House (1910) in Frankfort, Kentucky has a similar plan to the Evans house. Designed as a home for a Presbyterian minister, this two storey house is sited on a small city lot and it was constructed while Wright was in Europe. After a decade of development and refinement the quintessential example of the Prairie Style, the Robie House (1908–1910), was constructed in Chicago, Illinois. Designed as a family home, the three storey structure fills most of its corner site. Unlike many of Wright's other houses of the era, the Robie house features a façade of exposed Roman bricks with horizontal raked joints.

The five houses by Le Corbusier were completed between 1922 and 1928. Four of the five are in France and one is in Germany. The houses are significant because during the 1920s Le Corbusier developed five strategies for an architecture that would reflect the technological and social advances of its era. Initially published in the journal *L'Esprit Nouveau* and later collated in *Vers Une Architecture*, these strategies (*pilotis*, *plan libre*, *façade libre*, *fenêtre en longueur* and *toît jardin*) are found in their most refined form in the Villa Savoye.

The first house, the Maison-Atelier Ozenfant (1922–1923) in Paris, was designed as both a home and studio for Le Corbusier's close friend, the cubist painter Amédée Ozenfant. Set on a complex and steep corner site, in an urban area "already populated by artists' studio houses" (Curtis 1986: 56), this three storey, white rendered masonry building has large glazed areas and a saw-tooth roof. The Villa Cook (1926) in Boulogne-sur-Seine was described by Le Corbusier as "the

[4] Despite its designation, original working drawings record that this project was for Raymond W. Evans.

true cubic house" because, as Gans observes, "[p]lan, section and elevation all derive from the same square and in reference to one another" (Gans 2000: 66). The house is a three storey structure of white rendered masonry with a roof garden. The Villa Stein/de Monzie (1926–1928), also known as "Les Terrasses" or simply the Villa Stein, is set on a narrow block in the suburbs of Vaucresson. The unusual domestic brief was for a house and studio for Gabrielle de Monzie and her daughter, to be shared with Michael and Sarah Stein. Curtis describes the house as the "most monumental and luxurious of Le Corbusier's houses of the 1920's" (Curtis 1986: 79). In the late 1920s Le Corbusier and Pierre Jeanneret were invited to produce residential designs for the Second International Exposition of the *Deutscher Werkbund* at Stuttgart. The resulting Weissenhof-Siedlung Villa 13 (1927) is on a corner block in the outskirts of the city. Le Corbusier and Jeanneret designed Villa 13 as a prototype for suburban row housing. The Villa Savoye (1928), or "Les heures Claires", is sited in Poissy, France. This two storey, freestanding house in white rendered masonry, with an extensive roof-garden, is set in an open landscape. For Meier, "the Villa Savoye illustrates with extreme clarity and is perhaps the most faithful in its observation of (Le Corbusier's) 'Five Points of a New Architecture'" (Meier 1972: 2).

Discussion and Results

For each of the ten houses the *Archimage* and *Benoit* determinations of *D* were recorded for every elevation and plan. For example, for the north elevation of Wright's Henderson house the results are, $D_{(Archi)} = 1.59$ and $D_{(Benoit)} = 1.54$. Two variations on the average *D* value for the visual properties of each house were then produced, the first combining the plan and elevations and the second just the elevations. For the Henderson house, the first average values for the house were $D_{(\text{Plan+Elev},\ Archi)} = 1.55$ and $D_{(\text{Plan + Elev},\ Benoit)} = 1.51$. The second, holistic calculation, reliant solely on elevations, produced the following results, $D_{(\text{Elev},\ Archi)} = 1.56$ and $D_{(\text{Elev},\ Benoit)} = 1.53$. The final composite *D* result for each house was calculated using both *Archimage* and *Benoit* outcomes and just the elevations. The result for the Henderson house was $D_{(\text{Comp})} = 1.545$ (Table 69.1).[5] For Le Corbusier's Maison-Atelier Ozenfant, the average results for the elevations were $D_{(\text{Elev},\ Archi)} = 1.53$ and $D_{(\text{Elev},\ Benoit)} = 1.46$, leading to $D_{(\text{Comp})} = 1.495$ (Table 69.2). An analysis of just these two houses, the first in the sequence for each architect, suggests that while, on average, Wright's design exhibits a slightly higher degree of characteristic complexity (a difference of 0.05 or approximately

[5] *Archimage* and *Benoit* produce data to three or more decimal places but present it rounded to, respectively, two and three decimal points. In the present research the $D_{(Arch\ +\ Benoit)}$ values are rounded to two decimal places while $D_{(\text{Comp})}$ results are left at three (because they are an average of two results).

Table 69.1 *D* results for Wright's Henderson house

Henderson house	$D_{(Archi)}$	$D_{(Benoit)}$	$D_{(Comp)}$
Plan (Ground)	1.50	1.45	$D_{(Comp)} = 1.545$
Elevation (North)	1.59	1.54	
Elevation (South)	1.57	1.53	
Elevation (East)	1.56	1.53	
Elevation (West)	1.53	1.51	
$D_{(Plan + Elev.)}$	1.55	1.51	
$D_{(Elev.)}$	1.56	1.53	

Table 69.2 *D* results for Le Corbusier's Maison-Atelier Ozenfant

Maison-Atelier Ozenfant	$D_{(Archi)}$	$D_{(Benoit)}$	$D_{(Comp)}$
Plan (First)	1.56	1.56	$D_{(Comp)} = 1.495$
Elevation (1)	1.55	1.48	
Elevation (2)	1.57	1.51	
Elevation (3/4)	1.46	1.38	
$D_{(Plan + Elev.)}$	1.54	1.48	
$D_{(Elev.)}$	1.53	1.46	

Fig. 69.4 South elevation of the Henderson house ($D_{(Archi)} = 1.57$) (*left*); elevation (2) of the Maison-Atelier Ozenfant ($D_{(Archi)} = 1.57$) (*right*). Image: authors

3.5 %) there are multiple occasions when Wright's and Le Corbusier's façades have almost the same fractal dimension. Compare the south elevation of the Henderson House—$D_{(Archi)} = 1.57$ and $D_{(Benoit)} = 1.53$—with elevation (2) of the Maison-Atelier Ozenfant—$D_{(Archi)} = 1.57$ and $D_{(Benoit)} = 1.51$ (Fig. 69.4). While one elevation has a horizontal bias, the human eye can detect some similarities in visual detail. Furthermore, in this specific comparison, Wright's elevation has slightly more detail at the finer scale, while Le Corbusier's has a greater and more consistent spread of visual complexity across all scales.

The most visually complex and consistently scaled of all of the elevations considered were the front façade of the Villa Stein/de Monzie ($D_{(Benoit)} = 1.59$) and the south elevation of the Robie house ($D_{(Benoit)} = 1.53$). Again, in this pairing Le Corbusier's façade exhibits more levels of characteristic complexity than Wright's. The rear elevation of the Robie house has one of the lowest fractal

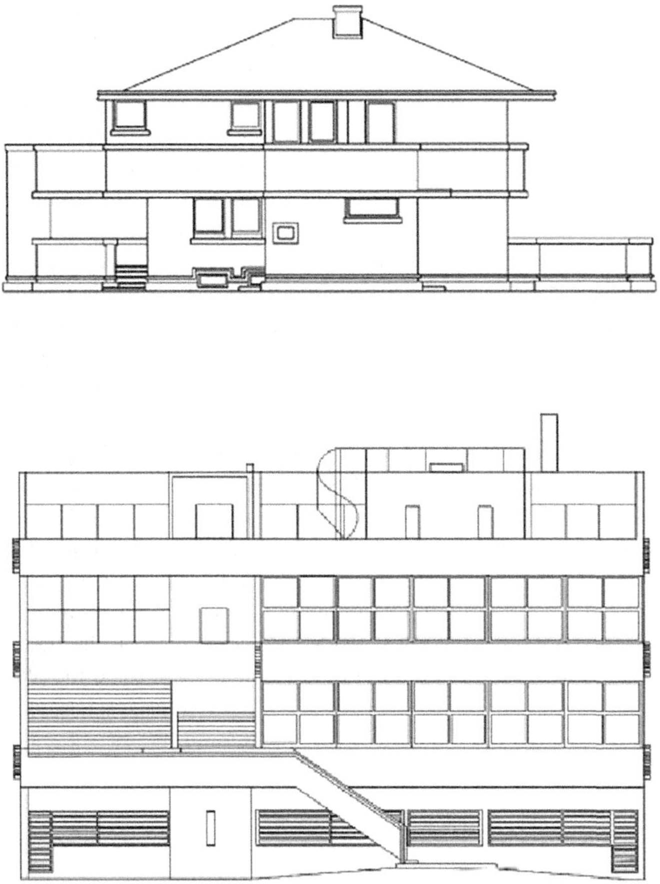

Fig. 69.5 North elevation of the Zeigler house ($D_{(Archi)} = 1.50$) (*above*); elevation 1 of the Villa Stein/de Monzie ($D_{(Archi)} = 1.67$) (*below*). Image: authors

dimensions of any of Wright's façades ($D_{(Benoit)} = 1.45$) along with most of the façades of the Zeigler house ($1.46 < D_{(Benoit)} < 1.49$). Setting aside the featureless party-walls, the lowest results for the Le Corbusier houses were typically found in the Villa Savoye ($1.41 < D_{(Benoit)} < 1.49$) (Fig. 69.5).

The combined results for Wright's houses ranged from $D_{(Comp)} = 1.505$ for the Zeigler House to $D_{(Comp)} = 1.580$ for the Evans house. The Robie house result was between these two extremes ($D_{(Comp)} = 1.550$). Overall, the aggregated result for all of Wright's five houses was $D_{(Agg)} = 1.543$ (Table 69.3). For Le Corbusier, the combined results ranged from the simplest building, the Weissenhof-Siedlung Villa 13, $D_{(Comp)} = 1.420$, to the most complex, the Villa Stein/de Monzie ($D_{(Comp)} = 1.515$). The aggregate result for all five of Le Corbusier's houses was $D_{(Agg)} = 1.481$ (Table 69.4).

Table 69.3 Results for Wright houses

Frank Lloyd Wright houses	$D_{(\text{Elev, } Archi)}$	$D_{(\text{Plan + Elev, } Archi)}$	$D_{(\text{Elev, } Benoit)}$	$D_{(\text{Plan + Elev, } Benoit)}$	$D_{(\text{Comp})}$
Henderson	1.56	1.55	1.53	1.51	1.545
Tomek	1.57	1.55	1.50	1.48	1.535
Zeigler	1.54	1.52	1.47	1.46	1.505
Evans	1.60	1.58	1.56	1.54	1.580
Robie	1.59	1.58	1.51	1.49	1.550
All houses	1.57	1.56	1.51	1.50	$D(\text{Agg}) = 1.543$

Table 69.4 Results for Le Corbusier houses

Le Corbusier houses	$D_{(\text{Elev, } Archi)}$	$D_{(\text{Plan + Elev, } Archi)}$	$D_{(\text{Elev, } Benoit)}$	$D_{(\text{Plan + Elev, } Benoit)}$	$D_{(\text{Comp})}$
Ozenfant	1.53	1.54	1.46	1.48	1.495
Cook	1.52	1.54	1.47	1.50	1.495
Stein/De Monzie	1.55	1.57	1.48	1.51	1.515
Weissenhof. S.	1.46	1.46	1.38	1.39	1.420
Savoye	1.52	1.55	1.44	1.49	1.480
All houses	1.52	1.53	1.45	1.47	$D(\text{Agg}) = 1.481$

The difference between the aggregated fractal dimensions of the houses of Wright and Le Corbusier was 0.062 (approx. 4 %). Despite this, many of Le Corbusier's façades and houses had higher fractal dimensions that Wright's. One reason the aggregated result for Wright's houses had a slightly higher D value is that several of Le Corbusier's houses have walls that are almost blank (future party walls but with one or two windows) whereas all of Wright's houses had elevations with a relatively consistent level of detail in each façade. If these future party walls were removed then the overall difference between Wright and Le Corbusier would be further reduced.

When all houses are analysed in this way some larger patterns are revealed. *Archimage* consistently produces higher results than *Benoit*, typically in the order of 4 %. Because this variation is relatively consistent, it has limited impact on the composite and aggregate results. For the five houses by Wright, the difference is in the order of $D_{(\text{Agg})} = 0.06$. Variations in results for the two programs for Le Corbusier's houses are marginally higher at $D_{(\text{Agg})} = 0.07$. Two further factors influence these discrepancies. As Lorenz warns, minor variations and inconsistencies in initial image framing result in a range of disparities in D of up to 2 % (Lorenz 2003: 63–65, 117–119). Furthermore, rounding inconsistencies (from three to two decimal places) also produce subtle variations. Given the differences in the way the two programs handle starting grid identification and divergent results, the variation between the two programs is neither substantial nor unexpected.

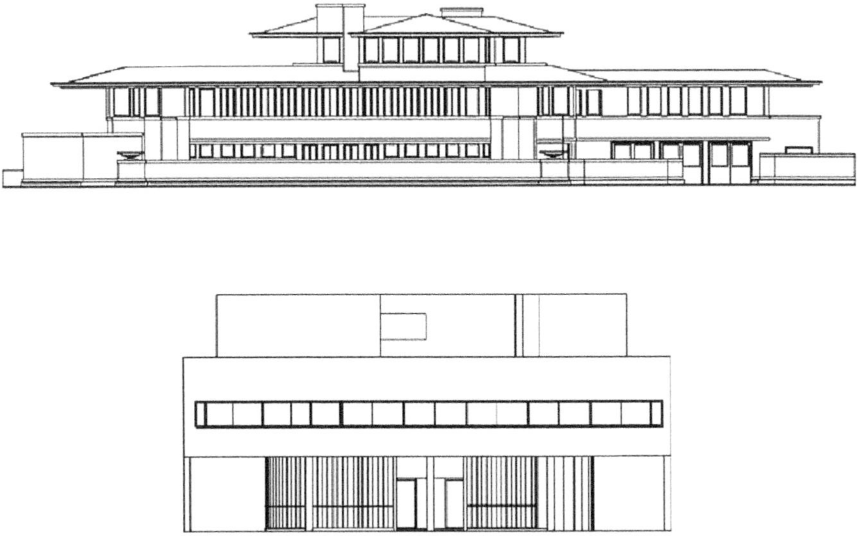

Fig. 69.6 South elevation of the Robie house ($D_{(Archi)} = 1.62$) (*above*); elevation 1 of the Villa Savoye ($D_{(Archi)} = 1.53$) (*below*). Image: authors

Conclusions

If each sequence of five houses, produced over a ten-year period by Wright and by Le Corbusier, is taken in its totality, then there is relatively little difference between the fractal dimension of each architect's works. Certainly, Wright's designs typically possess higher $D_{(\mathrm{Comp})}$ and $D_{(\mathrm{Agg})}$ values, but the variation is often marginal. Moreover, a number of Le Corbusier's façades not only have higher fractal dimensions than Wright's, but they are more consistent over a wider range of scales.

When Bovill analysed a single façade of a masterwork by Wright and one by Le Corbusier, he happened to choose Wright's most complex house façade from the era and one of Le Corbusier's simplest (Fig. 69.6). It is not surprising then that he found Wright's façade displayed a higher, and more consistent, level of characteristic visual complexity. However, when a much larger sample size is considered, the difference between the two architects' works is considerably reduced. The present analysis confirms Bovill's isolated comparative case, but it undermines the extrapolation of this result by other scholars to suggest that organic or regionalist architecture is more fractal than Modernist architecture.

Prior to commencing this research, it was anticipated that the inclusion of plans alongside elevations would not assist in the overall interpretation of the results. The reason for this assumption is the view that while plans indirectly reflect the aesthetic and spatial intentions of the architect, elevations provide a more thorough description of the shape and form of the building. The present research only partially supports this position. Most of the plans that were analysed possessed D values that were similar to the results for the associated elevations. Typically Wright's plans had lower $D_{(Archi/Benoit)}$ results while Le Corbusier's plans often had slightly higher results. But for plans to be conclusively used as part of such an analysis a range of decisions need to be made about which lines in a plan are significant for the analysis. This question will be the topic of future research.

A final observation arising from this research is that, in isolation, fractal dimension may not be the most useful way of differentiating the visual qualities of elevations. Other factors, including line length, line clustering, implied directionality and distribution of detail (not over multiple scales but in different zones) may be usefully combined with fractal dimension for a more complete analysis of architecture.

Biography Michael J. Ostwald is Professor and Dean of Architecture at the University of Newcastle (Australia) and a visiting Professor at RMIT University. He has previously been a Professorial Research Fellow at Victoria University Wellington, an Australian Research Council (ARC) Future Fellow at Newcastle and a visiting fellow at UCLA and MIT. He has a PhD in architectural history and theory and a DSc in design mathematics and computing. He completed post-doctoral research on baroque geometry at the CCA (Montreal) and at the Loeb Archives (Harvard). He is Co-editor in Chief of the *Nexus Network Journal* and on the editorial boards of *ARQ and* Architectural Theory Review. He has authored more than 300 scholarly publications including 20 books and his architectural designs have been published and exhibited internationally.

Josephine Vaughan is a research associate in the School of Architecture and Built Environment at the University of Newcastle and a member of CIBER, the architectural research group. Her research is focused on the fractal analysis of buildings and landscapes. Josephine directs the firm One Thousand Years, which specialises in sustainable design. Her works have been exhibited and installed regionally and nationally.

Chris Tucker is a lecturer in the School of Architecture and Built Environment at the University of Newcastle and a director of the architectural practice Herd. Chris has been awarded regional, state and international prizes for architecture, and his buildings and designs have been widely exhibited and published. His research interests revolve around the development of software tools to analyse the geometric and mathematical qualities of buildings.

References

ANDO, Tadao, ed. 2001. *Le Corbusier: Houses*. Tokyo: Toto Shuppan.

BATTY, Michael and Paul LONGLEY. 1994. *Fractal Cities: A Geometry of Form and Function*. New York: Academic Press.

BECHHOEFER, William and Marilyn APPLEBY. 1997. Fractals, Music and Vernacular Architecture: An Experiment in Contextual Design. In *Critical Methodologies in the Study of Traditional Environments* (Working Paper Series, vol. 97), Nezar Al Sayyad, ed. Berkeley: University of California at Berkeley. (Unpag.)

BOVILL, Carl. 1996. *Fractal Geometry in Architecture and Design*. Boston: Birkhäuser.

———. 1997. Fractal Calculations in Vernacular Design. In *Critical Methodologies in the Study of Traditional Environments* (Working Paper Series, vol. 97), Nezar Al Sayyad, ed. Berkeley: University of California at Berkeley. (Unpag.)

BURKLE-ELIZONDO, Gerardo, Nicoletta SALA and Ricardo David VALDEZ-CEPEDA. 2015. Geometric and Complex Analyses of Maya Architecture: Some Examples. Pp 113–126. in Kim Williams and Michael J. Ostwald eds. *Architecture and Mathematics from Antiquity to the Future: Volume I Antiquity to the 1500s*. Cham: Springer International Publishing.

COOPER, J. 2003. Fractal Assessment of Street-level Skylines: a Possible Means of Assessing and Comparing Character. *Urban Morphology: Journal of the International Seminar on Urban Form* **7**, 2: 73–82.

———. 2005. Assessing Urban Character: The Use of Fractal Analysis of Street Edges. *Urban Morphology: Journal of the International Seminar on Urban Form* 9, 2: 95–107.

CURTIS, William J. 1986. *Le Corbusier: Ideas and Forms*. New York: Rizzoli.

FEDER, Jens. 1988. *Fractals*. New York: Plenum Books.

FOROUTAN-POUR, K., P. DUTILLEUL and D. L. SMITH. 1999. Advances in the Implementation of the Box-counting Method of Fractal Dimension Estimation. *Applied Mathematics and Computation* **105**, 2: 195–210.

GANS, Deborah. 2000. *The Le Corbusier Guide*. Princeton: Princeton Architectural Press.

HILLIER, Bill. 1996. *Space is the Machine: A Configurational Theory of Architec*ture. Cambridge: Cambridge University Press.

KAKEI, Hidekazu and Setsuko MIZUNO. 1990. Fractal Analysis of Street Forms. *Journal of Architecture, Planning and Environmental Engineering* **8**, 414: 103–108.

LORENZ, Wolfgang. 2003. Fractals and Fractal Architecture. Masters Diss. Vienna University of Technology. http://www.iemar.tuwien.ac.at/modul23/Fractals/.

MAKHZOUMI, Jala and Gloria PUNGETTI. 1999. *Ecological Landscape Design and Planning: The Mediterranean Context*. London: E and FN Spon.

MANDELBROT, Benoit. 1977. *Fractals: Form, Chance, and Dimension*. San Francisco: W. H. Freeman and Company.

———. 1982. *The Fractal Geometry of Na*ture. New York: W. H. Freeman and Company.

MEIER, Richard. 1972. *Le Corbusier: Villa Savoye, Poissy, France, 1929–31*. Global Architecture, vol. 13. Tokyo: A.D.A. Edita.

OKU, T. 1990. On Visual Complexity on the Urban Skyline. *Journal of Planning, Architecture and Environmental Engineering* **8,** 412: 61–71.

OSTWALD, Michael J. 2001. Fractal Architecture: Late Twentieth Century Connections Between Architecture and Fractal Geometry. *Nexus Network Journal* **3**, 1: 73–84.

———. 2003. Fractal Architecture: The Philosophical Implications of an Iterative Design Process. *Communication and Cognition* **36**, 3/4: 263–295.

PFEIFFER, Bruce Brooks and Yukio FUTAGAWA. 1987. *Frank Lloyd Wright Monograph 1907–1913*, Vol. 3. Tokyo: A.D.A Edita.

RODIN, Vladimir and Elena RODINA. 2000. The Fractal Dimension of Tokyo's Streets. *Fractals* **8**, 4: 413–418.

SCHROEDER, Manfred. 1991. *Fractals, Chaos, Power Laws: Minutes from an Infinite Paradise*. New York: W. H. Freeman and Company.

STORRER, William Allin. 1993. *The Frank Lloyd Wright Companion*. Chicago: University of Chicago Press.

THOMSON, Iain. 1999. *Frank Lloyd Wright: A Visual Encyclopedia*. London: PRC.

YAMAGISHI, R., S. UCHIDA and S. KUGA. 1988. An Experimental Study of Complexity and Order of Street-Vista. *Journal of Architecture, Planning and Environmental Engineering* **2**, 384: 27–35.

Chapter 70
From Cosmic City to Esoteric Cinema: Pythagorean Mathematics and Design in Australia

Graham Pont and Peter Proudfoot

All things are exposed in time. Like the great fugues of Bach, Griffin's concepts expand into our growing consciousness. In moments of crisis new dimensions emerge, new signs, new energy

(Muller 1976).

Introduction

In his Walter Burley Griffin Memorial Lecture of 1976, Australian architect Peter Muller argued that the plan of Canberra is a "solid mandala… an intelligible, sensible three dimensional diagram packed full of significance and purpose" (Muller 1976). The first to explore the "esoteric world" of the Griffins' Canberra, Muller rightly emphasised the significance of concentric circles in the original proposals for the federal capital and its university. In both cases the radiating circles represented a hierarchy (one governmental and administrative, the other epistemological and encyclopaedic) proceeding "from the more essential to less essential, from generals to particulars". The mandala of concentric circles and radiating diameters, he concluded, was "inscribed into the plan of Canberra in the form of street patterns, appropriately named", thus defining the "physical, political and symbolic centre of our Nation".

In *The Secret Plan of Canberra*, Peter Proudfoot rediscovered a lost cosmic city and another mandala hidden at its ceremonial centre: a set of three intersecting circles that he called the "double Vesica". Proudfoot interpreted its role and significance in

First published as: Graham Pont and Peter Proudfoot, "From Cosmic City to Esoteric Cinema". Pp. 195–206 in *Nexus IV: Architecture and Mathematics*, Kim Williams and Jose Francisco Rodrigues, eds. Fucecchio (Florence): Kim Williams Books, 2002.

G. Pont (✉)
41 Forbes Street, Newtown, NSW 2042, Australia
e-mail: graham_pont@hotmail.com

P. Proudfoot
1 Ontario Avenue, Roseville, NSW 2069, Australia

K. Williams and M.J. Ostwald (eds.), *Architecture and Mathematics from Antiquity to the Future*, DOI 10.1007/978-3-319-00143-2_23,

the city plan through a wide-ranging survey of urban forms, esoteric traditions, sacred geometry and cosmological symbolism (Proudfoot 1994: Chaps. 1–2). More recently Graham Pont has pointed out that the three equal intersecting circles are an old Pythagorean symbol for the "Triad" and he has also identified Pythagorean forms, numbers and symbols in the Griffins' design for the Capitol Theatre, Melbourne.

Our latest research indicates that there is an esoteric rationale, also based on the circle, that connects the first plan of central Canberra, the Capitol Theatre, Newman College and several other Griffin designs. In all three of these major Australian works, the Griffins made a focal point or generative seed out of the same mandala—a circle quartered by a cross.

The Plan of Canberra

According to Marion Mahony Griffin, the plan of Canberra was produced during 9 weeks of "driving work" and submitted at the last possible moment. Although the prize was awarded to her husband, it is now recognised that Marion was also deeply involved in preparing the competition entry as a whole, and not just in her exquisitely rendered plans and perspective views of the envisaged city. Our primary focus is her beautiful *City and Environs* painting, which reveals the mathematical key to the "Secret Plan of Canberra" (Fig. 70.1).

The Griffins' plan is based on two axes suggested by the natural topography. The principal "Land Axis" runs south-southwest from Mount Ainslie to Bimberi Peak, some 30 miles away in the Brindabella Ranges. The minor and much shorter "Water Axis" runs at right angles from Black Mountain in the northwest towards an area known as "Mount Pleasant" (the prominence itself does not quite manage to provide the fourth natural terminus). The role of this formative cross, its plan, geometry and terminology (but none of its esoteric symbolism), are set out in "The Federal Capital; Report Explanatory of the Preliminary General Plan" (Melbourne, 1913). What the Griffins never revealed is how they generated the rest of their plan for central Canberra.

At the foot of the *City and Environs* plan there is a scale, the significance of which has not previously been explained. One unit of the scale equals a mile. If we adjust our compass to this unit we find that it exactly corresponds to the radius of a circle which, centred on the intersection of the Land and Water Axes, passes through the centre of Capital Hill (the ceremonial climax of Canberra) and the site of what was planned to be Canberra's second most important building, the people's Casino at the foot of Mount Ainslie. Thus the primary form or generative seed of the Griffins' Canberra is a mandala of a circle enclosing a *crux decussata* or St Andrew's Cross: this is also the Egyptian hieroglyph for a city or town. Less explicit versions of the same mandala lie at the centre of Griffith, New South Wales, a town planned by Griffin in 1914, and of his proposed Arsenal City at Tuggeranong (ACT), ca. 1916, as well as appearing in windows of Griffin houses at Castlecrag, New South Wales.

Fig. 70.1 Detail from Marion Mahony Griffin, *City and Environs* (1911). Image: courtesy National Archives of Australia. Image processing: Keith Lo Bue

The ancient pictograph concisely represents the fundamentals of urban form—the circular enclosure of consecrated space and its division by the crossing of the main streets into the four quarters traditionally assigned to the four castes or classes of society. The mandala of circle-and-cross is probably the oldest geometric representation of urban space. This symbol combines the paleolithic circle, one of the first rational forms known to humanity, and the neolithic rectilinear grid, representing the new geometry that was more convenient for the arrangement of towns, cities and their supporting farmlands. The deft impressing of the mandala—the essential mathematics of urbanity or civilisation—on a landscape the Griffins had not yet seen, was a touch of genius.

When our compass, still adjusted to the canonic radius, is centred on Capital Hill, we can retrace the second most important of Canberra's formative circles, the one enclosing the prominence which the Griffins planned to be the centre of Government and which is now the site of the new Parliament House. Here again, the esoteric circle is reflected in the surrounding street plan. A circle of the same radius would neatly enclose Mount Ainslie whose peak is slightly more than two canonic radii or one diameter from the intersection of the main axes. Despite all the subsequent tinkering, that larger circle, centred on the crossing of the primary axes in Lake Burley Griffin, still defines the ceremonial centre and spiritual heart of Canberra.

Here the main Land Axis links two natural prominences, Mount Ainslie and Capital Hill. The crucial importance of the intervening chord was emphasised by the contraposition of two isosceles triangles whose apexes are respectively Mount Ainslie and Capital Hill. The triangles share a common base that joins the intersecting points of two circles, thus bisecting a *vesica piscis*. This bisector forms the secondary "Municipal Axis" which runs parallel to the Water Axis, joining the proposed "Urban Administration" and "Urban Mercantile/Military" centres. In the original plan these nodal points (sites of the present Civic Centre and the defence complex at Russell) are both located exactly at the distance of one canonic radius from the central water-crossing. The geometry of the two isosceles triangles is also determined by the grid system that originates at the peak of Mount Ainslie. The grid module is the square of the canonic radius (one mile by one mile).

The canonic circle delimits the octagonal suburban centre to the north of the city and—somewhat less precisely—those to the west and southeast. The centres of the two suburban octagons just south of the dammed Molongo River are one diameter apart. The three centres of the southeastern suburbs form another isosceles triangle that almost replicates the triangles of the principal axis. The centre of the suburban octagon immediately southeast of the Capitol is located at exactly one diameter from the ceremonial centre. The east–west boulevard of the southeastern hexagon is three diameters south of Mount Ainslie. The centre of the northern octagonal suburb is almost, but not quite, one diameter from the peak of Mount Ainslie. The centre of the western "residential suburb" (the present Yarralumla) is slightly less than one diameter from the centre of Capital Hill. Some of these arbitrary departures from geometrical consistency appear to be the result of the Griffins' aim to make the natural prominences, particularly Capital Hill, the termini of vistas along the main axes and thoroughfares.

The original plan of Canberra was a brilliant synthesis of a priori geometry, classical precedent and ancient symbolism, on the one hand and, on the other, of a prophetically modern vision of natural landscape, indigenous character and ecological propriety. The Griffins' appreciation of Australian landscape—*a posteriori*—was immediate, joyous and definitive. After almost a century, their environmental sensitivity and insight can hardly be faulted.

According to the original geometry, Canberra's two most important monuments, the governmental complex on Capital Hill and the people's Casino at the foot of Mount Ainslie, were to be linked by three intersecting circles of equal radius and having a common diameter, thus forming a Pythagorean Triad. The neo-Platonist Iamblichus (ca. 250–325 A.D.) recorded that the Triad had the following "names" (or, as we might now say, "connotations"):

- Proportion (*analogia*)
- Harmonia
- Marrriage
- Knowledge (*gnosis*)
- Peace
- Every Thing
- Hecate (chthonic goddess of the cross-roads)
- Good Counsel
- Piety
- The Mean Between Two Extremes
- Oneness of Mind
- The All
- Friendship
- Purpose (Guthrie 1987: 22, 322).

How entirely appropriate for the ritual centre of a new democratic Commonwealth! As the grand Triad of the Land Axis symbolises the harmony that should exist between the people and their government, so the minor Triad on Capital Hill echoes an analogous harmony or balance of the legislative and executive powers—Parliament being positioned in the centre as the "mean" between the "extremes" of the democratically-elected Prime Minister and the imperially-appointed Governor-General. The Triad also symbolises the cardinal virtue of Temperance and all its resonances, moral, domestic, civil and political.

The Griffins fully understood the ancient "Idea of a Town"; their ideal Canberra was a true city, classically formed and cosmically oriented.[1] The city might almost

[1] The Griffins, who were both well read in the classics, were evidently familiar with the ancient traditions of town planning. They might have seen the summary of the Etruscan-Roman rituals in the article on "Augures" in Oskar Seyffert's *Dictionary of Classical Antiquities* (1891) an abridgment of which was published in New York in 1908.

have been ritually inaugurated too, according to the most venerable precedents: there could be much more to the secret magic that was so speedily wrought in 1911, in faraway Chicago. Australian artist Judith McConchie has recently discovered an intriguing congruence between the plan of Canberra and the constellation Crux or "Southern Cross".[2] Could it be that the generative mandala of Canberra was no other than a starry templum descried in the macrocosm and described on the world below? Later on, at Castlecrag, the Griffins were involved in the production of a community play called *Mirrabooka*—an Aboriginal name for the Southern Cross. If McConchie is right, Canberra must indeed be the last cosmic city, the last complete city—its mandala sealing a sublime marriage of Heaven and Earth.

Coming from a free and liberal democracy, the Griffins had expected to create a new democratic capital where the magic of Pythagorean geometry would inform and sustain the growth of a new and enlightened nation: their ideal—a modern Pythagorean polity, no less, a new Croton. On arriving in Australia, however, they soon had to cope with the mystifying constitution and colonial mentality of a country that was still officially a "Dominion" of the British Empire. Confronted as well by a ruthless, incompetent and incorrigible bureaucracy with its own agenda, the Griffins were forced to revise their plan several times (1913–1918) before losing control of the project altogether and moving to Melbourne, where they lived until 1924. The analogies between the cruel treatment of the Griffins and that later accorded to Jørn Utzon—the only other architectural genius to work in Australia—are too painful to dwell on.

The Griffins in Melbourne

Of the works from the Griffins' Melbourne sojourn, only two major buildings survive (both in Melbourne's principal thoroughfare, Swanston Street). They are Newman College, a students' residence at Melbourne University (1918) and the Capitol Theatre, an office building and cinema which opened in 1924. Designed contemporaneously, these important buildings both incorporate the circle-and-cross mandala previously employed at Canberra.

The proposals for Newman College were drastically revised and only partly realised but the formative mandala is clearly evident in the plan of the domed and spired refectory which commands the profile of the extant building. It is hardly surprising that the Griffins should adopt the classic plan of the Christian sanctuary for the communal focus of a religious institution. Their first plan (1915) shows two matching circle-and-cross structures terminating the principal façade. It is interesting to note that, while the plan is symmetrical, the elevation is asymmetrical, with the spire of the central chapel displaced to the left. The original plan obviously reflects the traditional form of ecclesiastical cloisters and

[2] "Canberra, city that reflects the stars", *Canberra Times*, *Panorama*, 10 March 2001, 2–3.

a
b
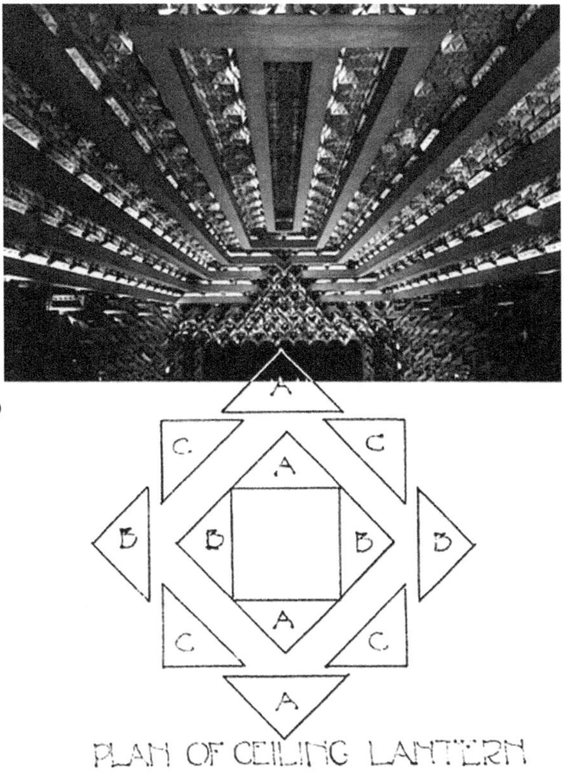

Fig. 70.2 Capitol Theatre, Melbourne: (**a**) (*above*) view of auditorium with "Crystal Tetraktys" above the proscenium. Image: Harold Paynting Collection, courtesy State Library of Victoria. (**b**) (*below*) part of Detail Sheet 82 (dated 1922.4.15). Image: courtesy Melbourne University Architectural Collection, State Library of Victoria

academic quadrangles; but is it too much to suggest that the grand axial opposition of the temple-like termini is a deliberate echo of Canberra's Land Axis and the way it joins the governmental complex on Capital Hill to the people's Casino below Mount Ainslie?

According to Marion's recollections, the design of the Capitol Theatre took fully 10 years from its original conception in 1914 to its spectacular opening in 1924. The cinema auditorium, with its famous illuminated ceiling intact, has been acquired by RMIT University and is now undergoing extensive refurbishment. On the occasion of the re-opening on 7 November 1999 (its 75th anniversary), the ceiling was ceremonially relighted. Suddenly, above the proscenium arch, there appeared the glowing cosmogram of the Pythagoreans, the Tetraktys (Fig. 70.2a). The symbolism of this mystical device and its significance for the Capitol Theatre are to be explored in a forthcoming article which argues for the following conclusions:

- The Capitol Theatre is a secular temple with an esoteric Pythagorean rationale;
- The dominant motif of the cinema space is a "Crystal Tetraktys";
- The decoration of the cinema is based mainly on the right-angled isosceles triangle of Pythagoras's theorem (the Griffins regarded the triangle as the symbol and form of light);

- The Capitol design employs the constituent numbers of the Tetraktys (1 + 2 + 3 + 4) and the whole-number ratios of harmonic proportion (1:1, 1:2, 2:3, 3:4);
- The building was deliberately "tuned" to simple harmonic proportions, including the major third (4:5, a modern harmony that falls outside the old Pythagorean canon).

The Griffins were well-informed musically and left numerous, though not always entirely explicit, allusions to their philosophy of music and its application to architectural design. Architecture, for them, was a kind of music—not frozen but full of life.

By the 1960s the Capitol Theatre had become unprofitable and in danger of demolition. Public opinion and professional objections, however, ensured the preservation of the fabric but at the cost of major internal alterations which included the replacing of the theatre lobby and stalls with a banal shopping arcade (1964).

It was probably the Griffins themselves who persuaded their Melbourne clients to call the proposed theatre and office development the "Capitol", for this was the name that had already been used for the monumental centrepiece of Canberra—the unbuilt ziggurat that was to crown Capital Hill. The common appellation is an important clue to the esoteric rationale of the theatre complex and its surprising relationships with the original Canberra plan.

One of the most striking continuities between the Capitol and the Capital is the exploitation of crystalline forms and imagery (Proudfoot 1994: Chap. 5). Marion was particularly keen on crystals and her enthusiasm is evident in her dreamy perspective views of Canberra and the astonishing interior decoration of the Capitol Theatre. It is said that she was entirely responsible for the design of the crystal ceiling and this is confirmed by several of the detail sheets with her personal annotations. Among the most interesting is her design for the theatre's "Crystal Lanterns" which were built up out of squares and isosceles triangles on the plan of a typical Pythagorean Tetrad (Fig. 70.2b). In much the same spirit is her "Icicle Lantern" for the foyer, which was designed on the plan of the eight-spoked "Cosmic wheel" (Detail Sheet 438). This was also the form of some of Canberra's projected suburbs.

We conclude with the latest and no less surprising discovery that the Capitol Theatre also incorporated the mandala of circle-and-cross in another temple-like edifice that reflected in miniature the Pythagorean philosophy, geometry and symbolism of Canberra.

Original plans and photographs taken before the alterations of the 1960s show how the cinema auditorium was approached through a lavishly-decorated lobby raised slightly above the street level. Here the circle and the cross became the dominant motifs of both plan and elevation (Fig. 70.3).

To the arriving patrons this elegant, if somewhat cramped, space might have given the impression of a domed pavilion or—to the more discerning, perhaps—a narthex or *tempietto*. Such impressions are confirmed by the plans for this room, especially Detail Sheets 159 and 269. The reflected ceiling plan shows how the octagonal columns and their matching "pilasters" form a square and support a space

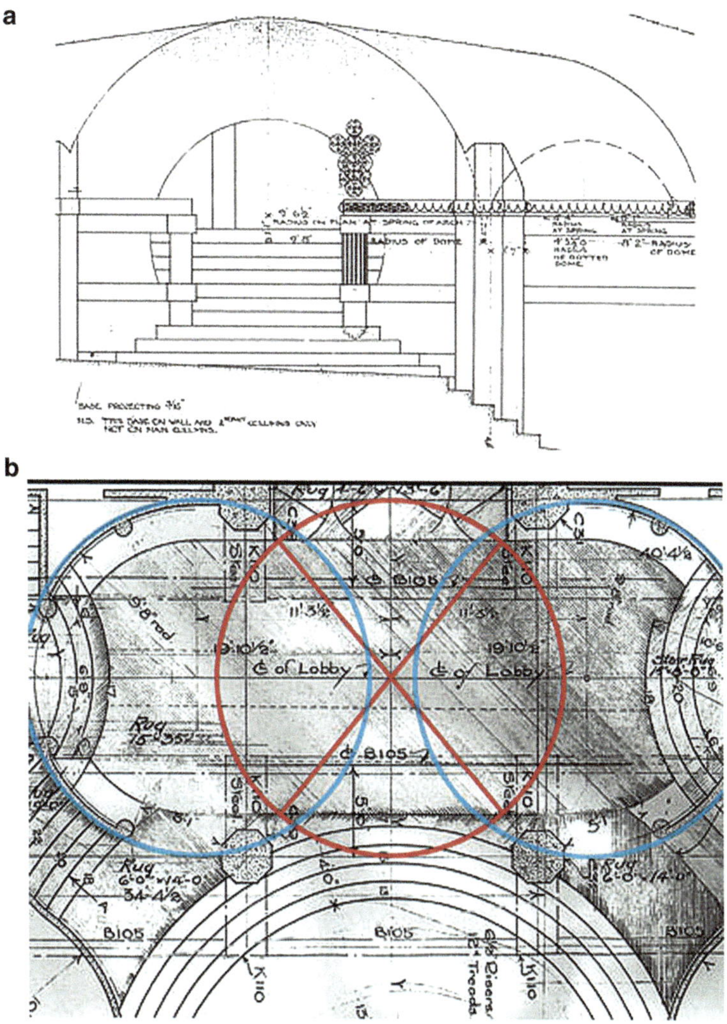

Fig. 70.3 Capitol theatre, Melbourne: (**a**) (*above*) part of Detail Sheet 270, "Lobby Ceiling" (last dated 1924.2.4). (**b**) (*below*) part of Detail Sheet 159, "Ground Floor Plan" (dated 1922.11.9). Images: courtesy Melbourne University Architectural Collection, State Library of Victoria. Image processing: Keith Lo Bue

of circular forms. The main foyer space is arranged symmetrically around a cross whose vertical and horizontal arms are on the central axes of the main building and the theatre lobby respectively. Detail Sheet 159 also indicates the diagonal arms that connect the centre of the lobby with its supporting columns Again, the formative circle, made manifest, completes the mandala that the Griffins employed in all their major Australian works.

Like Canberra and Newman College, the Capitol has—or had—a ceremonial centre planned on the circle and the cross. The continuity of form and symbolism is unmistakable and the recreation of the ancient cosmic temple form within the narrow site of the Capitol Theatre was both ingenious and highly significant. For the Griffins, the new theatre was to be not just a scene of idle diversion but a temple of the arts dedicated to public edification. Here again they revealed their determination to establish a spiritual ethos as part of that covert Pythagorean programme which was to achieve its finest expression in the development of the bushland suburb of Castlecrag on Sydney's magnificent Middle Harbour. These high intentions are confirmed by the choice of Cecil B. de Mille's silent epic *The Ten Commandments* for the opening of the Capitol, which was embellished with live music and drama as well as the magical illuminations of the crystal ceiling. The Capitol Theatre was a veritable *Gesamtkunstwerk*—a total art-work.

Conclusions

Essential to the Pythagorean polity is the all-important notion of harmony—harmony in nature, in art, in society and its habitat, in the heavens and in the human body and soul:

> Nature demands a unity in her ideal which embraces human nature and its expressions as well as all animals, vegetables and inanimate creations, and requires of each a contribution that shall fit into the great harmony.[3]

The Griffins saw their architecture, their life, as part of the "Great Chain of Being".

Their insistence on the human need for harmony, and its origin in the cosmos itself, is thoroughly Pythagorean, even though they never mention the Master's name. Their silence, of course, is entirely in accordance with the esoteric tradition.

Harmony is fundamental to the Griffins' world-view—their morality, polity and sensibility. The sophistication and sincerity of their beliefs are perfectly apparent in their writings, both published and unpublished; and these convictions are expressed in, and impressed upon, their architecture and landscape design with extraordinary passion, precision and determination:

> In town planning as in architecture there must be a vision. There must be a scheme the mind can grasp, and it must be expressed in the simplest terms possible. Just as music depends on simple mathematical relations so do Architecture and Town Planning.[4]

Their commitment to the Pythagorean ideal of harmony was reaffirmed in the lobby of the Capitol Theatre—which aspired to be no less than the narthex of a secular cathedral. Detail Sheets 159 and 269 show how, on the horizontal axis of

---

[3] Walter Burley Griffin, as reported in Griffin (1940: Vol. IV, 34).

[4] Walter Burley Griffin, in Griffin (1940: Vol. I, 361).

the foyer, the centre point is flanked by two almost equidistant centres of the semi-circular apsidal spaces that flow out from the central square. If we complete and make manifest these three hidden circles we finally discover that this lost anteroom to the Capitol Theatre also embodied the same Pythagorean Triad that had previously informed the site of Canberra: again, a plain and potent symbol of social harmony. There can be little doubt that the Griffins had knowingly contrived to deploy all the harmonic symbolism and spiritual power of the Triad in this new cosmopolitan setting.

The resemblances between Canberra and the Capitol Theatre are sometimes subtle but altogether too strong to have been accidental. They include, most conspicuously, the transcendental form of the ziggurat which reappeared in the crystal ceiling of the cinema (see Fig. 70.2a) and they probably extended to the remarkable similarity in plan and disposition between the domes over the stairways to the theatre lobby and the circular water basins of ceremonial Canberra.

Like Aeneas and his father, the Griffins fled the ruins of their capital and eventually founded a new "capitol" in a distant land. Whether or not they accepted the Virgilian version of the legend, the Griffins surely knew that the Roman colony or camp was technically a provincial microcosm of the great Urbs itself, formed on the same timeless geometry of circle-and-cross.

At Melbourne the essentials of Canberra's visionary theatre and its crowning monument, the proposed temple on Capital Hill, were boldly redistilled into a jewel of enlightened urbanity—a brilliant-cut crystal that is the very quintessence of the Griffins' world-view, religion and philosophy. Thus the new southern Capitol, Australia's first skyscraper, became a microcosm of world-harmony, a reaffirmation of moral, civic and spiritual values in the remote Antipodes and—in due course—a mutilated endnote to the destruction of the ideal Canberra.

Biography Graham Pont, a specialist in interdisciplinary studies, taught in the General Education programme at the University of New South Wales for 30 years, where he introduced the world's first undergraduate courses in Gastronomy. He was a founding convener of the Symposium of Australian Gastronomy (1984) and co-editor of *Landmarks of Australian Gastronomy* (1988). His last appointment was a visiting professorship in the School of Science and Technology Studies. Trained in philosophy, his principal research area has been history and philosophy of music but his interests have also extended to environmental studies, landscape, history of gardening, philosophy of technology, bio-acoustics and wine history. In 2000 he published the results of the first major computer analysis of Handel's musical notation, and he is completing a biography of Australia's first composer and musicologist, Isaac Nathan (1792–1864).

A Rome Scholar in architecture, Peter Proudfoot was Visiting Professor in the Faculty of the Built Environment, University of New South Wales, Australia. In articles and books he has written on the origin and history of Australian cities, the effect of seaport development on city growth and urban construction, the theory and practice of the Picturesque movement, ancient geomantic paradigms and their influence on modern urban planning, and architectural education theory. He was a consultant to the National Estate Division of the Australian Government on the conservation of historic structures, a consultant to the Commission of Inquiry into the Maritime Industry and the Australian Development Corporation.

References

GRIFFIN, M. M. 1940. *The Magic of America* (second version of her unpublished memoirs). New York: New York Historical Society.

GUTHRIE, K. S. 1987. *The Pythagorean Sourcebook and Library: An Anthology of Ancient Writings which Relate to Pythagoras and Pythagorean Philosophy*. D. R. Fideler, ed. Grand Rapids, MI: Phanes Press.

MULLER, P. 1976. *Walter Burley Griffin Memorial Lecture: The Esoteric Nature of Griffin's Design for Canberra*. National Library of Australia MS 7817, Box 23 (vii). http://www.pnmull.supanet.com/wbg1.htm (accessed 19 November 2013).

PROUDFOOT, P. R. 1994. *The Secret Plan of Canberra*. Kensington: University of New South Wales Press.

SEYFFER, O. 1891. *Dictionary of Classical Antiquities*. New York: Macmillan Company.

Chapter 71
The Ruled Geometries of Marcel Breuer

John Poros

Introduction

In the 1950s, architectural structures began to be built that seemed to have more in common with the incipient space race than traditional buildings. Thin shells with double curvatures, mostly with saddle-shaped forms, began to roof large and small buildings. These forms were characterized by their purely mathematical, abstract qualities. While the forms looked exotic, the ability to make these complex curved surfaces came from a simple generator, the ruled surface. Using these ruled surfaces, a complex curve can be made from straight lines, making the construction of these surfaces possible for the technology of that era.

A ruled surface is a surface defined by sweeping a line in space along two paths. The swept lines are known as rulings. The two paths at the endpoints of the lines are known as the base curve and the director curve. The simplest ruled surface is a plane, that is, a straight line swept along two parallel, straight line base and director paths. Two other simple ruled surfaces include cylinders, a line swept between the circular base and the director curves, and cones, a line swept along a circular base curve and held at a point at the other end of the ruled line (Weisstein n.d.).

While simple ruled surfaces such as planes, cylinders or cones have been used in architectural geometry for thousands of years, the doubly-ruled surfaces being explored in the 1950s were of a more complex nature. In a doubly-ruled surface, two lines sweep along paths that can be co-planar, defining a plane; non-coplanar, defining a hyperbolic paraboloid; or along curves, defining hyperbolic paraboloids

First published as: John Poros, "The Ruled Geometries of Marcel Breur", pp. 233–242 in *Nexus VII: Architecture and Mathematics*, Kim Williams, ed. Turin: Kim Williams Books, 2008.

J. Poros (✉)
College of Architecture, Art, and Design, Mississippi State University, P.O. Box AQ, Mississippi State, MS 39762, USA
e-mail: jporos@caad.msstate.edu

K. Williams and M.J. Ostwald (eds.), *Architecture and Mathematics from Antiquity to the Future*, DOI 10.1007/978-3-319-00143-2_24,

or hyperboloids. These surfaces have the great advantage of being able to be constructed from straight lines or identical curves, simplifying the construction process. The other advantage of these surfaces is structural: the curved surface acts as a membrane and, ideally, forces are purely compressive or tensile, leading to a highly efficient use of material. These shell structures have the capacity to be very thin, in many cases simply several centimetres thick spanning tens of metres.

For engineers of the time, the structural logic of these surfaces was paramount. Engineers such as Felix Candela argued against the use of these forms as simply new shapes:

> Space frames, hanging roofs and concrete shells are all legitimate prey in what is pronounced a move to humanize the arid, primitive idiom left to us by the pioneers (of modernism). 'Structuralism' is originality's new escape valve (Candela 1958: 191).

Yet these new forms in their geometrical purity were problematic as architecture. The surfaces were typically quite large, and had no real human scale to them as purely mathematical forms. The engineer Mario Salvadori understood this problem of scale and wrote "Architects and engineers who want to use the new shell forms effectively must understand the disturbing reactions they can evoke—and how to deal with them" (Salvadori and Raskin 1958: 112). The thinness of these shells gave them an insubstantial feeling that ran counter to the primeval need to feel the sheltering of a roof or the traditional weight of building.

The architect Marcel Breuer, like many of his contemporaries, worked with shells and the geometry of ruled surfaces in his work of the 1950s and 1960s. Unlike his contemporaries, Breuer's initial fascination with ruled surfaces to create thin, weightless shells gave way to using ruled surfaces to express weight and mass. In his best work, a ruled surface geometry is used to both create thin shell structures and visually ground the structure.

The Challenges of Ruled Surfaces

The use of ruled surfaces after World War II started with the work of structural engineers, but quickly caught the attention of architects as the new forms began to be published. "The hyperbolic paraboloid is now a project type in design offices and school workshops across the world," commented Felix Candela in an article for *Architectural Record* (1958: 191). Structural engineers such as Candela, Pier Luigi Nervi and Enrique Torroja had made this form of ruled surface popular, using it for factories, warehouses, churches and other building types throughout the 1950s. For structural engineers, structural efficiency was what recommended these forms; the ability to span large spaces with very thin shells was the reason for using a form such as a hyperbolic paraboloid. Candela reminds architects of this point in this article:

> It is forgotten that the paraboloid stemmed from purely functional and economic reasoning. I doubt very much that it can be the answer to any stylistic problems. But after the novelty of its shape has subsided perhaps it will be realized that the thin shell paraboloid has qualities as a building form that are far more persuasive than just esthetic considerations (Candela 1958: 191).

In an article in *Architectural Forum*, the engineer Mario Salvadori and architect Eugene Raskin defined some of the architectural problems inherent in the use of these mathematical surfaces. Salvadori and Raskin saw the mathematical and, to their minds, 'abstract' forms of shells as providing little sense of protection and no indication of scale, two traditionally important architectural qualities. Salvadori defined the first problem thus:

> One such basic assumption is that a roof is for safety and protection; it is an emotional as well as physical symbol of security. How can the shell provide reassurance on this point? It looks as though it had recently and temporarily alighted from a voyage on the wings of a breeze. To the eye, which seeks protection in the conventional terms of bulk and strength, the shell will seem unsatisfactory, forcing the designer to solve the human need for shelter by another approach (Salvadori and Raskin 1958: 112).

A separate but related problem Salvadori and Raskin identified with shells is defining the scale of a shell: "Shells are mathematical abstractions, just as well represented by paper models as in concrete over 300 foot spans" (Salvadori and Raskin 1958: 113). Salvadori and Raskin argued that the effect of these two problems was to create "tension, anxiety and stimulation" in people encountering these shell structures (1958: 113). Combined with the unfamiliarity of these shapes as buildings, Salvadori and Raskin argue that these structures have very different qualities to them than most traditional architecture, which telegraphs messages of stability and permanence. For Salvadori and Raskin, the key to the architectural use of these shell structures, besides their very weak suggestion of using terraces, plantings, steps and paving to introduce scale, is to use the tension and anxiety these structures produce in a positive way:

> The tension and anxiety that these unfamiliar shapes generate are not always undesirable, however. Games of skill, for example, produce tension and anxiety in a pleasurable sense; so does the reading of an adventure novel, or the watching of a suspense drama. There is no reason why these emotional states should not be favorably exploited architecturally as well, for example, in public buildings for sports, shopping, meeting, or in domestic spaces for entertaining. Actually, the argument could be made that serenity, calm, and restfulness are obsolete as expressions of our age, and that the shell is not only the outcome of a technological advance, but also the inevitable signal of a twentieth century aesthetic (1958: 113).

While ruled surface shells were used in the 1950s and 1960s for many building types, the issues that Salvadori and Raskin identified plagued most of these designs. In many cases, the shell structure is very symmetrical and balanced, trying to fight the inherent tension and anxiety of the shell's shape and thinness. In only a few cases, such as the work of Eladio Dieste, was the ability of the shell to cause a sublime sense of balanced tension exploited. In most of the work of the period there is an awkward sense of an alien object dropped in place.

Marcel Breuer and the Ruled Surface

Complex ruled surfaces began to appear in Marcel Breuer's work in the mid-1950s. While for most architects of the period ruled surfaces were only used to create shells and space-defining structures, Breuer's work shows ruled surfaces both as sculptural elements and as shells. While there is acknowledgment that the ruled surface can be used for its structural efficiency, Breuer's sculptural sensibility is clearly seen to be driving the form of these elements.

The first use by Breuer of simple ruled surfaces for sculptural elements is evident in the columns of the UNESCO Secretariat Building of 1955–1958 by Breuer and Bernard Zehrhaus, with structural engineering by Pier Luigi Nervi. The columns on the ground floor act as legs to create an open space underneath the Secretariat building. While the columns are sculpted to impart a sense of muscularity, much as are the legs of Le Corbusier's Unite d'Habitation completed 3 years earlier, the shaping of the columns also has a structural purpose. The legs of the UNESCO headquarters are created by ruled surfaces where the base curve and the director curve are lines askew to one another. These skewed lines allow the columns to be thinner at the bottom and fatter at the top in the transverse direction but fatter at the bottom and thinner at the top in the longitudinal direction. Besides the mass this imparts to the columns, the thickening at the bottom allows for greater rigidity of the column to resist forces in the longitudinal direction while the width at the top in the transverse direction allows for the column to act as part of a rigid frame bracing the building. According to press accounts of the time, the credit for the shaping of columns is assumed to be Nervi's, but there is no evidence one way or another in the project correspondence whether the idea came initially from Nervi, Breuer or Zehrhaus. Breuer himself saw the project as being a collaboration. In a letter to the *New York Herald Tribune* on October 8, 1952, Breuer corrects an article that mentions himself and Zehrhaus, but not Nervi in connection with the UNESCO building calling Nervi, "co-designer of our team in equal standing" (Breuer 2005: Series 8.3.5, Reel 5722, Frame 828).

While the UNESCO building used ruled surfaces in a building component, Breuer's first large work dominated by a ruled surface was for the Hunter College Library in the Bronx, New York City, built between 1957 and 1960. The basic structure consisted of six hyperbolic paraboloid shells placed on singular columns like inverted umbrellas and fused together. Such a configuration of shells was certainly not unprecedented. Candela's "High Life" Garment Factory in Mexico City of 1955 uses a similar pattern of square, inverted hyperbolic paraboloid shells to form a roof. A more direct influence is found in the work of Eduardo Catalano, an Argentinean architect, who is listed as a consultant on the Hunter College job. Catalano first became famous in the United States as the architect of his own house in Raleigh, North Carolina built in 1954. The Catalano House, a 1,700 sq. ft. house roofed by a 4,000 sq. ft hyperbolic paraboloid, was published extensively in the trade journals of the day. In an *Architectural Forum* article of 1955, the Catalano house is featured, as well as Catalano's own diagrams and models showing how hyperbolic paraboloids can be combined to create larger roofed structures (Candela 1955). While the multiple paraboloid combinations in the article do not match the Hunter College paraboloid configuration, the concept of combining the parabaloids is common to both.

In the design of the Hunter College Library, Breuer treats the hyperbolic paraboloid shell much as his contemporaries do, as a thin, weightless, abstract surface. The shell roof is hidden from exterior view by glass walls on all sides screened with chimney flue tiles. The glass walls are actually supported by the shell structure, but their placement on the edge gives the sense that the shell structure is lightweight and is supported by the glass. The shell is only revealed inside the main reading room of the library. The construction details give the ceiling a sail-like quality that belies the weight of the shell. The ceiling, while revealing the use of wooden formwork for the shell, with raised fillets where concrete was allowed to seep out between the wooden boards, is painted white, which negates the materiality of the concrete and further reduces the ceiling's sense of weight and mass. The raised fillets from the formwork are unbroken along the rulings of the hyperbolic paraboloid surface, emphasizing the mathematical geometry of the shell. Even the roof of the shell is striped with alternate colours, which again emphasizes the nature of the shell as a geometrical figure and not a mass.

Breuer continues his work with ruled surface columns in a number of office buildings, but usually not with the structural sophistication of the UNESCO building columns. Sculpted columns become a feature of Breuer office buildings, such as the IBM Building in La Gaude, France, or the HUD Headquarters in Washington D.C. These columns are created using ruled surfaces in the most basic sense; rectangular solids are cut at angles with planes to create the column shapes. The use of non-parallel base and director curves, as in the UNESCO building, is not used again by Breuer except for two projects: the Priory of the Annunciation in Bismarck, North Dakota, and St. Francis de Sales Church in Muskegon, Michigan.

At the Priory of the Annunciation, Breuer was commissioned to create a new priory and college in the plains of North Dakota. The Priory, the first piece of the campus to be completed, is composed of living and communal areas, a chapel, bell tower, and cloister. While the living and communal areas are rectilinear buildings, the chapel and bell tower use ruled surfaces in new ways that close the gap between the abstract shell of the Hunter College Library and the muscularity of the UNESCO columns. The Chapel has a shell roof that is in some ways like the shell of the Hunter College library. The difference between the Hunter College Library and the Chapel is that the Library has a complete "umbrella" shell supported by a column, where the Chapel is covered by four halves of an umbrella shell (Fig. 71.1).

The shells are split in half at the column, allowing the columns to be at the edge of the space rather than in the centre as with the Hunter College Library. While placing the columns at the edges allows the space in the centre to be column free, it does create a problem structurally. The lateral forces, which are balanced in the complete shell, are not balanced by the asymmetry of the half umbrella. The

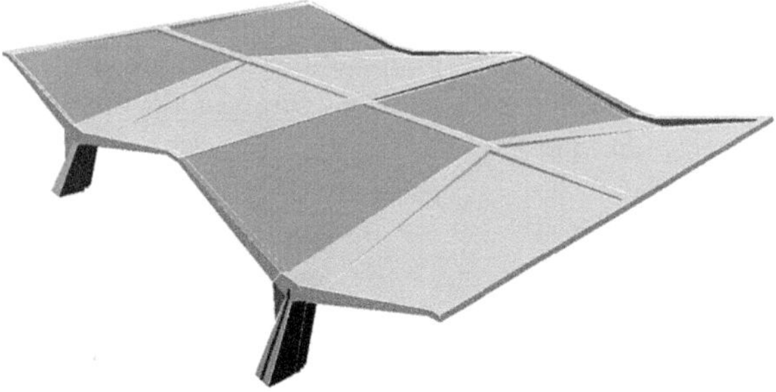

Fig. 71.1 Shell of annunciation priory chapel. Drawing: author

columns need to become buttresses, much in the same way as in a Gothic cathedral the flying buttresses need to brace the vaults. The columns achieve this through ruled surfaces which flare out at the bottom to act as buttresses and twist to run parallel to the wall at the top, much in the way the columns at the UNESCO building work.

The idea of the ruled surface as a way to brace a column is carried into the design of the bell banner at the Priory of the Annunciation as well (Fig. 71.2).

The bell banner is a 100 foot high tower that holds a banner surface with a cross cut into it and underneath the surface, bells. To resist wind pressures due to the large, flat banner surface, the base of the banner is oriented perpendicular to the banner. The ruled surface runs the entire 100 foot height of the tower and twists from the base orientation to the orientation of the banner. The discontinuity of the ruled surface makes it clear that the ruled surface is not a necessary structural element, but instead an expression of the forces that are being transferred by the steel reinforcing hidden within the concrete.

Critical to the development of the bell banner and the priory is the use of isonometric drawings (plan, section, elevation) to describe the geometry completely. The geometry of the ruled surface for the bell banner is described only through elevation, section and plan, with no three-dimensional representation. In the construction documents for the bell banner, the curving surface is not apparent, except as rulings between the straight line base and director curves. The priory roof is drawn in plan and section, but also in two axonometrics for the construction documents. Studies for the priory roof show a grid of 10 $\times$ 20 squares. This grid is organized into eight larger squares of 5 $\times$ 5 grid squares; each large square is then bisected by two diagonals. The diagonals plot the curvature of the roof, as shown in an elevation below, with one diagonal showing the maximum curvature and the other showing the minimum curvature of the shell. As shown in the drawing, the curvature of the shell is the result of setting the elevation of the ends of the grid line, the grid is the ruling of the surface.

The ability to show the constructional geometry in isonometric views is critical to the ability to construct these forms as well. For the bell banner, the construction drawings show plans at the level of each concrete lift. The formwork would be built so that the plan outline would be built at the top and bottom of the form. Planks would then be nailed between the two outlines creating the rulings of the surface

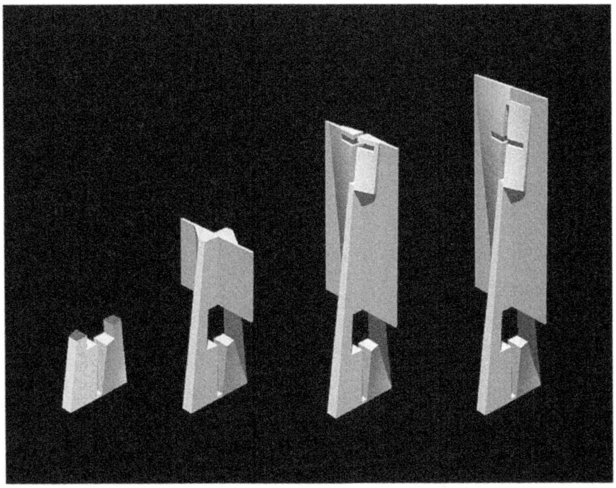

Fig. 71.2 Bell Banner geometry, priory of the annunciation. Digital model: author

and the sides of the form. Concrete would be poured and the process repeated for the next lift. The entire curved surface can thus be created by following a series of outlines.

The Priory roof shell was constructed in much the same way that the vaults for a medieval cathedral were constructed. From the construction photographs, it appears that the formwork for the ribs, the lines of the 5×5 grid squares, was constructed first. The ribs set up the system for planking, again the rulings of the surface, to be laid in straight lines infilling between the ribs. Again, the advantage of the ruled surface is that the entire surface can be formed by connecting points with a straight line.

The most dramatic use of a ruled surface in Breuer's work was at St. Francis de Sales Church in Muskegon, Michigan, completed in 1967. The side walls of the church twist upward in a hyperbolic paraboloid for 72 f. to meet beams that are twisted as well to meet the geometry of sloped, splaying front and end walls. The church from the exterior seems to be a large, trapezoidal solid with inward sloping east and west walls and twisting side walls which rise up to hold the roof (Fig. 71.3).

From the inside, a series of structural ribs and beams that hold the building together as a giant moment frame are exposed. These ribs run east-west and splay out to match the trapezoidal surface of the east wall and narrow to match the west wall. The result of this geometry at the roof level are beams that not only need to fan out from the west wall to the east wall to connect to their respective ribs, but must also twist in their cross section to connect to the sloping, splaying column ribs (Fig. 71.4). The side walls, as evidenced by the construction photos, which show the building of the east and west walls and roof first, are not structural but are only for enclosure.

Breuer spoke about the use of the hyperbolic paraboloid for St. Francis in an interview conducted by Shirley Rieff Howarth in 1977:

Fig. 71.3 Exterior geometry of St. Francis de Sales. Digital model: author

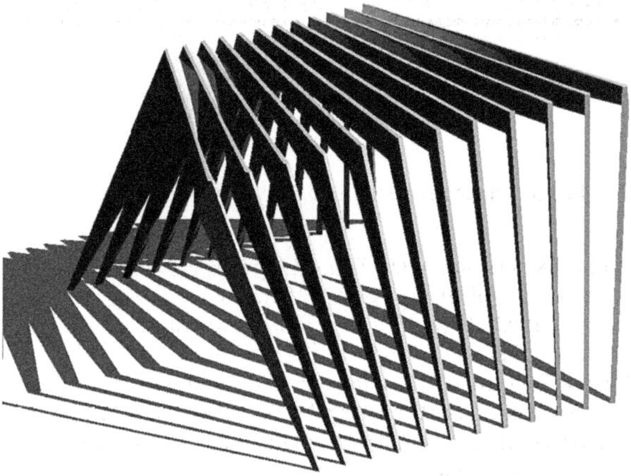

Fig. 71.4 Concrete structural frame of St. Francis de Sales. Digital model: author

> . . .the hyperbolic paraboloids are used in the St. Francis Church for enclosure and not for structural support. Now, this is the first time that they have been used for enclosure—they were never used in this manner before. However, hyperbolic paraboloids have been used in structure before. An example of this is in the Hunter College Library in New York. . . But in the St. Francis church, the hyperbolic paraboloids are side walls and they are really enclosures and purely for space and form reasons—this was not done before (Howarth 1979: 257).

Why was Breuer so concerned to state that his use of the hyperbolic paraboloid was non-structural? I believe that this had to do with his desire to bring mass and a sculptural sense to modern architecture. In his lecture "Material and Intrinsic Form," the Reed and Barton University Lecture on Design at the University of Michigan, March 6, 1963, Breuer stated:

> With the rebirth of solids next to glass walls, with supports which are substantial in material but not negligent in structural logic and practical requirements, a three dimensional

modulation of architecture is again in view; the brother or lover of our pure space. Although not resting on lions or acanthus leaves, space itself is again sculpture into which one enters (Breuer 2005: Series 6.1, Reel 5718, Frames 1092–1183).

Breuer saw a possibility in the use of these ruled surfaces that Salvadori and Raskin did not: ruled surfaces used to create mass, scale and connection to the earth. The ruled surface was not an abstract mathematical form for Breuer, but rather a method for creating sculptural form using the construction methodologies of the day. Breuer understood the structural implications of the ruled surface, but he was also able to look beyond Nervi's and many other structural engineers' view of the ruled surface as simply an efficient structural form. For Breuer, his own sense of the architectural and sculptural requirements came first in the design, and the structural implications were understood but did not overtake those requirements. Breuer's work was always grounded in the traditional qualities of building in many ways—mass, solidity and material. It may have been difficult for him to see the new possibility of expression in lightness and immateriality that Salvadori and Raskin pointed to. However, by reconciling this geometry with traditional architectural qualities, Breuer solved many of the problems of the shell seen by Salvadori and Raskin, bringing new expression to this geometry.

Biography John Poros is an Associate Professor in the School of Architecture at Mississippi State University. He is currently the director of the Carl Small Town Center, a community design and outreach component of the school. He is also the former director of the Educational Design Institute, a research center for improving school design. Before joining the faculty at Mississippi State, Prof. Poros had worked with the architecture firm of Kieran Timberlake Associates in Philadelphia for 7 years. Prof. Poros received his Masters of Architecture Degree from the Harvard Graduate School of Design and his Bachelor of Arts from Columbia University.

References

BREUER Marcel. 2005. *Marcel Breuer Papers Online*. Archives of American Art, Smithsonian Institution. http://www.aaa.si.edu/collectionsonline/breumarc/index.cfm.

CANDELA, Felix. 1955. A New Way to Span Space. *Architectural Forum* **103** (November 1955): 170–177.

———. 1958. Understanding the Hyperbolic Paraboloid. *Architectural Record* **123** (July 1958): 191–195.

HOWARTH, Shirley Rieff. 1979. Marcel Breuer On Religious Architecture. *Art Journal* **38**, 4: 257–260.

SALVADORI, Mario and Eugene RASKIN. 1958. The Psychology of Shells. *Architectural Forum* **109** (July 1958): 112–114.

WEISSTEIN, Eric W. N.d. Ruled Surface. From *MathWorld – A Wolfram Web Resource*. http://www.mathworld.wolfram.com/ RuledSurface.html. Accessed 19 November 2013.

Chapter 72
Conoids and Hyperbolic Paraboloids in Le Corbusier's Philips Pavilion

Alessandra Capanna

The Expo 1958

The Philips Pavilion at the Brussels World's Fair (Fig. 72.1) is the first of Le Corbusier's architectural works to connect the evolution of his mathematical thought on harmonic series and modular coordination with the idea of three-dimensional continuity (Capanna 2000).[1] This propitious circumstance was the consequence of his collaboration with Iannis Xenakis, the famous contemporary musician working at that time as engineer in Rue de Sèvres.[2] Xenakis's very deep interest in mathematical structures was improved on his becoming acquainted with the Modulor, while at the same time Le Corbusier encountered double ruled quadric surfaces.

At the beginning of 1956, Louis Kalff, the art director of Philips Industries, proposed to Le Corbusier a new kind of participation in the World's Fair: their intention was not to expose their products, but rather they wanted a bold show of sound and light effects, to illustrate what Philips's technical progress was able to lead to. "I won't design a pavilion with façades, I'll give to you a *Poème Électronic* and the bottle containing it!", answered Le Corbusier (Xenakis 1976). He designed a building that represented a real synthesis of arts: coloured lights, contemporary

First published as: Alessandra Capanna, "Conoids and Hyperbolic Paraboloids in Le Corbusier's Philips Pavilion", pp. 35–44 in *Nexus III: Architecture and Mathematics*, ed. Kim Williams, Ospedaletto (Pisa): Pacini Editore, 2000.

[1] Precisely the use of ruled surfaces was yet foreseen for enveloping Chandighar's assembly hall, designed during approximately the same period as the Philips Pavilion with the contribution of Xenakis, as one can learn reading the letters conserved at the Fondation Le Corbusier, Paris.

[2] For more about Xenakis, see (Capanna 2001).

A. Capanna (✉)
Dipartimento di Architettura e Progetto, Università di Roma "La Sapienza", Via Flaminia, 359 00196 Rome, Italy
e-mail: alessandra.capanna@uniroma1.it

K. Williams and M.J. Ostwald (eds.), *Architecture and Mathematics from Antiquity to the Future*, DOI 10.1007/978-3-319-00143-2_25,

Fig. 72.1 View of the pavilion. Photo: Wouter Hagens

music, the projection of enormous warped images in a space without architectonical quality. It could be, at the minimum, even a scaffold.

Minimal Surfaces in Three Space

The idea of a container without an aesthetic claim allowed Le Corbusier to think only about the show; in the meantime he entrusted Xenakis with a "mathematical translation" of his sketches, which represented the volume of a rounded bottle with a stomach-shaped plan (Xenakis 1976).

Synthetically, the form of the pavilion had to comply with the following principle: there was to be a maximum of free volume for a minimum of enclosing surface. The classical answer is the sphere, but the sphere, beautiful in itself, is bad for acoustics and is less perceptibly rich than some other double-curved, warped or skewed forms. In that large family of surfaces they had to single out those conforming self-loadbearing shells, easy to build in a common reinforced concrete yard.

Xenakis planned a first solution of the stomach-shaped volume defined by conoids and hyperbolic paraboloids (Fig. 72.2). The pavilion was enveloped by the conoid *E*, a composite surface formed by two conoids *A* and *D*, the hyperbolic paraboloids *K* and *G*, the connecting cone *L* and a couple of empty triangles for the entrance and the exit. Those warped surfaces culminated on three cusps, 17, 13 and 11 m high.

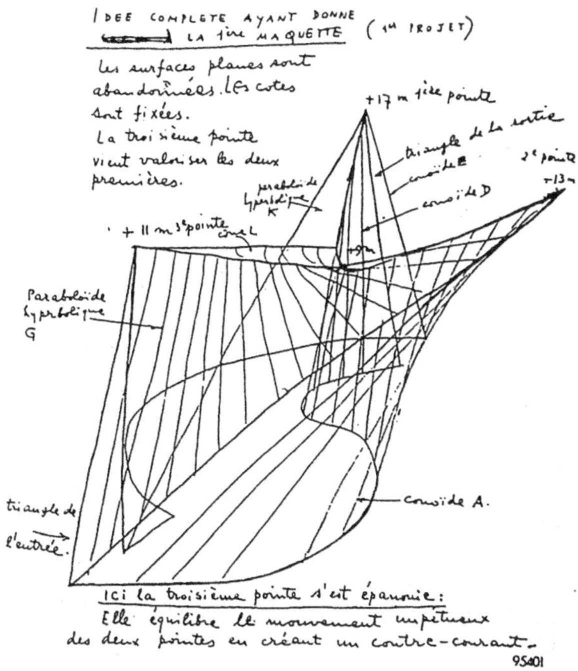

Fig. 72.2 First project. Sketch: I. Xenakis

Some constructive difficulties brought about work on a new project. First of all, it was necessary to change all conoids[3] into hyperbolic paraboloids, whose straight-lined generatrices made statics calculation easier. This second project was the fruit of a hybrid of analytic and descriptive geometry.

> As a matter of fact it is impossible to arrange all the surfaces of this kind only working with its algebrical functions. The refinement in curvature and surface dynamism cannot be imagined by studying its equation. The pavilion must be plastic, above all; for this reason in an infinity of possible curves we had to choose the best composition of warped surfaces (Xenakis 1958: 7).

Thus, the new height of the cusps and their consequent projections were established accordingly on the horizontal plane, so as to increase the dimension of the central cone *L*. The first cusp was fixed at 21 m from the ground, the second at 13 m and the third at 18. Afterwards, all paraboloids were modelled in accordance with the condition that the intersection with the horizontal plane conformed to the primitive scheme of a "stomach-shaped" plan.

Compared with the first one, this project introduced unchanged the hyperbolic paraboloids *G* and *K*, conforming the principal surfaces for filmed projections; it redefined and visibly widened cone *L*, and conoids *A*, E and *D* were transformed in five paraboloids: *A*, *E* and *B*, *N* and *D*. Moreover, the paraboloids *C* and *F* had been

[3] Conoids are those ruled surfaces defined by a curvilinear directrix and a couple of straight-lined directrices.

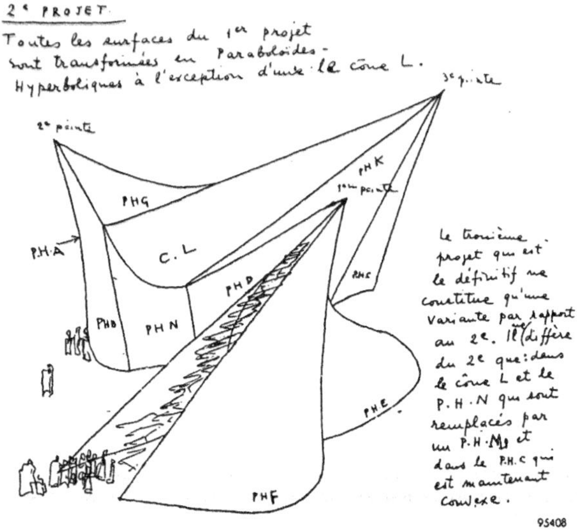

Fig. 72.3 Second project. Sketch: I. Xenakis

inserted. This last one, approached to the curved wall E, delimited the space necessary for the room of the automation of the show, technical spaces and services (Figs. 72.3 and 72.4).

This volume still required two pillars for the stability of the whole building. Xenakis later said that, with the assistance of Mr. Duyster, the Strabed engineer who suggested building the pavilion in pre-compressed concrete, he decided on some final modifications in order to abolish all the vertical supports:

> Mr. Duyster thought that the cone L and the hyperbolic paraboloid N formed the only hyperbolic paraboloid later on defined with the M. Geometrical clearness was so improved. . . . I suggested to him a further slight change to the hyperbolic paraboloids M and B. . . . Moreover I changed the concave paraboloid C into a convex one: the stability of the very steeply sloped third cuspid was so improved as well. Then I remoulded the hyperbolic paraboloids up to the edges of the exit-entrance triangles. The pavilion became completely self-loadbearing, lightweight and pillarless (Xenakis 1958: 10–11).

From this comes the configuration that makes use of hyperbolic paraboloids, thus shaping a kind of enveloping form, closed and opened to the world at the same time through the convergence of its geometrical construction (Fig. 72.5).

Correspondence of Geometrical and Mechanical Properties in Hyperbolic Paraboloid Shells

> In my opinion the fascination of forms derived by the minimal surfaces in design is based on several properties:

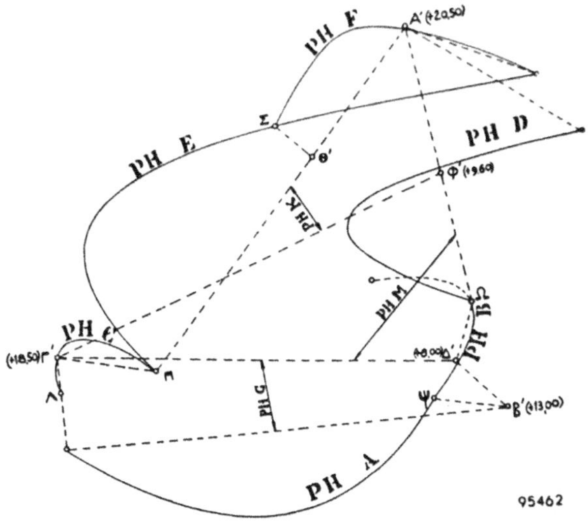

Fig. 72.4 General plan of the pavilion

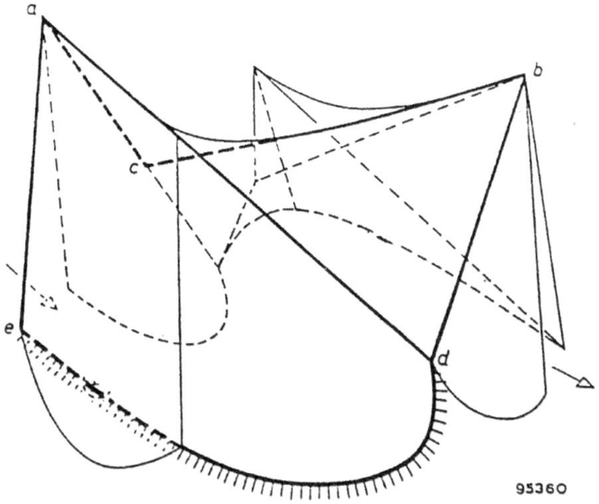

Fig. 72.5 Volume

1. the shapes of minimal surfaces can be astonishing from the aesthetical point of view;
2. the shapes of minimal surfaces allow the optimal use of materials;
3. the structural surfaces with a saddle shape are very stable and resistant;
4. the structures of minimal surfaces have natural geometric rigidity (Almgren 1982).

In 1935, B. Lafaille and E Aimond published their research on the distribution of forces in thin curved walls, vaults and shells defined by simple ruled quadric surfaces such as hyperbolic paraboloids and conoids. Mathematicians had long known those kinds of surfaces, but architects only began designing with this geometry after the accepted use of reinforced concrete to build modelled roofings. Frei Otto pointed out the influence of Bernard Lafaille in the

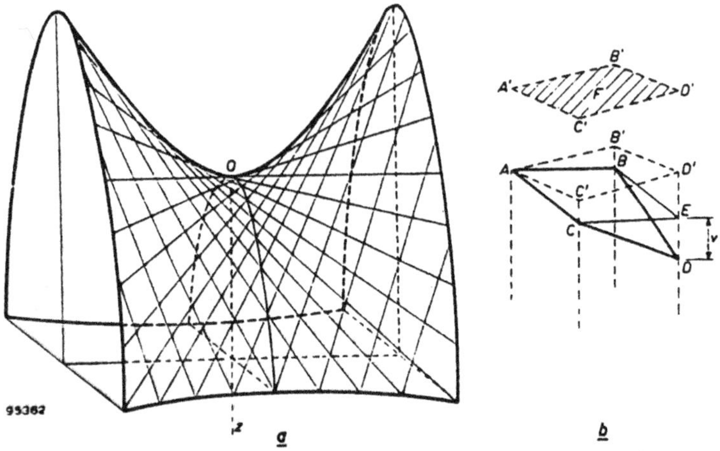

Fig. 72.6 (**a**, *left*) The double-system of generatrices of a hyperbolic paraboloid; (**b**, *right*) Part of the surface delimited by the intersection of four generatrices: AB, CD, AC, BD

development of lightweight roofing and tensile structures; he also remarked that at the Brussels World Fair about 20 pavilions made use of saddle surfaces positively corresponding to the aims of temporary buildings.

The coincidence of saddle surfaces with minimal surfaces was later denied.[4] However, C.G.J. Vreedenburgh, from the University of Technology of Deft, confirmed all the positive characteristics of such structures. He mathematically verified that hyperbolic paraboloids, rigidly fixed to the ribs, fulfilled the requirements of building[5]; he then provided some charge tests.

Before proceeding to the final construction in the assigned exposition area, two models were required: one at a scale of 1:25 made with plaster, which served for studying deformations due to accidental loads, proper weight and the eventuality of fire; another at a scale of 1:10, built to simulate the reality of setting up the thin shells; this one was also used to individualize the lines along which to put in tension the cables, that coincided with the generating geometry. Again, the spirit of geometry and mathematical theory adhere perfectly with constructive practice.

By testing the elementary deformations, it was possible to verify that in the enclosing hyperbolic paraboloid shapes, a uniform system of solicitations on the whole surface takes place: they were all funicular strains up to the edges, along which the loads are transferred to the foundations in the form of normal efforts.

Owing to the large dimension and to the steep slope of the surfaces, it was impossible to built them in one piece, therefore experiments with a sort of assembly

---

[4] See Emmer (2015 ('Periodical Minimal Surfaces')); compare with the unconditional statement taken from Otto (1973).

[5] For the detailed exposition of the demonstration coming down from the equation of hyperbolic paraboloids, see Vreedenburgh (1958).

system were undertaken. The big paraboloids were divided in portions of around a square metre, cut according to the irregular network constituted by the intersection of the generatrices (Figs. 72.6 and 72.7) and then temporarily fixed to a framework (Fig. 72.8).

Mathematics, Music, Architecture: The *Poème Électronic*

In any project such as this, scientific thought is a means with which to realize ideas that have been born of intuition or of some kind of vision, even if they are not of scientific origin.

In 1954 Xenakis was composing *Metastasis*.[6] He declared that he had some visual fantasy: straight lines representing the geometrical form of glissandi, which transformed themselves into an auditory fantasy.

> I started from a very simple serial problem: that is how to reach through the 12 sounds a different formulation of chords. I sketched some lines connecting every other single sound deriving from a chord of 12. Suddenly I thought of those lines as if they were glissando; this effect of sonorous continuity was linked together with my remote experiences with mass internal transformations (Restagno 1988).

Metastasis was the starting point for his compositional research in which science and mathematics are applied. Paradoxically, he declared that this work was not inspired by music itself or by some logical principle, but by the impressions gained during the Nazi occupation of Greece. He listened to the noise of the masses marching towards the centre of Athens, to the shouting of slogans, to the intermittent shooting of guns, superposing each other in a chaos. He never forgot the unforeseeable transformation of the regular, rhythmic noise of a hundred thousand people into some fantastic disorder.

From the beginning of the 1950s, Xenakis was interested in two compositional themes. First of all, he wanted to write a kind of dodecaphonic music with the help of computations, which builds the macroforms out of a few basic principles. In *Metastasis* he made computations based on the permutation of intervals, with the help of the axiomatic approach known in mathematics. In the second place, he was interested in the continuous change of chords. Let us take, for instance, 6 of 12 notes. We get a harmonic colour. Then let us take the complementary pairs of those six notes. Once again, we get a harmonic colour. The change between the two occurs without any transition, abruptly.

---

[6] The orchestra opus in which the Modulor defines a strict relationship between tempo and sound. The name derives from "meta" that means beyond, after and "stasis" that means immobility. The problem fascinated the ancient philosophers, beginning with Parmenide and Zenone Their paradox about Achille and the turtle illustrate the very problem about the contrast between movement and immobility. At the same time the whole word metastasis recalls the medical way to speak about the proliferation of carcinogenic cells, as if one can indicate that they are similar to the improvement of the mass density.

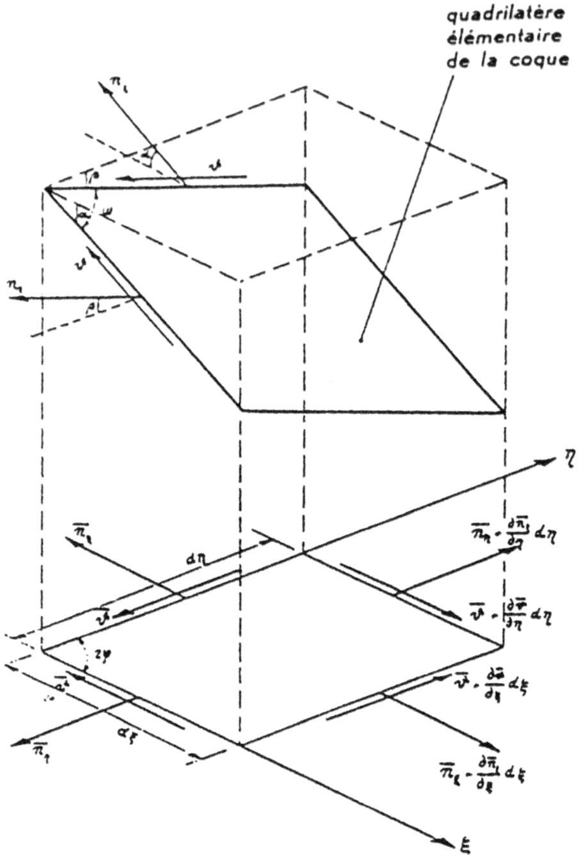

Fig. 72.7 Elementary quadrilateral

Coming from the theory of geometrical transformation and from the study upon sonorous masses—and its regular and irregular variations—the question is whether it was possible to get from one point to another without breaking the continuity.

In music, if we remain in the same scale, the only solution is a *glissando*. From the first six-note group toward the second one we can start in the direction of any note and in each case we get a different perceptive result. Moreover, if we have talked about the idea of continuous variation ("continuity charge"), in *Metastasis* the parallel development of the idea of discontinuous variation ("discontinuity charge"), represented by the permutation of intervals and the organisation of time based on the Golden Section, is very interesting.

In Le Corbusier's studio, Xenakis ran into fertile ground for his compositive obsessions. Le Corbusier's lessons on the mathematical spirit of the Modulor—about the opposition between the harmony of nature and the intellectualism of rules—joined forces with his researches on the Golden Section applied to dimensional changes of scales. Music becomes the image of a continuous movement produced by its own geometrical structure: a held note becomes a particular case of sonorous

Fig. 72.8 Shell's assemblage

curve (that is, the straight line) and, on the contrary, we find that the *glissando* makes perceptible a perfect continuity that can go on to infinity.

In *Metastasis* and in the Philips pavilion, Le Corbusier and Xenakis proved that the forms used in music and in architecture are closely linked: in the musical work, this problem led to *glissandi*, while in the pavilion, it resulted in the hyperbolic paraboloid shapes. The difference between physical and musical space is that the former is homogeneous: both dimensions are lengths and distances. In music the natures of two dimensions (pitch and time) are alien to one another, only connected by their ordering structure.

In the *Poème Électronic* the correspondence between music and architecture is not only a matter of geometry. It was projected by Le Corbusier as if it were an orchestral work in which lights, loudspeakers, film projections on curved surfaces, spectators' shadows and their expression of wonder, objects hanging from the ceiling and the containing space itself were all virtual instruments. Architecture played, at the same time, the role of orchestral instrument and of sound box, container and contents.

The *Son et Lumière* Show

The *Poème Électronic* is an alchemy of 10 min length. Going into the pavilion, one's sight was lost in a neutral and disquieting space delimited by screen-walls largely displaying their geometrical construction. Loudspeakers, electro-acoustics and lighting sets were combined with the strained cable network in a curved interior where architecture was losing its character to become an allegorical and apocalyptical show. The *Poème* was composed therefore from seven sequences, projected according to a rigorous script.

This subdivision was not, however, in some way perceptible: the show flowed instead without solution of continuity with the exception of two pauses during which obscurity and the total silence went down, followed by the apparition of the *object mathematique* and of a mannequin that were lit up with ultraviolet light. In the place of the synthesis of the arts, the organic world of the living subject and the rational, concerning the intellect, were represented separately. The 8 min of Le Corbusier's sequence of images was accompanied by the music of Edgar Varèse and preceded by a brief interlude, composed by Xenakis: 2 min of concrete music destined to accompany the projection of an introductory text in English, French and Dutch. The most intense moments of this brief composition are those in which the most rapid rhythm succeeds to mix a feeling of granularity and powderines of the sound with that of perceptive continuity. The organization of time and rhythm are introduced here as continuity in the discontinuous, as a mass of single discreet elements. The result is the presentation, also in this very inner detail of the poem, of the general theme of the work, that of continuity. The introduction of the acoustic phenomenon as the element of dominion of the forms confirms the theory according to which the geometric-mathematic rule constitutes the common base of the architectural and musical compositions.

The novelty of the twentieth century, expressed in the Philips Pavilion, is the overcoming of a geometry that governs the simple repetition of crystallographical elements, in order to adopt more complex systems with independent and not homogeneous variables in comparison to the unit of measure. The introduction of the fourth dimension in the built space, after having discovered it in the abstract but equally real realm of mathematics, is not an automatic operation. To the three dimensions of Euclidean geometry is generally added the unit of measure of the time to make perceptible the effect of movement. A pictorial representation, or at least a sculptural one, of architecture, such as in a work of Boccioni, cannot exhaust the search of new spatial quality in architecture.

For Le Corbusier, the expression of beauty would still seem to derive from exactness, from a strong resistance to discord and from an iron will to oppose randomness—as is apparent from his reflections on the form of houses and cities; but the creation of the Philips Pavilion sketches the advent of sceneries in which the rule can be also that of the random or that of light and sound diffusion. In fact, the *Poème Électronic* predicted the deconstruction of music and of architecture: the same necessity, more than 10 years after the demolition of Philips Pavilion, was demonstrated, for example by John Cage, in the refusal of both dodecaphonical tyranny and the rigidity of the roles inside the places appointed for the listening. In the same way the space of the pavilion was flexible and the public could assume different positions to follow the show. Music and noise were always on the point of trespassing the one into the other. Deformity not only represented an ugly alteration of the form, but also constituted a first idea in the studies on the modification of architectural geometry. Double ruled quadric surfaces had freed the plastics of the volume, and with it the sonorous materials, conducting composition to the limit of disorder, without yet going beyond it.

"The splendid result was the natural gift of numbers. The implaccable and magnificent play of mathematics" (Le Corbusier 1954: 55). Music and architecture are thus inserted once again in the historical-scientific tradition that from pre-Socratic philosophy leads to contemporary mathematics, assuming eternal forms and innovating subject matters.

Acknowledgment The images in Figs. 72.2, 72.3, 72.4, 72.5, 72.6, 72.7, and 72.8 are reproduced from the author's personal archives.

Biography Alessandra Capanna is an Italian architect living and working in Rome. She received a degree and a PhD in Architecture from University of Rome "La Sapienza". Among her published articles on mathematical principles both in music and in architecture are: "Una struttura matematica della composizione", remarking the idea of self-similarity in composition; "Music and Architecture. A cross between inspiration and method", about three architectures by Steven Holl, Peter Cook and Daniel Libeskind (*Nexus Network Journal* 11, 2 (2009); "Iannis Xenakis. Combinazioni compositive senza limiti", "Limited, Unlimited, Uncompleted. Towards The Space Of 4d-Architecture" and "Tesseract Houses", about the topic of conceiving higher dimension architectures. She is a *Ricercatore* at the Faculty of Architecture of Rome "La Sapienza". She is the author of *Le Corbusier. Padiglione Philips, Bruxelles* (Universale di Architettura 67, 2000), on the correspondence between the geometry of hyperbolic paraboloids and technical and acoustic needs, and its aesthetics consequences.

References

Almgren, F. 1982. Minimal Surfaces Forms. *The Mathematical Intelligencer* **4,** 4 (1982).

Capanna, A. 2000. *Le Corbusier, Padiglione Philips, Bruxelles.* Universale di Architettura. Torino: Testo & Immagine.

———. 2001. Iannis Xenakis: Architect of Light and Sound. *Nexus Network Journal* **3**, 1: 19-26.

Emmer, M. 2015. Architecture and Mathematics: Soap Bubbles and Soap Films. Pp 451–460. in Kim Williams and Michael J. Ostwald eds. *Architecture and Mathematics from Antiquity to the Future: Volume II The 1500s to the Future.* Cham: Springer International Publishing.

Le Corbusier. 1954. *The Modulor: A Harmonious Measure to the Human Scale Universally applicable to Architecture and Mechanics.* London: Faber and Faber.

Otto, Frei. 1973. *Tensile Structure: Design, Structure and Calculation of Building of Cables, Nets and Membranes*, Boston: MIT Press.

Restagno, E. 1988. *Iannis Xenakis*, Torino: EDT/Musica.

Vreedenburgh, C. G. J. 1958. The Hyperbolic Paraboloid and its Mechanical Properties. *Philips Technical Review* **20**, 1 (1958-1959): 9-16.

Xenakis, I. 1976. *Musique. Architecture.* Paris: Casterman.

———. 1958. Genese de I'Architecture du Pavilion. *Revue Technique Philips*, **1** (1958).

Chapter 73
Oscar Niemeyer Curved Lines: Few Words Many Sentences

Benamy Turkienicz and Rosirene Mayer

> *Purism imposed a very heavy limitation in that the free plan could occur only within the boundaries of pure geometric forms*
>
> (Zevi 1974; our trans).
>
> *I am attracted by free-flowing, sensuous curves. The curves I find on the mountains of my native land in the sinuous course of its rivers in the clouds in the sky, and on the body of the beloved woman*
>
> (Oscar Niemeyer, quoted in (Petit 1995)).

Introduction

This study is aimed at the description of the curves that identify the architecture of Oscar Niemeyer and it is based on the geometrical structure of the identity of buildings belonging to the so-called style of Niemeyer.

As opposed to the well-known discipline of Corbusier and Wright, Oscar Niemeyer is widely known for his inventive attitude and conceptual freedom, both associated with the use of curved surfaces, often called "free forms." While Corbusier and Wright postulated the importance of certain elements of control such as the Modulor (Corbusier) and the grid (Wright), Niemeyer preferred to leave the "theoretical" background of his designs at discrete levels, to the extent that his lack of explanation has occasionally been interpreted as the opposite, that is, that every individual building had been the result of a spontaneous act of will related to the so-called "architectural inspiration" (Turkienicz 1994).

First published as: Benamy Turkienicz and Rosirene Mayer, "Oscar Niemeyer Curved Lines: Few Words, Many Sentences", pp. 135–148 in *Nexus VI: Architecture and Mathematics*, Sylvie Duvernoy and Orietta Pedemonte, eds. Turin: Kim Williams Books, 2006.

B. Turkienicz (✉) • R. Mayer
Faculdade de Arquitetura, Universidade Federal do Rio Grande do Sul, Sarmento Leite, 320 SimmLab/sala 306 Centro, 90020-150 Porto Alegre, RS, Brazil
e-mail: benamy.turkienicz@gmail.com; rs_mayer@yahoo.com.br

K. Williams and M.J. Ostwald (eds.), *Architecture and Mathematics from Antiquity to the Future*, DOI 10.1007/978-3-319-00143-2_26,

The expression "free forms" raises two, more general questions related to the geometrical identity of Niemeyer's architecture:

1. that of the apparent paradox between the concept of free forms and individual style;
2. that of what elements and geometrical relations are determinant for the identification of an individual style.

The first question refers to the relationship between design freedom and shape control: the concept of design freedom associated with the recursive use of curves "free forms" may well suggest the absence of control, i.e., the idea of randomly generated curves. Paradoxically, the identification of individual styles and architectural languages does require a family resemblance between buildings, implying that Niemeyer's "free forms" are not as free as they may appear.

The second question addresses the rationale behind the identity between buildings: architects such as Kenzo Tange, Frank Lloyd Wright, Alvar Aalto, Eero Saarinen, Louis Kahn, Le Corbusier and the Uruguayan Eladio Dieste, among others, associated the circumference or the arc with orthogonal forms. However, the way Niemeyer selects and utilizes the curves substantially differs from these architects. His choice is clearly distinctive, in that he can create unique buildings out of a very limited range of curves without primarily associating them with orthogonal forms.

Amidst the variety of curvilinear volumes designed by Niemeyer, such as the Bobigny's Employments Office and the recent Contemporary Art Museum of Niteroi, Brazil, it is possible to trace a formal identity to the extent to which these buildings can be identified as belonging to an individual style or architectural language. The identity between these curves or "free forms" unequivocally unveil the main stylistic attributes of Niemeyer's architecture.[1]

Architectural critics have left unexamined the geometrical foundations of Niemeyer's architecture, and have thus failed to identify some important aspects related to sophisticated mechanisms of shape generation and shape control applied by the Brazilian architect in buildings with curved shapes. The present study will help demonstrate that Niemeyer not only systematically uses classical dimensional control mechanisms (rules of proportion) but is able to generate singular buildings out of a limited (and thus predictable) vocabulary of curves and a sequence of less predictable mathematical operations.

The study is not concerned with a formal approach. It deploys instead a graphical description of the relationships that characterize the language of Niemeyer's curves. It

---

[1] The identification of these attributes as parts of a language has lead to the basis of what can be described as a Niemeyer's first shape grammar. The construction of the grammar was based in the analysis of 20 buildings designed by Niemeyer between 1943 and 2003, all of them with curved surfaces. In the study a comprehensive vocabulary of curves is shown in detail in addition to the description of the dimensional control mechanisms pervading the architectural language of Niemeyer. See Mayer (2003), Mayer and Turkienicz (2005a, b).

should be understood as a preliminary step towards the depiction of the grammar that Niemeyer has used in his projects throughout his almost 70 years of professional life.

The chapter is divided into four parts. In the first part, the structure of Niemeyer's architectural grammar is briefly introduced. In the second part, the main argument of this chapter is developed through the description of Niemeyer's vocabulary of curves, followed by the third part, the representation of the dimensional control mechanisms (rules of proportion) observed in the buildings analyzed. The fourth part, the conclusion, refers to possible steps that might unfold as future developments for this work.

Generative Processes in Niemeyer's Style

To create a shape grammar for the description of Niemeyer's architecture, a model based on the idea of Shape Grammars (Stiny 1975; Gips 1975) was applied in the description of the volumetry of 20 buildings designed by Niemeyer between 1941 and 2003 (Mayer 2003) (Fig. 73.1).

These 20 buildings have in common the use of curved shapes (mainly conic curves, specially the parabola). Assuming that all 20 buildings were different from each other, the goal was to describe their volumes (according to the Shape Grammar model) using the smallest set of rules and operations. In order to assess their generative aspects, shape schematic building plans and facades were drawn. The drawings formed the basis for the correlation between the generatrix and the directrix with the depiction of the operations necessary to the generation of the surface of the volume of each building.

The generative elements of the volumes' surfaces (generatrix and directrix) correspond to the vocabulary of curves and line segment. Operations for the generation of the initial volume were rotation and translation. A reference structure was set so as to enable the description of the rules based upon axes for generative elements as well as for the operations. Generative rules for volumetry were described according to the relations between the generative element's axes (generatrix and directrix) and the axes of the generative operations: the position and the inclination angle referred to the generative element and to the direction of the operation related to the generatrix. Other determinant rules, as far as form is concerned, regard the selection of the generative elements (vocabulary) and the number and existence or not of similarities between the directrices.

After the generation of the initial volume the generative process evolves throughout a series of complementary operations including reflection, translation, scaling, intersection, addition and subtraction. Niemeyer apparently explores possibilities of the generated initial shapes with these complementary operations. The combination of the initial and the complementary operations, within Niemeyer's chosen vocabulary, gives rise to an immense set of alternative designs (Fig. 73.2).

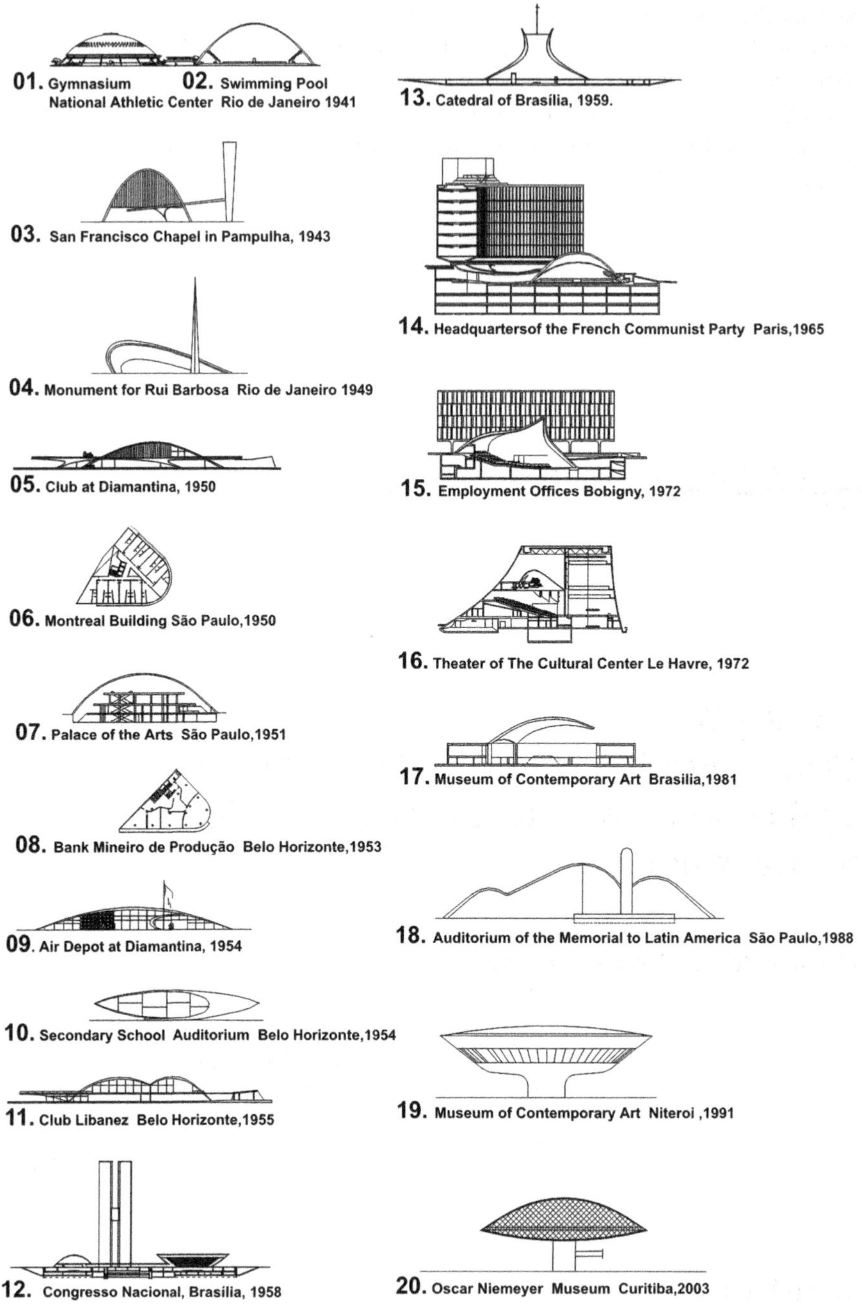

Fig. 73.1 Twenty buildings designed by Oscar Niemeyer between 1941 and 2003, that have *curved shapes* as a common feature. Image: authors

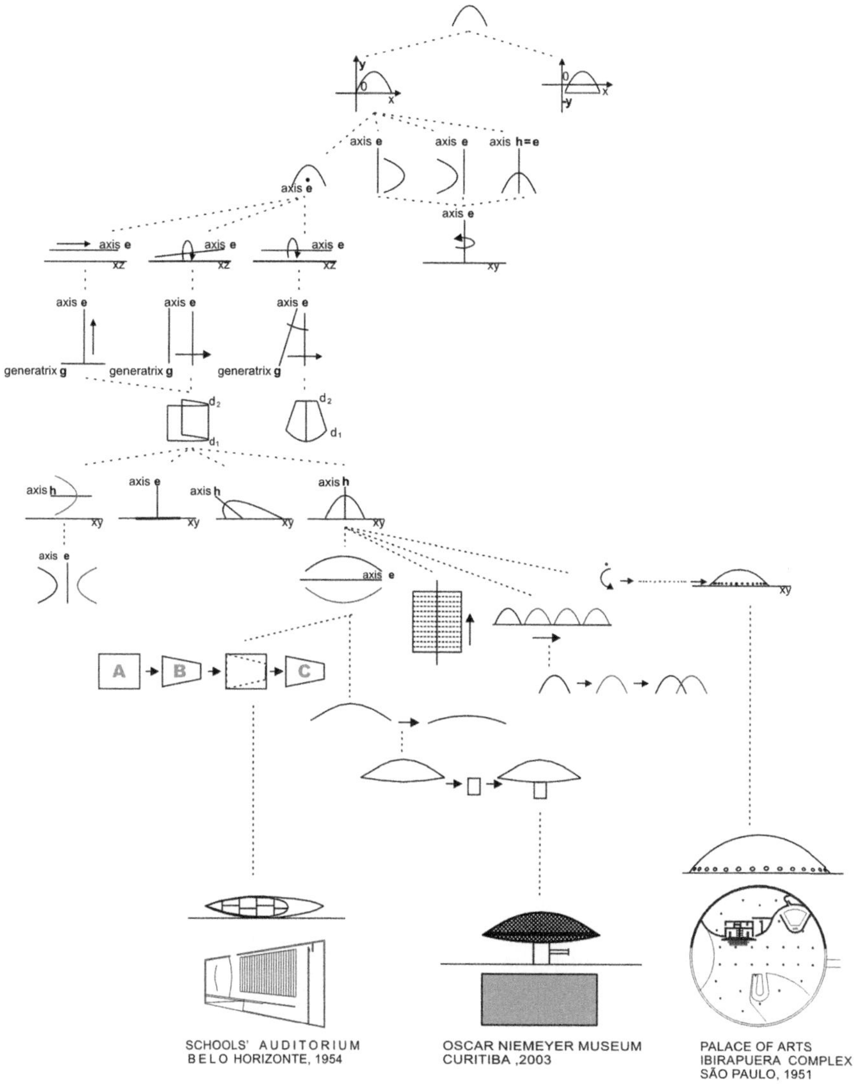

Fig. 73.2 Diagram showing the sequence of rules that generate the buildings selected for the sample. Image: authors

The study of Niemeyer's language and its generative steps has shown that cognitive aspects related for example, to the environment context (and its relations with the definition of the rules or with the generative steps) are relevant for the description of his language (Mayer and Turkienicz 2005b).

The next section explores some generative aspects related to Niemeyer's vocabulary of curves.

Curves in Niemeyer's Style

> In Pampulha my architecture's plastic vocabulary began to be defined, like an unexpected game of straight lines and curves (Niemeyer 2004: 153).

The description of Oscar Niemeyer's architectural language according to a shape grammar has unveiled the relative importance of the curve as a formal determinant generative element and hence for the perception of a kinship relation between buildings. In this study a detailed description of this shape grammar has been made, along with the description of the vocabulary and of the curve's grammar. The process is complementary to the characterization of the architect's generative process and had two goals:

(a) To characterize the kind of curved geometry identifying Niemeyer's architecture;
(b) To shed some light on the strategies involving different utilizations of the curve in the generation of elements and surfaces in Niemeyer's architecture.

Considering the curved line as Niemeyer's initial form, a preliminary classification of the curves present in the sample analyzed was made. In mathematical terms, he utilizes curves segments, polynomial and parametric. We have identified two types of recurrent curves: (1) conic curves, specially the parabola, and (2) the "composite" curve.

Conic Curves

The parabola is the most frequent of all curves found in the sample examined. Different parabolas in different scales were identified (parametric shapes) In order to examine the parametric shapes, an AutoCAD script using a Visual Basic application was developed for the computational generation of parabolas and to allow a parametric description of these curves:

The p parameter (distance from the focus to the directrix d) and the height (y coordinate) having the vortex positioned in (0, 0) may be specified through the parabola generator (Fig. 73.3). The line segment AB can also be specified through the coordinates A (x, y) and B(x, y).[2]

A comparative scheme was elaborated to demonstrate the transformations from the initial parabolic shape to the actual buildings where this curve is observed (Fig. 73.4). The scheme helped us to track transformations from the initial shape and, at the same time, to identify the basic elements common to all buildings by Niemeyer that were analyzed.

[2] Parabolic lines were utilized as generatrices or directrices combined with line segments or in a simultaneous manner as depicted by the grammar of Niemeyer's 20 buildings designed with curved shapes.

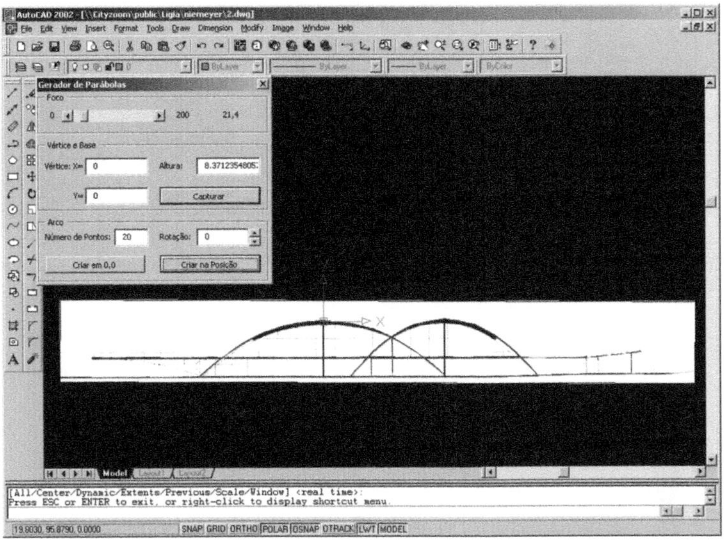

Fig. 73.3 Parabola generator: an AutoCAD script using a Visual Basic application. Screen shot: authors

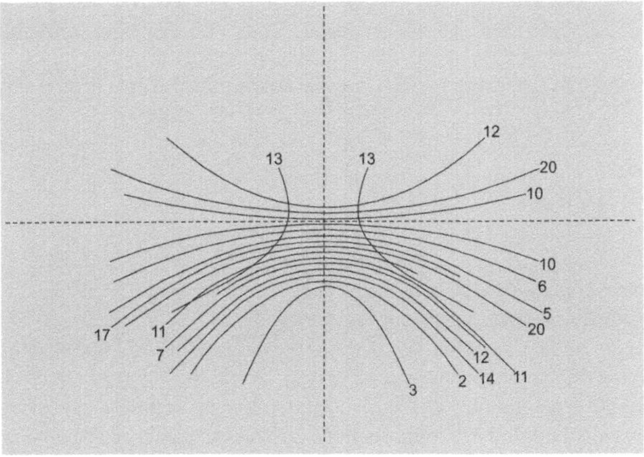

Fig. 73.4 Comparative scheme of the parabolic occurrence in different buildings designed by Niemeyer. *Numbers* indicate the correspondent building from the sample in Fig. 73.1. Image: authors

Hyperbolic and elliptical curves are used less frequently. For example, the columns of the Justice Palace may well be associated with the equilateral hyperbole; in the Cathedral of Brasilia, despite the probable association with the hyperbole, the curve fits better along a parabola; the shape of the Cathedral's baptistery is the resulting of the revolution of ellipses about its transversal axis (Fig. 73.5).

Fig. 73.5 Some *conic curves* in Niemeyer's buildings. Photos: authors

Fig. 73.6 The National Congress: *modifying curve* from the vertical profile. Photo with overlay: authors

Composite Curves

> ... when I decided to adopt the cupola at the National Congress building ... I have tried to work it in a plastic manner, modifying it, turning the curve upside down, trying to make it lighter ... In the Senate's cupola, overriding the self bearing characteristics offered by the primitive shape the hemisphere and departing from the supporting circle line I have designed straight lines aiming at the same sensation of lightness ... After inverting the National Congress cupola I explored its lines horizontally aiming at two goals: to increase the internal visibility angles and, at the same time, trying to give the sensation as it has simply "landed" at the Congress rooftop. I remember Joaquim Cardozo's telephone call: "Oscar, I have found the tangent which will allow the cupola to be set free, according to your will" (Niemeyer 1993: 12).

The transformations suggested by Niemeyer were performed on the vertical section, that is, from the curvilinear vertical profile. In his design for the Brazilian National Congress, Niemeyer departs from the hemispheric cupola, transforming its profile to achieve the configuration of a curve with the profile of a parabola (Fig. 73.6). This sort of transformation gives rise to the so-called composite curves which, in turn, in Niemeyer's buildings, fall into two categories: mixing of lines and composition of lines.

Composite curves are a designation adopted in this study for the composition of curves: considering the straight line a special case of curve (first degree polynomial), mixed lines constitute a special case of curve composition. In the following section operations and rules utilized in the generation of this kind of curve are described.

Mixed Lines

In Niemeyer's vocabulary "mixing of lines" designates those lines defined by the composition of arcs unified by line segments. Primitives for the mixing lines (or original forms of the vocabulary) are, consequently, the arc and the line segment. In the spatial relation "arc and line segment," the following rules were identified:

- In vertical profiles line segments are tangent to curves, and are therefore perpendicular to the circumference radius which bypasses the tangency point. Thus the tangent line's inclination angle with respect to the axis on the AO segment (height) of the arc is inversely proportional to this segment's dimension (Fig. 73.7);
- When the arc is a semi-circumference (plain arc) tangents bypassing the curves extreme parts are parallel to the AO segment;
- The plan analysis has allowed the identification of other spatial relations, beyond those found at vertical planes. These are:
  - The use of the arc over 180° in horizontal sections along with tangent lines or curves;
  - A clear composition of lines where Niemeyer "softens" the corners with the concordance of arcs (Fig. 73.8);
  - Mixed lines are generated along with operations such as rotation, translation, reflection, intersection, scale and addition. Figure 73.9 shows examples of these generative rules.

Composition of Curves

> I defined then the museum's profile. A line which springs from the ground, and without interruption grows and unfolds, sensually, up to the roof (Niemeyer 2000).

Composite curves are constituted by the composition of curved lines of different radii, alternating concave and convex lines (sinuous lines), or by a sequence of concave segments. The analysis has shown that Niemeyer frequently used a sequence of parabolas or arcs of circumference of different diameters (Fig. 73.10). The composition primitives the initial forms of the vocabulary are therefore the arc and the parabola.

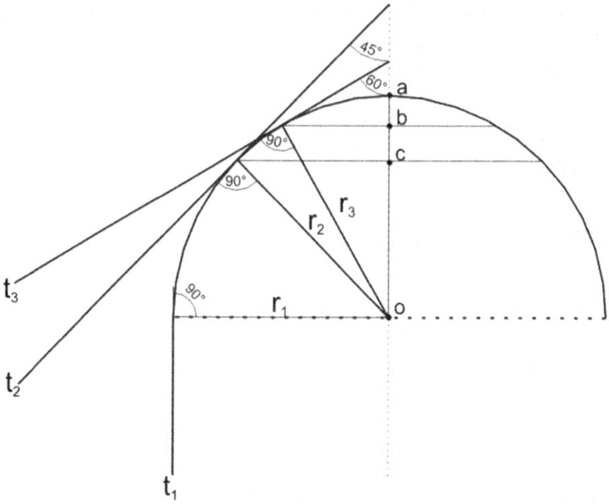

Fig. 73.7 A scheme of the tangency relation between *curve* and *line* segment in *mixed lines*. Image: authors

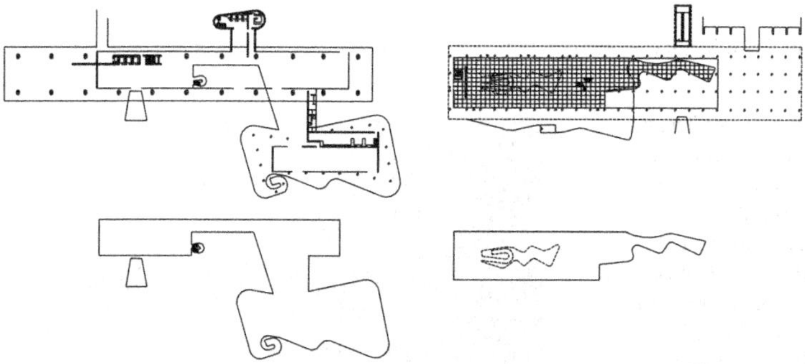

Fig. 73.8 In *straight lines* compositions Niemeyer "softens" the corners with the concordance of arcs. Image: authors

The arc establishes spatial relations with other curves as secant, non-secant and tangent (Fig.73.11).

As for the sample's distribution of the curve types, 70 % have their origin in the parabola; the composition of curves takes place predominantly in the horizontal plane.

The percentile refers to the number of buildings with the curve incidence; an overlapping of different types of curves occurs in many buildings in the sample analyzed:

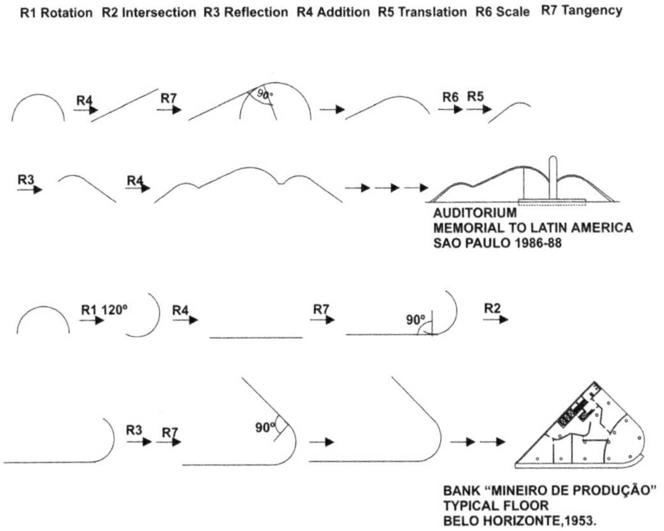

Fig. 73.9 Generation of designs based on *mixed lines*. Image: authors

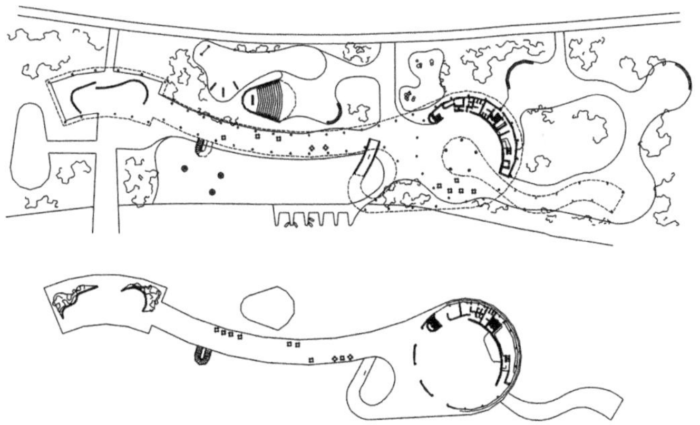

Fig. 73.10 Restaurant and boat house, Rio de Janeiro, 1944. Image: authors

- Mixed lines occur both in the vertical as in the horizontal planes (more alternatives are found in the horizontal plane);
- Composition of curved lines is predominant in the horizontal level;
- More frequent curves vary, in the average, between arcs of 45° to 180°.

Fig. 73.11 The arc establishes spatial relations with other curves: *secant*, *non-secant* and *tangent*. Image: authors

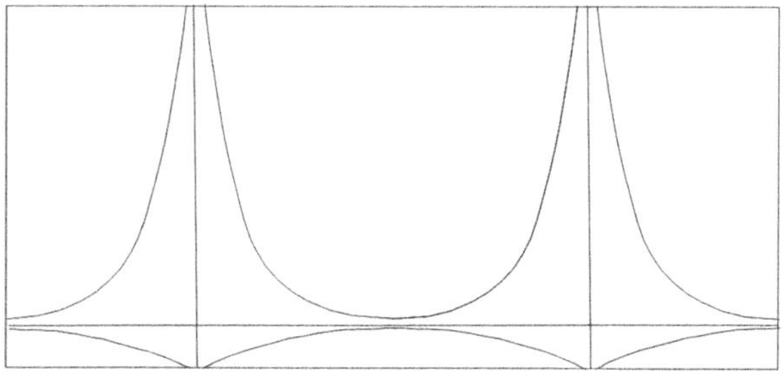

Fig. 73.12 The thin edge column: "... the palaces like just touching the ground ..." (Niemeyer 2004: 180). Image: authors

Geometrical Composition's Strategies

- In composite curves that constitute vertical profiles generatrices or directrices of surfaces the straight line or a parabolic line is generally used to create a transition between the curve and the ground level (with lines closer to the tangent to the curves);
- In vertical planes (such as pillars) the curve touches down with a proportionally thin section, i.e., the "thin edge" column (Mayer 2003: 33) (Fig. 73.12);
- The relation between the curve and the volume generated;
- Closed composite curves constitute horizontal planes (generally canopies) or directrices in the generation of one floor volumes situated aside or underneath buildings;
- A frequent composition associates a cylindrical form with a circular or elliptical base and a sinuous canopy (concave–convex curve), such as the restaurant and boat house in Rio de Janeiro, 1944 shown in Fig. 73.10.

Proportions in Niemeyer's Style

In the literature about Niemeyer's designs and his written work no relevant references are made concerning his attitude towards the use of proportions or dimensional control mechanisms. In his writings he does refer to the importance of linking plastic freedom to rules of proportion: "we guarantee absolute freedom of conception, within, of course, proportion rules that architecture has always required" (Corona 2001: 83).

In the Modulor, Le Corbusier describes the golden section and the Fibonacci series "as rhythms apparent to eye and clear in their relations with one another. And these rhythms are at the very root of human activities." Lucio Costa, Niemeyer's first mentor, insisted on the importance of the understanding of proportion for the achievement of harmony in designs: "...It is advisable for would be architects to have from the very beginning of their education a perfect notion of the meaning of proportion, modulation and *modinatura*..." (Costa 1995: 117).

Before developing the Modulor, Le Corbusier used in his designs a composition system based in rules of proportion (*tracés régulateurs*). As Niemeyer was strongly influenced by Costa and Corbusier, we have assessed the 20 buildings analyzed in search of proportions eventually used in the design process. The proportional analysis of the sample has revealed many relationships between parts of the building; all of them referred to the golden section.

Some early examples of these relations of proportion are found in his project for the airport in Diamantina (1954) (Fig. 73.13), where the (A′, a′) rectangle arises out of the relation between the building's total width (a′) and the building's total length (A′). Two golden rectangles (B′, b′) are shown demarcating the main access to the building and the front yard. The (C′, c′) rectangle is inferred from the airport's total length and is, subsequently, divided as to generate (D′, d′) which demarcates the stone wall. From D's subdivision also emerge (F′, f′), (G′, g′) and (H′, h′), (g′) corresponding to the building's maximum height and (h′) the distance between the tangent to the parabola and the building's geometrical center. The dimensions of the stone wall generate in turn the golden rectangle (I′, i′).

Conclusions

Some spatial relations, such as the connection of pair of arcs through lines or directrices composed by multiple curves (Fig. 73.14), are more frequent at the beginning of Niemeyer's career namely at the Pampulha project and others, becoming more rarefied from the 1960s (that is, after Brasilia). This strategy remains present from there on in the generation of horizontal planes (canopies) connecting volumes or ramps (parallel lines) or as directrices constituted by two or three lines.

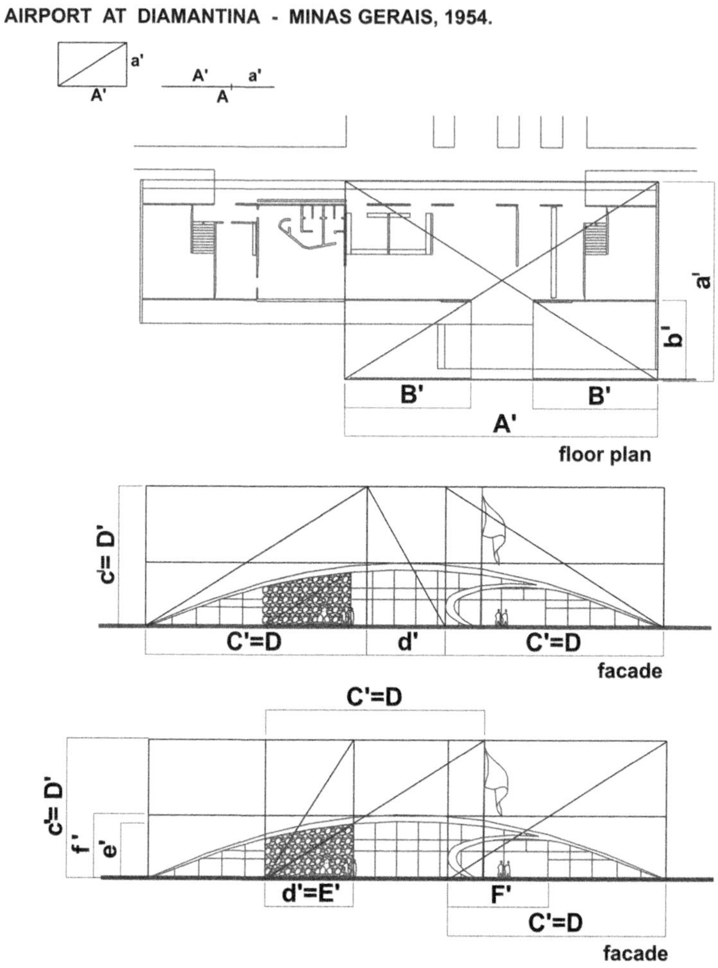

Fig. 73.13 The proportion analysis: regulating *lines* based on the golden section. Image: authors

It seems quite clear that the original steps of Niemeyer's strategy are based on two types of curves: in the parabola and in the arc of circumference added by the line segment. Departing from these two curves, he further develops operations such as rotation, translation, reflection, intersection, scaling (parametric curves) and addition. The parabola is, by far, the most recurrent curve in Niemeyer's buildings both in the vertical as well as in the horizontal planes.

However, in the horizontal planes we find the majority of spatial relations and consequently more freedom for new combinations within his vocabulary of forms. Some patterns will be found only in the horizontal planes.

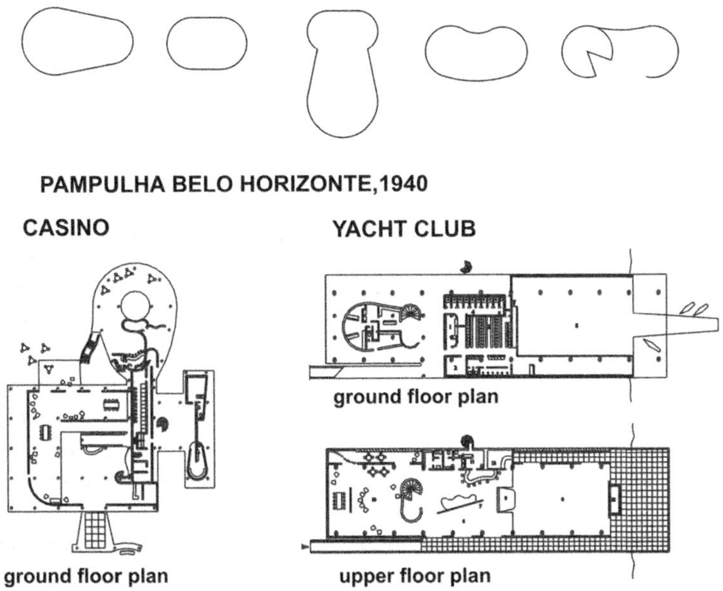

Fig. 73.14 Some compositions that were more frequent at the beginning of Niemeyer's career. Image: authors

The graphic description through schematic drawings has allowed the visualization of curve types and has enabled the depiction of their origin. The drawings, based in orthogonal projections, allowed the dissection of buildings into their geometrical primitives and the identification of the *tracés régulateurs* originated in the golden section.

It is impossible to know whether or not Niemeyer has used the golden section in his design process. The relevant issue here is that there exists enough evidence to postulate that Niemeyer used classical rules of proportion as an intrinsic feature of his work even in his very recent projects. The identification of the geometrical structure of Niemeyer's language is the key factor for the analysis and the syntactical description of his work. This description does not have as its goal the retrieval of Niemeyer's design processes but rather seeks to create the conditions for the generation of buildings according to Niemeyer's language. In this respect, the generation of buildings using Niemeyer's curves may be understood as an addition to the work developed, for example, by Koning and Eizenberg (1981) and Pinto Duarte (2004), who have generated buildings that could have been designed by Wright or Siza Vieira.

The knowledge that results from the study of Niemeyer's language may well be implemented as a pedagogical tool for the beginning years of architectural education. It is quite evident that the multiplicity of combinatorial possibilities based on spatial relations and simple geometrical transformations existent in

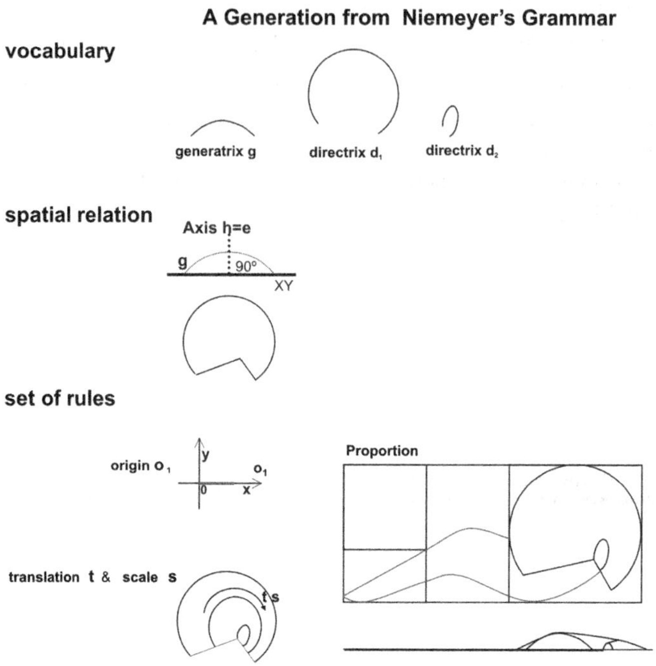

Fig. 73.15 A new design generated thanks to Niemeyer's shape grammar. Image: Authors

Niemeyer's language are easily learned and applied. A simple example of this application is given in Fig. 73.15, where a student was asked to design according to Niemeyer's style.

The example illustrates that once the geometric structure is made clear, it is possible to correlate formal control mechanisms without the usual restrictive character but as creative tools that are constantly fed by almost infinite combinatorial possibilities of the chosen language.

This study has analyzed only buildings generated from a curved profile or constituted by curved surfaces. The extension of the analysis towards other geometries also present in Niemeyer's work will certainly add other relevant elements to the knowledge about his language and architecture.

Biography Benamy Turkienicz graduated as an architect from the Universidade Federal do Rio Grande do Sul (UFRGS) in 1976. He has a MA in Urban Design, Joint Center for Urban Design—Oxford Polytechnic (1979), a M.Sc. from the Bartlett School of Architecture/University College London (1981), and a PhD in Urbanism, Chalmers University of Technology, Sweden (1982). He is full professor at the Department of Architecture of the Faculty of Architecture at UFRGS, and teaches in the program for Post-Graduation and Research in Design. Co-author of software for architectural and urban design, including "CityZoom", Prof. Turkienicz has given lectures in universities in Brazil and abroad and published in international and national periodicals and journals. He is the Director of the Laboratory for the Simulation and Modeling in Architecture and Urbanism, UFRGS, and for the NTU/UFRGS, a joint research group supporting municipalities and consulting firms in the planning and assessment of settlements and large-scale architectural and urban projects.

Rosirene Mayer is an architect and holds a Doctorate in Architecture (2012) from Federal University of Rio Grande do Sul—UFRGS, Brazil, and a Master in Architecture from the same institution. Currently she is a researcher at Simmlab—UFRGS (Laboratory for the simmulation and Modeling in Architecture and Urbanism). Since 1993 she has worked in the Municipal Housing Department of Porto Alegre on land tenure regularization and low income housing policies. Her research interests are in the fields of Architectural and Urban Design, Design Learning, Shape Grammar, and other generative systems for design exploration as well as Low Income Housing and Public Policies.

References

Corona, E. 2001. *Oscar Niemeyer: uma lição de arquitetura. Apontamentos de uma aula que perdura há sessenta anos*. São Paulo: FUPAM.

Costa, L. 1995. *Lúcio Costa: registro de uma vivência*. São Paulo: Empresa das Artes.

Gips, J. 1975. *Shape Grammars and their uses*. Basel: Birkhäuser.

Koning, H. and Eizenberg, J. 1981. The language of the prairie: Frank Lloyd Wright's prairie Houses. *Environment and Planning B: Planning and Design* **8:** 295-323.

Mayer, R. 2003. A linguagem de Oscar Niemeyer. MSc. Diss. Federal University of Rio Grande do Sul.

Mayer, R. and Turkienicz, B. 2005a. Generative processes in Niemeyer's style. In *Proceedings of the Dresden Symposium of Architecture 2004*, R. Weber and M. A. Amann, eds. Mammendorf: pro Literatur.

———. 2005b. Cognitive process styles and grammars: the cognitive approach applied to Niemeyer's free forms. In *Proceedings of the 23rd Conference on Education in Computer Aided Architectural Design in Europe*, J. Pinto Duarte, G. Ducla-Soares and A. Z. Sampaio, eds. Lisbon: Instituto Superior Técnico, Technical University of Lisbon.

Niemeyer, O. 1993. *Conversa de arquiteto*. Rio de Janeiro: Revan.

———. 2000. *Minha arquitetura*. Rio de Janeiro: Revan.

———. 2004. *My Architecture 1937-2004*. Rio de Janeiro: Revan.

Petit, J. 1995. *Niemeyer: poeta da arquitetura*. Lugano: Fidia Edizione d'arte.

Pinto Duarte, J. 2004. Towards the mass customization of housing: the grammar of Siza's houses at Malagueira. *Environment and Planning B: Planning and Design* **32,** 3: 347-380.

Stiny, G. 1975. *Pictorial and formal aspects of shape grammars: on computer generation of aesthetic objects*. Basel: Birkhäuser.

Turkienicz, B. 1994. Brasília: a arquitetura da crítica. *Arquitetura e urbanismo* **55** (Aug-Sept 1994): 53-56.

Zevi, B. 1974. *Il linguaggio moderno dell'architettura: guida al codice anticlassico*. Torino: Einaudi. Eng. trans: The Modern Language of Architecture, New York: Van Nostrand Reinhold, 1978.

Chapter 74
Dom Hans van der Laan and the Plastic Number

Richard Padovan

Introduction

The "plastic number" is the system of measure discovered in 1928 by the Dutch Benedictine architect Dom Hans van der Laan (1904–1991). In fact he regarded it not as a discovery but as the *rediscovery* of a forgotten system widely known in ancient times. Nevertheless, in my view it is the first fundamentally new system of architectural proportion discovered for at least 2,000 years. It is new both in its mathematics and in its philosophical underpinning. I will start by describing a problem with the older philosophies that van der Laan's theory sets out to overcome.

Collapse of the Ancient Cosmological Basis of Architectural Proportion

Up to and including the Renaissance it was believed that the natural world had been designed in accordance with a great mathematical order. We are all familiar with medieval illustrations of Genesis showing God as Architect creating the heaven and the earth with the aid of a huge compass; for God, the Bible tells us, "set a compass upon the face of the earth". And not only upon the face of the earth but also upon the human body. The most famous of all proportion diagrams is Leonardo's illustration of Vitruvius's statement that if a man lies on his back with arms and legs

First published as: Richard Padovan, "Dom Hans van Der Laan and the Plastic Number", pp. 181–193 in *Nexus IV: Architecture and Mathematics*, Kim Williams and Jose Francisco Rodrigues, eds. Fucecchio (Florence): Kim Williams Books, 2002.

R. Padovan (✉)
28 Petersham Road, Richmond upon Thames, Surrey TW10 6UW, UK
e-mail: kwb@kimwilliamsbooks.com

K. Williams and M.J. Ostwald (eds.), *Architecture and Mathematics from Antiquity to the Future*, DOI 10.1007/978-3-319-00143-2_27,

outstretched, and a compass placed with one point at his navel, the other will graze the tips of his fingers and the soles of his feet. Le Corbusier's "Modulor Man" is a recent descendant of this ancient idea. Now, if the universe and all it contains, including our own bodies, are governed by a divinely-ordained harmony, what could be more reasonable than to apply the same harmony to the things we add to it: architecture and other works of art?

The ancient belief in "a celestial and universally valid harmony" sustaining architectural proportion was undermined, according to Rudolf Wittkower, by the scientific revolution of the seventeenth century, which replaced it by a universe of mechanical laws, of iron necessity with no ulterior plan, "a universe in which the artist had to find his bearings by substituting purely subjective standards for the old super-personal ones" (Wittkower 1978: 117).

Paradoxically, the universe revealed by Galileo and Newton, and later by Einstein, Bohr and Schrödinger, is more completely mathematical than the ancient cosmos, not less, so one might expect it to have inspired a resurgence of mathematical proportion in art rather than to have killed it. But the new mathematical universe is no longer, like the old, an architectonic whole—a sort of house—hierarchically ordered and delimited, with mankind at its centre and the sky and fixed stars as a protective roof. There is a beautiful passage in one of Primo Levi's books which describes the contrast very well:

> Now, the sky which hangs over our head is no longer domestic. It becomes ever more intricate, unforeseen, violent and strange; its mystery grows instead of decreasing; every discovery, every answer to old questions, gives birth to thousands of new questions. Copernicus and Galileo had wrenched humanity from the centre of creation: it was only a change of location, yet many felt deposed and humiliated by it. Today we realize much more: that the imagination of the artificer of the universe does not have our limits, indeed has no limits, and our astonishment also becomes limitless. Not only are we not at the centre of the cosmos, but we are alien to it: we are a singularity. The universe is strange for us, we are strange in the universe (Levi 1991: 12).

The loss of the familiar "domestic" universe, and the absence of limits in the new, intangible universe revealed by science, are central to the philosophy underlying the plastic number. But we must first look briefly at other modern responses to the problem.

Since the seventeenth century, Wittkower writes, art has been driven by "an irrational creative urge" (1978: 117)—that is, by a reliance on the subjective judgement of the individual artist. But not all artists took this route. Since the mid-nineteenth century there have been continued attempts to construct a new objective foundation for proportion in art and architecture. The existence of the "Nexus: Relationships Between Architecture and Mathematics" conference series exemplifies this. These attempts have been most commonly associated with the cult of the golden section as an aesthetic paradigm.

Some have tried to give the golden section an objective justification by subjecting it to statistical testing. Gustav Fechner is the outstanding example, although he set out to disprove, not prove it. The trouble with this approach is that the more rigorous one wishes to make the test, the more one must eliminate all variables. The result, as Rudolf Arnheim has observed, is that

> "the more strictly investigators adhered to the criterion of preference, the more completely their results neglected everything that distinguishes the pleasure generated by a work of art from the pleasure generated by a dish of ice cream" (Arnheim 1986: 45).

The other route that has been followed is, in effect, to reconstruct the ancient cosmological argument, but now on the basis of the golden section. In the 1850s the polymath Adolf Zeising—mathematician, philosopher, poet, novelist and playwright—claimed that the golden section was the key to plant growth, the skeletons of animals, the proportions of chemical compounds, the geometry of crystals, the physics of light, sound and magnetism, the disposition of the heavenly bodies, and—not surprisingly—to that touchstone of ancient proportion theory, the human body. From Zeising to Le Corbusier, the body constantly reappears as the preferred example, but now—miraculously—we discover it to be subdivided according to an irrational number, the golden section, and no longer, as Vitruvius and Leonardo believed, rationally divided into aliquot parts. Thus much modern proportion theory is a thinly disguised version of the ancient religious belief in a God-designed universe, except that it is now based on the so-called Divine Proportion. A typical example is Le Corbusier. In *Towards a New Architecture* he argues that

> We say that a face is handsome when the precision of the modelling and the disposition of the features reveal proportions which we *feel to be harmonious* because they arouse, deep within us and beyond our senses, a resonance, a sort of sounding-board which begins to vibrate. An indefinable trace of the Absolute which lies in the depths of our being . . . This is indeed the axis on which man is organized in perfect accord with nature and probably with the universe . . .; this axis leads us to assume a unity of conduct in the universe and to admit a single will behind it . . .If the results of mathematical calculation appear satisfying and harmonious to us, it is because they proceed from the axis (Le Corbusier 1946: 187–193).

So the biblical Creator who "ordered all things in measure and number and weight" is still active behind the scenes, but now concealed in a sort of agnostic disguise as the "single will" that accounts for the unity of nature. This universal harmony, of which as products of nature we are supposed to be intuitively sensitive, is allegedly the underlying cause of our aesthetic response to the proportions of the things we see.

Van der Laan's Empiricism

Van der Laan's approach is altogether different. I shall outline the following ways in which it differs from both ancient proportion theory and such modern variants as Le Corbusier's:

1. It is entirely empirical, and not based on any metaphysical belief in a designed universe;

2. It is concerned primarily with scale rather than shape;
3. It is not concerned with precise measures but with "types of size" with a "play" between lower and upper limits. This play, which is determined by human psychology and not by any external datum, constitutes the basic proportion of the system;
4. The proportions do not extend indefinitely, but are contained within a limited number of "orders of size";
5. The system is intrinsically three-dimensional;
6. Proportion is for van der Laan the key to the formation of architectonic space;
7. Furthermore, it plays a formative role in human society.

The first difference from previous theories is that van der Laan's does not depend on any explicit or implicit quasi-religious belief in a divine creation or in the human body as a microcosm of the universe, such as one finds even in Le Corbusier's argument for the Modulor. Although as a Catholic priest van der Laan himself believed that God created the world, no such belief is demanded by his theory of proportion, which is entirely empirical. It is equally valid if you believe that nature is the product of blind chance. That is why in my book *Proportion* (Padovan 1999) I compare van der Laan to such eighteenth-century philosophers as Hume and Kant. The plastic number is not derived from the order of nature as such, but from our inability to discover that order except through the filter of our senses and intellect. To grasp the phenomena that nature presents to us, we must *impose* upon them our own human, intellectual limits. By accepting our limitations and making them its foundation, the plastic number seeks to overcome the alienating limitlessness of nature described by Primo Levi, in particular the limitless continuity of natural space.

Being empirical, the aim of the plastic number is genuinely *aesthetic*: not in the modern but in the original Greek meaning of the word, which concerns not beauty but perception. The word "beauty" hardly occurs in van der Laan's vocabulary. His argument for the plastic number is not based on the alleged beauty of a particular shape—the golden rectangle or any other—but on the limit of our ability to perceive differences of size. Its ground-ratio or basic proportion is that size-difference between two objects that is just enough to be instantly recognizable, even when the objects are seen separately and cannot be measured directly against each other. Psychologists estimate the smallest difference that can be distinguished by the eye when two sizes are compared directly as about 4 % of the sizes concerned. But the instantly recognizable difference that concerns van der Laan is much larger, as we shall see: about 25 %.

I am indebted to the Dutch architect Leo Tummers for an elegant illustration of this phenomenon. Many years ago, a visitor told van der Laan that a new length of cigarette had just been introduced, called "King Size". Van der Laan asked to compare examples of both types of cigarette. The difference turned out to be almost exactly the basic ratio of the plastic number. Asked how he knew this would be the case, he answered that it was inevitable. The manufacturer's choice of length must be governed by two parameters: on one hand the new must be recognizably longer than the regular one, but on the other he wants to give as little extra tobacco as he can get away with. Therefore the King Size has to be the first size just large enough to be obviously larger than the normal one.

Measuring and Counting

We cannot measure continua, only count individual things. We can count exactly the number of individuals in this room, because each of us is an indivisible unit, but if we want to measure our heights we must resort to some conventional unit such as the foot or metre. We can then measure our heights to any desired accuracy as multiples or submultiples of that arbitrary unit. Such measurement serves well enough in practical affairs, but architecture, for van der Laan, demands a kind of measure that recognizes the continuity of natural space and derives from that very continuity the limits necessary to make it tangible for us.

Van der Laan writes in *Architectonic Space* that the house (which, by the way, is his general synonym for all architecture: for him every kind of building, and indeed every city, is just a particular kind of house) is above all a means by which we give measure to the boundless space of nature. He draws an analogy with music, which gives measure to time in a similar way. It is not a matter of "measuring up", in the sense that we measure up space with a metre rule, or measure time with a clock, but of "measuring out": imposing measure on what is otherwise measureless:

> The house must therefore bring us into intellectual contact with the continuous quantity of the spatial datum. Hence the quantity presented to us in the space and form of the house must embody its own unit and number, just as both unit and number are given in the discrete quantity of the things we count (van der Laan 1983: 47).

In short, one can say that the plastic number differs from previous systems of proportion in that it is primarily concerned with *scale* rather than *shape*. Certain preferred shapes do arise from it, but these are secondary. The essence of the system is relative size.

Types of Size

Being empirical, the plastic number is not, like Le Corbusier's Modulor, a fixed set of mathematically exact measures, but a series of approximate relations derived from everyday experience. It starts out, as I said, from the observation that spatial quantity is a continuum, as opposed to the discrete quantity of individual things that we count. In order to quantify this continuum in some way, it breaks it down into a series of segments or "types of size" contained between certain limits (Fig. 74.1). The upper limit of a smaller type of size coincides with the lower limit of the next larger type. The succession of types still forms a continuum, like space itself, but

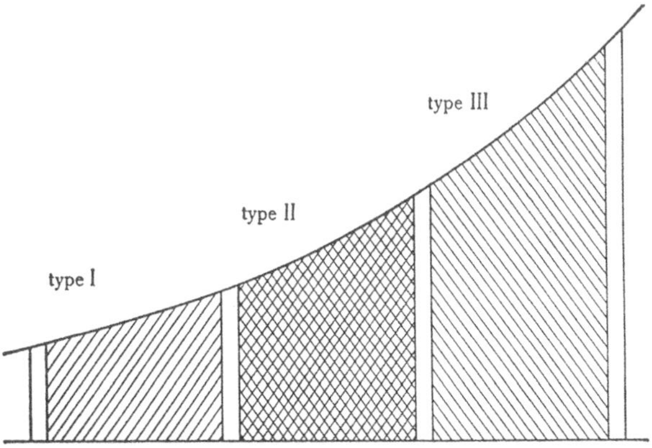

Fig. 74.1 A breakdown of a continuum into a series of segments or "types of size" contained between certain limits. Image: author

the continuum now comprises a sequence of discrete, and therefore perceptible, segments. The extent of the type constitutes the basic proportion or ground ratio of the plastic number.

This is how, in everyday life, according to van der Laan, we deal with the endless continuity of sizes we encounter in nature: by sorting them roughly into types of size. When it is hard to distinguish from one another a certain group of objects, we say that they are "of the same type of size". We do not bother with the size of each individual leaf on a tree or each pebble on a beach, but classify them as small, medium, large, and so on (a bit like the SMLXL used by clothing manufacturers). The first object that is unmistakably larger than those of a first type becomes the first representative of a first larger type. The King Size cigarette was just large enough to be unmistakably larger than the familiar one. Van der Laan demonstrated the grading process empirically by inviting his visitors to grade a collection of pebbles (Fig. 74.2). First the group of largest pebbles is picked out, until a pebble appears that is definitely too small to belong with the others. This becomes the first of a second type of size, and so on.

Orders of Size

As soon as we establish the point at which two sizes begin to differ just enough to belong to recognizably different *types*, we automatically determine also the limit of what van der Laan calls an *order* of size: that is, the limit beyond which two sizes cease to count for each other. The addition or subtraction that just begins to change the type of a given size is by definition the smallest portion of that size that still just counts for it. So like the type, the order is contained between two limits: a smallest and a largest size that still just count for each other—or just begin not to count, which amounts to the same thing. The order of size embraces seven consecutive types, contained between eight measures (Fig. 74.3).

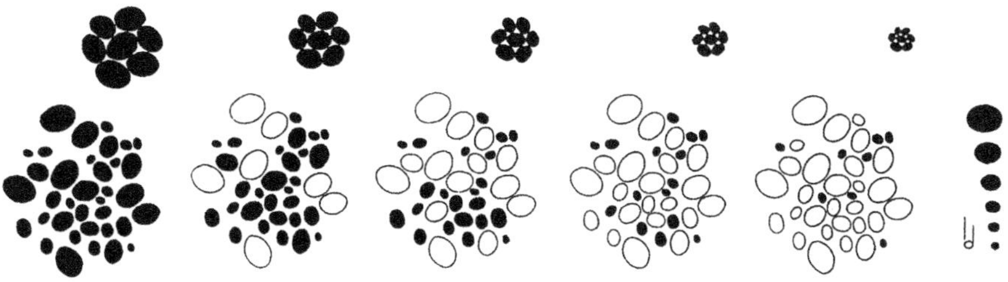

Fig. 74.2 The grading process of size applied to pebbles. Image: author

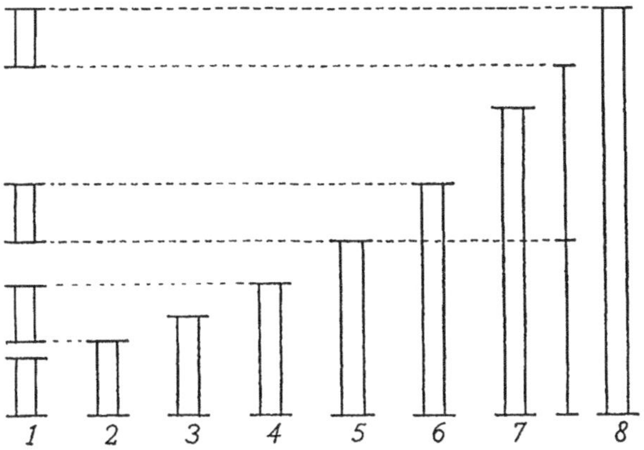

Fig. 74.3 The order of size embraces seven consecutive types contained between eight measures. Image: author

The system generates a limited range of eight clearly distinguishable shapes, which can be expressed by the simple numerical ratios 1:1, 4:3, 7:4, 7:3, 3:1, 4:1, 16:3 and 7:1 (Figs. 74.4 and 74.5). The extreme proportion allowed by the plastic number is thus 7:1. Here it contrasts with most other systems—in particular, with Le Corbusier's Modulor.

The proportions of the Modulor form a limitless geometric progression; they recede in infinite proliferation. Only the immediate relation between the whole and its two parts is readily graspable. But the eight proportions of the plastic number are held firmly between a lower and an upper limit, analogous to the octave in music. And just as the type and order of size are each defined by a lower and upper limit, so is proportion as such. The lower limit is provided by the extent of the type of size, the upper limit by the extent of the order of size. Van der Laan sums this up as follows:

> Within the limits of a type of size we call all concrete sizes identical; there is as yet no question of proportion.
>
> Within the limits of an order of size the types of size can be compared with each other; here it is a question of proportion.
>
> Beyond the limits of an order of size no relation is any more possible between types of size; there can no longer be any question of proportion (van der Laan 1967: 30).

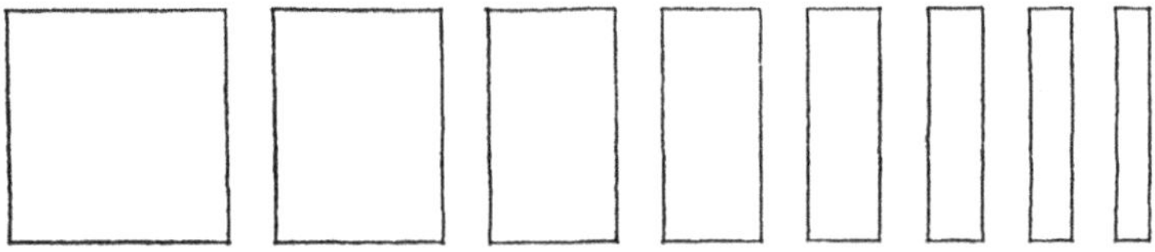

Fig. 74.4 The range of shapes in van der Laan's system are expressed by the numerical ratios 1:1, 4:3, 7:4, 3:1, 4:1, 16:3 and 7:1. Image: author

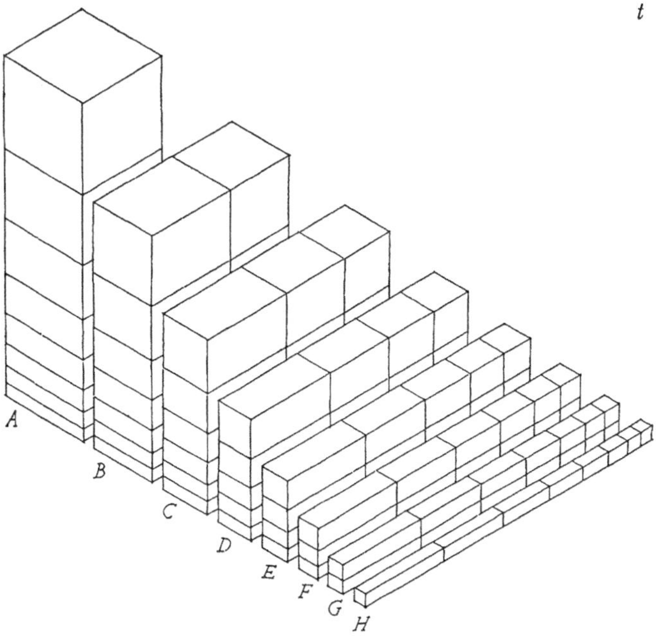

Fig. 74.5 The range of two-dimensional figures generates 120 three-dimensional forms. Image: author

A System Based on Three Dimensions

The obvious question is: how can the extent of the type and order of size be established objectively? The empirical demonstration using a set of pebbles is too subjective to provide a satisfactory basis. Different individuals will sort the pebbles differently and arrive at different proportions. Van der Laan employs a rather sophisticated mathematical argument based on the distinction between lines, planes and volumes. He argues that since we experience space in three dimensions the appropriate proportion must be derived from relations of volumes, and not (like the golden section and so far as I know all other systems) from the geometry of lines or planes. This is another respect in which van der Laan's system seems to be unique. From this argument, which I cannot describe here in more detail, he concludes that the only correct proportional system for architecture must comprise a geometrical progression in which the largest of any four consecutive terms equals the sum of the two smallest. Compare this with the golden section progression in which the largest of *three* consecutive terms equals the sum of the two smallest (Fig. 74.3).

I must confess that I remain unconvinced by van der Laan's argument, and cannot accept his dogmatic contention that the plastic number is the only valid system of proportion, although it is certainly a very interesting addition to the repertoire. In my view the real merit of van der Laan's theory lies not in the specifics of his system but rather in the fundamental concept of types and orders of size, which could be applied to any system. Proportion in art should not be bound by rigid algebraic formulae or geometrical constructions, but thought of in terms of broad perceptual categories.

Van der Laan's whole theory of proportion, and indeed of architecture, can be reduced to this: there are limits within which sizes can be related to each other, and limits beyond which this relation breaks down. Taking these approximate limits as a basis one can establish a chain of relationships by which the whole architectonic environment, from a brick to a city, can be connected together and made intelligible and humane. A certain piece of space is then no longer merely part of a measureless continuum but has become a delimited territory, marked off in clearly recognizable graded intervals. A piece of the unknown has become known.

Architectural Implications of the Theory

I earlier compared the order of size to the octave. And just as more than one octave is allowed in music, more than one order of size is allowed in architecture. But the orders are not arbitrary. Each corresponds to a particular level of architectural scale: there is an order appropriate to the wall, another to the house, a third to the district and a fourth to the city. The lower limit of the smallest order of size is determined by the wall-thickness, the basic keynote, so to speak, of the whole architectural composition. Here we touch on a central principle of van der Laan's theory: that architectonic space is *constituted* by a certain proportion of mass to distance, and specifically of a wall-thickness to the distance between two walls. I fear I cannot do justice to this theory on this occasion; but I will try to give you the gist of it.

Each of us, he says, relates a piece of space to ourselves: a zone of which we become the focus by our presence; "a space involved in our existence" (van der Laan 1983: 20, 5). This zone he calls "our neighbourhood". Architecture arises when we project this personal space-image onto other objects. Imagine you come across a large boulder standing in an open field. At a certain distance you become aware of being "in the neighbourhood of the boulder". However, this "neighbourhood" is not really a property of the boulder but the subjective projection of your own neighbourhood onto the boulder. Van der Laan describes the phenomenon as follows:

> ... at a greater distance we see the stone as completely isolated, and it is only when we get closer that at a certain point we find ourselves in its neighbourhood ... The neighbourhood of the stone is in fact wholly dependent on my presence. Actually when I approach the stone it comes gradually into my neighbourhood: into that part of the great space that I command by my presence (van der Laan 1960).

The temporary subjective perception can be made objective and permanent, however, by placing a second boulder a certain distance from the first:

> The two stones now stand in each other's neighbourhood. The neighbourhood of each stone, which first depended on my presence, now depends on the presence of the other stone, so that between the two a neighbourhood arises which has become independent of our presence (van der Laan 1960).

The mutual neighbourhood depends on the proportion between the size of the stones and their distance apart. In architecture, two parallel walls replace the two stones (Fig. 74.6).

An architectonic space arises when the proportion of the wall-thickness to the distance apart is 1:7, the limit of the order of size. Once this primary space-cell is constituted, the rest of architecture follows. Van der Laan expressed this concisely in a lecture on music and architecture in 1978:

> The space formed by the walls thereby acquires a certain autonomy, which enables us to relate other space to it in turn ...And this process can continue into the town, where streets and squares arise between the houses (see Fig. 74.7). The whole urban space can by this means be composed on the basis of the solid form of the walls, which are the only things that we determine directly by their dimensions and their placing with respect to each other ... The architectonic process can thus be summed up as follows: linear measures define the figures of surfaces which determine the form of the wall-mass, and it is this form that in turn defines the form of the delimited space, by which means an "inside" is established which can again determine new spaces with respect to the absolute "outside" of nature (van der Laan 1978).

Ordered Measure and Ordered Society

Van der Laan's proportion system is not intended as a mere aesthetic device, a sort of optional addition to architecture by which it is smartened up and given a final polish before being sent out into the world. As we have just seen, it is fundamental to architecture, and he even went so far as to say that the plastic number is not merely a *means*—something that contributes usefully to the making of buildings—but an *end*—the ultimate reason why we build at all (van der Laan 1967: 113). For even if we did not need shelter from the elements, we would still need architecture in order to give measure to the measureless space of nature.

Moreover, proportion has a social dimension, not just a purely architectural one. Even as a student in the 1920s van der Laan rejected both Functionalism and the anti-modernist movement known in Holland as the "Delft School". This movement, led by van der Laan's own teacher, M.J. Granpré Molière, played a crucial role in Dutch architecture in the years 1925–1950. But in van der Laan's view both

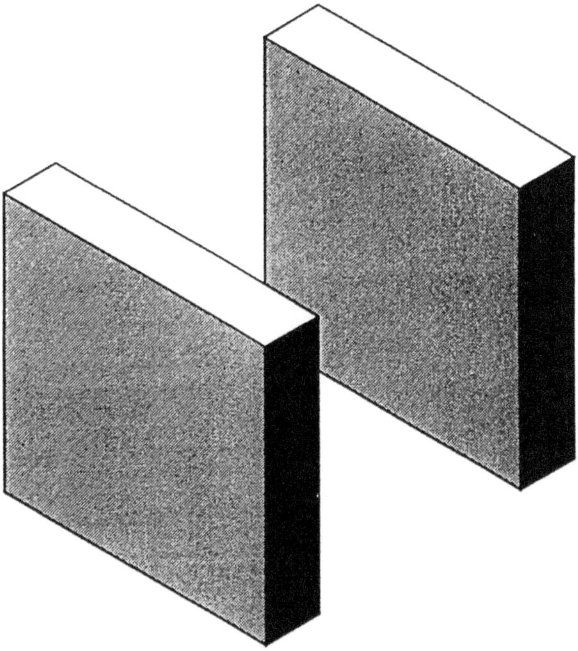

Fig. 74.6 In architecture, two parallel walls are in each other's "neighborhood". Image: author

Functionalism and Delft made the mistake of regarding architecture as the expression or reflection of social forces. Their difference lay only over their vision of the ideal society: for the Functionalists, one dominated by mechanization and urbanization, but for the Delft School one based on hand production and the small country town. Van der Laan rejected both: "The reason why both Molière and the Functionalists ground to a halt was that they directed their attention rather to the influence of society on architecture, than to the influence of architecture on society" (van der Laan 1972: 32).

The forms of architecture are not, according to van der Laan, dependent on society, but the forms of society dependent on architecture. Architecture, with its hierarchy of orders of size, is not just a background for social life, but the ordered framework necessary for society to arise. It is the condition that makes society possible. The same idea appears in Hermann Hesse's novel *The Glass Bead Game*, and both men illustrate it with an identical story, except that van der Laan places architecture, the ordering of space, alongside music, the ordering of time. Here is Hesse's version of the tale:

> We recall that in the legendary China of the Old Kings, music was accorded a dominant place in state and court. It was held that if music throve, all was well with culture and morality and with the kingdom itself. The music masters were required to be the strictest guardians of the original purity of the "venerable keys". If music decayed, that was taken as a sure sign of the downfall of the regime and the state (Hesse 1987: 28).

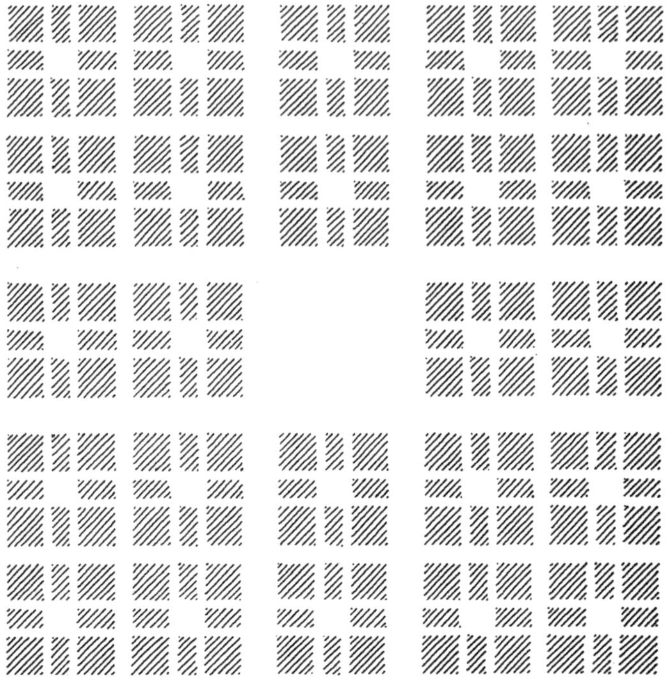

Fig. 74.7 The constitution of a primary space-cell is followed by the rest of architecture. Image: author

As I said, van der Laan believed that he had not discovered a new principle of architectural proportion but recovered an ancient, long forgotten one. Innovation, which inspires so much of modern art, held no interest for him. What concerned him was not innovation but *renewal*, by which he meant the stripping away of everything inessential to architecture and life. If you want to be truly original, he would say, go back to the origins.

Biography Richard Padovan studied architecture at the Architectural Association, London (1952–1957). Since then he has combined practice with teaching and writing on architecture. He believes, however, that his real architectural education began when in encountered the work and thought of the Dutch Benedictine architect Dom Hans van der Laan in 1974. His translation of van der Laan's treatise *Architectonic Space* appeared in 1983, followed by a monograph, *Dom Hans van der Laan, Modern Primitive,* in 1994. He is the author of *Proportion: Science, Philosophy, Architecture* (E & FN Spon 1999) and *Towards Universality: Le Corbusier, Mies and De Stijl* (Routledge 2002), which contrasts the grandiose philosophical ideals of European modernism with its failure to realize those aims, particularly in the building of cities.

References

Arnheim, R. 1986. *New Essays on the Psychology of Art*. Berkeley: University of California Press.
Hesse, H. 1987. *The Glass Bead Game*. London: Picador.
Le Corbusier. 1946. *Towards a New Architecture*. London: The Architectural Press.
Levi, P. 1991. *Other People's Trades*. London: Sphere Books.
Padovan, R. 1999. *Proportion: Science, Philosophy, Architecture*. London: E. & F. N. Spon.
Van der Laan, Dom H. 1960. Second lecture on architectonic disposition. *Den Bosch*, 2 April 1960.
———. 1967. *Het plastische getal*. Leiden: E. J. Brill.
———. 1972. Discussie over de betekenis van Granpré Molière. *Plan* **6**.
———. 1978. Lecture to music students, 27 April 1978.
———. 1983. *Architectonic Space*. Leiden: E.J. Brill.
Wittkower, R. 1978. The Changing Concept of Proportion. Chap. IV (pp. 109-123) in: *Idea and Image*. London: Thames & Hudson.

Chapter 75
Louis Kahn's Platonic Approach to Number and Geometry

Steven Fleming

Introduction

A debate in the *Nexus Network Journal* over the proportional aspects of Palladio's Villa Emo highlights a sticking point in the analysis of partially-documented ancient buildings. Where Lionel March (March 2001) finds no documentary evidence to warrant cloaking the Villa Emo in the gold of the golden proportion, Rachel Fletcher maintains that an accurate survey of the building as it was ultimately constructed, does reveal golden mean proportions, regardless of what the extant documentation suggests (Fletcher 2001). Doubts about on-site procedures, the relevance of surveys and certain historical evidence could fuel such a debate indefinitely. If, on the other hand, March and Fletcher were debating the proportions of a modern building, one for which dimensioned working drawings and complete office files were in existence, Fletcher would have fewer avenues to refute March's arithmetisation of geometry.

In his books on the works of Le Corbusier (Gast 2000) and Louis Kahn (Gast 1998), Klaus-Peter Gast provides a lively and scholarly commentary on the works of these two great architects, accompanied by some of the author's own very revealing photographs. The two books also present a number geometrical analyses which can be compared with dimensioned working drawings and complete office correspondence held in the respective archives of these two figures. The quantity of such evidence in each of Gast's books has already been remarked upon in book reviews published by the present author (Fleming and

First published as: Steven Fleming, "Louis Kahn's Platonic Approach to Number and Geometry", pp. 95–107 in *Nexus IV: Architecture and Mathematics*, Kim Williams and Jose Francisco Rodrigues, eds. Fucecchio (Florence): Kim Williams Books, 2002.

S. Fleming (✉)
School of Architecture and Design, University of Tasmania, Locked Bag 1323, Launceston 7250, TAS, Australia
e-mail: Steven.Fleming@utas.edu.au

K. Williams and M.J. Ostwald (eds.), *Architecture and Mathematics from Antiquity to the Future*, DOI 10.1007/978-3-319-00143-2_28,

Ostwald 2001; Fleming 1998). The current chapter is more specific in its critique, arithmetically testing some of Gast's claims about Kahn's buildings against those buildings' known dimensions. Discrepencies between Kahn's drawings and Gast's analysis prompt an alternative interpretation of mathematics within Kahn's work.

The discussion will focus primarily on Kahn's First Unitarian Church and School in Rochester, New York, since this building is a prime exemplar of what Kahn refers to as his "form and design" thesis. Later, it will be seen that this thesis could hold the key to Kahn's actual approach to number and geometry.

What Do the Drawings Say?

Before looking at Gast's analysis of Kahn's church in Rochester, consideration of an event that occurred late in the working drawings stage of this project will provide some insight into Kahn's interest—or disinterest—in proportion. Most unlike an architect who is concerned with mathematical proportions, Kahn allows the sectional proportions of this building to be altered late in its documentation stage, at the advice of acoustical consultants. The central ceiling of the auditorium is flattened out and the four light towers are made lower and wider, so that acoustically they will act as part of the whole auditorium space.[1] In the office correspondence pertaining to this issue, there is no suggestion that any previously calculated proportional system would be affected by such a late change, or that any new proportional system would need to be conceived to accommodate such dramatic alterations. The implications of this evidence for Gast's claims will be discussed in a moment.

According to Gast, many of Kahn's buildings emanate from what is referred to as a "Platonic Form," the square. One of Kahn's frequent sayings, "what will be has always been," is claimed to relate directly to Plato (Gast 1998: 185), thus establishing Kahn as a Platonist in word and in deed.

In his chapter on the church in Rochester, Gast argues that the plan of this building has a strict geometrical order, which derives from a "growth process" of geometrically dependant extensions. The geometrical reconstruction of the building begins from an imaginary central square, the corners of which are defined by the inner edges of the aforementioned light towers above the auditorium. The original square is twice bisected along the long and short axis of the space to form four squares. It is then claimed that the auditorium's width is determined by a "major golden section" growth of those imaginary quadrants across the auditorium (Fig. 75.1). In other words, measuring *across* the auditorium, as opposed to *along*

[1] See the file labeled, "UCRNY—Bolt, Beranek & Newman Acoustical Correspondence," Box L.I.K. 15, Louis I. Kahn Collection, University of Pennsylvania and Pennsylvania Historical and Museum Commission (hereafter cited as Kahn Collection). Refer specifically to "Notes from December 15th 1960 Conference with Bolt Beranek & Newman, Cambridge, Mass."

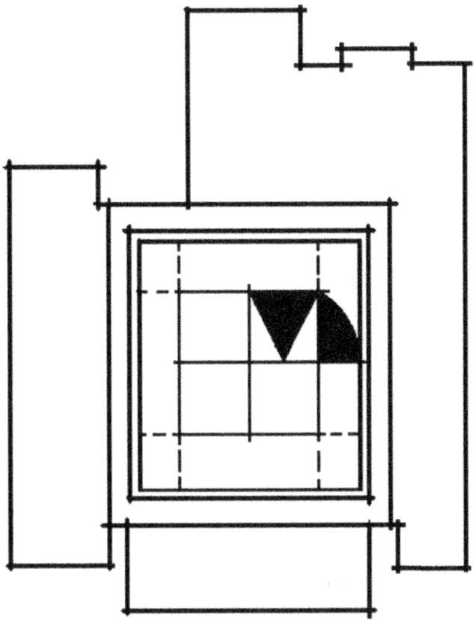

Fig. 75.1 Width of auditorium in relation to central square formed by inner edges of light towers, according to Gast. Drawing: author

it, a major golden section relationship exists between the distance from the center line of the auditorium to the inner edge of the light tower, and the distance from that same center line to the inner face of the auditorium wall. This line of reasoning is extended, with claims that the geometry of the whole building goes on growing, stemming always from this imaginary square formed by the inner edges of the four light towers.

If this is true, then upon accepting the acoustical engineers' recommendation to make those towers wider—this occurs in December 1960, just 6 months prior to the commencement of construction in June 1961—Kahn redesigns the entire building, applying a new geometrical system, which needs to originate from what is now a significantly smaller generating square. However the files for this project contain no evidence that a last-minute revision of this kind ever occurred. Between Kahn and his especially perspicacious clients, there are no letters to explain a further delay, or to explain changes to room sizes resulting from such a proportional revision. Faced with this evidence, it is hard to imagine such a major and time-consuming revision occurring at all.

In the conventional way, the First Unitarian Church and School in Rochester was built in accordance with scaled working drawings. According to common practice, these drawings bear an instruction to builders that they work from written dimensions, rather than scaled measurements. It is through these dimensions that Kahn can be considered to formally and legally communicate his intentions. Can claims regarding a hidden geometry underlying this building be supported by the dimensions on Kahn's working drawings?

According to the written dimensions on Kahn's working drawing entitled "A2: First Floor Plan," (Kahn Collection), the auditorium is 53′ wide, and 66′ long, and it

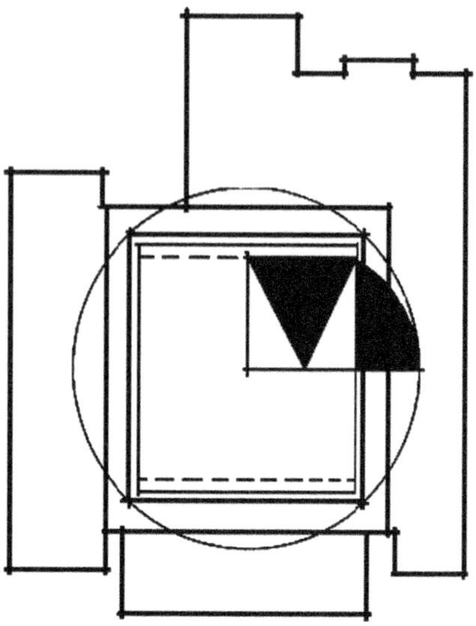

Fig. 75.2 Length of auditorium in relation to width, according to Gast. Drawing: author

is enclosed by a hollow blockwork wall with a uniform thickness of 2′. According to Gast, a major golden section growth based on half of the auditorium's width provides the radius for a large circle which touches the outer corners of the auditorium, thus determining the auditorium's length (Fig. 75.2). In other words, the distance from each of the auditorium's outer corners to its center (that distance will henceforth be referred to as x), should be 1.618 times greater than half of the auditorium's internal width (a distance henceforth referred to as y).

In order to check Gast's analysis against Kahn's working drawings, x will first be calculated for the diagonal distance from the outer corner of the auditorium to its center. According to the Pythagorean Theorem (in a right-angled triangle, the length of the hypotenuse squared is equal to the sum of the squares of the other two sides), x squared equals half of the auditorium's external width (that being 28′-6″) squared, plus half of its external length (that being 35′), squared.

$$x^2 = 28.5^2 + 35^2$$
$$x^2 = 2037' - 3''$$
$$x = 45' - 1''.$$

Meanwhile, an identical x value should be found by multiplying half of the auditorium's internal width (y) by 1.1618

$$x = 1.618y.$$

Since from Kahn's working drawings y is known to measure 26′-6″, then

$$x = 1.618 \text{ x } 26' - 6''$$
$$x = 42' - 11''.$$

This represents a discrepancy of 2′-2″ (or 4.8 %) against the diagonal calculated above. Perhaps this is negligible. Rachel Fletcher argued at the Nexus 2000 Round Table Discussion, that arithmetic precision of the kind that only the mind can behold, does not affect architecture, which "is meant to be experienced, it is meant to be lived in, it is meant to be perceived" (Watts et al. 2000: 112). However, the wall in question is constructed using 8″ cubic concrete blocks, and Gast's analysis is out by more than three block widths. Had Kahn intended to produce a palpable effect, surely he would have done so to within one block width.

The dimensions most affected by Gast's analysis, namely, the length and breadth of the auditorium, the thickness of its walls and the width of the ambulatory, are all measured in whole feet on Kahn's working drawings. They are 66′, 53′, 2′ and 6′ respectively. The odds against complex geometrical constructions producing so many lengths measurable in whole feet are literally impossible. Granted, such considerations may determine the width of the light towers, but such a minor calibration, if it is there at all, hardly rivals the fascination for proportions usually associated with geometrising architects. The dimensions of the auditorium and surrounding ambulatory would appear to be modulated according to whole feet for the most pedestrian of reasons. As stated above, the ambulatory wall is constructed using an 8 inch cubic blockwork module and every three blocks creates a dimension of two whole feet.

There is some doubt surrounding Gast's statements regarding many of Kahn's other buildings as well. For example, it is claimed that the distance by which the rectangular plan of the Kimbell Art Museum falls short of being a double square, determines the span of that building's concrete vaults. According to Kahn's working drawing A4 ("Upper Level Floor Plan," Kahn Collection) the Kimbell is 318′ wide (measuring from north to south) and 174′ deep (measuring from east to west). To be a double square, the building would need to be 348′ wide, that is, twice as wide as its depth of 174′. The difference between its actual width and the width it would be were it a double square is 30′ and this is the distance Gast refers to in his analysis as x, which should also be the span of the concrete vaults. However, 30′ is not the span of the concrete vaults. These only span 20′, or 22′ feet when measuring from the centers of the supporting columns. This represents a discrepancy of at least 8′ (or 40 %).

It is claimed that the separate living and bedroom sections of the Dr. and Mrs. Norman Fisher House are based on two squares sized according to the short width of the living section, that is 23′-6″. The distance by which the living section square is claimed to have stretched in one direction is meant to be exactly twice the distance by which the original bedroom section square grows in two directions. The living section square is 4′-6″ longer than 23′-6″ in one direction, whilst the bedroom section plan is 2′-6″ greater than 23′-6″ in both directions.[2] Twice 2′-6″ is 5′, where it should be 4′-6″ according to Gast's analysis. This represents a discrepancy of 6″ (or 12.5 %).

Gast analyses 14 more of Kahn's buildings by similar means. Unfortunately, many of these descriptions are difficult to follow since it is often unclear as to whether the analysis is referring to the center lines or the edges of columns or walls. Whilst not wishing to diminish Gast's greater contribution to the field of Kahn scholarship, the apparent discrepancies in three examples provided above suggest that there could be another way in which to approach the topic of number and geometry in Kahn's work.

What Does Kahn Say?

Alberto Pérez-Gómez argues that "intentions have to be understood in reference to their epistemological contexts" (Pérez-Gómez 1984:13). With this in mind, the remainder of this chapter proposes an approach to number and geometry that is intrinsic to Kahn's espoused metaphysics.

Consideration of Kahn's statements about proportion suggests that further geometrical analysis of his buildings in search of mathematical relationships may well be in vain. There is no record of Kahn advocating any interest in mathematical proportions as they apply to architectural composition. To the contrary, Kahn specifically states his preference for buildings *without* a clear sense of proportion. He states that,

> to make a thing deliberately beautiful is a dastardly act; it is an act of mesmerism which beclouds the entire issue. I do not believe that beauty can be created overnight. It must start with the archaic first. The archaic begins like Paestum. Paestum is beautiful to me because it is less beautiful than the Parthenon. It is because from it the Parthenon came. Paestum is dumpy—it has unsure, scared proportions. But it is infinitely more beautiful to me because to me it represents the beginning of architecture. It is a time when the walls parted and the columns became and when music entered architecture. It was a beautiful time and we are still living in it (Kahn 1986: 91).

In the context of this quotation, the phrase "to make a thing deliberately beautiful" refers to the application of sophisticated proportional systems to architectural compositions, since what differentiates Paestum from the Parthenon is its "unsure, scared proportions." Whilst not rejecting the use of proportions outright, Kahn subordinates this device to a sense of the "archaic." Paestum is championed for its chronological and, in a sense, its ontological proximity to architecture's mythical beginnings as a poetic discipline.

Many of Kahn's theoretical pronouncements make an ontological distinction between transcendent and terrestrial concerns. According to Kahn scholars

[2] See the drawing entitled: "A3: First Floor Plan," Kahn Collection.

including Charles Jencks,[3] Christian Norberg-Schulz,[4] Joseph Burton,[5] Vincent Scully,[6] David Brownlee,[7] David De Long,[8] Gerhard Auer,[9] and the present author (Fleming 1998), Kahn's theory resembles classical Platonism in this regard. Kahn's Platonism is most evident in his "form and design" thesis (Kahn 1961), developed at the time of his work in Rochester.[10] According to that thesis, particular buildings of the same type share an archetypal counterpart, or "form," which is seen in the mind's eye, or "psyche," as a vague idea that can only be represented by an *esquisse*-like diagram. When contemplating the ideal "form" for schools, Kahn claims that an architect must

> start right at the beginning, as though he were Socrates when he's talking about school. He must be this man [Socrates]. From it [the Socratic contemplation of typology] comes form, from the other considerations comes another duty called design (Khan 1960).

The contention of this chapter, that Kahn is simply pragmatic when it comes to making dimensions for his First Unitarian Church and School in Rochester, is entirely consistent with his "form and design" theory, in which metaphysical significance is primarily conferred on a building's planning strategy, or "form." "Design" meanwhile, is the pragmatic act of building in a circumstantial world. Kahn states that, "[d]esign is a material thing. It makes dimensions. It makes sizes", while "[f] orm," on the other hand, "is not design, not a shape, not a dimension. It is not a material thing" (Kahn 1991: 141). According to Kahn's "form and design" theory,

---

[3] Charles Jencks describes Kahn as the "major prophet" of the "metaphysical school" who built elements that seem to have "arrived perfected from Plato's ideal realm; see (Jencks 1973: 43–44, 232–233).

[4] According to Norberg-Schulz, "Kahn's philosophy evidently has Platonic origins. Thus he talks about *form* in the Platonic sense of *idea* .... He even uses the word "shadow" in connection with the concrete things of the world, as did Plato in his *Allegory of the Cave*. Kahn also subordinates the *existentia* to the *essentia,* and thus thinks within the tradition of Western metaphysics" (Norberg-Schulz 1979: 35).

[5] According to Burton, "Kahn's primary notion of [f]orm is like Plato's theory of the ideas, also known in English by the term 'Forms', as well as 'Ideas'" (Burton 1983: 76).

[6] Scully describes Kahn's "form and design" theory as, "a curious but very useful amalgam of Platonic Idealism and Pragmatic Realism" (Scully 1987).

[7] Brownlee argues that Kahn's "vocabulary was fortified by allusions to respected authority. Most fundamentally, the role played by light and shadow in differentiating the ideal world from the world of daily experience was an echo of the famous discussion of the same subject in Plato's *Republic*" (Brownlee 1991: 129).

[8] David De Long uses the word "Platonic" as an adjective to describe Kahn's notion of "form" (De Long 1991: 72).

[9] "Kahn's *form* is not a visible idea," Auer writes, "but a (Platonic) idea which has not yet materialised, the premonition of a slumbering archetype, an intuitive inspiration, at best formulated as an ideogram" (Auer 1992: 68–69).

[10] The first recorded public expression of this precise thesis is contained within a public address delivered at the Cooper Union entitled "The Scope of Architecture" on 20 January 1960, at the time of a stand off between Kahn and his clients in Rochester over their request that he develop a bi-nuclear scheme (Kahn 1960).

dimensions are related to the measurable, or sensible process called "design." They are totally unrelated to the "unmeasurable," or intelligible, concept of "form."

Meanwhile, Kahn argues that a great building

> must begin with the unmeasurable, must go through measurable means when it is being designed, and in the end must be unmeasurable. The only way you can build, the only way you can get the building into being, is through the measurable. You must follow the laws of nature and use quantities of brick, methods of construction, and engineering. But in the end, when the building becomes part of living, it evokes unmeasurable qualities, and the spirit of its existence takes over (Kahn 1979: 48).

If dimensions are slaves to earthly circumstances, by what other means could mathematics lend an "unmeasurable" quality to Kahn's buildings? Possible answers to this lie in the Platonic character of his theory.

The similarities between Kahn's theory and Plato's theory of Forms is most clearly articulated by the philosopher Arthur Danto. Danto argues that Kahn's notion of "form" is "exactly like its Platonic and Pauline counterparts, invisible and eternal" (Danto 1999: 187). Contemplating what most would expect a Platonic building to look like—an image to which Kahn's buildings do not conform, that of a composite of archetypal geometrical solids—Danto finds Kahn to be even "more in the spirit of Plato than architects whose buildings look like diagrams for geometric theorems" (Danto 1999: 188). It is not that Danto would deny Plato's famous love of geometry, but he does remind us that Plato is not particularly concerned with cosmetic appearances, dedicating his intellect instead to the correct deduction of eternal essences. According to Danto, both Plato and Kahn are concerned with the essential elements required of such things as political states, beds and Unitarian Churches, for these things to exist at all. So crucial and perplexing is this search for ideal and timeless archetypes, that the "look" of things, be that geometrical or otherwise, becomes a secondary issue. Where to Gast, Kahn's alleged geometry is central to his being a Platonising architect, Danto casts Kahn as a Platonist precisely because his buildings are not geometrical.

What Does Plato Say?

It will be apparent at this point that when Danto compares Kahn to Plato, he is not thinking of the relatively small number of passages from Plato's *Timaeus* that are typically featured in discussions of architecture and mathematics and that are, in some respects, peripheral to Plato's philosophy as a whole.[11] Neither is he thinking in terms of the Neoplatonic tradition in architecture and the inscription of

[11] As it investigates nature, Jowett points out that the *Timaeus* is not expressed through the mouth of Socrates, but by a Pythagorean, since in the *Phaedo* Socrates refuses to even discuss physics. Concerned by its undue influence on posterity, Jowett warns that the *Timaeus* is not central to Plato's philosophy, but is like "a detached building in a different style" (Jowett 1953: 633).

imperceptible proportions of religious significance. Rather, Danto is referring to the Plato we read in the middle period dialogues, of which the best known is *The Republic*. Whilst other Platonic dialogues also contain statements about mathematics,[12] *The Republic* presents a number of thoughts which seem to resonate very well with Kahn's thinking and which, in an embryonic way, suggest a new line of inquiry into the relationship between Platonism and architecture.

The central message of *The Republic* is well known: rulers and their citizens need to be liberated from their attachment to the world of sense experience (which is sensible but unintelligible), and turn their thoughts towards the realm of archetypal Forms (which is intelligible, but cannot be sensed). Within this context, *The Republic* ascribes two functions to mathematics. Firstly, it is a practical discipline, since it can be applied to the organisation of armies and the pitching of camps (*Republic* 525b and 526d). It also serves an epistemological purpose, since it tends "to draw the mind to the truth and direct the philosophers' reason upwards [towards the Forms], instead of downwards [towards sensible particulars]" (Plato 1987: 274). However, *The Republic* provides few specific instructions to "craftsmen" regarding the embodiment of mathematics within human artifacts, aside from various prohibitions. For example, it is stated that harmony and rhythm in both music and in architecture, as well as in every other kind of manufacture, should be treated with simplicity and restraint (400–401), so that people will not develop a taste for empirical fuss. Likewise, scene painters are criticised for exploiting people's vulnerability to optical illusions and other kinds of "deceptive semblances."[13] Aside from what should *not* occur, what active role might embodied mathematics play in leading viewer's minds upwards, bearing in mind that *The Republic,* like most of Plato's dialogues, speaks of an elementary kind of mathematics that simply deals with whole numerals?[14]

In Book 7, Socrates suggests that

> *there are some perceptions which don't call for any further exercise of thought, because sensation can judge them adequately, but others which demand the exercise of thought because sensation cannot give a trustworthy result.* (Plato 1987: 268)

---

[12] Other than passages from the *Timaeus*, one passage from *Philebus* (51c) venerating "straight lines and circles, and the plane or solid figures which are formed out of them by turning-lathes and rulers and measures of angles," has exerted considerable influence on architectural theory, its influence on twentieth-century architecture being discussed by Reyner Banham; see Banham (1988: 205).

[13] "The apparent size of an object, as you know, varies with its distance from the eye .... So also a stick will look bent if you put it in water, straight when you take it out, and deceptive difference of shading can make the same surface seem to the eye concave or convex; and our minds are clearly liable to all sorts of confusion of this kind. It is this natural weakness of ours that the scene-painter and conjuror and their fellows exploit with magical effect" (Plato 1987: 370).

[14] One passage in *Theaetetus* touches on the topic of irrational numbers, specifically $\sqrt{2}$; see *Theaetetus* (147d).

The sight of one's own fingers does not stimulate thought insofar as counting fingers is concerned, since "at no stage has sight presented the finger to [the mind] as being also the opposite of a finger" (Plato 1987: 269). "But what about the size of the fingers?" Socrates asks.

> Can sight distinguish properly whether they are large or small? Does it matter which one is in the middle or at the end? And can touch distinguish thickness and thinness or degrees of hardness and softness? Aren't all the senses in fact deficient in their perception of such qualities? (Plato 1987: 269).

As Julia Annas explains,

> [w]hat we see enables us to say that the finger is large, but also, and equally well, to say that it is small. So in these cases the mind is forced to reflect, and to come in to settle the problem. ... [W]hen the mind comes in to reason things out (524b) it declares that the contradiction is only an apparent one; 'large' and 'small' cannot really apply to the same thing, since what is really large is distinct from what is really small (524c). Thus we are moved to ask questions about what sort of thing it could be that could be really large or small, and to grasp that it cannot be the same as the largeness or smallness that we perceive with no effort in something like a finger. Rather, it is something 'intelligible', something that has to be worked out and grasped by the mind (Annas 1981: 218).

Plato's logic leads naturally to a theory of architectural embodiment, according to which repetitive elements in a building edify the viewer as fingers do. A viewer's sensory apprehension of a particular building may be adequate to say that it has ten windows, but consideration of which are large, and what *is* "large," would, according to Plato, engage the viewer in thoughts associated with the purely intelligible realm.

The reasoning behind his analogy concerning fingers is the basis of Plato's mandate that trainee philosopher kings must study arithmetic. If the perception of a unit

> is always combined with the perception of its opposite, and seems to involve plurality as much as much as unity, then it calls for the exercise of judgment and forces the mind into a quandary in which it must stir itself to think, and ask what unity itself is (Plato 1987: 270)

The perception of two hands, for example, involves both unity and plurality, insofar as each hand, representing one unit, also consists of five fingers, causing to viewer to ask "what is *The Unit Itself*?" In the same way, a building comprising multiple bays, each with multiple columns, could engage the viewer in the exercise of pure mathematical reasoning.

Once we understand how the sight of hands can aggravate the curious mind, and how buildings can have the same effect, we can extend this kind of thinking to two of the remaining disciplines that Plato's prescribes for potential philosopher kings: plane and solid geometry.[15] It follows that the sight of a building that is not quite

[15] It will be noted that Plato also prescribes studies in astronomy and harmonics (*Republic* 529–531). However, since he treats these as species of motion, their potential embodiment in a static medium such as architecture is greatly limited.

square, nor quite cubic, can be seen to draw the viewer into contemplating *The Square Itself* and *The Cube Itself*.

What Did Kahn Do?

Of course Kahn was not a philosopher and his own latent dualism is not likely to have led him to the kinds of deeply considered conclusions that Plato reaches. However, be it due to his intuition, or simply coincidence, there is one sense in which his "form and design" theory perfectly conduces buildings that edify viewers in the same way as hands are claimed to. In the realm of abstraction, 1 *Hand Itself* = 5 × *Finger Itself*. In the phenomenal realm, circumstantial factors affect the finger's size, or perceived size, thus making the viewer think about *Largeness Itself*. Similarly, in Kahn's realm of "form," *The Unitarian Church Itself* = 1 *Auditorium Itself*, surrounded by multiple instances of *Classroom Itself*. However, circumstances related to the "design" phase of Kahn's church in Rochester caused each classroom to be slightly different. It could be claimed that these differences prompt the viewer to ask whether each is a classroom and what *is* classroom, which is large and what *is* "large." Kahn's "form diagram" for the church indicates a square auditorium, whilst circumstances related to the "design" phase caused the built auditorium to be not quite square.[16] The radial distribution of light towers and entrance points about the auditorium lead those entering it to believe that it is square. Their discovery that it is not square after all, would, according to the logic of Plato's finger analogy, cause some mental distress leading to contemplation of *The Square Itself*.

In terms of Plato's finger analogy, the treatment of this building's façade (Fig. 75.3) is potentially edifying, since the perception of any type of element, to use Plato's words, "seems to involve plurality as much as unity" (Plato 1987: 270). The masonry alcoves suggest a single-storey building made up of six or more tall spaces, but when "the mind calls in reasoning and thought," (Plato 1987: 269) it realises that the building's sides are two storeys high and, typically, three classrooms long. Likewise, the Fisher House looks like two double-storey forms, but one of those forms has a double-height volume. Kahn takes a similar approach with the Phillips Exeter Academy Library. This building looks to be five storeys high, until reason enters in, and finds there are nine storeys above ground. Its outer form and inner atrium space both appear to be cubic, but neither is exactly.

Renzo Vallebuona observes that Kahn treats the site of his National Assembly complex in Dacca "as if it were possessed of no relevant dimensions" (Vallebuona

[16] Kahn's preliminary schemes all feature square auditoriums. The "circumstance" that deformed the shape of this space, was a letter from the building committee expressing their dislike of his scheme's "inherent squareness." Letter, Williams to Kahn, 28 February 1960, file labeled, "Building Committee Correspondence, April 1959 through December 1960," Box L.I.K. 15, Kahn Collection.

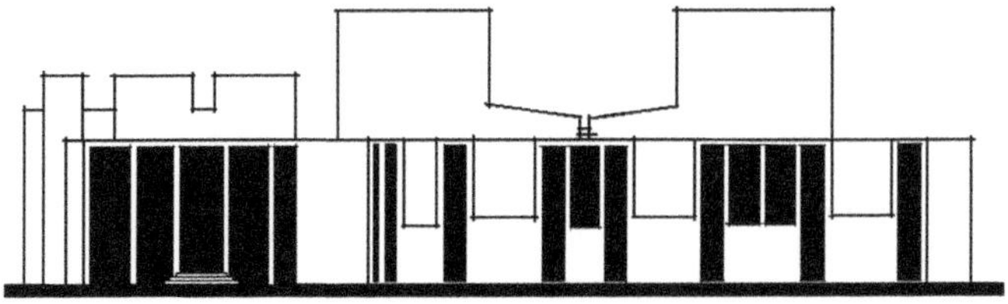

Fig. 75.3 The treatment of the façade of the First Unitarian Church. Drawing: author

1996: 65). Likewise, Vincent Scully links the "scalelessness and timelessness Kahn . . . build[s] in the housing at Dacca," to an earlier sketch by him of Siena's Campo in which "all details such as doors and windows that could suggest any particular scale or function were blotted out" (Scully 1991). According to the logic of Plato's finger analogy, the National Assembly building's ambiguity in scale might cause viewers to ask if the buildings are large, and what *is* large.

One expects the vaults of the Kimbell Art Museum to be elliptical, until reason enters in and finds them to be cycloidal. With the crescent-shaped glazing strips at the ends of each vault, Kahn makes it clear that he wants the viewer to think, to use reason. The upper chords of these windows are cycloidal, whilst the lower chords are defined by an ellipse.

Conclusion

That Plato's creation myth from the *Timaeus* should have had such an influence on Gothic and Renaissance architecture while dialogues such as *The Republic* were ignored, is an accident of history. An impression has been created that the admixtures, harmonic proportions, the golden mean, the square root of two and various elementary figures are central to Plato's thinking and that these can be viewed apart from Plato's central concerns, which were his epistemology and his metaphysics. Whilst the Neoplatonic tradition in architecture can be judged on its own terms, it has little to do with Plato.[17] Kahn can be forcibly enlisted into this tradition, or he can be approached on his own terms. It is doubtful that Kahn would have regulated his plans according to the golden mean when, by his own account, to do so would be to commit a "dastardly act," and the arithmetising of his working drawings confirms this.

Whilst the interpretation of mathematics in Kahn's work that has been presented here may seem obscure, it provides an explanation of Kahn's intentions that is intrinsic to his own espoused metaphysics. Whether speaking of the "measurable" and "the unmeasurable," "law and rule," "form and design," or "silence and light,"

[17] Jowett describes the neo-Platonism as "the feeble expression of an age which has lost its power not only of creating great works but of understanding them" (Jowett 1953: 631).

Kahn divided the universe into two realms, one a slave to circumstances and confusion, the other ideal. His discussions of his compromised Unitarian Church always make reference to the building's "form diagram," as Kahn entices his audience to contemplate the ideal, rather than the actual. To Kahn therefore, any device which could force viewers of his buildings to exercise reason over perception would be welcomed, since it would shift their attention from a compromised and circumstantial building and towards the unseen realm of "form." Being in the habit of thinking dualistically, Kahn may have delighted in the fact that visual apprehension alone does not reveal his buildings' height in storeys, their width in rooms, or their shape in terms of expected geometry. If these effects were indeed intended, then Kahn has stumbled part-way down a route once paved by Plato.

Biography Steven Fleming lectures in the history and theory of architecture at the University of Tasmania, Australia. He received his PhD in 2003 from the Department of Architecture at The University of Newcastle, with a thesis on Classical Platonism with respect to Louis I. Kahn's concept of "form". He has worked as a practicing architect in Australia and in Singapore. He is well-known among architects and urban designers working to promote bicycle transport, having announced cycling as an agenda for designers with his book, *Cycle Space: Architecture and Urban Design in the Age of the Bicycle* (Rotterdam: NAi010, 2012).

References

ANNAS, J. 1981. *An Introduction to Plato's Republic*. Oxford: Clarendon Press.

AUER, G. 1992. On the brick that wanted to become an arch: Louis I. Kahn's masonry structures. *Daidalos* **43**: 68-75.

BANHAM, R. 1988. *Theory and Design in the First Machine Age*. 2nd ed. London: Butterworth.

BROWNLEE, D. 1991. Light, the giver of all presences. In: *Louis I. Kahn: In the Realm of Architecture*, D. Brownlee and D. DeLong, eds. New York: Rizzoli.

BURTON, J. 1983. Notes from volume zero: Louis Kahn and the language of God. *Perspecta* **20**: pp. 69-90.

DANTO, A. C. 1999. *Philosophizing Art: Selected Essays*. Berkeley: University of California Press.

DE LONG, D. 1991. The mind opens to realizations: Conceiving a new architecture, 1951-61. Pp. 50-77 in D. Brownlee and D. DeLong, eds. *Louis I. Kahn: In the Realm of Architecture*. New York: Rizzoli.

FLEMING, S. and OSTWALD, M. J. 2001. Review of *Le Corbusier: Paris-Chandigarh. Nexus Network Journal* **3**, 1: 141-144.

FLEMING, S. 1998. Louis Kahn and Platonic mimesis: Kahn as artist or craftsman? *Architectural Theory Review* **3**, 1: pp. 88-103.

FLETCHER, R. 2001. Palladio's Villa Emo: The Golden Proportion Hypothesis Defended. *Nexus Network Journal* **3**, 2: 105-112.

GAST, K. P. 1998. *Louis I. Kahn: The Idea of Order*. M. Robinson, trans. Berlin: Birkhäuser.

———. 2000. *Le Corbusier: Paris—Chandigarh*. Basel: Birkhäuser.

JENCKS, C. 1973. *Modern Movements in Architecture*. Harmondsworth: Pelican Books.

JOWETT, B. 1953. Timaeus: Introduction. In: *The Dialogues of Plato*, 4th. ed. B. Jowett, trans. (vol. III, pp. 631-636). Oxford: Clarendon Press.

Kahn, L. 1960. "The Scope of Architecture at The Cooper Union Hall, 1-20-60." Cassette recording, Kahn Collection

———. 1961. Form and Design. *Architectural Design* **31**, 4 (April): 145-154.

———. 1979. Architecture. In: *Between Silence and Light: Spirit in the Architecture of Louis I. Kahn*, J. Lobell, ed. Boulder Colorado: Shambhala.

———. 1986. Wanting to be: The Philadelphia school. Pp. 89-92 in *What Will Be Has Always Been: The Words of Louis I. Kahn*, R. S. Wurman, ed. New York: Access Press Ltd. and Rizzoli.

———. 1991. The nature of nature. Pp. 141-144 in *Louis I. Kahn: Writings, Lectures, Interviews*, A. Latour, ed. New York: Rizzoli.

MARCH, L. 2001. Palladio's Villa Emo: The Golden Proportion Hypothesis Rebutted. *Nexus Network Journal* **3**, 2: 85-104.

NORBERG-SCHULZ, C. 1979. Kahn, Heidegger and the language of architecture. *Oppositions* **18**: pp. 29-47.

PÉREZ-GÓMEZ, A. 1984. *Architecture and the Crisis of Modern Science*. Cambridge, MA: MIT Press.

PLATO. 1987. *The Republic*, 2nd rev. ed. D. Lee, trans. London: Penguin Books Ltd.

SCULLY, V. 1991. Marvelous fountainheads: Louis I. Kahn: Travel drawings. *Lotus International* **68**: 48-63.

———. 1987. Introduction. Vol. I, pp. xvii-xxiii in *The Louis I. Kahn Archive: Personal Drawings: The Completely Illustrated Catalogue of the Drawings in the Louis I. Kahn Collection University of Pennsylvania Historical and Museum Commission*, A. Tzonis, ed. New York: Garland Publishing Inc.

VALLEBUONA, R. 1996. Louis Kahn in Wonderland. *Daidalos* **61**: pp. 62-67.

WATTS, C. M. et al. 2000. Methodology in Architecture and Mathematics: Nexus 2000 Round Table Discussion. *Nexus Network Journal* **2**: 105-130.

Chapter 76
The Salk: A Geometrical Analysis Supported by Historical Evidence

Steven Fleming and Mark A. Reynolds

Introduction

This is the second in a series of publications to marry the disciplines of geometrical analysis and architectural history, in order to present a holistic picture of geometry in Kahn's work and thinking; our earlier research analyzed Kahn's Kimbell Art Museum in Fort Worth, Texas (Fleming and Reynolds 2006). The publications are motivated by two overarching discontents: first, our discontent with geometrical analysis that pays no attention to historical documents that might suggest that proportions, no matter how accurate, are merely coincidental; second, our discontent with scholarship that would take an architect's silence on the topic of geometry as proof that no geometrical analysis of that architect's work is worth pursuing. Kahn said little to suggest an interest in the geometry that is buried everywhere in his work, making him the perfect subject for what we call historio-geometrical analysis.

In contrast to our earlier chapter, here we ruminate with relative liberty upon the possible meaning of the Salk's proportions. Of course, we are mindful of the arbitrary bond between signifiers and signified concepts,[1] but we also believe that the present context provides a fitting occasion for this kind of speculation.

First published as: Steven Fleming and Mark A. Reynolds, "The Salk: A Geometrical Analysis Supported by Historical Evidence", pp. 185–200 in *Nexus VII: Architecture and Mathematics*, Kim Williams, ed. Turin: Kim Williams Books, 2008.

[1] For an especially clear articulation of this argument see Saussure (1959).

S. Fleming (✉)
School of Architecture and Design, University of Tasmania, Locked Bag 1323, Launceston, TAS 7250, Australia
e-mail: Steven.Fleming@utas.edu.au

M.A. Reynolds
Academy of Art University, San Francisco, CA, USA
e-mail: marart@pacbell.net

K. Williams and M.J. Ostwald (eds.), *Architecture and Mathematics from Antiquity to the Future*, DOI 10.1007/978-3-319-00143-2_29,

Fig. 76.1 Salk Institute for Biological Studies, La Jolla, CA, Louis I. Kahn, architect. Photo: Steven Fleming

The Salk Institute for Biological Studies (Fig. 76.1) is a large building and much of the documentation for its construction was issued via addendums after work on site had commenced. We were therefore unable to defer to dimensions to confirm or refute all of our findings. Neither have we been able to arrange access to conduct in-situ measurements—though, in the case of the gardens shown in drawing LA4 ("Laboratory—middle level garden," dated 17 Jan. 1963), such measurements will remain an impossibility, as those gardens were never built. With less data at our disposal than we would have preferred, we were beholden to be especially accurate, thus we traced our analyses on high quality vellum, using only first generation copies, and we ensured our line weights matched Kahn's as well. For reasons of reprographics the illustrations shown here are small, redrawn versions of the actual analyses we used, and are therefore less accurate.

Our analysis was complicated by subtle complexities in the plan. Unlike Kahn's Kimbell Art Museum, which our earlier study found to be strictly regulated by a two-feet square grid, the Salk has many irregularities to belie the solemn aura of its famous plaza. Movable partitions mean interior ratios are elusive and variable, while the architect seems all too willing to abandon rhythms established along the length of a wing when end conditions dictate a change. In choosing what to measure, and what to ignore, we have been guided by Kahn's well known privileging of enduring elements, that is, those elements that are structural and therefore likely to outlive ephemeral fixtures when the Salk, in some mythical future, stands like the ruins that shaped Kahn's sensibility when he sketched them in Europe (Hochstim 1991). That is to say, we have measured the Salk's bones, not its flesh.

Superstructure

The Salk's bones start with a superstructure with column centers regulated by an array of 13 grid lines—running north–south like meridians—each 20′ apart. These grids lines are numbered 1 through 13 on sheet LA4, and they are crossed by four lettered grid lines (A,B,C and D) running east/west. As this grid serves as an invisible template for the project, we were interested to find the rectangle defining its extremities is a mere a 3.5′ short of being a 3:4 rectangle (it is 316.5′ × 240′).

That rectangle defines column centers. The rectangle that can be drawn about the outer edges of the columns is 319.5′ × 241.5′, and has proportions that deviate from the 3:4 rectangle by a mere 1.03 %.

In a diffuse but telling series of public statements, Kahn—who was also a pianist[2]—hinted at an interest in musical proportions, on one occasion attributing to an imagined space "a sound character alternating with the tones of the space" [Louis Kahn, cited in: (Robinson 1997: 12)], and on another claiming "a plan of a building should read like a harmony of spaces in light," just as "[t]o the musician a sheet of music is seeing from what he hears" (Kahn 1961: 149). We also know that Colin Rowe, Rudolph Wittkower's protégé, gave Kahn a copy of *Architectural Principles in the Age of Humanism* (Wittkower 1952), with a recommendation that in it, "I think that you will discover attitudes with which you are profoundly in sympathy."[3] The near 3:4 (or Major Fourth) proportions of the Salk's columnar grid tallies with Kahn's interest in musical proportions, and is consistent also with the musical proportions we discovered when we analyzed the Kimbell Art Museum that Kahn designed in Fort Worth, Texas (Fleming and Reynolds 2006).

Our argument, that the Major Fourth proportions of the Salk's superstructure is likely intended, is based on three tenets. First, the proportion applies to something that, because of his love of structure, Kahn is likely to have focused on. Second, it deviates from a Major Fourth by only 1.03 %. Finally, 3:4 is a kind of proportion that we know Kahn was interested in: it is a musical proportion. However, counter arguments could be advanced. Why, if he was laying out a proportioned columnar structure, didn't Kahn place those columns within a 240′ × 320′ rectangle, when nothing about reinforced concrete construction would have made that impossible?

[2] According to his daughter, Kahn helped support his family by playing the piano at a local cinema in his youth; see Tyng (1984).

[3] Letter from Colin Rowe to Louis I. Kahn, 7 February 1956, file labeled "Correspondence from Universities and Colleges", L.I.K. Box 65, Louis I. Kahn Collection, University of Pennsylvania and Pennsylvania Historical and Museum Commission (hereafter cited as Kahn Collection).

Also, why would Kahn have paid attention to a proportion that cannot be experienced from any point on the ground? Such arguments prevent us from calling the near 3:4 proportions of the Salk's overall structure a certain intention, despite the low percentage of deviation.

Overall Shape

Though the following claim may seem contradictory, we argue that the overall squareness of the Salk—including elements lying beyond the main columnar grid—is a much clearer intention, despite a percentage deviation comparable to the one indicated above. The complex as a whole measures $362.96' \times 372.6'$ and deviates from a true square by 0.972 %.

Why do we show extra leniency when it comes to squares? Because Kahn spoke of "square" buildings and spaces in his oeuvre that deviate from true squares by far more than 1 %. His thinking about squares is perhaps most evident in what he had to say about an earlier building, the First Unitarian Church in Rochester, New York. For over a year Kahn generated preliminary schemes for that church, all with perfectly square sanctuaries, before settling on a sanctuary that was $58'$ wide and $66'$ long. Though the sanctuary as eventually defined is hardly a precise square, he tells the congregation at their church's dedication, "[t]his building is ... a non-directional building"—this claim following a comparison to the Pantheon—"it's practically a square."[4] Like the $58' \times 66'$ sanctuary in Rochester, Kahn would have considered the $362.96' \times 372.6'$ footprint of the Salk to be practically a square, and symbolic of anything a perfect square might symbolize on that site.

Elevations and Sections

A more conventional approach to proportioning is evident in the elevations facing the Pacific, for which Kahn has used the Major Second ratio of 4:5, while the study towers flanking the laboratories in "Section 1 elevation" have 2:3, or Major Fifth proportions. Given the complexities of the Salk's program, with its highly serviced laboratory spaces, it is possible but unlikely that Kahn manipulated either the plan and/or the section to achieve this ratio, but programmatic constraints would not have prevented him from raising parapet heights to make elevations four fifths or two thirds higher than their respective widths.

We could debate the accuracy of these ratios, and whether or not they should be measured from the floor of the colonnade or else from the variable height of the

---

[4] Kahn quoted in the "Dedication of First Unitarian Church, Rochester, NY—Dec 2 1962", Historical Records of The First Unitarian Church, Rochester.

ground one step down, but in so doing we risk overlooking an unconventional aspect of the Salk, and one that calls for a less conventional examination of its walls. Questions surrounding the proportioning of elevations, we argue, are less important here than questions concerning the section. Height constraints related to aviation activity in the La Jolla area were a factor in Kahn's decision to bury one whole level (along with its attendant "pipe laboratory"), meaning the Salk's true sectional proportions, the proportions that mattered to Kahn, are likely to start below ground. Therefore it is especially interesting to find that the hidden, or buried Salk, has a $\sqrt{2}$ proportion. The significance of this proportion is that it is listed by Palladio—as recorded by Wittkower, who Kahn read—and that it also played a central role in Kahn's Kimbell Art Museum. For reasons known only to Kahn himself, it appears that he gave the Salk hidden—in every sense of the word—proportions; they are literally hidden beneath the ground.

Visitors to the Salk are more likely to see what we saw when we opened drawing LA11 ("Laboratory—garden elevations and sections," dated 7 Jan. 1963), the drawing from which we ascertained the Salk's sectional proportions, and that is an apparent grid of squares. Yet analysis reveals few of these are squares at all. This leads us to the view that Kahn had an aesthetic penchant for square-like proportions, without the fanaticism of one who would calibrate the thickness of every floor slab and wall in the service of a truly square grid (as, for example, the architects Giuseppe Terragni and Aldo Rossi have done). Qualifiedly, the Salk could be described as having a grid of squares, of the kind Kahn conceived in Rochester. They characterise the elevations and are evident in the remainder of the areas covered by our analysis, that is, the plaza, laboratories, studies and other office wings.

Plaza and Unbuilt Gardens

Drawing LA4 ("Laboratory—middle level garden," dated 17 Jan. 1963), the plan that was used to set out and commence construction work on the Salk, features sketchy details of garden beds where ultimately Kahn would make a plaza. Though an absence of dimensions for these unbuilt gardens[5] would restrict us to a purely geometric analysis, the fact that their location would have made them the physical and symbolic focus of the facility, calls for their close investigation. We invite readers to think of these gardens as the metaphorical keys that unlock a geometrical schema that could only have been conceived by an architect who was fascinated with music.

A musical theme that is announced by the Minor Third (5:6) ratio of the two smaller gardens at the western end of the plaza, would appear on first appraisal to be contradicted by the adjoining four gardens, all of which have proportions

[5] LA4 includes a note "see LSD 485" for details of the garden, though that particular drawing was not available for the purposes of this analysis.

approaching the golden section. But very close examination reveals those gardens have a ratio of 1.625:1, or 13:8, a fact that seems irrelevant until one considers there are thirteen notes in a musical scale, including the half-tones, and eight major notes when flats and sharps are excluded. It also happens that 8 and 13 are partners in the Fibonacci sequence: . . .3, 5, 8, 13, 21. . ..

At the eastern end of the plaza, a longer pair of garden beds have a ratio of 1.846:1. This is a compound ratio that includes the 5:6 ratio (the Minor Third) plus a square—the note. Effectively, Kahn has replicated the western-most rectangles in the East and added a square to each rectangle's long side.

In Fig. 76.2, AKMZ circumscribes the extremities of the two laboratories, while GNPR—which has a 2:3 ratio—roughly coincides with the unfinished gardens. In the hope of determining Kahn's way of laying out the gardens and pools we used the structural grid of the laboratories as the basis of a 12×16 grid of $10'$ squares, having three master squares: APRZ, EFHJ, and GKMN. Point W is the approximate center of the courtyard. Notwithstanding a few anomalies, the grid reveals order behind the seemingly random garden plan, such as the alignment of each garden's eastern inner edge to a north/south gridline.

Studies of Typical Plans

Geometric analyses of Kahn's drawings for studies of typical plans on sheet LA18 ("Studies—typical plans," dated 7 Oct. 1963) suggests he worked with the fractional parts of the square to generate musical ratios, similar to Palladio's approach of dimensioning lengths of simple integers related to music.

In Fig. 76.3, it appears that Kahn's basic plan is based on the 1:2 double square, the octave, AKMZ. In order to better understand the ratios in the plan, readers should note the ratios contained by the squares AKPR and PRMZ and then translate these to the square on the other side. Each of the individual squares of the double square was subdivided into fourths. For example, GU, UP, PV, and PN are fourths. This makes the interior rectangles, AKUh, AHmR, jVMZ, and RmJZ all 3:4 rectangles. These rectangles intersect within the square, and intersect along the diagonal at points, p1 and p2. These two points are critical to the width of the northern wing of the building, UV. LW marks the boundary of this wing as a double square, p1LWp2. When the partitioned space in the northern-most end of the building is included, this additional portion, LXYW, generates the 2:3 ratio, which contains two 3:4 ratios. (The opposite is also true: a 3:4 ratio will contain two 2:3 ratios.)

The 3:4 ratio is used for the width of the northern wing of the building, UV. UV is one fourth—the quarter note—of the double square's edge, KM, which then generates the 2:3 ratio. This 3:4 ratio was also used for the wall partitions inside the double square, along the line, HJ.

It can also be seen that the $\sqrt{2}$ ratio could also have been employed. Points p3 and p4 are generated where the side of each square passes through its diagonal, making

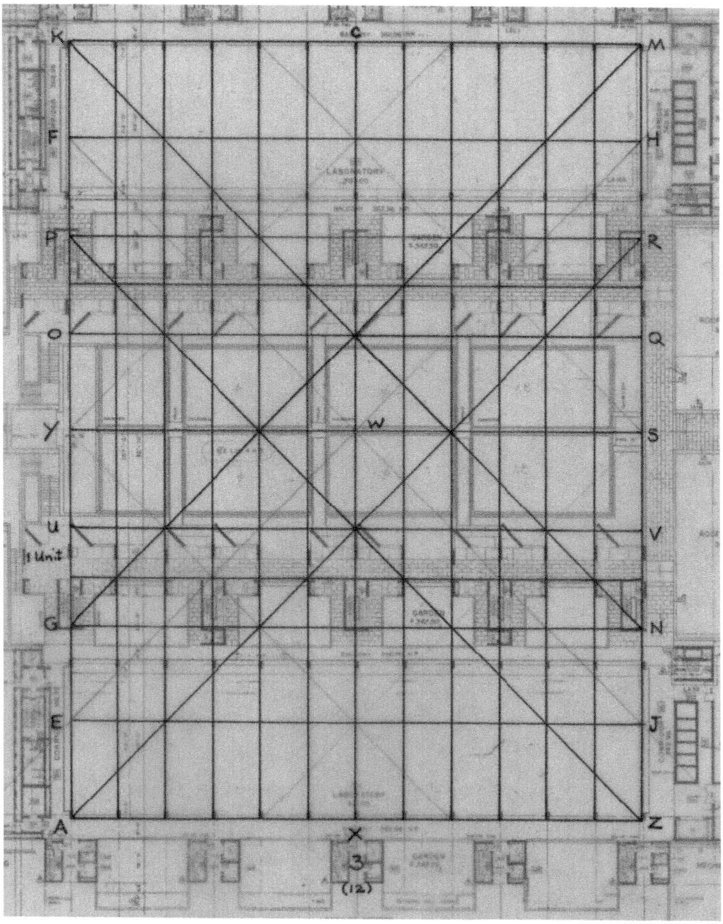

Fig. 76.2 Geometrical analysis overlaid on LA4. Drawing: Mark A. Reynolds

a reciprocal √2 (1:√2) in the base of each of the two squares. The top edge of each reciprocal rectangle runs along SB. Even though the √2 is not a musical ratio, it is on Palladio's list.

The splayed blades forming the walls at 45°, a and c, are in a 3:4, Major Fourth relationship, with a deviation of 2.7 %.

Laboratories

Each laboratory building is 3.87:1, being 240′ × 62′. The ratios of 3.75 × 1 to 4 × 1 are significant because the traditional canons of measure for a human body is a figure 7.5–8 heads tall by two heads wide at the shoulders. This can then be

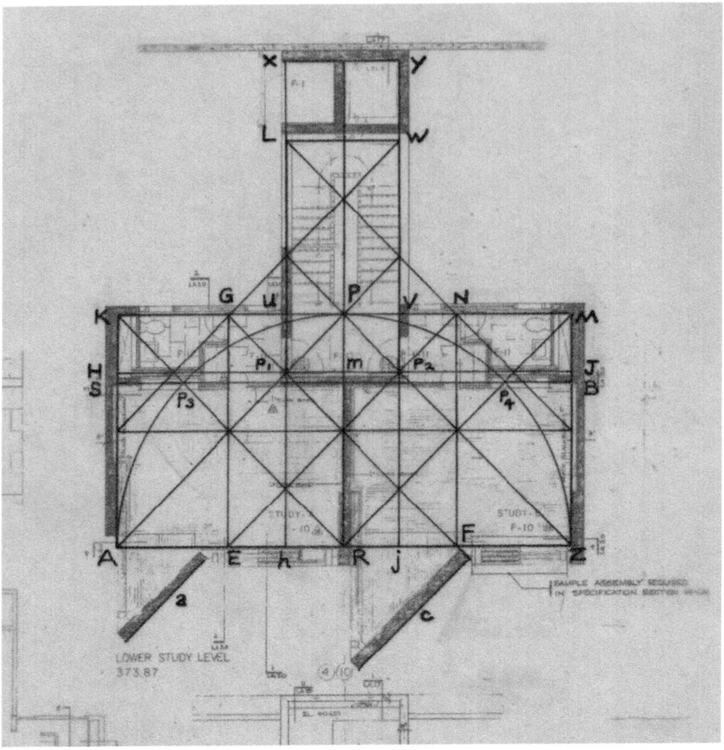

Fig. 76.3 Geometrical analysis overlaid on LA18. Drawing: Mark A. Reynolds

reduced to 3.75–4:1, the quadruple square. A human figure could stand within each or either of these 3.87:1 perimeters. As there are two laboratories, one could speculate that there is perhaps one for each view, front and back, or that one represents a man and the other a woman, as the space was intended for biological studies.

North Office Wing

Sheet LA47 ("Laboratory building—office wing elevations," dated 7 Oct. 1963) documents two elevations of the North Office Wing of the Institute. Figure 76.4 shows the geometric construction used to generate the ratio of the elevation, AKMZ. This is 5:6, the Minor Third. The base of the wall, AZ, was used to generate a square, APRZ; AZ = AP. This square was divided into fifths by the application of half-diagonals, AU and PF for example. Their intersection, at J, is one fifth. One of these fifths, NR, was added onto the top of the square with the length RM (NR = RM). Doing so makes the rectangle of 5:6 units.

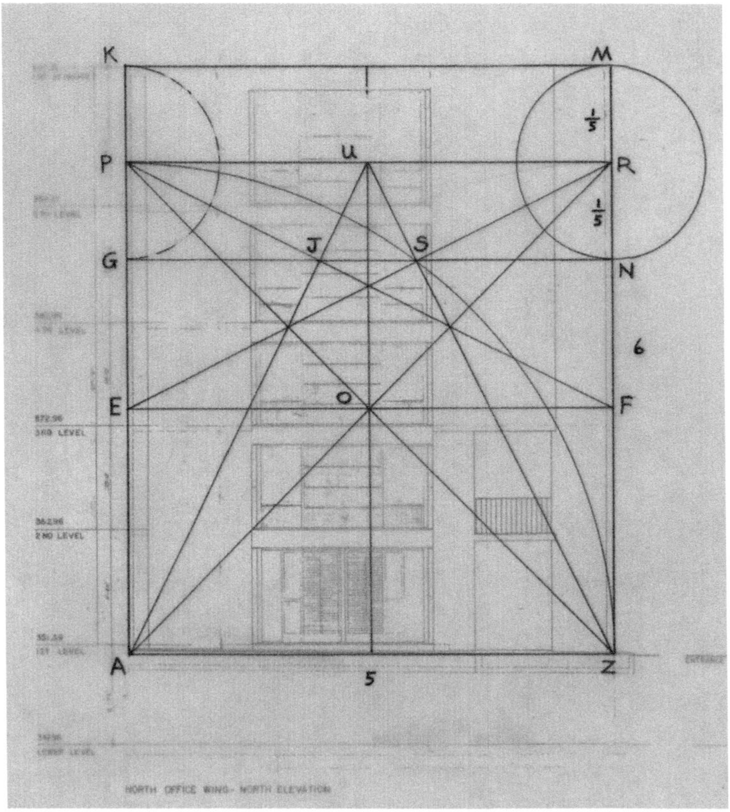

Fig. 76.4 Geometrical analysis overlaid on LA47 (elevation on the *left*). Drawing: Mark A. Reynolds

This technique is of great value when working with fractional parts and musical ratios when a square is employed. Most fractional parts are easily found when the diagonals and half-diagonals are drawn inside the square. Whatever fraction is applied, it converts the side of the square, considered to be unity (1), into that same number of units indicated in the denominator of the fraction. All fractions can be seen to be ratios, for example, generating fourths will change the value of the side of the square from 1 to 4. Taking one of these fourths away, we will have a 3:4 rectangle inside the square. Adding a fourth will make the new rectangle 5:4 units. The side of the square does not change length; only its assigned value does.

With regards to the other elevation on Sheet LA47, AKMZ (Fig. 76.5), has a ratio of 2:3, the Major Fifth. Here, the height of the building, AK, was used to make a square, AKPR. Then, half of this square, GNPR, was added to the right side to make RPMZ, which is now 50 % longer than it is high; AKNG = GNPR = RPMZ. The result is the 2:3 ratio.

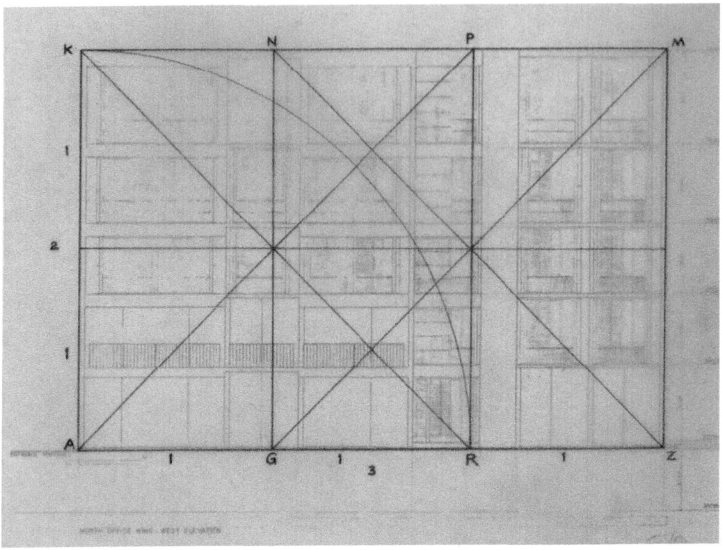

Fig. 76.5 Geometrical analysis overlaid on LA47 (elevation on the *right*). Drawing: Mark A. Reynolds

Conclusions

Do the plans for the Salk "read like a harmony of spaces" as "[t]o the musician a sheet of music is seeing from what he hears" (Kahn 1961: 149)? Because these words are quoted from the text Kahn would routinely send to those inquiring about his theory,[6] which those close to him would say embodied his theory better than any other,[7] it is hard to dismiss them as hollow, and tempting to read them in the light of what we know about Kahn's reading of Wittkower, who wrote about the importance of harmonic proportions to Palladio.

Our analysis raises two broad hypotheses. Hypothesis one: a complex array of musical proportions informed many aspects of the design but, because he was not a fanatic, Kahn was happy to alter dimensions to accommodate technical considerations—advice from structural or services engineers, for example. This hypothesis is consistent with the slight deviations from true ratios mentioned throughout this chapter. It also tallies with the central tenet of his article "Form and Design", that a building's fundamental planning arrangement is otherworldly,

[6] David De Long (1991) claims that those inquiring about Kahn's theory would be routinely sent a copy of "Form and Design".

[7] According to Tim Vreeland from Kahn's office, "Form and Design" embodies Kahn's thinking better than any previous text. See: Letter, Vreeland to Pidgeon, 11 January 1961 "Master File, November 1 through December 30, 1960," Box L.I.K. 9, Kahn Collection.

almost divine, while dimensions are subject to circumstantial factors such as budgets and building technology.

Hypothesis two: geometrical analysis is capable of finding proportions an architect would be surprised to learn he or she has even used. As it calls into question our entire endeavor, we will admit this hypothesis cannot be adequately explored in the context of the present chapter, where our aim has been to uncover whatever proportions the Salk might contain, but we would invite others to ask the hard questions. Why, if proportions meant anything to Kahn at all, do the dimensions found on his plans indicate no precise ratios? Should analysts of recent buildings, for which dimensioned working drawings are available, tolerate percent deviations, no matter how slim, when that practice was developed with ancient buildings in mind? Why would Kahn have used such a disparate variety of ratios, when conventional wisdom would have architects striving for cohesion through the reiteration of one, or a small family of ratios?

The analysis we have presented has, by no means, been exhaustive. Neither should it be seen as an abrogation of previous studies of its kind, most notably that of Klaus-Peter Gast (1998). Moreover, we hope it will be seen as the start of a discussion that could be advanced in any of the following ways. A simple computer program could be developed to digitally analyze the Autocad plan that the Salk Institute so kindly provided us; given any point on the plan as a center-point, and any related point on which to base a radius, such a program could automatically populate the plan with circles with radii proportional to the first radius, in search of precise ratios that might otherwise have eluded a human analyst working with his eyes and a compass. It would be fascinating also to have a musicologist examine the musical ratios we have identified—they might harmonize in ways we are not aware, or even be related to a known tune. As we indicated in our previous research on the Kimbell, we continue to publish in the hope that past associates of Kahn's will come forward with any anecdotal evidence of his interest, or disinterest, in proportions.

Acknowledgments We would like to thank William Whitaker of the Architectural Archives of the University of Pennsylvania, for his assistance during a 2004 visit to Philadelphia to access relevant correspondence, when he also provided us with the first generation reproductions used in our analysis; they were of sheets: LA4 "Laboratory—middle level plan, garden" (used as an underlay for Fig. 76.2; LA18 "Studies—typical plans" (used as an underlay for Fig. 76.3); LA11 "Laboratory—garden elevations and sections"; and LA47 "Laboratory building—office wing elevation" (used as an underlay for Figs. 76.4 and 76.5). Thanks also to the Salk Institute for Biological Studies for their daily tours of the facility—these were invaluable in collecting photographic records to supplement the above drawings—and to Bob Lizarraga for providing an electronic copy of the plan. Keith Duke assisted with the analysis.

Biography Steven Fleming lectures in the history and theory of architecture at the University of Tasmania, Australia. He received his Ph.D. in 2003 from the Department of Architecture at The University of Newcastle, with a thesis on Classical Platonism with respect to Louis I. Kahn's concept of "form". He has worked as a practicing architect in Australia and in Singapore. He is well-known among architects and urban designers working to promote bicycle transport, having announced cycling as an agenda for designers with his book, *Cycle Space: Architecture and Urban Design in the Age of the Bicycle (Rotterdam: NAi010, 2012).*

Mark Reynolds is a visual artist who works primarily in drawing, printmaking and mixed media. He received his Bachelor's and Master's Degrees in Art and Art Education at Towson University. He was awarded the Andelot Fellowship to do post-graduate work in drawing and printmaking at the University of Delaware. For years Mr. Reynolds has been at work on an extensive body of drawings, paintings and prints that incorporate and explore the ancient science of sacred, or contemplative, geometry. He is widely exhibited, showing his work in group competitions and one-person shows, especially in California. Mark's work is in corporate, public, and private collections. A born teacher, he teaches sacred geometry, linear perspective, drawing, and printmaking to both graduate and undergraduate students in various departments at the Academy of Art University in San Francisco, California. Additionally, Reynolds is a geometer, and his specialities in this field include doing geometric analyses of architecture, paintings, and design.

References

De Long, David. 1991. The mind opens to realizations: Conceiving a new architecture, 1951–61. Pp. 50–77 in *Louis I. Kahn: In the Realm of Architecture*, David Brownlee and David De Long, eds. New York: Rizzoli.

Fleming, Steven and Mark Reynolds. 2006. Timely timelessness: Traditional Proportions and Modern Practice in Kahn's Kimbell Art Museum. *Nexus Network Journal* **8**, 1: 33–52.

Gast, Klaus-Peter. 1998. *Louis I. Kahn: The Idea of Order*. Michael Robinson, trans. Berlin: Birkhäuser.

Hochstim, Jan. 1991. *The Paintings and Sketches of Louis I. Kahn*. New York: Rizzoli.

Kahn, Louis I. 1961. Form and Design. *Architectural Design* **31**, 4 (April 1961):145–154.

Robinson, Duncan. 1997. *The Yale Centre for British Art: A Tribute to the Genius of Louis I. Kahn*. New Haven: Yale University Press.

Saussure, Ferdinand de. 1959. *Course in General Linguistics*. London: Peter Owen Limited.

Tyng, Alexandra. 1984. *Beginnings: Louis I. Kahn's Philosophy of Architecture*. New York: Wiley Interscience Publications.

Wittkower, Rudolf. 1952. *Architectural Principles in the Age of Humanism*, 2nd. ed. London: Tiranti.

Part X
Contemporary Approaches to Design and Analysis

Chapter 77
Architecture and Mathematics: Soap Bubbles and Soap Films

Michele Emmer

Introduction

> I do not suppose that there is any one in this room who has not occasionally blown a common soap-bubble, and while admiring the perfection of this form, and the marvellous brilliancy of its colours, wondered how it is that such a magnificent object can be so easily produced.
>
> I hope that none of you are yet tired of playing with bubbles, because, as I hope we shall see, there is more in a common bubble that those who have only played with them generally imagine (Boys 1959: 13).

These are words by Charles V. Boys in the introduction to his famous book *Soap Bubbles, Their Colours and the Forces which Mould Them*. The book is based on three conferences that the author gave for a young public at the *London Institution* during December 1889 and January 1890. *The Society for Promoting Christian Knowledge* first published the book in 1902 and the author's revised and expanded edition was published in 1911.

The book quickly became a very popular and classic essay on the popularization of a scientific subject. Boys was presenting in the volume the results of new experiments on soap bubbles performed by scientists 30 years before the publication of his book. There is no doubt of the great importance of the nineteenth century for the study of what it is commonly called *The Geometry of Soap Bubbles and Soap Films*.

First published as: Michele Emmer, "Architecture and Mathematics: Soap Bubbles and Soap Films", pp. 53–65 in *Nexus I: Architecture and Mathematics*, ed. Kim Williams, Fucecchio (Florence): Edizioni dell'Erba, 1996.

M. Emmer (✉)
Dipartimento di Matematica, Sapienza Università di Roma, Piazzale A. Moro, 00185 Rome, Italy
e-mail: emmer@mat.uniroma1.it

K. Williams and M.J. Ostwald (eds.), *Architecture and Mathematics from Antiquity to the Future*, DOI 10.1007/978-3-319-00143-2_30,

The Geometry of Soap Bubbles and Soap Films

The Belgian physicist Joseph Plateau published the results of his experiments on soap bubbles and soap films in a two-volume treatise in 1873: *Statique expèrimentale et théorique des liquides soumis aux seules forces moléculaires* (Plateau 1873). In 1872, a year before Plateau's results were published, the famous impressionist painter Edouard Manet painted his *Les Bulles de savon*. If the geometrical history of soap films starts with Plateau, soap bubbles already had a long story of their own in literature and art (Emmer 1987, 1991, 2009).

Plateau had examined a vast number of small frames dipped into soapy water, obtaining completely novel shapes which he did not hesitate to define as full of *charme*, airy forms which in their essentiality are simply mathematical surfaces. The problem in mathematics which is linked to the name of Plateau consists in taking a generic curve in three space and then finding a surface with the least possible area bounded by the assigned curve. If a tridimensional model of the curve can be made, it can be dipped into the soapy water. When it is withdrawn, in very many cases a soapy surface is obtained which is the empirical solution to the problem. While this type of demonstration may be satisfactory to a physicist, a mathematician demands a rigorous demonstration of the existence of the solution, and seeks, where possible, concordance with the physical experiments. The general mathematical solution to Plateau's problem was to prove somewhat arduous (Fig. 77.1).

The reason soap films and soap bubbles are excellent tools with which to study *minimal surfaces* (surfaces of least area under some fixed conditions) is due to the presence of *surface tension*, connected with the forces of attraction which come into play between the molecules of a liquid. Surface tension is proportional to the surface area of the film. For a liquid to achieve equilibrium, the energy must be the least possible, and consequently the free surface area must be diminished as far as possible. Hence the denomination: minimal surfaces. When a soap bubble is blown, the soapy surface stretches and, when blowing ceases, the film will tend toward equilibrium in the form of a sphere. The sphere presents the least exterior surface area of all surfaces containing the same volume of in-blown air.

Besides investigating the solutions for different boundaries of the Plateau problem, the Belgian physicist spent time on the geometry of soap bubbles and soap films. Blowing with a small tube into soapy water, one notices that the more one blows, the more complex the agglomeration of films becomes. Thus one could assume that the way in which the various films meet together could give rise to infinite possible configurations. And here we have Plateau's discovery, incredible at first sight. However many soap films come into contact with one another, there can never be other than two types of configurations. The three empirical rules found by Plateau concerning soap films are:

1. a cluster of soap bubbles or soap films attached to a wire consists of flat or curved surfaces which intersect along curves with very smooth curvature.

Fig. 77.1 E. Bisignani, M. Emmer, "Soapy Hypercube", an illustration of the Plateau Problem for a cubic wire. Photo: author

2. the surfaces meet in only two ways: either three surfaces meet along a line, or six surfaces give rise to curves which meet at a vertex.
3. the intersection angles of the surfaces along a line, or of the intersection curves in a vertex are always equal: in the former case at 120°, in the latter at I09°28′. However many intersections there may be, the only types of angles which the soap films form among themselves are those found by Plateau! (Fig. 77.2).

Only in 1976 the American mathematician Jean E. Taylor was able to prove that the laws of Plateau are correct. With some degree of pride, she wrote:

> Although in the past 100 years, several mathematical models for area-minimizing surfaces, all called Plateau's problem, have been constructed, none of these models allowed the general kind of surfaces which arise in real soap films (Taylor 1976: 135–142).

That same year Taylor and Almgren published an article in the *Scientific American* with the results achieved on the geometry of soap films (Almgren and Taylor 1976: 82–93; Emmer 1980). The text was accompanied by a set of splendid photos. The article inspired the idea of a motion picture on soap films. The use of the camera has made possible in-depth investigation of the soap film structures revealing in slow motion a series of effects otherwise invisible to the naked eye.

Many mathematicians have worked on Plateau's problem and the *Minimal Surfaces Theory* in the last 120 years. The problem was solved in the 1960s by the Italian mathematician Ennio De Giorgi, and again by American mathematician Reifenberg in a completely independent way. In 1974 Italian mathematician Enrico Bombieri received the Fields Medal (equivalent to the Nobel prize for mathematicians) for his research on minimal surfaces theory.

So there is no doubt that the geometry of soap films and soap bubbles is a very important topic in modern mathematics. But is there really any relationship with architecture?

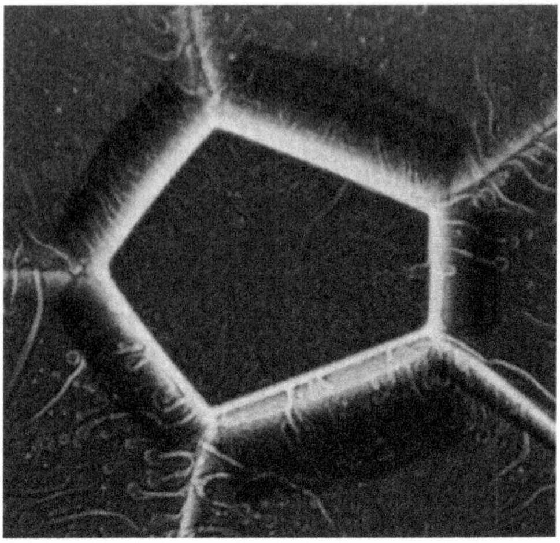

Fig. 77.2 Soap bubbles. Photo: author

Minimal Surfaces in Architecture

Soap Films and the Form of Towns

It is well known that among all closed plane curves of the same length, the circumference encloses the largest area, a property known as the *isoperimetric property*. The first one to consider the isoperimetric problem in mathematical terms was Pappus of Alexandria, in about 390 BC. In book V of his works on mathematics and physics, we find the word isoperimetric used in context of geometry (the book was written in Greek) (Hultsch 1933). It is possible to verify through very simple experiments with soap films that the property of the circumference is correct. It is enough to take a wire in the shape of a circumference and dip it in the soapy water and then dip it out: the soap film will form a circle spanning into the circumference, a soap film with the isoperimetric surface. An analogous formulation of the problem is the following: find a plane curve that surrounds a given area with the smallest length. The answer is again the circumference. In another experiment, one takes two threads, the first fixed to the wire (still in the form of a circumference) and the second fixed to the first. When, with a finger, one breaks the soap film between the two threads, they stretch out forming a perfect circumference, so that the space between them is the largest and the external one the smallest.

Mathematicians consider as very reasonable the hypothesis that those charged with the design of towns in ancient times knew of the isoperimetric property at least empirically, if not as a result of Pappus' work. An architect in the Middle Ages who wanted to construct town walls of the least possible length containing the largest inside area had to build the town in a circular form. Georg Gerster, in his book *La Terre de l'Homme: Vues Aériennes* (1975), has published photos of the earth taken from a airplane. With this technique it is possible to draw attention to the structures built by men. One of the chapters is dedicated to an *Archetype d'habitat: la cité en forme de cercle*. As Gerster points out, the most important motivation to build towns in a circular form, during various historical periods and among different civilizations, was probably not the isoperimetric property, but rather, "[t]he form of the circle was the symbol of harmony and perfection." Of course this does not mean that the only hypothesis for the circular form of towns is based on its symbolic meaning, excluding a priori the isoperimetric property. As Gerster points out, the circular plan allows the inhabitants whose houses are located on circumcentric circumferences to move easily towards the temple in the centre of the town or to the golf green outside the town. It is interesting to note that the presence of towns with circular plans is more significant in certain historical periods. In 1804 the French architect Claude Nicolas Ledoux (1736–1806) published the volume *L'Architecture Considérée sous le Rapport de l'Art, des Moeurs et de la Lé gislation* (1804). In his treatise Ledoux presented two projects: the first one for houses of workers, the second for the plan of a town. The first one was composed of a circle inscribed in a square. As a comment Ledoux wrote: "[i]ts form is pure" while for the second project: "the form is pure as the one the Sun describes in its movement." Moreover he added: "The circle, the square are the letters of the alphabet used by the architects for the structures of their best works."

Soap Films in Three Space: The Sphere

As the circle satisfies the isoperimetric property in the plane, the sphere satisfies the same property in three space: for the same external surface area, the solid which contains the maximum volume is the sphere, or if we assign a volume, the sphere is the solid which contains the assigned volume and has the minimal external surface area. When we blow to generate a bubble, we fix a volume of air (the air we blow), while the soapy water, forming the bubble that surrounds the volume of air, builds the spherical surface which solves the problem. As the isoperimetric property has probably been used in the foundation of towns, so the sphere has been (and still is) one of the favourite subjects in architecture.

One of the more utopian projects of Ledoux was related to the sphere. He was planning a house for agricultural guardians. It was a perfect sphere located down in a hollow and it was possible to reach it through four bridges. As Kaufmann has pointed out: "The project was a rare example of pure geometry" (Kaufmann 1978). For the *French Revolutionary Architects* (as he calls them), the sphere was particularly suitable for all the constructions related with the Death and the Eternity. Another French architect of the same period, Etienne-Louis Boullée (1728–1799), made the plan in 1784 for the cenotaph of Newton. Boullée was fascinated by the magnificent beauty of the spherical form, by the regularity of its transformation from shadow to light. So the soap bubble and its spherical form is both the allegoric symbol of the *Vanitas*, of the fragility of human life, and the symbol of perfection, still related to the idea of death.

Who knows if the Eskimos, when they build their igloos, are aware of solving an isoperimetric problem to have inside a construction based on a plane the greater possible volume for the same external surface. Of course there are static reasons too. From the mathematical point of view this problem has a constraint (the plane) in comparison with the *free* problem of blowing a simple soap bubble. The solution is the hemisphere, the igloo, based on a plane, as is very easily verified with a real soap bubble; when the bubble touches the plane, the phenomenon takes place so quickly that only with a slow-motion camera is it possible to see the precise moment when the bubble becomes a hemisphere.

Minimal Surfaces with Constraints in Three Space

The German architect Frei Otto has used soap bubbles and soap films as models to design and construct his well-known *tensile structures*. The work of Otto, as it is described in the volume *Tensile Structures* (Otto 1973), allows one to look to soap films from another point of view: that of the designer. Otto has used soap bubbles and films to study *pneumatic structures*, that is, structures which are under traction, like membranes. Otto points out that "every form which a soap bubble can assume can be obtained as pneumatic structure." Frei Otto considers as particularly important designs which can be obtained by experiments with soap bubbles.

> The knowledge of minimum surfaces is very important for the design of membrane and cable-net structures. However, minimum surfaces are not always the optimum structural shapes. A minimum surface defines only the surface of least area with a closed curve. A minimum surface is identical with a membrane everywhere uniformly stressed in all directions. The minimum surface is also the shape of the most economical surface support system. When additional loads act, the minimum surface is not always the optimum shape (Otto 1973: Vol. 2, 49).[1]

Otto adds:

> It is often necessary to change the form of the minimum surface in such a way that membrane zones requiring higher rigidity also have greater curvature. Of course any deviation from the minimum surface increases the total surface area. Determinations of the minimum surface is essential despite these major reservations (Otto 1973: Vol. 2, 50).

The essential problem for the designer is the measuring of the soap film models, a problem not so easy to solve. The equipe of Frei Otto set up, between 1959 and 1962, a technique to visualize soap film models in order to obtain a precise photogrammetric evaluation. During the years 1960–1964 they made many experiments to test the shape of soap films for different boundary curves. They have solved experimentally many different Plateau problems. In the construction of a membrane the problem is to find the minimum surface that touches the base support and goes up to reach the higher points. All the solutions obtained by Frei

[1] In mathematics the correct name is minimal surfaces instead of minimum surfaces.

Otto follow the rules of Plateau, but he was able to discover many new forms. To simulate these problems it is necessary to consider some fixed points in space (using small supports) and thin threads which connect them and then to dip all in the soapy water. After water has gone away, the soap film is stretched between the threads reaching the equilibrium position of the minimum area. Particularly interesting is the technique which uses the elasticity of soap film to make the film reach the vertices, which simulate the higher points in the membrane that is to be built. Otto has developed a technique to make the soap films get up through a very thin noose thread, called an *eye*. The Institute of Architecture of Otto at the University of Stuttgart was built along the lines of a model created by soap films using the *eye* technique. One of the more recent books of the Equipe of Frei Otto has the title *Seifenblasen—Forming Bubbles*.

The example of Otto is not unique in architecture. Due to their properties, soap bubbles and soap films are used in various fields of design and architecture. The American architect Peter Pearce in his volume *Structure in Nature as a Strategy for Design* (1979) considers soap bubbles as an archetype for any kind of modular structures.

Periodic Minimal Surfaces

In 1865 the German mathematician H.A. Schwarz was the first to solve the Plateau problem for a non-plane boundary, the quadrilateral obtained by choosing four vertices of a tetrahedron. Schwarz built several models of his solution. The surface is a minimal surface as the mean curvature is zero in each of its points, while the two principal curvatures have opposite directions: so the surface is of the type of a saddle surface, in each point concavity balances convexity. To visualize the surface, Hilbert and Cohn-Vossen suggest thinking of the pass of a mountain. An example of surface of this type is the hyperbolic paraboloid. The problem of Plateau solved by Schwarz has suggested many applications of soap bubbles in architecture. The idea is to consider a skew frame, made of rectilinear and curved tracts, like the one considered by Schwarz, to obtain the minimal surface that solves the Plateau problem. In the chapter 'Minimum Surfaces Stretched in Frames Subjected to Bending' of the volume *Tensile Structures*, Frei Otto wrote the following:

> Experiments were undertaken in which soap bubbles were stretched in different frames and measured photographically in order to determine their shapes. The first photos show soap films in skew frames consisting of rods of equal length. The experimental result that a minimum surface in a skew frame is not a hyperbolic paraboloid, was proved mathematically (Otto 1973: 52).

Pearce recalls an important observation on saddle type surfaces:

> Some areas on a saddle surface are flatter than others and, therefore, all points on the surface do not respond equally well to concentrated loads. When such saddle surfaces are associated in a periodic array, the physical interaction of one surface on another produces a compensatory effect which greatly increases their efficiency as structures. An isolated

> saddle polygon is not a fully stable structure. It requires the cooperative effect of associated saddle polygons if its advantages as a doubly curved surface are to be fully realized. This is true whether the surface is a tension system, as in the case of the soap film, or a surface of rigid material capable of resisting tension and compression forces (Pearce 1979: 14).

So it is necessary to put together saddle polygons or surfaces to obtain more stable structures. Schwarz was able to find rules which allow a suitable repetition of saddle-type minimal surfaces in order to obtain a new surface whose generating element is repeated many times. The procedure can be repeated, in principle, infinitely. The surfaces which can be obtained following the rules of Schwarz are called *Infinite Periodic Minimal Surfaces* (IPMS). A systematic description of frames to start the making of the most interesting IPMS can be find in the work of Anderson and Hyde *A Systematic Net Description of Saddle Polyhedra and Periodic Minimal Surfaces* (1984: 221–254).

IMPS without self-intersections can be characterized by the structures of tunnels which pass through them. Pearce has created labyrinthic structures starting from saddle polyedra for a very precise purpose: to amuse children.

> Moving from mathematical abstractions through small scale models to structures of the size used in the installation at the Brooklyn Children's Museum in New York requires considerable attention to details and to the subtleties of three dimensional space. The ideal minimal surface is a structure of zero thickness. The surface modules used in the present application have a typical wall thickness 0.190 inches. The translation of the zero thickness of the ideal minimal surface to a surface module of such a thickness is a technical problem of considerable complexity... Continuity of form is also a major consideration, since an important recreational intent of the curved space structures is to enable children to slide through the tunnel labyrinths. This requires joining systems without sharp edges and of compact physical dimension (Pearce 1979: 231).

There is the need for a labyrinthic structure to be closed in order for it to be stable. If we open it in order to provide an entryway for children and adults we lose some local stability. Moreover the openings made possible by the frames also help ventilate the labyrinth. These technical problems have been solved and in the Brooklyn Museum there are three labyrinthic structures which consist of 118 cells repeating themselves. The basic cell of all the systems is a saddle polyhedron. Children at the Brooklyn Museum make an important jump of quality: from playing with soap bubbles to entering in the labyrinthic structure of periodic minimal surfaces! (Fig. 77.3).

Final Remarks

The geometry of soap bubbles and soap films is of great importance in many fields of scientific research: mathematics, biology, chemistry, physics. Their structures are of great importance in architecture as well.

Not the least important element, as pointed out by Plateau, is the *charme* of soap bubbles and soap films. They are fascinating in the simplicity of their structures. It

Fig. 77.3 P. Pearce, "Labyrinthic Structure", Brooklyn Children's Museum, New York, ©Pearce Struct. Inc. Photo: author

is not by chance that soap bubbles, in addition to their importance in science, have a very long history in the visual arts.

> In my opinion the fascination of the forms derived by the minimal surfaces in design is based on several properties:
>
> 1. the shapes of minimal surfaces can be astonishing by the aesthetical point of view;
> 2. the shapes of minimal surfaces allow the optimal use of materials;
> 3. the structural surfaces with a saddle shape are very stable and resistant;
> 4. the structures of minimal surfaces have a natural geometric rigidity (Almgren 1982: 170).

The author of these words on the importance of Minimal Surfaces Forms in design is not an artist, but a mathematician. Fred Almgren is one of the most famous experts in the Minimal Surfaces Theory. Minimal surfaces is really an interdisciplinary field of great interest.

Biography Michele Emmer is full professor of mathematics at the University of Rome "La Sapienza". He was previously a professor at the Universities of Ferrara, Trento, Viterbo, L'Aquila, Sassari and Venice, and visiting professor, among others, at Princeton, Paris Orsay, Campinas Barcellona and in several Japanese universities. His areas of activity were PDE and minimal surfaces. He is also a filmmaker. His movies in the series "Art and Math" have been broadcast in Italy and abroad. He has organized many exhibitions and conferences on the topic of "Art and Mathematics", including the annual conference on "Mathematics and Culture" at the University of Venice, the exhibitions and conferences on M.C. Escher (1985 and 1998) at the University of Rome, and the section on Space at the Biennal of Venice (1986). He is the editor of *The Visual Mind* (MIT Press, 1993) and *Visual Mind II* (MIT Press, 2005). He is editor of the book series "Mathematics and Culture" and "Imagine Math (Springer)." His most recent book, *Bolle di sapone* (Turin: Bollati, 2009) was awarded a prize as best Italian non-fiction 2010.

References

ALMGREN, F. 1982. Minimal Surfaces Forms. *The Mathematical Intelligencer* **4**, 4 (1982): 164:172.

ALMGREN F. J. and J. E. TAYLOR 1976. The Geometry of Soap Bubbles and Soap Films. *Scientific American* (July 1976): 82–93.

ANDERSON, S. and S. T. HYDE. 1984. A Systematic Net Description of Saddle Polyhedra and Periodic Minimal Surfaces. *Zeitschrift für Kristallographie* **168** (1984): 221–254.

BOYS, C.V. 1959. *Soap Bubbles: Their Colours and the Forces Which Mould Them.* New York: Dover Publications Inc.

EMMER, Michele (with F. Almgren and J. Taylor). 1980. *Soap Bubbles.* Video, 27 minutes. Film and video in the series *Art and Mathematics.* Rome: Film 7 International.

———. 1987. Soap Bubbles in Art and Science: from the Past to the Future of Math Art. *Leonardo* **20**, 4: 327–334. Reprinted in *The Visual Mind: Art and Mathematics*, Michele Emmer, ed. Cambridge, MA.: MIT Press, 1993. pp. 135–142.

———. 1991. *Bolle di Sapone: un Viaggio Tra Arte, Scienza e Fantasia.* Firenze: La Nuova Italia.

———. 2009. *Bolle di sapone tra arte e matematica.* Torino: Bollati Boringhieri.

GERSTER, G. 1975. *La Terre de l'Homme: Vues Aériennes.* Zürich: Atlantis.

HULTSCH, F. 1933. *Pappi Alexandri Collectionis Quae Supersunt e libris Manuscriptis Editis.* Berlin: Weidmann.

KAUFMANN, E. 1978. *Trois Architectes Révolutionnaires: Boullée, Ledoux, Lequeu.* Paris: SADG.

LEDOUX, C. N. 1804. *L'Architecture Considérée sous le Rapport de l'Art, des Moeurs et de la Lé gislation.* Paris.

OTTO, F. 1973. *Tensile Structures: Design, Structure and Calculation of Buildings of Cables, Nets and Membranes.* Boston: The MIT press.

PEARCE, P. 1979. *Structure in Nature is a Strategy for Design.* Cambridge, USA: The MIT Press.

PLATEAU, Joseph. 1873. *Statique Expèrimentale et Théorique des Liquides Soumis aux Seules Forces Moléculaires* Paris: Gautiher-Villars.

TAYLOR, Jean E. 1976. The Structure of Singularities in Soap-bubbles-like and Soap-film-like Minimal Surfaces. *Annals of Mathematics* **103** (1976): 489–539.

Chapter 78
Aperiodic Tiling, Penrose Tiling and the Generation of Architectural Forms

Michael J. Ostwald

Introduction

In September of 1995 the Australian architectural practice Ashton Raggatt McDougall (ARM) invited the eminent mathematician Roger Penrose to open their soon to be completed refurbishment of the historic Storey Hall complex of buildings at the Royal Melbourne Institute of Technology. Penrose, who admitted that the design seemed "extremely exciting" (Penrose 1996), regretfully declined on the grounds that he was already overcommitted to too many projects to visit Australia at the required time. He concluded his response to the invitation with an enigmatic postscript which records that he is currently working on "the single tile problem" and recently "found a tile set consisting of one tile together with complicated matching rule that can be enforced with two small extra pieces" (Penrose 1996). This postscript contains the first clue to understanding the mysterious connection between Penrose and Storey Hall, between a scientist and a controversial, award-winning, building.

Storey Hall is significant for many reasons but only one prompted ARM to invite Penrose to open it. The newly completed Storey Hall is literally covered in a particular set of giant, aperiodic tiles that were discovered by Roger Penrose in the 1970s and have since become known as Penrose tiles. While architecture has, historically, always been closely associated with the crafts of tiling and patterning, Storey Hall represents a resurrection of that tradition.

First published as: Michael J. Ostwald, "Aperiodic Tiling, Penrose Tiling and the Generation of Architectural Forms", pp. 99–111 in *Nexus II: Architecture and Mathematics*, ed. Kim Williams, Fucecchio (Florence): Edizioni dell'Erba, 1998.

M.J. Ostwald (✉)
School of Architecture and Built Environment, The University of Newcastle, Callaghan, NSW 2308, Australia
e-mail: michael.ostwald@newcastle.edu.au

K. Williams and M.J. Ostwald (eds.), *Architecture and Mathematics from Antiquity to the Future*, DOI 10.1007/978-3-319-00143-2_31,

But what is Penrose tiling and what does it have to do with architecture in general and Storey Hall in particular? This chapter provides an overview of the special properties and characteristics of Penrose tilings before describing the way in which they are used in ARM's Storey Hall. The purpose of this binary analysis is not to critique Storey Hall or its use of aperiodic tiling but to use ARM's design as a catalyst for taking the first few steps in a greater analysis of Penrose tiling in the context of architectural form generation.

Periodic and Aperiodic Tilings

The geometers Grünbaum and Shephard record that the "art of tiling must have originated very early in the history of civilization" because, with the very first attempt to "use stones to cover the floors", humanity "could be said to have begun tiling" (Grünbaum and Shephard 1987: 1). Throughout history tiling has always been associated with architecture. From the moment the earliest primitive hut was floored with woven matting, walled with masonry, or carved with geometric patterns, it became, in mathematical terms, a tiled structure. The reason for this is that mathematical tiling is not defined by the craft of combining materials but by the repetitious creation of patterns formed through the application of a set of usually polygonal shapes. For this reason cut-stone mosaics are as much examples of mathematical tiling as painted frescos or carved Celtic knot-work. Any system of geometric patterning that covers or fills a surface using a finite set of shapes is considered tiling in a mathematical sense.[1] One of the reasons Grünbaum and Shephard suggest that the art of architecture has always been closely associated with the craft of tiling is that tiling or patterning adds richness, through ornamentation, to the surface of a building. Such ornamentation is not simply valued for aesthetic reasons but also for symbolic, practical and monetary purposes. At any point in history, Grünbaum and Shephard argue, "whatever kind of tiling was in favour, its art and technology always attracted skilful artisans, inventive practitioners and magnanimous patrons" (1987: 1). However, while many treatises have been written throughout history on the formation of architecture through geometric principles, such works rarely consider the relationship between tiling and architecture.[2] Further, despite Johannes Kepler's analysis of tiling patterns in his 1619 work *Harmonice Mundi*, a rigorous and scientific approach to understanding the properties of tilings has been formulated only in the last few decades.

[1] In theory it does not matter how large the set of tiles is. An infinite set of different shapes that fills a plane is still a form of tiling although an unconventional kind.

[2] Whereas minor or subtle references to the art of tiling may be discerned in various translations of the works of Vitruvious as well as in those of Alberti, Vignola and Serlio, even such minor references are increasingly rare in the treatises that followed; see Kruft (1994).

Charting the rise in enthusiasm for geometric tiling patterns to the works of Hao Wang in the early 1960s has become a veritable truism in mathematics (Penrose 1990: 174). In 1961 the philosopher Hao Wang became interested in questions of pattern recognition in the use of geometry as a tool for symbolic logic. Wang wanted to determine if, given a set of polygonal shapes, there is a procedure for determining whether or not they will tile a plane in such a way that they will necessarily repeat their configuration. Tilings that repeat their configuration or that display multiple lines of symmetry are usually called *periodic tilings*. The most recognisable periodic tilings are based on sets of squares, rectangles, trapezoids or parallelograms. In order to examine the question of whether or not a procedure exists for determining if a set of shapes will tile periodically, Wang developed a set of square tiles, each with different coloured edges. The edges of Wang tiles are only allowed to join other edges of identical colour—they may not be rotated or reflected, only translated. Wang conjectured that if *aperiodic tiles* (tiles that do not repeat their patterns) exist, then he could not derive a decision procedure whereby a given set of tiles will periodically tile a plane. Conversely, if a decision procedure could be determined, then there was no such thing as an aperiodic system of tiling (Rubinsteim 1996: 20–21).

In order to clarify Wang's agenda, it must be understood that there are two types of aperiodic tiles. There are sets of tiles that can fill a plane both periodically and non-periodically, and there are sets of tiles that only fill a plane non-periodically.[3] An example of the former is Gardner's set of quadrilaterals that tile both periodically and non-periodically; the choice is up to the person placing the tiles (Figs. 78.1 and 78.2). There are countless examples of geometric sets of this kind; Penrose frequently uses Marjorie Rice's 1976 single tile set to explain this idea (Fig. 78.3). Despite this, the sets of shapes that are conventionally referred to in mathematics as aperiodic are usually those that can fill a plane only non-periodically, or those that are necessarily non-periodic. This latter category of shapes is the one with which Wang was primarily concerned.

In 1965 Robert Berger developed Wang's thesis to prove that there is no decision procedure for tiling surfaces periodically and, thus, there must be a set of aperiodic tiles in existence. Following this realisation Berger set about finding the first set of aperiodic tiles. The tiling system he discovered, comprising a set of more than 20,000 different shapes, was exhibited the following year. However, Berger's tiling system was based on a peculiarity of logic, and in the following years a number of mathematicians produced increasingly less numerous sets of tiles that would fill a plane aperiodically. In 1967 Berger himself lowered the number of tiles from 20,426 to 104 and, in 1968, Donald Knuth further reduced the set to 92. Yet, in

[3] It must be noted that there is some confusion surrounding the terminology "aperiodic" and "non-periodic" as both terms are used interchangeably in popular mathematics and science. This chapter generally conforms to the wording used in Grünbaum and Shephard's encyclopedic work *Tilings and Patterns* (1987) and uses "aperiodic" as an accurate description of the properties of a tile set and "non-periodic" only when quoting from another work or when using the term as a broad, non-definitive, descriptor.

Fig. 78.1 Gardner's set of quadrilateral tiles (with a square period) used periodically. Image: author

Fig. 78.2 Gardner's set of quadrilateral tiles (still with a square period) used aperiodically. Image: author

1971, a more dramatic reduction occurred when Raphael Robinson both allowed the set of tiles to be rotated and reflected and then removed colour altogether from the tiles. Instead of colour, Robinson used a series of geometric additions and indentations to ensure that certain edges could be combined while others couldn't. In this way Robinson reduced the set of tiles, from Knuth's 92 to just 6 (Figs. 78.4 and 78.5). In essence, Robinson's tiles are still Wang tiles because they are still based on a square tiling pattern, and for this reason they represent the minimum possible set of aperiodic tiles founded on an underlying square period. However, while the square tiling period has a minimum limit of six tiles, Penrose proposed in 1973 that by using a parallelogram tiling period the set could be

Fig. 78.3 Rice's single polygon set tiling a surface periodically. Image: author

Fig. 78.4 Robinson's six tiles set. Image: author

reduced to just two tiles. Moreover, Penrose then proposed two sets of two tiles that each could be tiled only aperiodically.

The first set of "Penrose tiles", named by the mathematician John Conway the "darts and kites" set, is derived from a rhombus (or parallelogram) with four sides of equal length (*length* = ϕ) with obtuse angles of 108° and acute of 72°. A line is then drawn between the acute corners of the rhombus (bisecting each of these into two angles of 32°) and a length equal to the length of a typical side of the rhombus (i.e., ϕ) measured along this line (Figs. 78.6 and 78.7). The new point created in this way is connected to the remaining obtuse corners of the rhombus. The rhombus is then cut along these two lines creating a kite form (with angles of 72°, 72°, 72°, and 144°) and a dart form (with angles of 36°, 72°, 36° and 216°). Then, if the two

Fig. 78.5 Robinson's tiles displaying forced aperiodicity. Image: author

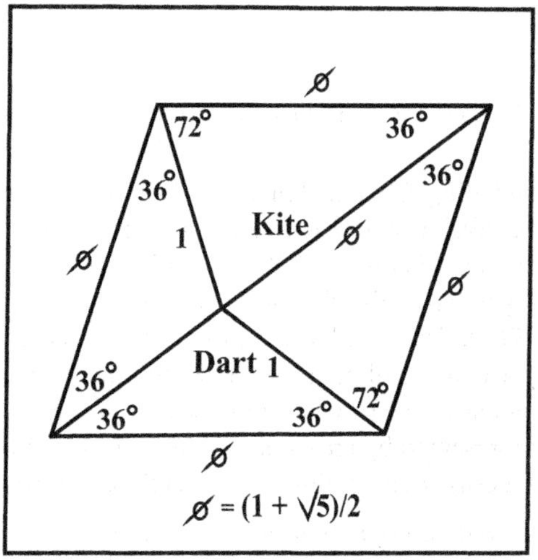

Fig. 78.6 Construction of Penrose dart and kite tiles. Image: author

cutting lines that connect the obtuse corners are of length 1, the other lengths created are reflections, proportionally, of the golden mean (i.e., $\phi = \frac{1+\sqrt{5}}{2}$). Finally, the two forms thus created, the kite and the dart, are coloured or indented to ensure that they may only be connected to certain other surfaces and thus tile only aperiodically.

An intriguing property of the "darts and kites" set of Penrose tiles is that if an infinitely large surface is to be tiled, $\frac{1+\sqrt{5}}{2}$ (approximately 1.618) times as many kites as darts is required; in other words, the ratio of darts to kites is the golden mean.

The first tile of the second Penrose set of aperiodic tiles is identical to the starting rhombus used to construct the first pair. That is, the first tile is a rhombus with sides

Fig. 78.7 Penrose dart and kite tiles displaying forced aperiodicity (note that the *coloured patches* and indents forcing the patterns are removed for clarity). Image: author

of equal length (*length* $=\phi$), with obtuse angles of 108° and acute of 72°. The second rhombus tile also has four sides of length equal to those of the first but with obtuse angles of 144° and acute of 36°. These are then modified with colours, shades or indentations to ensure that they tile aperiodically (Figs. 78.8 and 78.9). This second set similarly has proportions and ratios that reflect the characteristics of the golden mean. This close relationship between both sets of Penrose tiles and the golden mean may be more readily appreciated by closely examining a Pythagorean pentagram at multiple scales (Fig. 78.10).

One special characteristic of Penrose tiling patterns is that they exhibit *quasi-symmetry*. Normally any type of symmetry in a tiling pattern would render a set periodic, but Penrose tiles display partial, not complete, symmetry through rotation at 72°. For this reason Penrose tilings are said to be quasi-symmetrical. This characteristic is important because until 1984 it was believed that all crystalline materials must be based on lattices with conventional periodic symmetries. Crystals exhibit rotational symmetry at only 2, 3, 4 and 6 rotations. However, in 1984 Dany Schechtman discovered an aperiodic crystalline structure in aluminium-manganese by electron micrography. This crystalline structure, which was called a *quasicrystal*, almost possessed fivefold symmetry in much the same way that Penrose tiling patterns are almost symmetrical. Although, as Cracknell records, fivefold symmetry in crystals had been discovered as early as 1966, it was commonly believed that crystals could not have fivefold symmetry as such was unknown in morphological crystallography (Cracknell 1969). For Gardner, the discovery of the quasicrystal had great repercussions in science:

> Among physicists, chemists and crystallographers the effect of this discovery was explosive. Similar nonperiodic structures were soon being induced in other alloys, and dozens of papers began to appear. It became clear that solid matter could exhibit nonperiodic lattices with any kind of rotational symmetry. Wide varieties of solid tiles in sets of two or more

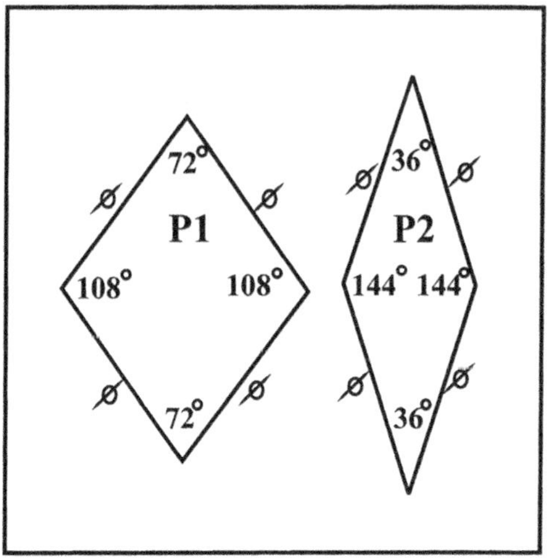

Fig. 78.8 Construction of Penrose twin parallelogram tiles. Image: author

Fig. 78.9 Penrose twin parallelogram tiles displaying forced aperiodicity (note that the coloured patches and indents forcing the patterns are removed for clarity). Image: author

were proposed as models, some forcing nonperiodicity, some merely allowing it (Gardner 1989: 25).

Fivefold quasi-symmetry defies the laws of crystalline restriction that suggests that crystalline lattices cannot posses fivefold symmetry. However, as Stewart and Golubitsky maintain, "quasicrystals 'almost repeat' their structure, and *can* have axes of fivefold symmetry" (Stewart and Golubitsky 1993: 95). Nevertheless, despite these recent developments in the geometry and mathematics of aperiodic tiling patterns, Penrose tiles are still widely regarded as examples of "recreational mathematics" (Gardner 1989). While Penrose tilings might possess some latent ability to describe crystal anomalies or "symmetry-breaking" in natural inorganic

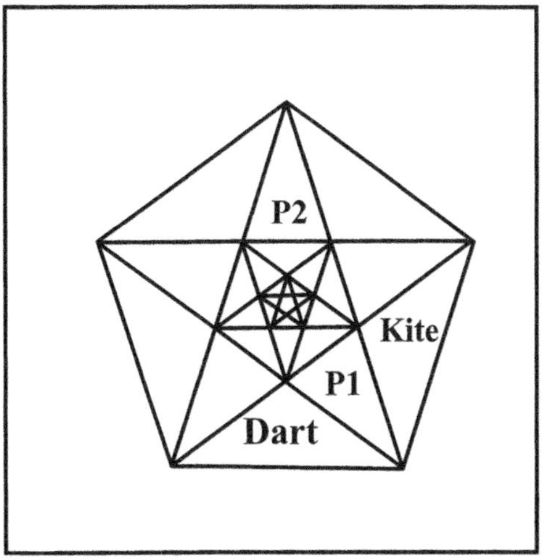

Fig. 78.10 The Pythagorean Pentagram displaying all four Penrose tiles at different scales. Image: author

forms (Stewart and Golubitsky 1993), they are still largely without clear application in any specific field.

Storey Hall

Storey Hall is a large auditorium with ancillary spaces that are built into the shell of an existing, historic structure (Figs. 78.11 and 78.12). The building was opened in late 1996 and has since been awarded a number of state and national design awards. The most contentious aspect of the design is the way in which ARM have carefully restored and retained parts of the original Victorian building, only to combine them with a glaringly modern, brightly coloured, geometric addition. The architectural designer and critic Norman Day describes the original Victorian detailing as being almost completely overwhelmed by a complex applique of "white, green, pink, purple and red panels, with inset green neon lights and great panels of white translucent and reflected light fittings" (Day 1995: 36). These tiles, which overlay both the façade of the building in Swanston street as well as most of the interior spaces, are all the second set of Penrose tiles (the twin parallelograms) marked in accordance with Penrose colouring (not Penrose indentations). In the words of ARM, on the exterior of the building the Penrose tiles are "shrouded in veil and drapery, folded sash, delicious lace, and strong rope lines, marking the inner boundaries of Penrose's mysterious geometry" (Ashton Raggatt McDougall 1996: 9). On the inside of the building, particularly in the auditorium space, "the tile is made to iterate its own pattern as if in multiple dimensions."

ARM assert that in the design of Storey Hall they are using Penrose tiles as part of a twofold strategy of transfiguring the dominant Euclidean geometry of the existing structure and as a symbol for the power of the "new sciences." For ARM

Fig. 78.11 Storey Hall, designed by ARM. Photo: courtesy John Gollings

Fig. 78.12 Storey Hall, designed by ARM. Photo: courtesy John Gollings

the use of Penrose tiles is not simply a reference to plane-filling geometry but to the greater conceptual shift associated with what they describe as the "new mathematics." This is a claim that various critics read as suggesting that there is a connection between Penrose tiling and fractal geometry. For example, Day claims that the Penrose tiles must be read in concert with the form of the facade itself, which is a "site-poured concrete wall" that is twisted "according to the new geometry of complexity" (Day 1995: 40). Inside the building the act of reading the meaning behind the giant Penrose tiles is further complicated because the tiles are layered with a range of other icons from science and geometry (Kohane 1996: 8–15). Norman Day suggests that amidst the geometry of Penrose there are also signs that lead to "chaos theory, Walter Burley-Griffin, urbanism, the sexual revolution, feminism, Einstein's grotto, Plato's cave, X and Y chromosomes, the Vault sculpture, paradoxes [and] contextualism" (Day 1995: 37).

But are the Penrose tiles significant in any way or are they simply icons for geometric knowledge? When the author of this chapter questioned Charles Jencks on the validity of applying mathematical forms to the exterior of an historic building, Jencks replied that the more "transformational" the use of iconography, the more interesting the outcome. For Jencks, while Storey Hall "doesn't really use the Penrose tiling pattern in section or plan ... it still uses it importantly in wall depth, in all-over ornament, and iconographically" (Ostwald et al. 1996: 30). All of these uses, Jencks argues, are appropriate means of creating architectonic forms. But doesn't this type of use border on simple "applique" or the "purely ornamental"?—seemingly not for Jencks. There is more happening in Storey Hall, according to Jencks, than simple conversion of Penrose tiles to architectural ornamentation. "Look at the lighting," he says, "and the way it relates to the old Victorian building. Its musculature is similar to the Victorian building and it has the same information density". If it were simply "applique, or applied ornament" it would be "less interesting than something that has greater organisational depth" (Ostwald et al. 1996: 30).

ARM's use of Penrose tiles as a generator of architectural form seems to reflect not so much a close reading of topographical Mathematics but an awareness of the role of these ideas in the understanding of quasicrystals and, less directly, as a connection to fractal geometry. Leon van Schaik (1996: 5) describes Storey Hall as "an architecture which works through the contemporary mathematics of surface". For van Schaik, the Penrose tiles refer not only to topographical mathematics but to an "unfolding symphony of forms that envelop us in an encounter with the spatiality of the new mathematics". When viewed in this way, the architectural forms generated through the use of Penrose tiles are simply extensions of the historic relationship between architecture and tiling, a relationship that otherwise seems, in recent years, to be lacking in creativity.

Penrose Tiling and Architecture

Despite their possible use in the interpretation of quasicrystals, Penrose tiles are still simply plane-filling patterns with a few unusual properties. Moreover, these unusual properties are not in any obvious way particularly relevant to architecture.[4] Ultimately, in the context of architecture, a periodic tile set is the same as an aperiodic set; the choice of using one or the other is simply aesthetic. Yet there are two recent developments in tiling geometry that have occurred as a result of Penrose's discoveries that seem potentially more profitable for the development

[4] Although Robbin's arguments to the contrary are very intriguing, this author remains sceptical; see Robbin (1990: 140–142).

of architectural forms. The reason these two developments seem more useful is that they promote the understanding of a spatial dimension in aperiodic tiling as well as a topological one. In 1976 the mathematician Robert Ammann proposed that a two-component set could be devised that would tile space aperiodically. This means that instead of being a "plane-filling" system, Ammann's tiles are "space-filling." Significantly, this same system was independently discovered at around the same time by the Japanese architect and geometer Koji Miyazaki (Miyazaki 1977). Ammann's aperiodic space-filling tiles are a pair of rhombohedra formed by creating two solids, each of which have six sides that are all the same as Penrose's starting rhombus for the formation of the dart and kite set. That is, each of the surfaces of space-filling tiles is a rhombus with sides of equal length and with obtuse angles of 108° and acute of 72°. The two solids produced in this way bear an uncanny resemblance to the basic geometry of Peter Eisenman's axonometric model for House X as well as his House El Even Odd project (Eisenman 1982, 1995). When coloured or modified in a certain way, Ammann's tiles will only fit together in three dimensions, aperiodically.

One final discovery in the geometry of tiling—a discovery that is rather more complex and is thus necessarily described in a very superficial manner here—concerns forced holes in tiled planes. Significantly, in order to describe the presence of forced holes in Penrose aperiodic tilings, mathematicians have resorted to the use of architectural metaphors (Ostwald and Moore 1995; Ostwald et al. 1997). For instance, John Conway describes the discovery of hole theory as akin to imagining "a vast temple with a floor tessellated by Penrose tiles and a circular column exactly in the centre. The tiles seem to go under the column. Actually, the column covers a hole that can't be tessellated" (Gardner 1989: 26–27). Certain combinations of Penrose tiles (and indeed any necessarily aperiodic tilings) can force areas that are unable to be tiled. Conventionally this type of error is rectified by removing a number of surrounding tiles and reworking the pattern until there are no holes. But if holes are formed, they impact on the greater pattern in many subtle and significant ways. Holes in tessellated planes, like space-filling aperiodic tiles, are emphatically spatial systems of geometry that broach many possible connections between architecture, Penrose tilings and other aperiodic tilings. These two aspects of aperiodic tiling warrant further investigation in architecture.

Biography Michael J. Ostwald is Professor and Dean of Architecture at the University of Newcastle (Australia) and a visiting Professor at RMIT University. He has previously been a Professorial Research Fellow at Victoria University Wellington, an Australian Research Council (ARC) Future Fellow at Newcastle and a visiting fellow at UCLA and MIT. He has a PhD in architectural history and theory and a DSc in design mathematics and computing. He completed post-doctoral research on baroque geometry at the CCA (Montreal) and at the Loeb Archives (Harvard). He is Co-Editor-in-Chief of the *Nexus Network Journal* and on the editorial boards of *ARQ* and *Architectural Theory Review*. He has authored more than 300 scholarly publications including 20 books and his architectural designs have been published and exhibited internationally.

References

ASHTON RAGGATT McDOUGALL. 1996. New Patronage. In Ashton Raggatt McDougall, eds. *RMIT Storey Hall*, Melbourne: Faculty of Environmental Design and Construction, RMIT.

CRACKNELL, Arthur. 1969. *Crystals and their Structure*. London: Pergamon.

DAY, Norman. 1995. Storey Hall. *Architecture Australia* **85**, 1 (1995): 36.

EISENMAN, Peter. 1982. *House X*. New York: Rizzoli.

———. 1995. *Eisenman Architects: Selected And Current Works*. Mulgrave: Images Publishing.

GARDNER, Martin. 1989. *Penrose Tiles To Trapdoor Ciphers*. New York: W. H. Freeman.

GRÜNBAUM, Branko and Geoffrey Colin SHEPHARD. 1987. *Tilings and Patterns*. New York: W. H. Freeman.

KOHANE, Peter. 1996. Ashton Raggatt and McDougall's Imitative Architecture: Real and Imaginary Paths through Storey Hall. *Transition* **52/53** (1996): 8–15.

KRUFT, Hanno Walter. 1994. *A History of Architectural Theory from Vitruvius to the Present*. In R. Taylor, E. Callander, and A. Wood, trans. New York: Princeton Architectural Press.

MIYAZAKI, Koji. 1977. *On Some Periodical and Non-Periodical Honeycombs*. Kobe: University Monograph.

OSTWALD, Michael J., and R. John MOORE. 1995. Mathematical Misreadings in Nonlinearity: Architecture as Accessory/Theory. In Mike Linzey, ed. *Accessory/Architecture*. Vol. 1. Auckland: University of Auckland. 1995.

———. 1997. Unravelling the Weave: an Analysis of Architectural Metaphors in Nonlinear Dynamics. *Interstices*. Vol. 4 (1997) CD-ROM.

OSTWALD, Michael J., Peter ZELLNER and Charles JENCKS. 1996. An Architecture Of Complexity. *Transition* **52/53** (1996): 30.

PENROSE, Roger. 1990. *The Emperor's New Mind: Concerning Computers, Minds, and the Laws of Physics*. London: Vintage.

———. 1996. Fax to Howard Raggatt. In Ashton Raggatt McDougall, eds. *RMIT Storey Hall*. Melbourne: Faculty of Environmental Design and Construction, RMIT.

RUBINSTEIM, Heim. 1996. Penrose Tiling. *Transition* **52/53** (1996): 20–21.

ROBBIN, Tony. 1990. Quasicrystals for Architecture: The Visual Properties of Three-Dimensional Penrose Tessellations. *Leonardo* **23**, 1: 140–141.

STEWART, Ian and Martin GOLUBITSKY. 1993. *Fearful Symmetry: Is God A Geometer?* London: Penguin.

VAN SCHAIK, Leon. 1996. Preface. In Ashton Raggatt McDougall, eds. *RMIT Storey Hall*. Melbourne: Faculty of Environmental Design and Construction, RMIT.

Chapter 79
Paving the Alexanderplatz Efficiently with a Quasi-Periodic Tiling

Ulrich Kortenkamp

Introduction

In this chapter we describe a quasi-periodic tiling that was entered in a 2004 competition for the landscaping re-design of the Alexanderplatz in Berlin. Although the tiling was only part of the whole design, we will concentrate here on the reasons for using it, its mathematical properties, and how it could be realized, and we will neglect all other aspects. The concept won one of the three first prizes in the competition, but was not chosen for realization.

The area that had to be redesigned is located in the heart of Berlin, a metropolis of 3.5 million inhabitants, which is still challenged by the process of the reunification of Germany. Its architectural history is quite interesting and probably well known to most readers.[1] A concise article that gives a brief historic summary can be found in the German *Wikipedia* at http://www.de.wikipedia.org/wiki/Alexanderplatz_(Berlin).

The most challenging issue in the landscaping is the wide open area of 26,000 m^2. One requirement postulated in the competition guidelines was that this area had to be maintained. However, as the area is neither rectangular nor quadrangular, but pentagonal (Fig. 79.1), it is not possible to use a customary orthogonal tiling without favoring one side of the area over the other. Also, for draining purposes tiling is preferred to a asphalt surface. Further, the area was to

First published as: Ulrich Kortenkamp, "Paving the Alexanderplatz Efficiently with a Quasi-Periodic Tiling", pp. 57–62 in *Nexus VI: Architecture and Mathematics*, Sylvie Duvernoy and Orietta Pedemonte, eds. Turin: Kim Williams Books, 2006.

[1] A concise article that gives a brief historic summary can be found on *Wikipedia*: http://www.en.wikipedia.org/wiki/Alexanderplatz

U. Kortenkamp (✉)
Martin-Luther-Universität, Institut für Mathematik, 06099 Halle (Saale), Germany
e-mail: ulrich.kortenkamp@mathematik.uni-halle.de

K. Williams and M.J. Ostwald (eds.), *Architecture and Mathematics from Antiquity to the Future*, DOI 10.1007/978-3-319-00143-2_32,

Fig. 79.1 Image: author

remain usable for small-wheeled vehicles such as baby carriages, scooters, inline skates and skateboards; this prohibits the use of small paving stones.

The landscape architects, Landschaftsarchitekturbüro Mettler of Berlin, who were consulting the mathematics department of the Technical University of Berlin, asked if we could provide a tiling that fulfils several properties:

1. All tiles should be 5-, 6- or 7-gons, each approx. 1 m^2 in size;
2. There should be no periodicity at all;
3. There should be only a few different types of tiles;
4. The layout should be easily executed by the construction team.

While the first two properties were due to the design concept of the architects, the others were owing to financial constraints.

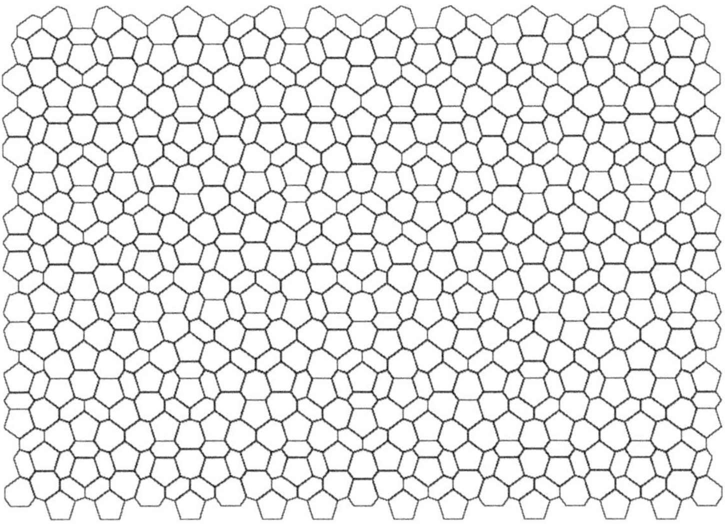

Fig. 79.2 Image: author

A Mathematical Solution

Of course, as soon as non-periodic, irregular, but automatically generated tilings are involved, a mathematician thinks of Penrose tilings and the like. Unfortunately, Penrose tilings are made from quadrangles, so we could not come up with a solution immediately, but we forwarded the request to Prof. Ludwig Danzer of Dortmund, who is a well-known expert in all sorts of tilings. He suggested several, and the landscaping architects chose one from them that seemed to be very well suited (Fig. 79.2). It consists of 5-, 6- and 7-gons, and uses only four types of tiles. It is non-periodic, and its flowery appearance is appealing (a television tower is next to the Alexanderplatz, so it would have been possible to view the tiling as a whole from 200 m above).

All seemed perfect, and there were still a few days left before the final plans had to be submitted. A student of Danzer started to draw a larger version of the tiling that could be included in the electronic version of the plan. A quick calculation shows that approximately 26,000 tiles are necessary for the whole area a mere number of 75,000 line segments to draw! Soon it became clear that it is not possible to finish this task by hand in time. However, we received no information on the rules that were behind the preliminary design. We could not find any publications that could help in analyzing the tiling, so we had to re-engineer it on our own.

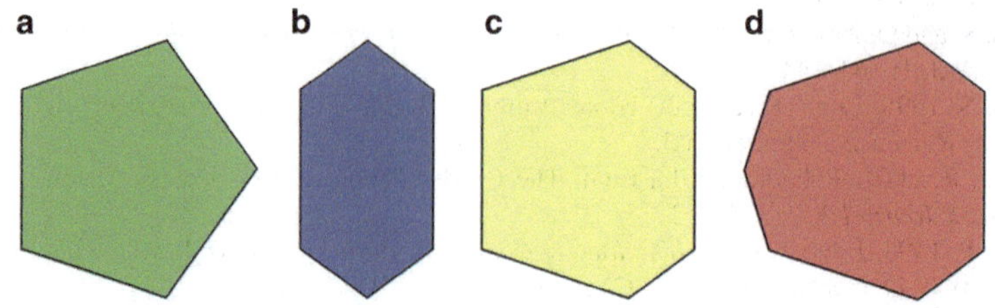

Fig. 79.3 Image: author

Re-Engineering the Tiling

Before we describe the re-engineering, we want to describe the four tiles that were used in the tiling.

(a) The Pentagon (Fig. 79.3a). The regular pentagon is reflecting the symmetry of the pattern and is responsible for its non-periodicity. The length *a* of its side can be chosen arbitrarily and thus sets the global scale of the tiling. The inner angle is 108°.

(b) The flat hexagon (Fig. 79.3b). This non-regular hexagon is used, for example, to surround a single regular pentagon. This description also shows that the inner angles are $360° - 108°/2 = 126°$. The long side *a* fits the regular pentagon, the short sides have length *b*, where *a*:*b* is ϕ (ϕ being the golden ratio).

(c) The hexagon (Fig. 79.3c). This hexagon can be created from the pentagon and the flat hexagon by using three sides of the pentagon and half of the flat hexagon. The lengths of sides and the inner angles are determined by this data. This tile is used rarely in the design, and is easily confused with the heptagon, tile D.

(d) The heptagon (Fig. 79.3d). Finally, the heptagon introduces a third inner angle, 144°. We can create the heptagon from the hexagon by placing to flat hexagons on it. These two create a small "roof" on top of the middle side that came from the pentagon. Again, we only use the two side lengths that were already used in the hexagons.

These four tiles are fun to play with, but it is far from trivial to create a large area without holes. It is definitely necessary to find the mathematical rules that create the tiling.

From Penroseto Danzer There is a striking similarity between dart/kite Penrose tilings and the Danzer tiling. Both have a kind of local fivefold rotational symmetry. So we started with a Penrose tiling (Fig. 79.4) and compared it to the Danzer tiling. Unfortunately, it was not straightforward to match these two this is not very surprising, because of the local-global structure of a quasi-periodic tilings. Quasi-periodic means that any local sub-configuration of a tiling can be found an infinite number of times in the (infinite) tiling of the plane, but the tiling as a whole is not periodic, i.e., it cannot be translated to match itself again.

Fig. 79.4 Image: author

In other words, even if we find a local part in one quasi-periodic tiling that matches another local part in another tiling, it is very likely that this matching does not extend beyond this local area. In our example, we can match a five-star in the Penrose tiling easily to a pentagonal flower in the Danzer tiling, but this will not tell us how to define rules that matches a full Penrose tiling to a full Danzer tiling.

We asked Danzer again, and he gave us a little bit more information: We "just" had to mark some points on each of the rhombic tiles, and this will give us the vertices of the Danzer tiling (Fig. 79.5). Then "it should be clear how to connect the dots." Actually, it was easy for a human to connect the dots on a printout, but it still was not clear how to do this automatically!

We managed to automate this step by marking the tiles with segments instead of only points (Fig. 79.6). Apparently the line segments are invariant of the placement of the tile. Using this observation we could generate drawings of 100,000 of tiles using the well-known recursive methods for Penrose tilings.

Direct Generation Still, we were not satisfied with the solution. We could create drawings, but the structure of the tiles was not represented at all. In particular, we could not count the number of tiles of each different type, and we had incomplete tiles along the border of the tiling. Also, we could not create lists from this data that could be used for placing the real tiles.

As we knew that there is a recursive scheme for creating the Penrose tiling, we were looking for a similar scheme for the Danzer tiling. The generation of Penrose tiling is usually based on triangles (both rhombi are made of two congruent triangles, which is also true for dart/kite-based Penrose tilings) that are subdivided into three smaller triangles in each recursion step. This method could

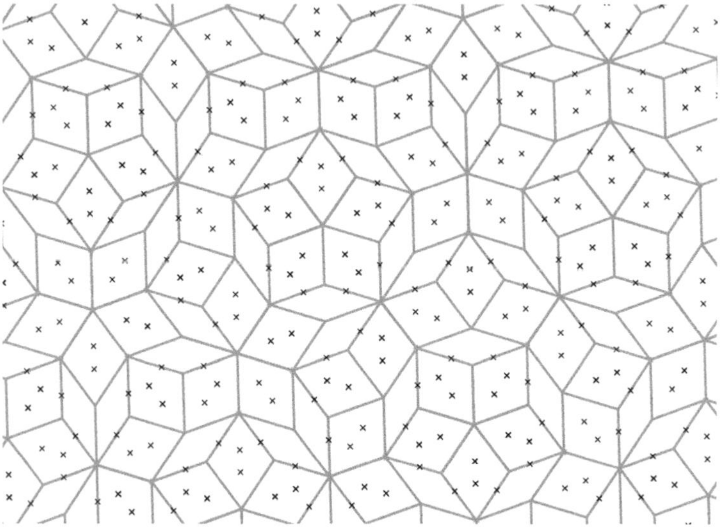

Fig. 79.5 Image: author

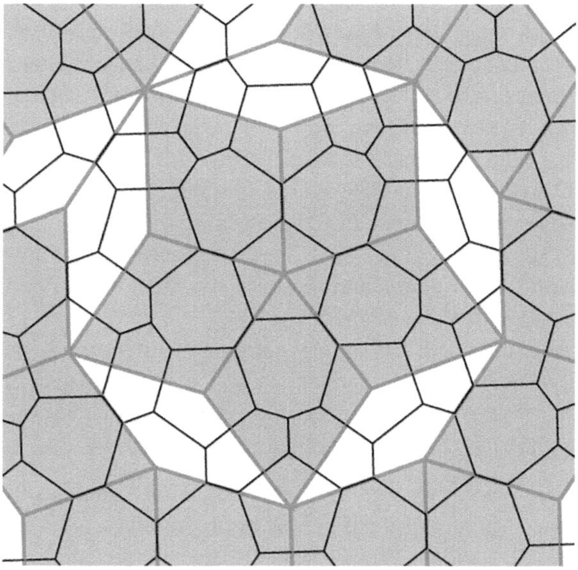

Fig. 79.6 Image: author

not be transferred: we could not find a simple subdivision into a few, say two to four, triangles that can be combined to create each of the four original tiles.

We used another method for identifying "subdivision" rules. Using the Penrose-based tiling software, we produced tilings that were produced using

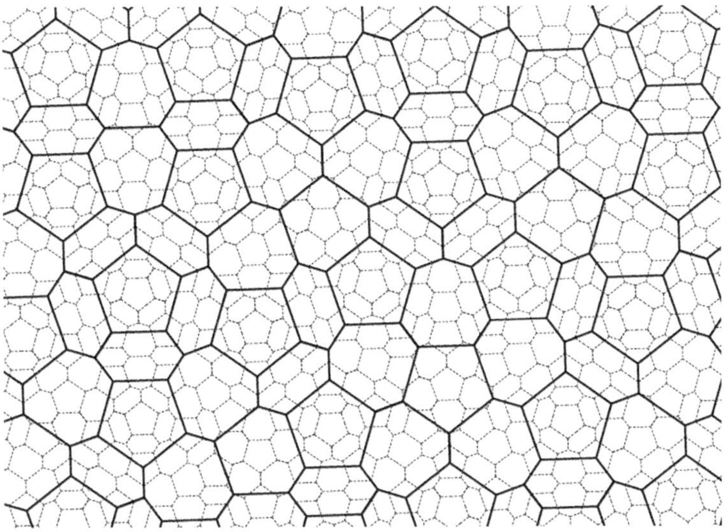

Fig. 79.7 Image: author

a certain number of recursion steps. By superimposing two drawings produced with n and $n+2$ recursion steps (Fig. 79.7), we could identify a first set of replacement rules for the Danzer tiles. Because applying the Penrose replacement rules once not only scales the tiles, but also translates the original tiling, it was easier to compare a tiling with the next one after. The translation is, in a way, undone.

Further investigation then showed that the four rules we read off the drawings were incomplete, or better: wrong. They worked for a few iterations, but then they produced holes in the tiling, and overlapping tiles. Comparing the tilings with the ones produced using the Penrose recursion showed that sometimes other rules had to be used for the pentagon and the heptagon. We ended up with six rules for six tiles, where two pairs of tiles are indistinguishable. The recursion rules are shown in Fig. 79.8.

Once we have found these rules, we can also easily calculate the expected number of tiles for each type, which is given as a non-negative eigenvector of the transition matrix below:

	P a	P b	Hex1	Hex2	Hep a	Hep b
P a	6	0	0	5	0	0
P b	0	1	0	0	5	0
Hex1	0	1	0	0	5	0
Hex2	0	0	0	2	1	1
Hep a	1	0	1	2	3	0
Hep b	0	1	0	0	5	0

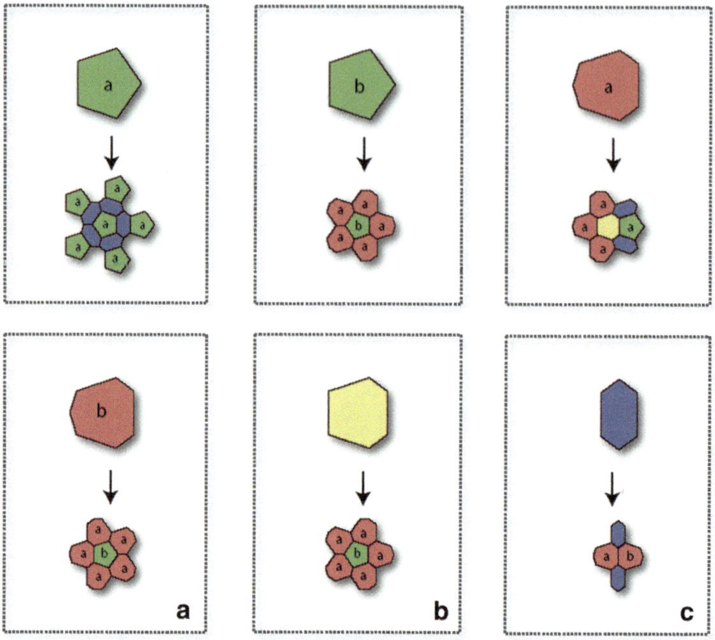

Fig. 79.8 Image: author

For the eigenvalue $\frac{7}{2}+\frac{3}{2}\sqrt{5}$ of this matrix we find the eigenvector

$$\left[9+4\sqrt{5}, 1, \sqrt{5}, 1/2 * \left(25+11\sqrt{5}\right), 1/2 * \left(15+7\sqrt{5}\right), 1/2 * \left(5+\sqrt{5}\right)\right].$$

In the limit case (i.e., for the whole plane), the ratios of tiles are thus:

(P1 + P2)	:	(Hex1)	:	(Hex2)	:	(Hep1 + Hep2)	=
10 + 4√5	:	√5	:	1/2*(25 + 11√5)	:	10 + 4√5	≈
18.94	:	2.24	:	24.80	:	18.94	

We see that the number of pentagons is equal to the number of peptagons (if we consider the whole plane, which is, of course, impossible). We also see that the "round" hexagons are very rare, only about 1/28 of the tiles are of this type.

Realization

Finally, we wanted to be able to create an easy description of the tiling that could be used by the construction team. Given the immense weight of the tiles (the landscape architects planned to use granite), it was clear that every tile had to be delivered by a truck near to its final place. The plan had to be easy to understand and execute.

As we had a combinatorial description of the whole tiling (i.e., for every tile we know all its neighbors), we could generate lists of adjacent tiles. Unfortunately, for a non-orthogonal tiling we cannot give a line-by-line description. Our solution was to remove the pentagonal tiles from the lists. Then, every "row" of tiles will be linear (i.e., we can place one tile directly next to the last one and adjacent to the previous row).

Not only does this "linearize" the tiling, but it also makes it possible to check the alignment of tiles using lasers. This is extremely important for such a large area which was to be built over several months.

Final Remarks

The tiling presented in this article is not new, and credit for it goes to Ludwig Danzer. Our contribution was to make it more accessible for automated generation using a recursive approach. We did not give rigorous proofs for the results in Sects. 79.3 and 79.4 as they are beyond the scope of this article.

Although the tiling was not used for the Alexanderplatz after all, we were able to show that it is feasible to use a quasi-periodic tiling for large areas. We hope that there will be another opportunity to use this work in the future. Please contact the author if you have an application for this or a similar tiling!

Biography Ulrich Kortenkamp works in mathematics and computer science. He currently teaches in the Institut für Mathematik at Martin-Luther-Universität in Halle, Germany. In his work in education he is always looking for topics that exhibit the beauty of Mathematics and the usefulness of computer science, which is almost always true for mathematically supported architectural themes. He is also co-author of the interactive geometry software Cinderella, which constitutes a user-friendly approach to geometry with a strong mathematical foundation.

Chapter 80
Generation of Architectural Forms Through Linear Algebra

Franca Caliò and Elena Marchetti

Introduction

This work has developed in a rather original environment, where a new approach to the relationships between mathematics and architecture is conceived. Namely, mathematics is not just considered as a useful calculation tool for structural problems, but is seen by some mathematicians and architects as an interpretative key to architectural forms. Under this aspect, it is capable of highlighting symmetries and harmonic relations among different parts, of making evident a structural logic, thus becoming a tool that lends itself to even critical and historical interpretation. In short, we can state that various aspects of mathematics are used as a technical language capable of speaking about architecture.

Obviously our efforts are aimed in this direction (Caliò et al. 1995; Caliò and Scarazzini 1997; Marchetti 1998). Namely, we are trying to apply to some significant classes of classical or modern architectural structures a mathematical taxonomy or, to be more precise, a geometrical model. In other words we want to describe them through mathematical formulae, even though it is very clear to us that such formulae have by no means influenced the creativeness of the designers. The purpose of our exercise is simply to better highlight the shape of the architectural object, to extract from it an inherent rule, to make evident its structural rigour.

The final result of this exercise is a geometrical three-dimension model, that is, a description of the geometrical object (a locus, or set of points) expressed through

First published as: Franca Caliò and Elena Marchetti, "Generation of Architectural Forms Through Linear Algebra", pp. 9–22 in *Nexus III: Architecture and Mathematics*, ed. Kim Williams, Ospedaletto (Pisa): Pacini Editore, 2000.

F. Caliò (✉) • E. Marchetti
Department of Mathematics, Politecnico di Milano, Piazza Leonardo da Vinci, 32, 21033 Milan, Italy
e-mail: franca.calio@polimi.it; elena.marchetti@polimi.it

K. Williams and M.J. Ostwald (eds.), *Architecture and Mathematics from Antiquity to the Future*, DOI 10.1007/978-3-319-00143-2_33,

geometric analytical formulae, and its subsequent display over normal bi-dimensional media, such as paper print or computer screen.

To achieve this result one could use classical methods based on geometrical properties of the described objects. However, our approach is a different one and, as such, it is useful also for shapes that at a first sight are not controlled by "classical" relationships. At the same time, is more readily suitable to a straightforward visualisation.

In this chapter we will illustrate the following approach: by establishing a few basic elementary shapes we build the "core" of the architectural objects; by imposing on them movements and deformation, we dynamically determine a final shape; finally, we will give its equations and supply the relevant graphical representations.

Describing the Technique in Use

This section deals with the mathematical concepts our method is based upon. It happens to be necessarily a rather technical section, even if we tried to limit the use of specific terminology and formulae to a minimum extent, perhaps at the expense of a rigorous dissertation. On the other hand we describe here *how* our method works, not *what* it can produce. For the latter one could directly go to the subsequent sections; however the reader would find cross-references to the present section which were impossible to eliminate completely.

There are two concepts basic to our method: the first one refers to the possibility of treating any surface as a set of vector points; the second one refers to the possibility of expressing in a matrix form some fundamental space motions. Do not be alarmed at this point by such obscure statements; we will attempt to clarify them immediately. Let us introduce, to describe the geometrical space, a three-dimensional Cartesian reference system $Oxyz$. Here O is the origin point, x y and z are the co-ordinate axes. We will call *Cartesian space* the space represented in such a way.

A three-component algebraic vector (i.e., an ordered set of three numbers) corresponds to a point in the Cartesian space, namely to the point having the three numbers as coordinates. Reciprocally, there is a unique algebraic vector corresponding to a given point. We have therefore a two-way correspondence between algebraic vectors and points in the Cartesian space. We will call *vector point* a point in the Cartesian space that corresponds to a vector.

Let us now assume that the components of a vector are not constant. Rather let us impose on them a continuous variation; consequently also the vector point will change and during this change it will describe a geometrical locus. If the components of the vector are dependent on a single parameter then the corresponding locus is what we call a curve, whose shape of course will depend on the mathematical relationship existing between the three vector coordinates and the parameter. Such a relationship is described by the *parametric equations* of the curve.

If instead the components of the vector depend on *two* parameters, then the geometrical locus is a family of curves (if we vary the first parameter we obtain a curve, if we subsequently vary the other parameter we obtain the curve family) which determines a surface. Hence, we conclude that one can generally express a surface through a set of vector points correspondent to two-parameter dependent algebraic vectors or, in other words, we can describe it through its parametric equations. It is important for the remainder of this discussion to keep in mind this concept, i.e., that a surface is "punctually" described by mathematical formulae.

As far as the second basic concept is concerned, let us make the following statement: "linear geometric transforms, among them isometric transforms scaling and similarities, can be expressed in matrix form". A clarification follows.

A geometric transform is an operation through which we go from an initial geometric locus (i.e. an initial shape) to another one. Since we describe shapes through vectors, in order to transform them we will act on vectors. If such a transform is applied to all vector points of a shape, a new shape is obtained. The previously quoted statement is then equivalent to saying that a linear geometric transform (a translation, a rotation or whatever else) can be expressed through the generic algebraic relation:

$$A\underline{v} + \underline{b} = \underline{w}.$$

In the previous expression A is a 3×3 matrix (i.e., a three-row three-column array, or a nine-value table) called *transform matrix*, $\underline{v}$ is the vector that must be transformed, $\underline{b}$ is the *translation vector* and $\underline{w}$ is the transformed vector. Such an algebraic transform acts on the vector points of a shape in the space, giving rise to an effect that can be interpreted as a movement. For instance a rotation (with no translation) of amplitude around the z-axis is given by:

$$\begin{bmatrix} \cos\theta & -\sin\theta & 0 \\ \sin\theta & \cos\theta & 0 \\ 0 & 0 & 1 \end{bmatrix} \begin{bmatrix} v_1 \\ v_2 \\ v_3 \end{bmatrix} + \begin{bmatrix} 0 \\ 0 \\ 0 \end{bmatrix} = \begin{bmatrix} w_1 \\ w_2 \\ w_{\ 3} \end{bmatrix}.$$

Obviously, in the last formula the translations vector could be omitted since it is the null vector; we have put it in evidence just to show how this is a particular application of the general formula. Let us conclude such short comments by remembering that the product of a matrix by a vector produces a vector, while the product of a matrix by a matrix gives a matrix. This property allows us to operate composite transforms, as in the following example where the vector $\underline{v}$ is firstly modified through the partial transform A, subsequently the partial transform B and translation b are applied to give the final result:

$$BA\underline{v} + \underline{b} = \underline{w}$$

Let us now go to the core of our method. Let us take a vector and apply to it subsequent translations such that all the translation vectors have the same direction but different magnitudes: the resulting transformed points lie on a single straight line. We can conclude from this property that we can describe a line by means of a point and a variable translation (i.e., a set of infinite translations) having the property of not affecting the direction of the translation vector but, depending on the values of a parameter, of continuously modifying its magnitude.

In a similar way we could think of a circumference as generated by a point that starting from an initial place, rotates around an axis orthogonal to the circumference's plane through a continuous modification of its rotation angle. We can then describe the circumference by applying a variable rotation matrix having a parametric rotation angle to a vector point.

In the two previous examples we started with a point, applied to it a *parametric transform* and obtained a curve. Now we take our curve and, through similar considerations, generate a surface as a composition of transforms describing the motion of our curve with respect to another curve.

This second series of transforms, which must be applied to all points of our curve, is described by a second parameter. For instance the well-known saddle surface can be obtained by translating a parable over another parable, as long as the two parables lie in two perpendicular planes. If for instance one parable belongs to the yz plane and the other to the xz plane we get:

$$\begin{bmatrix} 1 & 0 & 0 \\ 0 & 1 & 0 \\ 0 & 0 & 1 \end{bmatrix} \cdot \begin{bmatrix} 0 \\ u \\ u^2 \end{bmatrix} + \begin{bmatrix} v \\ 0 \\ -v^2 \end{bmatrix} = \begin{bmatrix} v \\ u \\ u^2 - v^2 \end{bmatrix} = \underline{w}$$

The vector point corresponding to the transformed vector $\underline{w}$ describes, as parameter u and parameter v vary, the saddle surface. The expression for $\underline{w}$ just given is the *parametric-vectorial equation* of our surface.

It is clear that using this technique we can describe through mathematical expressions very different shapes, even distant from the "classical" typologies, since their parametric equations are easily handled by a vectorial graphic system. On the other hand it is also true that our proposed method is nothing but a different way of mathematically describing a shape. Many alternative solutions aiming at the same target could be envisaged.

Application # 1: The Sogn Benedetg Chapel

Through the utilisation of the above described ideas we concentrated our attention on some architectural structures and tried to interpret them mathematically. The examined shapes have not been selected in a systematic way, as perhaps one could expect from mathematicians, rather their choice has been constrained by the available materials on the one hand and by our limited knowledge of the architectural field on the other.

Fig. 80.1 Peter Zumthor, Sogn Benedetg chapel, viewed from the rear. Photo: authors

The first example that caught our attention is a 1988 work by Peter and Annelise Zumthor: the small wooden church of Sogn Benedetg in Switzerland (Zumthor 1998; Furer 1990) (Figs. 80.1, 80.2, and 80.3).

The new church is a single-roomed building at the top of a mountain pass overlooking the village houses, and is named after the patron saint. This explains the importance of the building and suggests that the modern construction is superimposed on some older structure. For these reasons the building rests on a privileged spot of the local geography and is given a peculiar architectural shape in

Fig. 80.2 Peter Zumthor, Sogn Benedetg, viewed from the entrance. Photo: authors

order to be distinguished from the normal farmers' dwellings. In fact, what distinguishes it is actually its shape rather than the construction material, which is common wood, typical of the local buildings.

Shape is therefore the dominant element. In an essay on Peter Zumthor, Martin Steinmann (1989) states that the poetry intrinsic to Zumthor's work must be found basically in the simple shapes of its constructions. The simplicity is such that, by attracting the attention to the delimiting surfaces, it gives the impression of a calm

Fig. 80.3 Peter Zumthor, Sogn Benedetg, view of the ceiling. Photo: authors

soul. Steinmann again considers the Sogn Benedetg church as the most significant example of "right shape" among Zumthor's work: on the one hand, it is smooth enough to be receptive; on the other, it inserts itself so as to be protected from avalanches (Figs. 80.3, 80.4, and 80.5). It is this characteristic—apart from any aesthetical or critical considerations regarding the results—which attracted our mathematical attention. We therefore tried our method and using the previously described techniques generated a model simulating the church structure.

Let us start with the predominant importance of the shape: both the horizontal plan and the longitudinal section are based on a specific plane closed and symmetrical centred curve called a "lemniscate" A lemniscate can be generated

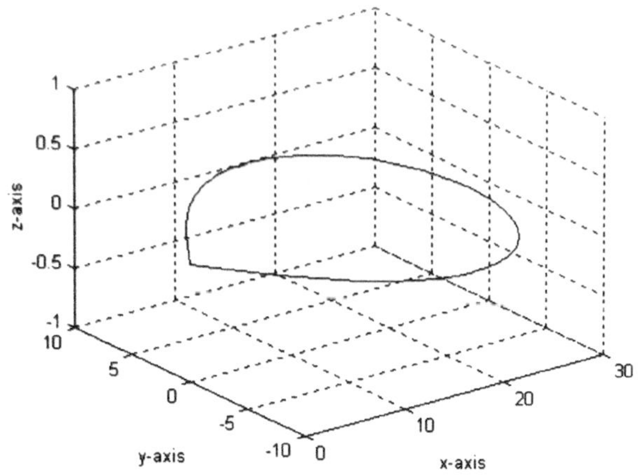

Fig. 80.4 A lemniscate on the *xy* plane with its origin at the centre. Image: authors

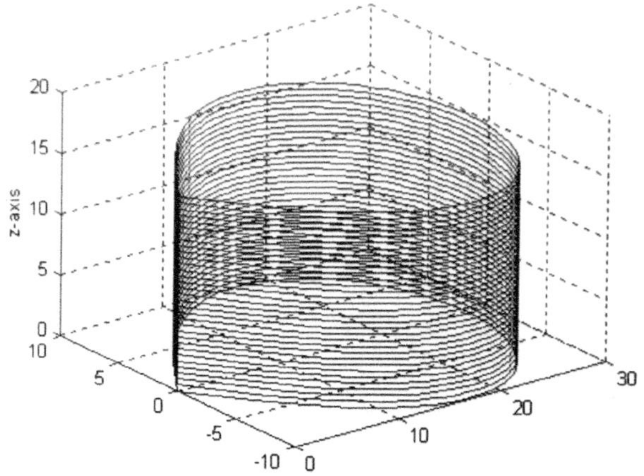

Fig. 80.5 Surface obtained by translation along the vertical *z* axis of the lemniscate in Fig. 80.6. Image: authors

starting from a vector point by applying to it a parametric rotation. Such a rotation modifies, during the rotation, the distance of the point from the centre.

Namely, we want to build a lemniscate lying on the *xy* plane and having as its centre the origin. We start from the vector point on the x-axis at unitary distance from the origin (its co-ordinates are 1,0,0). First we apply to it a scaling dependent on the rotation angle *v*; subsequently a complete rotation around the *z* axis is imposed, according to the algebraic expression:

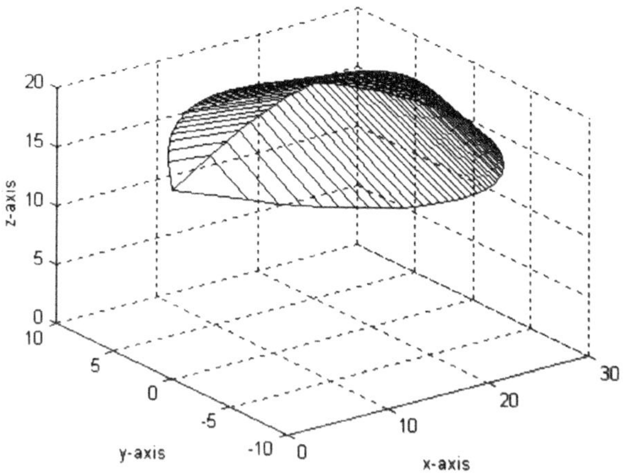

Fig. 80.6 The ruled surface obtained by moving a straight segment along two guiding curves. Image: authors

$$\underline{l}(v) = \begin{bmatrix} \cos v & -\sin v & 0 \\ \sin v & \cos v & 0 \\ 0 & 0 & 1 \end{bmatrix} \cdot \begin{bmatrix} a\sqrt{\cos 2v} & 0 & 0 \\ 0 & a\sqrt{\cos 2v} & 0 \\ 0 & 0 & a\sqrt{\cos 2v} \end{bmatrix} \cdot \begin{bmatrix} 1 \\ 0 \\ 0 \end{bmatrix} - \frac{\pi}{4} \leq v \leq \frac{\pi}{4}$$

where the a constant characterises the curvature of the lemniscate (Fig. 80.4).

The main structure can be considered as a $\underline{s}_1(u, v)$ surface obtained by translation along the vertical z axis of the lemniscate lying on the flat floor; it is therefore described by the expression:

$$\underline{s}_1(u, v) = \begin{bmatrix} 1 & 0 & 0 \\ 0 & 1 & 0 \\ 0 & 0 & 1 \end{bmatrix} \cdot \begin{bmatrix} a\cos v\sqrt{\cos 2v} \\ a\ \sin v\ \sqrt{\cos 2v} \\ 0 \end{bmatrix} \cdot \begin{bmatrix} 0 \\ 0 \\ u \end{bmatrix} - \frac{\pi}{4} \leq v \leq \frac{\pi}{4} 0 \leq u \leq b$$

where b is a constant. In our case we obtained through the analysis of the available data: $a = 28$, $b = 16$ (Fig. 80.5).

The chapel roof, being formed by beams, is straightforwardly represented by a ruled surface $\underline{s}_2(\mathrm{u}, \mathrm{v})$, i.e., a surface having the property that in any of its points there is a straight line completely contained within the surface. Our ruled surface is obtained by moving a straight segment, the extremities of which lie on two curves, called "guiding curves" (Fig. 80.6).

Mathematically, we determine two corresponding points on the guiding curves, i.e. two points obtained through the same parameter value, and impose that the straight line connecting them pertains to the surface. In our case the two guiding

Fig. 80.7 Image: authors

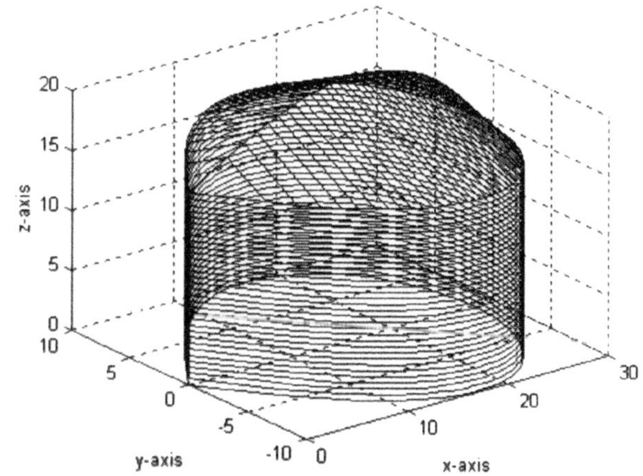

curves are the $\underline{l}_1(v)$ plant lemniscate after its elevation to the quote b of the roof and the lemniscate arc $\underline{l}_2(v)$, which is the longitudinal section of the roof.

Curve $\underline{l}_2(v)$ has been obtained as follows:

1. On the xz plane, which is normal to the base plane, is determined a lemniscate;
2. The lemniscate is turned by a suitable rotation ($\pi/12$ in our case) around the y axis;
3. A scaling whose constant parameter c depends on value a is applied;
4. Finally, the curve is vertically translated to the b height.

Results obtained through this procedure are illustrated in Fig. 80.7.

$$l_{1\_1}(v)\begin{bmatrix} a\cos \mathrm{v}\sqrt{\cos 2v} \\ a \sin v\sqrt{\cos 2v} \\ b \end{bmatrix} \quad -\frac{\pi}{4} \le v \le 0$$

with $\overline{v} = \nu + \frac{\pi}{4}$ and c a constant.

$$\underline{l}_2(v) = \begin{bmatrix} \cos\frac{\pi}{12} & 0 & \sin\frac{\pi}{12} \\ 0 & 1 & 0 \\ -\sin\frac{\pi}{12} & 0 & \cos\frac{\pi}{12} \end{bmatrix} \cdot \begin{bmatrix} 1 & 0 & 0 \\ 0 & 0 & -1 \\ 0 & 1 & 0 \end{bmatrix} \cdot \begin{bmatrix} c & 0 & 0 \\ 0 & c & 0 \\ 0 & 0 & c \end{bmatrix} \cdot \begin{bmatrix} \cos\overline{v}\sqrt{\cos 2\overline{v}} \\ \sin\overline{v}\sqrt{\cos 2\overline{v}} \\ 0 \end{bmatrix}$$

$$+\begin{bmatrix} 0 \\ 0 \\ b \end{bmatrix} - \frac{\pi}{4}$$

$$\le v \le 0$$

In our case $c = 30, \overline{v} = \frac{2}{3}v + \frac{\pi}{4}$; therefore the parametric-vectorial equation of our ruled surface of the roof results to be:

$$\underline{s}_2(u, v) = \underline{l}_1(v) + \left(\underline{l}_2(v) - \underline{l}_1(v)\right)u \quad -\frac{\pi}{4} \leq v \leq 0 \qquad 0 \leq u \leq 1$$

Application # 2: The American Air Museum

The other building we considered is completely different from the previous one both for its destination, dimensions, technology used, materials and the enclosing environment. However, in our opinion there is a common aspect they share, namely their purely essential forms.

The American Air Museum in Duxford, Cambridgeshire (UK), houses a collection of historic American combat aircraft in a building realised in 1995–1997 by the architect Sir Norman Foster (Foster and Partners, 1997; Jodidio 2001). Using the aircraft, other exhibits and supporting exhibitions, the American Air Museum tells the story of American air power and its effect on twentieth-century history from the Second World War to the Gulf War.

Apart from the advanced construction technique and the refined selection of building materials, shape is the most notable characteristic of this edifice. It is the shape that diverts the attention from its conspicuous dimensions, giving an impression of lightness, especially from a distance. In this case as well, the form, even if with purposes different from Zumthor's case, is a simple one enclosing the entire building into an essential shape.

This shape has been obtained as follows. The starting point is a toroidal surface having a horizontal rotation axis, a big radius of 277 m and a minor one of 63 m. Of this surface only a section is used, obtained by intersecting it with a horizontal plane sufficiently distant from the axis so that a complete circular section is never obtained.

Let us make some comments regarding the aesthetics of this choice. The resulting shape gives the effect of something coming out from the ground and fluently reaching towards the sky with the emerging part, while hiding the rest under the grass In fact, both the selection of materials and the fact that the dividing line between the metallic structure and the terrain is not simply horizontal augments this effect.

Now let us describe the mathematical treatment. The basic torus can be obtained by rotation of a circumference of radius r (the minor radius) around an axis lying in the circumference plane and having a distance R (the major radius) from the centre

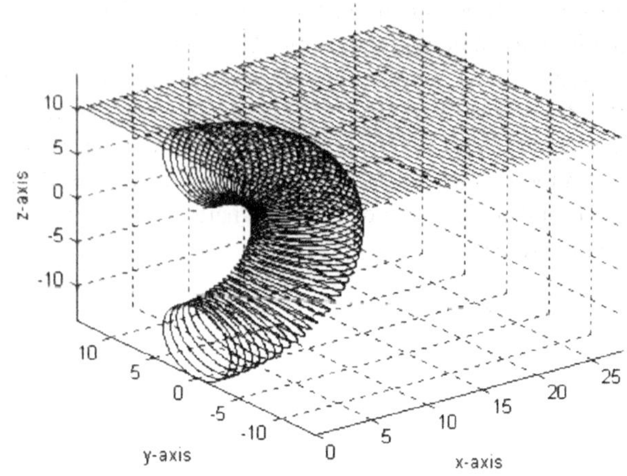

Fig. 80.8 Image: authors

of the circumference. A circumference, as we have already seen, is obtained by rotation of a vector point around a centred axis, normal to its plane.

In our case we have taken the x-axis as the torus axis; r and R are obtained from the available data. The mathematical expression is as follows:

$$\begin{bmatrix} 1 & 0 & 0 \\ 0 & \cos v & -\sin v \\ 0 & \sin v & \cos v \end{bmatrix} \cdot \left(\begin{bmatrix} \cos u & 0 & -\sin u \\ 0 & 1 & 0 \\ \sin u & 0 & \cos u \end{bmatrix} \cdot \begin{bmatrix} r \\ 0 \\ 0 \end{bmatrix} + \begin{bmatrix} 0 \\ 0 \\ R \end{bmatrix} \right)$$

$$= \begin{bmatrix} r\cos u \\ -\sin v(R + r\sin u) \\ \cos v(R + r\sin u) \end{bmatrix} 0 \leq u, v \leq 2\pi$$

We consider only a portion by limiting the rotation angle to a half revolution. Moreover, the geometric relationship between the variation limits of angle u and angle v due to intersection of the horizontal plane is the following (Fig. 80.8):

$$\arcsin\left(\frac{R\cos v - b}{r\cos v}\right) \leq u \leq \pi - \arcsin\left(\frac{R\cos v - b}{r\cos v}\right).$$

The final result corresponding to the object that stimulated our analysis is illustrated in Fig. 80.9.

Fig. 80.9 Image: authors

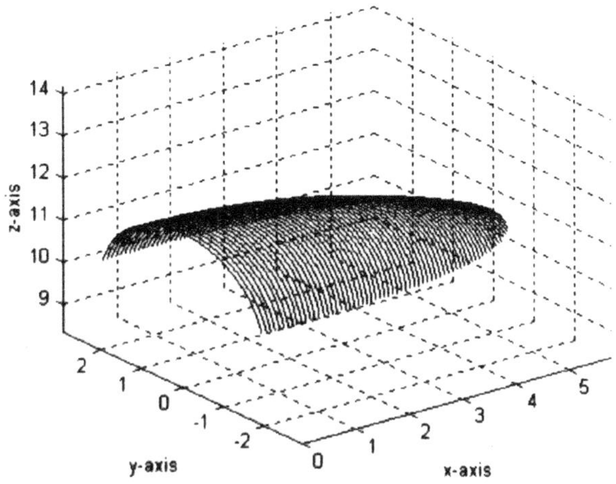

Conclusion

As a concluding remark we would note that the shapes used in our examples are still sufficiently "classical" so that alternative mathematical methods could be employed in order to describe them. However, they have been used on the one hand because of their intrinsic beauty, on the other with the simple purpose of suggesting an idea.

If this idea demonstrates itself as a useful tool and finds acceptance, the next step will be a systematic analysis of more complex shapes, for which alternative mathematical approaches are inadequate. Secondary—and perhaps more ambitious—outcomes could be given by historical and critical evaluations tied to the possible interpretation of the examined shapes in mathematical terms, according to the dynamic method we have envisaged and are proposing.

Biography Franca Caliò is Full Professor in Numerical Analysis in the Department of Mathematics of the Politecnico di Milano. She has taught courses in the area of numerical computation for engineering students and since 1988 she has taught mathematics in the Architecture and Design Schools of the Politecnico di Milano. She has published numerous papers in Italian and international scientific journals on numeric analysis, dealing with integral models of different kinds. Moreover, her field of interest is geometrical modelling in computer graphics. She has published various scientific papers and innovative didactics books on this topic.

Elena Marchetti, Associate Professor, has taught mathematics courses to architecture students at the Faculty of Architecture of the Politecnico di Milano since 1988. She has produced numerous publications in Italian and international scientific journals in the area of numerical integration. She has published many papers about the applications of mathematics to architecture and arts. The experience gained through intense years of teaching courses to architecture students led her to collaborate in some books dedicated to this topic, with multimedia support packages.

References

Caliò, Franca, Elena Marchetti and E. Scarazzini. 1995. *Operazioni e Trasformazioni su Vettori* (software). Milan: Città Studi Edizioni.

Caliò, Franca and E. Scarazzini. 1997. *Metodi Matematici per la Generazione di Curve e Superfici*. Milan: Città Studi Edizioni.

Foster and Partners. 1997. *Norman Foster: Selected and Current Works*. Mulgrave: Images Publishing.

Furer, R. 1990. Untergewichtig, aber Hochrangig (Zumthors Kapelle in Sogn Benedetg). *Architese* **6** (1990): 29–33.

Jodidio, P. 2001. *Foster*. Cologne:TaschenVerlag.

Marchetti, Elena. 1998. *Appunti di Istituzioni di Matematiche (Linee e superfici)*. Milan: Città Studi Edizioni.

Steinmann, M. 1989. On the Work of Peter Zumthor. *Domus* **710** (Nov. 1989): 44–53.

Zumthor, P. 1998. *Peter Zumthor Works*. Baden: Lars Muller.

Chapter 81
The Praxis of Roman Geometrical Ordering in the Design of a New American Prairie House

Donald J. Watts

Introduction

A new strategy for studying the properties and processes of ancient geometrical architectural design occurred in 1988 with the design and construction of the Watts house in Manhattan, KS, U.S.A. At the time, the author, together with his wife and colleague Carol Martin Watts, had been studying the geometric ordering of classical Roman architecture for nearly a decade and had learned many geometric design properties previously unknown to us and today's architectural profession (Watts 2014a, b). We realized that while much important knowledge can be learned from the analysis of historic structures, other important lessons could only be learned through applying these geometric systems to the process of a new design. After all, these historic geometric patterns were used as part of a design and construction process at the site of ancient buildings. The Watts house (Fig. 81.1) therefore became an important extension of our ongoing research of geometric design processes in architectural design.

First published as: Donald J. Watts, "The Praxis of Roman Geometrical Ordering in the Design of a New American Prairie House", pp. 183–192 in *Nexus I: Architecture and Mathematics*, ed. Kim Williams, Fucecchio (Florence): Edizioni dell'Erba, 1996.

D.J. Watts (✉)
The College of Architecture, Planning and Design, Kansas State University, Seaton Hall 211, Manhattan, KS 66506-2902, USA
e-mail: wattsd@ksu.edu

K. Williams and M.J. Ostwald (eds.), *Architecture and Mathematics from Antiquity to the Future*, DOI 10.1007/978-3-319-00143-2_34,

Fig. 81.1 The Watts house, Manhattan, KS, USA. Photo: author

The Site and Its Plan

Manhattan, KS is located in a region known as the Flint Hills, which lies at the edge of the great midwestern prairie of the United States. The small rolling hills afford visually enclosed lower ravines combined with hilltops having expansive distant views to the horizon. As with the architecture of ancient Romans, we chose to build upon a site that embodied the qualities of the larger landscape. The building site is located just below a hilltop and is encircled by native cedar trees that visually enclose the site at ground level. As one rises to higher elevations upon the site, the distant horizon becomes dramatically visible.

Like most cities in the midwestern United States, Manhattan, KS, is planned upon a one-mile grid and has minor deviations from this order to accommodate the major topographic features of the town. The building lot is located at the intersection of Crescent Drive and Hudson Avenue. Crescent Drive is a smaller residential street that encircles a hilltop while Hudson Avenue is a major neighbourhood street derived from the rural one-mile road grid of the region.

The major strategy for the design of the house was to clarify and order the inherent qualities of the site. In so doing, it was hoped that the house design would promote a strong connection of our lives to the neighbourhood, the city and the world beyond. A series of geometric diagrams will be useful in explaining the design responses to the building site. The place in which to build upon the site began by recognizing the locations of the many existing trees and the natural open spaces of the lot. This natural order was clarified through the creation of a datum square (Fig. 81.2), placed inside the property lines of the building site. The datum square reinforces the layers of enclosure occurring at the site by preserving the trees at the site edges. The datum square is rotated from the cardinal orientations of the

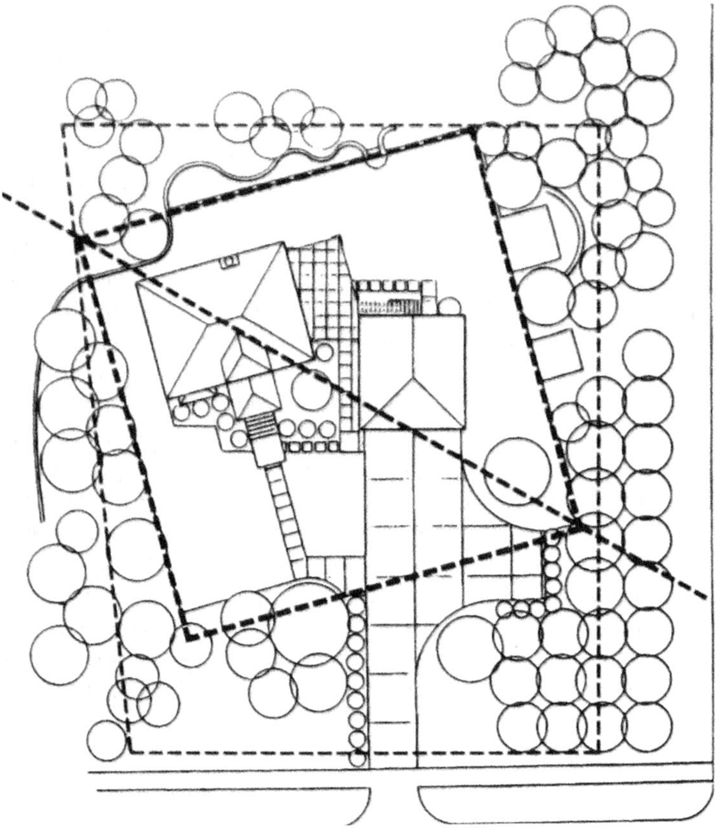

Fig. 81.2 Site plan datum *square* with the solar solstice diagonal. Image: author

city street grid by 15°. By such a rotation, the diagonal through its southeast and northwest corners aligns with the orientations of the winter solar solstice sunrise and summer solar solstice sunset. Sunrises and sunsets are particularly important times of the day for hilltop locations.

The topography of the site, the location of existing trees and the views from the site to the larger landscape all gave a direction for positioning the house. The particular location of the largest existing tree together with an adjacent clearing established the natural connection of the site to the street. The large tree became a reference point in the subsequent site design. A sacred cut of the datum square (Fig. 81.3) was found to closely relate to all of the above criteria and to establish a centre for the house.[1] Whereas the house was aligned with the datum square and the distant views of the prairie, the garage (Fig. 81.4) was given an independent

[1] Brunès (1967: vol. 2). The sacred cut is a major thesis of this book.

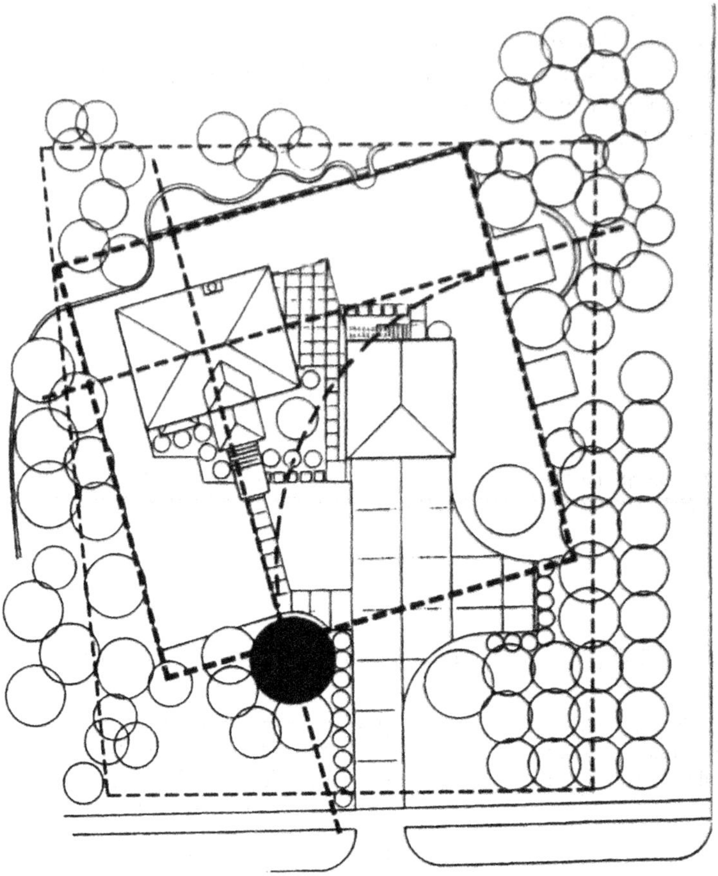

Fig. 81.3 The sacred cut proportioning of the site plan datum square. Image: author

massing and an orientation that aligns with the cardinal directions of the grid of the city.

The cross-axes of the house were conceived as visual axes that would connect the house interior to the world beyond the site. It was therefore important that these cross-axes extend beyond the confines of the datum square of the site. As shown in Fig. 81.5, planting beds in the form of semicircles were designed as focal points and visual extensions of the cross-axes of the house. A fireplace massing is located within the house to serve as an internal focal point in counterbalance to the external views. The fireplace axis occurs in parallel with the short axis of the house. These two axes are later integrated into the stair tower element of the house.

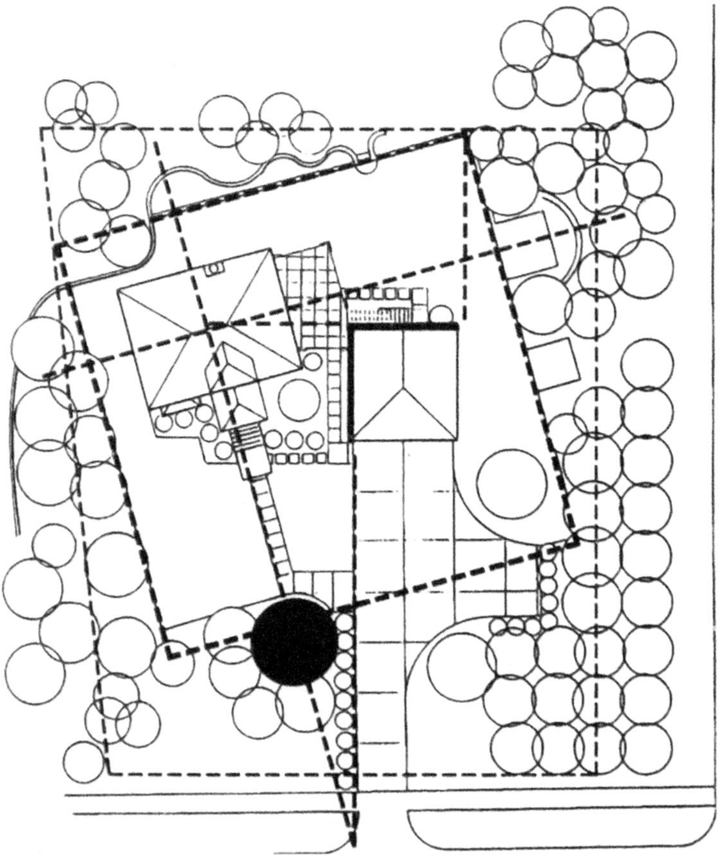

Fig. 81.4 Site plan datum *square* linkages to the city grid. Image: author

The House Massing and the Principal Plan

The massing of the house began as a nominal cube with a side dimension being the sacred cut square of the datum square of the site. The house was conceived as having three levels above ground; the first level consists of the children's sleeping and living areas, the second level is the *piano nobile* of the house with formal living and dining accompanied by kitchen and laundry, and the third level contains the parents' sleeping, study and bathing areas. Required spaces coupled with desired views to the southeast and limited solar exposure to the west encouraged the house proportions to elongate in an east/west direction. The sacred cut was used to proportion the house into the desired rectilinear proportion and to define cross-axes of the house. The positioning of the house within the larger site was accomplished through the alignment of the sacred cut cross-axes of the site with the cross-axes of the house. The initial regulation square of the house, derived from the

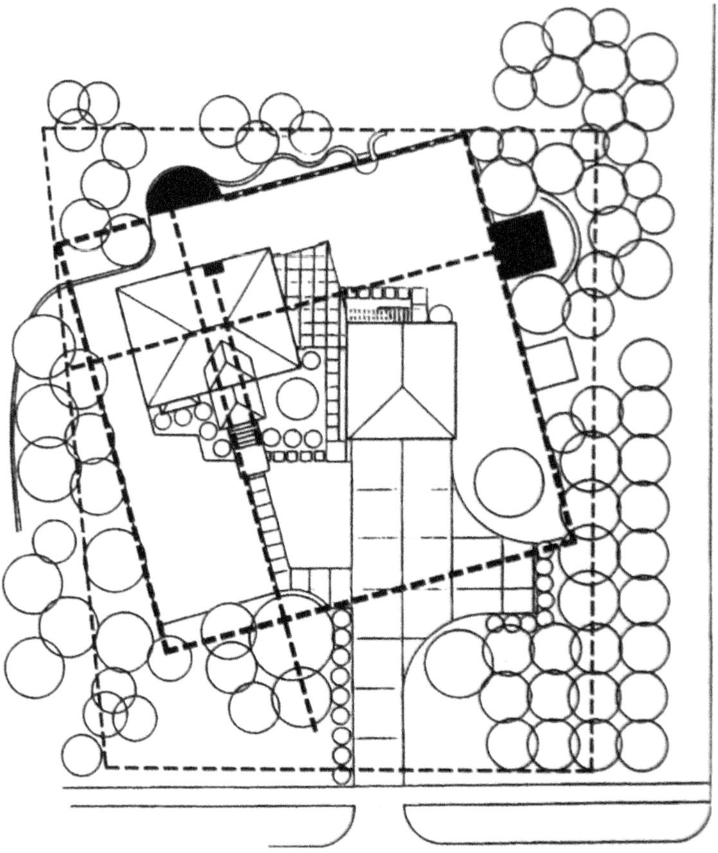

Fig. 81.5 Extensions of the sacred cut axes through creation of exhedra beyond the boundary of the site plan datum square. Image: author

site plan, was used to define a regulating cube with sides the size of the regulating square. Originally 35′–8″ on each edge, necessary wall thicknesses and room sizes required the cube to expand to 37′. The 37′ regulating cube encloses all of the house volume, including the roof with an overhang of 2′. Therefore, by subtracting the roof overhang from the regulating cube, the enclosed volume of the actual house fits within a smaller regulation dimension of 33′. The interior spatial composition of the house can be seen in plan as derived from a 33′ datum square.

The spatial development of the second floor piano nobile best illustrates the proportioning processes of the house plan. Following the dictates of the cross axes, the interior spaces are ordered in the manner of the four quarters of the whole. This plan is zoned such that the honorific spaces of formal living and dining take the

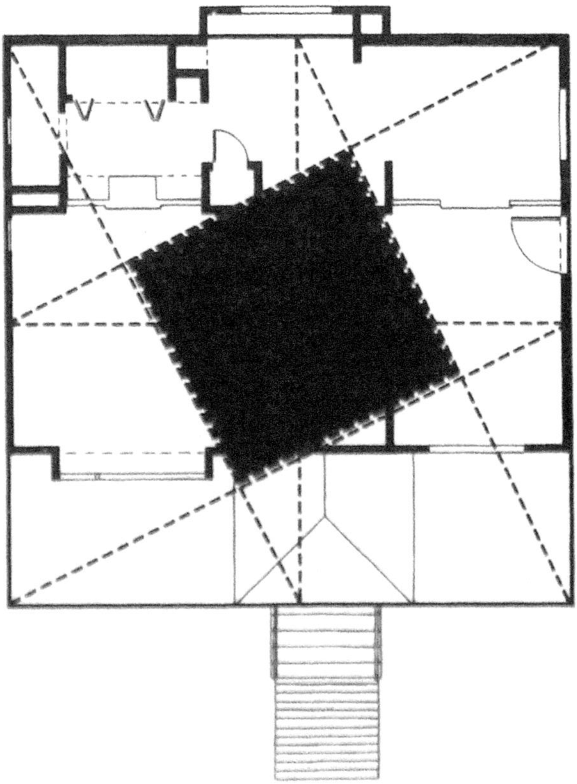

Fig. 81.6 One-fifth the area of the house plan datum square. Image: author

frontal position within the house. Service activities of kitchen, laundry and bath occur at the rear. The living room is the principal room of the second floor plan. Its size was determined not by simply dividing the datum square plan into one-fourth areas but rather by a one fifth proportioning strategy we discovered in our research of Roman Gerasa (Fig. 81.6).[2] The difference between one-fourth and one-fifth areas (5 % of the total area of the second floor), is allocated to interior wall thicknesses and other required masses within the plan. One-fifth the area of the 33′ datum square is a square of approximately 14′–9″ on each side. The 9″ were allocated to the thickness of the exterior walls of the house thereby leaving a 14′ square of open interior space for the living room.

As shown in Fig. 81.7, the 14′ living room is located in the southwest corner of the plan and its centre is aligned with the centre line of the overall house datum square. An *ad quadratum* pattern of successively smaller squares within the 14′

[2] Watts and Watts (1992). The construction of the one-fifth area subdivision of the square is a variation on the capabilities of the sacred cut which we discovered in the research of Gerasa.

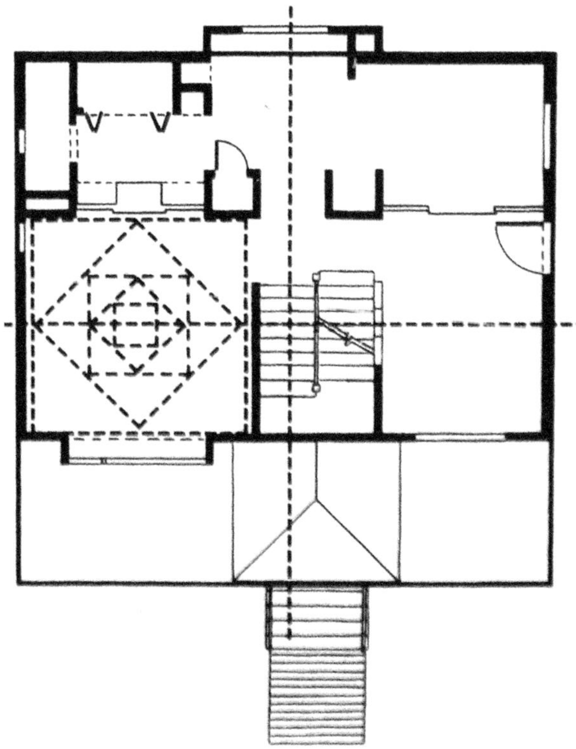

Fig. 81.7 *Ad quadratum* template of areas used in the plan of the *piano nobile*. Image: author

square defined squares of 10′, 7′, 5′ and 3′–6″ respectively. The ad quadratum subdivision of the living room square became a template for the sizing of the other principal spaces of the piano nobile. The sizes of the various spaces within the house is similar to that found in our study of the classical Roman domus (Watts and Watts 1986: 132–139).

The positioning of the ad quadratum series of squares within the plan is shown in Fig. 81.8. The 10′ square is used to size the dining room and kitchen. The 7′ square was used for the stair tower proportioning and as the nucleus of the breakfast space located on the north end of the major entry axis of the house. The 5′ square was used as a module for the open space of the kitchen and the 3′–6″ square was used to define the width of the secondary corridor of the laundry area.

Using principles analogous to a Roman urban street, the east–west axis of the piano nobile is defined by a freestanding datum wall adjacent to a full length walkway. The only opening for walking through this datum wall occurs at the cross-axial centre of the house and is marked by a central archway. Necessary

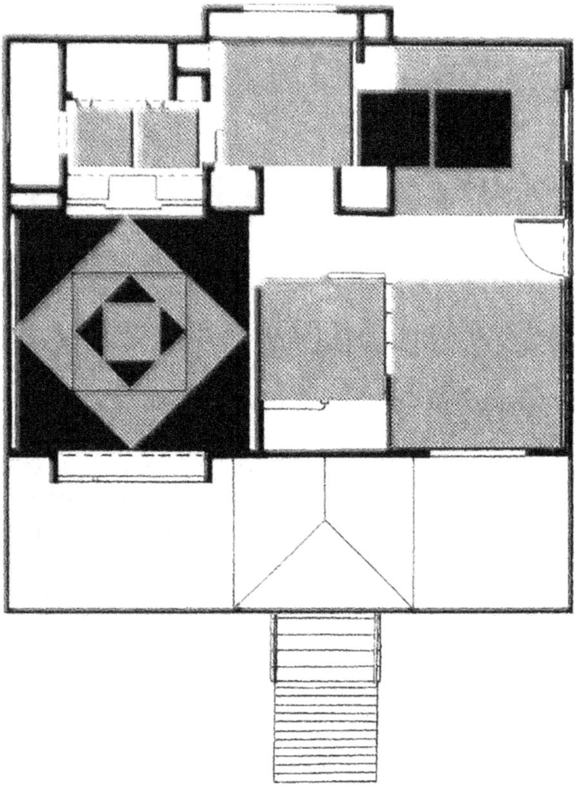

Fig. 81.8 Composition of the *ad aquadratum* areas within the design of the plan of the *piano nobile*. Image: author

machinery and storage areas are incorporated within the massing of the central archway. The archway massing, together with the massing for a small bath (Fig. 81.9) were positioned in such a manner as to form local symmetries for the house main entry axis and the living room. This process also experimented with allowing a massing element, the western jamb of the central arch, to participate in the local symmetries of both the entry axis and the living room elevation.

The House Massing and the Principal Elevation

The principal elevation of the overall 37′ regulating cube is a 37′ regulating square. A sacred cut arc (Fig. 81.10) is used to identify the position of the floor construction of the piano nobile. As seen in the floor plan geometry, the centreline of the regulating square coincides with the centreline of the western flight of stairs in

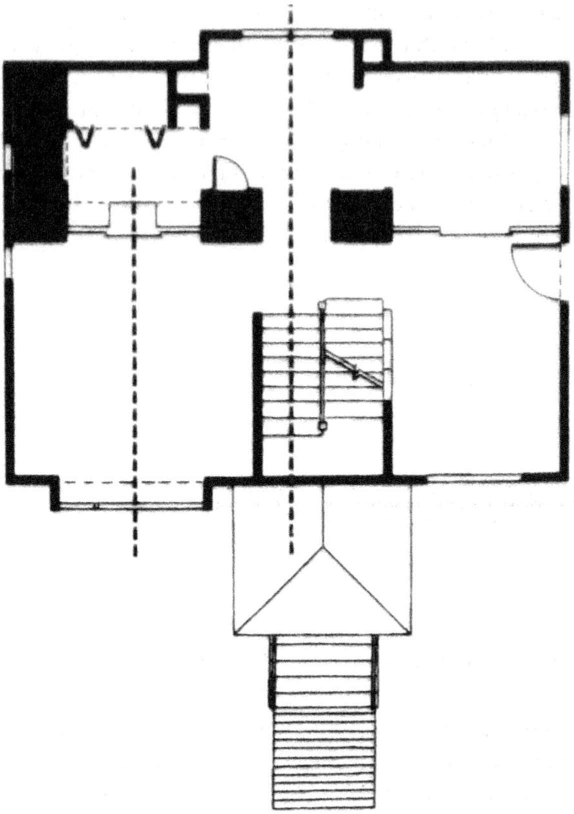

Fig. 81.9 Local symmetries developed within the *piano nobile*. Image: author

the stair tower. In the diagram of the elevation, the vertical centreline of the overall datum square is seen to coincide with the centreline of the western flight of stairs of the stair tower. As will be shown later, the horizontal centreline of the regulating square serves as an important datum in the design of the elevation of the piano nobile.

The determination of the position of the third floor construction atop the piano nobile is shown in Fig. 81.11. The centreline of the eastern flight of stairs of the stairtower is added to Fig. 81.10 and, by means of an arc passing through the centre of the datum square, the distance between these two stair flight centrelines is transferred to an equal dimension above the horizontal centreline of the datum square. This upper horizontal line positions the third floor construction. The separate rectangular box drawn within the confines of the floor and ceiling datums identifies the actual floor to ceiling space of the piano nobile Its height is one-fourth the regulating square, or 9′–3″.

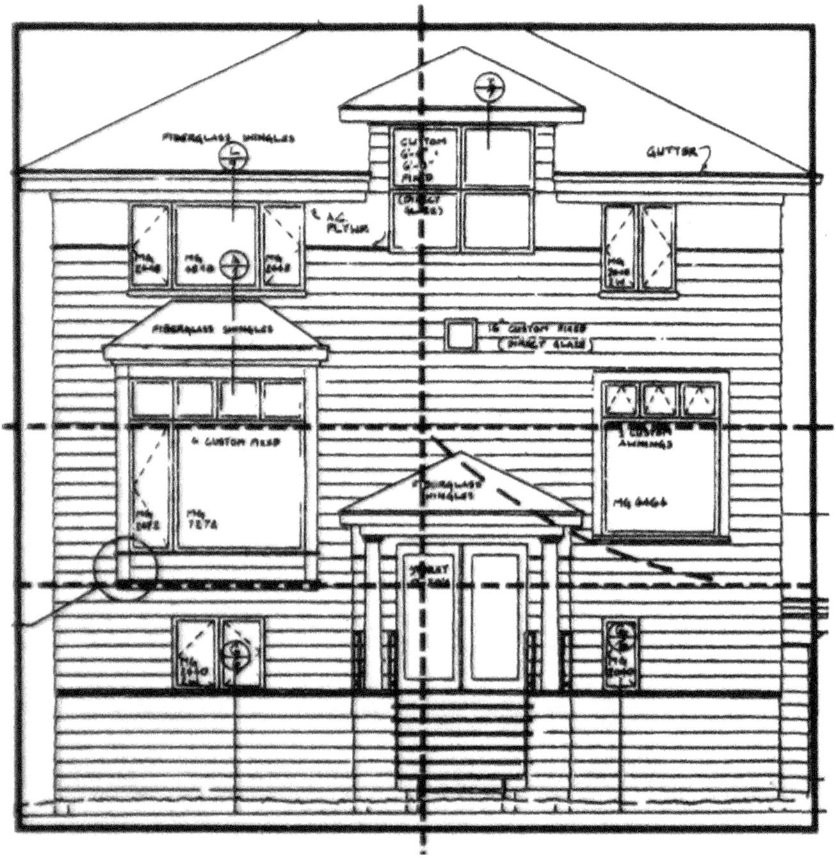

Fig. 81.10 Sacred cut of the primary elevation datum *square* determines the floor plane of the *piano nobile*. Image: author

The division of the regulating square by one-half, one-fourth and one-sixth increments (Fig. 81.12), begins to establish datum lines for the further development of the fenestration of the elevation. The lowest one-sixth horizontal subdivision marks the floor elevation of the entry porch and the top of an implied rustication band that remembers the plinth construction of the earlier stone buildings of this Kansas region. The upper one-sixth divisions are used to position strings of window sills upon the third floor elevation. A one-fourth vertical subdivision of the datum square is used for centring the fenestration of the living room side of the house elevation. The horizontal one-half subdivision of the datum square is used to position the upper level windows of the piano nobile.

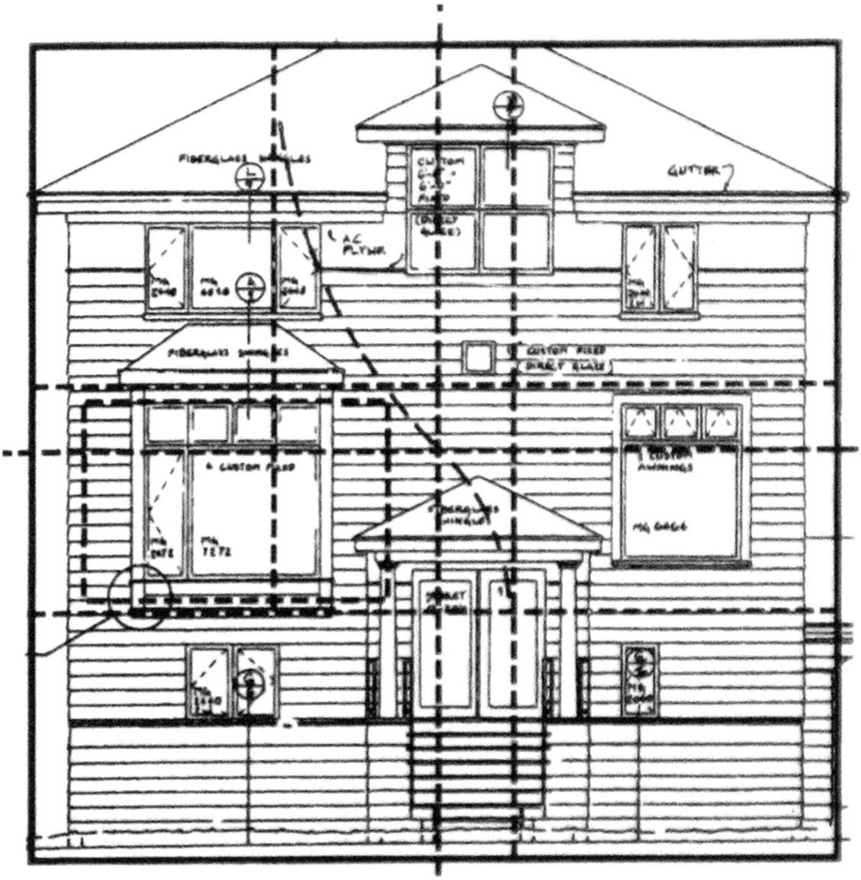

Fig. 81.11 Determining the height of the *piano nobile* within the primary elevation. Image: author

The datums of the previous figures, together with the interior room geometries of the plans, are interrelated through the final composition of the fenestration of the principal elevation, Fig. 81.13. The groupings of windows for the various rooms are composed in such a manner as to reflect the geometric composition of the rooms they serve. It is important to note that the fenestration is the most obvious connection between exterior and interior, and as such, its geometric design should be commensurate with both exterior and interior systems. It is a vital link between the world without and the world within.

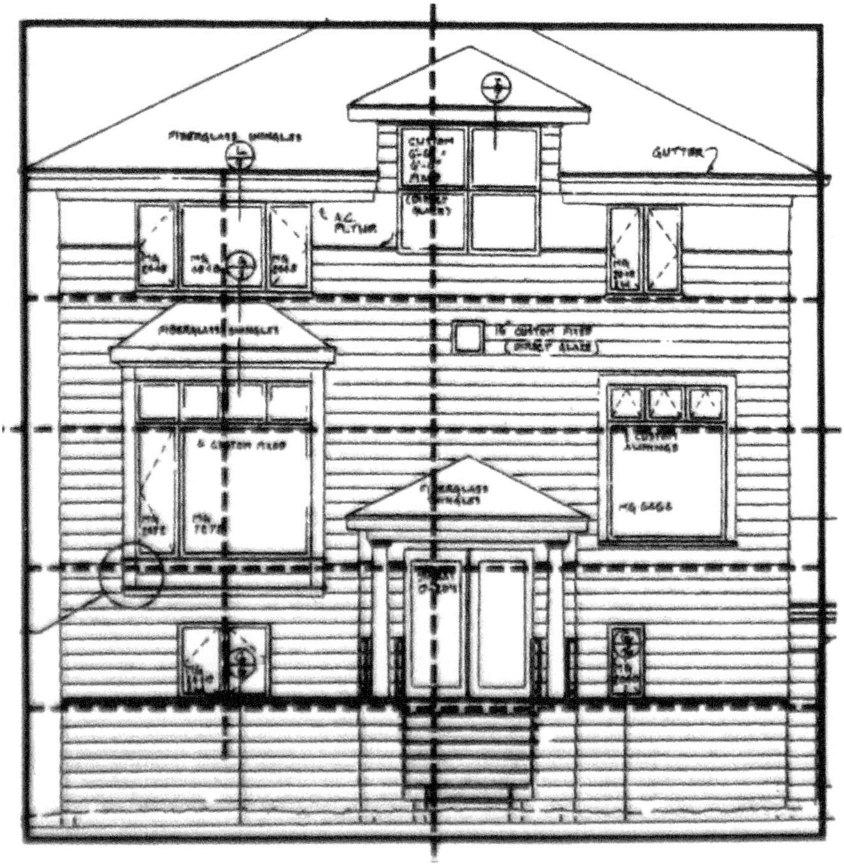

Fig. 81.12 One-half, one-fourth and one-sixth subdivisions of the primary elevation datum square. Image: author

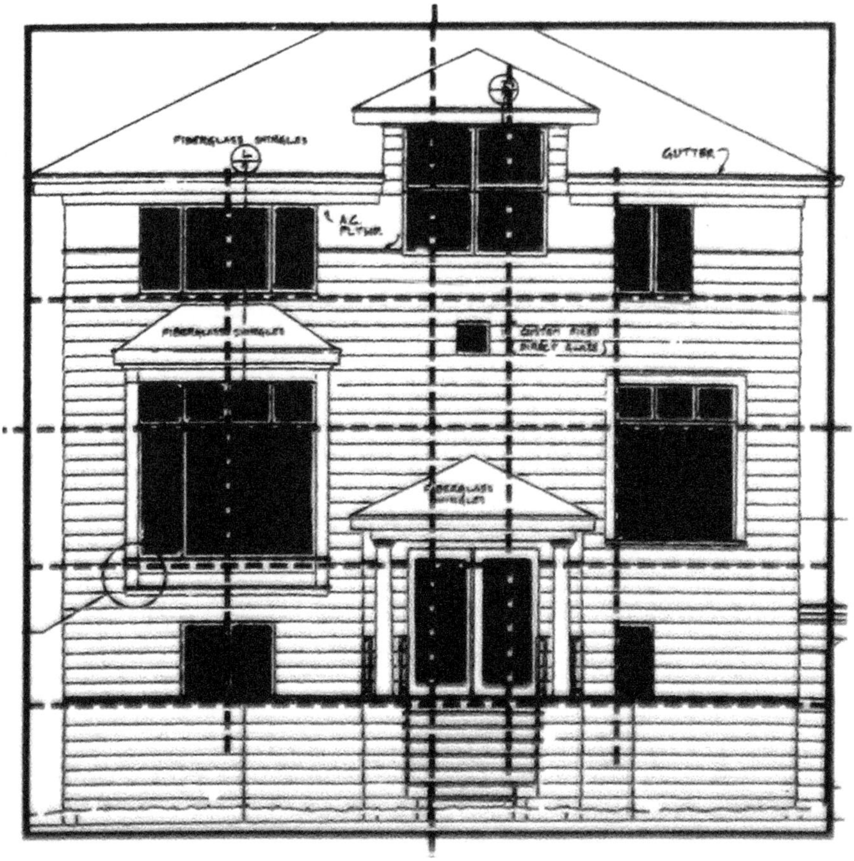

Fig. 81.13 Overall composition of the fenestration of the primary elevation. Image: author

Conclusions

Utilization of these geometric properties in an actual design application yielded a number of valuable insights. The practice of the process revealed its basic pragmatic virtues in terms of directly setting out datums and proceeding to investigate alternative commensurate subdivisions of the datum frame. This directness of application extends to the actual construction process whereby the ongoing implementation of the design can utilize some of the same geometrical processes done at full scale. Such a practice recalls that of classical times where a courtyard floor or building templum shows itself as the layout table and palimpsest of the construction process of the building.

Such a directness of application provides a ready feedback of correctness in the field. It can also provide the opportunity for unforeseen minor adjustments in the original design. In doing so, this process allows for improving the fit between the original design intention and its final implementation.

Applying these geometric processes to today's construction technology and materials obviously created some new conflicts. Constructing such geometric compositions is much easier when using building materials that can be formed to any particular proportion and dimension. Such a process was an integral part of the ancient construction on site process described above. The restrictions upon the designer of using standardized building components required some flexibility in the final construction process. The issues just described are no doubt why the architect Le Corbusier argued in the late 1940s for a modular system of industrialized building components. The history of why his idea never materialized is the source of another paper but let it be said that while his proposed modular was derived from human proportions, other scales derived from man's machines have increasingly become the dominant standards of measure.

These new standards of measure are reflections of new perceptions of existence in today's world. The cosmological meanings underlying the geometric systems of the ancient builders are past paradigms of perception and understanding. And yet, like past paradigms of science, the paradigm of geometric design processes retains its structure and applicability within its own frame of reference. While today's designer does not utilize the geometric systems in a ritualistic process that is seen as the recreation of the cosmos, the process none-the-less retains all its mathematical properties and their pragmatic attributes.

As an architect and an educator, I see this system of design as valuable "lost knowledge". It helps designers perceive their environment in new (old) ways and to discover new relationships. It has helped me see some of the ways this manner of thinking and working has been carried into today's homebuilding in the United States. Lastly, it may also yield new approaches to the conception and utilization of new technologies of design and construction.

Biography Donald J. Watts is an architect and professor in the Department of Architecture at Kansas State University, Manhattan, KS, U.S.A. His interest in the geometry of architectural design began with studies of vernacular Afghan architecture while being a Peace Corps architect at Kabul University. Subsequent research occurred at the Roman classical city of Gerasa, in Jordan while being a Senior Fulbright Professor at Yarmouk University. This introduction to the geometry of Roman architecture led to extensive research with his partner Professor Carol Martin Watts at various locations in Italy. His geometrical research has been published in sources including *Scientific American* and *The Journal of the Society of Architectural Historians.*

References

Brunès, Tons. 1967. *The Secrets of Ancient Geometry and Its Use*, 2 vols. Copenhagen: Rhodos.

Watts, Carol Martin. 2014a. The Geometry of the Master Plan of Roman Florence and its Surroundings. Chap. 12 in Vol. I of this present publication.

———. 2014b. The Square and the Roman House: Architecture and Decoration at Pompeii and Herculaneum. Chap. 14 in Vol. I of this present publication.

Watts, Donald J. and Carol Martin Watts. 1986. A Roman Apartment Complex. *Scientific American* **255**, 6 (December 1986).

———. 1992. The Role of Monuments in the Geometrical Ordering of the Roman Master Plan of Gerasa. *Journal of the Society of Architectural Historians*, **51**, 3 (September 1992).

Chapter 82
Exploring Architectural Form in Perspective: A Fractal Hypercube-Building

Tomás García-Salgado

Introduction

What would be more interesting for our presentation's purpose? To analyze how math has been used in the past or to attempt new applications for the present? This is the question being put forward here. I believe that form in architecture can be explored directly in three dimensions if one can manage an accurate and versatile perspective method.

2D drawings and 3D models have been used either to represent or build architecture since ancient times. These are usually made at any scale, whereas perspective cannot be scaled, it allows the truthful rendering of a form in proportions. Perspective mainly allows the designer to visualize a building within an imaginary space.

Architects frequently use perspective when the design process is almost at an end, to get an idea of how a building would be seen in reality. However the customary use of 2D projections during the design process compels designers to visualize forms fragmentarily. In contrast, the architectural example we are about to present here was not based on 2D projections or a preconceived form; it was rather the result of exploring form directly in perspective, of using mathematics to visualize form beyond 2D. The *Città Ideale* perspectives, attributed to Piero della Francesca, are one of the finest examples we have of architectural forms created in perspective.

First published as: Tomás García-Salgado, "Exploring Architectural Form in Perspective: A Fractal Hypercube-Building", pp. 35–46 in *Nexus VI: Architecture and Mathematics*, Sylvie Duvernoy and Orietta Pedemonte, eds. Turin: Kim Williams Books, 2006.

T. García-Salgado (✉)
Facultad de Arquitectura, UNAM, Ciudad Universitaria, Coyoacan, Mexico
e-mail: tgsalgado@perspectivegeometry.com

K. Williams and M.J. Ostwald (eds.), *Architecture and Mathematics from Antiquity to the Future*, DOI 10.1007/978-3-319-00143-2_35,

Form as Structure

Conventionally architects use drawings to communicate ideas in the same manner that writers use words to tell stories. Both languages, forms and words, encompass underlying structures, as *La Grande Arche* in Paris or Dan Brown's thriller *Deception Point*. The main issue in this thriller, a 'discovery' made by NASA and its implication during a presidential campaign in USA, becomes a nested sequence of events directing the plot line through an invisible narrative structure while the reader imagines all the places and characters until the story comes to an end.

Likewise, elsewhere I have used a geometric network as a visual environment for the design of the stained glass windows for a church (García-Salgado 2005: 457). Thus, the process of formal synthesis was developed in a visually organized environment in which the thematic form for each window were tried out until the desired results were achieved. Even though the windows are separated one from one another, the geometric network is perceived virtually as a continuous invisible structure that gives them unity. Thus, the support structure should be broadly understood as an organized system, since ultimately such a structure, visible or invisible, can organize stories, spaces, themes and anything else.

The perspective sketch by von Spreckelsen submitted for the international competition for *La Grande Arche* (Monnier 1988: 19) is a fine example of this concept. The sketch is a simple drawing, made up of squares combined with freehand lines, and needs nothing else to enhance its formal value. No other sketch whatsoever could have been so convincing of the building's formal novelty: an organized spatial system within a hypercube, showing eight cells, five of which make up the building's mass while the central void-cells relate to the city, addressing the visual axis of the Champs-Élysées from the distance and allowing it to pass through until it vanishes. Just one simple sketch was capable of convincing the competition's jury in favor of its originality.

Naturally, von Spreckelsen's sketches in plan and elevation were indispensable for analyzing the building requirements and fulfill the program. It seems however, that two towers linked by an extended roof was his initial idea, as one of these sketches suggests. He was probably inspired to give the arch its square form when he drew it in perspective. Certainly, perspective can improve the designer's visual thinking.

Sketch as Visual Thinking Approach

Why are perspective drawings so much more persuasive than other projections? We already know that perspective can anticipate how a building may look in reality, but why other projections cannot do that is the issue here. Plans, elevations and sections are mainly used to rationalize the building design until both spatial needs and constructive system are satisfied, but these projections can describe a building only fragmentarily, whereas perspective is par excellence the language of representing architectural ideas as a whole.

I believe that sketching in perspective, either freehand or with an accurate outlining method, not only serves to win international competitions, it also enriches the designer's visual thinking. Computers cannot do that because they are made to process, transform, simulate and store. The dreams of artificially emulating human intelligence are still just dreams, since creativity as an individual human manifestation involves the so-called "flexible thinking" (Caldwell 1999), which is beyond computers.

Before shooting the freeway chase sequence in the film *The Matrix Reloaded*, the scene was visualized as a storyboard by the artist Steve Scroce. A variety of images from quite different vantage points anticipated the dynamic sequence of events. Nothing more than sketches and imagination were needed to start creating. Later, a vast team of actors, expert drivers, mechanics, stuntmen, cameramen, remote-control cameras, contractors, and extras, were involved in playing out each event until the storyboard was converted into a spectacular crash course.

Genuine and Preconceived Forms

Years ago at a "Generative Art" conference at Milan, I said, "... the customary design process is born in 2D sketches, matures either way, and gets old in 3D. So, our perspective's gospel suggests for it to be born in 3D at once" (García-Salgado 2002). With this idea I presented a building embodied within a cross-form, structurally solved. It was an insight of a preconceived form without more options to explore (Fig. 82.1). However, a question arose after my presentation: how can a form be freely explored before it has to compromise with either a preconceived form or a structural system?

Frank Gehry modeled the Guggenheim Museum in Bilbao by hand, first shaping its forms manually and only then defining its load-bearing structure by means of a computer program (Di Cristina 2005). Historically, forms in architecture obey constructive principles derived from the properties and applications of the materials whether or not they were preconceived or modeled. The so-called constructive truth, rational or intuitional, has traditionally been the cornerstone in architectural design. This time, without expecting a lucky insight, I started outlining a hypercube in perspective with the aim of exploring it geometrically until an architectonic form was created. I knew where to begin but not where I would end.

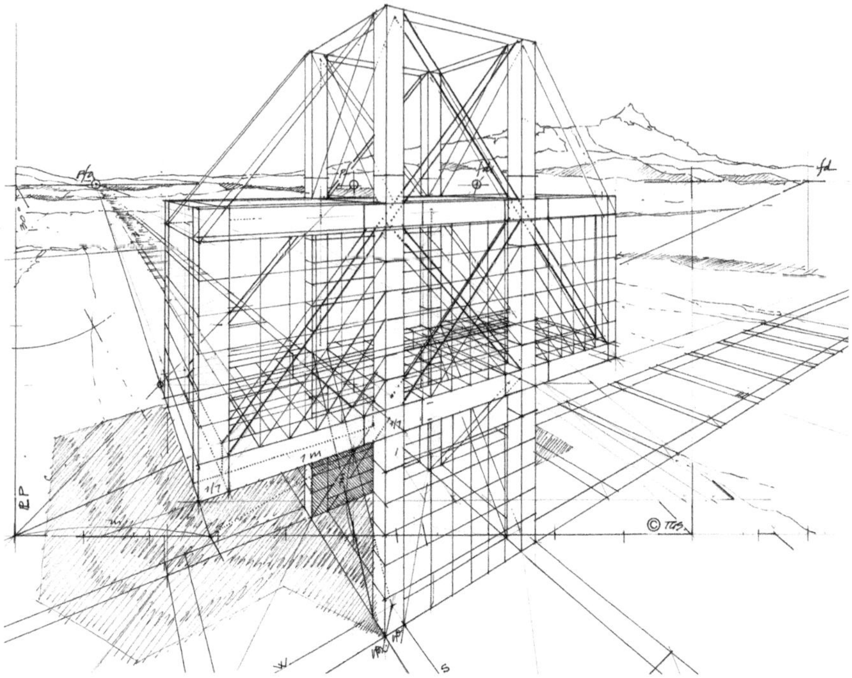

Fig. 82.1 The Cubic Church at the crossroads. Image: author

Exploring a Hypercube in Perspective

As we know, a hypercube can be geometrically represented in many ways as long it has 8 cubic cells, 24 square faces, 32 edges and 16 vertices. François Lo Jacomo gives a comprehensive description of the construction of a hypercube by using two hyper-planes, each one containing a square and a cube (Lo Jacomo 2002: 31–33). Thus aligning one hyper-plane underneath the other vertically, the fourth dimension (4D) is obtained for each one of the figures. As a result, the square becomes a cube and the cube becomes a hypercube. In general, a line is a plane, a plane is a cube and a cube is a hypercube in 4D space.

I will explain now how a hypercube can be rendered in Modular Perspective. Here I will use two practical rules of my method without going deeper into its theory.[1] The first rule is applied to foreshorten a cube in perspective, and the second one to deduce its vertical modulation.

According to Euclid, Book III proposition 31 (attributed to Thales of Miletus about 300 years earlier): *An angle inscribed in a semicircle is a right angle*. In my own words: in a 2D plane, two lines intersecting at a point on the circumference of a

[1] A full description of my method can be consulted in García-Salgado (1982, 1988).

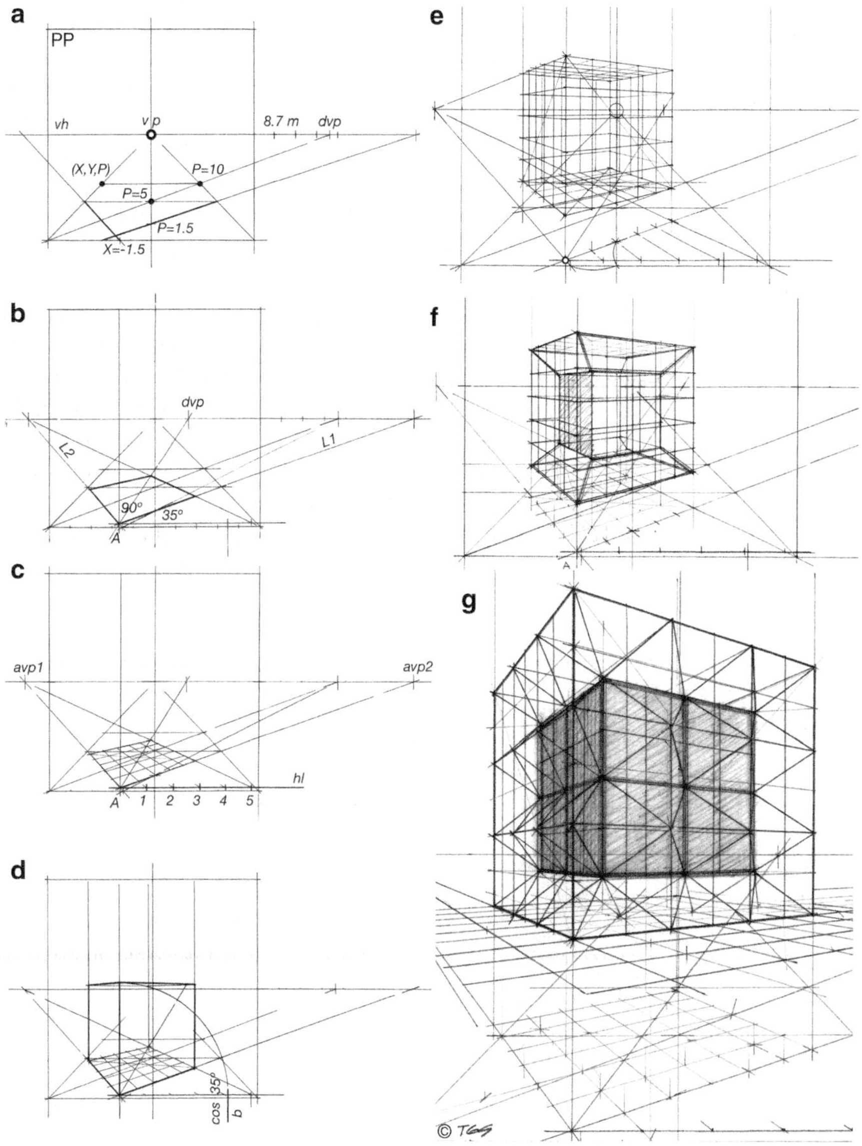

Fig. 82.2 Outlining the process of modular perspective: from the ground to the hypercube. Image: author

circle form a right angle if and only if both lines cross its diameter endpoints at the same time. This is quite a long statement but fits better with the first rule, since such lines are constructed differently within the perspective plane (PP). Let us commence step by step through Fig. 82.2.

First draw the PP as a square and locate the vanishing point (*vp*) at its center. Thus, coordinates X, Y, P, of a point within the observer's visual field can be rendered directly on this plane (I changed the customary 'Z' for 'P' because it stands for the first letter of the Spanish *profundidad*, meaning depth).

Run the visual horizon (*vh*) at the height of the *vp* and then carry out the base-diagonals of the PP to the *vp*.

Locate the distance vanishing point (*dvp*) at the right side of the *vh*, in this case at 8.7 *m* from the *vp*, therefore the *vp dvp* is the interval from the observer's distance to the PP. Next, run a line from the *dvp* to the left corner of the PP′ base and where it crosses with the central vertical line, $P = 5\,m$ is obtained. In the same manner, where this line crosses with the right diagonal, P = 10 m is obtained. Then you have on plane $Y = -5\,m$ a $10 \times 10\,m$ square floor (Fig. 82.2a).

Now using plane $Y = -5\,m$, we have Line L1 defined by points ($X = 0\,m$, $P = 1.5\,m$) and ($X = 5\,m$, $P = 5\,m$), and line L2 is defined by points ($X = -1.5\,m$, $P = 0\,m$) and ($X = -5\,m$, $P = 5\,m$). Notice that the second coordinates of both lines correspond to the circle's diameter $= |10\,m|$, whereas in this case lines L1 and L2 depicts an approximately 35° angle on the PP.

Thus, where L1 and L2 intersect the circle at point *A*, an accurate 90° angle is found since both lines cross the circle's diameter endpoints. Even more interesting is to bisect this angle in perspective by carrying out a line from point *A* to point ($X = 0$, $P = 10$).

By extending this line until it crosses the *vh*, a new vanishing point is found. This new point is called the diagonal vanishing point (*dvp*) and will be very useful for the hypercube construction. In general, this short cut allows us to intersect points on a circle in any radial direction we want (Fig. 82.2b).

Now, to divide the cube's surfaces into five equal modules we first proceed by dividing it horizontally and then vertically. Draw a horizontal line (*hl*) through point *A* and then divide it into five equal parts. By carrying out lines from each mark to the asymmetric vanishing point 1 (*avp1*), the cube's base diminishing is obtained where they cross the vanishing line *A-avp2* (Fig. 82.2c). Notice that a cube always has three *avp* (García-Salgado 2003: 36). Then to find out the true length of the cube's right base along the same *hl*, we must apply the formula $b = c \times \cos a$. Here b represents both the true length of the cube's side and its height in perspective. Once its height has been deduced it can be placed anywhere along the vertical line *A* (Fig. 82.2d).

If the $hl = 6$ cm wide, then it would diminish as $b = 6 \times \cos 35^\circ = 4.91$ cm. With practice this procedure becomes much easier than it looks; you just go directly in perspective with the squares and imagination. For instance, it took me just over 20 min to fully draw Fig. 82.2e. An additional advantage you may have following this procedure is the freedom to draw at any scale you want since all dimensions are modular.

The odd modulation of Fig. 82.2f made evident an obvious problem between the 5′s and the 3′s when attempting to connect the hypercube's outer shape and the inner-cube by diagonal lines. So one module was added in the subsequent sketch obtaining a $6\,m$ $4\,m$ hypercube. So doing, the outer shape can be related evenly with the inner cube at its center and extreme points (Fig. 82.2g). Here, instead of using the hypercube's external shells as the building mass not unlike *La Grande Arche* I used them to conform to an exoskeleton structure supporting the inner cube across its six faces.

Since we now have useable spaces and a preliminary structure, even though the exploration process has not yet been concluded, we need to somehow hypothesize as to the program. Like the children's toy "Meccano", the exoskeleton structure could be assembled and disassembled to move from one place to another, a facility that suggests the idea of a movable pavilion devoted, for instance, to a traveling Art and Math exhibition. Usually exhibits of this sort that travel from one museum to another must adapt to the available spaces. In this case the pavilion's transportability eliminates this problem.

Fractalizing the Hypercube-Building

A fractal is a self-similar object, a concept that is not new in architecture. Some Gothic buildings show recursive patterns with up to three iterations in windows, such as in Metz Cathedral, for example (Scott 1999: 27). Surely the reader will recall other, different examples since there are many. My curiosity to explore fractal geometry in perspective was basically a challenge because I had never done it before.

According to Mario Livio, "For a [three-dimensional] cube, a division into cubes of half the edge-length ($f = 1/2$) produces $8 = 23$ cubes, and one-third the length ($f = 1/3$) produces $27 = 33$ cubes" (Livio 2002: 218). Coincidentally our design combines both divisions since the exoskeleton's outer shape is divided in thirds while the inner cube's shape is in halves. Notice that one half of the inner cube is equivalent to one third of the exoskeleton's outer shape.

The exoskeleton structure is confirmed by six pyramidal stumps with all its vertices lying on the inner-cube surfaces. The resulting configuration is a system of bars connecting all the vertices of the eight sub-cubes diagonally. Afterwards the exoskeleton structure was fractalized, about one-third and a half and no further iterations were performed (Fig. 82.3). Notice how one of the eight sub-cubes of the inner cube is missing, since it was absorbed through the first iteration.

Once a computer program starts running it does not stop and think, as I did after the first iteration. So my hypercube could end resembling the Menger-Sierpinsky sponge if a computer program were to produce it, making it impossible to manage utilitarian spaces. Mathematics can help architectural design but can never rule its decisions. In Villa Rotonda, for instance, Palladio created new aesthetic rules for the proportions of the columns of the porticoes. "The designer creates the rules whereas math provides the tools," occurs to me here as an appropriate aphorism.

Up to now we have been working in modules without caring about the building's dimensions. Before converting modules (m) to dimensions (m) we must establish a module unit as a reference. Let this unit be 1/27 of the hypercube's volume, since any of its diagonal planes contains all the structural bars in their true dimension. As this plane measures $1\,m \times \sqrt{2}\,m$ and its diagonals $\sqrt{3}\,m$, the exoskeleton bars

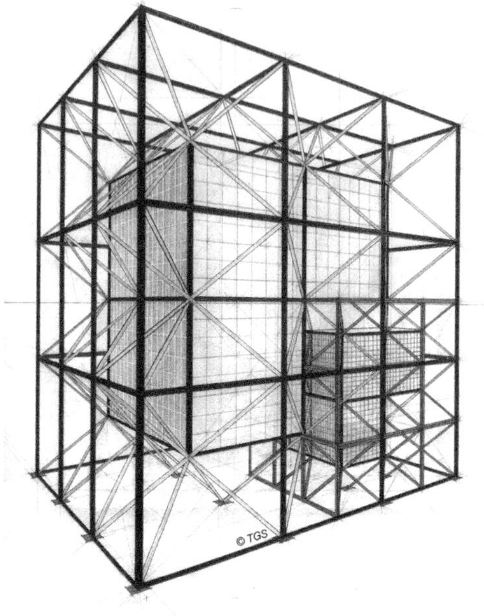

Fig. 82.3 Fractalizing the hypercube. Image: author

node-to-node would measure $\sqrt{3} \div 2$. In general, all possible lines joining the unit vertices are $\sqrt{2}$ and $\sqrt{3}$, which is the beauty of a cube whose side $= 1\ m$.

Consequently, if the module unit *m* were equivalent to 20 m then the outer shape of the hypercube would measure 60 m and its bars 17.32 m. In a similar manner the fractal hypercube dimensions were deduced, giving it now an equivalence of 10 m. Therefore its bars are 8.66 m. Observe in Figs. 82.3, 82.5, and 82.6 how the visual complexity of the exoskeleton structure increases after the first iteration.

Composition

Formal complexity in composition has, among other properties, three important aspects: the number of elements employed both structurally and architectonically; the way they are organized; and their dimensions. It is appropriate here to mention the Vitruvian triad of *firmitas*, *utilitas*, *venustas* to recall the timeless principles of composition, faithfully illustrated by Palladio in Villa Rotonda, for example. When one observes the porticoes of the Rotonda, the proportions appear well-balanced and dynamic due the subtle increase in distance between its middle columns. The magic formula for *venustas* was only the increase of the spacing by one column's half-diameter, enough to give it a unique appearance. Palladio's architecture has been widely imitated around the world but the imitations often fail to include such important refinements.

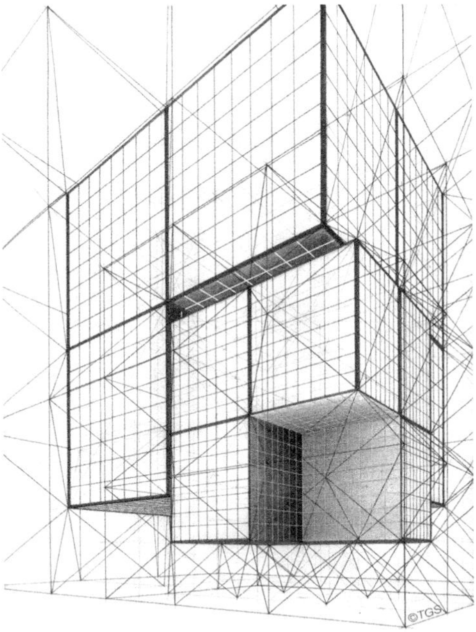

Fig. 82.4 The inner-cube and its fractal. Image: author

Recalling these principles raises a question regarding the composition of the hypercube: what options do we have for resolving both the inner cube and its fractal in relationship to the whole? What would be the formula for *venustas* in this case? To investigate some options I drew a detailed view of the fractalized corner to closely see how these cubes interrelate to each other. As can be seen in Fig. 82.4, the missing scaled fractal enhances the corner composition in such a way that it clearly defines a natural entrance for the building.

In the Rotonda, it is the arrangement of the columns that emphasizes the entrance, whereas here is a missing part. Now, two more aspects can be gained by framing each of the seven sub-cubes: a manifestation of its structure and a reinforcement of its scaled proportions. Intentionally I drew Fig. 82.5 at the observer's eye level in order to explore these features, otherwise they could hardly be perceived. As a result we have seven fractal sub-cubes, of 10 m × 10 m, totally independent of the seven inner-cubes' sub-cubes, of 20 m × 20 m, both sets floating within the exoskeleton structure.

Afterwards, several tiling patterns were attempted to encompass all the cube's surfaces in glass, a square grid being selected as best for dressing them. Thus both the inner-cube faces and those of the fractals were divided in 1/8 *m* according to its scaled proportions (see Fig. 82.4).

Of course there are many other design components not yet explored, such as stairways, elevators, floors, the exhibit design itself, all the systems (mechanical, electrical, passive, security, etc.), and so on. If this building were real all of them would have to be analyzed according to their functional and formal requirements

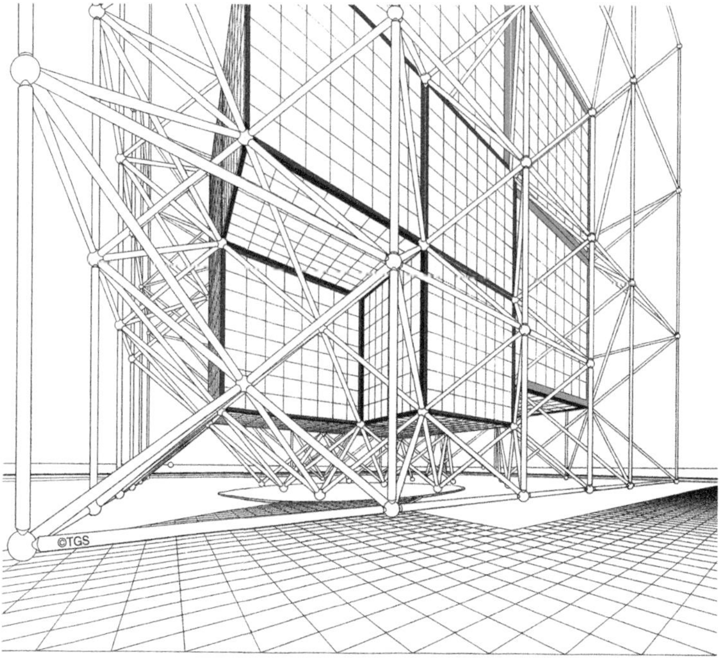

Fig. 82.5 Close-up view of the exoskeleton structure (not a computer drawing). Image: author

within the project. In this case as well our perspective model could help us to assess formal solutions.

The same is true for the exoskeleton structure, since I have assumed it can be built but have not yet explained how it could be done. For instance, the use of Mero's systems for the exoskeleton bars, flooring, spider fixing for the glassed surfaces, space frames, girders, and many other components might be suggested. To go deeper into each one of these possibilities would necessarily take us to another level of the process, which is beyond the scope of this presentation, in which I have wished to emphasize that improving the designer's visual thinking is the ultimate goal of exploring form in perspective.

To conclude, the idea of an exoskeleton structure supporting the glassed-enclosed cubes seems very suggestive. Just imagine this building, appearing to have no foundation, challenging the fundamental principle that a building must appear rooted in the ground, floating in the air like an object from outer space, bathed in sunlight that reflects its image on the water, or dazzling at night with multimedia screens and hologram projections on the glass surfaces. Perhaps you will agree with me (Fig. 82.6).

Of course my initial question is it more interesting to analyze how math has been used in the past or to attempt new applications for the present is a riddle. Simply, the present cannot be improved by ignoring what has been done in the past, as without the past there is no present. Thus our search for a new design approach can be

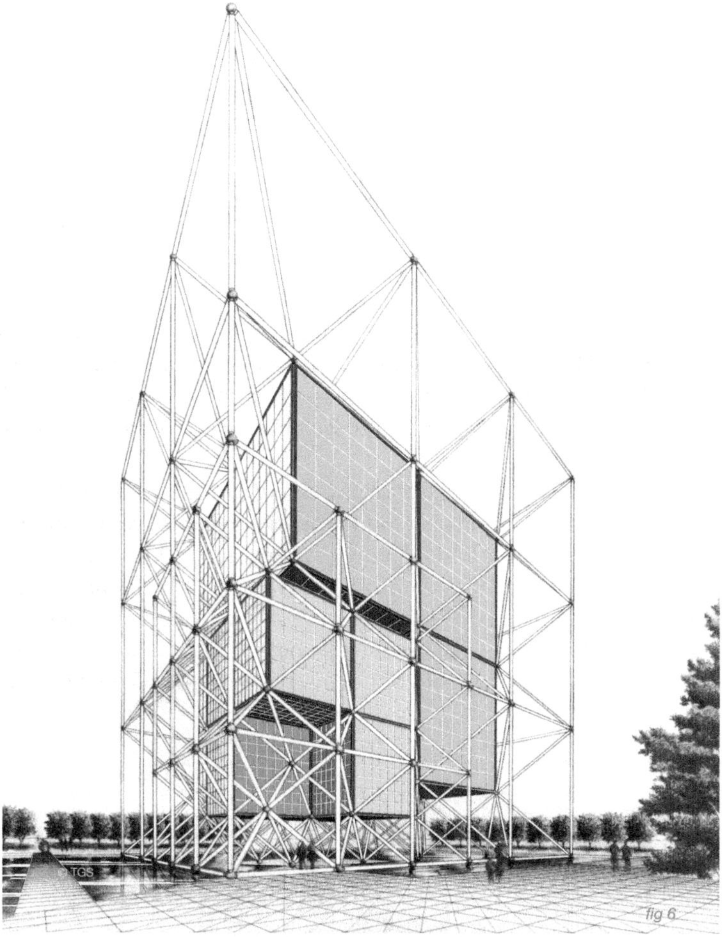

Fig. 82.6 The hypercube-building seen from the access. Image: author

summed up by saying, "if you can imagine it, you can draw it in perspective". This has been my intention here.

Biography Tomás García-Salgado received his professional degree in architecture (1968), Master's degree, and Ph.D. (1982–1984). He is a formal researcher in the Faculty of Architecture of the UNAM (México), and owns the distinction as National Researcher, at level III. Since the late 1960s, he has devoted his time in perspective geometry research, being his main achievement the Modular Perspective theory. He also has several works of art, architecture, and urban design.

References

CALDWELL, B. and Denis DAKE. 1999. Visual Thinking . http://www.design.iastate.edu/NAB/about/thinkingskills/visualbasics/visualthinkingbasics.html. Accessed 25 November 2013.

DI CRISTINA, G. 2005. Poetry in Curves: The Guggenheim Museum in Bilbao. Pp. 159–186 in M. Emmer, ed. *The Visual Mind II*. Cambridge, Massachusetts: MIT Press.

GARCÍA-SALGADO, T. 1982. *Perspectiva Modular Aplicada al Diseño Arquitectónico*. Vol. II. México: UNAM.

———. 1988. A Modular Network Perspective Model vs. Vectorial Models. *Leonardo*, **21**, 3: pp. 277–284. Oxford: Oxford University Press.

———. 2002. Modular Perspective as a Method for Generative Design. Pp. 6.1-6.10 in *Proceedings of the 5th Generative Art Conference*. Milan: Generative Design lab of Politecnico di Milano University.

———. 2003. Distance to the Perspective Plane. *Nexus Network Journal*, **5**, 1 (Spring): pp. 22–48: Basel: Springer.

———. 2005. Rational Design versus Artistic Intuition in Stained-Glass Art. Pp. 449–468, in M. Emmer, ed. *The Visual Mind II*. Cambridge, Massachusetts: MIT Press.

LIVIO, M. 2002. *The Golden Ratio*. New York: Broadway Books.

LO JACOMO, F. 2002. *Visualiser la Quatrième Dimension*. Paris: Vuibert.

MONNIER, G. 1988. Hommage à l'Architecte de La Grande Arche. *Techniques & Architecture*, **376**: pp. 16–24. Paris: Editions Jean Michel Place.

SCOTT, A. 1999. *Developing a Fractal Architecture*. Master thesis, Miami University, Oxford, Ohio.

Chapter 83
The Compass, the Ruler and the Computer: An Analysis of the Design of the Amphitheatre of Pompeii

Sylvie Duvernoy and Paul L. Rosin

Introduction

The purpose of research on design problems in historic architecture is usually to understand and to reveal the hidden theoretical sciences that the architects applied in their projects. The research may be oriented and supported by some written documents: architectural treatises from the epoch that provide the researcher with the basic information about the main theoretical design principles. Renaissance and early Baroque are periods in which written evidence of existing mature design theories are abundant. On the other hand, when studying ancient architecture from classical antiquity, apart from a few minor texts that incidentally mention various topics related to architecture, scholars have only the treatise *De Architectura Libri Decem* by Viruvius to help them understand the design principles from a theoretical standpoint. However, among the traditional typologies of Roman monuments, the amphitheatre is not among the case studies discussed by Vitruvius. This absence of scientific concern and of cultural recognition for this kind of building recently engaged the curiosity of modern scholars, who for a few years now have been trying to write, in Vitruvius's style, the missing chapters about amphitheatres.

First published as: Sylvie Duvernoy and Paul Rosin, "The compass, the ruler and the computer", pp. 21–34 in *Nexus VI: Architecture and Mathematics*, Sylvie Duvernoy and Orietta Pedemonte, eds. Torino: Kim Williams Books, 2006.

S. Duvernoy (✉)
Politecnico di Milano, Milan, Italy
e-mail: syld@kimwilliamsbooks.com

P.L. Rosin
School of Computer Science and Informatics, Cardiff University, 5 The Parade, Roath, Cardiff CF24 3AA, UK
e-mail: Paul.Rosin@cs.cf.ac.uk

K. Williams and M.J. Ostwald (eds.), *Architecture and Mathematics from Antiquity to the Future*, DOI 10.1007/978-3-319-00143-2_36,

The present research is a further attempt at revealing the intentional geometrical and arithmetical order that underlies the design layout of some Roman amphitheatres. The presence of this order may be proved either by using the same procedures that ancient designers probably used themselves geometrical diagrams drawn with some kind of manual graphic device, relying on simple arithmetic calculations or by applying modern analytical procedures provided by contemporary mathematics and computing.

In this chapter both approaches will be discussed and will prove to be complementary.

Of course, while applying modern mathematical tools, we do not assume that the ancient designers were aware of them; the objective is to demonstrate a knowledge, but not the knowledge of the tool used for the analysis itself. On the contrary, while applying a more traditional approach, which makes use of the basics of classical geometry and arithmetic, we consciously and deliberately imitate the supposed research methodologies of ancient designers, being careful to respect the limits of their contemporary knowledge, as much as this is possible.

The modern approach is based on the use of computer technology and specific software related to advanced mathematical theories, while the traditional approach is based on the manipulation of the compass, the ruler and the natural integers.

Here, two different analyses of the same monument Pompeii's amphitheatre were conducted in an alternating rather than in a parallel way. The results that each kind of analysis suggested then oriented and enriched the other aspect of the study. The conclusions do not come from the mere comparison of two independent researches but from the discussion of hypotheses that emerged alternatively from one or another aspect of the inquiry. Both analyses are based on the same data coming from an accurate survey of the monument that was conducted in 2001, according to the "polar methodology", i.e., by means of a single electronic theodolite, located in the approximate centre of the arena.

This study is the continuation of research presented in another chapter in this present publication (Duvernoy 2014).

Analysis Through Computer Technology

Modern computing and mathematical facilities provide us with two extremely useful tools for the analysis of survey data: the first is a means of precisely fitting geometric models to the data and accurately measuring discrepancies between the model and data, and the second is statistical testing procedures to determine whether a given model is appropriate to the data. We will demonstrate the application of these tools to analyse the Pompeian amphitheatre.

Serlio's Oval Constructions We already mentioned the lack of information about amphitheatres in Vitruvius's treatise. The main literary source for this kind of monument is Sebastiano Serlio's treatise on architecture (Serlio 1545; Hart and

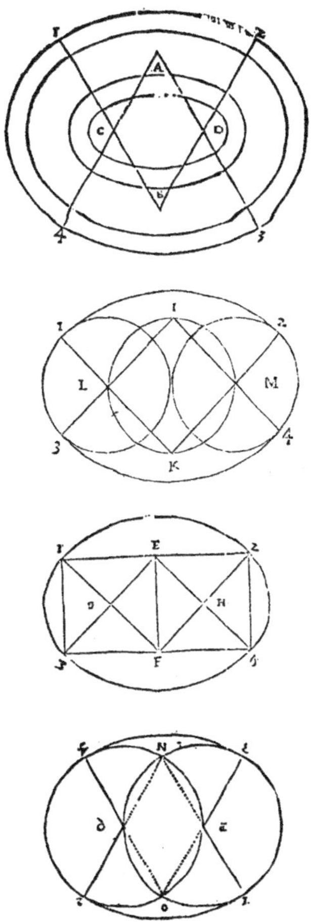

Fig. 83.1 Sebastiano Serlio's oval constructions (Serlio 1545: 14–15)

Hicks 1996). Serlio assumes that amphitheatres were oval and he discusses oval shapes at some length (Rosin 2001). His four methods of construction are illustrated in Fig. 83.1; these have been used extensively in architecture over the years.

Only the first construction allows a variable aspect ratio; the other three are all fixed. It is the former we shall use in the following analysis. The oval consists of four circular arcs with centres ($\pm h$, 0) and (0, $\pm k$) and radii $a-h$ and $b+k$ respectively (Fig. 83.2).

It has the attractive property that the arcs join smoothly with tangent continuity. This geometric constraint can be expressed algebraically as:

$$h = \frac{k - \frac{a-b}{2}}{\frac{k}{a-b} - 1} \tag{83.1}$$

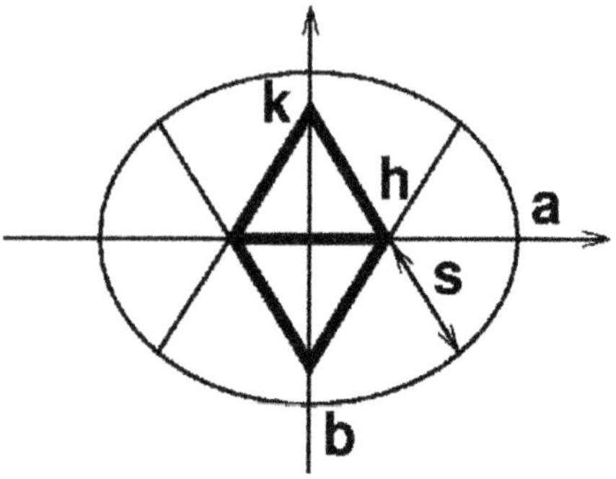

Fig. 83.2 Serlio's construction method for variable aspect ratio ovals. Image: authors

The construction can be generated conveniently using two equilateral triangles whose bases are centered on the origin. Their intersections with the axes determine h and k, which can be expressed as:

$$h = \frac{a-b}{\sqrt{3}-1}; k = \frac{\sqrt{3}(a-b)}{\sqrt{3}-1}.$$

The lengths of the circular arcs are specified by extending the triangles' diagonal sides by a length s. When s is increased both a and b are increased, and so it is not straightforward to choose geometrically correct values of h and s so as to achieve specified values of a and b. In fact, the ratio is given by

$$\frac{a}{b} = \frac{h+s}{h(2-\sqrt{3})+s}.$$

Not all aspect ratios can be achieved with this method since two arcs per quadrant are only drawn when $s > 0$ such that $\frac{a}{b} \leq \frac{1}{2-\sqrt{3}} \approx 3.732$.

The ratio of the radii of the two circular arcs also varies, and can be shown to be

$$\frac{2}{2h+s}.$$

Fitting the Model to Noisy Data The majority of the papers in the literature analyzing the layout of amphitheatres tend to overlay (often manually) the proposed shape (ellipse or oval) and then visually evaluate its appropriateness. While this might be feasible with perfect data (and perfect manual capability) the Roman amphitheatres have been subjected to damage over the years. On top of this, it is quite natural that inaccuracies and local adjustments would occur in construction. Thus, the positions of the walls no longer perfectly correspond to either an ellipse or an oval composed of circular segments.

This problem is aggravated by the similarity between an ellipse and an oval. Depending on the number of circular arcs used, an oval can be found that is a very good approximation of an ellipse (Rosin 1999), and so the difference between a four-centered oval and an ellipse only amounts to a few centimeters. Given such subtle differences, visual evaluation is inadequate, and a more objective approach is required. This has led researchers to perform a fit between the proposed models (i.e., ellipse, polycentric ovals) and use the fitting error as a criterion for selecting the most appropriate model (Trevisan 2000).

In this chapter we also perform a best fit between the candidate models and the data. Rather than use a traditional least squares error term we minimize the mean absolute error using Powell's method (Press et al. 1990). This is less sensitive to outliers than least squares. The error at each data point is taken as the shortest distance to the oval or ellipse. For the former the appropriate arc needs to be selected, making the fitting a non-linear, iterative procedure. In the case of the ellipse, computing the distances requires solving a quartic equation and choosing the shortest of the four solutions, and so it is also a non-linear process. (Note that to ensure proper testing of the ellipse we do not use the approximation to the distance to the ellipse developed earlier (Rosin 1998). However, we found that the approximation was in fact very accurate, resulting in values in the following tables very close to those obtained using the true distance.) In addition to Serlio's oval, the best fitting four-centered oval that satisfies the tangent continuity constraint (Eq. 83.1) is found. Since the oval and ellipse fitting is iterative an initial estimate is required; this is obtained by first directly fitting a conic with a non-linear constraint to guarantee an ellipse (Fitzgibbon et al. 1999).

Amphitheatre Data Figure 83.3 shows the data acquired from the amphitheatre; the three nearly complete rings will be used for model fitting. Moving from the innermost to the outermost, they contain 99, 69, and 80 points.

The ellipse and the two oval models were fitted to the data as described above, and the fits are illustrated in Fig. 83.4. It can be seen that all curves provide a very good fit to the data.

The mean absolute residual errors are given in Table 83.1.

The best fits (lowest errors) are highlighted and it can be seen that the ellipse provides the best match to the data in all cases. However, despite the error values providing such objective and quantitative information, interpretation of these results must still be done with care. It is not sufficient to identify the best fitting model and declare that this was the one used in the planning and construction of the amphitheatre. For any set of data one of the models is bound to give a lower error than the remaining models. However, the question is whether that improvement in error of fit between one model and another is *statistically* significant.

Statistical Analysis To analyze the data the first step is to determine what statistical model is appropriate. The fact that the different models have different degrees of freedom makes directly comparing them based on their mean fitting errors problematic. Instead we shall compare the models via their distributions of errors, and test if two error distributions are significantly different or not.

The simplest and most common model assumes normal distributions. Histograms of the *signed* residual errors for two of the rings are given in Fig. 83.5. While some histograms look reasonably normal others are doubtful,

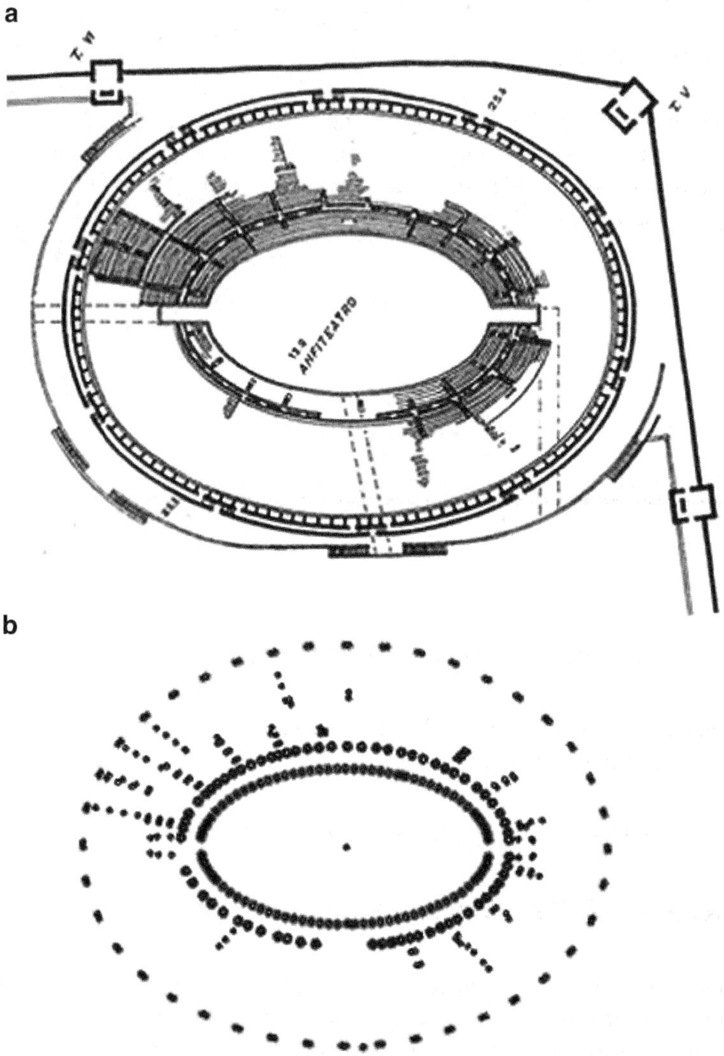

Fig. 83.3 (**a**) Plan of the ruins of the monument, (**b**) measured data from the amphitheatre. Image: authors

moreover their appearance is somewhat dependent on the histogram bin width. To check the hypothesis of normality more thoroughly the Shapiro-Wilk test (Siegal and Castellan 1988) is applied to the signed residuals (Table 83.2). If $p < 0.05$ then the null hypothesis of normality is rejected, and thus few fits (indicated by the highlighted values) can be considered to have normally distributed residuals.

The above means that many standard statistical tests for comparing hypotheses are not applicable to this data. A non-parametric test is more appropriate since it does not require the assumption of a normal distribution. However, even when

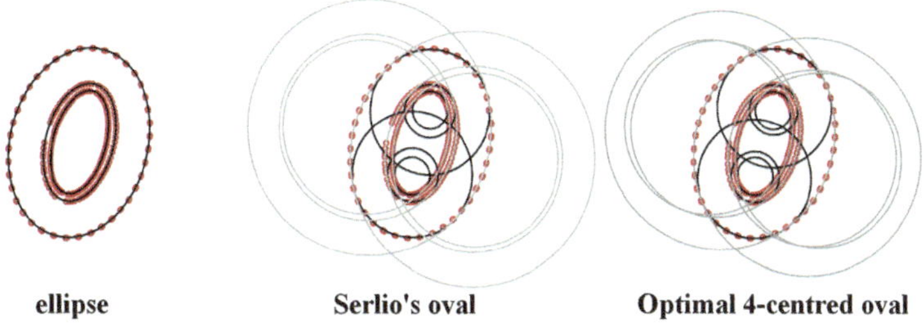

Fig. 83.4 Best fit curves overlaid on the amphitheatre data. Image: authors

Table 83.1 Mean absolute errors of fitted models to the amphitheatre data

Ring	Inner	Middle	Outer
Ellipse	0.043	0.078	0.166
Serlio's oval	0.119	0.162	0.289
Optimal oval	0.119	0.118	0.172

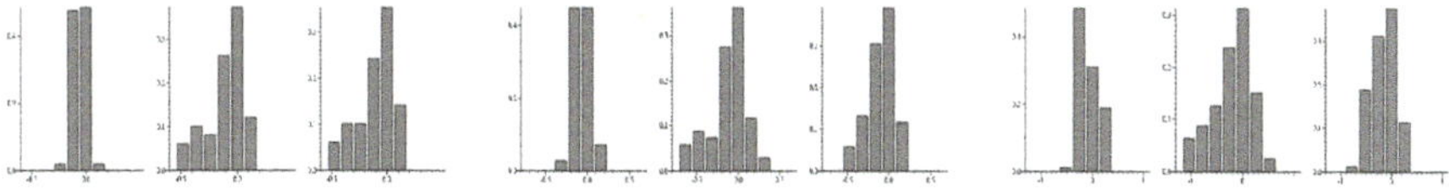

Fig. 83.5 Histograms of residuals for the three rings of points; each group of three shows the residuals for the fitted ellipse, Serlio's oval, and the optimal oval

Table 83.2 Normal test computed by the Shapiro-Wilk statistic; only the highlighted values can be considered normal

Ring	Inner	Middle	Outer
Ellipse	0.211	0.493	0.003
Serlio's oval	0.000	0.000	0.001
Optimal oval	0.000	0.018	0.030

considering non-parametric tests, most still make some assumptions on the two groups of data (error distributions in our case) such as:

1. they have the same standard deviations;
2. they have symmetric distributions;
3. the distributions have the same shape.

There are few tests that do not assume any of the above, and so we are forced to use the median test, which tests if two independent groups differ in central tendencies (the alternative hypothesis) (Siegal and Castellan 1988). The limitation of the median test is that since it makes so few assumptions it inevitably has a low statistical power, i.e., it needs more data points and/or lower noise levels to reach similar confidence levels than other comparable tests that make more assumptions (especially parametric tests).

Table 83.3 c2 values computed by the median test and applied to detect significant differences between the residuals from the ellipse fit and either of the two oval fits; only the highlighted values can be considered statistically significantly different

Ring	Inner	Middle	Outer
Serlio's oval	7.733	1.378	0.461
Optimal oval	6.761	0.000	1.842

The results of the median test are shown in Table 83.3. If $\chi^2 \geq 6.64$ then the null hypothesis of no difference between the distributions (at a significance level $\alpha = 0.01$) is rejected. In the case of the inner ring there is sufficient information in the data to allow the median test to discriminate between the ellipse and ovals. For the other two rings, although the ellipse model also fits the data better than the four-centered ovals, it is not possible to prove that this is statistically significant.

It is of interest to compare a recently published analysis of a similar problem: testing whether patterns in prehistoric wall paintings correspond to geometric models (ellipses, spirals, polygons) (Papaodysseus et al. 2005). In this case analysis was helped by two factors not available in our case of the Pompeii amphitheatre. First, the residual errors were shown to be normally distributed, and thus the more powerful *t*-test could be used to accept or reject the hypothesis. The second difference is that multiple instances of various patterns exist in the paintings. Assuming that the ones that appear to be circles were really intended to be painted as accurate circles enabled the painter's residual errors to be estimated. Consequently, it is straightforward to test if the residual errors incurred when fitting the other models (e.g., the ellipse) to patterns is greater than predicted by the error model, and should therefore be rejected.

Area and Perimeter Comparisons Rather than concentrate on residual errors calculated from the data points Kimberling took another approach when analyzing the grassy square in Washington DC called the "Ellipse" (Kimberling 2004). He compared values of area and perimeter measured from the data against the corresponding values determined from both an ellipse and tangent continuous oval. The ellipse was not fitted to all the data points, but rather its axis lengths were set to the same values as the diameter and width of the data. Likewise, the axis lengths of the oval were copied from the data. Two instances of the oval were then chosen by determining the values of h and k to match either the area or perimeter values measured from the data.

Tables 83.4 and 83.5 show the values we have obtained for the amphitheatre. The "polygon" row indicates values derived from the original data, with adjacent points being connected to form a polygon. Calculating the remaining values is straightforward except for the perimeters of the ellipses, as no closed form solution is available; therefore this was computed numerically using Mathematica. It can be seen that in all cases the area and perimeter values measured from the data are most closely matched by the ellipses' values. However, there is insufficient information available to perform statistical analysis.

Table 83.4 Estimated values of area obtained from the models fitted to the amphitheatre data (in m^2)

Ring	Inner	Middle	Outer
Polygon	*1862.47*	*2721.04*	*8746.07*
Ellipse	1865.41	2729.57	8772.99
Serlio's oval	1870.11	2738.37	8797.88
Optimal oval	1871.09	2734.41	8778.08

Table 83.5 Estimated values of perimeter obtained from the models fitted to the amphitheatre data (in m)

Ring	Inner	Middle	Outer
Polygon	*165.363*	*194.983*	*337.255*
Ellipse	165.293	194.986	337.455
Serlio's oval	165.785	195.377	337.854
Optimal oval	165.800	195.274	337.663

Analysis Through Ancient Mathematics

By the means of these various computations we have made an analysis of ellipse and four-centered oval fits to the concentric walls of the Pompeii amphitheatre. We have shown that both models provide accurate matches to the data, but the ellipse always fits better. However, these differences are relatively small and subtle.

In order to come closer to a definite conclusion, the problem must be approached from a practical architectural standpoint and no longer from a purely theoretical point of view.

The Geometry of the Building Pompeii is the oldest of the extant Roman amphitheatres (Fig. 83.6). It belongs to a building type that relies mainly on ground modeling, rather than on heavy masonry technology. Its construction involved partly excavating and partly raising the natural ground. The level of the arena lies below the exterior ground level, while the upper part of the *cavea*, the tiered seating area, is higher (Fig. 83.7).

The stone seats for the spectators used to lie directly on the ground, pre-modeled to provide a suitable slope running all around the central void of the arena. Therefore, tracing an oval curve for the arena would have meant that two out of the four centres would have been placed high in the middle of the slope of the raised ground, much above the level of the arena itself, while tracing ellipses using "the gardener's method" (two poles planted in the ground, and a rope) meant working on a horizontal surface, and dealing with two "centres" only: the two focal points of the curve located symmetrically on its major axis, inside its perimeter.

The superimposition of the diagrams, either of the ovals and the ellipses, on the plan of the monument shows the hypothetical position of the centres of the circles and the foci of the ellipses inside the monument itself (Fig. 83.8).

While discussing the advantages and disadvantages of each kind of curve, scholars always point out the fact that parallel ellipses cannot be traced from the same foci, while concentric oval curves can be traced from the same centres, and this very point obviously lead later architects to prefer the use of oval shapes for the design of later amphitheatres. But Pompeii's monument, being among the earliest

Fig. 83.6 View from the outside façade. Photo: authors

Fig. 83.7 View of the cavea from the arena. Photo: authors

buildings of this kind, has to be considered as innovative architecture, one that experimented with new geometrical patterns, and therefore attempting to reveal an utterly perfect diagram embedded in its composition might not be an appropriate goal [its builders did not even name it *amphitheatre*, but rather *spectacula*; the word "amphitheatre" itself appeared later on (Golvin 1988)]. Figures 83.4 and 83.10 show that the hypothetical arcs of the ovals of Pompeii would not be concentric anyway and their centres would be so distant from one another that this could not be

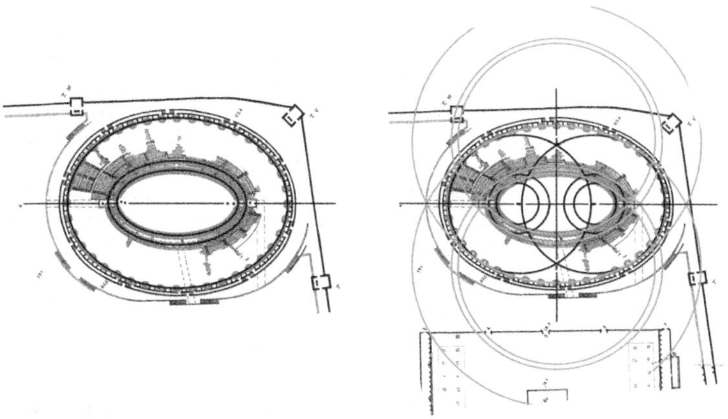

Fig. 83.8 The amphitheatre is built in the corner of the city wall. The superimposition of the diagrams on the plan of the monument shows that the centres of the larger *oval curves* would be in an awkward position for an easy layout. Image: authors

considered as errors due to historical traumas and subsequent permanent deformations.

In the late Roman Republic conic curves were part of common mathematical knowledge. We know that conic curves were first discovered by Manaechmus while searching for the solution to the Delian problem around 350 B.C. He is credited for having obtained them from the sectioning of acute-angled, right-angled and obtuse-angled cones. The curves were further studied by Aristaeus and Euclid. According to Sir Thomas Heath (1981), in Euclid's day some kind of focus-directrix property was already known. In any case, for tracing an ellipse only the knowledge of the existence of the foci is necessary (independent of the directrix). By the time Pompeii was built Archimedes and Apollonius of Perga had further investigated the properties of the conics. The ellipse was also known to Archimedes as the section of a cylinder; Apollonius even suggested that the planets had elliptic orbits of which the sun occupied one focus. In his treatise entitled "Conics", Apollonius examines various properties of the lines drawn from the foci to points on the curve or to specific points on straight lines tangent to the curve. And finally, Book III, proposition 52 shows that if, in an ellipse, straight lines are deflected from the "*points resulting from the application*" (i.e., the foci) to any point on the curve, the sum of the distances will be equal to the main axis (Densmore 1998). This well-known property of the ellipse, which has been extensively applied to drawing elliptic curves by the so-called "gardener's method", because of its simplicity, is not emphasized by the author and no corollary of the proposition is provided to point out a reverse practical aspect.

Archimedes and Apollonius were the greatest mathematicians to dedicate themselves to the study of conics, at the close of the golden age of Greek geometry. They died respectively around 212 and 192 B.C., about a century and a half before the building of Pompeii's amphitheatre, which was not the first to appear

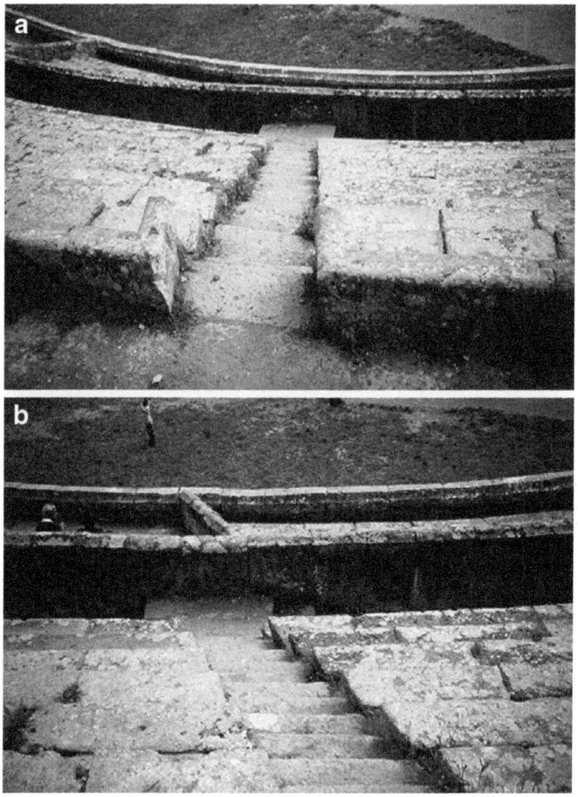

Fig. 83.9 Two views of some of the remaining stairs that used to divide the cavea into 40 different *cunei*, wedges. Photo: authors

in south Italy. We may be forgiven for thinking that we have here the proof of the interaction between science and art, and mathematics and architecture in the early Roman World.

Some attention must be paid as well to another aspect of the geometrical pattern of Pompeii, i.e. the radii that divide the cavea in several wedges (Fig. 83.9).

Some awkward attempts have already been made to determine the position of some converging points that would have facilitated their layout (Duvernoy 2014). But in the context of an elliptical pattern and a non-oval shape, the wedges of the cavea must have been drawn in some other way than radial division from a centre that does not exist. The upper perimeter of the cavea is pierced by 40 gates from which 40 stairs used to go down to the *ima cavea*. The upper ellipse of the monument was thus divided into 40 segments approximating the curve by an irregular 40 sided polygon. If we do the same thing with the lower ellipse of the cavea, and if the vertices of the two polygons are connected by rays, the resulting diagram quite perfectly matches the data coming from the survey (Fig. 83.10).

The inscription of polygons inside ellipses and circles is the methodology that Archimedes used to apply in order to find out simple ratios between the areas of ellipses and areas of circles having common "diameters". Archimedes is legendary

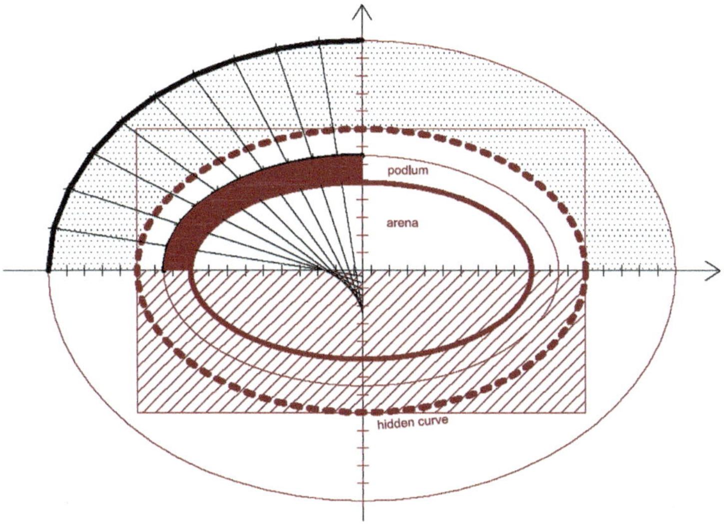

Fig. 83.10 Geometric diagram of Pompeii's amphitheatre. Highlighted areas (in *grey*) are of the same magnitude. Image: authors

for having attempted squaring the circle, but his efforts on squaring conics the ellipse and especially the parabola are less famous. Nevertheless, he stated that the area of a circle whose diameter is equal to the greater axis of an ellipse is in the same ratio to the area of this ellipse as its diameter is to the shorter axis of the ellipse (*Conoids and Spheroids*, prop. 4). Further propositions compare the areas of the ellipses to the areas of the rectangles that circumscribe them, and to circles and their circumscribing squares, etc. Since Archimedes also determined that the value of π can be approximated to $3 + 1/7$, we may assume that the calculations about perimeters and areas of the various curves of Pompeii's amphitheatre, required for building construction and financial purchases (estimating material quantities and costs), were possible.

The Arithmetic of the Building: Dimensions and Numbers of Pompeii The search for a module acting as the major common divisor of all dimensions (Duvernoy 2014) has shown that the axes of the three measured curves are respectively 10–19, 13–22 and 26–35 modules. Applying Archimedes' method for calculating the area, $S = (3 + 1/7)ab$, we find out that the area of the arena is exactly two thirds of the area of the curve closing the podium. In other words, the elliptic ring of the podium is exactly half the area of the arena. Because of the irrational value of π, numbers cannot be round for both the perimeter and the area of each curve, but ratios and proportions are precise.

Table 83.6 includes numbers about the "hidden" curve that could not be measured since it lies beneath the cavea, but its position can be precisely assumed from the entrances to the *ambulacrum*, or passageway, behind the podium.

Table 83.6 Examples of computation in modules and square modules

	"Diameters"		"Radii"			
Curve	A	B	a = A/2	b = B/2	Perimeter	Area
Arena	10	19	5	9.5	≈45.5	≈150
Podium	13	22	6.5	11	55	≈225
"Hidden"	16	25	8	12.5	≈64.5	100 (3 + 1/7)
Amphitheater	25	36	12.5	18	≈96	715

A "module" is equal to 12 Roman feet (roughly 3.50 m) and a "square module" is 144 square Roman feet (roughly 12.25 m^2). Comparison between Tables 83.4, 83.5 and 83.6 must be done considering the approximate value for π

This curve is the perimeter of the supporting wall of the outer part of the amphitheatre. It marks the division between the ground that was lowered and the ground that was raised: the "void" and the "solid". Therefore it must have been traced as a first step in the building process, since the supporting wall needed to be erected before starting any other operation. The area enclosed by the supporting wall, whose axes measure 16 and 25 modules, is exactly 314.2857 [=100(3 + 1/7)]. Consequently, the area of the raised cavea around the central void is

$$715 - 314.2857 \approx 400,$$

which corresponds to the area of the rectangle circumscribing the "hidden" ellipse (16×25). In other words, the ratio between the "void" and the "solid" is equal to the ratio of the area of the ellipse and its circumscribing rectangle. The elliptic ring around the void has thus been "squared": 400 is not only a rectangular number but also a square one. The "squaring" of irregular areas has always been one of the main concerns of ancient mathematics since Egyptian times, and in the case of Pompeii's amphitheatre, the squaring of the elliptic ring has been achieved thanks to the clever choice of specific numbers for modular dimensions. Calculations and computations are therefore easy. Each of the 40 wedges of the cavea has an area of 10 square modules, etc. (Fig. 83.11).

Conclusion

The analysis of the amphitheatre of Pompeii by the means of ancient mathematics was thus accomplished from two different standpoints. First, by noting the curves' shape, their centres, and the tracing of the radii, we discussed the geometry (i.e., the manipulation of the classical drawing tools, straightedge and compass mainly), and then, by carefully interpreting the dimensions of the monument, thanks to our knowledge of ancient metrology, we discussed the arithmetic (i.e., the manipulation of the natural integers). The geometrical analysis and the arithmetical analysis both converge to the same conclusion. Furthermore they corroborate the conclusions suggested by the numerical analysis with modern

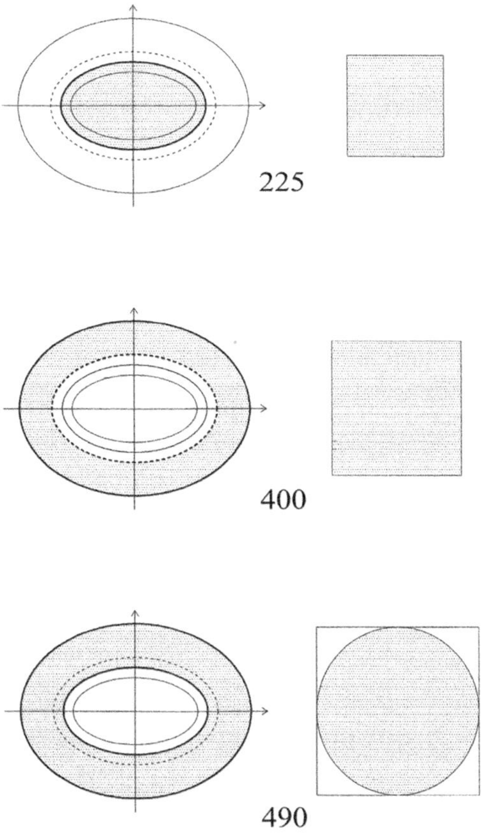

Fig. 83.11 The squaring of elliptic areas, and the figurate numbers of Pompeii's amphitheatre. Image: authors

mathematics (i.e., the manipulation of computer science). Therefore, the coherence of the results coming from our different approaches allows us to assert that the geometrical pattern of Pompeii's amphitheatre is a rare example of elliptic shape in architecture. Furthermore, its geometry and dimensions also show some of the finest evidence of direct application of the latest discoveries in mathematical knowledge and science in architectural design in classic antiquity.

And, most important, the influence of mathematical research on architectural formalism survived the geometric improvements or changes in successive amphitheatre design since the difference between an oval shape and an elliptic one is imperceptible to the observer.

Acknowledgment All images and photographs in this chapter are by the authors.

Biography Sylvie Duvernoy is an architect, graduated from Paris University in 1982. She was awarded the Italian degree of "Dottore di Ricerca" in 1998. After having taught architectural drawing for several years at the engineering and architecture faculties of Florence University, she recently began teaching at the

Politecnico di Milano. Her research mainly focuses on the reciprocal influences between graphic mathematics and architecture. The results of her studies were published and communicated in several international meetings and journals. In addition to research and teaching, she has always maintained a private professional practice. She is the author of *Elementi di disegni, 12 lezioni di disegno di Architettura* (Florence, Le Lettere, 2011).

Paul L. Rosin is professor at the School of Computer Science and Informatics, Cardiff University. His research interests include the representation, segmentation, and grouping of curves, knowledge-based vision systems, early image representations, low level image processing, machine vision approaches to remote sensing, methods for evaluation of approximations, algorithms, etc., medical and biological image analysis, mesh processing, non-photorealistic rendering, and the analysis of shape in art and architecture.

References

Densmore, D., ed. 1998. *Apollonius of Perga. Conics, Books I-III.* Santa Fe, NM: Green Lions Press.

Duvernoy, S. 2014. Architecture and Mathematics in Roman Amphitheaters. Chap. 13 in vol. 1 of this present publication.

Fitzgibbon, A., Pilu, M. & Fisher, R. 1999. Direct least-square fitting of ellipses. *IEEE Transactions on Pattern Analysis and Machine Intelligence* **21**(5): 476–480.

Golvin, J. C. 1988. *L'Amphithéâtre Romain.* Paris: ed. Pierre Paris.

Hart, V. & Hicks, P., eds. 1996. *Sebastiano Serlio on Architecture: Books I-V of "Tutte L'Opere D'Architettura et Prospetiva"*. New Haven: Yale University Press.

Heath, T. 1981. *A History of Greek Mathematics.* New York: Dover.

Kimberling, C. 2004. The shape and history of the ellipse in Washington, D.C. In *Proceedings of Bridges Conference*, R. Sarhangi and C. Séquin, eds. http://www.archive.bridgesmathart.org/2004/bridges2004-231.pdf. Accessed 25 November 2013.

Papaodysseus, C. et al. 2005. Identification of geometrical shapes in paintings and its application to demonstrate the foundations of geometry in 1650 B.C. *IEEE Transactions on Image Processing* **14** (7): 862–873.

Press, W. H., Flannery, B. P., Teukolsky, S. A. and Vettering, W.T. 1990. *Numerical Recipes in C.* Cambridge: Cambridge University Press.

Rosin, P. L. 1998. Ellipse fitting using orthogonal hyperbolae and Stirling's oval. *Graphical Models and Image Processing* **60** (3): 209–213.

———. 1999. A survey and comparison of traditional piecewise circular approximations to the ellipse. *Computer Aided Geometric Design* **16** (4): 269–286.

———. 2001. On Serlio's constructions of ovals. *Mathematical Intelligencer* **23** (1): 58–69.

Serlio, S. 1545. *Il primo libro d'architetettura di Sebastiano Serlio.* Venice: Francesco de' Franceshi Senese.

Siegal, S. & Castellan, N. J. 1988. *Non-Parametric Statistics for the Behavioural Sciences.* New York: McGraw Hill.

Trevisan, C. 2000. Sullo schema geometrico costruttivo degli anfiteatri romani: gli esempi del Colosseo e dell'arena di Verona. *Disegnare Idee Immagini* **18–19**: 117–132.

Chapter 84
Correlation of Laser-Scan Surveys of Irish Classical Architecture with Historic Documentation from Architectural Pattern Books

Maurice Murphy, Sara Pavia, and Eugene McGovern

Introduction

Traditional methods used for surveying and recording historic structures are (a) based on manual measurement systems using tapes and levels, (b) instrument based, using theodolite and level, and (c) imaged-based, using rectified photography or photogrammetry. More recently, digital technologies have automated the process of both the collection and the processing of measurement data. These technologies are based on the use of a laser to detect range and geometry, digital photo-modelling for supplying geometry and texture and, finally, software platforms to compute and model virtual buildings and environments. The automated measuring systems for recording the historic structures examined in this chapter are terrestrial laser scanning, digital photo-modelling and a combination of both systems. The product of the laser scan survey is described as a point cloud (Fig. 84.1.2) which represents the x-, y-, z-coordinates of a scanned object. The point cloud can then be textured from image data to create a virtual 3D model of a structure or object. Architectural pattern books which were published in the eighteenth century in Great Britain and the colonies are correlated with the laser scan survey data to facilitate a more comprehensive analysis of historic construction techniques used in this period.

First published as: Maurice Murphy, Sara Pavia and Eugene McGovern, "Correlation of Laser-scan Surveys of Irish Classical Architecture with Historic Documentation from Architectural Pattern Books", pp. 23–32 in *Nexus VII: Architecture and Mathematics*, Kim Williams, ed. Turin: Kim Williams Books, 2008.

M. Murphy (✉) • E. McGovern
Dublin Institute of Technology, Bolton Street, Dublin 1, Ireland
e-mail: maurice.murphy@dit.ie; eugene.mcgovern@dit.ie

S. Pavia
Trinity College, Dublin, Ireland
e-mail: pavias@tcd.ie

K. Williams and M.J. Ostwald (eds.), *Architecture and Mathematics from Antiquity to the Future*, DOI 10.1007/978-3-319-00143-2_37,

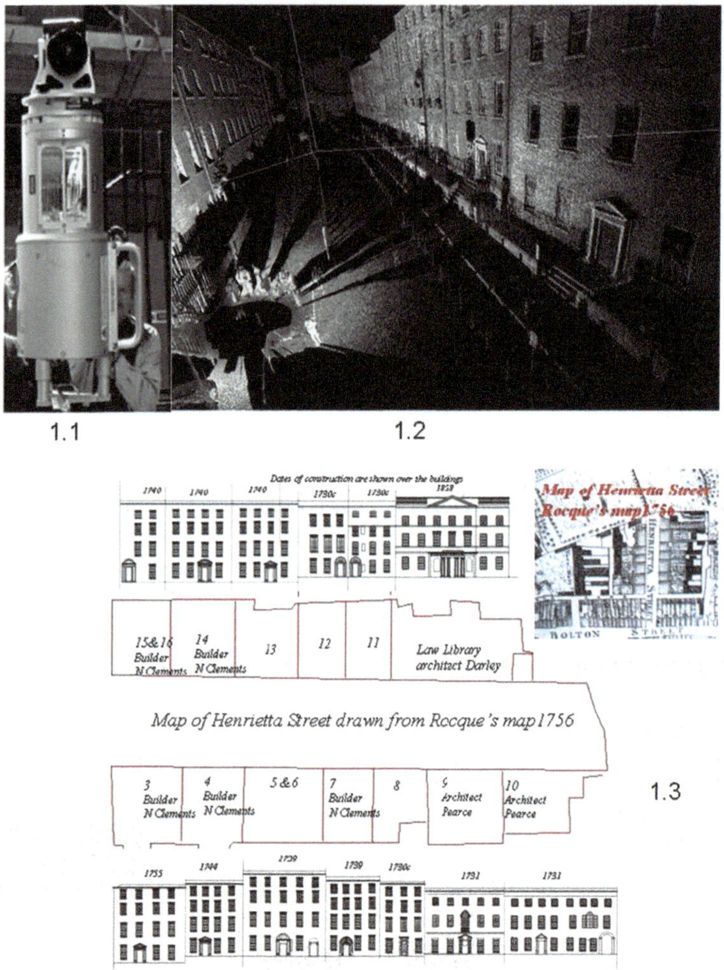

Fig. 84.1 Laser scan survey and detail geometry of façades of Henrietta Street. Image: authors

This research focuses on the recording and analysis of historic construction techniques used in post-medieval (from 1700 to 1830) structures in Ireland. For the purpose of this chapter, a case study is used which is confined to the front elevation façade and street fabric of Henrietta Street. The street is located near to the centre of Dublin City and is one of the earliest of the Georgian streets to be developed in the city. Henrietta Street was chosen as a case study because it represents the beginning of Dublin's great classical building period of the eighteenth century and can be traced back to its initial development in 1724 (Crimmins 1987).

Recording System-Terrestrial Laser Scanners

The survey of the front elevation façade and street fabric of Henrietta Street was carried out using a terrestrial laser scanner (*RIEGL* LMS-Z420i illustrated in Fig. 84.1.1). The terrestrial laser scanner is a device that automatically measures the three dimensional coordinates of a given region of an object's surface, in a systematic pattern and at a high capture rate near real time. The laser ranger is directed towards an object by reflective surfaces that are encoded so that their angular orientation can be determined for each range measurement. The entire instrument and/or the recorded object are rotated to achieve, where possible, complete 3D point coverage (Mills and Barber 2004). There are three types of scanners suitable for metric surveys for cultural heritage: triangulation, phase comparison, and time of flight. Triangulation scanners calculate 3D coordinate measurements by triangulation of the spot or stripe of a laser beam on an object's surface, which is recorded by one or more CCD (charge-coupled device) cameras. Phase comparison systems calculate range based on the difference in phase between emitted and returning wavelengths. Time of flight scanners calculate range, or distance, from the time taken for a laser pulse to travel from its source to an object and be reflected back to a receiving detector (Boehler et al. 2001). Time of flight scanners are most suitable for the metric survey of historic structures because of their long range, which is between 2 m to 2,500 m as opposed to triangulation and phase comparison systems, which are more suitable for recording smaller objects (Blais 2004).

All scanning systems are fitted with a CCD camera and the image data can be used to colour the product of the laser scan survey data, which is described as a point cloud. The point cloud represents the x, y, and z coordinates of a scanned object (Fig. 84.1.2). The RGB colour data from a digital camera can be mapped onto the point cloud by taking account of point translation, instrument rotation and perspective projection (Fig. 84.2.1). Both the laser and camera must be correctly calibrated geometrically (Abmayr, et al. 2005). The mounting position and orientation of the accompanying camera is defined in respect to the scanner's coordinate system, with every image representing a calibrated and registered image. High-resolution colour images can be precisely mapped onto a geometric model represented by a point-cloud, provided that the camera position and orientation are known in the coordinate system of the geometric model (Beraldin 2004) (Fig. 84.2.1).

Processing Laser and Image Survey Data

The point cloud (Fig. 84.1.2) requires cleaning, sorting and combining of different sets of point cloud data before processing takes place. Re-sampling by reducing the density of the data for overly dense point clouds can reduce the amount of

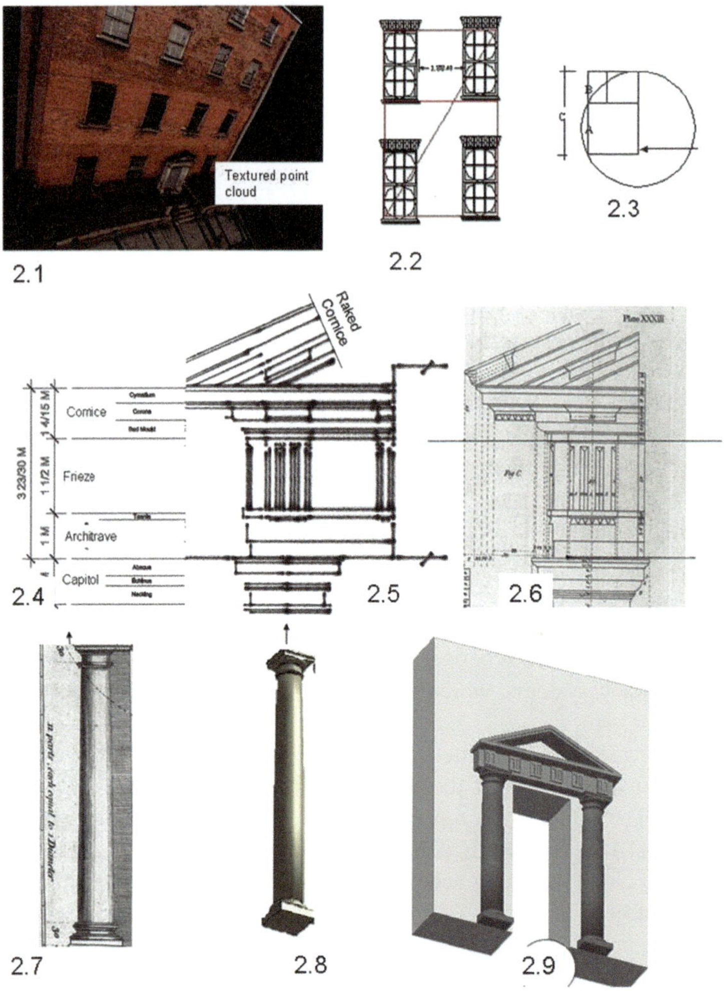

Fig. 84.2 Correlation of a laser scan survey and historic data. Image: authors

unnecessary data and file size. Software platforms, based on organisation of all data in an octree structure (a data structure useful for visibility of 3D data) permits the data to be sorted into an even point density on the scanned surface of the object (Remondino and El-Hakim 2006). Registration is the combination of several point clouds taken from different observation points or the referencing of the scanned object in a global or project coordinate system. This is achieved through the use of tie and control points that are either features of the object (e.g., corners) or special targets (spheres, flat targets with high reflectivity), which are identifiable in the point cloud at the processing stage. Software for registering point clouds usually facilitates registration by special targets, by overlapping point clouds, or by a combination of both.

The range data in the form of the point cloud from the laser scanner can be considered as a skeletal framework for recording the geometry of the historic structure. This geometric framework is then mapped with the associated image data allowing for more precise identification of the structure's texture and features. Polygonal surface meshing is the initial process which creates a surface on a point cloud; the created surface is made up of triangles connecting the data points into a consistent polygonal model (Remondino and El-Hakim 2006). Following the creation of a triangular mesh the results are then textured from associated image information (Fig. 84.2.1). The creation of an ortho-image from point cloud data allows for all of the image and geometric data to be exported for visualisation or further processing in CAD, VRML or other modelling platforms. In the processing of laser scanning data ortho-images are photo realistic 3D models containing the width, breadth and height of an object. The ortho-image represents the data for a particular plane on the x, y, and z axes; this can therefore represent elevation, plan, or section of an object.

Historic Data: Sources

The arrival of classical architectural styles in Ireland in the early 1700s marked a change in technology and the beginning of the modern construction industry. Before the arrival of classical architectural styles, the buildings were mainly medieval in the form of fortified structures (Ryan 1994). The style and construction of Irish classical architecture is based on fractal like components (Capo 2006), geometric proportion and a limited range of material and texture. There is sufficient historic data from this period concerning the historic construction techniques, and when combined with the field survey scan data, a more complete re-engineering survey emerges.

The architecture of the Renaissance introduced new and more scientific rules for the production of drawings and surveys. The most significant were the rules of perspective defined by Brunelleschi and Alberti in the mid-fifteenth century. Perspective rules introduced the concept of geometrical accuracy to represent and visualise architectural forms before they are built (Wyeld and Allan 2006). In the following century Palladio's 1570 *Quattro Libri dell' Architettura* introduced concise documentation of the rules of classical architecture (Jokilehto 1986). Palladio's principles and documentation of classical architecture were further developed by British architects in the seventeenth and eighteenth century. These later publications inspired the architectural pattern books which were published in the eighteenth century. The pattern books were based on the publications of seventeenth-century British architects rather than directly based on Palladio's work, creating a British and colonial Palladian style referred to as Georgian. The pattern books contained historic construction techniques used in the eighteenth century such as geometry and principles of the external and internal structure and fabric, construction, the positioning of openings, proportional relationship of the building's elements, and classical detailing.

Correlation of Laser-Scan Surveys and Historic Data

Orthographic elevations representing the street façade are the first results as shown in Fig. 84.1.3; these were plotted from the laser scan survey. Rocque's street map of 1756 was rectified, scaled, and then plotted to show the building line in 1756 (Fig. 84.1.3). The street morphology can be identified from comparing the historic map of 1756 and the façade geometry. From this comparison, it is evident that the street is practically intact, with the exception of demolition of part of 15 and 16, and the construction of the Law Library in 1828. Additional historic data, such as the identification of builder and architect, assist in identifying the evolving patterns of construction of the street. For example, Edward Lovett Pearce, who was one of the most prominent designers in Dublin in the early 1700s (Fearon O'Neill Rooney 2003), is accredited with the design of nos. 9 and 10 Henrietta Street. Pearce studied Renaissance manuscripts and in particular the Palladian styles that were popular in the early 1700s in Ireland and Great Britain, and subsequently this influenced his designs for Henrietta Street. There are no surviving design drawings for Henrietta Street. The façade elevations, plotted historic map and construction dates of the houses are shown in Fig. 84.1.3.

Detailed Geometric Façade Proportions

The rectangles in Fig. 84.2.2, 3 represent the Golden Section, a rectangular proportion that was believed to embody aesthetic qualities and used to position bay sizes and openings. The section was used to set out different combinations of window openings and bay sizes and the opening sizes of doors or windows and can be found in pattern book geometry. The proportional relationship between window openings and the bay width is determined by variations in the use of the golden section. The result is a modular system which can establish the dimensions of the bays in the façade walls, the sizes of window components, and the widths of supporting lintels (Ching 1979). The construction of the golden section in Fig. 84.2.2, which determines the bay width, was plotted from the point cloud. The section is formed initially by constructing a square, inscribing a circle from the centre of one of the sides of the square, then side A is extended by the distance B to meet the tangent, and the rectangle is completed (Fig. 84.2.3).

Façade Construction

The components and materials used in the façades of the buildings in Henrietta Street were identified and analysed from the point cloud and ortho-images. The main elements, which make up the façades of the houses on Henrietta Street, are the brickwork and stone in the external walls, the sliding sash windows, and the door-cases, doors and fanlights. Some of the original elements of the buildings are still in place and date back to about 1730 at the earliest, or have evolved with additions over the centuries. The external walls in Henrietta Street are mainly constructed in brick and laid in a Flemish Bond using a mortar mix of lime and sand. Flemish and English were the principle bonds used in the seventeenth and eighteenth centuries. Flemish bond was more economic when using facing bricks in the thick walls of the early 1700s, as the proportion of stretchers is greater (Innocent 1916). The arrangement for header followed by stretcher for each course in Flemish bond is more complex to build for thicker walls than English bond (Fidler 1875: 17–22), which is built with one course of headers and a next course of stretchers. The brick joints for houses nos. 16 and 17 are tuck-pointed. Introduced in the early nineteenth century, tuck pointing involved painting the existing brickwork and mortar which was then re-pointed creating an impression of a more uniform brick with regular mortar joints. Most of the original brickwork survives in the street with minor alterations (Fearon O'Neill Rooney 2003). A sample of brick dimensions measured from the scan survey is 9″ in length by 2 1/2″ high. The minimum dimensions specified in the building act of 1729 (Roundtree 2002: 34–35) (introduced by the architect Pearce who was also a member of parliament) were 9″ in length by 4 1/4″ in depth by 2 1/4″ high. More than likely locally manufactured bricks were used, as there were brickfields in Gardeners Mall and Moore Street nearby to Henrietta Street in the 1700s. The colours of bricks varied from red, purple, or grey in the late seventeenth century and up until about 1730 (Lynch 1993). Imported bricks from Holland (Nicholson 1823) were grey in colour. The scan survey indicates that the predominant colour of the facing brick in Henrietta Street is red to reddish-brown, further evidence that these bricks were manufactured locally.

Stone

The scan survey illustrates that both limestone and granite are found in the façades of the houses. Calp, a local limestone (Lewis 1837), is used in the walls of some of the basements and is mainly covered with a render. Local granite is used in plinths, cornices, steps, sills and copings. Imported Portland stone is used in the stone door cases. Fredric Darley (Craig 1980), the architect of the Law Library, was a member of the Darley family, which was involved in granite quarrying and building from the early 1700s in Dublin, Wicklow and Meath (Ryan 1992: 36–37); local granite was more than likely used for the construction of the façade of the Law Library.

The Law Library was built later than the other buildings in the street; the building methods were greatly influenced by the neo-classical style. Neo-classical detail and proportion in the façade of the building is component-based but demanded a greater detail in geometric accuracy in the carving and forming of the stone than did the Palladian-influenced buildings constructed much earlier. The precision of construction detail is applied across the variations of the classical components in the building from the columns to balustrades to the dressed ashlar stone.

Windows and Doors

The sliding sash window, the traditional window used in Georgian houses in Dublin in the 1700s, is used in all of the buildings in the street. From the early part of the eighteenth century, stone door cases in a classical style were used. Constructed with pediments supported by columns, the door cases were further enriched with carved mouldings. Figure 84.2.6 is an example of a part illustration from a pattern book showing Pain's 1788 drawings of elements of a Doric pediments, including the column capitals which were commonly used for building entrances.

Creation of 3D Objects Based on Historic Detail

Door-cases in particular were copied from the pattern books because of the detail required in reproducing the classical orders. The pediment of the door case which is plotted from Pain's 1788 drawings (Pain 1788: Pl. XXXIII) as illustrated in Fig. 84.2.5 as a 2D vector representation of its classical decoration and proportioning. This modular relationship as detailed in Fig. 84.2.5 is based on Pain's 1788 interpretation of the Doric classical orders. The Classical proportioning which is used is related to a series of modular relationships based on the diameter at the base of the column. Figure 84.2.5 sets out the modular relationships, all based on 1 module, which is the diameter of the base of the column. The pediment is made up of the entablature in the centre with the capital of the column supporting it and two raked cornices over it. These ratios establish the depth and thickness of materials in the pediment and columns, identifying their construction details behind the surface of the scan survey.

The 3D objects detailed in Fig. 84.2.8, i were constructed using the software platform ArchiCAD. Initially the pediment is modelled as a 2D vector object, as detailed in Fig. 84.2.5. The historic data taken from Pain's 1788 pattern book is used to build the profile of the 2D vector object. The main modular components, the architraves and entablature (based on the diameter of the base of the column), are used to model the object, the level of detail can be added to at later stages. The column is built from Pain's column detail (Pain 1788: Pl. XXXIII) again as a 3D model. The pediment and column are combined with the wall section and mapped to fit the geometry of the laser survey data (the point cloud; see Fig. 84.2.1). It is important to note that the 3D objects are modelled on the historic data first and later edited to fit the point cloud surveys. The 3D object (Fig. 84.2.9) represents a door case which contains details of direction, *x*-, *y*-, *z*-coordinate values, and construction components (i.e., the capital, base and column can be separated). It can be modified to suit accurately the geometry, scale and rotation of the point cloud survey.

Conclusion

The modelling process detailed in Fig. 84.2 illustrates a representation of some of the historic elements which make up the buildings of Henrietta street. It is not possible to model all of the components within the space of this chapter. In Fig. 84.2, a historic building information model (HBIM) (Eastman 2006) is proposed which correlates laser scan survey data with historic data in order to analyse the building techniques of this period. The documentation and recovery of historic methods of construction is critical in order to maintain, conserve and restore the existing building stock. The parallel process of augmenting the results of the laser scan survey with relevant historic data can be improved through increasing the number of sources of archival information such as original maps, drawings and related text. The most relevant sources of information concerning the construction of Henrietta Street are the architectural pattern books of the early 1700s. While the pattern books of the eighteenth century contain detail of building geometry, they are limited in analysis of historic methods of construction. In contrast, technical manuals from the middle of the nineteenth century were very comprehensive in describing the science of building materials and varied approaches to building technology. This is best illustrated in such texts as Nicholson (1823) and Fidler (1875), which were published a century later but can assist in expanding the technological and scientific detail. The additional detail of construction techniques such as material specifications did not appear in pattern books but was passed on to builders from previous generations as empirical knowledge.

Biography Maurice Murphy is a lecturer and researcher in the area of building conservation in the School of Construction in the Faculty of the Built Environment, Dublin Institute of Technology. His research interests include building conservation and new technology applications, application of CD ROM and WEB design for educational courseware in building technology, Training and education in building conservation, Laser Scanning of existing buildings, BIM.

Sara Pavia is a lecturer and researcher on civil engineering materials, construction technology, applied conservation and building repair in the Department of Civil Engineering, Trinity College Dublin. She is a member of the European Committee for standarisation of limes (CEN, WG11, TG1) with responsibility for the evaluation and validation of the physical testing methods and chemical analysis of limes.

Eugene McGovern is a lecturer and researcher in the Department of Spatial Information Sciences in the Faculty of the Built Environment, Dublin Institute of Technology. He is currently leading a number of research projects in the area of laser scanning applied to both cultural heritage and engineering.

References

Abmayr T., F. Hortl, M. Reinkoster and C. Frohlich. 2005. Terrestrial Laser Scanning – Applications in Cultural Heritage Conservation and Civil Engineering. In: *Proceedings of the ISPRS Working Group v/4 Workshop 3d-arch 2005*: "Virtual Reconstruction and Visualization of Complex Architectures". S. El-Hakim, ed. Mestre-Venice.

Beraldin, J. A. 2004. Integration of Laser Scanning and Close-Range Photogrammetry – the Last Decade and Beyond. Pp. 972–983 in *XXth International Society for Photogrammetry and Remote Sensing (ISPRS) Congress, Commission VII*, Istanbul, Turkey.

Blais, F. 2004. Review of 20 Years of Range Sensor Development. *Journal of Electronic Imaging* 13, 1 (January 2004): 231–240.

Boehler W., G. Heinz and A. Marbs. 2001. *The Potential of Non-Contact Close Range Laser Scanners for Cultural Heritage Recording*. CIPA International Symposium. Potsdam, Germany.

Capo, D. 2006. The Fractal Nature of the Architectural Orders. *Nexus Network Journal* **6**, 1: 30–40.

Ching, F. D. K. 1979. *Architecture: Form, Space and Order*. New York: Van Nostrand Reinhold.

Craig, M. 1980. *Dublin 1600–1860*. Dublin: A. Figgis.

Crimmins, C. 1987. *Henrietta Street: A Conservation Study*. Master's thesis in Architecture, Dublin.

Eastman, C. 2006. *Report on Integrated Practice University and Industry Research in Support of BIM*. Georgia Institute of Technology, American Institute of Architects, ed. 2006.

Fidler, H. 1875. *Notes on Building Construction*. London: Rivingtons.

Innocent, C. F. 1916. *The Development of English Building Construction*. Rpt. 1999, Shaftesbury, UK: Donhead Publishing.

Jokilehto, J. 1986. *A History of Architectural Conservation*. York: Institute of Advanced Architectural Studies.

Lewis, S. 1837. *Topographical Dictionary of Ireland*. London.

Lynch, G. 1993. Brickwork: The Historic Development. *The Building Conservation Directory*. http://www.buildingconservation.com/articles/brick/brickwork.html

Mills, J. P. and D. M. Barber. 2004. *Geomatics Techniques for Structural Surveying Engineering* **130**, 2: 56–64.

Nicholson, P. 1823. *The New Practical Builder*. London.

Fearon O'Neill Rooney. 2003. *Surveys and Inventories of Henrietta Street*. Dublin.

Pain, W. 1788. *Pain's British Palladio, or The Builder's General Assistant*. London: I. and J. Taylor.

Remondino, F. and S.F. El-Hakim. 2006. Image-Based 3D Modeling: A Review. *The Photogrammetric Record Journal* **21**, 115 (September 2006): 269–291.

Roundtree, S. 2002. *A History of Clay. Brick as a Building Material in Ireland*. Master's thesis, Department of Art History, Trinity College Dublin.

Ryan, M. 1994. *Irish Archaeology Illustrated*. Dublin: Country House.

Ryan, N. 1992. *Sparkling Granite*. Dublin: Stone Publishing.

Wyeld, T.G. and A. Allan. 2006. The Virtual City: Perspectives on the Dystopic Cybercity. *The Journal of Architecture* **11**, 5: 613–620.

Part XI
Theories and Applications of Computer Sciences

Chapter 85
Mathematics and Architecture Since 1960

Lionel March

"Mathematics and Architecture since 1960"—an impossibly tall order! In a short chapter, I can only sketch an outline of work that I have had direct involvement with since the 1960s. I will touch upon some of my own work and work by some of my closest colleagues in architecture and urban studies.[1] Much of this work is recorded in the academic research journal *Environment and Planning B*, of which I was appointed founding editor in 1974, and which, under the banner *Planning and Design*, is now in its 29th year. Many *Environment and Planning B* contributors have been colleagues "at a distance" whom I may have met on occasions, or not.[2]

First published as: Lionel March, "Architecture and Mathematics Since 1960", pp. 9–33 in *Nexus IV: Architecture and Mathematics*, Kim Williams and Jose Francisco Rodrigues, eds. Fucecchio (Florence): Kim Williams Books, 2002.

[1] First in architecture at Cambridge University, for two years in urban studies at Harvard and MIT, back at Cambridge as Director of the Centre for Land Use and Built Form Studies, then in systems engineering at the University of Waterloo, in design technology at The Open University, later at the Royal College of Art, London, and—for the last two decades—in architecture and urban design at the University of California, Los Angeles, with a 6-month stop-over as a consultant at The National University of Singapore.

[2] For example, I co-authored a paper with the graph theorists Frank Harary in the USA and R. W. Robinson in Australia in 1978 entirely by correspondence. I met Harary some years later at a cocktail party in Cambridge, but Robinson, never.

L. March (✉)
The Martin Centre, University of Cambridge, 1-5 Scroope Terrace, Cambridge CB2 1PX, UK
e-mail: lmarch@ucla.edu

K. Williams and M.J. Ostwald (eds.), *Architecture and Mathematics from Antiquity to the Future*, DOI 10.1007/978-3-319-00143-2_38,

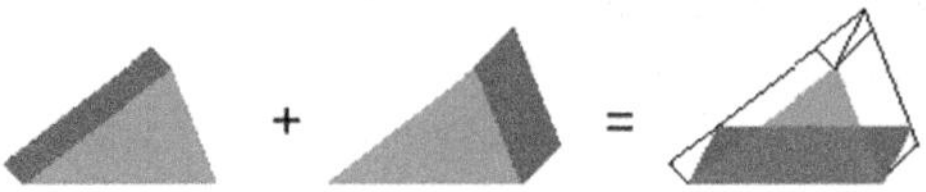

Fig. 85.1 Pappus's generalization of the Pythagoras Theorem where, for any triangle, the areas of the arbitrary *black parallelograms on the left* sum to the area of the appropriately constructed *black parallelogram on the right*. Image: author

Introduction

> A crude, but useful, distinction between mathematics and architecture is that the former tends towards abstract generalizations, whereas the latter is concretely particular.

In mathematics, take the simple rule of the so-called Pythagorean triangle—in a right triangle the squares on the sides sum to the square on the hypotenuse. Typical of the mathematical enterprise, the mathematician Pappus of Alexandria extended the rule to any triangle with any parallelograms on two sides (Fig. 85.1):

It is not difficult to see that Pythagoras's Theorem is just a very special case of Pappus. It is only a step further to arrive at what is known today as the cosine law.

Architecture tends to work the other way round. Given very general notions about buildings, the architect is asked to tease out a specific design to satisfy certain performance expectations. Mathematics can play a part at each of these three level: generic knowledge, specific design, and the prediction of performance. Mathematics has traditionally been used to predict physical performance such as structural, acoustic, thermal, lighting and so on. Here, accepted laws of physics are applied to specific geometries and materials to derive expected results which do, or do not, satisfy requirements. Armed with this information, the architects may modify their designs to seek improved performance. Similarly, there is economic modeling in which various costs and benefits are assessed. I shall not dwell on this analytical use of mathematical modeling which has nevertheless developed at a pace over the past 40 years. Instead, I shall concentrate on the role that mathematical thought plays in furthering generic knowledge in architecture, and then touch upon its potential use in generating specific designs.

I mean by "generic knowledge" notions generally entertained by architects and the public: the cultural fix, the conventional wisdom. In the 1950s, in the era of reconstruction following WWII, it was generally accepted that tall buildings would make better use of land than the low structures they were to replace. In the early 1960s, I was invited by Sir Leslie Martin to assist in producing a plan for such a reconstruction of Whitehall, the national and government center in London.[3] The government architects had already built a prototypical office complex some mile or so away. It sported three twenty-story towers over a three-story podium. It seemed that the government's intention was to pull down several Victorian structures

[3] Professor of Architecture, University of Cambridge. Best known in Portugal for his work at the Gulbenkian Foundation in Lisbon, but more widely for the Royal Festival Hall, London.

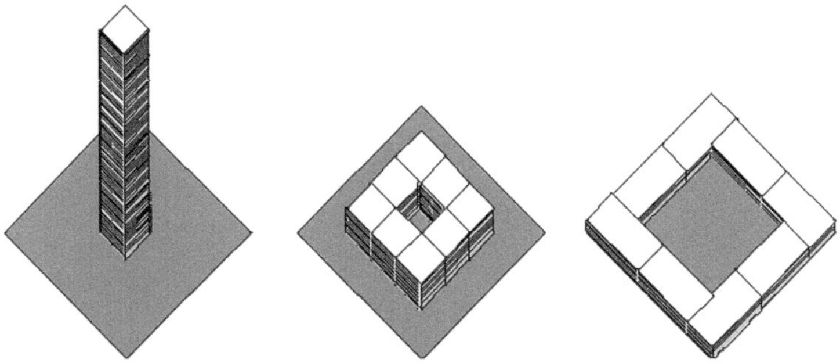

Fig. 85.2 Froebelian demonstration showing a given volume distributed in three different ways suggestive of tower and court forms. Image: author

(including Gilbert Scott's "battle of the styles" Foreign Office) and insert new buildings of this type next to Westminster Abbey, Inigo Jones's Banqueting Hall, William Kent's Horse Guards, and Norman Shaw's New Scotland Yard.

Consider an exercise using Friedrich Froebel's third gift from the wooden construction toys made famous among architects in Frank Lloyd Wright's *An Autobiography*.[4] A square table marked with 25 squares is the site. Placing all eight cubes on the center square creates a high-rise tower; placing them on the eight squares of the next ring creates a low-rise court form. Now the fourth gift has the same volume as the third, but is made from eight bricks with sides in the proportion 4:2:1. These eight bricks may be arranged around the sixteen perimeter squares at half the height of the last ring and one-sixteenth the height of the central tower (Fig. 85.2).

Buildings were arranged around courtyards in traditional development in, and projects for, the Whitehall area. London itself is renowned for its Georgian squares—examples of the courtyard form in urban design. I developed a program to explore the possibilities of arranging the built forms (as our very abstract representations of buildings were known) in three distinct arrays: the pavilion form (or isolated structure); the street form; and the court form.

The government was looking to put a certain number of civil servants to work in Whitehall. In building terms this translated into a gross building volume that was to be distributed over the site with a variety of options regarding conservation. The result of computations convinced us that the whole scheme could be arranged in courts surrounded by buildings no higher than the existing Victorian and Edwardian

[4] Friedrich Froebel (1782–1852) pioneered the modern study of form. He employed the three Aristotelian categories of quantity, quality and relation in structuring the educational content of his "gifts" which ran the gamut from solid, plane, line to point in descending order of concreteness, and ascending order of abstraction. His categories were "forms of knowledge" in which quantitative aspects are studied; "forms of beauty" in which the qualities of spatial transformations and symmetries come into play; and "forms of life" in which the forms are related to actual objects—a house, a bath, a chair, and so on. I use these categories to organize this chapter.

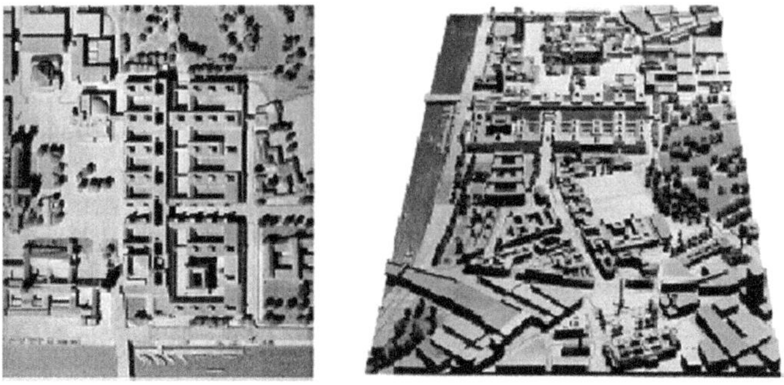

Fig. 85.3 Model of Leslie Martin's National and Government Centre, Whitehall, London, 1964. Image: author

office buildings in the area. Tall towers were not necessary, although even after the plan was accepted the government's architect pushed on with plans to build an office tower that would have dwarfed Big Ben in its bulk. It seemed that building something to blend in with the existing fabric was not appreciated at that time as a means to express the power of the "modern" state (Fig. 85.3).

These conclusions had been reached by number-crunching on an early card-hungry computer, and represent one of the first uses of electronic computing in architectural design. Shortly thereafter Leslie Martin established the research centre for Land Use and Built Form Studies (LUBFS) initially with funding from the Gulbenkian Foundation, and with myself as Director.[5]

Forms of Knowledge: Quantitative Studies

One of my first pieces of research was to formalize the results of the Whitehall study. Using no more than high school calculus, the new models demonstrated the superior effectiveness of low-rise courtyard development over street and pavilion forms. These results were then compared to the famous Heilgenthal-Gropius result that underwrote the internationalists' images of a city of residential slabs and towers. It was concluded that a more discerning appreciation of the mathematical model's 'structure' might have convinced Gropius that the greatest gains were to be found in *flachbau*, low-rise housing (March 1972). Of course, every mathematical model abstracts from actuality and only deals with a limited number of factors and assumptions. In my view, such models are useful in questioning our prejudices and sharpening our understanding as long as the limitations are taken fully into account.

A more general principle was extracted from such studies which we named after Fresnel, the French physicist, who had used such a pattern for light refraction gratings. Subsequently we learned that both Leonardo da Vinci and Albrecht

[5] Engineer Sr. Luis Lobato was an enthusiastic protagonist of this work.

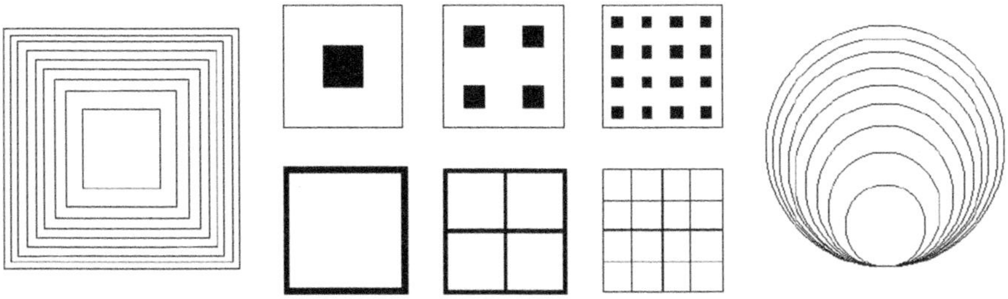

Fig. 85.4 *Left*, the Fresnel square. *Center*, different spatial distributions of the same content (the density of *black* against *white*) from *left* to *right*, concentrated to dispersed. *Center top row*, nucleated distribution. *Center bottom row*, reticulated distribution. *Right*, Leonardo da Vinci's diagram reconstructed from *Codex Atlanticus* 221v-b in which each crescent has the same area as the *full circle*. Image: author

Dürer had drawn related diagrams in their graphic studies of isoperimetric problems. Generically, the diagram shows a set of similar plane figures with areas equal to 1, 2, 3, 4, ... respectively. Fresnel and Leonardo used circles, we used squares. When the squares are placed about a single center, the diagram shows a series of concentric rings each of which is of unit area and therefore equal to the area of the central square (Fig. 85.4).

In this diagram, the unit square in the center occupies one-ninth of the largest square. The side lengths are thus proportionately $\sqrt{1} = 1{:}\sqrt{2}{:}\sqrt{3}$, $\sqrt{4} = 2{:}\sqrt{5}{:}\sqrt{6}{:}\sqrt{7}{:}\sqrt{8}{:}\sqrt{9} = 3$. The illustration shows this same unit area distributed in two ways—nucleated at the center and reticulated around the perimeter. Then, retaining the overall content, the area is fractured into smaller parts, but retaining the overall content. This diagram illustrates a problem with urban designers' use of the term density, and its subsequent architectural implication that high density implies tall buildings. Spatially there are four distinctions to be made: concentrated-nuclear, concentrated-linear, dispersed-nuclear, dispersed-linear. All have the same density, but exhibit quite distinct spatial distributions that carry quite different architectural and urban design implications.

Much of twentieth-century architecture has a basis in density calculations. Think only of the work of CIAM members in the pre-WWII years and the application of their doctrines in the years of reconstruction that followed. There was much "scientific" hood-winking concerning persons per square kilometer, bed-spaces per hectare and similar quantifications. This was the language of land-owners, public and private, who sought to make efficient or profitable use of land. The individual occupant is not put first. We very rarely hear about land per person, per individual.

If land were divided up equally, how much land do different people have around the world? Politically, this is a key question, because others have decided, or are deciding, how to parcel up that land. In Singapore, the Peoples Action Party clearly has decided to give over a substantial part of each individual's entitlement of land to public open space rather than private patios and gardens. Persons per hectare disguise this reality, whereas starting from the premise that each individual Singaporean—man, woman and child—might count the land of a single's tennis court as their own, the question might be asked, as a family looks out from its tenth-story apartment, how has most of it been used, for whom and by whom?

In the world today, each individual's portion of the land surface of the globe (excepting Antarctica) is equivalent to $150 \times 150 = 22{,}500\ m^2$. Estimates suggest that at the very worst this might fall to 13,500 m^2 by 2050, but is far more likely to have levelled out at 19,500 m^2, or the size of two soccer fields. Does this individual know that some two thirds of this land is already consumed in permanent pasture, forest, crop and arable land? Or that only two percent of the land surface is actually urbanized? Or that this urban use may barely increase to three percent of the land surface by 2050? Nevertheless, it does beg the question: how will this additional one percent—1,360,000 km^2—be urbanized, a land mass equivalent to the Northern Territory, Australia? Or, if everyone of the estimated seven billion persons in 2050 were to live at the density of sprawling Los Angeles, that no more land than in use today would be required for urban uses, and that an area the size of Japan might be returned back to non-urban uses? On average, each and every person in Los Angeles enjoys, appropriately, a basketball court of land. Yet, each individual making up the world's urban population (some 2.82 billion) average two and two-thirds times the land an Angeleno currently has (Fig. 85.5).[6]

I trust these simple arithmetical manipulations and interpretations of publicly-available data challenge the conventional wisdom. Part of the problem is to visualize the arithmetic in sensible ways, and part is that the numbers arise from the shape of urban and architectural artifacts. 2,750 persons per km^2 is an abstract concept: one person standing in an otherwise empty basketball court is concrete. Yet, they are equivalent.

> Architecture, in its applications, demands the concretisation of abstract mathematical statements.

A study prepared for an The Open University television program in urban geography is suggestive. I posed the problem of the compact city form and showed that without using more land other forms were possible. Take, for example, a nine block section of a theoretical compact city. The nine blocks might be reconfigured spatially as annulated form in which the depth of "service" from the roads is the same as in the compact city; or, as a cruciform. In each of these two reconfigurations the length of road is reduced by a factor of 55 %, which means that utility runs (water, sewers, electricity, telephone and cable) that normally follow roads are also reduced by this same amount. The maximum trip length across the compact section is 120 % of the same distances across the annular and cruciform configurations. Moreover, both these latter forms show lower expected mean distances than the compact city form: The cruciform shows that trips are

[6] Information extracted from AAAS (2000a, b).

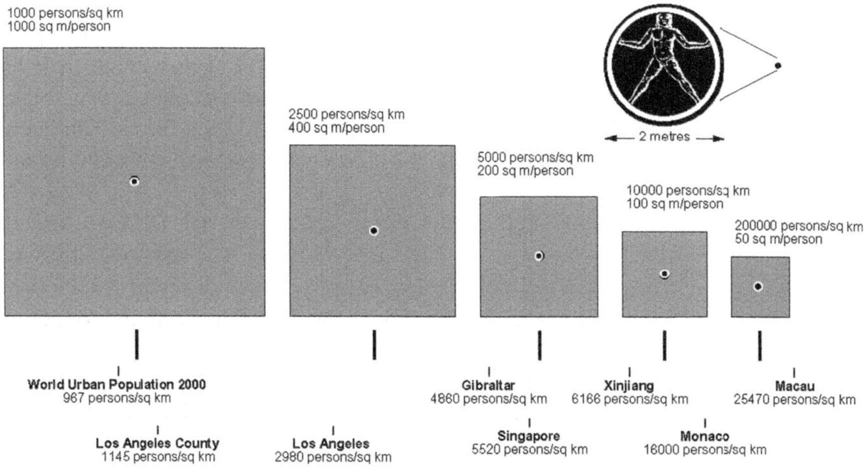

Fig. 85.5 Distribution of land in different communities and groups in the world population, 2000. The *circle* with Franceso di Giorgio's Vitruvian man marks the scale. Image: author

likely to be 95 % of the compact from, while the ring is best with 92 %.[7] Reductions this order are significant in energy terms as well as in convenience. The study suggests compelling counter-examples to the conventional wisdom about city forms (Fig. 85.6).

Suppose now that a new urban area is to be planned. Conventional surveys would pick out a piece of land and define a closed boundary around the urban area. But why not mark the protected open areas and let the urban form fill the interstitial spaces? The second two configurations suggest such an approach (Fig. 85.7).

There is no variety in these examples such as central city, market towns and villages which make up many urban regions. Another diagram indicates a "line" alternative to conventional "blob" thinking (Fig. 85.8).[8]

In the nineteenth century, the Spanish engineer, Ildefons Cerdà (1815–1876), set out the first modern theory of *urbanización* with the compelling maxim "Rurizad lo urbanos, urbanizad lo rural ... *replete terram*".[9] There is an 1861 sketch of his that illustrates the fundamental urban problem with compact city forms: that traffic requires more land as centers are approached at the same time that buildings require

[7] The unit of distance taken in this example is half a block length. The mathematical question of mean distances is addressed in Baglivo and Graver (1983: 98–111). An architectural investigation of built forms is found in Tabor (1971). On the assumption that all trips are equally likely, it should be noted that the compact city-section favors short trips over the other two configurations, whereas the annular form has an even distribution. In practice, trips are not likely to be equiprobable, and this needs to be factored in according to expectations.

[8] For a practical application of these ideas to the Central Region of Chile, see Echenique (1994).

[9] "Ruralize the urban, urbanize the rural. . .*fill the earth.*" I was first introduced to the works of Ildefons Cerdà by Dr. Marcial Echenique who had joined LUBFS from Barcelona to direct the urban systems study. A useful summary is Soria y Puig (1999).

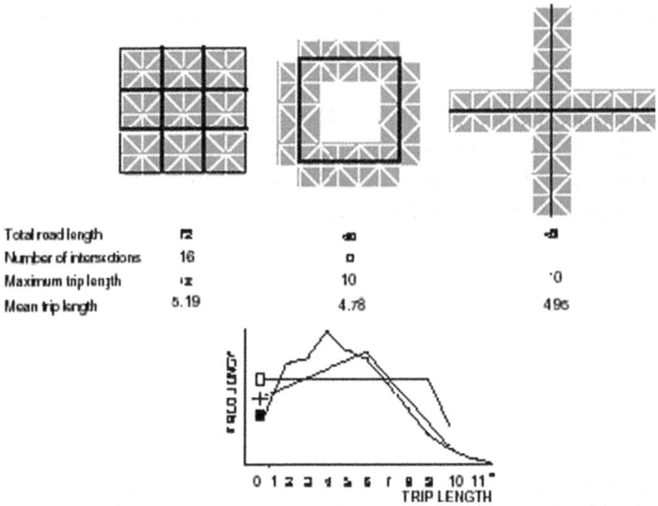

Fig. 85.6 *Top left*, a nine block section of a theoretical compact city. *Top center*, the same land area reconfigured as an annulated form. *Top right*, the same again reconfigured as a cruciform. *Below*, the frequency distribution of trip lengths. Image: author

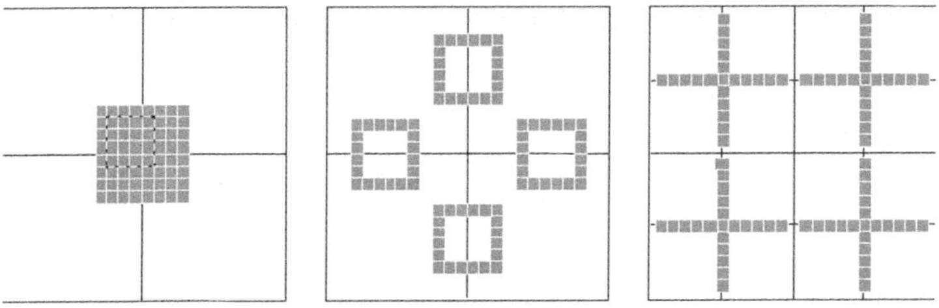

Fig. 85.7 *Left*, a theoretical compact city surrounded by a green belt. *Center*, four ring cities occupying the same amount of urban land. *Right*, four equivalent cruciform cities unite to create a reticular pattern. Image: author

more space. In 1961—quite unknowingly at the time—I illustrated this problem with two figures. The first showed the expected distribution of floor space in a theoretical city and the second the expected distribution of traffic. The two requirements are in mutual conflict: central blocks rise skywards as central roads become congested (Fig. 85.9).

The next exercise looks at this problem. Many downtown areas make use of small blocks established at a much earlier period. The city of Singapore is a good example in which the image of Manhattan has been adopted on sites so small that the footprints of the tower blocks are often less than 25 % the area of a typical New York high-rise. This has led to an exciting sky-line of elegant, but essentially grossly inefficient, buildings. Take again a theoretical model of a downtown: first with every one of $5 \times 5 = 25$ blocks filled by built forms eight floors high

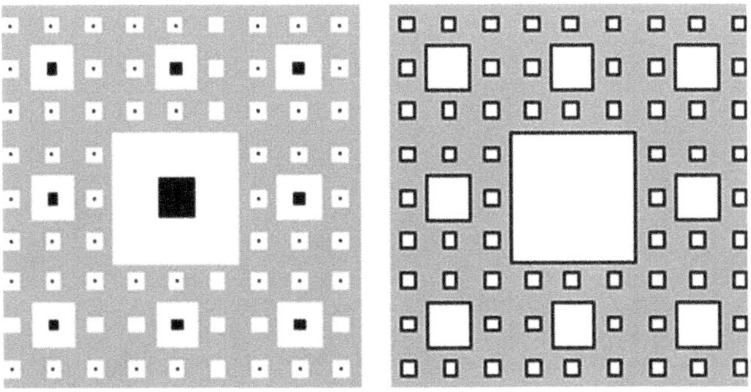

Fig. 85.8 *Left*, a theoretical urban region comprising a central city surrounded by a greenbelt, satellite market towns with their greenbelts, and necklaces of villages around each of these. *Right*, an alternative possibility: a regional park defined in the *center*, satellite community parks, and smaller village greens. The urbanization is linearized around these protected open areas, occupying the same amount of land as in the conventional configuration. Image: author

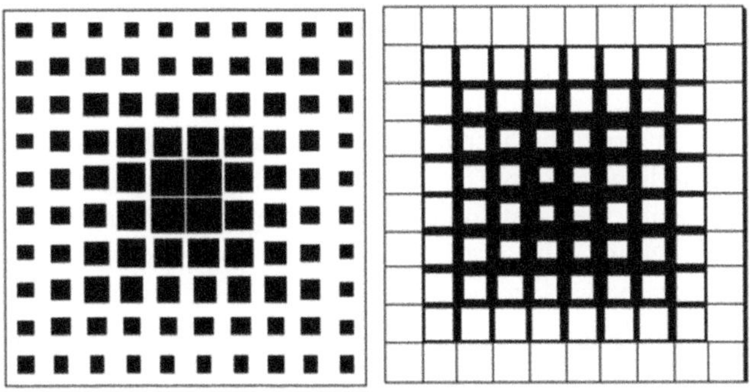

Fig. 85.9 A theoretical city. *Left*, distribution of floor space. *Right*, distribution of traffic. Image: author

($25 \times 8 = 200$ units). Now suppose that this floor space is reconfigured to look more like the arrangement with which we are familiar: each annuli from the center has the same amount of floor space so that the 16 outer blocks have built forms four floors high ($4 \times 16 = 64$ units); the next annuli of eight blocks retains the eight story high built forms ($8 \times 8 = 64$ units). While the center 64 units of floor space (at 32:1 ratio) occupy just one block to make 192 units in all annuli. There is a discrepancy of 8 units between the two arrays, but this of no consequence to the argument. The "exponential" growth in building height mirrors, but exaggerates, reality. The same roads serve the buildings, but it is clear that the central area will suffer acute congestion.

The whole floor space may be reconfigured around a single courtyard. The surrounding road consumes the same amount of land as before, but now there are just four roads of much higher capacity than the 12 roads of the original scheme. There is a ninefold reduction in intersections improving pedestrian flows, reducing

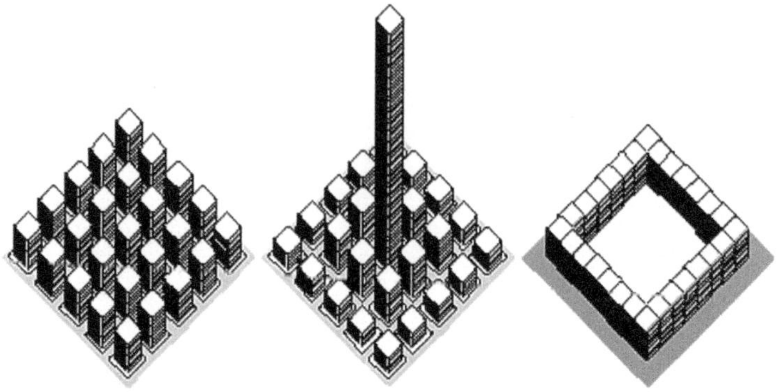

Fig. 85.10 A theoretical $5 \times 5 = 25$ block urban *center*: (**a**) An array of identical eight story blocks; (**b**) The same floor space reconfigured to provide equal floor space in each annuli from the *center*; (**c**) The same floor space reassembled in a seven story court form. Image: author

the risk of accident and the pollution of waiting vehicles. The most surprising result is that much of the land of the narrow and traffic-vulnerable sidewalks is collected together in a protected, traffic-free, courtyard the size of nine former blocks. The building is seven floors high containing 196 units of floor space (Fig. 85.10). Efficient tall buildings tend to have large footprints because of the service core. It is possible to retain the court form while halving the building depth for natural ventilation and lighting. This is achieved in a four court array.

Even so, for every two lanes of traffic per road in the original layout, the single court provides five lanes per road and the four courts provide three lanes per road: always holding the area of land devoted to transportation constant. The lane capacity of roads increases with the number of lanes. Keeping the road surface constant, suppose the 25 blocks are regrouped into four blocks, two-lane roads are replaced by three-lane roads with a lane capacity one and a quarter times higher, or into a single block surrounded by four-lane roads with a lane capacity just over double the two-lane scheme. Cerdá's problem finds a solution in such a courtyard arrangement, and this is precisely the form he chose as a type for Barcelona (Figs. 85.11 and 85.12).

These studies of built forms and the distinct ways they use land are based on the simple criterion of a sky ratio, or the angle at which the base of one built form makes with the parapet of a form immediately opposite. It will be observed that this angle is steep in the case of the pavilion forms, but much shallower in the comparable court forms.

The implication is that the ground floor occupant in the court forms are likely to see more sky than those in the tower configurations. A criticism that I make myself is that the sky view from a ground floor window is not as simple as the model assumes. In each individual case, a computer simulation will be needed to compute a more realistic assessment. The tower forms suffer from not taking into account the

Fig. 85.11 Three road networks showing the increase in traffic flows by increasing the lanes per route, but maintaining the same total road surface. Image: author

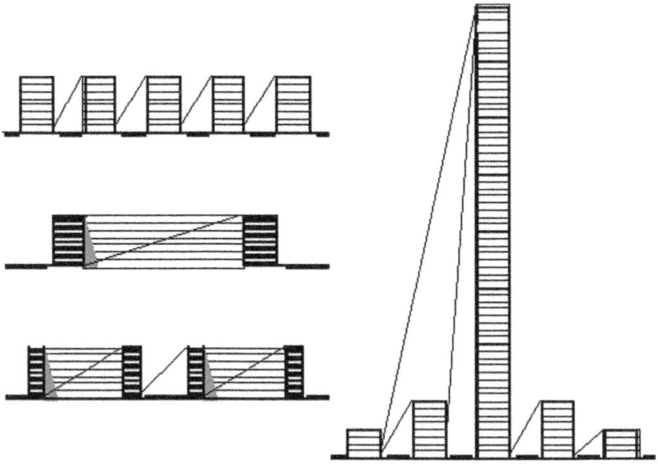

Fig. 85.12 Sky angles in the four configurations in Fig. 85.11 together with the cross-section of a four-court configuration, *bottom left*. Image: author

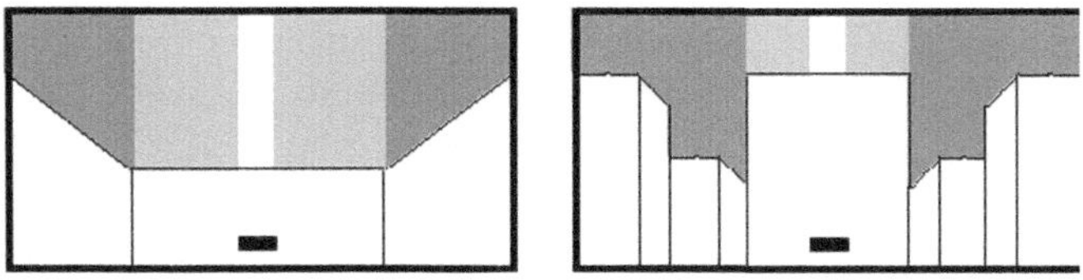

Fig. 85.13 *Left*, a view through a window in a court. *Right*, a view from a window in an array of towers. The *white* area is the sky accounted for in the model from a ground floor window. The middle *gray* area, together with the *white*, is the sky assumed to be seen from the rooms on the façade facing the opposite building. The *dark gray* is the additional sky not accounted for by the mathematical model. Image: author

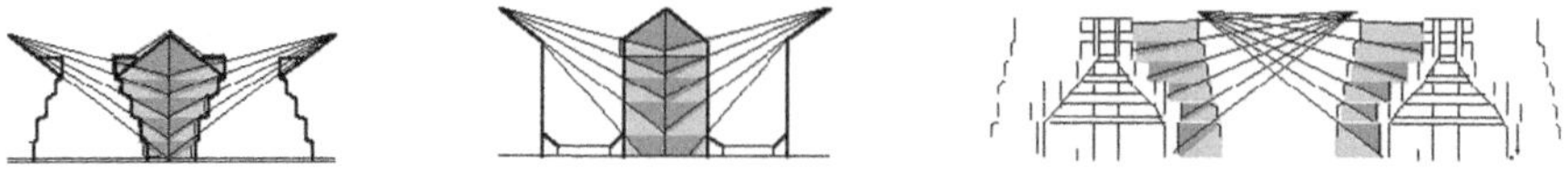

Fig. 85.14 *Left*, the sky angles for a typical Tudor street. *Center*, the sky angles for a typical post-Fire street. *Right*, the sky angles for the cross-section of the proposed Whitehall court form. Image: author

views around the sides of the built forms opposite, whereas the court only has internal views of the buildings surrounding the court (Fig. 85.13).

A practical example of the effect of sky angles on architectural design is shown in the considerations that led to Leslie Martin's project for new office buildings around courts in Whitehall, London. Historically, the early Tudor streets in the area, with their overhangs, may be seen as an attempt to have light penetrate to the center line of each floor. The later post-Fire (1666) terraces solved this problem by varying the story heights (Fig. 85.14). The Whitehall project proposed a stepped section on either side of an enclosed galleria for public use. The room depths now vary, but this variation accommodated the variety of room sizes in the program. Research on atrium offices since has confirmed that such a strategy has potential for energy saving in the British climate. The stepped-back elevations to the court provide a greater sense of openness in the courts themselves as well as in the offices on opposite sides.

Mathematical processes of thought have been used here to provide counter-examples to the conventional wisdom that society and professions have a habit of promoting. Alfred North Whitehead has described this mode of thought as "speculative"; it is profoundly radical in that it attempts to go back to basics. The alternative to centralized development gained the name "perimeter development". Some London boroughs made use of this alternate concept in their public housing, especially following a devastating gas explosion in one high-rise residential tower. The award-winning work of Richard McCormac is notable in this regard, as is a scheme in Battersea in which a proposal for 22-story towers was replaced by 4-story town houses and walk-up apartments. Later New Towns in the UK such as Livingstone and Milton Keynes adopted perimeter development. Richard McCormac realized that it was possible to increase the perimeter length around a site by something akin to crenelations, or more generally by foldings. This corresponds to increasing the available frontage for the homes (Fig. 85.15).

Phoenix, Arizona, plans to expand urban development north of the city, into the Sonoran Desert for some 300,000 persons. The desert is of great ecological significance.[10] The Nature Conservancy and partner organizations have identified some 100 landscape-scale conservation sites and some 30 smaller areas. Urban developments may flood around these conserved "islands". The lines of desert cities

[10] The Sonoran Desert covers about 222,700 km^2. in California and Arizona in the United States, and Baja California and Sonora in Mexico. It is the subject of a case study by the Nature Conservancy in AAAS (2000a: 188–89). In 1995 the regional population was 5.5 million, growing at the rate of 3.0 % per year.

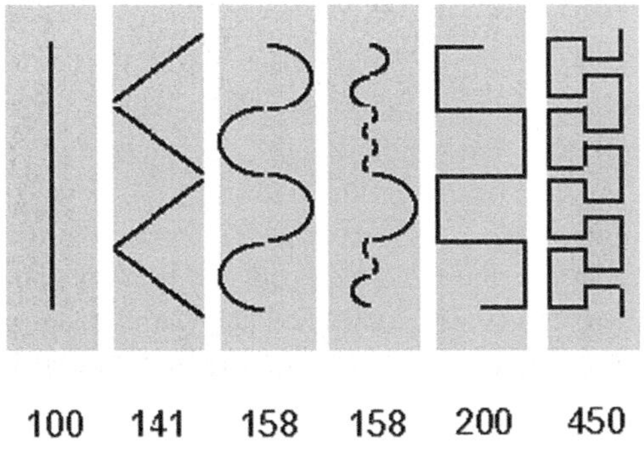

Fig. 85.15 The effective length of a perimeter may be increased by folds and crenelations. The numbers compare the increase to the straight line as 100 units. Image: author

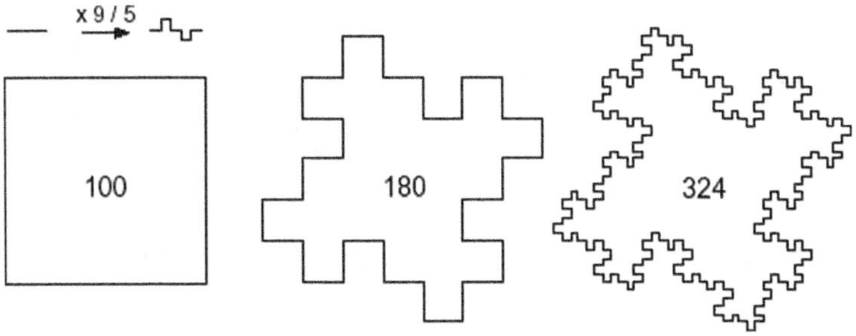

Fig. 85.16 Fractal development in which the perimeter between one use and another is increased while holding area constant (fractal dimension 1.36). Image: author

that fan out from Palm Springs, California, for some 50 km north and south of Joshua Tree National Monument provide an example of this open-space centered development. In opening a workshop on the Phoenix-Sonoran development, I questioned the drawing of closed boundaries around future urban uses. All parties agreed that people were migrating to the desert because they wanted to live the desert life, yet the planners and architects were visualizing gated and walled communities little different from other exurban developments elsewhere. It seemed to me that a desirable goal would be to maximize the boundary between the desert and the homes. I presented a simple fractal demonstration of how the urban area allocated might remain constant, but the perimeter might be increased threefold over the closed boundary favored by surveyors, or as much as required depending on scale. This is not dissimilar to the convolutions of inlet and isthmus that developers create in new oceanside communities to maximize access from the homes to water. Curiously, and sympathetic to my prior arguments, the original square tends towards a cruciform after a few iterations of the crenelating rule (Fig. 85.16).

Forms of Beauty: Qualitative Studies

For Froebel, forms of beauty were abstract designs that exhibited strong symmetries. Interestingly, he would break the symmetry of one design and, through a series of moves, transform the design into a novel one.[11] Symmetry is unavoidable, in modern terms, since asymmetry, called the identity, acts like the "one" of multiplication in common arithmetic, and thus it counts as a unique form of symmetry. An object has symmetry if there are spatial transformations that allow the object to move, and yet end up occupying the initial space. For example, if I take a paper square and place a pin in the center, I can rotate the square through any multiple of the right angle and each time its position will coincide with the original. I could also turn the square over along any one of four axes through the center—the horizontal, vertical, and the two diagonal—to achieve the same result. In the modern group theory of symmetry, it is possible to precisely say whether an object has more or less symmetry than another, and to know just how many subsymmetries of an object there may be to exploit in design. Louis I. Kahn's Assembly building in Dacca, Bangladesh, provides an excellent lesson. The ceiling of the main chamber is based on a regular 16-gon (the number 16 being a symbol of wholeness in Indo-Arabic cultures). The order of symmetry of this shape is 32. According to Lagrange,[12] the order of subsymmetries must be factors of this number: 16, 8, 4, 2 and 1, the identity. The facets of the ceiling mark out the symmetry of the octagon (order 16), the mosque is based on the symmetry of the square (order 8), the double squares of the offices follow the symmetry of the rectangle (order 4), while their internal planning is bilateral (order 2). Finally, the whole design has no overall symmetry (order 1): a globally asymmetrical composition that is replete with local symmetries. Quite deliberately, the powerful axis of the entrance lobbies, crossing that of the assembly chamber itself, is symbolically broken by the mosque, which adjusts its orientation to Mecca.

While symmetry may be described mathematically, it is not the conventional mathematics of solving numerical equations—typical of quantitative studies. The geometry of Euclid, for example, is about determining the measurements of lines, areas and volumes and comparing these properties for different figures such as the five Platonic solids. There is absolutely no explicit discussion in *The Elements* of the symmetry of these forms, yet it is symmetry that provides their compelling, foundational importance today (Baglivo and Graver 1983: 246–250).[13]

---

[11] Froebel had trained as an architect in Frankfurt, but became an assistant to Christian Samuel Weiss, one of the founders of the modern science of crystallography.

[12] Joseph-Louis Lagrange (1736–1813) was the founding Professor of Mathematics at L'Ecole Polytecnique, Paris, during the tenure of the architect J. N. L Durand in the stereotomy department under Abbé Haüy—Weiss's French rival in the architecture of crystal forms.

[13] There are just seven non-infinite spherical groups in space, two each associated with the Platonic duals, the cube/octahedron and the dodecahedron/icosahedron, and three related to the self-dual tetrahedron.

A simple example of qualitative study of form is given by examining the set of 12 pentominoes—two-dimensional figures in which five identical squares are joined in all possible ways edge to edge. Early configurational studies in architecture made use of such objects as synthetic, impersonal populations of designs. The 12 configurations may be grouped according to their symmetries: five are asymmetrical (identity, of order 1); one exhibits half-turn symmetry (the cyclic group, C_2, of order 2), two are bilaterally symmetrical in the orthogonal direction, two along the diagonal (all four are examples of the dihedral group, D_1, of order 2), one shows bilateral symmetry in two directions (the dihedral group, D_2, of order 4), and the regular cross has the full symmetry of the square (the dihedral group, D_4, of order 8). Preserving the orthogonal orientation, the identity forms may assume eight distinct positions; the C_2 form, four positions including left- and right-handed versions; the D_1 forms have four positions; the D_2 form two distinct positions, and the regular cross (say, the exterior of Palladio's Villa La Rotonda)[14] has only one position (Park 1996).

The population of pentominoes may also be classified topologically in terms of connectivities between adjacent squares (Fig. 85.17). A linear, planar graph represents this where a vertex of the graph stands for a square and the line joining two points marks a shared edge. The 12 designs fall under just four equivalence classes: seven are topologically equivalent to a simple "path", three to a "tree" with two branches; one to a "tree" with three branches; and one with a "cycle" where four squares make mutual contact around a common point. Another representation is the planar map, which may be thought of a rubber-sheet transformation of a pentomino itself. I shall return to the importance of this later.

Polyominoes may be used to define the footprint of a built form. Suppose the task is to accommodate 18 cubic units. One built form, 18 cubic units high, would achieve this. There is also just one way with a footprint of two units—a nine cubic unit high configuration. With a footprint of three units there are two distinct plan possibilities at six cubic units high: a simple slab, or an L-shape. With three floors, the footprint is six units. There are 35 hexominoes. Thirteen have the topology of a "path"—a simple string of six units on each floor, no matter how this is folded. Every cubic unit has at least two exposed sides. When the configurations have the topology of a "tree" with branches then some units will have only one exposed face, or possibly none. Twenty-two of the configurations are like this. A vertex with three branches means that that unit has $4–3=1$ exposed face. A vertex of degree 4 represents a unit with $4–4=0$ exposed faces. That is to say, the unit is completely surrounded on the interior. Eight of the scheme include a "cycle" in their topology. These are the most compact schemes. The mean distance measure within each floor of the configuration ranges from 1 for the nine-story slab; 1.33 for both six-story designs, and from 2.33 to 1.67 for the 35 three-story schemes (Figs. 85.18 and 85.19).

[14] The interior of La Rotonda has the subsymmetry D_2.

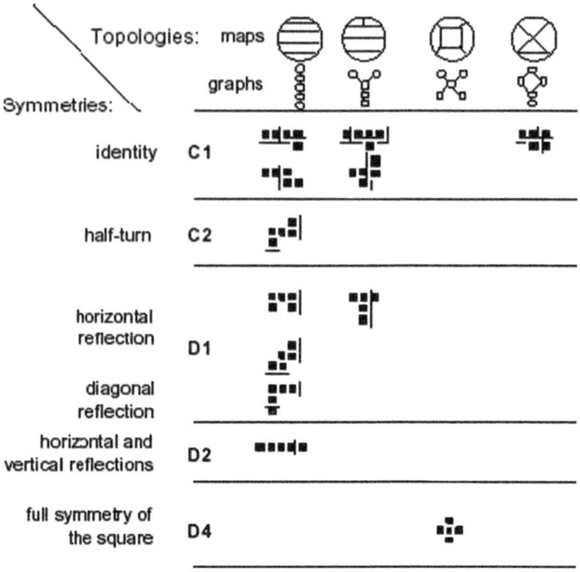

Fig. 85.17 The population of 12 pentominoes classified by symmetry and incidence represented by linear graphs and planar maps. Image: author

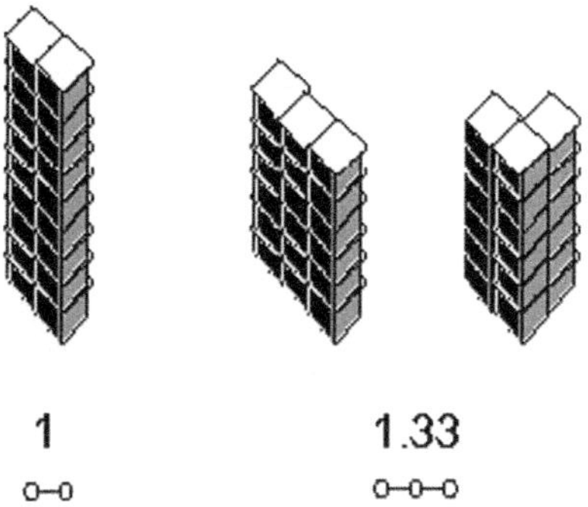

Fig. 85.18 *Left*, 18 spatial units assembled in one nine-story tower based on the unique domino plan. *Right*, the two six-story towers generated from the two triomino plans. The connectivity graphs and mean graph distance are given for the configurations. Image: author

An indication of the combinatorial explosion with this approach is demonstrated by the 1285 distinct two-story 9-omino designs, and the 192,622,052 one-story 18-omino configurations. Almost all of these schemes are asymmetrical.

If the quantitative studies have the musty scent of Renaissance isoperimetric and proportional problems, then these particular qualitative studies exude the antiseptic smell of Durand's polytechnic rationalism. Discrete elements are combined in tinker-toy fashion. Even if three-dimensional polycubes were used instead of the prismatic extension of two-dimensional polyominoes into three dimensions, such as in the above example, nothing will have changed—we are still in tinker-toy land. But until recently, this has been the cost of the abstraction necessary for the mathematization of synthetic design in architecture (Fig. 85.20).

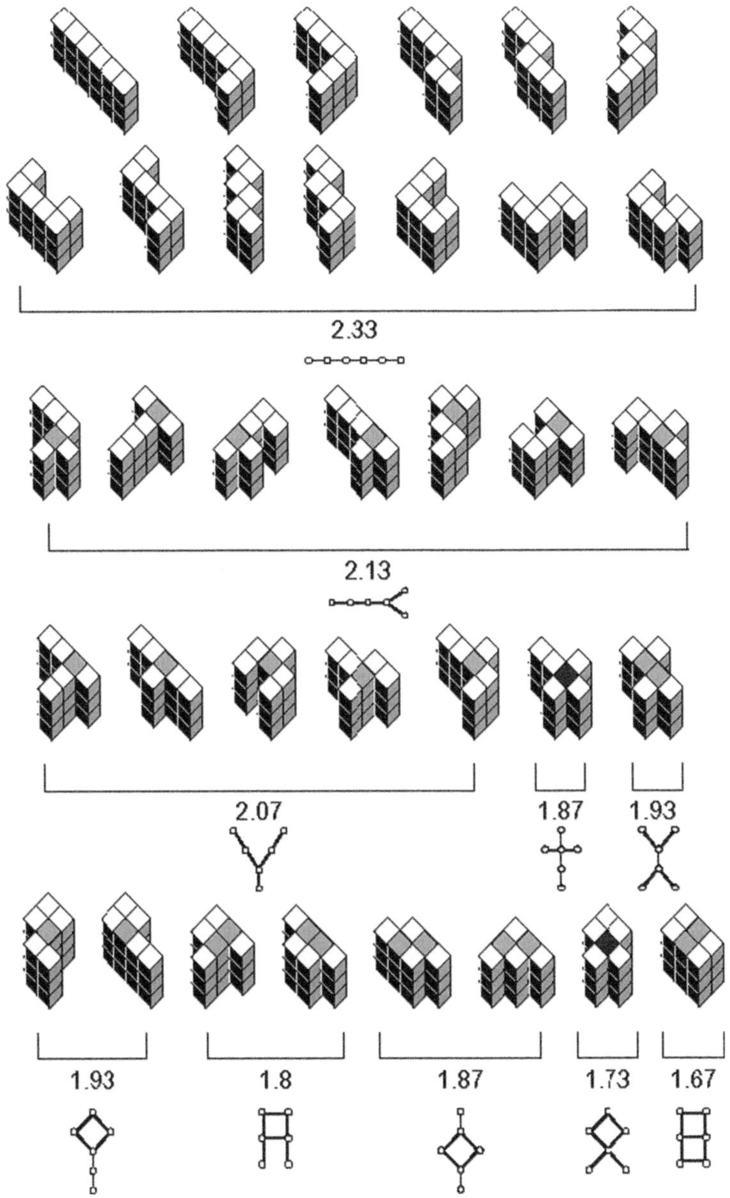

Fig. 85.19 The population of three-story built forms based on the prismatic development of the 35 hexomino plans. The *lightly tinted* areas show spaces with only one external face, and the *dark tint* indicates a space that is entirely interior. Image: author

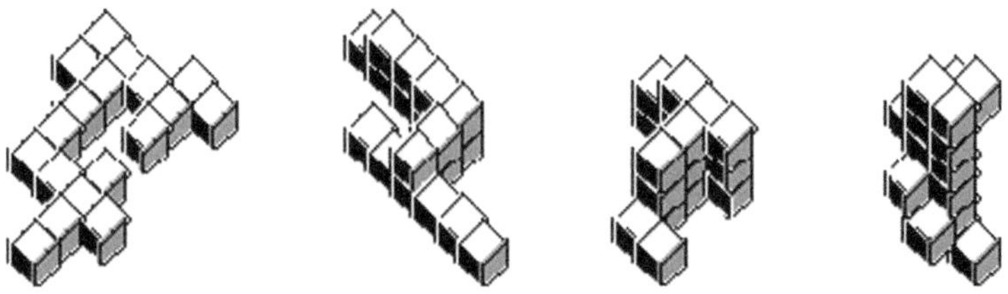

Fig. 85.20 From left to right: a 18-polyomino single story composition, and then three 18-polycube compositions—two-story, three-story and six-story—which are not prismatic projections of two-dimensional polyominoes. Image: author

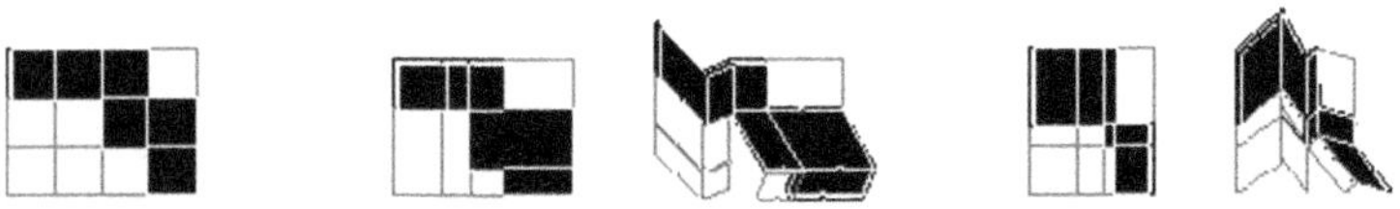

Fig. 85.21 *Left*, a dimensionless grating. *Center and right*, some dimensional transformations, shears and cuts, of the same grating. Image: author

To generate more realistic schemes, several authors, including myself, toyed with the idea of the "dimensionless grating".[15] This is best illustrated with one of the hexominoes (Fig. 85.21).

It will be seen that it sits in 3×4 grid, or grating. This configuration may be described using a Boolean—0, 1—code for computational purposes:

$$1110 \mid 0011 \mid 0001$$

reading from left-top to bottom-right of the grating.[16] Dimensions may then be applied to the grating to transform the configuration, and the grating may be "torn" to produce a yet more general transformation, while making the straight lines curvilinear would produce "morphs" similar to those found in D'Arcy Thompson's *On Growth and Form*. Thompson's book has been an influential source in the development of contemporary morphological studies in architecture.[17]

Let me turn now to an example from my stay in Singapore. I had used the earlier arguments to suggest that Singapore's largely high-rise, UN Habitat award-winning, New Towns had never been necessary. Those arguments had been introduced to at least one Singaporean architect, Tay Kheng Soon, who for his criticism found himself

---

[15] This concept arose initially from Newman (1964). Newman had directed the effort to build the world's first programmable, electronic computer at the University of Manchester, 21 June 1948. He is especially regarded for having recruited Alan Turing—of Turing Machine fame—to his staff.

[16] This Boolean description of a building appears in March and Steadman (1971: 121–144). It was subsequently elaborated in March (1976).

[17] See, for example, Steadman, P. (1983). *Architectural Morphology: An Introduction to the Geometry of Building Plans*. London: Pion Ltd.

exiled in Malaysia planning and designing high-density, low-rise housing.[18] My own contribution was deliberately contentious. I showed that Singaporeans could have been housed in one-story dwellings on no more urban land than currently used. The form of housing that achieved this result was the courtyard, which has a long tradition in the Chinese, Indian and Islamic cultures that meld together in Singaporean society, in contradistinction to the high-rise slab and tower of twentieth-century internationalism (March 1992).

Here, I use the example to illustrate an application of the modern mathematics of symmetry. One type of site that was considered was the familiar nine-square pattern. The site was walled and the dwelling could occupy at most 5/9 of the land within. For the sake of argument, I envisaged five units somewhat after the open manner of Kahn's Trenton bath-houses, leaving the enclosures to individual owners. The question was to maximize variety within these parameters. This reduces to the mathematical question: how many ways are there of "coloring" five of the squares in a 3×3 square. The colored squares would then represent a possible house plan and the uncolored squares would represent the open spaces, or courtyards. A simple answer would be the mathematical expression "9 choose 5" which is the factorial expression 9!/5!4! equal to 126. Many of these plans are the same under symmetry: right- and left-handed, or rotated. How many distinct plans is given by one of the most elegant pieces of modern combinatorial theory—George Pólya's Enumeration Theorem.[19] I can only hint at its power here. First, I give the answer diagrammatically (Fig. 85.22).

The result that there are just 23 distinct configurations under symmetry is derived from the group of Pólya figures which indicate the symmetry operations on the nine squares. The central square always remains the central square. Corner squares may exchange only with corner squares, and the four side squares with side squares (Fig. 85.23).

The information contained in the eight figures is captured in the cycle index:

$$\frac{1}{8}\left(f_1^9 + 2f_1f_4^2 + f_1f_2^4 + 4f_1^3f_2^3\right),$$

which is short-hand for "in eight figures we may choose a figure with nine 1-cycles *or* 2 figures with a 1-cycle and two 4-cycles *or* a figure with a 1-cycle and four 2-cycles *or* 4 figures with three 1-cycles and three 2-cycles". The next step is to make the following substitutions:

[18] "Meng Ker introduced me to the book—(March 1972)—and from there we did some morphological studies together and he did his thesis on Woodlands, one neighbourhood of Woodlands. And he redesigned a neighbourhood, the typical HDB [Housing Development Board] neighbourhood which was 10, 12, storeys at that time. He redesigned it to 5, 6 storeys, with one-third of the dwellings having gardens. So, to answer your question, are there alternatives to HDB, there are plenty of alternatives" (Bay 1998).

[19] See Pólya et al. (1983), where the lively "voice" of the nonagenarian Pólya can be heard still teaching at Stanford University, California.

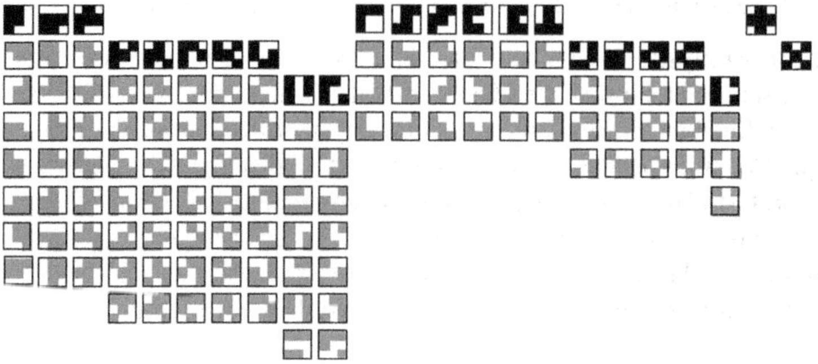

Fig. 85.22 The catalogue of the 126 courtyard houses in a 3 × 3 square in which four of the nine squares are open spaces. The 23 *black figures* are the configurations enumerated by Pólya's theorem, the *gray figures* show symmetrically-equivalent configurations in the vertical columns. The *indents* show configurations in which all five units are fully connected, then those in which some contact is corner to corner, and finally configurations in which the house is divided in two or more parts across a courtyard. Image: author

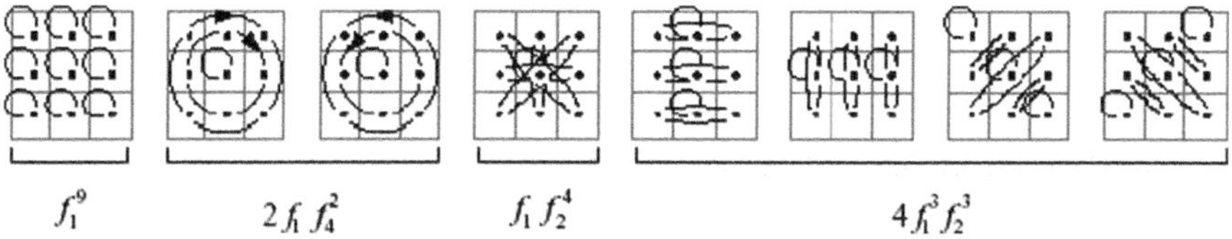

Fig. 85.23 The group of eight Pólya figures for the nine square problem. From *left* to *right*, if the nine-square is not moved, then all nine squares also remain stationary. "Nine 1-cycles" is symbolically represented by the expression below. Next, the figures show rotations through 90°, both clockwise and anticlockwise, to give "two figures with one 1-cycle and two 4-cycles". *Center*, is the exchange resulting from a half-turn, 180° rotation. This gives one figure with "one 1-cycle and four 2-cycles". *Right*, four figures in which reflection is indicated in the vertical axis, the horizontal, the leading and trailing diagonals. The symbolic notation reads "4 figures with three 1-cycles and three 2-cycles", the 1-cycles occurring along the axis of symmetry in each case. Image: author

$$f_1 \to 1 + x, \quad f_2 \to f + x^2, \quad f_4 \to 1 + x^4,$$

which are examples of what Pólya calls the "figure inventory". The first basically reads "in a 1-cycle we may choose to either not to color an element, 1, or to color it, x". The second reads "in a 2-cycle, if we color one element, then the other element of the pair must carry the same color—hence the term x^2". The third reads "in a 4-cycle, if one element is colored then so must the other three be colored likewise under symmetry—hence x^4". After the substitution, the cycle index looks like

$$\frac{1}{8}\left((1+x)^9 + 2(1+x)\left(1+x^4\right)^2 + (1+x)\left(1+x^2\right)^4 + 4(1+x)^3\left(1+x^2\right)^3\right),$$

which, when expanded, simplifies to the "counting polynomial"

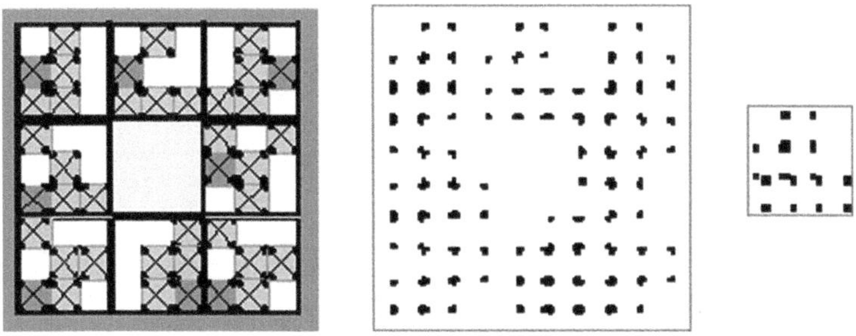

Fig. 85.24 *Left*, a group of nine-square homes arranged around a common open space. The *darker tint* indicates a service area. *Center*, the *poché* of the columns creates its own aleatoric pattern, although strictly based on symmetry considerations. *Right*, the *poché* of columns in Kahn's project for the Adler House, 1954–1955. Image: author

$$1 + 3x + 8x^2 + 16x^3 + 23x^4 + 23x^5 + 16x^6 + 8x^7 + 3x^8 + x^9.$$

The powers of x represent those designs with that number of colored elements. There is one configuration with no colored elements ($1 = x^0$), as there is just one in which all nine elements are colored (x^9). The number we are looking for is the coefficient of x^5 which tells us there are 23 distinct designs of the nine square homes under symmetry. These are the ones illustrated in black above. The coefficient of x^4 is the same, since this could be read as coloring the four open spaces instead of the five enclosures.

An elaboration of this example includes a second color that stands for the service unit to the served spaces. The inventory then looks like

$$f_1 \rightarrow 1 + x + y,$$

and so on for the 2-cycles and 4-cycles. The coefficient of the term x^4y counts the number of configurations with four served spaces, x^4, and one service space, y. That number is 89 (Fig. 85.24).[20]

Finally, in this section on qualitative studies, I return to the representation of architectural plans by planar maps. It turns out that the number of potential architectural plans can be enumerated and classified. There are only a certain number of planar maps with n-faces, or the compartition of the plane into n-rooms. These in turn may be derived from trivalent planar maps through "ornamentation", and every trivalent map in the plane is a stereographic projection of a three-dimensional trivalent polytopes. In other words, Plato had the right idea.

Christopher Earl and I wrote,

[20] For an excellent introduction to Polyá's work in the design context, I recommend (Economou 1999).

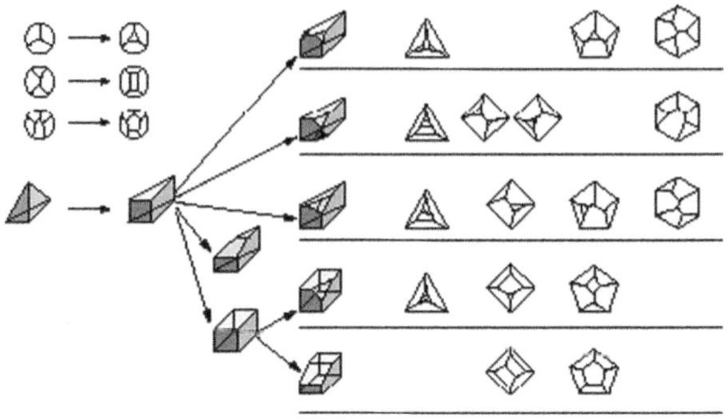

Fig. 85.25 *Top left*, the three necessary and sufficient rules to enumerate all trivalent 3-polytopes starting with the tetrahedron. The first rule shaves a 3-vertex to create a new triangular face, three new lines, and two new points; the second cuts a line to produce a rectangular face, three new lines and two new points; and the third takes three contiguous vertices and makes a cut which produces a pentagonal face. Starting with a four-faced tetrahedron, the next trivalent polytope, like a slice of cheese, has five faces; then there are two distinct six-faced polytopes including the cube; and five trivalent polytopes with seven faces. *Right*, stereographic projections through faces of the polytopes produce a countable number of trivalent planar maps, in this case with seven faces including the exterior. Image: author

> Essentially, the catalogue of trivalent 3-polytopes constitutes the "periodic table" for the "chemistry" of room formations. The catalogue is exhaustive. Such forms pre-exist in the recursive sense. They are produced by a simple rule system. At root we are saying that room formation is not itself a design problem, whereas ornamentation is. Immanent structures for each and every room formation are finite in number and are known aprioristically: architectural design is pre-eminently a matter of selection and the appropriate physical and material *transformation* of one of these fundamental plans (Fig. 85.25) (March and Earl 1977; Earl and March 1979).

It is also a mathematical fact that every such trivalent map can be represented by an orthogonal arrangement of lines—that "free-forms" are no freer than the right-angle. It may be of interest to note that there are just three "perfect" plans corresponding to the tetrahedron, the cube and the dodecahedron. In the first, all four spaces connect to three others; in the second, all six spaces connect to four others; and in the third, all twelve space connect to five others (including the exterior in all cases) (Fig. 85.26).

Forms of Life: Relations

We leave behind classical and neoclassical approaches such as those I have described in the previous two sections with the statement by George Stiny that "a design is an element in a *n*-ary relation among drawings, other kinds of descriptions, and correlative devices", and that "a relation containing designs is defined recursively in an algebra that is the Cartesian product of other algebras"

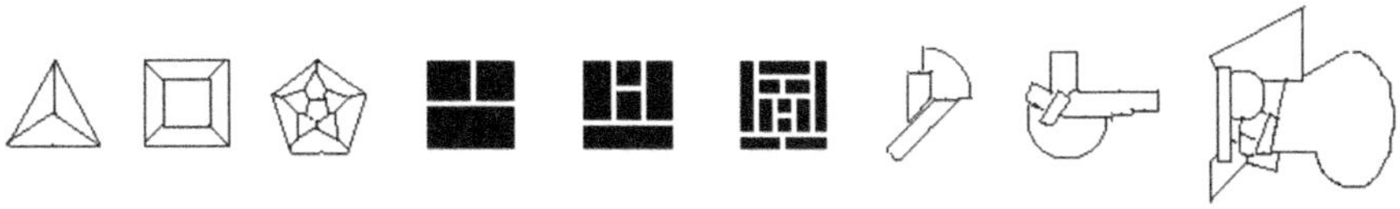

Fig. 85.26 Perfect plans. *Left*, the planar maps of the tetrahedron, the cube and the dodecahedron. *Center*, orthogonal presentation of these maps as rectangular dissections. *Right*, free-forms satisfying the same topologies. Image: author

(Stiny 1990). This is the language of modern mathematics and computation, not—unfortunately—of contemporary computing such as to be found in commercially available computer-aided design software. I draw a strong distinction between "computation" and "computing". Designers who compute using certain programs unwittingly become prisoners of encoded modes of thought, which, despite the flashiness of three-dimensional display, remain one-dimensional in origin. Almost all current systems rely on representations of lines, planes and solids as lattice point sets. Except for the square, it is rarely appreciated that a computer cannot draw an equilateral triangle, or a regular pentagon, or any other regular polygon. It has to fudge and fool the user into suggesting it can. Nor can a computer scale up a square to double, or triple its area, and draw it. It looks as if it is capable, but the task is actually beyond its digital powers. Digitization has its terrible limits, and it exacts a frightening intellectual price.

The shape grammar formalism provides an escape in its approach to shape computation. I have no space here to go into details. It is perhaps worth remarking that the Turing Machine is the standard theoretical construct for modern computation, and that a shape computation satisfies the criteria for a Turing Machine using shapes in place of symbol strings, or other tinker-toy sets in which there are primitives, or fundamental units. I will say that the repercussions of "seeing" shapes in a calculation liberates us from the norms of thinking with symbols, words and numbers (Stiny 2001).

Briefly, I will mention three examples of shape computation that the reader may care to follow up at more leisure. First, the shape grammar formalism has been successfully applied in examining questions of style in architecture and the visual arts. Some of the best work in this area is by Terry Knight, whose examination of the transformation in style for Frank Lloyd Wright's Prairie houses to the Usonians is particularly revealing (Knight 1994). My second case is recently completed work by Dr. José Pinto Duarte at MIT, who wrote a shape grammar interpreter to emulate Álvaro Siza's housing at Malagueira, Portugal. The 1,200 home project started in 1977 and continues today. Duarte has enjoyed the support of Siza in this exercise.[21] Finally, anyone who has had the misfortune to see a Frank Gehry building under construction, before it is dressed in its sexy metallic integument, will have been gravely disappointed by the massive clumsiness of the angled standard framing.

[21] See Duarte (2005a, b). Stiny and Knight were among Duarte's Ph.D. advisors.

Skin and skeleton are no more integrated than in Eiffel's Statue of Liberty. At the Engineering Design Centre, University of Cambridge, Dr. Kristina Shea is using the shape grammar formalism to take up the "next challenge in free-form structural design", which she sees as the "simultaneous design of intriguing surfaces and their corresponding structure" (Shea 2000; Papalambros and Shea 2001).

Concluding Remarks

In the late 1960s, I was told that the architects Alison and Peter Smithson found themselves at a loss with their son's schoolwork involving the "new maths". Apparently, they brought this generational gap problem to the attention of the Royal Institute of British Architects (RIBA) Library Committee, which in turn invited me to write a book for architects that would illustrate the potential of the "new maths" in their field. I invited Philip Steadman to join me in this task. *The Geometry of Environment* was published by the RIBA in 1971, and subsequently by Methuen and the MIT Press. Hungarian and Italian editions followed. The Smithsons had been impressed earlier by Rudolf Wittkower's *Architectural Principles in the Age of Humanism* (1949) and, in the debate on that subject at the RIBA, Peter Smithson had expressed an interest in the relationship of mathematics to architecture at mid-twentieth century as a parallel to the Renaissance. I have not mentioned my own *Architectonics of Humanism*, published in 1999 as a fiftieth anniversary companion to Wittkower, although its subject is probably closer to current interests among Nexus readers than the matters I have described. I must confess that my main motivation in writing *Architetonics* was not the proportional analysis of architectural works, but the origins of the modern arithmetization of geometry, essentially the digitization of shape. There is no map, there are no shapes, in Alberti's *Discriptio Urbis Romae*, only lists of coordinates—it is pure digitization. The primacy of number has dogged us architects ever since, and shape has been put on the back-burner. Yet, surely, giving shape to the human environment is the architect's primary task. The craze for digital imaging only serves to obscure centuries of neglect in the study of shape itself.

In *The Geometry of Environment*, Philip Steadman and I introduced to the architectural profession and schools the mathematical concepts of relations and mappings; set theory and Boolean algebra; group theory and symmetry; spatial transformations and matrix representation; and graph theory. We also updated proportional theory to cover modular co-ordination. Today, shape grammars use an extension of Boolean algebra called a Stone algebra that allows for a null shape, but not for an infinite shape; they also employ Euclidean spatial transformations that include scaling, translations, rotations and reflections. Symmetry plays a key role in local shape rule applications. Graph theory provides another description. Lattice theory provides the means to partially order decompositions of shape, which with each shape rule application may change—there are no fixed parts, only creative ambiguities. We can know what constitutes a shape after the final computation in its generation, and even then it depends on what rules we adopt to view it!

I have not discussed space syntax, which has proved to be popular in South American countries and especially in Brazil.[22] Its recent architectural applications include predicting pedestrian movements in buildings and the visual comprehension of interior spaces (isovist studies). The architect Norman Foster has employed these techniques in some projects (Hillier and Hanson 1984).

It would be amiss not to mention my old Cambridge student companion, Christopher Alexander, and the stimulus that his *Notes Towards a Synthesis of Form* (1964) gave to the mathematical treatment of architectural topics. Even though he came to reject his original thesis, the subject of decomposing complex programmes into manageable sub-problems persists. My problem in writing on mathematics and architecture since 1960 remains the utter unmanageability of the topic. In this, the Queen of the Sciences has forsaken me.

Acknowledgment I dedicate this chapter to the memory of Dr. John Ashby (1928–1999), a biochemist, and who, as a publisher with Pion Limited, London, had the foresight and courage to nurture and promote the journal *Environment and Planning B*.

Biography Lionel March holds a BA (Hons) in Mathematics and Architecture, Diploma in Architecture, MA, and Doctor of Science (ScD) from the University of Cambridge. He was co-editor with Sir Leslie Martin of the 12-volume *Cambridge Architectural and Urban Studies. He was founding editor in 1967 of* Environment and Planning B, now *Planning and Design*. He is currently Visiting Scholar, the Martin Centre for Architectural and Urban Studies at University of Cambridge, and emeritus member, Center for Medieval and Renaissance Studies, University of California, Los Angeles. His publications include *The Geometry of Environment (*with Philip Steadman, RIBA, 1971)*, Urban Space and Structures* (with Leslie Martin, Cambridge University Press, 1975)*, The Architecture of Form (*Cambridge University Press, 2nd ed., 2010), *R. M. Schindler: Composition and Construction* (with Judith Scheine, Academy, 1995) and *Architectonics of Humanism (*Academy, 1998). His most recent book is *The Mathematical Works of Leon Battista Alberti* (with Kim Williams and Stephen R. Wassell, Birkhäuser, 2010).

References

[AAAS] American Association for the Advancement of Science. 2000a. *AAAS Atlas of Population and Environment*. Berkeley: University of California Press.

———. 2000b. *Millennium World Atlas*. New York: Rand McNally.

[22] The work of Dr. Mário Júlio Teixeira Krüger, University of Coimbra, relates the space syntax tradition with the LUBFS approach. Dr. Nuno Portas was an important link with the Cambridge school following the 1974 Portuguese Revolution, as was Manuel Solá-Morales Rubó in Barcelona, who steered through the 1975 Spanish edition of *Urban Space and Structures*.

ALEXANDER, C. 1964. *Notes Towards a Synthesis of Form*. Cambridge, MA: Harvard University Press.

BAGLIVO, J. A. and GRAVER, J. E. 1983. *Incidence and Symmetry in Design and Architecture*. Cambridge: Cambridge University Press.

BAY, P. 1998. Interview with Philip Bay and Architecture Students from NUS on SPUR, transcribed by Dinesh Naidu. http://www.rubanisation.org/index.php?option=com_content&view=article&id=287:interview-with-philip-bay-and-architecture-students-from-nus-on-spur-transcribed-by-dinesh-naidu&catid=40:articles&Itemid=56. Accessed 27 November 2013.

Duarte, J. P. 2005a. A Discursive Grammar for Customizing Mass Housing: the case of Siza's houses at Malagueira. *Automation in Construction* **14**, 2 (March 2005): 265–275..

———. 2005b. Towards the Mass Customization of Housing: the grammar of Siza's houses at Malagueira. *Environment and Planning B: Planning and Design* **32**, 3 (May 2005): 347–380.

EARL, C. F. AND MARCH, L. 1979. Architectural applications of graph theory. In R. J. Wilson and L. W. Beineke, eds. *Applications of Graph Theory*. London: Academic Press.

ECHENIQUE, M. 1994. 'Let's build in lines' revisited. *Planning and Design*, **21**, 7: pp. s95-s105. London: Pion Ltd.

ECONOMOU, A. 1999. The symmetry lessons from Froebel building gifts. *Planning and Design*, **26**, 13: pp. 75–90. London: Pion Ltd.

HILLIER, B. and HANSON, J. 1984. *The Social Logic of Space*. Cambridge: Cambridge University Press.

KNIGHT, T. W. 1994. *Transformations in Design*. Cambridge: Cambridge University Press.

MARCH, L. and EARL, C. F. 1977. On counting architectural plans. *Environment and Planning B*, **4**: pp. 57–80. London: Pion Ltd.

MARCH, L. and STEADMAN, P. 1971. *The Geometry of Environment*. London: RIBA Publications.

MARCH, L. 1972. Elementary Models of Built Forms. Pp. 55–96 in L. Martin and L. March, *Urban Space and Structures*. Cambridge: Cambridge University Press.

———. 1976. A Boolean description of a class of built forms. Pp. 41–73 in L. March, ed. *The Architecture of Form*. Cambridge: Cambridge University Press.

———. 1992. From sky-scraping to land-hugging. Commentary. *Journal of The National University of Singapore Society*, **10** (December): pp. 12–19. Singapore: National University.

NEWMAN, M. H. A. 1964. *Elements of the Topology of Plane Sets of Points*. Cambridge: Cambridge University Press.

PAPALAMBROS, P. Y. and SHEA, K. 2001. Creating structural configurations. Pp. 93–125 in E. K. Antonsson and J. Cagan, eds. *Formal Engineering Design Synthesis*. Cambridge: Cambridge University Press.

PARK, J. 1996. Schindler, symmetry and the Free Public Library, 1920. *Architectural Research Quarterly*, **2** (Winter): pp. 72–83. Cambridge: Cambridge University Press.

PÓLYA, G., TARJAN, R. E. and WOODS, D. R. 1983. *Notes on Introductory Combinatorics*. Boston: Birkhaüser.

SHEA, K. 2000. Generating rational free-form structures. Pp. 118–128 in *Proceedings: Digital Creativity Symposium*. London: University of Greenwich.

SORIA Y, PUIG, A., ed. 1999. *Cerdà: The Five Bases of the General Theory of Urbanization*. Madrid: Electa Espanã.

STEADMAN, P. 1983. *Architectural Morphology: An Introduction to the Geometry of Building Plans*. London: Pion Ltd.

STINY, G. 1990. What is a design? *Planning and Design*, **17**: pp. 97–103. London: Pion Ltd.

———. 2001. How to calculate with shapes. Pp. 20–64 in E. K. Antonsson and J. Cagan, eds. *Formal Engineering Design Synthesis*. Cambridge: Cambridge University Press.

TABOR, P. 1971. Traffic in Buildings. Ph.D. Diss., University of Cambridge.

Chapter 86
BiOrganic Design: A New Method for Architecture and the City

Alessandra Capanna

Introduction

Over the last 20 years, many architects have proclaimed a new design philosophy based on the emergent condition of complex-systems science, which opposes conventional analytical methodology, or reductionism, and non-linear processes including computer aided design.

There are those who claim that with the support of computers, entirely new forms of design have become possible and others who believe that computers have even modified the creative processes and design theory. In this sense, architects are involved in scientific investigations of artificial life, genetic algorithms and neural network programs.

Artificial Intelligence supporting the development of digital systems, both those produced for self-generated architectures as well as those for drawing topological transformations in Euclidean space is evolving faster than human intelligence and it has often been speculated that it is only a matter of time before our machines become smarter than us.

It is true that the use of digital systems for animation, on which programs such as ALIAS and MAYA are based, has had a liberating and cathartic effect on architects, enabling them to draw and control unusual shapes with high levels of complexity.

First published as: Alessandra Capanna, "BiOrganic Design. A New Method for Architecture and the City". Pp. 11–20 in *Nexus VI: Architecture and Mathematics,* Sylvie Duvernoy and Orietta Pedemonte, eds. Turin: Kim Williams Books, 2006.

A. Capanna (✉)
Dipartimento di Architettura e Progetto, Università di Roma "La Sapienza", Via Flaminia, 359 00196 Rome, Italy
e-mail: alessandra.capanna@uniroma1.it

K. Williams and M.J. Ostwald (eds.), *Architecture and Mathematics from Antiquity to the Future*, DOI 10.1007/978-3-319-00143-2_39,

> The idea is not to automate design. It is not about being able to complete a design with a click of the mouse . . . It is about higher quality, not more efficiency. We want it to be better, not faster. It is also not about having the computer create a large number of proposals from which to choose. It is not about using computers to create unusual forms. When used like that, a computer would be nothing more than an extension of the pen in the hand. It is about using computers to think, as an extension of the brain (Watanabe 2002: 7).

Artificial Intelligence: A Game Played in the Field of the Architects

In 1950 Alan Turing wrote a very prophetical and provocative paper on Artificial Intelligence, entitled "Computing Machinery and Intelligence" (Turing 1950).[1] Turing was convinced that if a computer could do all mathematical operations, it could also do everything a person is capable of, a still highly controversial opinion. There are theoretical objections many like to believe that Man, in some subtle way, is superior to the rest of creation and mathematical objections. According to a large number of scientists, computer science will be able to create a machine capable of simulating a fraction of the human minds capacity, at the most.

The English mathematician Roger Penrose is sure that he can demonstrate mathematically that software cannot in any case produce a copy of human intelligence (Penrose 1990). He is convinced that human mind does not work like a computer even those based on neural network analogy following common physical laws, because the rules of intelligence and the mechanism generating thoughts are written in a quantum theory still to be acknowledged.

However, 50 years ago, skipping all the preliminaries, Turing started his paper directly with the question: "Can machines think?" and then went on in less than 30 pages to describe all the hopes and the fears about the possibility of attributing to the machines some specific "human" quality.

He called his method, played in the form of a dialog, the "imitation game." The test that enables us to distinguish a man from a machine is commonly known as Turing test. The Turing test says: "If a computer is found to give answers to questions that cannot be distinguished from answers given by a person, it must be concluded that the computer can think."

Q: Please write me a sonnet on the subject of the Forth Bridge.
A: Count me out on this one. I never could write poetry.
Q: Add 34,957 to 70,764.
A: (Pause about 30 s and then give as answer) 105,621.

---

[1] Alan Mathison Turing (London 1912–Manchester 1954), Ph.D. mathematics at Princeton in 1938, was a pioneer in computational theory. With David Champernown he wrote the first chess-playing program for computers. He is probably the first to imagine the possibility of machines really thinking. The term "Turing machine" was introduced by Alonzo Church in his review on Turing's paper "On computable numbers" (Church 1937).

Q: Do you play chess?
A: Yes.
Q: I have K at my K1, and no other pieces. You have only K at K6 and R at R1. It is your move. What do you play?
A: (After a pause of 15 s) R-R8 mate.

The question and answer method seems to be suitable for introducing almost any one of the fields of human endeavor that we wish to include (Turing 1950: 434).

Now, let us examine this well-known part of "Computing Machinery and Intelligence" as an exploration of the relationship between Architecture and Artificial Intelligence.

Douglas R. Hofstadter (1979) points out that few people realize that the solution of the arithmetical problem took too long and, above all, the sum is wrong! The fact would be natural if a human being was answering; it would be just a mistake. However, it being a machine, we can think about several possibilities:

- a hardware random error (one that never occurs again);
- a accidental hardware or coding error, which repeated arithmetical mistakes;
- an intentional fault generated by a program code to make the machine wrong in order to introduce incidental arithmetical mistakes and thus deceive the questioner;
- an unexpected phenomenon: the computer program finds it difficult "to think" in the abstract and simply made a "genuine mistake", which it will probably never repeat;
- a joke: the machine has intentionally teased the questioner… because is playing the imitation game.

However we can tell that the answerer is a machine because of its incapacity to work out "aesthetic thoughts". People willingly accept a machine able to understand and translate texts, to manipulate symbols (playing chess), to solve algorithms of course, even to have sense of humor. If we admit it is able to joke, it must be unable to shine in beauty competitions.

In fact the idea that Artificial Intelligence is unable to compose poetry supports the argument of the impossibility of producing thinking hardware.

In the beginning of the information science revolution, architects, fascinated by new technologies as well as by biological models, had begun to think along the lines of virtual architecture and the architecture of artificial life. Virtual architecture is more concerned with non-Euclidean geometries and includes architects-video-artists, the trend being to design liquid spaces made with digital materials, moving and transforming images that existing only in the personal computer. The architecture of artificial life follows the new organic paradigm, which is born from the union of auto-generative software and bio-morphic architecture.

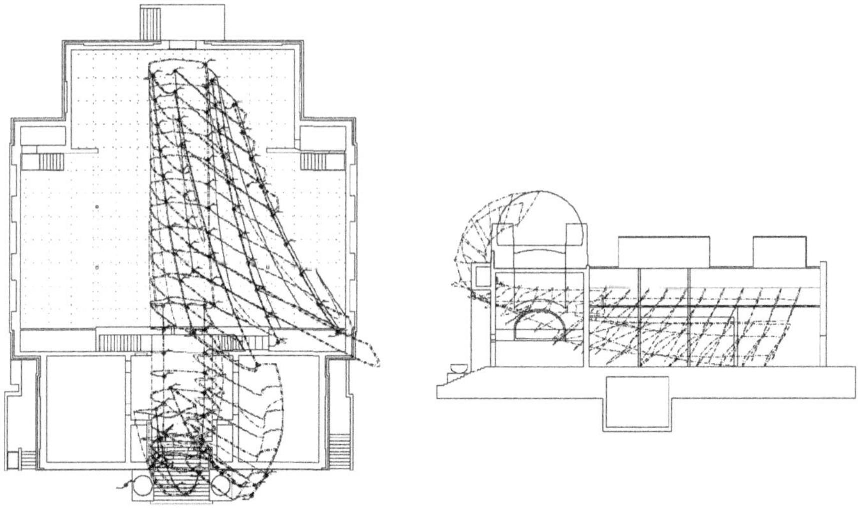

Fig. 86.1 Greg Lynn, The Tingler at the secession house, Vienna, 1999: Plan and section. Image: Courtesy of Greg Lynn FORM

"The Tingler," a painting/architecture hybrid performed in Vienna in the summer 1999, is an example of the architecture of artificial life. Greg Lynn,[2] together with the Argentine painter Fabian Marcaccio, who now lives in New York, transformed the interiors of Olbrich's Secession House into a disquieting vertebrate, using Alias/Wavefront, Maya and Microstation (Figs. 86.1, 86.2, 86.3, 86.4, and 86.5).

The architectural structure from which Marcaccio's paintings were suspended grew out of the golden dome into the Secession House just like the parasite protagonist of William Castle's movie "The Tingler," which was the inspiration for the exhibit. In the film, the bothersome organism was living inside the human backbone and grew bigger as one's fears grew. Analogously in the exhibit, the parasite designed by Lynn/Marcaccio penetrated inside the building, transforming it into an animate form that seemed to have grown so enormous that the walls could hardly contain it (Brizzi 1999).

In other projects by Lynn, animation software and computer-assisted manufacturing are combined to build previously unimaginable, "living" architectural forms. The ideas of the designer are supported by the amazing performances of the machine.

[2] Greg Lynn heads up the architecture firm FORM based in Venice, California, where he uses computers to design his wavy structures. His works, according to an article by Mark Dery in *Artbyte*, "are made possible by the computer's ability to generate warped or fluid forms. A typical Lynn creation is a monstrous hybrid of architectural theory and cyberpunk science fiction" (Dery 2000) Lynn teaches at UCLA and at the University of Applied Arts Vienna.

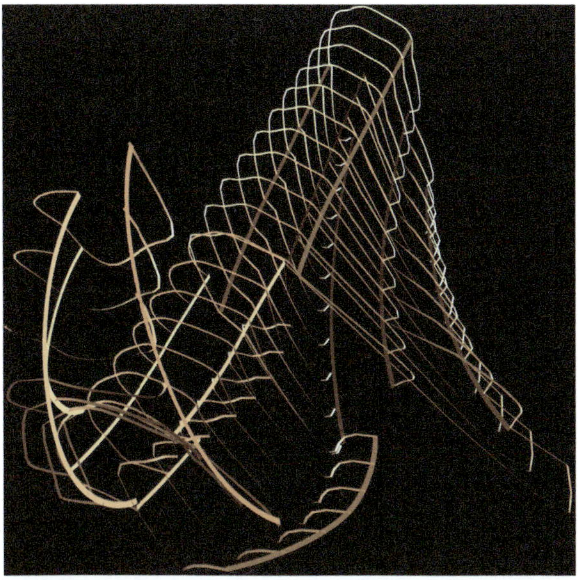

Fig. 86.2 The Tingler. Digital sketch. Image: Courtesy of Greg Lynn FORM

Fig. 86.3 The Tingler, exterior view. Image: Courtesy of Greg Lynn FORM

About 30 years ago, studies concerning pattern language analyzed thinking processes in order to derive a coded method for automated design. The greatest shortcoming of this system was probably only that it was conceived before the age of personal computers, essential instruments for verifying the design processes and acquiring the issues for possible new applications.

Fig. 86.4 The Tingler, exterior view. Image: Courtesy of Greg Lynn FORM

Fig. 86.5 The Tingler, interior view. Image: Courtesy of Greg Lynn FORM

On Growth of Form

It is evident that we are introducing a sort of mechanized creativity, as if artificial minds could master an automatic artistic power in order to develop complex systems design.

The Induction Cities[3] project by architect Makoto Sei Watanabe began in 1994 as research into Program Aided Design, which is capable of proposing solutions to complex building settlements through the use of self-generated computerized programs.

Its procedures are akin to the studies on pattern language. Both are based on the discovery of hidden structures in urban and social systems. Complexity science has discovered simple principles hidden within what appears to be utter disorder. Consequently, instead of imitating forms, it is the process that contributes to design.

The purpose is not to build a grid-pattern city or a labyrinthine town on a whim. It is to present the potentiality of a methodology for creating a city that will be as natural as a spontaneously generated living organism, and will even satisfy certain conditions. From 1994 to 1997 Watanabe made tests of the program, developing a series of different "inducted cities": the Self- and Other-Determining City, the Sun-God City, the On Demand City, the City of Distorted Space, the Comparative District City and the City of Correlative Wave Motion.

Each project uses the computer, combined selected output results, and computer simulation to depict patterns of influence amongst local services, etc., in order to find a fluid order of housing design, accessibility and pleasure.

Unlike conventional design, Induction City is a method for inducing results that meet selected conditions. In planning residential quarters for multi-unit housing architecture in Japan, for instance, access to maximum sunshine for each unit is given top priority. The result of analytical methods, which often tries to decide everything, is sometime monotonous box-like housing complexes. Sun-God City is a program that performs this task automatically: the software combines units randomly and sends sunlight to each. The process is repeated until all the units are optimally arranged. This method can also be applied to create automatic programs for other important conditions such as privacy and access.

On Demand City to take another example is a program created to optimize the location of urban facilities and their relationships.

Because a city is a constantly moving and growing much like a living organism it can be planned by an Evolutionary Design Program, a step forward as regards the programs for City Generative. Again, it is a matter of programs exploiting the power of computers to perform a huge number of procedures quickly and efficiently in order to memorize them, to compare and to evaluate all the potential results. This is just how "Deep Blue," the machine able to play chess and win against the world champion Garry Kasparof, works.

Going beyond the principle of randomness, Makoto Sei Watanabe designed the Iidabashi Subway Station, developing, with his team, the Web Frame project,[4]

---

[3] See the website http://www.makoto-architect.com/idc2000/.

[4] See http://www.makoto-architect.com/idc2000/syb_e2.htm.

Fig. 86.6 Makoto Sei Watanabe. Iidabashi subway station, entrance at street level. The steel is realized straight from the same data files and can be considered the next higher level of generation. Photo: Courtesy Andrew I-kang Li

whose aim is to move forward from the first phase of Induction Cities into the field of "aesthetic" evaluation.[5] Watanabe has explained that they intend the program to satisfy "fuzzy" criteria such as "enjoyable" or "dynamic:"

> At this point, we have to return to our earlier question, what is a "good" thing? In City of the Sun Goddess, we chose as an index for evaluation exposure to sunlight, and in On Demand City, our index was distance. . . . In both cases, that is, some aspect of "naturalness" showed up. Naturalness is something that everyone can understand.[6]

He continues the same web page, "Why not introduce some principles of nature for example, the laws of motion governing the movements of waves that give such a sense of pleasure?"

The Iidabashi Subway Station is on the Ōedo line of the Tokyo underground (Fig. 86.6). It is the first architecture to be completely carried out by a computer. Completed in October 2000, Watanabe designed it in 1991 after winning a

[5] See http://www.makoto-architect.com/WF_II/WF_II_e1.html.

[6] Quoted from http://www.makoto-architect.com/subway/subway_2e.html.

Fig. 86.7 Makoto Sei Watanabe. Iidabashi Subway Station. The electrical equipment network. Photo: courtesy Andrew I-kang Li

competition held by the Tokyo Metropolitan Subway Construction Corporation. He applied the so-called "Induction Subway Project" system associating the city planning with the "Induction cities".

The principles on which the subway station are based are the same as in two other architectural designs by Watanabe, the "K-Museum," rising in the very heart of an incipient city, and "Fluid Cites Fiber Waves", designed just like a living organism to interact with the visitors of the 2000 Venice Biennale. Both are conceived "to make visible the invisible."

Iidabashi's subway passages and flow patterns constitute a submerged space that delineates physical and virtual routes which are no more complex, however, than the intense maze of the city's nameless streets above ground. Both the city and the underground accommodate the daily transit of people (whose level of stress and feeling of disorientation is even greater underground), electrical systems, lighting circuits, rainwater conducts (Figs. 86.7 and 86.8).

The design is anticipated by Tokyo's urban configuration: roads which appear to follow a regular right angled grid pattern imperceptibly deviate from the orthogonal; the disorientation is accentuated by urban blocks not having a building number sequence, thus only by knowing the neighbourhood it is possible to reach any address. Consequently, the aid of clear, outstanding informative signs indicating physical reference landmarks becomes indispensable so as to find one's way amidst the city's urban layout, both above the ground and below.

Watanabe followed his guideline theory by inserting multi-sensory signalling icons in the underground area. Large installations and small figurative works run across its multimedia routes so that "to see is to touch."

Fig. 86.8 Makoto Sei Watanabe. Iidabashi Subway Station. The electrical equipment network. Photo: courtesy Andrew I-kang Li

Hence Watanabe's architectural design comes very close to the objective of being a total communication *media* governing the theoretical as well the actual physical space, constantly mediating between the mind's logical processes with its sensory systems.

The exposed networks of old and new tube routes that were visible during the excavation to construct this underground extension were especially suggestive. The 13 subway lines that wind their ways beneath the surface of Tokyo delineate a topological system resembling fossils suspended "in mid-air" (or should we say "in mid-ground"?). Visible only for a short time before being buried below the earth once more, the construction project truly made the invisible visible, in terms of urban structure, its beauty, and indeed its character.

Trans-sensorial clues reveal directions and destinations from the inside as well. A portion of a wall is treated with a Braille-like texture to enhance the juxtaposition of visual and tactile feelings (Fig. 86.9).

If one "reads" with one's fingertips of the fingers the phrase embossed on a metallic surface, "WOODEN SURFACE TREATMENT," two sensations became superimposed, a tactile feeling and a visual image. The knowledge we have of the different sensory experiences translates in our mind as messages that begin a

Fig. 86.9 Makoto Sei Watanabe. Iidabashi subway station. Braille wall. Photo: Courtesy Andrew I-kang Li

process of transferring information from a state of self-consciousness to one of subconscious feelings and vice versa: a sort of perceptual oxymoron.

The whirl of the staircase handrails and of connecting routes between Iidabashi station and the other underground levels make evident, like a collection of superimposed layers, the topological structure of the site.

The network of underground conduits in this part of the Tokyo subway is organised through three adjacent parallel tunnels, the central one being the station proper, the lateral ones where the rail lines run.

In coherence with the initial premise making visible the invisible the central tunnel's ceiling was substituted by juxtaposed technological conduits lying in a cavity aligned longitudinally with the pedestrian walkway. Finally, from the depths, 35 m below the city, a steel flower blooms and germinates: this is "Wing," the ventilation tower housing all the technological equipment for the entire subway station (Fig. 86.10).

The mechanism of auto-generation seeks more water, more light, just like "Elsie" (Electro Light Sensitive Internal External) and "Elmer" (Electro Mechanical Robot) built in 1948 by Walter Gray Walter, an English neurophysiologist who amused himself by fabricating cybernetic beings.[7]

[7] The intriguing history of Artificial Intelligence, from the first self-acting machines up to modern artificial life, is told by Castelfranchi and Stock (2000).

Fig. 86.10 Makoto Sei Watanabe. Iidabashi subway station: the Wing. Photo: Courtesy Andrew I-kang Li

Conclusions

All digital computers now operate according to sequences of rules that the machine follows step by step. A methodical exploration of computing theory brought Penrose to criticize one of its philosophical corner-stones, the Turing test. However, many computer scientists accept the test as a valid way of distinguishing an intelligent program from a non-intelligent one, and a machine from a human being.

After all, whether the program happens to be executed by a man or by a computer makes no difference. It is not even a matter of believing whether it is possible or not to achieve Artificial Intelligence, because some of the consequences of the first attempts of computer science to create a huge thinking machine are already useful design instruments.

In conclusion, in order for software developed for architecture to pass the Turing Test, it must not only pretend to be human but also show an amazing capacity for optimizing design choices and generating architectural organisms comparable with those produced by humans, as well as and above all demonstrate its capacity to generate aesthetically definable figures. (Our purpose here is not to argue the beauty of digital architecture!) To the first question of the test, "Please write me a sonnet on the subject of the Forth Bridge," or if we like, "Please draw me a folded building or an optimized new town," the machine will not answer, "Count me out on this one. I never could write poetry or draw architecture," but rather, "Yes, you can choose amongst the following solutions...", thus winning the imitation game.

Acknowledgment I wish to thank Andrew I-kang Li for taking the wonderful photographs of the Iidabashi Subway Station.

Biography Alessandra Capanna is an Italian architect living and working in Rome. She received a degree and a PhD in Architecture from University of Rome "La Sapienza". Among her published articles on mathematical principles both in music and in architecture are: "Una struttura matematica della composizione", remarking the idea of self-similarity in composition; "Music and Architecture. A cross between inspiration and method", about three architectures by Steven Holl, Peter Cook and Daniel Libeskind (*Nexus Network Journal* 11, 2 (2009); "Iannis Xenakis. Combinazioni compositive senza limiti", "Limited, Unlimited, Uncompleted. Towards The Space Of 4d-Architecture" and "Tesseract Houses", about the topic of conceiving higher dimension architectures. She is a *Ricercatore* at the Faculty of Architecture of Rome "La Sapienza". She is the author of *Le Corbusier. Padiglione Philips, Bruxelles* (Universale di Architettura 67, 2000), on the correspondence between the geometry of hyperbolic paraboloids and technical and acoustic needs, and its aesthetics consequences.

References

Brizzi. M. 1999. Greg Lynn. Diagrammi d'una esposizione. *Arch'it – IN A BIT*, http://architettura.it/inabit/19990930/.

Castelfranchi, Y. & Stock, O. 2000. *Macchine come noi. La scommessa dell'Intelligenza Artificiale*. Rome-Bari: Ed. Laterza.

Church, Alonzo. 1937. Review: A. M. Turing, On Computable Numbers, with an Application to the Entscheidungsproblem. *Journal of Symbolic Logic* **2**, 1: 42-43.

Dery, Mark. 2000. Soft House: Home Grown. *Artbyte*, November-December 2000. http://www.artbyte.com/mag/nov_dec_00/lynn_content.shtml.

Hofstadter, D. R. 1979. *Gödel, Escher, Bach: an Eternal Golden Braid*. New York: Basic Books Inc.

Penrose, R. 1990. *The Emperors New Mind*. Oxford: Oxford University Press.

Turing, A. M. 1950. Computing Machinery and Intelligence. *Mind* **59**: 433-460.

Watanabe, Makoto Sei. 2002. *Induction Design: A Method for Evolutionary Design*. Basel: Birkhäuser.

Chapter 87
Formal Mutations: Variation, Constraint, Selection

Andrzej Zarzycki

Novelty in human clocks requires independent acts of creation. Novelty in biological clocks seems more suited to iterative modification from a common origin.
M. Kirschner and J. Gerhart (2005: 7).

Origins and Parallels

We all are familiar with writers' or artists' blocks when faced with a clean sheet of paper, white canvas or blank computer screen. As designers, we are full of expectations and desires to create…, but the real question is how to begin. Many painters would break through this initial moment by smearing a canvas with abstract and meaningless scribbles. This breaking moment often helps us to forget the difficulty involved in starting a new project. This quickly and randomly chosen context for beginning the act of creation puts us on a specific path where we start thinking in terms of transformations, changes and adaptation, and not in terms of defining something from nothing. The idea of change, adaptation or inhabiting the pre-existing context, seems to be a nature-like process that is intrinsically gradual and as such, less threatening for the artist or creator. M. Kirschner and J. Gerhart parallel this observation in the opening quote for this chapter.

While we often find the ideas of building on the work of predecessors or preexisting conditions comforting, the question may be raised whether this approach predetermines and limits possible outcomes. After all, there are examples of great ideas emerging from questioning the basis of our assumptions about the

---

First published as: Andrejz Zarzycki, "Formal Mutations: Variation, Constraint, Selection", pp. 33–46 in *Nexus VII: Architecture and Mathematics,* Kim Williams, ed. Turin: Kim Williams Books, 2008.

A. Zarzycki (✉)
New Jersey Institute of Technology, College of Architecture and Design, University Heights, Newark, NJ 07102, USA
e-mail: zarzycki@njit.edu

K. Williams and M.J. Ostwald (eds.), *Architecture and Mathematics from Antiquity to the Future*, DOI 10.1007/978-3-319-00143-2_40,

nature of the world and reality. This was true with Einstein's theory of relativity and the constancy of the speed of light as well as many others human advancements.

Starting with an already predefined canvas may put the creative process into a particular trajectory, resulting in a certain class of solutions. However, this possible pre-determination does not limit the chosen method's creative capacity. Some might argue that it actually increases the creative challenge, resulting in more interesting solutions. In *Poetics of Music* (1993), Igor Stravinsky talks about the necessity of restraints and limitations in achieving creative outcomes. Working against the hard edge of design limitations and imposed boundaries is what makes solutions innovative and unique.

My interest in studying tectonic evolutions and simulating form mutations in design comes from the observation that these operations are natural ways to manipulate data and models within digital environments. It builds upon the observation that editing already existing data is more native to digital environments than inputting new data. Architecturally this could mean that transforming already existing forms is a potent and effective way to derive new forms, ideas, and designs. Finally, creating new ideas from scratch is almost always more difficult than arriving at new ideas through gradual transformations of the old.

Towards an Augmented Design Process

Traditionally, we assume that the design process is a linear, gradual and creative development of products arriving at the finality of a completed design project. This means that if we were to choose continually the best scenario, we would end up with the most successful design. However this static, somehow optimistic, approach to the design process often misses many opportunities that, while part of some possible scenarios, may be obscured by local inconsistencies. It may miss possibilities that behave like many natural processes, where sometimes a series of uninteresting or inferior solutions will precede a highly innovative form. The reference to nature and biology is just one of many similarities. If we consider the case of the caterpillar and the butterfly, we see that a caterpillar does not visually imply a butterfly, or in other words, a butterfly is not an obvious or 'rational' consequence of an evolution of a caterpillar.

We assume that if we always do the logical thing we end up with the best possible scenario. However, everyday life—not to mention advanced design simulations—does not support this conviction. While experience leads us to this conclusion in real life, we are just starting to realize alternate possibilities in design with digital technology and mathematically based 3D simulations.

Nothing prevents the traditional process from exploring a multiplicity of possibilities. The significant difference lies in the ease with which digital technology permits studying multiple alternatives and pursuing parallel scenarios. One common realization is that the copy command does not cost anything in digital environments.

This is particularly true when we consider the parametric and procedural nature of digital design. In these ramifications it is relatively easy to alter various parameters and test alternatives. This ease with which alterations can be made may on occasions result in designers looping themselves in the possibilities and having difficulty breaking away from what appears an infinite number of possibilities. However the same characteristics enables digital design to create ample variations that are prerequisite for any evolutionary system.

The design process begins with assumptions that are often arbitrary, but it results in logical and unique solutions. It is judged by its logic and consistency in the context of its starting assumptions. While one can certainly question assumptions made, oftentimes there is no definite way to determine whether they are appropriate or not. These assumptions can only be judged against their potential for creating a broad number of design possibilities that have the ability to lead to innovative ideas without the introduction of self-contradicting elements. In this sense, the design process resembles the discipline of math, where we start with assumed fundamental truths called axioms, and we build a self-consistent and integral set of knowledge.

This consistency and design integrity means that with each step in the design process the number of the possible solutions is being reduced, slowly converging on a final design. We could illustrate it as a design decision tree; with each step forward, towards the resolution, we advance to the higher branch; thus, are left with fewer choices that are consistent with our past decisions. Unlike when climbing a tree, where we always can see other branches and understand our past climbing choices, in design the further we progress the more difficult it is to see other possibilities that are not a part of our present design trajectory, also called a design horizon. This continuously narrowing focus brings many benefits in decision-making, but also makes us miss design possibilities that may be more suitable for our intentions.

"Seeing other branches" is especially critical in situations where we are faced with the decision of choosing the less-than-perfect scenario. At that point, the simple method of elimination of less desired solutions does not lead to the best results. A weighted average of the possible scenarios and an understanding of their ultimate potentials is the best approach to designing.

Types of Transformations

Since change and transformation have become the norm and the basic building elements in the creative process, a new set of instructions is necessary to direct these design agents. These instructions may involve simple form transformations as well as topological changes, including object discontinuities.

The design is executed by applying simple rules and behaviours to the original form. Each of these rules represents limited vocabulary and produces very recognizable effects, such as the 'bend' transformation. However, by compounding even a small number of simple transformations, the forms' complexity and design

Fig. 87.1 An object with no transformations applied. Image: author

possibilities grow exponentially and escape predictable visual patterns. The phenotypic results of a single transformation may often appear not to change its resultant form qualitatively, but the transformation is still present in its genotypic definition of an object waiting to emerge. This dormant transformation may be responsible later for a rapid emergence of the form/design once other transformations are applied, leading to complex and sophisticated forms. This rapid emergence of form results from narrowing the difference between the phenotype and genotype potentials (Figs. 87.1, 87.2, and 87.3).

In most cases, the order of applied transformations is critical. Different orders will produce different products, in the same way as the compounding of mathematical functions F(G(x)) will, in most cases, produce different results than G(F(x)). This is consistent with the non-linear nature of these transformations (Figs. 87.4 and 87.5).

With the transformational model of design we can distinguish a number of factors that influence form outcomes. These factors can be classified as:

- the type of transformation;
- the way a transformation is applied;
- the internal structure of a transformed object;
- the spatial properties of where the transformation occurs.

These factors reflect the critical component of any real life form development: formal expression, what it is made of, and a context it exists in.

Fig. 87.2 The original object with one transformation applied. Image: author

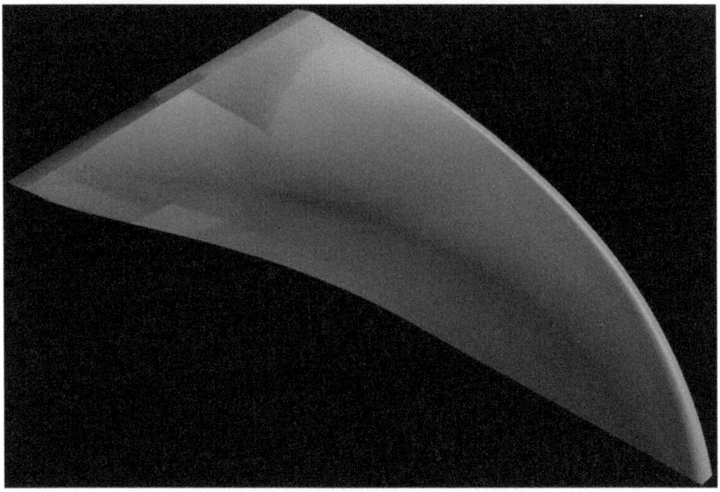

Fig. 87.3 The original object with a different transformation applied. Image: author

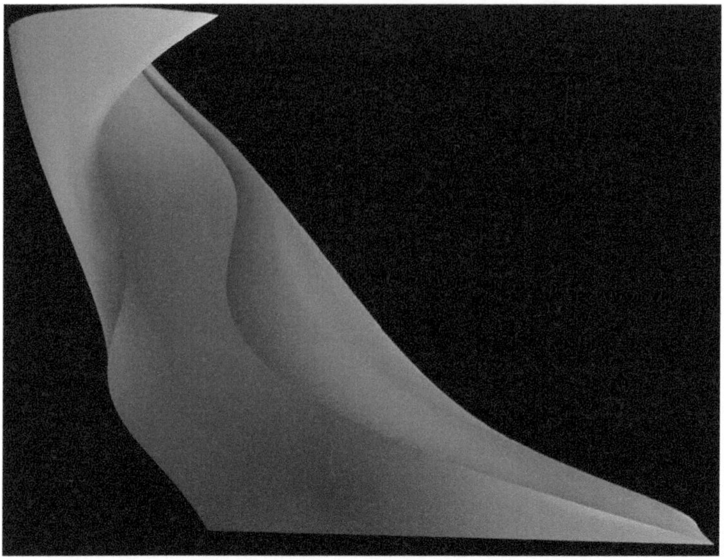

Fig. 87.4 The original object with both transformations applied. Image: author

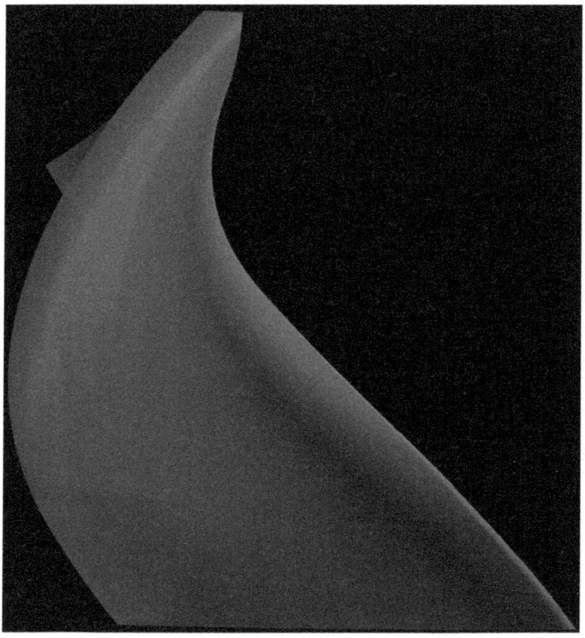

Fig. 87.5 The same object transformations applied in a different order. Image: author

Additionally, the form transformation factor can be divided into three categories of transformations:

- continuous;
- destructive;
- cohesive.

The way a transformation is applied—the relationship between the transformational "gizmo" and the object's axes of symmetry—would result in visually different outcomes. Although all the outcomes would be consistent with the mathematical definition of a particular transformation, they may not be obvious and would be seen as distinct forms.

An internal structure of a transformed object is critical in expressing its resultant form. For example, bending of a meshed object is dictated by its segmentation. Since individual faces do not bend and are the smallest building blocks of a meshed form, the size and number of segments may drastically change the result of applied transformation. The difference between shapes like letters "V" and "U" lies in an internal segmentation, not necessarily in a difference of transformation applied to the letter "I" or a character "-". In those situations, segmentation can be seen as an object's transformational degree of freedom, which defines a number of pivotal points controlling facets and curvatures.

Transformation may result in a new form but also in a change in the internal definition of a form. These changes, when continuous, result in a *texturizing* of an object, creating an interesting relationship between a form and its texture (facture).

Spatial properties, often achieved through space deformers and warps, are similar to object transformations; they are, however, a property of a space, not an object. This means that they affect various parts of an object or various objects in space differently based on their location.

The critical difference between spatial and form deformations is that an object changes depending on its position in space—emergence of a design context. I will be discussing spatial deformations later in this chapter in relationship to form animations.

The following are three transformation categories. These categories are organized around the object identity, not necessarily around topological commonalities. They allow for disappearance and emergence of forms through changes in objects' identities. This is a critical characteristic in evolutionary design systems.

Continuous transformations preserve an object's topological identity and continuity while deforming it. Examples are functions such as bend, twist, or smooth with NURBs. These transformations, on occasions, may interfere with sub-object topological levels but will not affect the cohesiveness of an object as a whole (Fig. 87.6). In some cases the topology of an object can be disturbed while preserving its identity. I would still call those transformations continuous although they do involve some points of discontinuity. An example of such a condition is a transformation from a sphere to a torus.

Destructive or populating transformations break an object's physical identity resulting in multiple new objects. This is achieved through object fragmentation,

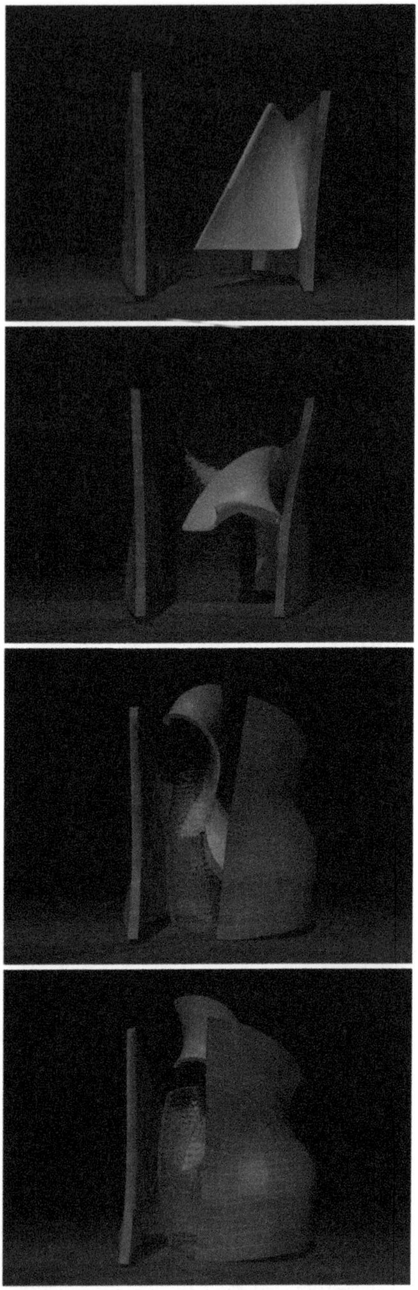

Fig. 87.6 Stages of a form evolution using exclusively continuous transformations. Image: author

not copying. Common examples are modelling functions such as subdivide, explode, and shatter, with each of them having slightly different properties or addressing different topological levels. Explode results in multiple three-dimensional objects,

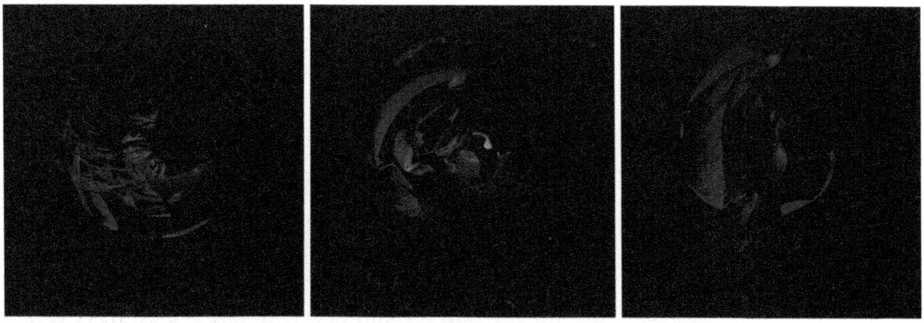

Fig. 87.7 Destructive transformations in conjunction with continuous transformations applied to a single object result in a rich visual landscape. Three stages of an evolution. Image: author

while shatter results in subdivisions of surfaces and planes. The rate of population can be controlled by transformation parameters, but also by the object's surface subdivisions. The surface subdivision can further control the shape or proportions of resulting objects (Fig. 87.7).

Cohesive transformations result in merging multiple objects into one larger entity. This can be achieved by attaching 'adhesive' properties into objects, but also by capturing these objects in a space bubble through the use of space warps. An example is a metaball or meta-object that behaves similarly to mercury, a liquid with strong cohesive forces. Also its molecules seek to minimize surface tension. Space warps are another means of forced cohesion. They are particularly effective with objects using dynamics and with particles (Figs. 87.8 and 87.9).

This transformational model of design expresses a form in relative terms, as related to other forms, and does not need to rely on the absolute definitions of forms. This shift from an absolute to a relative reading means that an event or existence is no longer predetermined by its initial conditions, but is rather primarily defined by the local circumstances—having the capacity to behave in a non-linear manner.

Animation—Interpolation and Extrapolation of Static States

While using form transformations is a new and effective way to derive designs, even greater design possibilities are achieved by animating those initial, static forms with the use of space warps, morphs and form modifiers. Since most transformations are parameter-based, it is easy to animate numeric values of these parameters and study evolution of forms. This is usually executed by defining critical static states of a form, also called keyframes, as an analogy to the traditional animation process, and interpolating these values, as well as spatial positions and properties, into in-between forms. While morphing forms we identify moments in an animation that have interesting design opportunities and we can retrieve the parameters that define the transformation's particular states for further refinement. We can also register the

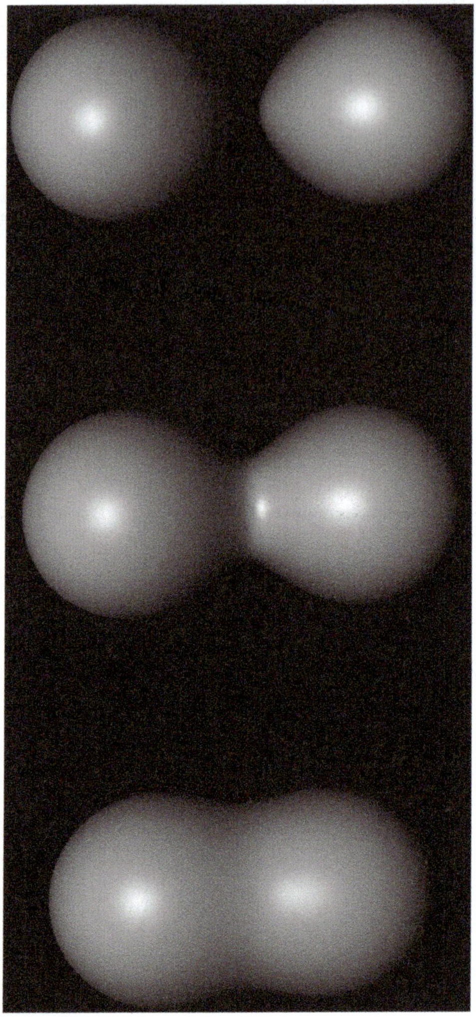

Fig. 87.8 Behaviour of two metaballs. Image: author

Fig. 87.9 Metaball particle system trapped in a space bubble. Image: author

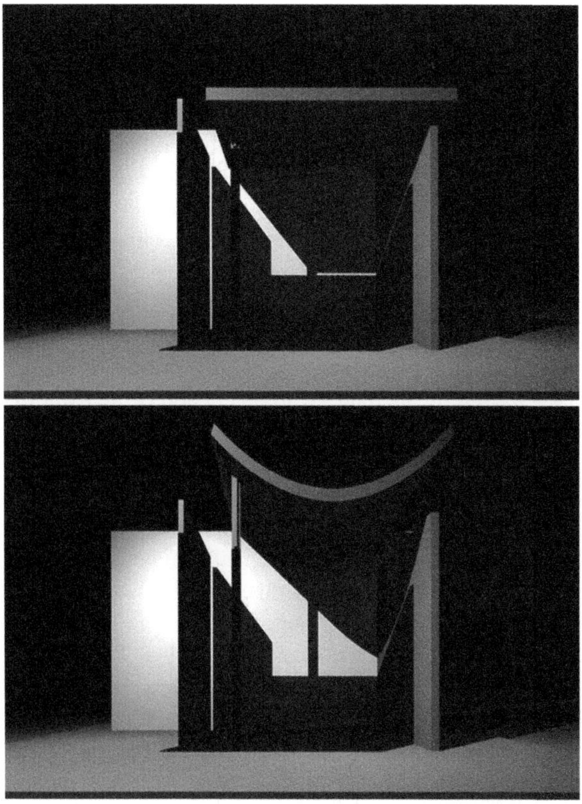

Fig. 87.10 Animation of a building envelope allows for in-depth lighting studies. Image: author

object's particular state and output as a static, transformed form that no longer relies on changing parameters. This newly-shaped form becomes a seed for another process. Its simpler existence as a 'flattened' transformation stack, reduces the complexity of the object's definition, but is not critical in terms of the object's visual definition.

Since many of these parameter driven functions behave non-linearly, the results of these animations, as well as the in-between stages, are often unpredictable, even though no random values are introduced. This is particularly evident in transformations that introduce singularities and discontinuities into space. This is perhaps the strongest element in this approach since it allows for the creative leap—mutation—to occur. It also introduces strong and effective explorative components into architecture in a way that is similar to how explorations of a physical model can bring surprising new discoveries. However, in this case it happens in a much more pronounced way and with greater intensity.

Tectonic animations can also be used as study tools. While it is common to employ digital technology in performing light and shadow studies for a static architectural space or a building, with this approach, we can animate the envelope of a building with changing window apertures to arrive at the most desirable lighting scenarios (Fig. 87.10). This effectively repositions the question from what is the best lighting scenario for a particular design, to what is the design that uses existing lighting possibilities most effectively.

This compounding strategy can be brought to another level of design thinking where any form can be subsequently deformed and used as a seed for another design. Consequently, through the parallel processing of ideas and designs, we often talk about a class of all possible solutions or about the results the solutions are tending towards, and less about geometric absolutes, which are seen as static and finite design solutions, as well as designs that start obeying probabilistic rules rather than definite and predictable patterns.

With an introduction of animation into design, two classes of transformations emerge: form and space deformers. Form deformers change the object's geometry, which is a permanent change even if it only exists for a short period of time. This new form is an attribute of an object and is not location-dependant. Form deformers react only with particular objects and do not interfere with other objects that are in the same locality.

Space deformers, also called space warps, are properties of space and affect any object that is within a space unless specifically excluded from the operation. They allow transformations that are only relevant to space or context, not a particular object. Furthermore, their influence is location-in-space related, which means that the form of an object is dependant on the location within a space warp and will change if the object is moved (Fig. 87.11). This distinction between form and space deformers is particularly applicable to architecture since space deformers can be seen as the design context or environment. Ability to assign properties to space, not much different than in real life, allows for global treatment of design. It also creates favourable conditions for simulations of form mutations and dynamic systems.

Language of Mutations

The concept behind the project *Formal Mutations* brought this transformative design methodology a step further, where the process of change is paralleled to other processes like those found in nature and evolution. As a result, design methodology has to account for the creative error—a mutation which helps a designer to break away from the obvious and predictable while setting the design on unexpected but meaningful trajectories. This can be achieved by introducing chaotically behaving functions into design or by compounding multiple simple rules that behave like switches enabling individual transformations.

Formal mutations, an example of a non-linear design process, relies on generating new design forms from previously created forms. If established as a part of a generative process, an element of reiteration is introduced into design. A present design state can only be seen in the context of the immediately preceding state. As a form evolves from generation to generation there is an opportunity to introduce elements of noise or imperfections that can push designs in unexpected directions.

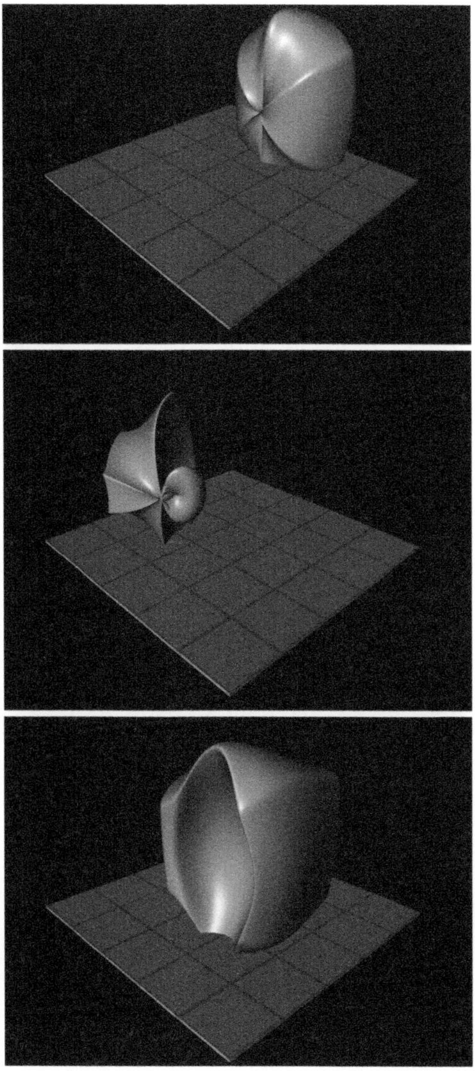

Fig. 87.11 Behaviour of an object as the location-in-space changes. Image: author

Even in a simplified model of behaviour, where no mutation or contamination is introduced, we can observe the development of great variations in forms.

Non-linear processes, especially those employing dynamics such as cloth deformations or particles, are defined by their immediately preceding states. As such, they tend to carry some residual values in discontinuities or deformations, called here *traces*, from iteration to iteration. Traces, such as flexion, often result from the inertia present in the material's physical properties. In such situations, the speed changes in the dynamics system can proceed faster than the material's ability to react to the change, leaving a discreet trace from the action. These traces can

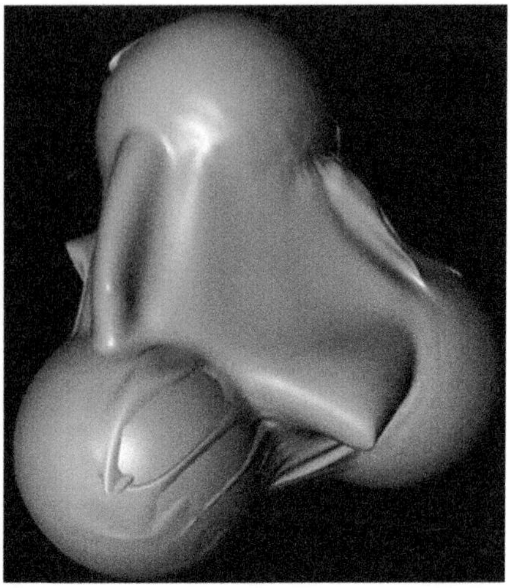

Fig. 87.12 Dynamic cloth deformations register traces resulting from object collisions. Visible tearing and folding of surfaces. Image: author

manifest themselves as tears, folds, or other surface imperfections (Fig. 87.12). Another example of a trace is inconsistencies in the spatial distribution of particles. These inconsistencies are carried from generation to generation by the dynamic interaction between particles.

Particle systems are, in many ways, what the idea of Formal Mutations suggests. Particles are objects that, once created, become freely behaving agents that can only be controlled with space deformations or other deformations that affect the entire population of objects, not just individual instances. The two control areas over particle systems are the initial conditions and global deformations.

A new set of design criteria is emerging. Terms such as contaminations, traces, seeds, thresholds, attractors, etc., are becoming building blocks in the design process. In this new world, chaotic functions become contaminants; residual elements and values from previous states of existence are seen as traces. Any form can be used as a seed for another architectural form while trajectories of individual evolutions/mutations disobey a simple causality.

In some aspects, this iterative process is what design has always been about. A process of continuous refinement, one layer of trace paper over another, is present in both traditional and digital approaches to design. The difference is that digital simulations allow for more parallel processing in design through a co-development of several trajectories or multithreading.

These formal and tectonic possibilities are not always immediately recognizable. Often they emerge from obscure landscapes through the process of spatial mutations, and they are only noticeable when other components become activated. Since they are often interdependent, they may remain dormant while waiting for a spatial activator. These situations are particularly visible through the use of space warps and dynamics, where objects have an ability to interact and acknowledge each other.

Conclusions

This chapter identifies three levels of computational design as they relate to the concept of formal mutation, with higher levels being a generalization of lower levels.

Level one, is a *simple transformation* of forms within the computational environment through the use of transformations. This traditional-yet-digital method brings great potential to design. It is fully interactive and enables users with a limited knowledge of computational concepts and software to engage digital design on the 'user' basis.

Level two, *formal transformations*, has all the benefits of the previous level, plus the ability to morph form transformations with animation tools that bring a new class of design possibilities. The design process is still fully interactive, with design results usually deviating from preliminary expectations.

Level three, *formal mutations*, introduces randomly-behaving functions into level two transformations. It relies more on particle and dynamic systems that are designed to obey the laws of physics. The role of a designer shifts from being clearly interactive into a system manager who controls naturally evolving processes through arranging various starting conditions.

In this new paradigm, a designer can retrace design steps for future revisions and reconsiderations. This goes beyond the 'undo' button and helps us not only to create new designs but also, and more importantly, to study the design process itself. This design methodology allows for better scanning of potential design possibilities, bringing them from the realm of the possible to the probable to the real. The second critical advantage is that it enables us to understand, explain, and produce complex designs with a set of simple rules or transformations. It is important to add that the complexity of digital designs is not seen as an aim in itself, but rather it acknowledges the nature of reality. These computational methods are looking for ways to address this complexity, as well as to explain complex ideas and forms with the simplest language possible.

Biography Andrzej Zarzycki is a designer and educator who uses digital tools to create experiential architectural spaces. His research focuses on media-based environments and on validation methodologies of generative design through building performance analysis and simulation tools. He earned his master of architecture from the Technical University, Gdansk, Poland, and his master of science in architecture studies from MIT. He is a co-winner of the SHIFTboston Ideas Competition 2010 and a co-founder of TUTS (Tremont Underground Theatre Space) (the-tuts.org), a design initiative focusing on innovative adaptations of infrastructure into contemporary public spaces and on the integration of digital (ubiquities) technologies into urban life. He has taught digital design studios and courses in the interior architecture department at Rhode Island School of Design. He is currently Assistant Professor at the College of Architecture and Design, New Jersey Institute of Technology and partner in the architectural firm TechnoMorphic.

References

KIRSCHNER, Marc W. and John C. GERHART. 2005. *The Plausibility of Life: Resolving Darwin's Dilemma*. New Haven and London: Yale University Press.

STRAVINSKY, Igor. 1993. *Poetics of Music in the Form of Six Lessons* (The Charles Eliot Norton Lectures). Cambridge, Massachusetts: Harvard University Press.

Chapter 88
The Role of Mathematics in the Design Process Under the Influence of Computational and Information Technologies

Arzu Gönenç Sorguç

Introduction

The aim of this chapter is to discuss the evolving role of mathematics in the design process under the influence of Computational and Information Technologies. Mathematics is not just a tool for understanding form, order, harmony and more, but it changes our minds and the traditional way of thinking into 'algorithmic' thinking, which provides the means to manage today's multi-dimensional design problems.

The rapid changes and developments encountered in all the disciplines from science, technology to art, humanities, politics and others not only influence our daily lives but also force us to redefine many fields of interest and disciplines accordingly. In particular, the involvement of information and computational technologies makes harder to propose rigid definitions for any field of interest.

Among many other disciplines, architecture, which can be defined as the art and science of designing building and structures, has always been an interdisciplinary field involving mathematics, science and technology, history, art, philosophy, sociology, politics, etc. Therefore, it is possible to claim that architecture has been more exposed to all these changes and consequently its definition, tasks, tools and even the perception of 'space' have been continuously redefined. But it

---

First published as: Arzu Gönenç Sorguç, "The Role of Mathematics in the Design Process under the Influence of Computational and Information Technologies", pp. 47–56 in *Nexus VI: Architecture and Mathematics*, Sylvie Duvernoy and Orietta Pedemonte, eds. Turin: Kim Williams Books, 2006.

A. Gönenç Sorguç (✉)
Department of Architecture, Middle East Technical University, 06531 Ankara, Turkey
e-mail: arzug@arch.metu.edu.tr

K. Williams and M.J. Ostwald (eds.), *Architecture and Mathematics from Antiquity to the Future*, DOI 10.1007/978-3-319-00143-2_41,

is also possible to claim that in all these evolutions, mathematics has always been a tool for architects and today may be even being more important than before.

Mathematics, which is defined by *Cambridge Dictionaries* as "the study of numbers, shapes, and space using reason and usually a system of symbols and rules for organizing them", helps architects in the search for order and beauty by providing valid structural solutions to design 'sub-spaces' in harmony with the universe. However, the impact of IT and computational technologies and the resulting new design media, which are closely associated with mathematics, forces architects to question the role of mathematics once again. Hence, in this chapter the changing role of mathematics in architectural design will be discussed not only in terms of design in digital media but also by questioning what changing our minds to algorithmic thinking to solve complicated 'multi-dimensional' architectural design problems means.

The Role of Mathematics

In Book I, Chap. III of *De Architectura* Vitruvius proposed three criteria that must be balanced and coordinated in order to achieve "good buildings": "beauty", "firmness" and "utility" (Vitruvius 2009: 19). For this, mathematics and especially geometry was indispensable for providing order and harmony in the visual elements and, to some extent, to understand structural behaviors. It has been argued that the golden mean or golden rectangle proposed by the Greeks was the primary guide in planning. In Islamic architecture the ratio of $1{:}\pi$ was employed in plan and section in the organization of the buildings. In India and Egypt complicated mathematical models were used to map the movements of stars and planets to understand cosmic harmony. The use of symmetry in Renaissance architecture, the complicated curvilinear forms of the Baroque, the use of Euclidian geometry and, more recently, non-Euclidian geometry employed mostly by architects who participated in the Deconstructivist movement in the beginning of twentieth century also show how mathematics has played an important role in the history of architecture.

However, it should be pointed out that in today's architecture mathematics has become more important than ever before, not only by providing geometrical relations or ordering principles but providing a new way of thinking to solve complicated "multi-dimensional" design problems comprising solutions from many other disciplines in new design media. Igor Verner and Sarah Maor (2003) considered the interaction between mathematics and architecture in three different ways: the geometrical analysis of architectural forms and objects regarding dimensions, proportions and spatial transformations; formal description and interpretation of architectural concepts and symbols such as infinity and multi-scaling; and mathematical background concerning science and technology in design and construction.

Here, besides the well known interventions between mathematics and architecture briefly discussed above, the new and evolving role of mathematics will be discussed, first by considering "design process" regardless of whether the design media is either real or virtual, and second, by considering "design" in digital media. Algorithmic thinking, which is a kind of mathematical thinking, must be considered the keystone in this evolution, associating different domains of knowledge in order to solve highly complicated today's "multi-dimensional" architectural design problems, allowing architects to fully explore new "design media" and "space".

The Evolving Role of Mathematics and Possible Role of Algorithmic Thinking in Architectural Design

What Does Algorithmic Thinking Mean in Architecture?

Involvement of several disciplines and demands of "clients/users" result in "multi-dimensional" design problems for which architects must satisfy not only the requirements of the design itself but also find ways to bring optimized solutions fed by these different disciplines and satisfy the "client". Yet architecture and all the disciplines associated with it are continuously changing and developing, so achieving "good designs" and "good buildings" is getting harder. When the complexity of design problems are considered, then we must inevitably reconsider design as a "process", continuously evolving, from the "problem definition" through the end of construction and even afterwards (Gönenç Sorguç 2005a, b).

At this point, it is possible to define design process with very basic mathematical terms borrowed from set concepts, functions and mappings. In that way the universal set of design can be thought as the collection of different domains of knowledge involved in the design process and design itself; then the ideas of architects with all those contributions of these domains of knowledge can be mapped to a "state of form" in physical or a digital media, which can be perceived by the "clients". Although this model of design process appears to be very simple, the implementation of it requires a different way of thinking. At this point, I believe that the architects and designers in general should be more familiar with so-called "algorithmic thinking".

Although many people think that the word "algorithm" belongs to the computer age, it actually dates back to the ninth century and comes from the name of the Persian mathematician Al-Khwarizmi. At that time, algorithm was used to mean arithmetic operations with Arabic numerals. Today, algorithm is used to define systematic approaches having definite input, output and procedures to solve problems and to fulfill complicated tasks. Consequently, algorithmic thinking is a way of mathematical thinking that can be defined as the mental capacity to develop algorithms (i.e., well-defined systems and methods) for problem solving. Thus, it is possible to say that not only in computer sciences and programming but in any field, algorithms and algorithmic thinking can be employed to define problems, constraints and the procedures for solutions in a controlled and optimized way, no matter how complicated the problem is. Actually, due to the involvement of IT and computational technologies every discipline has reexamined its way of thinking and new models have been proposed to cooperate with other disciplines; one such study discussed engineering design thinking and structured that thinking process in a rigorous way (Dym et al. 2005).

At this point, it is possible to ask whether or not such a structured way of thinking might limit the creativity of the designer. In order to answer this question, it is necessary to consider the design as a process according to the model proposed above, in which algorithmic thinking helps architects first to clearly identify and rigorously define the design problem and its constraints, and then to define the universal set of the design, the processes that include different domains of knowledge as well as the processes of how mapping will be achieved between different sets and how the design will be mapped in any media as the output of algorithmic thinking. Since the architect or designer is free to define the universal set and the algorithms, I believe that algorithmic thinking does not impose limits on creativity, but to the contrary, with a well defined problem approach, the output that is, the design and its process can be optimized according to the set of constraints imposed by the designer. Hence, I claim that the new and evolving role of mathematics goes beyond the conventionally recognised role of mathematics in architecture and now means "a change in thinking".

Algorithmic Thinking and Impact of IT in Architectural Design

The impact of computers, computational technologies and IT in general, and their close association with mathematics and algorithmic thinking, has forced architects to question not only the design process but also different design media and their different interpretations of design as well as in order to find ways to take advantage of all the potential for design solutions offered by these by the new tools. Therefore, architects should not only develop algorithmic thinking for the design process but they should also learn to cope with these rapidly developing tools and media. Today digital media are considered not only as new presentation media but also as new "space" in which the architectural design has found new "meanings".

Several CAD and CAM software applications allow architects to map their ideas in an easily perceived way in 2D and 3D models. In this modeling process architects sometimes force the limits of these software applications or even develop their own interfaces and algorithms according to their needs. Besides, these models change the role of architectural design to "a left and right unique" set on which the whole domain of knowledge interacts through that model in the design and construction process and even afterwards (Gönenç Sorguç 2005a).

Thomas Fischer and Christiane Herr (2001) discussed "generative thinking", which became more plausible with the introduction of computer-aided design and computer-aided manufacturing. In their article, generative design is considered a design methodology in which "the designer does not interact with products and materials in a direct way but via a generative system which is a set-up based on abstract definitions of possible design variations". As can be understood, in generative design the designer should also provide "algorithms" not just for the creation of the product but for the creation of process as well.

Luísa Gama Caldas and Leslie Norford (2001) investigated ways to "map" architectural design intentions to a generative design system, taking as an example the School of Architecture in Oporto, Portugal. Their aim was to illustrate that Generative Systems (GS) are very important tools for evaluating possible problems and enhance the design accordingly. GS can then be considered as a feedback system that is used to encode the design through algorithms.

Although these examples and the field of applications can be multiplied not only in architectural design but in many other disciplines in which design is involved, here it is intended to illustrate the how digital media and algorithmic thinking can effect the architectural design process. It is important to note that algorithmic thinking and digital media co-exist in any design executed in digital media both as a way of design methodology and as a way to take advantage of the potential of these media.

Moreover, these 2D and 3D models, which can also be easily be employed by other disciplines for analysis and computation, allow architects to explore more challenging geometries and combinations of materials. Frank Gehry's ground-breaking design for the Guggenheim Museum in Bilbao can be considered as one of the most remarkable example of this tendency. Gehry and his team obtained a 3D model of the museum with CATIA software, which was originally developed for the aerospace industry. In that way, the complicated geometry of the building and the structural system to support that geometry could be investigated and implemented. In Gehry's design, digital media was not merely a presentation media. Rather, it was employed as a common interface between architectural design and the other domains of knowledge involved in it. Although only one example is included in this chapter, today CAD software and other 3D visualization techniques are common practice in any design process and architects have chances to experiment interactively with different geometries, configurations, structural systems for very complicated organizations.

As mentioned above, another direct result of IT and designing in digital media is the introduction of the concepts of "cyberspace" and "virtual reality", which are encountered very frequently even in daily life. There exist hundreds of articles written about these new concepts in various fields of interests. It can be said that the common element of all these studies is the effort to understand the nature of this new "space" and to explore its potential and, of course, to understand the possible impacts on different fields. Several definitions have been proposed for cyberspace and virtual reality in these studies. Here, in order to discuss these concepts within the realm of mathematics, I prefer to use the following definitions from the *Oxford English Dictionary*: cyberspace is defined as a "notional environment within which electronic communication occurs, especially when represented as the inside of the computer system; space perceived as such by an observer but generated by computer system and having no real existence"; the definition of virtual reality is "the notional image or environment generated by computer software with which a user can interact realistically".

It can be seen in these definitions that for cyberspace to exist or to have virtual reality the "designer" in any discipline should be capable of not only using computer software but, more importantly, he or she must be capable of questioning this rather new and continuously changing media and its tools in order to reveal all the potential and consequently the nature of "space" in that media. Mathematics, which actually underlies everything "constructing" these new media and space, also provides the basic tools for this exploring, learning and designing process.

Steve Ferrar (2001) questioned non-physical space and architects' perception of it. One of the important conclusions of that study is that multi-dimensional thinking and the collection and management of information are inevitable in understanding and creating non-physical spaces. In accordance with that study, I believe that mathematics with its tools to collect and process information and algorithmic thinking as a way of multi-dimensional thinking to manage all these various domains of knowledge are ideal for understanding and design non-physical spaces. In that sense, Marcos Novak's attitude towards design, which comprises physical and virtual spaces, and his perception of virtual space as an information space, plays an important role in attempts of understanding non-physical space and its direct relevance with mathematics. Nowadays, inquiries into non-physical space through designs, movies, or books result in further inquiries and questions.

A Mathematics Course and Digital Design Studio Held in the Department of Architecture of Middle East Technical University

The concept of design and the design process and the role of mathematics discussed above resulted in the proposal of a new course in the Department of Architecture at Middle East Technical University entitled "Mathematics in Architecture", which I taught, and the "Digital Design Studio", which was taught by myself and my colleague, Dr. Sebnem Yalýnay. "Mathematics in Architecture" is aimed at increasing awareness in the minds of students about the very close association between mathematics and mathematical thinking and architectural design, aside from geometry, which they are familiar with (Gönenç Sorguç 2005b).

In the "Digital Design Studio" this intention has been stated explicitly. In these studios, the design medium is considered as the digital space and the subjects are proposed accordingly. The students are asked to question this "new medium" by changing minds to algorithmic thinking. It is important to make clear that it is not expected that students will develop algorithms or codes for their designs; instead they are expected to develop an awareness of the multi-dimensional design process

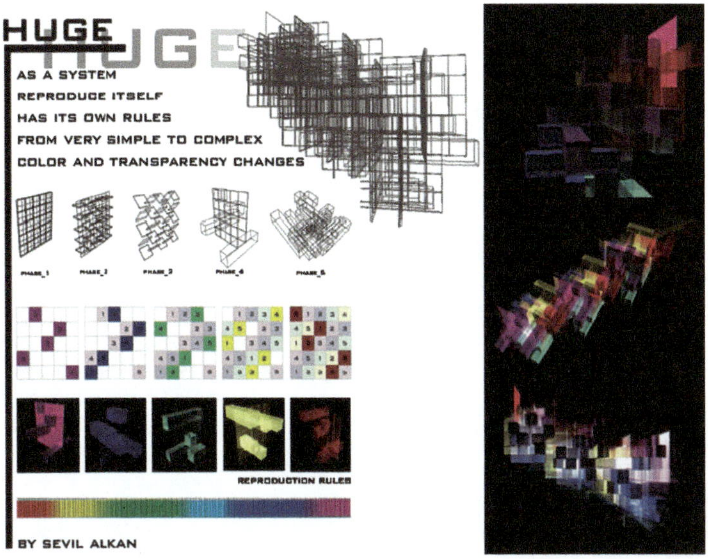

Fig. 88.1 "Huge", Arch. 470 digital design studio project by Sevil Alkan, 4th-year student

through mathematics and algorithmic thinking without losing sight of architectural design. In order to provoke students into making these design inquiries, the design subjects are chosen as abstract concepts which are thought to be more generative in the exploration of the digital medium. Some of the design topics were "huge", "a scalarity", "one as many, many as one" and "the game", for which students are free to interpret the concept as they like in the very beginning of the studio, then as the studio progressed, it was shown through criticism and lectures that digital medium was much more than a presentation medium and that algorithmic thinking and mathematics were essential for mastering that medium and its potential for the design, which was actually a process in digital medium. Figures 88.1 and 88.2 show the design projects that were accompanied by 2-min animations by students Sevil Alkan and Onur Yuce Gun, aimed at explaining their interpretation of designing in digital medium and algorithmic thinking.

Conclusion

In this study I have briefly discussed the evolving role of mathematics and algorithmic thinking in today's multi-dimensional architectural design under the influence of computational and information technologies. I tried to avoid giving names and examples which are actually acknowledged by all those who are interested in the design process, the ones actually referred to here are intended to reveal the existence of mathematics and algorithmic thinking in the center of the

Fig. 88.2 "Huge", Arch. 470 digital design studio project by Onur Yuce Gun, 4th-year student

design process. Aesthetics, design movements and the social and psychological impacts are not argued; instead I concentrated on different implementations.

I wished to show that design has become a complicated process interacting with various disciplines that influence the process in a direct or indirect way, resulting in multi-dimensional design problems. Architects have several means for achieving good designs and yet the amount of information or different domains of knowledge makes this task harder. Mathematics and its ways of thinking especially algorithmic thinking can provide ways to handle these complicated problems without posing any limits on creativity or design insight.

Biography Arzu Gönenç Sorguç earned a degree and PhD in Mechanical Engineering at Middle East Technical University. She went to Tokyo Institute of Technology for her post-doctoral studies. She is currently an associate professor at Department of Architecture at METU and the head of International Joint Graduate Program (TU Delft and METU) for 'Computational Design and Fabrication Technologies in Architecture'. Her research centres on acoustics, software technologies, environmental control and performance based design, structural systems, mathematics in architecture, computational design and technologies, computer-aided engineering and computer-aided manufacturing. She is the author of *A Parametric Approach to Biomimesis: A Proposal for a Non-Dimensional*

Parametric Interface Design in Architecture (with Semra Arslan, Lambert Academic Publishing, 2010). She was co-director with Kim Williams of "Nexus 2014: Relationships Between Architecture and Mathematics" (METU, Ankara, June 2014).

References

CALDAS, L. and NORFORD, L. 2001. Architectural Constraints in Generative Design System: Interpreting Energy Consumption Levels. Pp. 1397-1404 in *Proceedings of the 7th International IBPSA Conference*. R. Lamberts, C.O.R. Negrao and J. Hensen, eds. College Station, Tx: Organizing Committee of Building Simulation.

DYM, C., AGOGINO, A., ERIS, O., FREY, D. and LEIFER, L. 2005. Engineering Design Thinking, Teaching and Learning. *Journal of Engineering Education* **94**, 1: 103-120.

FERRAR, S. 2001. The Nature of Non-Physical Space' – Or how I learned to love cyberspace wherever it may be. Pp. 208-213 in *Architectural Information Management: 19th eCAADe Conference Proceedings: Education & Curricula Design Theory*. Helsinki: Helsinki University of Technology (HUT).

FISCHER, T. AND HERR, C. 2001. Teaching Generative Design. In: The Proceedings of the Fourth International Conference on Generative Art, C. Soddu, ed. http://generativedesign.eu/on/cic / ga2001_PDF/fischer.pdf (accessed 28 November 2013).

GÖNENÇ SORGUÇ, A. 2005a. Role of System Architecture in Architecture in Developing New Drafting Tools. "Mechanical Systems, Machine Elements, and Manufacturing, Special Issue on the Latest Frontiers of CAD/CAE/CG". *Japan Society of Mechanical Engineers JSME International Journal*, **28**, 2: 244-250.

———. 2005b. Teaching Mathematics in Architecture. *Nexus Network Journal* **7**, 1: 119-124.

VERNER, I. AND MAOR, S. 2003. The Effect of Integrating Design Problems on Learning Mathematics in Architectural Design College. *Nexus Network Journal* **5**, 2: 103-115.

VITRUVIUS. 2009. *On Architecture*. Richard Schofield, trans. London: Penguin Classics.

Chapter 89
Generative Design Grammars: An Intelligent Approach Towards Dynamic and Autonomous Design

Ning Gu

Introduction

Generative design research, particularly the theories and applications of design grammars, have formed a very important part of design research. The generative concept and the rule-based structure have heavily influenced the latest development of Computer-aided Architectural Design (CAAD) technologies, for examples, parametric design tools. Through the early mathematical models and the more recent computer implementations, they have contributed to the formal understandings of design analysis and design generation. By analysing existing designs of a known style, design grammars can formally describe this design style and generate other designs that also share the style. For nearly four decades, design grammars have been refined and examined across a wide range of design disciplines including architectural design, product design, engineering design and so on. Grammars as design formalisms are advanced in a number of ways: They are able to apply simple rules to produce designs with rich descriptions; and they enable different designs that share a similar style to emerge by alternating the sequence of the rule application.

With the adoption of computer technologies, design grammar research has transformed from the early focus of mathematical models to the current computational implementations, and the technologies have provided new ways to further the design grammar research. Under such an influence, this chapter revisits the grammatical approach to generative design, and presents the conceptual

An earlier version of this paper was published as: Ning Gu, "A Grammar for Dynamic and Autonomous Design in 3D Virtual Environments", pp. 11–22 in *Nexus VII: Architecture and Mathematics,* Kim Williams, ed. Turin: Kim Williams Books, 2008.

N. Gu (✉)
School of Architecture and Built Environment, University of Newcastle, Callaghan, NSW 2308, Australia
e-mail: Ning.Gu@newcastle.edu.au

K. Williams and M.J. Ostwald (eds.), *Architecture and Mathematics from Antiquity to the Future*, DOI 10.1007/978-3-319-00143-2_42,

framework of Generative Design Grammars (GDG). The framework outlines the general structure of GDG and the general structure of the basic components—design rules. Integrated with a computational agent model (Russell and Norvig 1995) and a 3D virtual environment as the simulation engine, GDG can support an intelligent approach towards dynamic and autonomous design. For demonstration, a scenario of designing a virtual gallery is presented, where a computational design agent reasons, dynamically and autonomously generates, simulates and modifies designs of a virtual gallery in the 3D virtual environment, through the application of a GDG example.

This research has enhanced the design grammar research by presenting an intelligent approach to automating and optimising the generative design process, through an integrated GDG framework supported with powerful agent reasoning and design simulation. This approach to utilising computer technologies in design implies both opportunities and challenges for designers.

Background

Design Grammars

The concept and development of GDG for this research are inspired by the notions of shape grammars (Stiny and Gips 1972). The inspiration comes directly from shape grammars as a design formalism for describing and generating designs in general. In architectural design, there are many successful shape grammar applications; for example, the Palladian grammar (Stiny and Mitchell 1978), the Mughul Gardens grammar (Stiny and Mitchell 1980), the Prairie Houses grammar (Koning and Eizenberg 1981) and the Siza Houses grammar (Duarte 1999).

Knight (2000) summarizes that a shape grammar is a set of shape rules, which can be applied in a step-by-step manner to generate a set, or language, of designs. The nature of shape grammars is both descriptive and generative:

- Shapes (points, lines, planes or volumes) as the basic components of shape rules are descriptions of the designs that the grammars generate.
- The applications of the shape rules generate designs via shape operations and spatial transformations.

Inheriting the descriptive and generative nature of shape grammars, our GDG are capable of describing designs in 3D virtual environments using components of the design rules, and generating designs via rule applications. The descriptive and generative qualities of GDG well serve the purposes of designing in 3D virtual environments.

3D Virtual Environments

The terms "virtual environments", "virtual worlds", or "cyberspace" can be understood as networked environments designed using the place metaphor. The place metaphor provides a consistent context for people to browse digital information, interact with the environment and communicate with each other. The applications of 3D virtual environments have expanded from the original internet gaming and military simulation to provide supports for other activities such as electronic institutions, virtual museums, distant education, virtual design studios, and so on. They have become an important part of the holistic living environments we inhabit supporting everyday economic, cultural, educational and other human activities. As CAAD tools, 3D virtual environments have shown promising potentials in areas such as design simulation, distant team works as well as interdisciplinary design collaboration.

Technologies for designing virtual environments have developed over the years supporting multi-user text-based, 2D graphical and 3D virtual environments. Nowadays, most virtual environments are visualised using 3D models. Platforms for designing 3D virtual environments include Active Worlds (http://www.activeworlds.com), Virtools (http://www.virtools.com), Second Life (http://www.secondlife.com), and others that have been developed from gaming engines such as Quake (http://www.idsoftware.com). Maher and Simoff (2000) first characterise the design activities in 3D virtual environments as "Designing within the Design". Unlike in most CAAD systems, designers are also represented within the virtual design. They are called avatars (animated characters). "Designing within the Design" lately become the main idea for exploring and enhancing remote team collaboration in design practice. For global design teams, 3D virtual environments provide an integral platform that utilises digital communication, design representation, and collaborative modelling. Figure 89.1 illustrates selected designs in 3D virtual environments from our recent research and teaching.

Except for the input and output devices, 3D virtual environments are implemented entirely in the computer environments. After all, they only comprise of assemblies of computing entities, which can be flexibly programmed and configured. This flexibility makes it possible to consider designing in 3D virtual environments in terms of dynamics and autonomy. However, the current designs in 3D virtual environments often limits to static simulations. Our research challenges this conventional use of 3D virtual environments in design; and presents GDG for dynamic and autonomous design in 3D virtual environments.

Fig. 89.1 Designs in 3D virtual environments. Image: author

The Conceptual Framework of Generative Design Grammars

This section presents a conceptual framework that provides guidelines and strategies for developing GDG by defining the general structure of a grammar and the general structure of its basic components: design rules. Using this framework, designers define grammars that produce different design languages for 3D virtual environments, rather than predefine every detail of all possible designs.

Generative Design Grammar Framework

A GDG is comprised of design rules R, an initial design Di, and a final state of the design Df.

$$\mathrm{GDG} = \{\mathrm{R}, \mathrm{Di}, \mathrm{Df}\} \tag{89.1}$$

The basic components of a GDG are design rules R. The general structure of a GDG comprises of four sets of design rules: layout rules Ra, object placement rules Rb, navigation rules Rc, and interaction rules Rd.

$$\mathrm{R} = \{\mathrm{Ra}, \mathrm{Rb}, \mathrm{Rc}, \mathrm{Rd}\} \tag{89.2}$$

The structure of GDG is determined by the four general phases of designing in 3D virtual environments. They are:

- To layout places/areas of the design: Each place/area has a purpose that accommodates certain intended activities;

- To configure the places/areas of the design: Each place/area is then configured with certain objects, which provide visual boundaries of the place/area and visual cues for supporting the intended activities;
- To specify navigation methods: Navigation in the virtual design can be facilitated to consider the use of way finding aids for assisting the designers and visitors' in exploring the design in the 3D virtual environment;
- To establish interactions: In general, this is a process of ascribing behaviors to selected objects in each place/area of the virtual design so that physical interactions of the design can be simulated or the designers and visitors can interact with the virtual design.

The four set of design rules—layout rules, object placement rules, navigation rules and interaction rules—address the above four design phases accordingly. The generative design grammar framework is illustrated in Fig. 89.2.

The firing sequence of the design rules follows the order of layout rules, object placement rules, navigation rules and finally interaction rules. Integrated with relevant design and domain knowledge, GDG can be developed by following this general structure. The stylistic characterisations of the generated designs—in terms of the syntax (visualization, layout and object placement) and in terms of the semantics (navigation and interaction)—are defined accordingly in these four sets of rules.

Design Rules

The basic components of GDG are design rules. The general structure of design rules is similar to the general structure of shape rules. In shape grammars, a shape rule can be defined as:

$$\mathrm{LHS} \rightarrow \mathrm{RHS} \tag{89.3}$$

which specifies that when a left-hand-side shape (LHS) is found in the design, it will be replaced by a right-hand-side shape (RHS). The replacement of shapes is usually applied under a set of shape operations or spatial transformations. The shapes are labeled (the use of spatial labels and state labels) for controlling the shape rule applications.

Similarly, a design rule of GDG is defined as:

$$\mathrm{LHO} \quad + \quad \mathrm{sL} \quad \rightarrow \quad \mathrm{RHO} \tag{89.4}$$

which specifies that when a left-hand-side object (LHO) is found in the 3D virtual environment, and the state labels sL are matched, the LHO will be replaced by a right-hand-side object (RHO). The term "object" used here can refer to a virtual object, a set of virtual objects or virtual object properties. Virtual objects are visualised as 3D models in 3D virtual environments. Like shape grammars, GDG

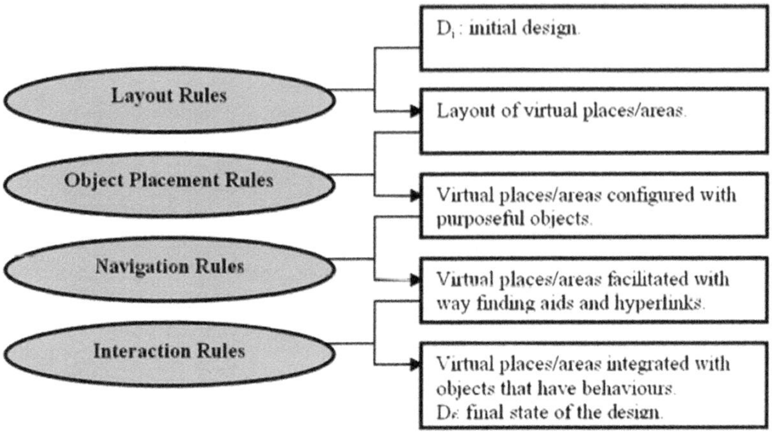

Fig. 89.2 Generative design grammar framework. Image: author

also use spatial labels and state labels to control the application of design rules. The original use of state labels in a shape grammar is to control the sequence of shape rule applications. In GDG, on one hand, this original purpose is maintained so that the designs rules can be applied in the sequence of layout rules, object placement rules, navigation rules and interaction rules. On the other hand, a special set of state labels are also developed as discussed below. The general structure of design rules implies the following two aspects:

- State labels are singled out and expressed explicitly as sL in the structure. The use of state labels is essential to the application of GDG as they direct the grammar application to ensure that the generated design satisfies the current design goals. Each design rule is associated with certain state labels representing specific design contexts that can relate to different design goals. In order for a design rule to be fired, a virtual object, a set of virtual objects or virtual object properties need to be found in the 3D virtual environment that match the LHO of the design rule, and the design contexts represented by the sL of the design rule need to be related to the current design goals;
- The basic components of design rules are objects and their properties, not shapes. Therefore, they are not entirely visual/spatial. For the interaction rules and parts of the navigation rules, the replacement of LHO with RHO is applied under a set of general transformations.

Layout rules are the first set of design rules to be fired in the application of a GDG. They are visual/spatial rules that generate the layout of places/areas according to the kinds of activities to be supported, as outlined in the design requirements. Figure 89.3 illustrates two example layout rules taken from a GDG for a gallery design. By applying the first rule, the design of the gallery will be expanded by adding an additional area. By applying the second rule, the design of the gallery will be changed by subtracting an area.

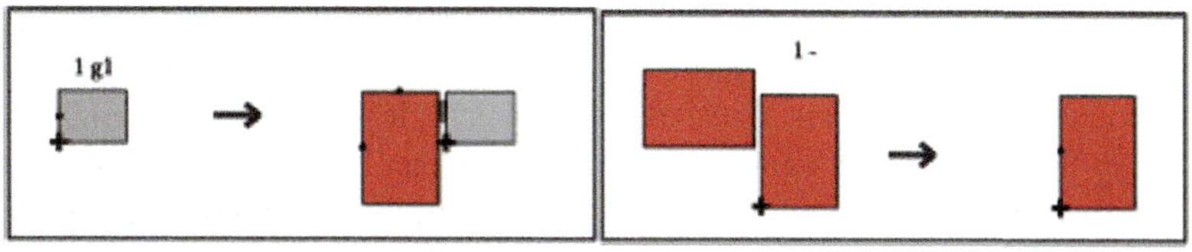

Fig. 89.3 Two examples of layout rules. Image: author

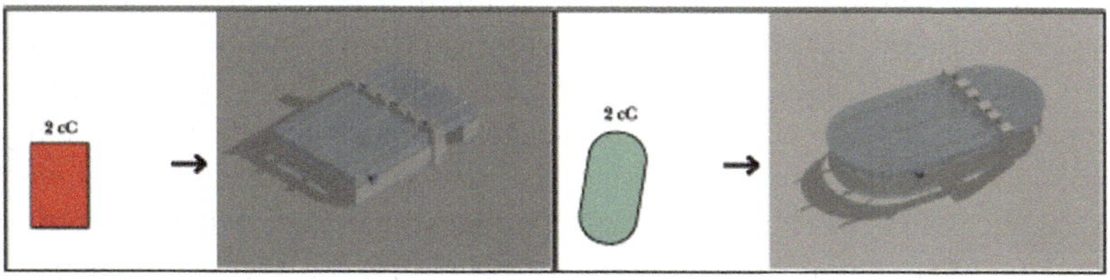

Fig. 89.4 Two example layout rules that generate visual boundaries for different areas of a gallery design. Image: author

Object placement rules are fired after layout rules, they are also visual/spatial rules. After a layout is produced, object placement rules further configure each place/area to provide visual boundaries of the place/area and visual cues for supporting the intended activities, through object placements. Figure 89.4 shows two example object placement rules that generate the visual boundaries for two different areas in a gallery design. Figure 89.5 shows an example object placement rule that arranges the interior of a display area for exhibition.

Navigation rules are fired next in the application of a GDG, after layout rules and object placement rules. Navigation rules provide way finding aids in the generated places/areas to assist the designers and visitors' navigation. Way finding aids can be simulated in 3D virtual environments with direct references to those in built environments (Vinson 1999; Darken and Sibert 1993, 1996). There are at least two kinds of way finding aids we use in built environments:

- The use of spatial elements, for example, paths, openings, hallways, stairs, intersections, landmarks, maps, signs and so on;
- The use of social elements, for example, the assistance gained from guides or other people.

Besides these way finding aids originating from built environments, 3D virtual environments also have their unique forms of navigation since virtual places/areas are hyper-linked. Most 3D virtual environments allow people to move directly between any two locations using hyperlinks. Hyperlinks are not parts of the actual design as they cannot be reproduced in built environments. However they are important for designs in 3D virtual environments as they often enable designers and visitors to navigate and explore more efficiently.

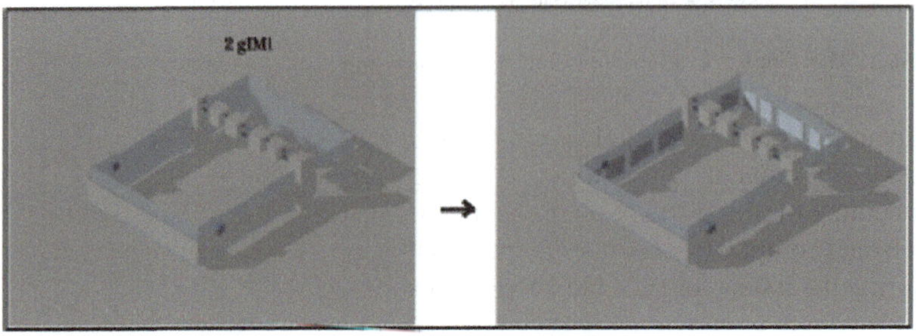

Fig. 89.5 An example layout rule that arranges the interior of a display area for exhibition. Image: author

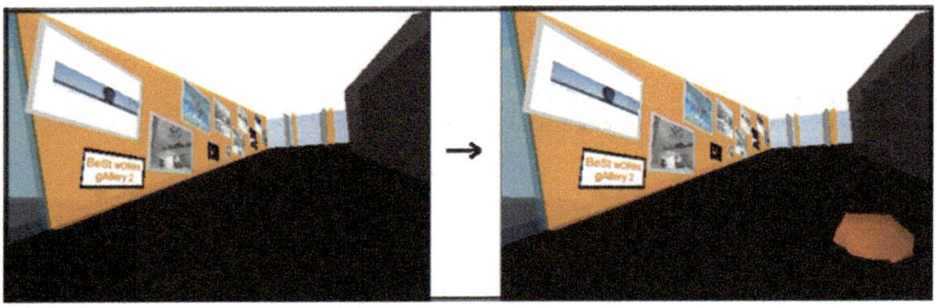

Fig. 89.6 The effect of an example navigation rule. Image: author

Navigation rules are not entirely visual/spatial. The application of the rules indeed involves object placements for defining way finding aids in the generated places/areas. However, before these object placements are made, navigation rules are mainly about recognising the connections among these generated places/areas and finding appropriate navigation methods for the designers and visitors to access these places/areas. Figure 89.6 shows the effect of an example navigation rule. The left-hand-side image is the interior of a display area in a gallery design. The right-hand-side image shows that a hyperlink is created and appears as a color stone on the floor. After appropriate behaviors are ascribed, the link will take the designers and visitors to a different display area when it is "stepped" on, if they require immediate exit from the current area.

Interaction rules are the final set of design rules to be fired in the application of a GDG. The application of interaction rules ascribes appropriate behaviors to selected objects in each generated places/areas. Therefore, physical interactions can be simulated or the designers and visitors can interact with the virtual design.

Interaction rules are non-visual/spatial rules that recognise selected objects in the virtual design and ascribe appropriate behaviours to these objects. There are two different types of interaction rules. One supplements object placement rules and the other supplements navigation rules. Object placement rules define visual boundaries for each generated place/area and place purposeful objects in the place/area. The first type of the interaction rule ascribes behaviours to relevant objects in

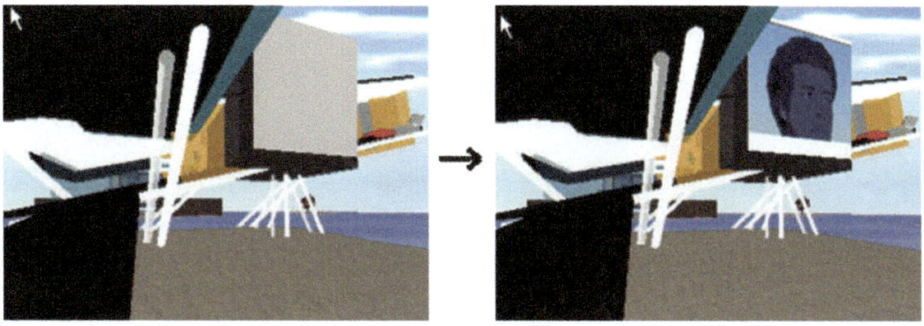

Fig. 89.7 The effect of an example interaction rule. Image: author

order to simulate certain physical interactions in these places/areas. The other type of interaction rule looks for way finding aids and hyperlinks generated by navigation rules and ascribe appropriate behaviours to activate them.

Because interaction rules do not operate on a visual/spatial level they are not appropriate to be expressed using illustrations. In this study, interactions rules are expressed in the form of "IF... THEN..." Without getting into the technical details of ascribing behaviours to objects in 3D virtual environments, Fig. 89.7 shows the effect of an example interaction rule of the first type for supplementing object placement rules. The left-hand-side image is the exterior of a gallery design with an empty advertisement board. The right-hand-side image shows the same advertisement board displaying digital images in an animated sequence, after the interaction rule is fired, which configures the object properties of the advertisement board object using a scripting language to enable the animation to be shown.

Designing a Virtual Gallery

This section presents a scenario of designing a virtual gallery. The scenario aims to demonstrate the application of GDG, and their effectiveness for dynamic and autonomous design in 3D virtual environments. GDG as generative design systems can be manually applied by human designers, or as demonstrated in this gallery design scenario, dynamically and automatically applied by computational agents. In the context of computer science, agents as intentional software systems operate independently and rationally, seeking to achieve goals by interacting with their environment (Wooldridge and Jennings 1995). Unlike most computational objects, agents have goals and beliefs and execute actions based on these goals and beliefs (Russell and Norvig 1995). The agents used in this demonstration are Generative Design Agents (GDA) (Gu and Maher 2005). GDAs are rational design agents specified to have five computational processes: Sensation, interpretation, hypothesising, designing and action. These processes provide a basis that allows design and other domain knowledge to be integrated into GDA, which together support reasoning and designing in 3D virtual environments.

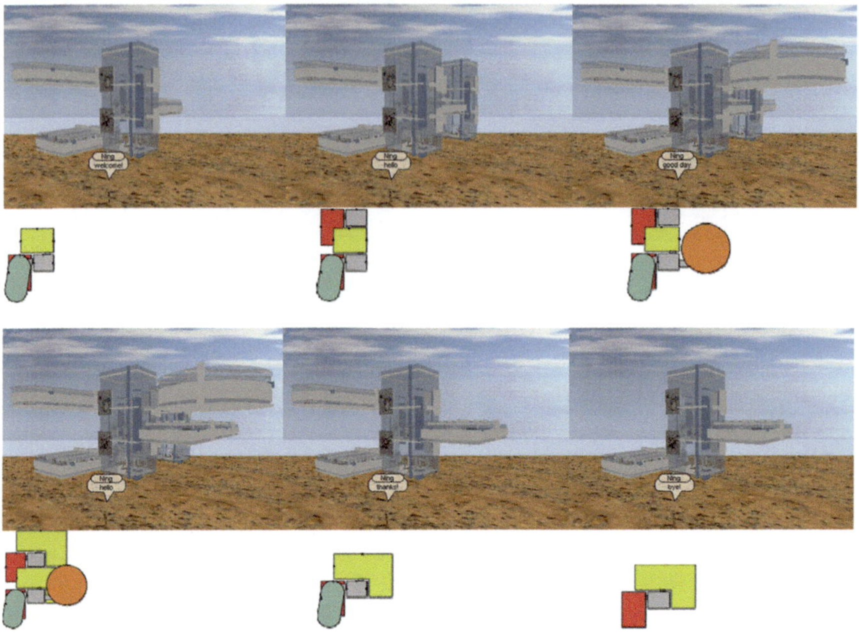

Fig. 89.8 3D models and Plans of a virtual gallery generated for stages 1–3 (*above*) and stages 4–6 (*below*) of the design scenario. Image: author

The design scenario consists of six different stages. The different stages present various changes of design requirements during the designing process, for example, changes of activities, changes of exhibition requirements, changes of gallery capacities, and so on. The scenario shows that a GDA senses these changes of design requirements, either being simulated in the 3D virtual environment or entered directly by the human designer. By altering the sequence of the design rule application in the GDG example, the GDA dynamically and autonomously generates, simulates and modifies different designs in the 3D virtual environment to address different changes of design requirements.

In terms of the technical implementation, the GDA is implemented using Java. The scenario is implemented in a 3D virtual environment developed using the Active Worlds platform. The design rules of the GDG example, and a general rule base for supporting the GDA's reasoning, are written using Jess (http://herzberg.ca.sandia.gov), a rule-based scripting language (Friedman-Hill 2003). Figure 89.8 illustrates different designs of the gallery, both 3D models and plans, generated by the GDA through the application of the GDG example, for the six stages of the scenario.

In the cases when there is more than one design rule that match the current design context, a control mechanism is needed to resolve the conflict. In general, there are three main methods for controlling the generative design grammar application. They are random selection, human designer intervention, and agent learning mechanism. This scenario uses the human designer intervention method.

- The random selection method allows the system to randomly select one design rule from the set of rules that meet the conditions.
- The human designer intervention method allows the system to turn to human designers for instructions once such a conflict occurs.
- The agent learning mechanism provide a more dynamic but complex approach to allow the system to resolve the conflict based on the agent's past design experience.

Conclusion

This chapter presents GDG: an intelligent approach towards dynamic and autonomous design. The GDG approach extends design grammar research by providing an integrated framework with powerful agent reasoning and design simulation that can automate and optimise the generative design process.

- The use of 3D virtual environments provides a simulation tool for visualising designs that are generated by GDG; and more importantly serves as a platform for representing design requirements as well as simulating contextual information that may affect the generation of the designs;
- Design automation is supported by the use of GDA. Computational agents as intentional software systems can operate independently and rationally, seeking to achieve goals by interacting with their environments (Wooldridge and Jennings 1995). The use of GDA in conjunction with GDG presents a robust approach to design reasoning and automation for executing the design grammar.

The effectiveness of both GDG and GDA has been demonstrated through a virtual gallery design scenario. Although the design scenario is constructed with a specific kind of virtual gallery design and a specific platform of a 3D virtual environment in mind, it demonstrates the general effectiveness of GDG and GDA both for autonomous designs in 3D virtual environments. Integrated with different design and domain knowledge, they can be adapted for dynamic and autonomous design in 3D virtual environments for other purposes.

This grammatical approach to dynamic and autonomous design in 3D virtual environments also implies the following opportunities and challenges for design, design research as well as designers.

For design, highlighted with GDG's generative design capabilities, the use of GDG provides a formal framework to describe and generate designs in 3D virtual environments with certain stylistic considerations. The use of GDA presents a robust approach to design reasoning, generation and automation. These design agents actively seek to satisfy their design goals to meet the changing design requirements, by interacting with the 3D virtual environment.

For generative design research, the GDG framework presented in Sect. 89.3 can serve as a base for developing GDG for different design styles that suit different purposes. The GDG approach is a highly integrated and interdisciplinary one. Combining with the use of agent technologies, it introduces dynamics and autonomy to designing in 3D virtual environments. 3D virtual environments integrated with GDG can go beyond the conventional purpose of design communication and static simulation to support design generation and automation. The GDG framework also provides a foundation to formally study the styles of 3D virtual environments. Compared to other novice designs, 3D virtual environments designed with a specific style in mind will achieve better consistency in terms of visualisation, navigation and interaction, and this consistency provides a strong base to assist the designers and visitors' orientations and interactions in these virtual environments.

For designers, on one hand, GDG provides the generative force for dynamic and autonomous design. On the other hand, each grammar defines coherent stylistic characterisations shared by the designs it generates. Designers do not repetitively produce individual designs. Instead, they specify GDG incorporating with their own design styles, for design agents such as GDA. The application of GDG is then applied by GDA on behalf of the human designers, and directed by the actual design requirements, to dynamically and automatically generate individual design instances in 3D virtual environments as needed. This will require designers to gain a different set of skills in order to understand and accommodate the grammatical approach to design. The emergence of generative design tools such as GDG offers cutting-edge technologies for designers to exploit alternative ways of designing, but at the same time imposes new challenges redefining their roles in design.

Biography Ning Gu is a senior lecturer at the School of Architecture and Built Environment, University of Newcastle in Australia, where he was awarded the Pro Vice-Chancellor's Award for Research Excellence in 2010. His research interests are virtual architecture, generative design, design cognition, computer-supported collaborative design and Building Information Modelling (BIM). He has published extensively in international refereed journals and conferences. His research has been funded by competitive research grants from the Australian Research Council (ARC) and Australian Cooperative Research Centre for Construction Innovation (CRC-CI). As a digital architect, he has published conceptual designs of virtual architecture, and has designed and implemented virtual environments for a wide range of purposes including design collaboration, e-commerce, online learning, and so on. He has been a visiting scholar at MIT, Columbia University and Eindhoven University of Technology.

References

Darken, R. P., and Sibert, J. L. 1993. A Toolset for Navigation in Virtual Environments. Pp. 157-165 in *Proceedings of the 6th Annual ACM Symposium on User Interface Software and Technology*. Atlanta,.

———. 1996. Way Finding Strategies and Behaviours in Large Virtual Worlds. Pp. 142-149 in *Proceedings of ACM SIGCHI'96*. New York: ACM.

Duarte, J. P. 1999. Democratized Architecture: Grammars and Computers for Siza's Mass Housing. *Proceedings of the International Conference on Enhancement and Promotion of Computational Methods in Engineering and Science*. Macau: Elsevier Press.

Friedman-Hill, E. 2003. *Jess in Action*. Greenwich, CT: Manning Publications.

Gu, N. and Maher, M. L. 2005. Dynamic Designs of Virtual Worlds Using Generative Design Agents. Pp. 239-248 in *Proceedings of CAAD Futures 2005*. Dordrecht: Springer.

Knight, T. W. 2000. Shape Grammars in Education and Practice: History and Prospects. http://www.mit.edu/~tknight/IJDC/ (accessed 28 November 2013).

Koning H. and Eizenberg J. 1981. "The Language of the Prairie: Frank Lloyd Wright's Prairie Houses." *Environment and Planning B: Planning and Design* 8, 3: 295-323.

Maher, M. L., and Simoff, S. 2000. Collaboratively Designing within the Design. Pp 391-399 in *Collaborative Design: Proceedings of Co-Designing 2000*, S. A. R. Scrivener, L. J. Ball and A. Woodcock, eds. London: Springer.

Russell, S., and Norvig, P. 1995. *Artificial Intelligence: A Modern Approach*. Englewood Cliffs, NJ: Prentice Hall.

Stiny, G., and Gips, J. 1972. Shape Grammars and the Generative Specification of Painting and Sculpture. Pp. 1460-1465 in *Proceedings of Information Processing 71*. Amsterdam: North Holland Publishing.

———. 1980. The Grammar of Paradise: on the Generation of Mughul Gardens. *Environment and Planning B: Planning and Design* **7**, 2: 209-226.

Stiny, G. and Mitchell, W. J. 1978. The Palladian Grammar. *Environment and Planning B: Planning and Design* **5**, 1: 5-18.

Vinson, N.G. 1999. Design Guidelines for Landmarks to Support Navigation in Virtual Environments. Pp. 278-285 in *Proceedings of the ACM CHI 99 Human Factors in Computing Systems Conference*, M. W. Altom and M. G. Williams, eds. Pittsburgh: ACM Press.

Wooldridge, M., and Jennings, N. R. 1995. Intelligent Agents: Theory and Practice. *Knowledge Engineering Review* **10**, 2: 115–152.

Chapter 90
Ethics and Geometry: Computational Transformations and the Curved Surface in Architecture

Michael J. Ostwald

Introduction

Is it possible to criticise architectural form, the shape of a building, from an ethical perspective? Since the nineteenth century a growing number of scholars have maintained that not only is this conceivable but is indeed almost inevitable. While the validity of this position is still being questioned to the present day, a presupposition founding the question has rarely, if ever, been adequately considered. If the shape of a building can be criticised from an ethical perspective, and the production of architectural form is reliant on a range of underlying conceptual armatures, then such devices must also be complicit in architecture's ethical status. Moreover, while a wide range of concerns influence the built environment, it is the application of geometry that most consistently enables the production of architectural form. Given this situation, is it feasible to criticise or interpret architectural form from a moral perspective without first considering the ethics of geometry?

The research contained in this chapter is broadly concerned with the way in which a designer constructs architectural form from geometry and more specifically with whether such constructions can legitimately be considered from an ethical perspective. Because this topic is too extensive for a single chapter, the present research focuses on a specific case wherein geometry and architectural form are inextricably connected. This situation is found in the production of complex or

First published as: Michael J. Ostwald, "Ethics and Geometry: Computational Transformations and the Curved Surface in Architecture", pp. 77–92 in *Nexus VI: Architecture and Mathematics*, Sylvie Duvernoy and Orietta Pedemonte, eds. Turin: Kim Williams Books, 2006.

M.J. Ostwald (✉)
School of Architecture and Built Environment, University of Newcastle, Newcastle, NSW 2308, Australia
e-mail: michael.ostwald@newcastle.edu.au

K. Williams and M.J. Ostwald (eds.), *Architecture and Mathematics from Antiquity to the Future*, DOI 10.1007/978-3-319-00143-2_43,

compound curvilinear forms. Historically the creation of such forms has involved a demonstrable awareness of geometry (or at a minimum its practical application) whereas other architectural forms, especially rectilinear ones, do not have the same overt reliance on geometric construction. While a possible connection between curvilinear form and geometry is apparent in historic examples, more recently the development of computational design tools have embedded explicit geometric constructions into the creation of architectural form. One design strategy in particular, algorithmic transformation, perfectly illustrates the way in which a geometric construction can produce a curvilinear or organic architectural form. For this reason, the present chapter analyses the computational transformation of an object as a means of producing and representing architectural form. The analysis is not concerned with the mechanics of the deformation the actual algorithm or mathematical formula that underlies the geometric construction. The designer is almost never aware of the mathematics governing the software. He or she is only conscious of the geometric boundaries and rules which they can manipulate to produce form. Instead the analysis is focussed on whether or not the transformative operation has ethical implications for the resultant architectural form.

One of the reasons that the question concerning the role of geometry in architecture's moral dialectic is so complex is that classical philosophy holds, and for good reason, that the question should not be asked at all. Within a neo-Platonic or Aristotelian philosophical framework, architectural and ethical systems are mutually exclusive. From this classical perspective the *actions* of a person may be ethical (or not) but the objects they use or create are free of innate ethical bias even if they possess functions that lend themselves to use in potentially unethical ways. Thus, a gun is without intrinsic ethical bias because it may be used to assist human survival (potentially ethically sound) in the same way that it might be used to injure or kill (potentially ethically unsound). The actions of the person using the gun are open to ethical interpretation, not the gun itself. Similarly, in the case of geometry and mathematics, in *Rhetoric* Aristotle argues that "mathematical speeches have no *éthé*, since they do not involve any resolute choice" (Lachterman 1989: xi). Here Aristotle affirms the view that like architecture, mathematics and geometry are objects in a philosophical framework; they do not, in any connotative sense, result in a particular response. While it is difficult to categorically refute the classical position, it has not hindered the production of an extensive body of discourse dedicated to the ethics of architecture. In contrast, the Aristotelian position seems to have largely overshadowed any propositions concerning the ethics of geometry. One reason for this difference between architecture and geometry from an ethical perspective is related to ubiquity.

Objects in classical philosophy are typically aligned with ideal or singular uses. Thus, for Plato a chair has the same terrestrial function regardless of whether it is made of wood or stone, or is decorated or plain. Even though the physical chair is a universal object, all chairs are variations of an idealised or transcendent support mechanism that the artisan is striving to imitate. While geometry occupies a different role in Platonic metaphysics it is sufficiently omnipresent that it does not control, in a moral sense, comportment or conduct (Elden 2001). Jacques

Derrida describes the role played by geometry in metaphysics as being one of a number of "ideal objects of science" (1989: 25), things that exist outside of human desires, motives, or behaviours. In contrast to such ideal objects, architecture is less ubiquitous precisely because it is shaped by an individual or group with an agenda that is necessarily social or human. This is why a wider definition of ethics is normally adopted in architectural discourse than in the philosophy of geometry.

Architectural ethicists typically postulate that some objects are sufficiently complex that they have either: (1) a capacity to shape the attitudes or behaviours of their users; or (2) some ability to communicate the designer's intentions to the user. In both instances the object, in isolation, has an impact on an individual or group and is therefore vulnerable to ethical criticism both of itself and its creator. The former proposal (1) is certainly open to dispute but at the very least scholars agree that architecture provides shelter, a function that directly shapes inhabitation and facilitates community (Levine et al. 2004). Barry Wasserman, Patrick Sullivan and Gregory Palermo argue that architecture is fundamentally concerned with "the design and construction of places for human affairs and human inhabitation" (2000: 36). This means that from the moment humankind constructed "a shelter for protection in the natural wild landscape ... architecture's import for well-being" became manifest (idem.). In the latter proposition (2) set out above, that some objects can transmit a designer's intent, the evidence is equally contentious. First, can an artist's or architect's creative agenda be communicated through their work to an observer? Second, can this communication occur in such a way that it influences the user's actions? Mary Devereaux (2001) and Berys Gaut (2001) independently assert that the answer to both of these questions is yes and each focuses on objects (art, film or architecture) that function as aestheticised propaganda. For example, Albert Speer's vast neo-classical plan for Adolf Hitler's reformation of Berlin could not be viewed as anything other than a sign of totalitarian power (Dovey 1999). However, while Speer's Berlin may evoke a subconscious recognition of fascism, autocracy or opulence, a person who is unaware of history would be unlikely to intuit overtones of xenophobic nationalism in its classical entablature and trabeated colonnades. This line of argument, when taken to its logical conclusion, confirms that architecture is at best a poor communication tool for complex concepts and at worst is, in a linguistic sense, unable to sustain the connection between signified and signifier (Eco 1980). Because geometry in itself has a further diminished capacity to communicate, in ethical theory its role is relegated to that of an ideal object. The present chapter acknowledges these philosophical limitations, but accepts the possibility that architecture is open to ethical criticism and that, by inference, the geometry that constructs architecture is also morally complicit in its production. The chapter then tests this proposition in a case wherein architecture, geometry, human experience and implied meaning are closely connected. That case is the curved form in architecture and explicitly its production through computational transformation.

One final clarification that is necessary before progressing is that in philosophical enquiry the phrase "the construction of geometry" encompasses both the physical drafting of shapes (tracing arcs, parallel lines and right angles to establish more complex configurations) as well as the logical description of a sequence of actions (or calculations) which produce a spatial outcome. The concept of geometric construction also assumes that it is a conscious endeavour; nature does not construct geometry, humans do. Importantly, the phrase does not refer to the physical erection of a building using bricks and mortar.

The present chapter is structured in four sections. This introduction has set out the field of research, the approach taken to the specific case and some methodological limitations. The second section reconstructs a short history of ethical theory in philosophy and in architecture with the intention of drawing out aspects which are most closely connected to geometry. This section relies on the small number of philosophical theories that have been proposed for interpreting geometry in an ethical framework. The third section describes the creation of curved, irregular or organic surfaces in architecture, focusing on the connection between geometric construction and architectural form. This penultimate section includes three examples of computational transformations that result in a non-orthogonal architecture. Finally the chapter considers computer-generated architecture and the specific case of the transformative process of creating and representing a design. This section concludes by questioning the design practice from an ethical perspective.

Ethics, Geometry and Architectural Form

Ethics is defined as the study of the moral standards that shape an individual's conduct. Ethical discourse is typically relied upon to provide advice on the right (in a moral rather than legal sense) course of action in a situation. Ethics, like geometry, does not rely on empirical observation but rather on formal argumentation or construction; an ethical statement or position relies on a logical proposition or presupposition (Beauchamp 2001).

The branch of ethics that provides guidance about deeds or actions is called normative ethics. Normative ethics develops logical arguments that are intended to assist moral decision making. Most writings on architectural ethics participate in the normative approach because they are written with the intention of providing advice on ways to approach the problems of architectural practice or design. There are three primary normative schools of thought in Western ethical philosophy. First there is the view, attributed to Aristotle, that a personal value system benefits both an individual and society; this theory is called virtue ethics. The second is reliant on the argument that people must learn duties and be taught values that lead, according to proponents like Immanuel Kant, to the creation of an ethical society. This school is known as deontology: the theory of ethics which develops universal laws governing decisions regardless of local consequences. The third school of thought says that the needs of the community are the benchmark of ethical thinking and that the conduct of an individual should be shaped by these greater needs and values. This theory of ethics is known as either utilitarianism or as teleological ethics. The critical study of these three schools of normative ethics, known as metaethics, challenges the assumptions that are made as part of the ethical decision-making process. Only a limited number of architectural examples exist of metaethical theory. Because architecture is not, in a classical sense, a legitimate field of ethical enquiry, its claims are rarely framed in such a way that they may be interrogated by metaethical theorists. However, virtue theories of architectural ethics have been common for many hundreds of years and in the last few decades the deontological position has become increasingly prevalent.

From the earliest treatises, architecture has been presented as serving, either in a functional or symbolic sense, the good of humanity. The Vitruvian triad of "firmness, commodity and delight" clearly articulates architecture's aspirational role as central to the protection of humanity, the formation of community and the maintenance of higher goals. The similarly archaic foundation myth of architecture in the construction of the primitive hut also confirms that architecture's role is to shelter humanity, shape its practices and sustain its communal vision (Rykwert 1987). Wasserman, Sullivan and Palermo support both assertions when they note that in the historical treatises architecture's role is "providing 'a good' for society through building" (Wasserman et al. 2000: 29). In the seminal architectural treatise of the Renaissance, Alberti, like Vitruvius before him, argued for the "well-being of society through architecture", a claim that is concerned with "architecture's ethical potential" (idem). The Vitruvian moral argument, developed by Alberti, is essentially a form of virtue ethics, as Spector (2001) observes. Importantly, in these early works geometry is considered essential for the production of an ethical architecture. For example, Vitruvius identified a system of geometry circles, squares and associated Phileban solids as being an abstraction of human form, which is itself a reflection of the image of God. The Vitruvian position has strong parallels to the argument Plato offers that artefacts and nature may be beautiful but they are both poor reflections of their ideal forms. Natural beauty is a type of terrestrial diversion whereas geometric purity is more closely aligned to the form realm, as it raises the mind's focus to transcendent ideals. Although there are serious differences between the Vitruvian and Platonic positions, both affirm the connection between geometry and higher virtues. Each also supports the second proposition which opened this present chapter that if architectural form is susceptible to ethical criticism because it influences behaviour then the underlying geometric construction must also be exposed in this way.

In the *Ethical Function of Architecture* Karsten Harries asserts that geometric construction is central to the creation of architectural form which expresses the spirit of the age; this is for Harries the pivotal ethical consideration for architecture. A primary theme in his argument is what he calls "perennial Platonism"; the connection between local objects (such as architecture) and universal ones (the virtues and ideals of the form realm) by way of geometric and aesthetic rules which are ethically framed. For example, Harries's reading of Platonism leads to a proposition about the possible relationship between geometry, form and the ethics of timeliness associated with the view that architectural form should express *Zeitgeist*, the spirit of the age. For Harries geometric construction shapes the ethics of form precisely because the "appeal of the language of geometry transcends particularities of time and of place" (Harries 1997: 255): the geometric construction is the action that imbues architectural form with an innate ethical quality. Architectural form cannot simply be an object, in the classical sense advocated by moral philosophy, because the underlying geometry has, in its specific process of construction, a range of qualities or virtues. There are two related arguments in philosophy, from Hobbes and Lachterman, that support Harries's view.

In the seventeenth century the British philosopher Thomas Hobbes applied the construction methods adopted in geometry and mathematics to political science to create a logical method of social formation and, by inference, a moral community (Grant 1990). For Hobbes, "geometry is said to deal with motion" (Sacksteder 1980: 103); this is because geometry cannot exist without action. Hobbes maintained that it is the conscious process of description or motion (what the present chapter calls construction) that gives geometry its innate rightness; a proposition replete with ethical overtones. In the most famous twentieth-century argument about the ethics of geometry David Lachterman (1989) develops a related proposition.

Lachterman's definition of ethics is derived from the Aristotelian *ta éthé*, which refers to the common values that define the way in which a society operates. Like the majority of other examples in this section, it is derived from virtue theory; its focus is primarily on motivation and comportment rather than actions and their consequences. Lachterman's argument is that a close reading of the explication and application of geometry throughout history reveals the motives and virtues of the geometer and the resultant form or theory:

> Accordingly, in speaking of a radical difference between the ethos of Euclid and the ethos of Descartes I am not suggesting anything like a moral discrimination between persons; rather, I have in view the disparate ways (mores) and styles in which the Euclidean and the Cartesian geometer do geometry, comport themselves as mathematicians both toward their students and toward the very nature of those learnable items . . . from which their disciplined deeds take their name (Lachterman 1989: xi).

Thus one geometer may, in the process of construction, display a disdain for their audience (of students) and their subject (geometry) whereas another's method of construction might be more respectful. In this way Lachterman identifies in the geometric construction of Descartes "compelling evidence that [Descartes] wants his 'samples' of method to display those virtues of absolute originality and self-sufficiency by which genuine learning is distinguished from simple *historia*" (Lachterman 1989: 149). Lachterman suggests that geometry, in its construction (in modern philosophy) or axiomatic method of geometrical demonstration (in classical philosophy) retains an ethical trace that reflects the attitudes and virtues of its creator.

The arguments of Lachterman and Harries each affirm the close relationship between the attitudes of the designer, the geometric construction they are employing and the outcome it produces. In the nineteenth century John Ruskin advanced an approach to the ethics of architecture that also accepts that geometry, as it is expressed in architectural form, is ethically culpable. At the heart of Ruskin's polemic is the suggestion that aesthetics, like ethics, must have an underlying and supportable series of suppositions. If the aesthetics of a design rely on misdirection or hidden structure then the design is no longer true or right (Moore and Ostwald 1997). Ruskin argued that mechanically manufactured curvilinear forms were lifeless and therefore lacking a suitable symbolic connection to nature, a connection which is assumed to be their fundamental purpose for existing on the façade of a building. For Ruskin, the "proper material of ornament" is "whatever God has created" (Ruskin in Cook and Wedderburn 1903: vol. 9, 265). In practice, Ruskinian "right line" is a hand drawn curve that provides an indelible connection between the human body (the stone mason), the construction of geometry (by hand) and the carving of form, "their universal property being that of ever-varying curvature in the most subtle and subdued transitions" (Cook and Wedderburn 1903: vol. 9, 267). Only the construction of a right line elevates the viewer's mind to higher moral thoughts.

In the examples outlined in this section there is the suggestion that geometry, through its construction, can be ethically coloured and that the architectural form that results from such a construction will bear evidence of this taint. There is also specific agreement in the examples that it is the construction process that reflects the moral standards of the geometer or designer, not the resultant geometry. Intriguingly, all of the arguments concerning the ethics of geometry are closely related to the first of the normative schools of thought, virtue theory. Virtue theory does not tell people how to distinguish between moral or immoral actions, rather it argues that if a designer or geometer has the right personal qualities they will not be motivated in immoral ways.

The Construction of Curvilinear Form

Historically, the production of curvilinear architectural forms has occurred in an environment that is aware of the cultural, political or symbolic significance of such surfaces and the geometry that underlies them. In semiotic theory this environment is said to possess a "social contract" wherein words and forms have some degree of consistent meaning, and that this meaning results from the construction of language or architecture (de Saussure 1959; Broadbent 1970). For example, in medieval art the vertical aureole form, the mandorla or *vesica piscis*, repeatedly frames images of religious significance. The same almond-shape is also present in the curved vaults and the window forms of religious structures of the era. Such arched forms are a direct result of the underlying geometric armature and the symbolic significance of its generation.

During the Renaissance circular forms were central to many religious structures, and the treatises of Vitruvius and Alberti ensured that such constrained curves were understood, at least by *cognoscenti*, as derived from the proportions of the human body which, in turn, have a more numinous origin (Kruft 1994). The archetypal Baroque compound curve found in the façade of Borromini's *S. Carlo alle Quattro Fontane* has both regular sinusoidal flowing surfaces along with more dynamic syncopated curves constructed of broken oval segments. Such curves responded to the social, symbolic and phenomenological needs of the counter-reformation (Norberg-Schulz 1971, 1980). Their construction, especially at the hands of Guarino Guarini, was reliant on a rigorous geometric interpretation of the connection between experiential or secular qualities (especially perspective) and symbolic or profane characteristics (Meek 1988). The rise of Modernism, initially in French Enlightenment architecture and later in Russian Constructivism, saw the return of strong geometric curves. In such cases, ranging from Ledoux and Boullée's spherical buildings to Melnikov's and Chernikhov's intersecting cylinders, architectural form was derived from clearly articulated geometric constructions. Moreover, while geometry in such works was no longer theologically nuanced it nevertheless had a symbolic role to play in evoking the *Zeitgeist*. In all of these examples the curved architectural form has resulted from a conscious understanding that certain geometric constructions possess properties that lend themselves to particular semiotic (associated with meaning) or phenomenological (associated with experience) applications. However, in recent years the methods for producing complex curvilinear architectural forms have become increasingly reliant on the computer and the process of transformation.

In the interpretive systems developed to analyse the philosophical or ontological implications of classical geometry the persistent influence of an operation on a spatial system is called a *transformation*. An example of a simple geometric transformation is found in D'Arcy Thompson's famous book of 1942 *On Growth and Form* (1961). In this work Thompson presents a series of line drawings showing how one species of fish can be seen to resemble another when they are visually stretched according to a series of geometric rules (Fig. 90.1). William Mitchell (1990) defines this process as a continuous geometric deformation and compares it to Albrecht Dürer's explanation of human physiognomy and proportion using similar continuous deformations (Fig. 90.2).

Mitchell classifies a series of different types of linear deformations, including "stretch", "shear" and "perspectival"; so named because of the dominant geometric distinction between a regular two-dimensional grid placed over an object and its post-deformation version. Mitchell also provides examples of non-continuous deformations where certain relational properties of a shape are retained while others are altered. Importantly, in all of these cases the geometry that underlies the transformation is both visible and controlling. Yet, in computer aided design software the geometric construction is typically invisible. An informative example of such a computational transformation is found in the work of architects Solan Kolatin and Bill Macdonald.

In the last decade Kolatin and Macdonald have achieved international recognition for their designs that have been produced through the application of generative algorithms (2001). Their widely published Chimerical Housing project takes as its starting point a found object; a "normative three bedroom, two-and-a-half bathroom colonial house" (Kolatin and Macdonald 2000: 71), which came pre-packaged as a library part with the modelling software they use. This found object is then subjected to an iterative process that transfigures its form: a type of

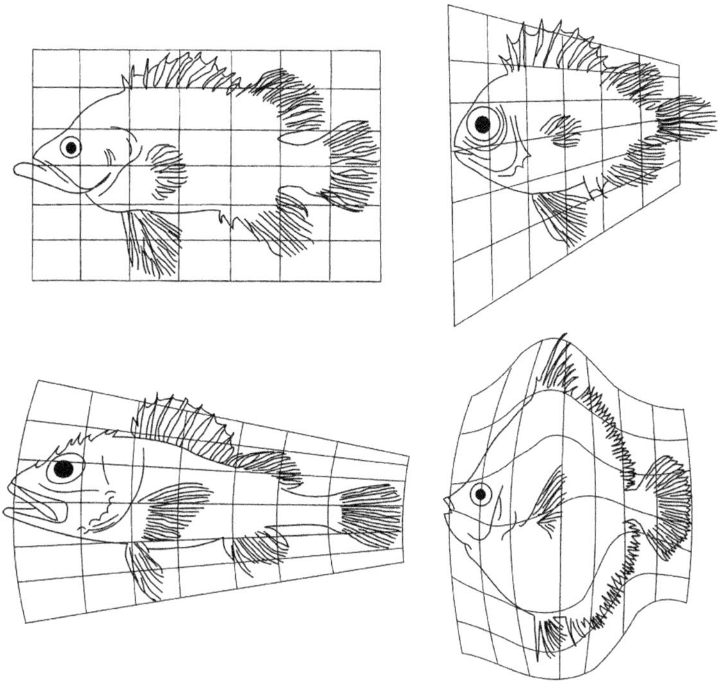

Fig. 90.1 Continuous deformation as a strategy for creating different species of fish. Image: courtesy Tessa Morrison, after D'Arcy Thompson

extreme weathering that attacks the edges of the object, slowly liquefying it until it has been converted into an amorphous puddle. Kolatin and Macdonald's intention with the Chimerical Housing project is to determine the point at which the form has become most visually arresting. They describe the transformative process as consisting of three stages. In the first stage the transformation creates what they call "useful" genetic aberrations: forms which still foster some level of semiotic recognition. In the second stage the transformation produces "monsters": mutations where the form has become sufficiently attenuated or distorted that it may no longer be a spatially useful outcome. Finally, when the algorithmic regression has reduced the house to a flattened mass, they describe the form as "noise" or visual static. Kolatin and Macdonald see their role in this process as deciding the stage at which the transformation is halted. They are not explicitly interested in either the starting point or the end point, nor is the geometric construction that underlines the transformation ever mentioned. This method is in stark contrast to the historical process of deploying geometric constructs to produce an architectural form that possesses characteristic properties derived from the underlying geometry.

In 1995 I suggested an experimental case for challenging students to think about the relationship between architectural form, geometry and the computer. Again a found object is the starting point: most computer-aided design tools rely on found or

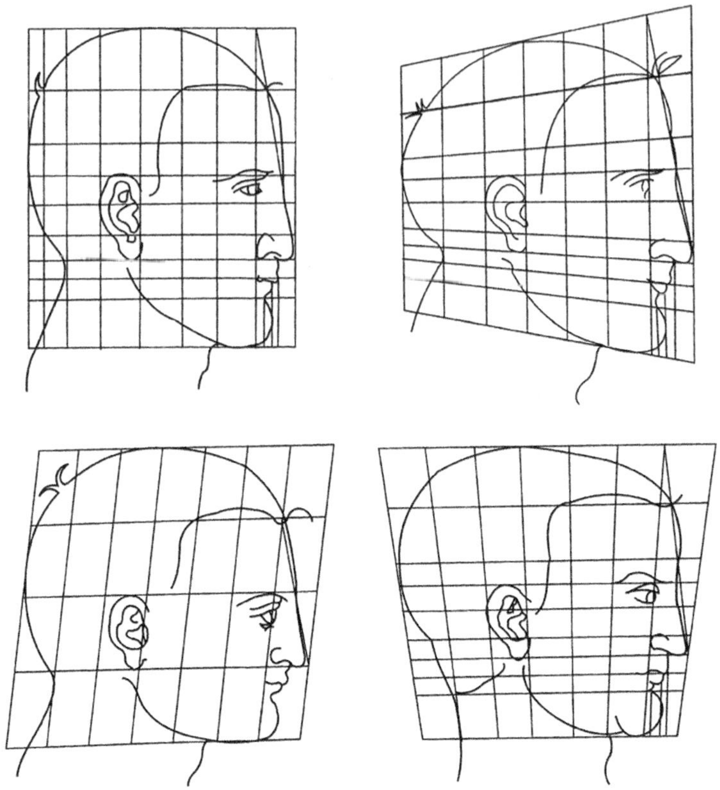

Fig. 90.2 Geometric Transformation as a strategy for creating different faces. Image: courtesy Tessa Morrison, after Dürer

pre-packaged objects. In the majority of software modelling packages, walls, surfaces and doors are pre-prepared and are simply modified by the designer. The starting object in this case is Le Corbusier's Villa Savoye, a building that has been repeatedly used as a basis for different architectural investigations associated with space and form. This experimental design, the Villa Sphere, has been modelled not just as an overall form, but also individually so that every element can be contoured, textured or surfaced. Moreover, each element is also fixed in certain distinct relationships columns and beams are instructed to remain connected regardless of what happens to the model. Then the starting object is selected and a single item is chosen by the designer from a "drop-down" menu. The building's form is gradually transformed as the entire structure is "spherized" a standard operation from the transformation palette of many modelling programs. Like Kolatin and Macdonald's Chimerical Housing, Le Corbusier's seminal work soon buckles, unfolds and inflates like a balloon, although all elements retain their necessary relationship (Figs. 90.3 and 90.4). If left long enough, the transformation will stretch the

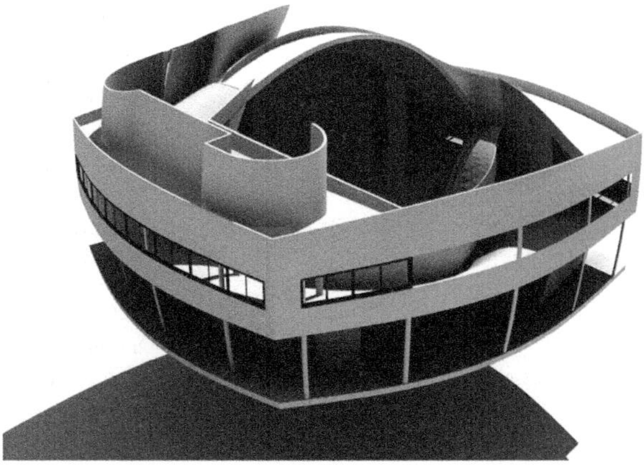

Fig. 90.3 Villa Sphere: "Stage N + 14 transformation". Image: Michael Ostwald and Rod Halligan

building into a contorted globe with beams as meridians, columns as lines of latitude and floors as continents.

This process of digital transformation is strikingly similar in concept to that employed in creating many architectural forms in recent years. For example, it is conceptually akin to Greg Lynn's development of "blobs" isomorphic polysurface structures from disconnected primitives and the application of a transformative tool, the "metaball" function (Lynn 1995: 39–44; Lynn 1999).

In these three computational examples, Chimerical Housing, the Villa Sphere and blobs, architectural form results from a process of transformation that is disconnected from the geometric construction which underlies the operation. If then the proposition that commenced this chapter that architectural form is open to ethical criticism is tentatively accepted and the associated presupposition is also accepted that geometric construction is complicit in the ethical framing of form then what does ethical discourse say about these computational examples?

Conclusion

Throughout the chapter various arguments have been offered that the geometric construction of architectural form has ethical implications. These ethical arguments, which are primarily drawn from virtue theory,[1] suggest that in the case of geometry:

[1] A problem with virtue theory, and its architectural application, is that it is asynchronous with many contemporary values. Virtue theory tends to be derived from traditional or normative western thought. Virtue theory should not be dismissed for this reason alone but it does have more in common with historic approaches to architecture than contemporary computational ones.

Fig. 90.4 Villa Sphere: "Stage N + 20 transformation". Image: Michael Ostwald and Rod Halligan

1. the process of construction must be a conscious act for it to be open to ethical criticism. Virtue theory maintains that a conscious act should demonstrate a degree of forethought, logic and rigour;
2. the process of construction reflects the respect the geometer or architect shows for their audience and the subject. Virtue theory suggests that a person possessing the right moral characteristics will comport themselves in an appropriate way and that this behaviour will be reflected in their works.

In addition to the ethical characteristics of geometric construction in itself, the architectural forms that result from such operations also possess potential moral dimensions. The geometric construction, as it is expressed in architectural form, has:

3. *phenomenological qualities* that condition the way in which an individual or crowd reacts in a space. This experiential quality, an indirect result of geometric construction, is a traditional architectural value directly associated with firmness and commodity;
4. *semiotic qualities* that shape the messages communicated to an individual or a community. Whether or not such qualities make architecture ethical depends on the message being propagated.

In the remainder of this section these four types of ethical criticism of the geometric construction are analysed in terms of computational transformation and the Villa Sphere.

Consider the process of computational transformation. Even the most banal computer-aided design software can generate complex three-dimensional forms without the architect having to possess any knowledge of geometry. In the process of computational transformation, geometry, as a system of construction, is hidden behind iconic "tools" or "functions", optimised operations prepared by software programmers to allow architects to avoid geometric construction altogether. While it is historically true that architects have been able to create form without a detailed understanding of the underlying geometric construction, never before has this been able to occur without a visible construction process. It is possible to argue that by making the construction process invisible the computer has freed the architect from the burdens of geometric construction and allowed them to focus exclusively on form. Yet, historically it is the very process of geometric construction that provides architectural form with meaning and governs its capacity to guide experience (Pérez-Gómez 1992; Evans 1995; Pérez-Gómez and Pelletier 1997). Before considering the semiotic and phenomenal implications of the transformative approach, questions concerning consciousness and comportment need to be considered; if the process of construction is invisible, has it occurred at all?

If an architect sets in motion a form-making procedure, regardless of whether it is reliant on geometric construction or computational transformation, he or she is acting consciously. However, if virtue theory is adopted as a framework then the process is only ethical if the conscious act is at the very least logical, rigorous or planned; Nigel Taylor calls this an ethics of "care" (Taylor 2000). Consider the Villa Sphere example in this context and ignore its discursive intent. The act of selecting a starting object and then subjecting it to transformation is undeniably conscious and by inference the design must be open to the ethical criticism of its geometric construction. The process of creating the object (the Villa Savoye) also displays a certain care, but the transformation does not. The transformative process may be stopped and analysed by the architect, again showing a limited form of interest, but the end point is chosen without foresight or demonstrable logic. This suggests that the design of the Villa Sphere may be ethically unsound. The second ethical question associated with geometric construction is concerned with comportment. Lachterman's theory of the ethical virtue of geometric construction may be extrapolated to the case of the Villa Sphere with two possible readings. First, in the computer program the construction of geometry is without comportment (expression of individual virtue or lack thereof), at least insofar as the designer using the software has no personal expression in the geometric construction. Alternatively, the comportment of the software programmer may be embedded in the resultant form because the past construction (hidden in the programming code) is enacted every time the software operates. In this example there is a separation between construction and form a separation that either disconnects architecture from the ethics of geometry or provides a diminished connection. In either case, the architect's disinterest in the underlying construction process is a strong indicator of his lack of respect for potential inhabitants and for the resultant architecture.

The invisibility of the construction process is also ethically contentious from the point of view of both semiotics and phenomenology. This chapter has previously argued that the visible construction of form from geometry is central to the historic "social contract"; the communal agreement that, in this case, architectural form has the capacity for communication in and of itself. While all form has some capacity for evoking meaning, without the social contract every person will interpret form in their own way. Thus, from an ethical perspective, a potential problem with the loss of connection between construction and form is that the resultant architecture can only communicate with an individual, not with a community and ethical rules are typically mediated by communal needs.[2] Another way of understanding this dilemma is to realise that without an appreciation of geometric construction the historic rigour of form creation has been replaced with experimental and intuitive form generation (neither of which display due care). Thus, regardless of whether or not the social contract is important, the rigour of form generation is lost when the construction process is invisible. In the case of the Villa Sphere, the geometric construction lacks a human hand; the selection of a function in a software menu like "spherize" or "metaball" may repeat the general rules of geometric construction but they lack the spirit, or intentionality, which is presumed to give the resultant form meaning.

While the present chapter cannot categorically support the opening proposition that the process of geometric construction is complicit in the ethics of the resultant architectural form it does identify several important approaches to this argument. Moreover, in the specific case of the computational transformation there are evidently serious limitations from the perspective of moral philosophy that have to be considered by architects creating complex curvilinear forms in this way.

Biography Michael J. Ostwald is Professor and Dean of Architecture at the University of Newcastle (Australia) and a visiting Professor at RMIT University. He has previously been a Professorial Research Fellow at Victoria University Wellington, an Australian Research Council (ARC) Future Fellow at Newcastle and a visiting fellow at UCLA and MIT. He has a PhD in architectural history and theory and a DSc in design mathematics and computing. He completed post-doctoral research on baroque geometry at the CCA (Montreal) and at the Loeb Archives (Harvard). He is Co-Editor-in-Chief of the *Nexus Network Journal* and on the editorial boards of *ARQ* and *Architectural Theory Review*. He has authored more than 300 scholarly publications including 20 books and his architectural designs have been published and exhibited internationally.

References

BEAUCHAMP, Tom L. 2001. *Philosophical Ethics: an Introduction to Moral Philosophy*. Boston: McGraw-Hill.

BROADBENT, Geoffrey. 1970. Meaning into Architecture. Pp. 50-75 in C. Jencks and G. Baird, eds. *Meaning in Architecture*. New York: George Braziller.

COOK, E. T. and WEDDERBURN, A. eds. 1903-12. *The Works of John Ruskin* (Library Edition, 39 vols.). London: George Allen.

DE SAUSSURE, Ferdinand. 1959. *Course in General Linguistics*. London: Peter Owen Limited.

[2] A counter argument would be that, historically the social contract was never as widely accepted as architectural scholars would like to believe.

DERRIDA, Jacques. 1989. *An Introduction to Edmund Husserl's Origin of Geometry*. Lincoln: University of Nebraska Press.

DEVEREAUX, Mary. 2001. Beauty and Evil: The Case of Leni Riefenstahl's Triumph of the Will. Pp 227-256 in J. Levinson, ed. *Aesthetics and Ethics: Essays at the Intersection*. Cambridge: Cambridge University Press.

DOVEY, Kim. 1999. *Framing Places: Mediating Power in Built Form*. London: Routledge.

ECO, Umberto. 1980. Function and Sign: The Semiotics of Architecture. Pp. 56-65 in G. Broadbent and C. Jencks, eds. *Signs, Symbols and Architecture*. New York: Wiley.

ELDEN, Stuart. 2001. The Place of Geometry: Heidegger's Mathematical Excursus on Aristotle. *The Heythrop Journal* **42**, 1: 311-328.

EVANS, Robin. 1995. *The Projective Cast: Architecture and its Three Geometries*. Cambridge, Mass.: MIT Press.

GAUT, Berys. 2001. The Ethical Criticism of Art. Pp 182-203 in J. Levinson, ed. *Aesthetics and Ethics: Essays at the Intersection*. Cambridge: Cambridge University Press.

GRANT, Hardy. 1990. Geometry and Politics: Mathematics in the Thought of Thomas Hobbes. *Mathematics Magazine* **63**, 3: 147-154.

HARRIES, Karsten. 1997. *The Ethical Function of Architecture*. Cambridge: MIT Press.

KOLATIN, Solan and MACDONALD, Bill. 2000. Excursus Chimera? *Architectural Design* **70**, 3: 71-77.

———. 2001. Chimerical Housings: Mass Customised Housing. Pp. 24-29 in M. Burry, ed. *Cyberspace, The World of Digital Architecture*. Melbourne: Images Publishing.

KRUFT, Hanno W. 1994. *A History of Architectural Theory from Vitruvius to the Present*. New York: Princeton Architectural Press.

LACHTERMAN, David R. 1989. *The Ethics of Geometry: A Genealogy of Modernity*. New York: Routledge.

LEVINE, Michale P., MILLER, Kristine and TAYLOR, William. 2004. Ethics and Architecture. *The Philosophical Forum* 35, 2: 103-115.

LYNN, Greg. 1995. Blobs. *Journal of Philosophy and the Visual Arts* **6**: 39-44.

———, ed. 1999. *Animate Form*. New York: Princeton Architectural Press.

MEEK, Harold A. 1988. Guarino Guarini and His Architecture. New Haven: Yale University Press.

MITCHELL, William J. 1990. *The Logic of Architecture: Design Computation and Cognition*. Cambridge, Mass.: MIT Press.

MOORE, R. John and OSTWALD, Michael J. 1997. Colubrine Chains and Right Lines: A Dædalic Analysis of Ruskin's 'Living Waves'." Pp. 436-439 in *Architecture: Material and Imagined*. Washington DC: Association of Collegiate Schools of Architecture.

NORBERG-SCHULZ, Christian. 1971. *Baroque Architecture*. New York: Harry N. Abrams.

———. 1980. *Late Baroque and Rococo Architecture*. Milan: Electra.

PÉREZ-GÓMEZ, Alberto. 1992. *Architecture and the Crisis of Modern Science*. Cambridge, Mass.: MIT Press.

PÉREZ-GÓMEZ, Alberto and PELLETIER, Louise. 1997. *Architectural Representation and the Perspective Hinge*. Cambridge, Mass.: MIT Press.

RYKWERT, Joseph. 1987. *The First Moderns: The Architects of the Eighteenth Century*. Cambridge, Mass.: MIT Press.

SACKSTEDER, Williams. 1980. Hobbes: The Art of the Geometricians. *Journal of the History of Philosophy* 18: 131-146.

SPECTOR, Tom. 2001. *The Ethical Architect: The Dilemma of Contemporary Practice*. New York: Princeton Architectural Press.

TAYLOR, Nigel. 2000. Ethical Arguments about the Aesthetics of Architecture. Pp 193-206 in W. Fox, ed. *Ethics and the Built Environment*. London: Routledge.

THOMPSON, D'Arcy W. 1961. *On Growth and Form* (1942) London: Cambridge University Press.

WASSERMAN, Barry, SULLIVAN, Patrick and PALERMO, Gregory. 2000. Ethics and the Practice of Architecture. New York: John Wiley & Sons Ltd.

Chapter 91
Equiangular Numbers

Henry Crapo and Claude Le Conte De Poly-Barbut

Some mathematical problems are resolutely geometric. No matter what you do to them, subjecting them to different sorts of manipulations and calculations, their 'geometric content' persists even in the tiniest parts of what remains, even in the numbers used to express their solution, like the parts of an image residing 'everywhere' in a hologram, or like the smile of a Cheshire cat. We want to tell you of one such problem, and of a delightful series of real numbers starting with 0,1, ... and tending toward 2, that does its best to recall the struggles along its path into existence. We maintain that it is because of these ancient struggles (which are bound to recur when one tries to 'construct' them) that these numbers are of architectural and artistic significance.

σ_2	$\sigma 3$	σ_4	σ_5	σ_6	σ_7	σ_8	$\ldots \sigma_8$
0	1	1.41421	1.61803	1.73205	1.80194	1.84776	... 2

You will recognize the first few even in this inappropriate form, rounded off to five decimal places: (σ_4 is $\surd 2$, while σ_5 is τ, the Golden Mean, and σ_6 is $\surd 3$). We call the sequence $\{\sigma_n\}$ the *equiangular numbers*.

The story begins with one of Donald Coxeter's masterpieces, his algebraic characterization of groups generated by reflections (Coxeter 1935). His formulation is simple: you insist that your group be generated by a finite set of

First published as: Henry Crapo and Claude le Conte de Poly Barbut, "Equiangular Numbers", pp. 9–21 in *Nexus II: Architecture and Mathematics*, ed. Kim Williams, Fucecchio (Florence): Edizioni dell'Erba, 1998.

H. Crapo (✉)
Centre de Recherche Les Moutons Matheux, 28 Grand'ave, La Vacquerie 34520, France
e-mail: henry.crapo@ehess.fr; moutons.matheux@gmail.com

C. Le Conte De Poly-Barbut
Centre d'analyse et de mathématique sociales, EHESS, Paris
e-mail: claude.barbut@ehess.fr

K. Williams and M.J. Ostwald (eds.), *Architecture and Mathematics from Antiquity to the Future*, DOI 10.1007/978-3-319-00143-2_44,

elements of order 2, say $\{s_i \mid i=1\ldots n\}$, and that the defining relations be all of the form

$$(s_i s_j)^{c_{ij}} = \varepsilon$$

for extended integer values c_{ij}, where $i<j$ and $2 \le c_{ij} \le \infty$, and ε is the identity element of the group. These values are recorded in a *graph* whose vertices are the generators, and where the edges ij are labelled c_{ij} whenever this value is at least 3. (By $c_{ij}=\infty$ we mean simply that there is no corresponding relation imposed; all the powers $(s_i s_j)^n$ are distinct.) For instance the graph $\circ\overset{m}{-}\circ\overset{n}{-}\circ$ denotes the group with three generators, and relations

$$s_1^2 = s_2^2 = s_3^2 = (s_1 s_2)^m = (s_2 s_3)^n = (s_1 s_3)^2 = \varepsilon.$$

The hard part was then to show that every such group is geometrically representable as a group generated by reflections.

Say you have a group generated by reflections in n mirrors, which we call the *generators*, surrounding a fundamental region in a space. These generators are reflected in each other to form virtual reflectors, which we call *mirrors*; algebraically they are conjugates xsx^{-1} of a generator s by an element x of the group. The space is divided up into *cells*, each an image of the fundamental region under a succession of reflections, and representing the element of the group that carries the fundamental region to that location (Fig. 91.1). Since each element x of the group is expressible as the product of a *word* in generators, it has a *length* $\ell(x)$, equal to the minimum length of a word $s_1 \ldots s_n$ with product $\pi(s_1 \ldots s_n) = x$ x. A word of this length is called a *short* word for x. Geometrically, the length is the number of mirrors you have to cross in order to get from the identity (fundamental region) to the cell representing the element x. Every mirror is a conjugate xsx^{-1} of a generator s by an element x; take this element x to be of minimum length among such expressions, and choose any short expression for x. You find a short *palindrome* $s_1 \ldots s_n \ldots s_1$ for the mirror.

Under the partial order

$x \le y$ if and only if some short word for x
is a prefix of some short word for y

the group becomes a *semilattice* (to be precise: a complete meet-semilattice: every subset of the group has a greatest lower bound), or simply a *lattice*, if the group is finite. Each *step*, or covering pair $[x,y]$, where $x<y$ and $\ell(x)+1=\ell(y)$, has an associated generator $s=x^{-1}y$ and an associated mirror $m=yx^{-1}$, which we call the *generator label* and *mirror label*, respectively, of the step. It is a nice surprise to find that there are consistent drawings in which *steps with the same mirror label are drawn parallel*. Perhaps even more surprisingly, if these vector directions x_m, one for each mirror m, are *very* carefully chosen in n-dimensional space, the resulting

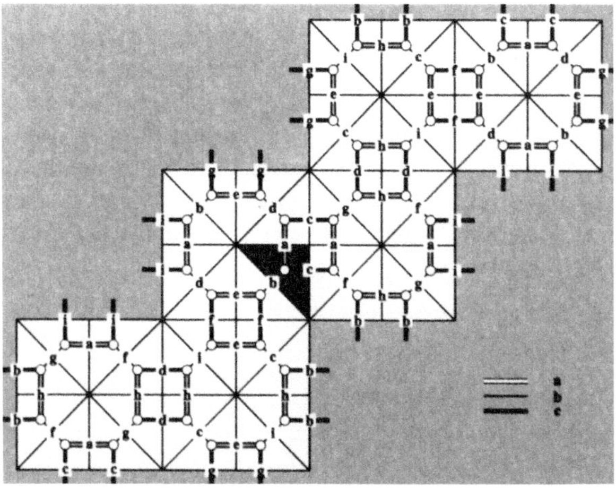

Fig. 91.1 The group $\circ\overset{4}{—}\circ—\circ$ generated by three reflections of the cube

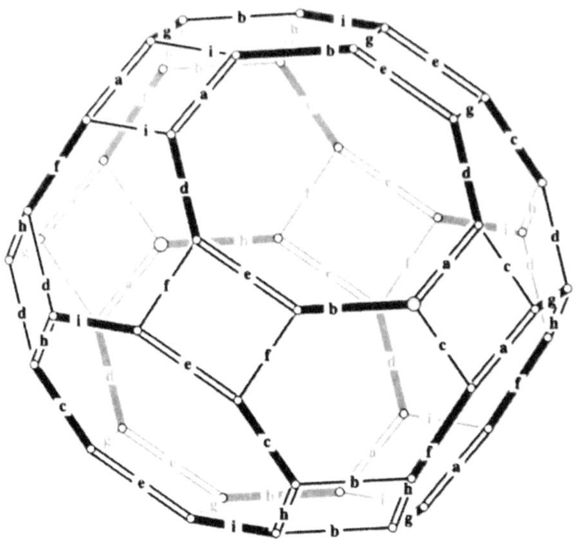

Fig. 91.2 The group $\circ\overset{4}{—}\circ—\circ$ as zonohedron

figure becomes the 1-skeleton of a *zonotope* (Fig. 91.2), the convex figure formed as the Minkowski sum of the line segments $[0, x_m]$, or of a zonotopal tiling.

The question remains, for a Coxeter group, given, say, by its graph of generators and relations, how do we choose the vectors x_m in order correctly to draw the corresponding zonotope or zonotopal tiling? In his charming article on zonotopes (1962), Coxeter showed how it suffices to cut all the vectors by a hyperplane, and so

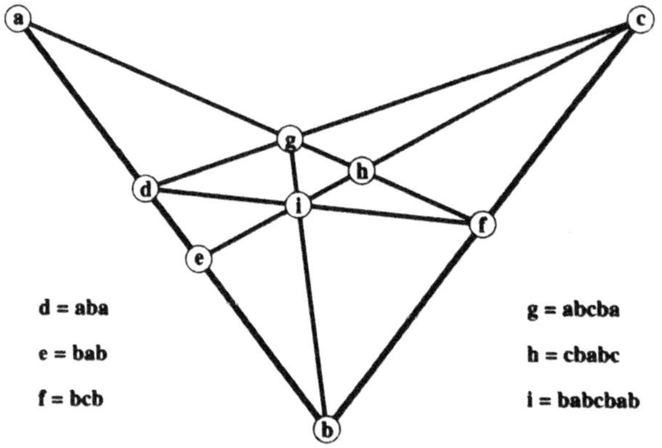

Fig. 91.3 The mirror configuration for ○–4–○—○

work with figures of projective points. This will be our approach, to construct these *mirror diagrams*.

Look at Fig. 91.1 in detail. There is a cell for each element of the group ○–4–○—○ generated by reflections of the cube, with fundamental region shown in grey. The three types of edges encode the corresponding generator labels, while the letters are the mirror labels. The outer edges should be identified in pairs so that the sheet forms a polyhedral surface (here, the cube). Figure 91.2 is a correct projection of the corresponding zonotope, with flat octagonal, hexagonal, and square faces in 3-space. Figure 91.3 shows the mirror diagram for this group. Note that for any pair xy of mirrors, their successive conjugates

$$x, xyx, xyxyx, xyxyxyx, \ldots$$

are collinear, and that the mirrors at the ends of each line are of minimal lengths for that line. Here are nine mirrors, arranged on seven major lines. This simple figure already possesses a non-trivial *projective property*. Since the four mirrors a, $d = aba$, $e = bab$, b lie at the four points of intersection of a line ab with the six edges of a plane tetrahedron (vertices $cfgi$), these four mirrors are *harmonic*. If we assign coordinates (1, 0, 0) to a, (0, 1, 0) to b, and place the points d and e symmetrically relative to the midpoint (1, 1, 0), then the point d will have coordinates $(\sqrt{2}, 1, 0) = (\sigma_4, 1, 0)$.

Mirror diagrams for the groups ○—○—○ ··· ○—○ of permutations of an n-element set are generalized Desargues configurations, as in Fig. 91.4, formed by the intersection of n hyperplanes in general position in a space of dimension $n-1$. For S_4 this is a complete quadrilateral in the plane, for S_5, the usual Desargues configuration in 3-space.

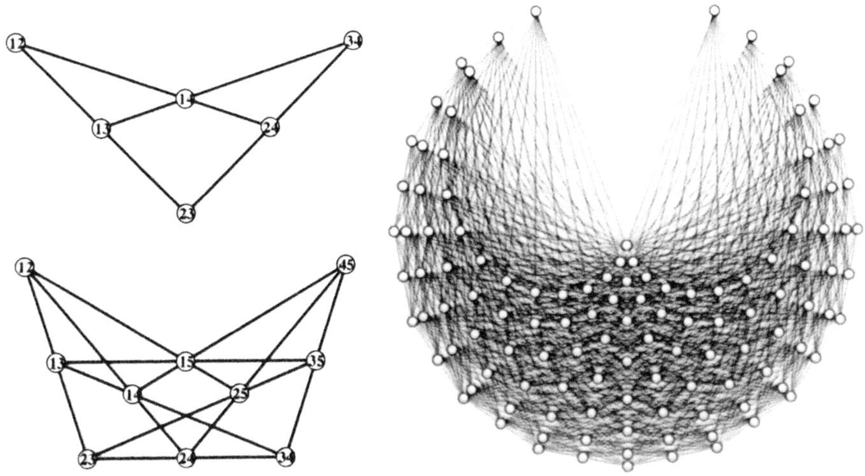

Fig. 91.4 Mirror diagrams for the symmetric S_4, S_5, S_{16}

For the group $\circ\overset{5}{—}\circ—\circ$ generated by reflections of the icosohedron or dodecahedron, the mirrors configuration (Fig. 91.5) already has *no possible projective construction*! Any such construction would be a projective construction of the golden mean, which is known to be impossible using straightedge only. Try drawing this figure, just looking at a list of sets of points that are supposed to be collinear. You will quickly see why we are making a fuss about equiangular points! (You can get it right quickly by trial and error, but trial and error has no standing as a projective construction.) We have shaded some triangular regions of the diagram in order to emphasize that this is a *projective regular pentagon chgij*, with inner and outer stars: take the line *ab* to be the line at infinity.

The problem of drawing mirror diagrams for Coxeter groups has a simple general solution if we are willing to impose appropriate restrictions on the positions of those mirrors on lines joining pairs of generators. If these choices are made in a natural way, there is a *straightforward* construction of the remaining positions; everything just falls into place. We must take a closer look at the case of two generators.

For two generators *a*,*b*, the simplest such group is that for $(ab)^\infty = \varepsilon$. This is the group you see in the barbershop (Fig. 91.6) with parallel mirrors on opposite walls. You see not only infinitely many chairs, but infinitely many mirrors, each making its own faithful reflected image of the entire infinite scene. The generators are the two mirrors bounding region ε, with real silvered glass.

If the two generators are not quite parallel, the series of images will bend along a circular path of large diameter. Whenever the angle between them is a rational multiple of π, the images will pile up in a finite number of distinct positions. For mirrors at an angle of $\frac{\pi}{n}$ we find $2n$ images, one for each element of the dihedral group D_n, and n concurrent mirrors at equal angles (Fig. 91.7).

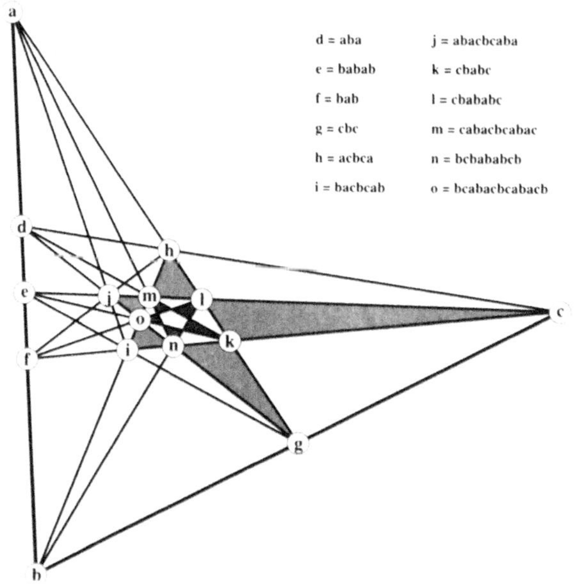

Fig. 91.5 Mirror diagram for the group $\circ\overset{5}{-}\circ-\circ$ of the icosahedron

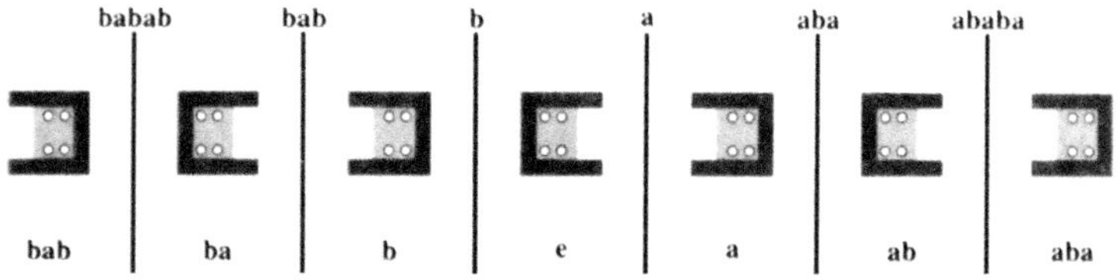

Fig. 91.6 A portion of the group $\circ\overset{\infty}{-}\circ$

Consider a family P of n coplanar and concurrent equiangular lines $L_1, \ldots, L_n$ through a point c in the plane. Intersect this family of lines with a line L parallel to the bisector of a pair of consecutive lines, say of L_1, and L_n. By $\wedge$ and $\vee$ we denote the operators join (of a pair of points, to form a line) and meet (of a pair of lines, to form a point), respectively, in the projective plane. Let $p_i = L_i \wedge L$, for $i = 1 \ldots n$. We call such a set $E_n = \{p_1, \ldots, p_n\}$ a centrally symmetric set of n *equiangular* points (Fig. 91.8).

Without loss of generality we may select homogeneous (projective) coordinates

$$p_1 \to (1,0)$$
$$p_n \to (0,1)p_n \to (0,1)$$
$$\text{midpoint of the segment} \quad [p_0, p_n] \to (1,1).$$

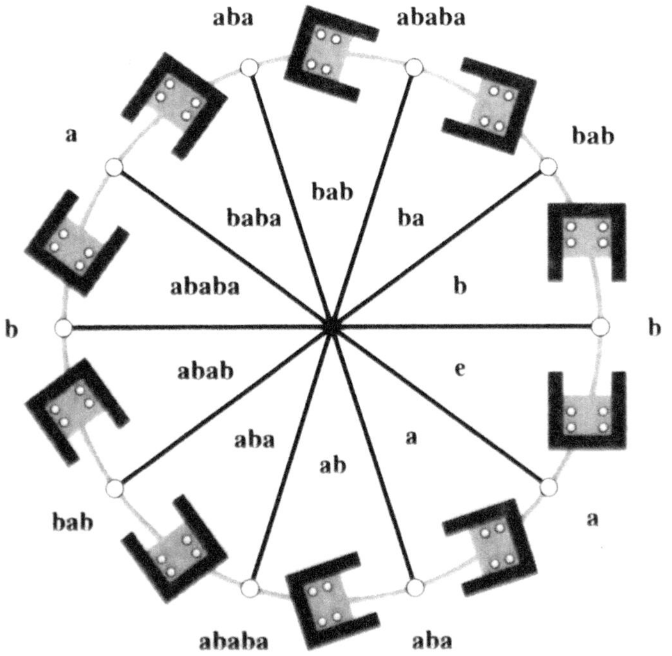

Fig. 91.7 The group $\circ\overset{5}{\text{---}}\circ$ (barbershop quintet)

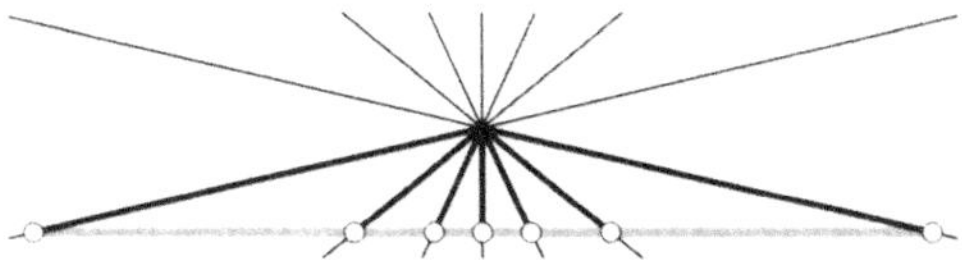

Fig. 91.8 A line of seven equiangular points

Then the projective coordinates of all points p_i are determined, each up to a non-zero scalar multiple. Let $(\sigma_n, 1)$ be the coordinates of p_2, given n equiangular lines.

These values σ_n can be computed as roots of a sequence of polynomials, as follows. Let r_1 be reflection of the plane in mirror L_1. The mapping

$$A : p \to L \wedge \left(r_1\left(p \bigvee c\right)\right)$$

is a projective map, an involution of the line L fixing the point p_1 and inducing the permutation

$$(p_1)(p_2p_n)(p_2p_{n-1})\cdots(p_{k+1}) \quad \text{if } n = 2k$$
$$(p_1)(p_2p_n)(p_2p_{n-1})\cdots(p_kp_{k+1}) \quad \text{if } n = 2k-1$$

of the points p_i. This mapping A can be expressed as right-multiplication by the (2×2)-matrix

$$\begin{pmatrix} -1 & 0 \\ \sigma & 1 \end{pmatrix}$$

since this linear transformation and its non-zero scalar multiples are the only linear maps that send (1,0) to a scalar multiple of itself, exchanging (σ,1) and (0,1) with scalar multiples of each other.

Our symmetric choice of projective coordinates (1,0) for p_1 and (0,1) for p_n, permits us to express the central symmetry D of the figure by the linear transformation with matrix

$$\begin{pmatrix} 0 & 1 \\ 1 & 0 \end{pmatrix}.$$

This transformation induces the permutation

$$(p_1p_n)(p_2p_{n-1})\cdots(p_kp_{k+1}) \quad \text{if } n = 2k$$
$$(p_1p_n)(p_2p_{n-1})\cdots(p_k) \quad \text{if } n = 2k-1.$$

Composing the maps D, then A, we obtain a map that induces the cyclic permutation ($p_1\, p_2 \ldots p_n$) which advances the points along the line (a turn by one of the $2n$ cogs of the wheel), and has matrix

$$\begin{pmatrix} 0 & 1 \\ 1 & 0 \end{pmatrix}\begin{pmatrix} -1 & 0 \\ \sigma & 1 \end{pmatrix} = \begin{pmatrix} \sigma & 1 \\ -1 & 0 \end{pmatrix}.$$

Using multiplication by this matrix DA to compute the coordinates of the successive points p_i, we find

$$\begin{aligned} p_1 &= (1,0) \\ p_2 &= (\sigma, 1) \\ p_3 &= (\sigma_2, -1, \sigma) \\ p_4 &= (\sigma_2 - 2\sigma, \sigma_2 - 1) \\ p_5 &= (\sigma_4 - 3\sigma_2 + 1, \sigma_2 - 2\sigma) \\ &\vdots \\ p_m &= (f_m(\sigma), f_{m-1}(\sigma)), \\ &\vdots \end{aligned}$$

for $m = 1, \ldots, n$, where the f_m form a sequence of polynomials

$$\begin{aligned} f_0 &= 0 \\ f_1 &= 1 \\ f_2 &= x \\ f_3 &= x^2 - 1 \\ f_4 &= x^3 - 2x \\ f_5 &= x^4 - 3x^2 + 1 \\ f_6 &= x^5 + 4x^3 + 3x \\ f_7 &= x^6 - 5x^4 + 6x^2 - 1 \\ &\vdots \end{aligned}$$

determined by initial values $f_0(x) = 0, f_1(x) = 1$ and the simple recursion

$$f_m = xf_m - 1 - f_m - 2,$$

for all $m \geq 2$. In closed form:

$$f_m(x) = \sum_{i=0}^{\lfloor (m-1)/2 \rfloor} (-1)^i \binom{m-i-1}{i} x^{m-2i-1}$$

The terminal condition $f_n = 0$ applies, and permits us to determine the correct value of σ_n, the largest positive root of f_n. Factorizations of these polynomials, with integer coefficients, and exact expressions for their roots in terms of radicals, begin as follows:

0:	0	
1:	1	
2:	x	0
3:	$(x-1)(x+1)$	± 1
4:	$x(x^2-2)$	$0, \pm\sqrt{2}$
5:	$(x^2-x-1)(x^2+x-1)$	$(\pm 1 \pm \sqrt{5})/2$
6:	$x(x^2-3)(x^2-1)$	$0, \pm 1, \sqrt{3}$
7:	$(x^3-x^2-2x+1)(x^3+x^2-2x-1)$.	

Our attempts to use computer algebra systems to solve the polynomial equations for $f_n = 0$ yielded useful results only for $n \leq 6$, a difficult expression in radicals for $n = 7$, and no results at all for $n > 7$.

A trigonometric solution, however, to the equations $f_n(x) = 0$ exists, and takes the form:

$$2\cos\frac{k\pi}{n} \quad \text{for } k = 1, \ldots, n-1,$$

with largest positive root

$$\sigma_n = 2\cos\frac{\pi}{n}.$$

Using these values of σ_n we can construct a linear representation of the group. Each generator s_i is given by standard unit vector,

$$s_i = (0, \ldots, 1 \ldots 0),$$

while the linear operator 'conjugation by s_i' is given by the matrix

$$\begin{pmatrix} 1 & 0 & \cdots & \sigma_{1i} & \cdots & 0 & 0 \\ 0 & 1 & \cdots & \sigma_{2i} & \cdots & 0 & 0 \\ \vdots & \vdots & & \vdots & & \vdots & \vdots \\ 0 & 0 & \cdots & -1 & \cdots & 0 & 0 \\ \vdots & \vdots & & \vdots & & \vdots & \vdots \\ 0 & 0 & \cdots & \sigma_{n-1,i} & \cdots & 1 & 0 \\ 0 & 0 & \cdots & \sigma_{ni} & \cdots & 0 & 1 \end{pmatrix}$$

Extend this matrix representation multiplicatively, first representing each element x in the group as a product of generators, then 'conjugation by x' as the corresponding product of matrices. It is now an easy matter to compute projective coordinates for all the mirrors, since each is the conjugate of a generator by an element of the group, and is thus the image of a standard basis vector under multiplication by one of these product matrices. For instance, each mirror in the symmetric group , being a permutation with cycle structure (ij), gets coordinates (0,. . .,0,1,. . .,1,0,. . .,0), where the 1s are in positions i through j–1.

In the limit, with $\sigma = 2$, the 'translation' map DA has matrix

$$\begin{pmatrix} 2 & 1 \\ -1 & 0 \end{pmatrix}$$

and creates an infinite sequence of points

$$p_1 = (1, 0), p_2 = (2, 1), p_3 = (3, 2), \ldots p_k = (k, k-1) \ldots$$

reaching a projective limit at the midpoint (1,1) (Fig. 91.9). Mirror reflection in the line L_1 the linear map A, permutes pairs of points on opposite sides of this midpoint (Fig. 91.10):

$$(k, k-1)\begin{pmatrix} -1 & 0 \\ 2 & 1 \end{pmatrix} = (k-2, k-1).$$

In closing, we should notice that a geometric situation gave rise to a difficult (yea, impossible) construction problem in projective geometry, then to a problem in

Fig. 91.9 The limiting case of infinitely many equiangular points (Venice harbour entrance)

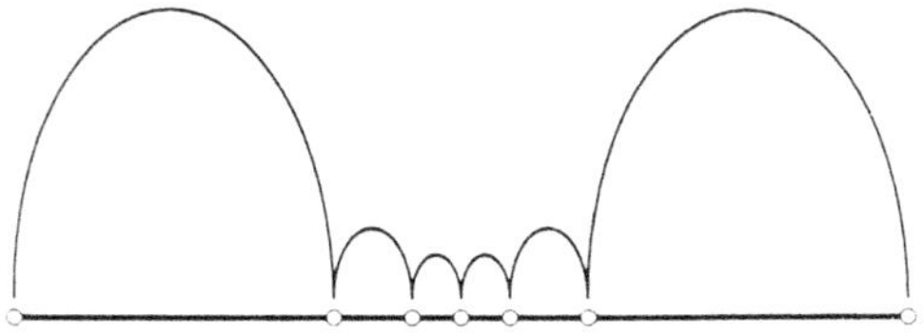

Fig. 91.10 Pairs of points on opposite sides of the midpoint

polynomial algebra that taxes the powers of the best modern computer algebra systems, but which had a simple solution in terms of trigonometry. It is fair to ask whether these further values of σ_n, for $n = 7,8,\ldots$ occur already in nature, for the simple reason that they are the natural coordinates of *equiangular points*. Finally, since the merits of the golden mean are well recognized in artistic matters (planning of paintings, design of building façades, or choice of relative dimensions for European paper stock), where the aspect of 5-equiangularity is thoroughly disguised, surely the subsequent values of s_n for $n > 5$ can give rise to analogous aesthetic feelings in similar situations. Can our readers point to any instances of the use of s_7 in ancient or contemporary architecture?

Acknowledgment All images in this chapter are by the authors.

Biography Henry Crapo has a Ph.D. in mathematics from MIT (1964). He was a professor at the University of Waterloo in Ontario, Canada. He carried out his research in geometry at INRIA-Rocquencourt, France and at Centre d'analyse et de mathématique sociales (CAMS), EHESS in Paris.
Claude Le Conte De Poly-Barbut earned her diplôme d'études approfondies mathematiques (DEA) from the Universié di Paris (2001). She taught at the University of Paris V, and was affiliated with the Centre d'analyse et de mathématique sociales (CAMS), EHESS in Paris.

References

Coxeter, H.S.M. 1935. The Complete Enumeration of Finite Groups of the Form $R_1^2 = (R_iR_j)^{k_{ij}} = 1$. *Journal of the London Mathematical Society* **10**: 21–25.

———. 1962. The Classification of Zonohedra by Means of Projective Diagrams. *Journal de Mathematiques Pures et Appliquées*. **41**: 137–156.

Chapter 92
Architecture as Verb and the Ethics of Making

Alberto Pérez-Gómez

Introduction

Recent architectural production, considered at the leading edge of design, celebrates the use of computers to generate impossibly novel forms and "spaces," with a complete disregard for historical precedent and seemingly oblivious of humanity's embodied consciousness, i.e., the oriented spatiality of the user. This application of computer technology to design seems to have left behind its simple utilitarian justification as a tool that might improve the efficiency of architectural production, to be driven by the claim of the tool's capacity to generate "new forms," totally "other" from our traditional "orthogonal" building practices. Indeed, recent powerful software packages are now capable of treating surface as the primary element in design, allowing for unimaginable geometric configurations that are at once structurally sound and open up an infinity of formal possibilities. Theoretical arguments behind these practices often argue an ethical intention, a desire to overcome the wrongs of an architecture traditionally associated with repressive power, artistic self-indulgence, and political control. These forms, exhibiting "similarity" in their generative patterns to *both* cultural artifacts and natural phenomena, are arguably capable of transcending many of our old dualistic assumptions, particularly the opposition between scientific rationality and irrationality, yet in their novelty they seem incapable of offering a place for the collective imagination.

First published as: Alberto Pérez-Gómez, "Architecture as Verb and the Ethics of Making", pp. 35–46 in *Nexus IV: Architecture and Mathematics,* Kim Williams and Jose Francisco Rodrigues, eds. Fucecchio (Florence): Kim Williams Books, 2002.

A. Pérez-Gómez (✉)
School of Architecture, McGill University, 815 Sherbrooke Street West, Montreal, QC, Canada H3A 2K6
e-mail: alberto.perez-gomez@mcgill.ca

K. Williams and M.J. Ostwald (eds.), *Architecture and Mathematics from Antiquity to the Future*, DOI 10.1007/978-3-319-00143-2_45,

Indeed, while it is imperative that we keep searching for more appropriate poetic forms for the new "programs" of human life in our electronic age, it is simplistic to imagine that formal extrapolations of "democratic/chaotic" structures may be the way of the future, an instrumental vehicle to transcend our stylistic conundrums and make a better dwelling place for humanity. These instrumental processes are necessarily dependent on mathematical models, themselves "designed" by computer engineers with the interests of electronic corporations in mind, and when extrapolated to architecture often become an empty exercise in formal acrobatics. Engaging these processes, architects tend to forget the importance of verticality (our spatial engagement with the world that defines our humanity, including our capacity for thought), our historicity (we are, effectively, what we have been), and gravity (the "real world" of bodily experience into which we are born, and that includes our sensuous bond to all that which is *not* human). At best an excuse for decontextualized novelty, this rhetorical instrumentality typically results in new forms of self-referential, structural determinism. The result is an architecture both oblivious of its specific cultural context and of the experiencing body, hardly attuned to its intended programmes and paradoxically, disengaged from its ethical imperatives.

Cultural critics have declared that modernity's blind faith in novelty for its own sake should no longer be taken for granted. At a more personal level, we recognize that spending all our time in front of a computer screen has rewards, but we suspect that it also carries a price. We lose other kinds of knowledge, usually related to the body and the senses, other potential sources of wisdom. This epic of "loss" is already an old story, a transaction that has been going on for many years, accelerating after the Scientific Revolution of the seventeenth century. Yet, in our professional capacity, we *must* act, avoiding nostalgic escapades and engaging in some way the increasingly more powerful instrumental devices that through their rational transparency augment our power of transforming and controlling the world.

As architects we start to recognize that the reality of our discipline is infinitely complex, both shifting with history and culture, and also remaining the same, analogous to the human condition which demands that we continually address the same basic questions to come to terms with mortality and the possibility of transcendence opened up by language, while expecting diverse answers that are appropriate to specific times and places. Architecture possesses its own "universe of discourse," and over the centuries has seemed capable of offering humanity far more than a technical solution to pragmatic necessity. Our technological world is often skeptical about architecture having any meaning *at all* (other than providing for shelter). Yet, our dreams are always set *in place*, and our understanding (of others and ourselves) could simply not *be* without architecture. We know architecture allows us to think and to imagine, it opens up the "space of desire" that allows us to be "at home" while remaining always "incomplete" and open to our personal death, this being our most durable human characteristic. Even cyberspace could not "appear" if we were not first and foremost mortal, self-conscious bodies *already* engaged with the world through direction and gravity. We don't merely "have" a body, we "are" our bodies. It suffices to try to think in a totally dark room for more than a few minutes to convince oneself of the "reality" of this unarticulated, pre-conceptual "ground" of being which includes "architecture" as the external, visible order.

Thus I have often argued that architecture as an art communicates to us the possibility of *recognizing* ourselves as complete, in order to dwell poetically on earth and thus be wholly human. The products of architecture have been manifold, ranging from the *daidala* of classical antiquity to the gnomons, *machinae* and buildings which Vitruvius names as the three manifestations of the discipline, from the gardens and ephemeral architecture of the Baroque period to the built and unbuilt "architecture of resistance" of modernity such as Le Corbusier's La Tourette, Gaudí's Casa Batlo, or Hejduk's "masques." This *recognition* is not merely one of semantic equivalence, rather it occurs in experience, and like in a poem, its "meaning" is inseparable from the experience of the poem itself. As an "erotic" event, it overflows any reductive paraphrasing, overwhelms the spectator-participant, and has the capacity of changing one's life. In order to propitiate such events the architect must necessarily engage language. The main concern of any generative theory of architecture is therefore, in my view, to find appropriate language (in the form of stories) capable of modulating intended actions (projects) in view of ethical imperatives, always specific to each task at hand. The practice that emerges from such a theory can never be an instrumental application, but rather appears as a VERB, as a process that is never neutral and should be valorized, *a process that in fact erodes the boundaries between the artistic disciplines concerned with space*.

This has been the story of an architecture of resistance since Piranesi, passing through Hejduk, Libeskind and Peter Greenaway. From the moment when the traditional divisions among the fine arts were subverted, first in epistemology and eventually in practice., between the eighteenth and the early twentieth centuries, the most significant works of art that "construct" a mysterious depth, a significant spatiality, belong within this story.[1]

Disembodied computer generated processes usually claim that instrumentality allows designers to bypass the questions of cultural specificity and ground design in "scientific natural principles," finally "closing the distance" between theory and practice. This is arguably in response to cultural critics whose interest in the vindication of minority rights has led them to condemn the artistic imagination and its abuses as a vehicle for power, oppression and exploitation. Respect for the other and political correctness, however, is hardly assured by instrumental processes that operate in a historical and cultural vacuum. Richard Kearney has convincingly shown, on the other hand, that only by engaging our *own* imagination (with its inescapable horizon of language, and *despite* its dangers) can we be truly compassionate (Kearney 1988). It is our imaginative faculty that allows us to identify with the other, and truly understand her suffering. This entails a very real, yet *opaque* connection between words and deeds. Thus valorizing the work

[1] This is the story I tell in my own *Polyphilo or the Dark Forest Revisited* (1994).

as process, as embodied making, is not only a means of formal discovery, it is also a vehicle for ethical production. This form of relationship between theory and practice, between words and process, is obviously not unprecedented in art, but is less prevalent in architecture. In this essay I would like to sketch two examples, from the fifteenth and the twentieth centuries, that illustrate this relationship. The examples are intentionally far apart chronologically, and my use of them is unusual in contemporary scholarship. My aim is to draw a map of the vicissitudes of *architecture as verb* during the modernization of Europe.

Luca Pacioli's *Divina Proportione*

Part Two of Luca Pacioli's *Divina Proportione* (1509) is dedicated to the masons, stone-cutters and sculptors from his home town who, according to Pacioli, had asked him to provide guidelines for architecture based on arithmetic and geometry. Unlike Vitruvius and other contemporary writers on the subject, Pacioli defines the realm of architecture exclusively as buildings. This assumption, which would become the norm for modern architecture, was indeed a novelty in the early sixteenth century. Pacioli's architectural theory has a distinct "technical" emphasis. Yet, despite his interest in stone-cutting and other practical issues, Pacioli's understanding of theory and of the architect's tools of representation are unique. As a Franciscan mathematician, interested in teaching theology through geometry, his characterization of the role of architecture as a true "middle realm" between the human and the divine challenges our conceptions of Renaissance artistic culture as a monolithic prelude to a scientific modernity. Rather, Pacioli's understanding of the relationship between thinking and making as moments of an embodied process, differs significantly from both medieval practice, and from the most generalized modern concepts. His is a non-instrumental relationship, unlike that which we have come to take for granted for architecture after the Scientific and Industrial revolutions.

Pacioli was primarily a professor of theology. He is portrayed in his Franciscan habit by Jacopo de'Barbari[2] and addresses us as a teacher prepared to demonstrate, with his various mathematical and geometrical implements, the wonders of revealed Truth. On the table lies a beautifully bound volume with initials identifying it as a book by Pacioli himself. On top of the book is a wooden dodecahedron, described by Pacioli as the symbol of the "quintessence" because its construction subsumes the other four (the tetrahedron, cube, octahedron and icosahedron) and because it must be constructed from the "divine proportion," the golden-section ratio that is inherent in the pentagonal faces of the solid. With his left hand Pacioli points to the words "LIBER XIII" in an open book, while the

[2] The responsibility of Pacioli in the design of his own portrait has been well established; see Daly Davies (1977: 74–76).

pointer in his right hand is directed toward the geometric diagram on a slate with "EUCLIDES" inscribed on the side of its frame. Pacioli is demonstrating Proposition 8 of the 13th (and last) book of Euclid's *Elements*, where Euclid discusses the regular bodies. This theorem is crucial for nesting regular bodies into a sphere. It is also the beginning of speculation about the "squaring of the circle," the attempt to construct a square whose perimeter would be equal to the circumference of a circle inscribed in the square (a problem that was recognized as impossible only in the nineteenth century, when the irrational constant π was understood). In other words, this theorem was believed to be the geometrical key to the potential "solution" of duality into unity. It was a significant reference in the discourse of logical reason for architects, alchemists, mathematicians, and theologians until the late eighteenth century.

The most striking feature in the painting, however, is the floating, shimmering *corpo transparente* on Pacioli's right. This crystalline icosahexahedron (26-faced body) seems to be half-filled with a transparent elixir, and appears both solid and hollow. It is reminiscent of the engravings (by Leonardo da Vinci) of regular and space-filling bodies that illustrate the *Divina proportione*, likely an allusion to the "ungraspable true nature" of the primordial substance/space of the universe that is described by Plato in *Timaeus*, the *prima materia* which is both the *substance* of human artifacts (such as art and architecture), and the geometric space which is the *place* of human culture. In the painting all 18 squares and 8 equilateral triangles are perfectly and simultaneously visible, illuminated by an unseen source of light that makes the vessel appear to radiate from within.

What is remarkable about Pacioli's theoretical work is the explicit desire to demonstrate a non-instrumental, "opaque" relationship between the most abstract and the most concrete. His many books provide a comprehensive examination of practical applications of arithmetic and geometry, *as well as* mystical numerology. Architecture he believed to be the most propitious site for "experiencing" this dark, "irrational" continuity. In his *Divina proportione* Pacioli's mystical discussion of the golden section synthesizes Pythagorean and Platonic themes with Christian theology, culminating in a section on "practical" aspects of architecture. Pacioli evidently believed that architecture could fulfill the human quest for spiritual unity that underlies the mathematical demonstrations in his treatise.

The first four chapters of his book discuss theory and *mathemata* in general, including their relevance for painting, sculpture, music, perspective, and architecture. He emphasizes the origins of theory in vision, based on the wonder that likely accompanied the experience of cosmic phenomena such as a lunar eclipse (Pacioli 1509: 63). He insists that nothing can be grasped by the intellect unless it has been previously offered to perception in some way, emphasizing the distance existing between the world of perfect ideas and that of human existence. In other words, Pacioli insists that a geometrical point or a line *is not* the real point or line traced by a drafting instrument. The most noble sense is sight because it enables the intellect to "understand and taste." This "theory" is always *in* and *of* the world, in accordance with the Greco-Roman understanding of *theoria* as a contemplation of truth that also "saves the phenomena." *Such theory is always "discovered"; it never dictates to the hands of the artist "how" or "what" to do, yet its epiphanies are corroborated through enlightened human action.* The psychosomatic unity of human consciousness (as opposed to post-Cartesian concepts) remains here a primary assumption. This "traditional" theory could never be an imaginary (scientific) construction of the world (like Copernicus's cosmology, for instance) understood from some godly point of view. For Pacioli (like in Plato) only in works of art and craftsmanship do humans experience (without understanding clearly) a "dark coincidence" between Being and becoming.

After this important preamble, Pacioli explains why the golden section merits the attribute of "divine." He claims that this proportion resembles God himself through five major correspondences, and reveals Christian truth through 13 important properties that correspond to Christ and his 12 disciples.[3] It is crucial to note that "proportions" for Pacioli refer to the "practice" of architecture, the actual stereotomy and stereometry of stone masonry, rather than to the design of Albertian *lineamenti* in the mind or the architect's drawings of plans and elevations. This, and not Vitruvian theory, is indeed Pacioli's true concern: the capacity of the "craft" of architecture to reconcile duality into unity through geometric operations. Speaking about "columns with sides" (pillars or "square-based columns"), he refers to the "difficult problem of proportioning the circle to the square using the science of *quadratura circuli.*" He then speculates that

[3] The first of the five major correspondences is with the absolute uniqueness of this proportion, "the supreme epithet of God Himself"; there is simply no other proportion (a:b::c:d) with the same characteristics. The second correspondence is with the Holy Trinity; this proportion demonstrates unity with only three terms, and with three terms alone. It is defined by one mean and two extremes (*proportio habens medium et duo extrema*), and therefore is analogous to the Trinity's one substance in three distinct persons. The third correspondence concerns the impossibility of defining God in human terms; this proportion cannot be constructed with "intelligible numbers," remaining always "occult and secret... irrational in the words of the mathematicians" (the "golden number," is approximately 1.618). The fourth correspondence is with the immutable essence of God; the "divine proportion" is invariable and "continuous," it arises as the constant factorial relationship between consecutive terms of the Fibonacci series (mentioned above). The fifth correspondence is an analogy to the quintessence or "celestial virtue." In this case, Pacioli identifies the Christian God with the Platonic demiurge, and Creation with the cosmogony described in *Timaeus*. Pacioli argues that God himself created the quintessence, and from it, the four elements that compose the universe: earth, water, air and fire. The generative function of the quintessence (God's heaven), identified here with a *prima materia*, is analogous "to our holy proportion that provides formal being (according to Plato in his *Timaeus*) to the heaven itself, attributing to it the figure of the dodecahedron... the body made of twelve pentagons that cannot be formed without our proportion." Consequently, this proportion functions as a "continuous quantity" that assigns its respective forms to the four elements: the tetrahedron to fire, the cube or hexahedron to earth, the octahedron to air, and the icosahedron to water; the quintessential dodecahedron completes the set of five regular bodies. "And through them, our proportion gives form to an infinite number of dependent bodies" ("space-filling" or irregular polyhedra), that provide the complex richness we normally encounter in our experience of the world. Most importantly, Pacioli concludes, without the divine proportion it is impossible to establish the geometric relationship among these bodies and to demonstrate how can they be circumscribed by a sphere and thus reconciled with a primordial unity (Pacioli 1509: 69–70).

the wise philosopher who is capable of finally solving the problem may have been born already, "as for me I can demonstrate [its truth] *palpabiliter* [in a palpable manner, through tactile intuition], to anyone who may question it" (Pacioli 1509: 170–171). This "perceptual knowledge" of wholeness is precisely the province of architecture.[4]

In conclusion, Pacioli's concept of architecture was unique, perhaps similar only to the equally "unorthodox" understanding of architecture in the "pagan" and "alchemical" *Hypnerotomachia Poliphili* (1499), arguably the work of another Frate, Francesco Colonna.[5] Pacioli's concern is "technical" rather than "humanistic," recalling the alchemist's search for "gold" as a mineral "*sol*," that is distinct from the true sun in heaven. Indeed, alchemy insists that the quintessence is a "mortal heaven" that is not identical to God's heaven. Analogously, Pacioli's mathematics were never truly "of this world," but must be grasped through the senses. The aim of the architect/craftsman was not to render the ideal world as a concrete physical presence; this would be an absurd impossibility. Only close to a century later, Juan Bautista Villalpando legitimized the obsession to actualize the Temple of Jerusalem in Philip II's Palace/Monastery of El Escorial, inaugurating a modern tradition that would continue to our own time, passing through Walter Gropius and Peter Eisenman. Consequently, Pacioli was not interested in the ability of an architect/author to produce "pictures" of a future building. The *pictura* of the Temple of Solomon in his treatise remained "otherworldly." Although he was familiar with the new power of art (particularly the perspectival epiphanies of painting), his emphasis on craft distinguishes him from most contemporary writers on architecture. More significantly, although architecture may be a craft, it need not be devoid of a "philosophical" component. On the contrary: its discourse is mathematical and theological.

Pacioli's portrait, in which nothing is superfluous or accidental, reveals not only his Euclidean teaching and its allusion to divine proportion, but also his interest in a stereotomic glass architecture. The "philosophical" work of the architect was modeled on the architecture of the Platonic cosmos that was echoed in God's design for the otherworldly Temple of Jerusalem. Stereotomy and stereometry offer techniques, but also a philosophical understanding of what may be revealed in the process: the "ephemeral gold" that must be recognized in the unending process of transmutation which is a human work. Architecture, a "mediating art" *par excellence*, emerges from humble materials, from the earth itself, like the glass that forms Pacioli's floating icosahexahedron is generated artificially from lowly

---

[4] We might recall here the "wondrous demonstration" of the squaring of the circle when a beam of sunlight is projected into a dark chamber through a square orifice, and the projection turns out to be a circle. This phenomenon remained a source of wonder during the sixteenth and seventeenth centuries, well after Kepler's demonstration of the "pin-hole" principle, according to which an aperture of *any* shape will project the sun as a circle.

[5] The authorship of the *Hypnerotomachia* has been the subject of much recent debate. My reading of Pacioli's Franciscan understanding of architecture tends to corroborate, albeit indirectly, the original authorship attribution; see Pérez-Gómez (1994: xi–xvii).

ashes and sand, an alchemical symbol of rebirth and salvation. Following Pacioli's alchemical path, architecture could be construed as a virtuous and ethical craft, truly a form of meditation, capable of transmuting matter (the earth) and liberating it from gravity, and enabling humanity (humus) to recognize its spiritual wholeness.[6]

Le Corbusier's *Poème*

Operating in a very different world than Pacioli, in the wake of Nietzsche, but within the same European tradition, Le Corbusier provides an excellent example for the value of process work and the productive non-instrumental relationship between thinking and making which we seek. The relationships that emerge in this historical sketch of a *longue duree* are illuminating.

Le Corbusier admired Choisy's axonometric images as an example of "modern space," objectified and determined precise measurement, a 3-D matrix governed by scale. In this sense, he appropriated them for his arguments in his early definition of *L'Esprit Nouveau*,[7] perceiving the appropriateness of this objectified space to his interest in technique as a determinant of form. He also understood quite early the relationship between axonometric space and the new space of painting.

It is well known that Le Corbusier sought the integration of the arts. He painted all his life, seeking at first public exposure. After 1927, however, he decided to make his painting a private activity, and devoted to it every morning of the week. He believed that his painting was crucial for his understanding of architecture, the issue being not one of formal analogies, but rather the activity of *making* itself, i.e., an inquiry into the world of appearances coupled with the careful construction and realization of projects.[8] Critics and historians, incapable or unwilling to think beyond the Renaissance categories of the Fine Arts, have often underplayed Le Corbusier's architectural quest through painting, or have dismissed it as a propaganda ploy to have his architecture valorized as "art." On the contrary, Le Corbusier's rhetorical plea is crucial to understand his legacy. As a reader of Friedrich Nietzsche's *Zarathustra* and admirer of Alfred Jarry, Le Corbusier struggled all his life to find ways to translate into architecture, into the pragmatic world of embodied experience, the new fascinating depth and temporality first fabricated in canvas and paint, or sketched for sculpture or tapestry.

[6] It is worth recalling the Franciscan tradition of seeking self-realization through making, one that was never free of controversy. The ex-communicated Brother Elias, second General of the Franciscans (1226–1284), was thought to be the author of various alchemical treatises during the fourteenth and the fifteenth centuries.

[7] The magazine *L'Esprit Nouveau* was launched in 1920 and ceased publication in 1925. Le Corbusier was co-editor, with Amédée Ozenfant and Paul Dermée.

[8] He described his work in these terms in a catalogue for an exhibition of his paintings at the Galerie Balay et Carré, Paris, 1938.

Le Corbusier's abundant sketches and graphic meditations demonstrate a pervasive, sharp self-consciousness of the difficulties involved in art as will to power. Drawing for him was an act of reconciliation between the artist and a pre-existing reality. He was not naive about the polarities between nature and culture, geometry and life, or the city and the country. It is true that early in his life, during his "purist" period, he stressed geometry, and possibly never abandoned a progressive view of history. His painting during the early 1920s paid attention to the intricate formal relationships between objects. Transformed into planes rather than merely "flattened," the superimposed objects seem to push out from the picture surface, "squeezing out" space, i.e., common perspectival depth (Green 1987: 113). By the late 1920s he became friends with Fernand Léger. He had in his possession a catalogue of a major exhibition of Giorgio de Chirico's work (1928), and his own paintings focused on erotic themes. In the 1930s, Le Corbusier came to the realization that the space of human significance had to be articulated as an erotic distance. He considered the potential polarity of man against nature no longer operative; hereafter the manufactured world and the natural one were to be accepted equally, without compromising either.

A careful examination of Le Corbusier's sketchbooks (1981) from 1914 to 1964 reveals a consistent use of perspectival views to visualize his ideas. There are often diagrammatic plans and elevations, but never objectified drawings. Never does one find an axonometric as the initial idea for a building. There are, of course, some "air views," but these are always contextual and perspectival. The only generative "axonometric" spaces we find in the sketchbook are ideas for his paintings. We may recall here how Le Corbusier emphasized the importance of patience in creative work, expressing notions that echoed Filarete when it came to describing the process of gestation of an idea. After being entrusted a task, he would "place it in the interior of [his] memory." He would sketch no more and let the problem "float, simmer, ferment" until one day "out of the spontaneous initiative of the inner being" one takes a pencil, charcoal, color pencils, and "one gives birth on the paper: the idea comes out – the child is delivered" (quoted in Eardley 1981: 13).

The synthetic, comprehensive nature of architectural design generated in the matrix of axonometric space is incompatible with this conception. Indeed, this observation is consistent with the total disappearance of axonometric representation from Le Corbusier's mature work. Only in the period preceding 1929 (volume one of the *Œuvre complète*) was there a frequent use of axonometric drawing. Significantly, after the early 1930s and coinciding with his new realizations in painting, there are almost no axonometric drawings in the remaining six volumes of the *Œuvre complète*. An examination of the vast *Le Corbusier Archives* (1982–1985) yields a similar picture.

Le Corbusier's concern with this "modern space" in painting was, therefore, not simply transplanted to architecture. His work can hardly be reduced to the application of the matrix of axonometry as a place for modernist syntax, as has been often assumed. Relatively early in his career his true struggle was to find equivalent modes of presenting in the visual/erotic space of architectural situations. His experimentation with projection in the space of representation was a life-long passion which included an interest in film, and that culminated in an awareness of artistic discovery as the unveiling of unexpected relationships between objects of the environing world, emerging from the new contiguities construed in the work. In 1938 he wrote, in terms that recall a surrealist understanding of collage, that the difference between everyday, prosaic spoken language and painting consisted in a different way of denoting things. While the former names things narrowly and specifically, the latter is concerned with the quality of things, bringing them together freely. Thus it is the unexpected relationships, i.e., the space of metaphoric tension, that the artist discovers; this "is what the poet proclaims, that which the inspired being creates" (Green 1987: 117). A decade later he wrote again about art in *New World of Space* (1948). He explained the genesis of his *Ubu* sculptures, deliberately named in honor of Alfred Jarry, the founder of pataphysics: "Stones and pieces of wood led me on involuntarily to draw beings who became a species of monster or god." Le Corbusier was aware of how the process offered much to his architecture precisely because it revealed "new things," "unexpected" and "unknown." He concluded:

> When the inexplicable appears in human work, that is, when our spirit is projected far from the narrow relation of cause and effect . . . to the cosmic phenomenon in time, in space, in the intangible . . . then the inexplicable is the mystery of art . . . (Corbusier 1987: 246).

Articulating this experience in poetic language, Le Corbusier's *Poème de l'Angle Droit* (1955) remains his most comprehensive, rather misunderstood theoretical statement about architecture. Almost every important aspect of the architect's thought finds a place in the verses and images of the *Poème*. The *Iconostase,* the prescribed arrangement for the color lithographs accompanying the text, is deployed between the tool of the architect (the right angle) at the center bottom, and the realm of the "cosmos" at the top. The form of the work itself, in the wake of many other discursive writings, establishes a tension between words and images, demonstrating Le Corbusier's awareness of the importance of the poetic word and of the space of collage to fully express his thoughts. The "opaque" relationship between image and words communicates the possibility to reconcile the ordering imagination of the architect and his tools, i.e., the right angle, with the pre-given order that the architect encounters "already there."

This issue, a fundamental question for human making and architecture, is not resolved naively through some kind of theological formulation. Like Nietzsche, Le Corbusier looked at the unquestionably significant traces of human history, at the presence of genius, and at our ability to grasp the outlines of destiny, to fill the void left by God's demise. Echoing, perhaps unknowingly, the late writings of Nietzsche in *The Will to Power*, he seemed to identify the power of artistic creation with the erotic drive that characterizes the *Übermann*, leading the way for humanity to discover a new form of spirituality through a reconciliation of will to power and *amor fati*. Le Corbusier wrote: "A man who searches for harmony has a sense of the sacred, the secret which is in every being, a great limitless void where you may place your own notion of the sacred – individual, completely individual" (Le Corbusier and Jeanneret 1995: IV, 170). The issue then is to reconcile extreme individuality, the work of the creator's imagination, with a given world, *both* natural *and* constructed, in the absence of a positive theology or cosmology. The result is much more than a simple reiteration of the old Romantic themes of transcendence through art, if by art we understand the production of "aesthetic objects" placed in the homogeneous space of a universal museum.

The ultimate ground and foundation of the *Iconostase* (G.3) is the hand of the architect, drawing a right angle within a *bounded* space, the space of the human horizon. The *Poème* fails as "theory" if the expectation is to find a logical, universally applicable structure for action, or if one still expects the architecture it evokes to "signify" following a semiotic model. Le Corbusier sought the revelation of coincidences, bringing together a perceptual faith and the implementation of the architect's tools. While respecting the primacy of our technological world—we must emphasize that he was never naive about an unchanging nature as the ground of meaning—the issue was the possible revelation of the poetic, the only kind of human "truth." This *poésie* could not be something imposed through a fabricated cosmology or imported from another time. It had to emerge from our world of experience, without resentment, embracing all its contradictions.

Some commentators have put forward an alchemical interpretation of the text and images. At times this tends to rationalize the unresolved tensions in the work and has been used as an argument to demonstrate Le Corbusier's "symbolic" preoccupations, beyond the "formal" and "functional" (Moore 1980). Used in this way, the analogy cold be problematic. True symbolization occurs only when the work is made "of" its own world, rather than construed through some alien construct. In fact, Le Corbusier never expressed himself in the language of alchemy or Gnosticism. The alchemical interpretation of his pervasive concern with dualities and their engagement in the work of architecture is only adequate *after* arguing that alchemy and mythology embody the traces of a collective unconscious, rendering his quest in terms similar to those I outlined for Pacioli. The work does disclose a "rare gold" in the gaps between fragments and words, in the spaces unveiled by bringing together disparate objects on the same "playing field" of the *Iconostase*. There is, of course, a significant affinity between the techniques of collage and the search for the ineffable unity in alchemy, and the issue of substantial transmutation (a self-transformation) through the work on the *opus* is as present in the arcane science as it is desired by Le Corbusier from his work. Yet, the issue is still the specific disclosure of truth in the particular embodiments appropriate to our world, through its forms of representation.

The *Poème de l'Angle Droit* is, more specifically, a pataphysical text, where individuality and universality are reversed and reciprocal. The issue of self-transformation is indeed present, and is accessible to us through Jarry's exemplary work. This is indeed the fundamental question for the architect, as it is for the participants (not mere observers) in his work. Truth must be sought in the unique coincidences disclosed in each artistic work, a wondrous truth that cannot be repeated or universalized. Like pataphysics, a modern incarnation of alchemical theory, the *Poème* is a search for the exceptional, which is the universal, and architecture is therefore construed as a "science" of imaginary solutions. This discourse is posited as our only possible "cosmology," the "ground" for artistic action. The *Poème* is therefore much more than a celebration of the *événement plastique*.[9] It is clear that Le Corbusier identified a sense of the sacred with a harmonious relationship (*accord*) to the cosmos, attainable through art and architecture. His work, however, is not simply a reiteration of the Romantic program of man in a secular "nature;" his (our) nature is *also* art, without for this reason being a "mere" construction. Le Corbusier must have been conscious of this Nietzschean "paradox."

Le Corbusier's "patient search" from purism through surrealism led to an awareness that architecture could not be conceived or perceived in "aesthetic" terms, that its meaning had to be disclosed in a temporal medium, distinct from that of the "aesthetic objects" of the modernist tradition. This, I would argue, is most explicit in Le Corbusier's late buildings, particularly in La Tourette, so different in this respect from the early purist work, so respectful of the internal historicity of the "type" and yet so revolutionary in its transformation. In this late work, Le Corbusier recognizes that the "re-writing" of the program as a human situation is a crucial aspect of architectural work, the construing (in words) of a choreography beyond formal manipulations. The result is a building almost impenetrable to prosaic human function, inhabitable yet "uncomfortable" and demanding for human spirituality. While this topic is beyond the scope of this article, it is worth noting how Le Corbusier became intentionally detached from the construction process in this case, and "let go" formal aspects of the building that he would have been obsessed with in his earlier work. La Tourette celebrates "faulty" craftsmanship and "mistakes" in the translation process between drawings and building; it profits from the space of unpredictability in the construction process and transforms it into the place of meaning, the poetic instant that reveals, in a flash, the coincidence between sublime work and human mortality.

Biography Alberto Pérez-Gómez earned his undergraduate degree in architecture and engineering in Mexico City, did postgraduate work at Cornell University, and was awarded a Master of Arts and a Ph.D. by the University of Essex in England. He has taught at universities in Mexico City, Houston, Syracuse, and Toronto, at the Architectural Association in London, and was Director of the Carleton University School of Architecture (1983–1986). He has lectured extensively worldwide. He is currently Saidye Rosner Bronfman Professor of the History of Architecture at McGill University. He is the author of *Polyphilo or The Dark Forest Revisited* (MIT Press, 1992), and co-editor of a book series entitled *CHORA: Intervals in the Philosophy of Architecture (*McGill-Queen's University Press). His most recent book, *Built Upon Love: Architectural Longing after Ethics and Aesthetics* (MIT Press, 2006), examines points of convergence between ethics and poetics in architectural history and philosophy, drawing important conclusions for contemporary practice.

---

[9] I take issue with this crucial point made in the otherwise fascinating interpretation of the *Poème* by Peter Carl (1988).

References

CARL, Peter. 1988. Le Corbusier's Penthouse in Paris, 24 Rue Nungesser-et-Coli. *Daidalos* **28**: 65-75.

DALY DAVIS, Margaret. 1977. *Piero della Francesca's Mathematical Treatises*. Ravenna: Longo Editore.

EARDLEY, Anthony. 1981. Grandeur is the Intention. Pp. 4-23 in *Le Corbusier's Firminy Church*. Kenneth Frampton and Sylvia Kolbowski, eds. New York: Rizzoli.

GREEN, Christopher. 1987. The Architect as Artist. Pp. 113-117 in *Le Corbusier, Architect of the Century*. London: Arts Council of Great Britain.

KEARNEY, Richard. 1988. *The Wake of the Imagination*. London: Century Hutchinson Ltd.

LE CORBUSIER. 1948. *New World of Space*. New York: Paul Rosenberg Gallery.

———. 1955. *Poème de l'Angle Droit*. Paris: Éditions Verve.

———. 1981. *Le Corbusier Sketchbooks*. 4 vols. Cambridge: MIT Press.

———. 1982-1985. *Le Corbusier Archives*. 15 vols. New York: Garland Press.

———. 1987. *Le Corbusier, Architect of the Century*. London: Arts Council of Great Britain.

LE CORBUSIER and JEANNERET, Pierre. 1995. *Œuvre complete*. 8 vols. (1929-1970). Basel: Birkhäuser.

MOORE, RICHARD A. 1980. Alchemical and Mythical Themes in the 'Poéme de l'Angle Droit' 1947-1965. *Oppositions* **19/20**: 110-141.

PACIOLI, Luca. 1509. *De Divina Proportione*. Venezia: Paganino dè Paganini.

PÉREZ-GÓMEZ, Alberto. 1994. *Polyphilo or the Dark Forest Revisited: An Erotic Epiphany of Architecture*. Cambridge MA: MIT Press.

Index for Volume II[1]

[1] Note: Individual buildings may generally be found listed under one of the following classifications:
Amphitheatres
Apartment Buildings
Archaeological Sites
Baptisteries
Basilicas
Bridges
Castles
Cathedrals
Chapels
Churches
Government Buildings
Houses
Libraries
Mausoleums
Monuments and Memorials
Mosques
Museums
Palaces
Public Spaces / Squares
Pyramids
Shrines
Temples
Theatres / Entertainment
Towers
Villas

K. Williams and M.J. Ostwald (eds.), *Architecture and Mathematics from Antiquity to the Future*, DOI 10.1007/978-3-319-00143-2,

Index for Volume I

K. Williams and M.J. Ostwald (eds.), *Architecture and Mathematics from Antiquity to the Future*, DOI 10.1007/978-3-319-00143-2,

D

E

F

GPSR Compliance

The European Union's (EU) General Product Safety Regulation (GPSR) is a set of rules that requires consumer products to be safe and our obligations to ensure this.

If you have any concerns about our products, you can contact us on ProductSafety@springernature.com

In case Publisher is established outside the EU, the EU authorized representative is:

Springer Nature Customer Service Center GmbH
Europaplatz 3
69115 Heidelberg, Germany

Batch number: 10739969

Printed by Printforce, the Netherlands